Lecture Notes in Networks and Systems

Volume 226

The series "Lecture Notes in Networks and Systems" publishes the latest developments in Networks and Systems—quickly, informally and with high quality. Original research reported in proceedings and post-proceedings represents the core of LNNS.

Volumes published in LNNS embrace all aspects and subfields of, as well as new challenges in, Networks and Systems.

The series contains proceedings and edited volumes in systems and networks, spanning the areas of Cyber-Physical Systems, Autonomous Systems, Sensor Networks, Control Systems, Energy Systems, Automotive Systems, Biological Systems, Vehicular Networking and Connected Vehicles, Aerospace Systems, Automation, Manufacturing, Smart Grids, Nonlinear Systems, Power Systems, Robotics, Social Systems, Economic Systems and other. Of particular value to both the contributors and the readership are the short publication timeframe and the world-wide distribution and exposure which enable both a wide and rapid dissemination of research output.

The series covers the theory, applications, and perspectives on the state of the art and future developments relevant to systems and networks, decision making, control, complex processes and related areas, as embedded in the fields of interdisciplinary and applied sciences, engineering, computer science, physics, economics, social, and life sciences, as well as the paradigms and methodologies behind them.

Indexed by SCOPUS, INSPEC, WTI Frankfurt eG, zbMATH, SCImago.

All books published in the series are submitted for consideration in Web of Science.

More information about this series at http://www.springer.com/series/15179

Leonard Barolli · Isaac Woungang ·
Tomoya Enokido
Editors

Advanced Information Networking and Applications

Proceedings of the 35th International
Conference on Advanced Information
Networking and Applications (AINA-2021),
Volume 2

 Springer

Editors
Leonard Barolli
Department of Information
and Communication Engineering
Fukuoka Institute of Technology
Fukuoka, Japan

Isaac Woungang
Department of Computer Science
Ryerson University
Toronto, ON, Canada

Tomoya Enokido
Faculty of Business Administration
Rissho University
Tokyo, Japan

ISSN 2367-3370   ISSN 2367-3389   (electronic)
Lecture Notes in Networks and Systems
ISBN 978-3-030-75074-9   ISBN 978-3-030-75075-6   (eBook)
https://doi.org/10.1007/978-3-030-75075-6

This Springer imprint is published by the registered company Springer Nature Switzerland AG
The registered company address is: Gewerbestrasse 11, 6330 Cham, Switzerland

Welcome Message from AINA-2021 Organizers

Welcome to the 35th International Conference on Advanced Information Networking and Applications (AINA-2021). On behalf of AINA-2021 Organizing Committee, we would like to express to all participants our cordial welcome and high respect.

AINA is an international forum, where scientists and researchers from academia and industry working in various scientific and technical areas of networking and distributed computing systems can demonstrate new ideas and solutions in distributed computing systems. AINA was born in Asia, but it is now an international conference with high quality thanks to the great help and cooperation of many international friendly volunteers. AINA is a very open society and is always welcoming international volunteers from any country and any area in the world.

AINA international conference is a forum for sharing ideas and research work in the emerging areas of information networking and their applications. The area of advanced networking has grown very rapidly, and the applications have experienced an explosive growth especially in the areas of pervasive and mobile applications, wireless sensor networks, wireless ad hoc networks, vehicular networks, multimedia computing and social networking, semantic collaborative systems, as well as Grid, P2P, IoT, Big Data and Cloud Computing. This advanced networking revolution is transforming the way people live, work and interact with each other and is impacting the way business, education, entertainment and health care are operating. The papers included in the proceedings cover theory, design and application of computer networks, distributed computing and information systems.

Each year AINA receives a lot of paper submissions from all around the world. It has maintained high-quality accepted papers and is aspiring to be one of the main international conferences on the information networking in the world.

We are very proud and honored to have two distinguished keynote talks by Dr. Flora Amato, University of Naples "Federico II", Italy and Prof. Shahrokh Valaee, University of Toronto, Canada, who will present their recent work and will give new insights and ideas to the conference participants.

An international conference of this size requires the support and help of many people. A lot of people have helped and worked hard to produce a successful AINA-2021 technical program and conference proceedings. First, we would like to thank all authors for submitting their papers, the session chairs and distinguished keynote speakers. We are indebted to program track co-chairs, program committee members and reviewers, who carried out the most difficult work of carefully evaluating the submitted papers.

We would like to thank AINA-2021 general co-chairs, PC co-chairs, workshop co-chairs for their great efforts to make AINA-2021 a very successful event. We have special thanks to the finance chair and web administrator co-chairs.

We do hope that you will enjoy the conference proceedings and readings.

Leonard Barolli
Makoto Takizawa
AINA Steering Committee Co-chairs

Isaac Woungang
Markus Aleksy
Farookh Hussain
AINA-2021 General Co-chairs

Glaucio Carvalho
Tomoya Enokido
Flora Amato
AINA-2021 Program Committee Co-chairs

AINA-2021 Organizing Committee

General Co-chairs

Isaac Woungang	Ryerson University, Canada
Markus Aleksy	ABB Corporate Research Center, Germany
Farookh Hussain	University of Technology, Sydney, Australia

Program Committee Co-chairs

Glaucio Carvalho	Sheridan College, Canada
Tomoya Enokido	Rissho University, Japan
Flora Amato	University of Naples "Federico II", Italy

Workshops Co-chairs

Kin Fun Li	University of Victoria, Canada
Omid Ameri Sianaki	Victoria University, Australia
Yi-Jen Su	Shu-Te University, Taiwan

International Journals Special Issues Co-chairs

Fatos Xhafa	Technical University of Catalonia, Spain
David Taniar	Monash University, Australia

Award Co-chairs

Marek Ogiela	AGH University of Science and Technology, Poland
Arjan Durresi	Indiana University Purdue University in Indianapolis (IUPUI), USA
Fang-Yie Leu	Tunghai University, Taiwan

Publicity Co-chairs

Lidia Ogiela   Pedagogical University of Cracow, Poland
Minoru Uehara   Toyo University, Japan
Hsing-Chung Chen   Asia University, Taiwan

International Liaison Co-chairs

Akio Koyama   Yamagata University, Japan
Nadeem Javaid   COMSATS University Islamabad, Pakistan
Wenny Rahayu   La Trobe University, Australia

Local Arrangement Co-chairs

Mehrdad Tirandazian   Ryerson University, Canada
Glaucio Carvalho   Sheridan College, Canada

Finance Chair

Makoto Ikeda   Fukuoka Institute of Technology, Japan

Web Co-chairs

Phudit Ampririt   Fukuoka Institute of Technology, Japan
Kevin Bylykbashi   Fukuoka Institute of Technology, Japan
Ermioni Qafzezi   Fukuoka Institute of Technology, Japan

Steering Committee Chairs

Leonard Barolli   Fukuoka Institute of Technology, Japan
Makoto Takizawa   Hosei University, Japan

Tracks and Program Committee Members

1. Network Protocols and Applications
Track Co-chairs

Makoto Ikeda   Fukuoka Institute of Technology, Japan
Sanjay Kumar Dhurandher   Netaji Subhas University of Technology,
                                       New Delhi, India
Bhed Bahadur Bista   Iwate Prefectural University, Japan

TPC Members

Elis Kulla	Okayama University of Science, Japan
Keita Matsuo	Fukuoka Institute of Technology, Japan
Shinji Sakamoto	Seikei University, Japan
Akio Koyama	Yamagata University, Japan
Evjola Spaho	Polytechnic University of Tirana, Albania
Jiahong Wang	Iwate Prefectural University, Japan
Shigetomo Kimura	University of Tsukuba, Japan
Chotipat Pornavalai	King Mongkut's Institute of Technology Ladkrabang, Thailand
Danda B. Rawat	Howard University, USA
Akio Koyama	Yamagata University, Japan
Amita Malik	Deenbandhu Chhotu Ram University of Science and Technology, India
R. K. Pateriya	Maulana Azad National Institute of Technology, India
Vinesh Kumar	University of Delhi, India
Petros Nicopolitidis	Aristotle University of Thessaloniki, Greece
Satya Jyoti Borah	North Eastern Regional Institute of Science and Technology, India

2. Next Generation Wireless Networks

Track Co-chairs

Christos J. Bouras	University of Patras, Greece
Tales Heimfarth	Universidade Federal de Lavras, Brazil
Leonardo Mostarda	University of Camerino, Italy

TPC Members

Fadi Al-Turjman	Near East University, Nicosia, Cyprus
Alfredo Navarra	University of Perugia, Italy
Purav Shah	Middlesex University London, UK
Enver Ever	Middle East Technical University, Northern Cyprus Campus, Cyprus
Rosario Culmone	University of Camerino, Camerino, Italy
Antonio Alfredo F. Loureiro	Federal University of Minas Gerais, Brazil
Holger Karl	University of Paderborn, Germany
Daniel Ludovico Guidoni	Federal University of São João Del-Rei, Brazil
João Paulo Carvalho Lustosa da Costa	Hamm-Lippstadt University of Applied Sciences, Germany
Jorge Sá Silva	University of Coimbra, Portugal
Apostolos Gkamas	University Ecclesiastical Academy of Vella, Ioannina, Greece

Zoubir Mammeri   University Paul Sabatier, France
Eirini Eleni Tsiropoulou   University of New Mexico, USA
Raouf Hamzaoui   De Montfort University, UK
Miroslav Voznak   University of Ostrava, Czech Republic

3. Multimedia Systems and Applications
Track Co-chairs

Markus Aleksy   ABB Corporate Research Center, Germany
Francesco Orciuoli   University of Salerno, Italy
Tomoyuki Ishida   Fukuoka Institute of Technology, Japan

TPC Members

Tetsuro Ogi   Keio University, Japan
Yasuo Ebara   Osaka Electro-Communication University, Japan
Hideo Miyachi   Tokyo City University, Japan
Kaoru Sugita   Fukuoka Institute of Technology, Japan
Akio Doi   Iwate Prefectural University, Japan
Hadil Abukwaik   ABB Corporate Research Center, Germany
Monique Duengen   Robert Bosch GmbH, Germany
Thomas Preuss   Brandenburg University of Applied Sciences,
                                      Germany
Peter M. Rost   NOKIA Bell Labs, Germany
Lukasz Wisniewski   inIT, Germany
Hadil Abukwaik   ABB Corporate Research Center, Germany
Monique Duengen   Robert Bosch GmbH, Germany
Peter M. Rost   NOKIA Bell Labs, Germany
Angelo Gaeta   University of Salerno, Italy
Graziano Fuccio   University of Salerno, Italy
Giuseppe Fenza   University of Salerno, Italy
Maria Cristina   University of Salerno, Italy
Alberto Volpe   University of Salerno, Italy

4. Pervasive and Ubiquitous Computing
Track Co-chairs

Chih-Lin Hu   National Central University, Taiwan
Vamsi Paruchuri   University of Central Arkansas, USA
Winston Seah   Victoria University of Wellington, New Zealand

TPC Members

Hong Va Leong	Hong Kong Polytechnic University, Hong Kong
Ling-Jyh Chen	Academia Sinica, Taiwan
Jiun-Yu Tu	Southern Taiwan University of Science and Technology, Taiwan
Jiun-Long Huang	National Chiao Tung University, Taiwan
Thitinan Tantidham	Mahidol University, Thailand
Tanapat Anusas-amornkul	King Mongkut's University of Technology North Bangkok, Thailand
Xin-Mao Huang	Aletheia University, Taiwan
Hui Lin	Tamkang University, Taiwan
Eugen Dedu	Universite de Franche-Comte, France
Peng Huang	Sichuan Agricultural University, China
Wuyungerile Li	Inner Mongolia University, China
Adrian Pekar	Budapest University of Technology and Economics, Hungary
Jyoti Sahni	Victoria University of Technology, New Zealand
Normalia Samian	Universiti Putra Malaysia, Malaysia
Sriram Chellappan	University of South Florida, USA
Yu Sun	University of Central Arkansas, USA
Qiang Duan	Penn State University, USA
Han-Chieh Wei	Dallas Baptist University, USA

5. Web-Based and E-Learning Systems

Track Co-chairs

Santi Caballe	Open University of Catalonia, Spain
Kin Fun Li	University of Victoria, Canada
Nobuo Funabiki	Okayama University, Japan

TPC Members

Jordi Conesa	Open University of Catalonia, Spain
Joan Casas	Open University of Catalonia, Spain
David Gañán	Open University of Catalonia, Spain
Nicola Capuano	University of Basilicata, Italy
Antonio Sarasa	Complutense University of Madrid, Spain
Chih-Peng Fan	National Chung Hsing University, Taiwan
Nobuya Ishihara	Okayama University, Japan
Sho Yamamoto	Kindai University, Japan
Khin Khin Zaw	Yangon Technical University, Myanmar
Kaoru Fujioka	Fukuoka Women's University, Japan
Kosuke Takano	Kanagawa Institute of Technology, Japan

Shengrui Wang   University of Sherbrooke, Canada
Darshika Perera   University of Colorado at Colorado Spring, USA
Carson Leung   University of Manitoba, Canada

6. Distributed and Parallel Computing
Track Co-chairs

Naohiro Hayashibara   Kyoto Sangyo University, Japan
Minoru Uehara   Toyo University, Japan
Tomoya Enokido   Rissho University, Japan

TPC Members

Eric Pardede   La Trobe University, Australia
Lidia Ogiela   Pedagogical University of Cracow, Poland
Evjola Spaho   Polytechnic University of Tirana, Albania
Akio Koyama   Yamagata University, Japan
Omar Hussain   University of New South Wales, Australia
Hideharu Amano   Keio University, Japan
Ryuji Shioya   Toyo University, Japan
Ji Zhang   The University of Southern Queensland
Lucian Prodan   Universitatea Politehnica Timisoara, Romania
Ragib Hasan   The University of Alabama at Birmingham, USA
Young-Hoon Park   Sookmyung Women's University, Korea

7. Data Mining, Big Data Analytics and Social Networks
Track Co-chairs

Eric Pardede   La Trobe University, Australia
Alex Thomo   University of Victoria, Canada
Flora Amato   University of Naples "Frederico II", Italy

TPC Members

Ji Zhang   University of Southern Queensland, Australia
Salimur Choudhury   Lakehead University, Canada
Xiaofeng Ding   Huazhong University of Science
                                   and Technology, China
Ronaldo dos Santos Mello   Universidade Federal de Santa Catarina, Brasil
Irena Holubova   Charles University, Czech Republic
Lucian Prodan   Universitatea Politehnica Timisoara, Romania
Alex Tomy   La Trobe University, Australia
Dhomas Hatta Fudholi   Universitas Islam Indonesia, Indonesia
Saqib Ali   Sultan Qaboos University, Oman

Ahmad Alqarni	Al Baha University, Saudi Arabia
Alessandra Amato	University of Naples "Frederico II", Italy
Luigi Coppolino	Parthenope University, Italy
Giovanni Cozzolino	University of Naples "Frederico II", Italy
Giovanni Mazzeo	Parthenope University, Italy
Francesco Mercaldo	Italian National Research Council, Italy
Francesco Moscato	University of Salerno, Italy
Vincenzo Moscato	University of Naples "Frederico II", Italy
Francesco Piccialli	University of Naples "Frederico II", Italy

8. Internet of Things and Cyber-Physical Systems
Track Co-chairs

Euripides G. M. Petrakis	Technical University of Crete (TUC), Greece
Tomoki Yoshihisa	Osaka University, Japan
Mario Dantas	Federal University of Juiz de Fora (UFJF), Brazil

TPC Members

Akihiro Fujimoto	Wakayama University, Japan
Akimitsu Kanzaki	Shimane University, Japan
Kawakami Tomoya	University of Fukui, Japan
Lei Shu	University of Lincoln, UK
Naoyuki Morimoto	Mie University, Japan
Yusuke Gotoh	Okayama University, Japan
Vasilis Samolada	Technical University of Crete (TUC), Greece
Konstantinos Tsakos	Technical University of Crete (TUC), Greece
Aimilios Tzavaras	Technical University of Crete (TUC), Greece
Spanakis Manolis	Foundation for Research and Technology Hellas (FORTH), Greece
Katerina Doka	National Technical University of Athens (NTUA), Greece
Giorgos Vasiliadis	Foundation for Research and Technology Hellas (FORTH), Greece
Stefan Covaci	Technische Universität Berlin, Berlin (TUB), Germany
Stelios Sotiriadis	University of London, UK
Stefano Chessa	University of Pisa, Italy
Jean-Francois Méhaut	Université Grenoble Alpes, France
Michael Bauer	University of Western Ontario, Canada

9. Intelligent Computing and Machine Learning
Track Co-chairs

Takahiro Uchiya,	Nagoya Institute of Technology, Japan
Omar Hussain	UNSW, Australia
Nadeem Javaid	COMSATS University Islamabad, Pakistan

TPC Members

Morteza Saberi	University of Technology, Sydney, Australia
Abderrahmane Leshob	University of Quebec in Montreal, Canada
Adil Hammadi	Curtin University, Australia
Naeem Janjua	Edith Cowan University, Australia
Sazia Parvin	Melbourne Polytechnic, Australia
Kazuto Sasai	Ibaraki University, Japan
Shigeru Fujita	Chiba Institute of Technology, Japan
Yuki Kaeri	Mejiro University, Japan
Zahoor Ali Khan	HCT, UAE
Muhammad Imran	King Saud University, Saudi Arabia
Ashfaq Ahmad	The University of Newcastle, Australia
Syed Hassan Ahmad	JMA Wireless, USA
Safdar Hussain Bouk	Daegu Gyeongbuk Institute of Science and Technology, Korea
Jolanta Mizera-Pietraszko	Military University of Land Forces, Poland

10. Cloud and Services Computing
Track Co-chairs

Asm Kayes	La Trobe University, Australia
Salvatore Venticinque	University of Campania "Luigi Vanvitelli", Italy
Baojiang Cui	Beijing University of Posts and Telecommunications, China

TPC Members

Shahriar Badsha	University of Nevada, USA
Abdur Rahman Bin Shahid	Concord University, USA
Iqbal H. Sarker	Chittagong University of Engineering and Technology, Bangladesh
Jabed Morshed Chowdhury	La Trobe University, Australia
Alex Ng	La Trobe University, Australia
Indika Kumara	Jheronimus Academy of Data Science, Netherlands
Tarique Anwar	Macquarie University and CSIRO's Data61, Australia
Giancarlo Fortino	University of Calabria, Italy

Massimiliano Rak	University of Campania "Luigi Vanvitelli", Italy
Jason J. Jung	Chung-Ang University, Korea
Dimosthenis Kyriazis	University of Piraeus, Greece
Geir Horn	University of Oslo, Norway
Gang Wang	Nankai University, China
Shaozhang Niu	Beijing University of Posts and Telecommunications, China
Jianxin Wang	Beijing Forestry University, China
Jie Cheng	Shandong University, China
Shaoyin Cheng	University of Science and Technology of China, China

11. Security, Privacy and Trust Computing
Track Co-chairs

Hiroaki Kikuchi	Meiji University, Japan
Xu An Wang	Engineering University of PAP, China
Lidia Ogiela	Pedagogical University of Cracow, Poland

TPC Members

Takamichi Saito	Meiji University, Japan
Kouichi Sakurai	Kyushu University, Japan
Kazumasa Omote	University of Tsukuba, Japan
Shou-Hsuan Stephen Huang	University of Houston, USA
Masakatsu Nishigaki	Shizuoka University, Japan
Mingwu Zhang	Hubei University of Technology, China
Caiquan Xiong	Hubei University of Technology, China
Wei Ren	China University of Geosciences, China
Peng Li	Nanjing University of Posts and Telecommunications, China
Guangquan Xu	Tianjing University, China
Urszula Ogiela	Pedagogical University of Cracow, Poland
Hoon Ko	Chosun University, Korea
Goreti Marreiros	Institute of Engineering of Polytechnic of Porto, Portugal
Chang Choi	Gachon University, Korea
Libor Měsíček	J.E. Purkyně University, Czech Republic

12. Software-Defined Networking and Network Virtualization

Track Co-chairs

Flavio de Oliveira Silva	Federal University of Uberlândia, Brazil
Ashutosh Bhatia	Birla Institute of Technology and Science, Pilani, India
Alaa Allakany	Kyushu University, Japan

TPC Members

Yaokai Feng	Kyushu University, Japan
Chengming Li	Chinese Academy of Science (CAS), China
Othman Othman	An-Najah National University (ANNU), Palestine
Nor-masri Bin-sahri	University Technology of MARA, Malaysia
Sanouphab Phomkeona	National University of Laos, Laos
Haribabu K.	BITS Pilani, India
Shekhavat, Virendra	BITS Pilani, India
Makoto Ikeda	Fukuoka Institute of Technology, Japan
Farookh Hussain	University of Technology Sydney, Australia
Keita Matsuo	Fukuoka Institute of Technology, Japan

AINA-2021 Reviewers

Admir Barolli
Adrian Pekar
Ahmed Elmokashfi
Akihiro Fujihara
Akihiro Fujimoto
Akimitsu Kanzaki
Akio Koyama
Alaa Allakany
Alberto Volpe
Alex Ng
Alex Thomo
Alfredo Navarra
Aneta Poniszewska-Maranda
Angelo Gaeta
Anne Kayem
Antonio Loureiro
Apostolos Gkamas
Arjan Durresi
Ashfaq Ahmad
Ashutosh Bhatia
Asm Kayes
Baojiang Cui
Beniamino Di Martino
Bhed Bista
Carson Leung
Christos Bouras
Danda Rawat
Darshika Perera
David Taniar
Dimitris Apostolou
Dimosthenis Kyriazis
Eirini Eleni Tsiropoulou
Emmanouil Spanakis
Enver Ever
Eric Pardede
Ernst Gran
Eugen Dedu
Euripides Petrakis
Fadi Al-Turjman
Farhad Daneshgar
Farookh Hussain
Fatos Xhafa
Feilong Tang

Feroz Zahid
Flavio Silva
Flora Amato
Francesco Orciuoli
Francesco Piccialli
Gang Wang
Geir Horn
Giancarlo Fortino
Giorgos Vasiliadis
Giuseppe Fenza
Guangquan Xu
Hadil Abukwaik
Hideharu Amano
Hiroaki Kikuchi
Hiroshi Maeda
Hiroyuki Fujioka
Holger Karl
Hong Va Leong
Huey-Ing Liu
Hyunhee Park
Indika Kumara
Isaac Woungang
Jabed Chowdhury
Jana Nowaková
Jason Jung
Jawwad Shamsi
Jesús Escudero-Sahuquillo
Ji Zhang
Jiun-Long Huang
Jolanta Mizera-Pietraszko
Jordi Conesa
Jörg Domaschka
Jorge Sá Silva
Juggapong Natwichai
Jyoti Sahni
K Haribabu
Katerina Doka
Kazumasa Omote
Kazuto Sasai
Keita Matsuo
Kin Fun Li
Kiyotaka Fujisaki
Konstantinos Tsakos

Kyriakos Kritikos
Lei Shu
Leonard Barolli
Leonardo Mostarda
Libor Mesicek
Lidia Ogiela
Lin Hui
Ling-Jyh Chen
Lucian Prodan
Makoto Ikeda
Makoto Takizawa
Marek Ogiela
Mario Dantas
Markus Aleksy
Masakatsu Nishigaki
Masaki Kohana
Massimiliano Rak
Massimo Ficco
Michael Bauer
Minoru Uehara
Morteza Saberi
Nadeem Javaid
Naeem Janjua
Naohiro Hayashibara
Nicola Capuano
Nobuo Funabiki
Omar Hussain
Omid Ameri Sianaki
Paresh Saxena
Purav Shah
Qiang Duan
Quentin Jacquemart
Rajesh Pateriya
Ricardo Rodríguez Jorge
Ronaldo Mello
Rosario Culmone
Ryuji Shioya
Safdar Hussain Bouk

Salimur Choudhury
Salvatore Venticinque
Sanjay Dhurandher
Santi Caballé
Shahriar Badsha
Shigeru Fujita
Shigetomo Kimura
Sriram Chellappan
Stefan Covaci
Stefano Chessa
Stelios Sotiriadis
Stephane Maag
Takahiro Uchiya
Takamichi Saito
Tarique Anwar
Thitinan Tantidham
Thomas Dreibholz
Thomas Preuss
Tomoki Yoshihisa
Tomoya Enokido
Tomoyuki Ishida
Vamsi Paruchuri
Vasilis Samoladas
Vinesh Kumar
Virendra Shekhawat
Wang Xu An
Wei Ren
Wenny Rahayu
Wuyungerile Li
Xin-Mao Huang
Xing Zhou
Yaokai Feng
Yiannis Verginadis
Yoshihiro Okada
Yusuke Gotoh
Zahoor Khan
Zia Ullah
Zoubir Mammeri

AINA-2021 Keynote Talks

The Role of Artificial Intelligence in the Industry 4.0

Flora Amato

University of Naples "Federico II", Naples, Italy

Abstract. Artificial intelligence (AI) deals with the ability of machines to simulate human mental competences. The AI can effectively boost the manufacturing sector, changing the strategies used to implement and tune productive processes by exploiting information acquired at real time. Industry 4.0 integrates critical technologies of control and computing. In this talk is discussed the integration of knowledge representation, ontology modeling with deep learning technology with the aim of optimizing orchestration and dynamic management of resources. We review AI techniques used in Industry 4.0 and show an adaptable and extensible contextual model for creating context-aware computing infrastructures in Internet of Things (IoT). We also address deep learning techniques for optimizing manufacturing resources, assets management and dynamic scheduling. The application of this model ranges from small embedded devices to high-end service platforms. The presented deep learning techniques are designed to solve numerous critical challenges in industrial and IoT intelligence, such as application adaptation, interoperability, automatic code verification and generation of a device-specific intelligent interface.

Localization in 6G

Shahrokh Valaee

University of Toronto, Toronto, Canada

Abstract. The next generation of wireless systems will employ networking equipment mounted on mobile platforms, unmanned air vehicles (UAVs) and low-orbit satellites. As a result, the topology of the sixth-generation (6G) wireless technology will extend to the three-dimensional (3D) vertical networking. With its extended service, 6G will also give rise to new challenges which include the introduction of intelligent reflective surfaces (IRS), the mmWave spectrum, the employment of massive MIMO systems and the agility of networks. Along with the advancement in networking technology, the user devices are also evolving rapidly with the emergence of highly capable cellphones, smart IoT equipments and wearable devices. One of the key elements of 6G technology is the need for accurate positioning information. The accuracy of today's positioning systems is not acceptable for many applications of future, especially in smart environments. In this talk, we will discuss how positioning can be a key enabler of 6G and what challenges the next generation of localization technology will face when integrated within the new wireless networks.

Contents

Contents

RADAR - Regression Based Energy-Aware DAta Reduction in WSN: Application to Smart Grids

Bashar Chreim$^{(\boxtimes)}$, Jad Nassar, and Carol Habib

JUNIA, Lille, France
{Bashar.chreim,Jad.nassar,Carol.habib}@junia.com

Abstract. The evolution towards Smart Grids (SGs) represents an important opportunity for the energy industry. It is characterized by the integration of renewable and alternative energy resources into the existing power grids while ensuring a fine-grained control for the different measuring points. Therefore, this evolution requires the ability to send a maximum of data over the network in real time while controlling the grid. A Wireless Sensor Network (WSN) deployed across the grid is a potent solution to achieve this task. However, sensor nodes have limited energy and computation resources especially the battery powered ones. For that, reducing transmission is an essential priority in order to increase the lifetime of the network. Data prediction is a widely used, yet effective, solution in literature to accomplish this task. In this paper, we propose a Quality of Service (QoS) aware algorithm based on time series prediction and linear regression for data prediction in WSN. We test our approach in a SG context on real data traces of photo-voltaic cells. Our algorithm takes into consideration the diversity of applications of SGs with different requirements while being energy efficient. Our results show that our proposal provides satisfactory results compared to literature solutions in terms of data reduction percentage, Root Mean Square Error (RMSE) and energy consumption.

Keywords: Smart Grids · Wireless sensor networks · Data reduction · Data prediction · Linear regression · Linear correlation

1 Introduction

The evolution towards Smart Grids (SGs) represents an important opportunity to shift the energy industry into a new era of reliability, availability and efficiency [1]. This transformation will offer huge advantages for the stake holders by giving them a broader vision, management and control of the grid while lowering the costs, as well as to the customers, making their daily life more comfortable and convenient [2]. This evolution is manifested by (1) the integration of renewable energy resources all over the grid, (2) a two-way communication between the utility and the customers and (3) automated decisions of the smart

© The Author(s), under exclusive license to Springer Nature Switzerland AG 2021
L. Barolli et al. (Eds.): AINA 2021, LNNS 226, pp. 1–14, 2021.
https://doi.org/10.1007/978-3-030-75075-6_1

connected devices. Therefore, these changes require the ability to transmit in real time a maximum of data over the network, in order to monitor and control the different heterogeneous decentralized energy resources. A WSN deployed all over the grid on the different measuring and control points, is a potential and plausible solution to be used with SGs [1]. Moreover, in a SG, electricity and energy do exist, but connecting sensor nodes to such high voltage with intermittent and ill-adapted energy levels is sometimes inappropriate or physically impossible. For that, battery-powered sensor nodes must be deployed all over the grid, in order to ensure data transmission from source to destination node.

In a WSN, monitoring, processing and transmitting are the main tasks accomplished by sensor nodes. These sensors have limited energy and computation resources [3]. Each time interval, source nodes perform data sampling and transmission to the destination, via a set of sensor nodes distributed across the network. However, most of the time, sensed values do not change significantly between consecutive readings. This is all true to some SG applications as well (i.e., photo-voltaic cells monitoring). Sending these samples periodically will cause exhaustion of the batteries of sensor nodes (knowing that wireless communication is considered the major energy consumer [3]) and information redundancy at the destination. For that, and in order to maximize sensor nodes lifetime, data reduction has proven to be a potent solution. This is done by reducing the data transmission rate or aggregating data packets within the network.

In literature, data reduction techniques have been widely used for WSN applications [4]. However these techniques are limited to specific applications. This is mainly due to their parameters (e.g., filter length, step size) that need to be tuned to adapt to a particular data type (e.g., temperature, humidity). Therefore, these techniques need specific customization before being used in SG applications which are characterized by their heterogeneity in terms of QoS requirements and data types [5]. For the rest of the paper, we will refer to data type by variable.

In this paper, we propose *RADAR*, an energy efficient data reduction algorithm based on data prediction. It uses time series prediction and linear regression models. We note that in a previous work [6], we presented the concept of our algorithm. More precisely, our algorithm exploits correlations among different variables collected from photo-voltaic cells and creates automatically a prediction model for each variable. A time series prediction model is used to predict one of these variables, and the rest of them are predicted using linear regression models. To the best of our knowledge, our work is the first effort to apply linear regression in data reduction for WSNs combined with a time series prediction model.

RADAR is tested on real data traces obtained from photo-voltaic cells. Simulation results show that our approach provides satisfactory results compared to literature solutions in terms of energy efficiency and *RMSE*, while reducing data transmission in the wireless network.

The rest of the paper is organized as follows: Sect. 2 presents a summary of related work. Section 3 describes our proposed solution. Section 4 shows the simulation setup and environment used to validate our proposition. Section 5 describes the performance evaluation of our approach and remaining issues are discussed in Sect. 6. Finally, Sect. 7 concludes the paper.

2   Related Work

Data reduction techniques can be divided into three main categories [4]: data compression, In-network processing and data prediction (Fig. 1). The main idea behind data compression is that processing data consumes much less power than transmitting it [7]. For that, data is compressed/aggregated before leaving the source node using compression and aggregation techniques [7,8]. Unlike data compression, the process of aggregation in In-network processing is applied at intermediate nodes between the source and the destination [9–11]. Therefore, the amount of data is reduced over the network while traversing towards the destination. In data prediction, the amount of data transmitted by the sensor nodes is reduced by predicting the sensed values using specific models. Usually, the network maintain two instances of each prediction model, one residing at the source node and the other at the destination node, this is called Dual Prediction Scheme (*DPS*) [12]. It consists of running an instance of the model on both the source and destination nodes. The destination node will start answering queries using the values predicted by its model, without any communication with the source node. However, if the difference between the predicted and the sensed value is greater than a certain threshold, the source node will send its sensed value to the destination. In the current work, data prediction is the point of our interest.

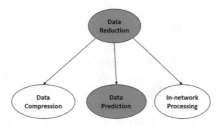

Fig. 1. Classification of energy saving in sensor networks

In literature, numerous studies focused on developing data prediction techniques in WSNs. In [13], the authors used Auto Regressive Integrated Moving Average (*ARIMA*) model for the information collection scheme. In their approach, the sink node uses historical data to build up an appropriate time series prediction model for each sensor node and send back its parameters to build the same model. In [14], the authors proposed a prediction scheme using Seasonal

ARIMA (*SARIMA*) model for short term prediction that can predict using only limited input data. This model is used to predict traffic flow. The main drawback with these methods is their requirement of high memory and computational overhead to initially build the model and to re-compute it when outdated. In [15], Least Mean Square (*LMS*) algorithm is used for data prediction in WSN. It consists of running two instances of the model at the sensor and the sink node, applying dual prediction scheme (*DPS*). The main complexity when using *LMS* is the task of choosing the best parameters to fit to a specific data type. In [16], the authors proposed a modification for the *LMS* algorithm by adding a phase of initialization and parameters determination. They varied the algorithm parameters within specific intervals, and selected those that minimize the *RMSE*. All described techniques have been successfully used in WSN applications, but it is important to note that for each type of application, the parameters must be computed. In addition, for the same variable, parameters may not fit for all different QoS necessities. For example, for different temperature based applications (e.g., solar power forecasting, solar irradiance prediction), the determined model parameters (e.g., filter length, step size) may not fit to all these applications. In fact, this is mainly due to having several difference thresholds between sensed and predicted values based on the application needs [16].

Moreover, other work focused on linear regression for data prediction [17–19]. In [18], the authors used multiple linear regression (*MLR*) to predict solar intensity from a set of weather metrics. In [19], the authors determined the most influential features for predicting photo-voltaic production, using linear regression. However, in these work, all input variables need to be available all the time in order to predict a single output. In other words, sensor nodes that are responsible of collecting the value of these variables need to execute the transmission task all the time, which may exhaust the batteries of the sensor nodes after a period of time.

3   Proposed Solution

The purpose of our contribution is to create an autonomous prediction algorithm (*RADAR*) for heterogeneous applications (e.g., photo-voltaic cells monitoring, electric vehicles control) in WSN. It generates simultaneous prediction models for all variables by exploiting linear correlation among them. This is done using two types of models: *time series prediction* and *linear regression*. These models provide satisfactory and accurate results while being simple and lightweight which is the most beneficial in energy limited WSNs.

The idea is to predict the first selected variable (from a set of pre-identified variables) using a time series prediction model. Here, the prediction algorithm could be any time series model from the state of the art. In our case we used *SARIMA* as the benchmark model. After that, this variable will allow us to predict the remaining ones using linear regression models (simple linear regression (*SLR*) for the first iteration followed by *MLR* for the remainders). All of that is based on the linear correlation between every couple of variables. As already mentioned, by variable we mean the different data types (e.g., temperature, humidity) that we used in our simulation. In the rest of this section, we will detail the steps of *RADAR* that is represented in Algorithm 1. We will consider an example of variables from Table 1 to support our explanation.

Table 1. Correlation matrix showing correlations between variables

	Temperature	Humidity	Irradiance	Current BP
Temperature	1	−0.8154	+0.7748	+0.6791
Humidity	−0.8154	1	−0.7906	−0.6673
Irradiance	+0.7748	−0.7906	1	+0.7464
Current BP	+0.6791	+0.6673	+0.7464	1

- Our algorithm takes as input a dataset with a random number of variables (e.g., temperature, humidity).
- In the first step, the correlation matrix (*CM*) is built (line 2 in Algorithm 1). It represents the correlation coefficient of each couple of variables (Table 1).
- After that, the negative values are replaced by their absolute ones in the *CM* (since the positive and negative values have the same impact on the correlation coefficients).
- In the next step, the maximum value (*Max*) in the matrix must be identified (line 8 in Algorithm 1). It represents the correlation coefficient of the most correlated couple of variables (e.g., the temperature and the humidity in Table 1 with the value −0.8154).
- A time series prediction model (Model 1 in Table 2) is created using *SARIMA* [14] for one variable of the couple, randomly selected (the temperature per example in this case). This variable will be predicted at the destination based on the *SARIMA* model.
- The output of this time series model is used as input to create an *SLR* model (line 11 in Algorithm 1), using the formula presented in [20]. It predicts the second variable previously identified. Since we chose the temperature as a random variable in the previous step, the humidity will then be the output of the current model (Model 2 in Table 2).
- The next model (Model 3) should be created now by identifying the next maximum value in the matrix (in Table 1). It is an *MLR* model, and takes as input the outputs of the previous created models (line 17 in Algorithm 1).
- This final step is repeated until prediction models are created for all variables (Model 4 in our example is the last model). Our algorithm returns a matrix that represents the input/s and output variable/s of each created model.
- Once the execution of the algorithm finishes, all the prediction models are created. The prediction process can now take place.

Algorithm 1: Data prediction algorithm

Require: Dataset
Ensure: Prediction Models
 1: $Count \leftarrow NbOfVariables(dataset)$ //Number of needed models
 2: $CM \leftarrow BuildCorrelationMatrix()$
 3: **for** each value in CM **do**
 4:     **if** $value < 0$ **then**
 5:         $value \leftarrow |value|$
 6:     **end if**
 7: **end for**
 8: $Max \leftarrow CM.MaximumValue()$
 9: $X, Y \leftarrow Max.IndexInCM()$
10: $SARIMA(X)$
11: $SLR(X, Y)$
12: $Count \leftarrow Count - 2$
13: $Max \leftarrow 0$ //Replace maximum by zero in the matrix
14: **repeat**
15:     $Max \leftarrow CM.MaximumValue()$
16:     $X, Y \leftarrow Max.IndexInCM()$
17:     $MLR(X \text{ and all previous outputs}, Y)$
18:     $Max \leftarrow 0$
19:     $Count \leftarrow Count - 1$
20: **until** $Count == 0$ //Each variable has its prediction model

---

In a nutshell, *RADAR* works as follows: a time series prediction model will predict the upcoming value of one of the variables of the applications. Next, the predicted value is used as input by an *SLR* model to predict the value of the second variable. Then, *MLR* models are executed simultaneously by taking the predicted values to predict the value of the next corresponding variable, and so on. *DPS* is applied by implementing the algorithm on both the source and the destination node in a WSN. It uses the predicted values from the models to answer queries, without any communication with the sensor node. The source node continue sensing the data and predicting the values without sending them over the network. The sensed value is sent to the destination only when there is a difference between this value and the predicted value, that is greater than a certain threshold. Indeed, the stronger the correlation among variables, the higher the accuracy of the predicted values.

4   Simulation Setup

In order to evaluate our proposition and compare it to existing approaches, we use real sensor values from NREL National Wind Technology Center [21]. We consider the temperature, humidity, global horizontal irradiance[1] and current black photon[2] between 01/06/2019 and 31/07/2019 with a fixed sampling rate of one sample every minute. For each variable, we consider five different thresholds. We note that these thresholds were chosen randomly and can be adjusted for specific QoS needs.

Our approach is compared to *LMS_MOD* [16] and *SARIMA* [13]. These algorithms have different parameters that must be computed. For that, we applied the same methodologies used in the corresponding papers for the tuning process. We would like to point out that the tuning process is repeated five times for each variable based on the different thresholds.

We consider a one hop communication environment with no loss in order to prove the efficiency of our proposal in an optimal case scenario. Simulations were conducted using Python language, and executed in a Raspbian environment, installed on a Raspberry Pi 3. It has a 1.4 GHz Arm Cortex-A53 quad-core CPU and 1GB of RAM. The main idea of using Raspberry Pi for simulation is to be able to measure the power consumed by the algorithm (the process of measurement will be detailed in Sect. 5.3).

4.1 Data Preparation

Our proposed solution is based on linear regression. This method uses least squares method in order to calculate its best parameters. However, this method is very sensitive to outliers [22]. For that, before running our algorithm, we cleaned up our dataset by eliminating abnormal observations (i.e., negative values, erroneous values). We note that the dataset is then splitted between training and testing (75% and 25% respectively).

4.2 Building Models

Our algorithm presented in Sect. 3 is executed taking as input the pre-processed dataset. Table 1 shows the correlation coefficients for the existing variables in the dataset using Pearson correlation coefficient. Based on the values of this matrix, the algorithm creates a prediction model for each variable. Table 2 shows the output of the algorithm. Each row of this table represents a prediction model for a specific variable. The first model is a time series prediction model for temperature. The second one uses an *SLR* model, and takes as input the temperature and predicts the humidity. The third and fourth models use the *MLR*, and take as inputs and output the variables denoted by "Input" and "Output" simultaneously in the table.

---

[1] Global horizontal irradiance represents the total solar radiation incident on a horizontal surface.

[2] Current black photon represents the current generated by photo-voltaic cells.

Table 2. Models structure table showing input/s and output for each model created

	Temperature	Humidity	Irradiance	Current BP
Prediction model 1	Output	–	–	–
Prediction model 2	Input	Output	–	–
Prediction model 3	Input	Input	Output	–
Prediction model 4	Input	Input	Input	Output

5   Performance Evaluation

The performance of our algorithm is evaluated using different metrics: Data reduction percentage, *RMSE* and energy consumption. Each of these will be discussed in the following subsections. We note that the temperature variable is not mentioned below in the metrics, since it is predicted in *RADAR* using *SARIMA* model as already detailed in Sect. 3 (It is the first variable selected in *RADAR* steps).

5.1   Data Reduction Percentage

Data reduction percentage corresponds to the number of packets whose predicted values fall within the range of the chosen threshold, and therefore not transmitted to the destination node. Figures 2, 3 and 4 show the data reduction percentage achieved for our approach (*RADAR*), *LMS_MOD* and *SARIMA*. We can see that *RADAR* presents higher reduction percentage than *LMS_MOD* for the humidity and current black photon. Between 5 and 14% for humidity and between 23 and 34% for current black photon. Concerning the global horizontal irradiance, the data reduction percentage is close between *RADAR* and *LMS_MOD* with a slight improvement for our proposal. This is mainly due to the parameters chosen for each variable in *LMS_MOD*, and more precisely the size of the filter[3] that may impact the data reduction percentage when the filter size is bigger. However, the percentage of data reduction obtained with *SARIMA* is higher than our proposition, and that because its forecasts are based on previous values, that make the predicted values close to the real ones. While in *RADAR*, the prediction is based on regression models which may affect the data reduction percentage in some cases. However, the gain achieved in *SARIMA* in terms of data reduction percentage comes at the cost of a higher error rate and a more consequent energy consumption percentage, which we will detail in the following subsections.

---

[3] It indicates the number of packets to be transmitted to the destination node, when the model is in the training phase or outdated.

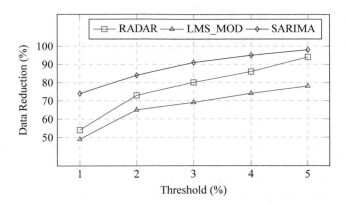

Fig. 2. Data reduction for humidity

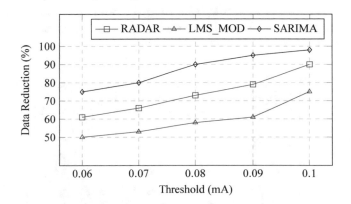

Fig. 3. Data reduction for current black photon

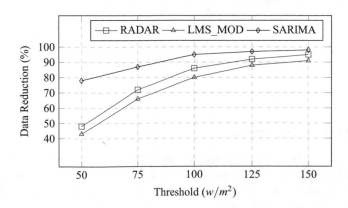

Fig. 4. Data reduction for global horizontal irradiance

5.2   Root Mean Square Error

In order to identify how close the predicted results are from the real ones, we compute the *RMSE*. It is calculated using the following formula:

$$RMSE = \sqrt{\frac{1}{n}\sum_{i=1}^{n}(y[n] - u[n])^2} \tag{1}$$

Where n is the number of samples, u and y are two vectors representing the set of real and predicted values respectively.

Figures 5, 6 and 7 show the *RMSE* for the three methods, and for the different variables (temperature, irradiance and current black photon). We observe that for the humidity, global horizontal irradiance and current black photon, our proposition has a lower *RMSE* than *LMS_MOD* and *SARIMA*. This is mainly due to the cumulative error that appear in time series models, because of the dependency of a predicted value from the past ones. However, in our model, each value is independent from others, and a prediction error does not affect the upcoming predictions.

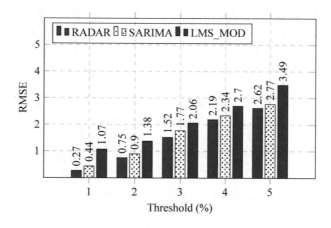

Fig. 5. RMSE for humidity

5.3   Energy Consumption

In order to estimate the energy consumed by the Raspberry Pi to execute each algorithm, we used a USB power and energy meter module. It measures the energy consumption during a specific period of time. It is connected to the Raspberry Pi via the USB port.

First, we measure the energy consumption for a period of five minutes, where no tasks are carried out by the Raspberry Pi. Afterwards, the same test for five

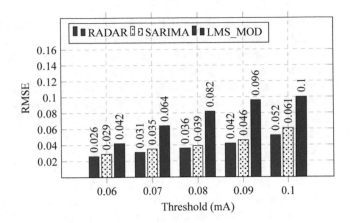

Fig. 6. RMSE for current BlackPhoton

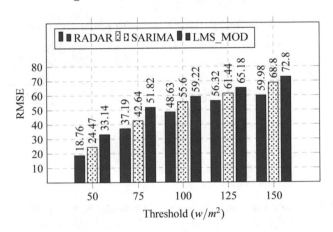

Fig. 7. RMSE for current global horizontal irradiance

minutes is repeated on the Raspberry for the different algorithms respectively (*RADAR*, *SARIMA* and *LMS_MOD*). The energy consumed by the algorithm is computed using the formula below:

$$E\_Consumed = E\_Algorithm - E\_Idle \qquad (2)$$

Where *E_Consumed* is the net energy while executing the algorithm, *E_Algorithm* and *E_Idle* represent the energy consumed by the Raspberry Pi when it's executing an algorithm and when it's not, respectively.

Table 3 shows the energy consumption and the processing time needed by our algorithm (*RADAR*), *LMS_MOD*, and *SARIMA*, to predict the set of variables. We notice that *RADAR* consumes less energy than *LMS_MOD* (10.8 J for *RADAR* vs 43.2 J for *LMS_MOD*), this is because *LMS_MOD* includes the previous values in the prediction phase, and thereby increases the time required for the prediction task which will consume more energy. Moreover, *RADAR* presents a huge improvement in terms of energy reduction compared to *SARIMA* (10.8 J for *RADAR* vs 579.6 J for *SARIMA*). This is mainly due to *SARIMA* training and testing phase that require recreating instances of the model so often which will result in a high energy consumption [13]. We would like to point out that in order to predict a set of variables, each one of these requires an instance of a prediction model. Therefore, to predict X variables based on *SARIMA* [14] or *LMS_MOD* [16] approaches, X time series instances of these models must be created. In *RADAR*, the prediction of X variables requires the use of only one instance of the time series model. In other words, in this example, *SARIMA* and *LMS_MOD* use 4 instances each, while *RADAR* uses only one instance of *SARIMA*, thus reducing the energy consumption significantly. The processing time is also more advantageous for *RADAR* compared to *SARIMA* and *LMS_MOD* for the same reasons stated above.

Table 3. Energy consumption and processing time

	RADAR	LMS_MOD	SARIMA
Energy consumption (Joules)	10.8	43.2	579.6
Processing time (Milliseconds)	27.1	29.8	153790

6   Discussion

Before coming to our conclusions, we discuss some relevant issues in our approach. Despite having shown efficiency for different variables, by reducing *RMSE*, energy consumption and ensuring a high reduction percentage, some minor potential issues should be highlighted. Knowing that our approach is based on linear regression models, strong relationships/correlations between variables are required to obtain accurate predictions. For that, low relationships can affect negatively the performance of our algorithm. In this case, a correlation threshold should then be fixed in order to apply our algorithm. Furthermore, our simulations were conducted on Python considering a no loss scenario of packets over the network. In a real sensor network our model could rise a reliability issue; in a real WSN with interference and losses, if the message containing the reading message transmitted by the sensor is lost, the model at the destination will go apart and the algorithm will predict erroneous values. This should be carefully handled by sending regular control messages per example in order to maintain the synchronization between the sink and the sensor nodes. This issue was not addressed in this work since its main focus was to test the performance of the algorithm in the best cast scenario.

7   Conclusion and Future Work

In this paper, we presented a data prediction correlation based approach, that automatically generates prediction models for different heterogeneous variables. The main advantage of our approach is the ability to adapt to different applications with different requirements as per a SG environment while being energy efficient (if the correlation requirements are satisfied). We tested our approach with real data traces for photo-voltaic cells and performed simulations considering one hop communication networks. *RADAR* provides satisfactory results compared to primary literature solutions. It shows a better performance compared to *LMS_MOD* in terms of *RMSE*, data reduction percentage and energy consumption. Compared to *SARIMA*, it offers a lower error percentage, while consuming much less energy. As future work, we will continue testing our approach on different applications. Later on, we will implement our algorithm in a real WSN scenario to evaluate its performance.

Acknowledgments. This work was partially funded by a grant from the SoMel SoConnected project. This project involves the MEL (Métropole Européenne de Lille), Enedis, EDF, Dalkia, Intent, the Lille Economie-Management Laboratory, HEI - Yncréa HdF and the Faculties of the Catholic University of Lille.

References

1. Nassar, J.: Ubiquitous networks for smart grids. Ph.D. dissertation, Universite des Sciences et Technologies de Lille (2018)
2. Ilic, D., Da Silva, P.G., Karnouskos, S., Griesemer, M.: An energy market for trading electricity in smart grid neighbourhoods. In: 6th IEEE International Conference on Digital Ecosystems and Technologies (DEST), pp. 1–6. IEEE (2012)
3. Raza, U., Camerra, A., Murphy, A.L., Palpanas, T., Picco, G.P.: What does model-driven data acquisition really achieve in wireless sensor networks? In: IEEE International Conference on Pervasive Computing and Communications, pp. 85–94. IEEE (2012)
4. Anastasi, G., Conti, M., Di Francesco, M., Passarella, A.: Energy conservation in wireless sensor networks: a survey. Ad Hoc Netw. **7**(3), 537–568 (2009)
5. Nassar, J., Berthomé, M., Dubrulle, J., Gouvy, N., Mitton, N., Quoitin, B.: Multiple instances QoS routing in RPL: application to smart grids. Sensors **18**(8), 2472 (2018)
6. Chreim, B., Nassar, J., Habib, C.: Regression-based data reduction algorithm for smart grids. In: IEEE 18th Annual Consumer Communications & Networking Conference (CCNC) (2021)
7. Kimura, N., Latifi, S.: A survey on data compression in wireless sensor networks. In: International Conference on Information Technology: Coding and Computing (ITCC 2005), vol. 2, pp. 8–13. IEEE (2005)
8. Grabisch, M., Marichal, J.-L., Mesiar, R., Pap, E.: Aggregation functions: means. Inf. Sci. **181**(1), 1–22 (2011)
9. Skordylis, A., Guitton, A., Trigoni, N.: Correlation-based data dissemination in traffic monitoring sensor networks. In: Proceedings of the 2006 ACM CoNEXT Conference, pp. 1–2 (2006)

10. Matos, T.B., Brayner, A., Maia, J.E.B.: Towards in-network data prediction in wireless sensor networks. In: Proceedings of the 2010 ACM Symposium on Applied Computing, pp. 592–596 (2010)
11. Bahi, J.M., Makhoul, A., Medlej, M.: An optimized in-network aggregation scheme for data collection in periodic sensor networks. In: International Conference on Ad-Hoc Networks and Wireless, pp. 153–166. Springer (2012)
12. Dias, G.M., Bellalta, B., Oechsner, S.: The impact of dual prediction schemes on the reduction of the number of transmissions in sensor networks. Comput. Commun. **112**, 58–72 (2017)
13. Liu, C., Wu, K., Tsao, M.: Energy efficient information collection with the ARIMA model in wireless sensor networks. In: IEEE Global Telecommunications Conference, GLOBECOM 2005, vol. 5, p. 5–pp. IEEE (2005)
14. Kumar, S.V., Vanajakshi, L.: Short-term traffic flow prediction using seasonal ARIMA model with limited input data. Eur. Transp. Res. Rev. **7**(3), 21 (2015)
15. Santini, S., Romer, K.: An adaptive strategy for quality-based data reduction in wireless sensor networks. In: Proceedings of the 3rd International Conference on Networked Sensing Systems (INSS 2006), pp. 29–36. TRF Chicago, IL (2006)
16. Nassar, J., Miranda, K., Gouvy, N., Mitton, N.: Heterogeneous data reduction in wsn: Application to smart grids. In: Proceedings of the 4th ACM MobiHoc Workshop on Experiences with the Design and Implementation of Smart Objects, pp. 1–6 (2018)
17. de Carvalho, C.G.N., Gomes, D.G., de Souza, J.N., Agoulmine, N.: Multiple linear regression to improve prediction accuracy in WSN data reduction. In: 7th Latin American Network Operations and Management Symposium, pp. 1–8. IEEE (2011)
18. Sharma, N., Sharma, P., Irwin, D., Shenoy, P.: Predicting solar generation from weather forecasts using machine learning. In: IEE International Conference on Smart Grid Communications (SmartGridComm), pp. 528–533. IEEE (2011)
19. Kumar, S.: Solar energy prediction using machine learning. Technical report (2015)
20. Zou, K.H., Tuncali, K., Silverman, S.G.: Correlation and simple linear regression. Radiology **227**(3), 617–628 (2003)
21. Jager, D., Andreas, A.: NREL national wind technology center (NWTC): M2 tower; boulder, colorado (data). National Renewable Energy Lab.(NREL), Golden, CO (United States), Technical report DA-5500-56489 (1996). http://dx.doi.org/10.5439/1052222
22. Hope, T.M.: Linear regression. In: Machine Learning, pp. 67–81. Elsevier (2020)

A Probabilistic Preamble Sampling Anycast Protocol for Low-Power IoT

Tales Heimfarth[1(✉)] and João Carlos Giacomin[2]

[1] Department of Applied Computing, Universidade Federal de Lavras, Lavras, Brazil
tales@ufla.br
[2] Department of Computer Science, Universidade Federal de Lavras, Lavras, Brazil
giacomin@ufla.br

Abstract. This paper presents AGAp-MAC, an asynchronous anycast protocol for low rate IoT, which waives the assurance of a quick communication establishment in favor of energy reduction. Developing multi-hop communication protocols with high scalability and low energy consumption is a challenging task in IoT. AGAp-MAC employs anycast communication pattern to reduce the latency due to sleep-delay, while duty-cycling is the main strategy to reduce energy consumption. Preamble sampling technique is used to assure communication between nodes. A series of strobed preambles is sent by a transmitter node until an answer is received from a forwarder candidate. AGAp-MAC is an improvement over AGA-MAC, our previous work. It achieves energy reduction using short awake periods, with a certain probability of hearing the preamble. This protocol was compared with others from literature using a network simulator. Energy consumption was reduced in at least 50% in comparison to the other tested protocols, maintaining compatible latency.

1 Introduction

Over the past 20 years, the availability of low-cost, low-power electronic devices has grown a large data communication market, making information from around the world easily accessible to people. At the beginning of the 21th century, smart-phones have expanded the ability to exchange information for all types of spoken and written communication. More broadly, the Internet of Things (IoT) technology involves the use of autonomous electronic devices with the ability to interact with the environment, collecting and sending data to a processing center. This network connects everything to the Internet, with the purpose of extracting information [2, 7].

A Low-Rate Wireless Personal Area Network (LR-WPAN) provides communication services on networks covering restricted areas. This kind of IoT is denominated micro-IoT [5], they employ small and low-power devices with short-range communication capabilities and low data rate, as those specified in IEEE 802.15.4 [1] reference model. Wireless Sensor Network (WSN) is a LR-WPAN widely used in environment monitoring. Its main challenge is to assure network operation for long periods, since the sensor nodes have limited energy resources,

L. Barolli et al. (Eds.): AINA 2021, LNNS 226, pp. 15–27, 2021.
https://doi.org/10.1007/978-3-030-75075-6_2

usually small batteries [8]. The communication is made by multi hop with low-power radios.

Medium Access Control (MAC) protocols are responsible for ensuring exclusive access to shared media at any given time. Another function of MAC protocols in LR-WPANs is to control the periods of radio operation. In order to save energy, the nodes employ duty-cycle [12]. The sensor nodes are kept in a low-power state, with radios off, most of the time. They are periodically reactivated to perform measurements, processing and communication. This method can extend network life-time for more than a year, however the sleep-delay problem arises as an untoward consequence. When a sensor node has a message to send through the network, it must wait for the active period of a forwarder, in order to establish communication. Latency reduction in duty-cycled WSNs has been a challenging task for many years.

Some works dedicated to classify the large number of MAC protocols proposed for WSNs were presented in the last decade. Kumar *et al.* [14] separate these MAC protocols in *contention-based* and *scheduled-based*. Contention-based are the most suitable protocols for WSNs, since they are more scalable and more flexible to topology changes, and they do not require centralized control. Most contention-based MACs are asynchronous, which use preamble sampling to establish communication between the sender and receiver nodes [3]. Before data packet the sender transmits a long preamble to be heard by the receiver when it enters the active period.

Asynchronous MAC protocols can mitigate the sleep-delay, exploring the redundancy of paths available on the network, if an anycast communication pattern is employed [6]. In this method, a group of candidates for the next hop, denominated *Forwarding Candidate Set* (FCS), is defined. The elements of this set are selected according to some routing metrics, defined in the network layer. The sender node transmits a series of short preambles in order to advise the FCS members its intention to communicate. The MAC protocol manages the preambles in a strobed manner, interleaving them with small hearing intervals, in order for the sender to wait for an answer. The final decision about the next forwarder is taken in the MAC layer, according to some conditions observed at the moment [11]. This procedure reduces the latency caused by the lack of synchronism between the nodes. The use of strobed preambles shortens the time needed to establish communication with the next relay node, when compared to protocols that send only one large preamble, like B-MAC [17]. However, more energy is spent in channel sampling, since sampling duration must be larger than the small listening interval of the sender node [12].

In this work, we propose AGAp-MAC (Adaptive Geographic Anycast Probabilistic MAC), an asynchronous protocol for wireless sensor networks, which uses strobed preambles to search for a forwarder node, but with a small period for channel sampling. All the nodes employ duty-cycles to save energy, but differently from previous works, the sampling interval is smaller than that separating two consecutive short preambles. There exists a probability of a FCS member to sample the channel during this interval, missing the opportunity to receive the message from the sending node. However we argue that a larger number of members in FSC gives a greater chance of establishing communication. In this way, the increment in end-to-end latency can be compensated with an enlargement in FCS. A mechanism is provided by AGAp-MAC to ensure end-to-end communication.

2   Related Work

In low-power IoT's, as the WSNs, a message sent by a source (S) node addressed to a destination (D) node propagates through the network by multi-hop due to the limited radio communication range. Several intermediate nodes, denominated forwarders (or relays), receive and forward data packets, establishing a route from S to D.

Medium access and routing protocols receive great attention in multi-hop communication development, since they are the main responsible for energy efficiency of sensor nodes and network life-time enlargement. MAC protocols coordinate access to the wireless medium maintaining very low energy consumption [3, 14]. On its turn, the main objective of network protocols is to reduce end-to-end latency [9]. A source node, with a message to be sent, must select one of its neighbors to relay the message ahead, considering as neighbor any node in its radio range. As stated in the Sect. 1, contention-based protocols are best suited to low-power IoT's and WSNs due to their scalability and flexibility to topology changes, and their decentralized topology control. Anycast communication pattern is employed in asynchronous contention-based protocol in order to mitigate the sleep-delay.

The GeRaF (Geographic Random Forwarding) protocol [19] was a pioneering asynchronous protocol in the use of anycast communication. GeRaF employs a Geographic Routing method to select the FCS's member. In order to take part in FCS, the neighbor has to shorten the distance of the message relative to destination. GeRaF divides the FCS members in groups of priority, those with larger advances are given more priority in answering the sender and becoming the next forwarder. Advance is the shortening in the distance to node D given by a FCS member.

CMAC [16] protocol is based on the same principles as GeRaF. It defines a minimum advance (r_0) towards the *sink* node as a criterion for selecting FCS members. The sender's neighbors must provide an advance to the destination greater than r_0 to take part in the FCS. r_0 depends on traffic load and network density. The first FCS member to wake up responds to the sender node and becomes the next message forwarder. CMAC proposes to reduce even more energy consumption employing dual preamble sampling. Two very short medium sampling are executed by the nodes in order to verify the presence of a short preamble. The interval between both the samples must be larger than the listen period (when the sender node waits for an answer) and smaller than a preamble transmitting time, in order to assure that one of the samples will sense the preamble signal. However, the switching times of the radios limits the efficiency of this method [3], since the interval between both the samples must accommodate the radio resting time and also its warm-up time.

AGA-MAC [10] follows the same principle as CMAC, imposing a minimum advance to the destination in each hop. In AGA-MAC, this threshold value (τ) depends on the size of the data message. For a short data packet τ is set to a small value in order to reduce the time spent to find a forwarder node. If data payload is long, τ is enlarged in order to reduce the hop count until destination. The authors demonstrated that for each data length there is an optimum threshold value that provides the smallest end-to-end latency in the network. This was pioneering work considering the length of the data to make decisions about routing in anycast protocols.

The AGAp-MAC, proposed here, improves the energy performance over the AGA-MAC [10], while it do not make a long preamble sampling to assure communication between sender and receiver node. With a short channel sampling, as that used in Low Power Listening (LPL) [17], rendezvous will be achieved with a certain probability. This probability is a function of the number of members in FCS, the preamble duration and the listening interval between preambles.

3 AGAp-MAC Protocol

AGAp-MAC is a cross-layer duty-cycled protocol for WSNs, which includes functionalities of the network (NWT) and medium access control (MAC) layers. Sensor nodes use preamble sampling to establish communication. Periodically each node wakes up, turns its radio on and checks whether the communication channel is busy, indicating that a neighboring node intends to send a message. It is an asynchronous protocol, since the active schedule of the nodes are independent, even if they use the same cycle time (the interval between two consecutive active periods). A RTS/CTS handshake is employed to achieve rendezvous between sender and receiver node. When a node is willing to send a data packet, it starts checking the channel for ongoing communications, in order to avoid collisions. If the channel is free, the node sends a series of RTS (Request To Send) packets, acting as a preamble, with a small hearing interval between two consecutive RTSs, waiting for a possible ongoing CTS (Clear-To-Send) packet, which is used by the receiver node to inform it is awake and ready to receive data.

Conventional duty-cycled protocols based on channel probing, commonly use a large sampling period in order to assure the reception of a preamble if a neighboring node is initiating a transmission. Differently from other state-of-the-art protocols based on preamble sampling, AGAp-MAC does carrier sense for a very short period of time, with a certain probability of hearing a RTS signal. This lack of assurance in receiving a RTS is compensated by its anycast communication pattern.

A problem that arises from duty-cycling is the sleep-delay. Every time a node has a message to send, it must wait until the receiver enters its active period to receive a data message, increasing latency. In order to reduce this problem, anycast communication pattern is employed, as stated in Sect. 1. Suppose that

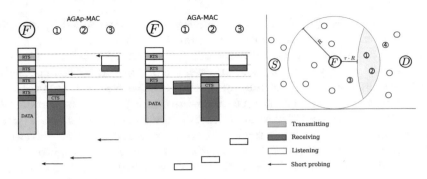

Fig. 1. Example of FCS formation and communication timings for AGAp-MAC and AGA-MAC

a forwarder node F has a message to send towards the destination node D, as depicted in Fig. 1. Before initiating transmission, the sender probes the channel during a contention window period and thereafter starts the series of RTS. The *Forwarding Candidate Set* (FCS) is a group of potential relays for the message which members are selected according to their proximity to D. In Fig. 1, nodes 1, 2 and 3 fulfill this premise. Node 4 is already out of range, being not included in the FCS. In our protocol, however, we have an additional requirement: nodes must bring a minimum advance which is given by $\tau \cdot R$, where $\tau \in [0, 1]$ is called threshold and R is the radio range. In our picture, only nodes 1 and 2 meet the τ requirement, composing the FCS.

When a FCS member wakes up and detects a RTS, it sends back a CTS packet signaling to the sender it is ready to receive data packet. A non-FCS member detecting a RTS just turns its radio off to save energy. After receiving a CTS, the sender node transmits the data packet to the elected forwarder. Using a set of nodes as possible next hop may reduce the advance towards the destination since the selected node may not bring the largest advance. However, a well selected FCS cardinality gives a substantial gain in time, balancing latency enlargement due to shorter hops.

All the nodes are aware of their positions. Decisions to route are made locally, following the principles of geographical routing. Sender node informs in the preamble the eligibility parameters to take part in FCS and neighboring nodes decide by themselves. These parameters are sender and destination positions and the minimum advance calculated by $\tau \cdot R$, as already stated. Adequate value to these parameters are used to determine the cardinality of the FCS. The first member to wake up and answer to the sender becomes the next forwarder. The more members are in the FCS, the sooner a rendezvous between the sender and a relay node is achieved. However, large FCS cardinalities lead to increasing number of hops to destination (since each hop brings, in average, lower advancement). Therefore, for a given data packet size, a different FCS cardinality will be optimal in the sense of minimum latency. Given a certain network density, it is possible to drive the average FCS cardinality by means of the threshold parameter. The best threshold for each packet size is determined by empirical experiments as presented in the Experimental Results section.

Before presenting a mechanism introduced in our protocol to reduce the global network energy consumption, an example of AGA-MAC communication will be presented. Figure 1 presents two possible communications: the traditional mechanism, employed in, for example, AGA-MAC and also the one introduced in this article. In the AGA-MAC communication, forwarder node F starts to send a series of RTS packets. Node 3 is the first to wake up and hear it. Since it does not fulfill the minimum desired advance (represented as $\tau \cdot R$ in the figure), it does not respond and goes back to sleep. Node 2 is the next to wake up, responding promptly with a CTS and receiving the data packet. Node 1 wakes up later, it does not receive RTS, and returns to sleep mode. The process is started again with node 2 acting as the new sender, until the packet is transported from source S to destination D.

It is important to remark here that, for most traditional asynchronous low power listening protocols based on preambles, when a node wakes up, it must probe the communication channel for a relatively large interval (carrier sense), to be sure of not missing the preamble. In fact, the hearing time must be greater than the slot reserved for the CTS packet by the sender. This is demonstrated for the AGA-MAC in Fig. 1. In opposition to this fact, AGAp-MAC employs a very short (minimum) carrier sense in order to save energy of all network nodes during duty-cycle, in the same way as Low-Power-Listening (LPL) process, as described in [17] for B-MAC protocol. Although very efficient at reducing energy consumption, this has a drawback: if the short carrier sense is carried out during a CTS time (while no RTS is being transmitted), the awaken node can not detect the presence of the sender and returns to sleep missing the chance of forwarding the message. Each node of the FCS will be able to hear the preamble with a given probability. The redundancy inside the FCS is used to increase the probability that some node will respond to the sender. The mechanism will increase the average response time, leading to a larger latency when compared with AGA-MAC. However the advantage here is the energy savings, since all network nodes are periodically doing carrier sense.

Figure 1 presents also an example of AGAp-MAC operation. Node 3 wakes up, does its short carrier sense and notices the presence of communication, staying active until receiving a complete RTS. Since it is not allowed to take part in FCS, it returns to sleep mode. After this, node 2 wakes up and performs its channel sampling. Unfortunately, this procedure is realized during a CTS waiting period. Differently from AGA-MAC, nothing is heard and the node returns to sleep. In the next moment, node 1 wakes up and performs its carrier sense, detecting traffic and receiving a RTS. After the contention period (checking channel for ongoing communications), it answers with CTS, signaling it is ready to receive the data message.

There is a probability of ending the RTS without any answer (all nodes in the FCS wake up at CTS waiting period or there is no node in the FCS), our protocol has different phases to assure communication. After sending RTS for a complete cycle time, the protocol goes to phase 2. In this phase, the sender reduces the threshold to $\tau = 0$ (accepting any node which brings advance towards destination) and sends the RTS sequence again. If no answer is received at this phase, phase 3 is started. It consists of an additional cycle of preambles. Between the phases, a time displacement equivalent to a RTS sending interval is introduced. The objective here is to break off the correlation between RTS times of consecutive phases. If a node probes the channel during a CTS time in one phase, it will hear the RTS in the next phase.

4   Experimental Results

In this section, we present the evaluation of our protocol in terms of energy consumption and latency. In order to assess the performance of our protocol, the Grubix Wireless Network Simulator, an extension of ShoX [15], was used. The simulator incorporates the protocol stack for an IEEE 802.15.4 [1] compliant radio. Our protocol was compared with the others well-known AGA-MAC [10], GeRaF [19] and X-MAC [4]. GeRaF is also an asynchronous anycast protocol whereas X-MAC is an standard asynchronous protocol which, as the others, uses preamble sampling to achieve rendezvous with the next relay node. AGA-MAC is our previous work, and it has good latency performance when compared with GeRaF and X-MAC. Nevertheless, none of these protocols employs the short channel sampling as AGAp-MAC.

Setup

For our simulations, we adopted a radio model based on the IEEE 802.15.4, bit rate of 250 kbps, fixed transmission power, bidirectional links and Free Space propagation model for isotropic point source in an ideal medium. Sensor nodes were deployed randomly in the simulation arena and their locations form a Poisson process. Each node has its own position information. The proposed MAC is combined with a geographic routing based on GPSR [13]. For all simulated protocols, the same parameters were set (see Table 1). The GeRaF protocol was adapted to become similar to CMAC [16], avoiding the use of continue packets which would decrease its performance. The waiting time for a CTS was adjusted to match other protocols, i.e., t_{CTS}. The adaptations were realized to enable fair comparison.

Each simulation consists of the transmission of a data packet from a source S to a final destination D, which are 2080 m apart. Data packet length was configured in order that its transmission time ranges from 5% to 80% of the cycle time. In order to obtain statistically relevant results, each simulation scenario was repeated 60 times. The confidence level $\alpha = 0.05$ was employed to calculate the confidence interval.

Table 1. Parameters employed in simulation experiments.

Parameter	Values	Parameter	Values
Source-Destination distance	2080 m	Data packet interval (%cycle time)	5%, 20%, 40%, 80%
Radio range (R)	40 m	Thresholds (τ)	0.1, 0.2, 0.3, 0.4, 0.5, 0.6, 0.7
Number of nodes	4000	Carrier Sense interval t_{cs}	1.84×10^{-3} s (AGAp-MAC), 3.85×10^{-3} s (Others)
Node density η	$6.25 \times 10^{-3}\,\frac{node}{m^2}$	RTS interval (t_{RTS})	2.0×10^{-3} s
Cycle time (t_{cycle})	0.10 s	CTS interval (t_{CTS})	1.5×10^{-3} s

Results

In order to determine the better threshold to transmit a packet in each situation, we first realized experiments to explore different values combined with diverse data packet sizes. The idea here is to find, for each packet size, the ideal threshold (τ) to minimize latency. Figure 2 presents the average latency obtained for each configuration. It is possible to notice, as expected, that latency its heavily influenced by packet size. Moreover, different threshold selection results in diverse total transmission latency. For each data packet size, a given threshold can be selected for optimal results. For example, for a data length of 80%, the threshold of $\tau = 0.5$ results in shorter transmission time. The optimal threshold for different packet sizes is stored in each node for production use.

It is possible to notice that for small packet lengths, lower thresholds yield lower total latency than higher thresholds. The conversely can be visualized for large data packets. This can be explained as the following. Large packets have a high transmission time. If a large FCS (small threshold) is employed, the average hop distance is reduced, increasing the total number of hops. This is not favorable for high transmission times. For small data packets, the reduction of the preamble times provided by large FCS (small threshold) pays off the increased number of hops.

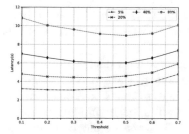
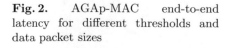
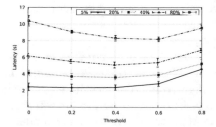

Fig. 2. AGAp-MAC end-to-end latency for different thresholds and data packet sizes

Fig. 3. AGA-MAC end-to-end latency, extracted from [10]

For the sake of comparison, Fig. 3 presents similar results of the AGA-MAC [10] protocol. As expected, the best threshold for all packet length yields lower latency when compared with AGAp-MAC. This means that the new protocol has slightly higher latency than the original AGA-MAC. This is expected, and it is a price to pay for much lower energy consumption, that will be verified in the next section.

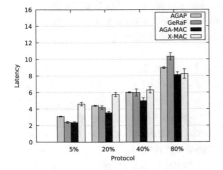

Fig. 4. End-to-end latency of AGAp-MAC compared to the others well-known MAC protocols

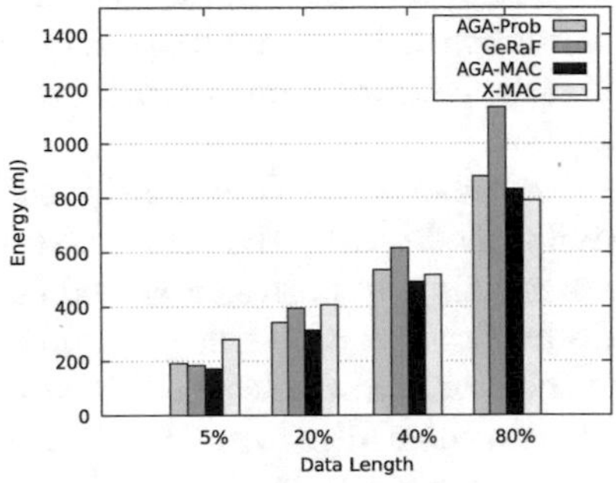

Fig. 5. Energy of a single communication of AGAp-MAC compared to the others well-known MAC protocols

Figure 4 depicts latency, in seconds for a multi-hop packet delivery of AGAp-MAC using optimal thresholds, compared with AGA-MAC, GeRaF and the standard X-MAC using geographic routing. The picture also presents confidence intervals. The results show that our protocol, due to its adaptive behavior, produced good results for all packet sizes evaluated. As expected, the pure AGA-MAC, due its long carrier sense, produces lower latency, since when a node of the FCS wakes up, it will always hear the RTS, differently from our protocol. In latter case, the awake FCS member will hear the RTS signal with probability $\frac{t_{RTS}}{t_{RTS}+t_{CTS}}$, where t_{RTS} is RTS transmission time, and t_{CTS} is CTS transmission time.

A very important aspect of the protocol is its energy expenditure, since a sensor is energy constrained. Our protocol aims at reducing the total network energy. Sensor nodes are programmed to wake up periodically, and if no other task is pending, the channel must be probed for incoming transmissions. As already stated, in traditional asynchronous protocols, as X-MAC or GeRaF, this carrier sense is long, in order to ensure a response. This means that even when the node wakes up in the middle of a waiting CTS period, it should continue to probe the channel, until the next preamble is sent. Our protocol, in opposition, has a short carrier sense period.

Before analyzing the global energy consumption, the energy expenditure in a single communication from S to D by different protocols is presented in Fig. 5. Only the impact of the multi-hop transmissions is visualized, not considering the energy spent by the rest of the network in duty-cycling. Energy spent is given by:

$$E_T = P \sum_{h=1}^{H} \left(t_{cw} + p_h(t_{RTS} + t_{CTS}) + \frac{1}{2}(t_{RTS} + t_{CTS}) + t_{RTS} + t_{CTS} + 2 \cdot t_{data} \right)$$

(1)

where P is the radio power (60 mW), p_h is the number of RTS in hop h, H is the number of hops and t_{data} represents the time spent to send a data packet. The radio consumption is considered the same in idle listening, transmission and reception.

The radio power P is multiplied by the sum of the total time of transceiver activity, modeled by the rest of the equation. This is the sum of the transceiver operating time in both sender and receiver modes for each hop. For hop h, sender node performs a contention window ($t_{cw} = t_{RTS} + t_{CTS}$) and sends p_h RTS's. The total time of consecutive RTS is calculated by the sum of the RTS sending time with the waiting time for CTS ($t_{RTS} + t_{CTS}$). After p_h preambles, the next awaken relay waits in average $\frac{1}{2}(t_{RTS} + t_{CTS})$ until start receiving a complete RTS packet (t_{RTS}) and then sends back a CTS packet (t_{CTS}). Data packet is then transmitted by the sender and received by the next relay. In energy computation, both events are considered ($2 \cdot tdata$).

As expected, in Fig. 5 it is possible to notice that a transmission for the AGA-MAC has lower energy consumption when compared to AGAp-MAC since more RTS packets have to be sent until rendezvous is achieved. For small packets, our protocol employs less energy than X-MAC, which suffers by its unicast nature. For large packets, our protocol expends less energy than GeRaF, since the latter has an increased number of hops to deliver the packet, while our protocol is adaptive to packet size. In general, the energy necessary for a transmission, in AGAp-MAC, was compatible with other well-known protocols.

For a real network, its life-time is a very important metric since it may enable or not different applications. Much more important than the energy spent by a single multi-hop transmission, is the overall energy expenditure through time. This is specially true for networks with limited traffic, which is the case for several WSN applications. In order to evaluate this aspect, we have simulated a time period of $t = 100$ s and included $m = 10$ transmissions during this time interval.

The calculation of global energy with the settings described can be done using:

$$E_G = N \cdot P \cdot t_{cs} \cdot \frac{t}{t_{cycle}} + \sum_{k=1}^{m} E_{Tk}$$

(2)

where E_G is the global energy, m is the number of transmissions, E_{Tk} is the energy spent in the transmission k, N is the number of network nodes (in our case 4000), t_{cs} is the carrier sense interval and $\frac{t}{t_{cycle}}$ returns the number of carrier senses per node during the period ($t = 100$ s). The cycle time was set to $t_{cycle} = 0.10$ s.

Table 2 presents the total energy consumption of the network under these conditions for the simulated total period. It is possible to see that, for every data packet length, AGAp-MAC has an energy expenditure at least 50% less than any of the others state-of-the-art protocols, for the used parameters.

Table 2. Global network energy consumption (in J) for 10 transmission during a period of 100 s.

Data length	AGA-Prob	GeRaF	AGA-MAC	X-MAC
5%	444	926	926	927
20%	445	928	927	928
40%	447	930	929	929
80%	450	935	932	932

As already explained, this energy saving comes from the size of the channel probing (t_{cs}), which is included in every cycle by all nodes. For the simulations, we considered that during inactivity time, sensor nodes are kept in power down (PD) mode, with the crystal oscillator turned off but with voltage regulator working. In order to probe the channel, the radio must be in reception mode. Taking as example the parameters of a CC2420 [18], its transition time from Power Down (PD) mode to receiving mode is 1.20 ms. For a trustfully channel probing, 5 samples are necessary [3], which accounts to another 0.64 ms. Our protocol needs a total of 1.84 ms for the whole operation. The other protocols need the same 1.20 ms, but in addition they need to sample the CTS entire period plus 9 samples (since a node may wake up at the end of a preamble, getting 4 samples and missing one of this preamble). Since the CTS period comprises 1.50 ms, the total sampling period adds to 3.85 ms. The values are presented in Table 1. Considering that every node is sampling at every cycle, our protocol has achieved a very large energy saving.

5   Conclusion

Wireless sensor networks employ duty-cycles to save energy. In asynchronous MAC protocols, a series of request-to-send packets (RTS) is employed as preambles to establish communication. Each RTS transmission is followed by an interval long enough to receive a possible clear-to-send (CTS) packet from the selected next hop, which signalize it is ready to receive the data packet from the sender. In order to detect possible transmissions, each node of the network realizes a carrier sense during its activity period. Traditional asynchronous MAC protocols carry out a long period of carrier sense, in order to assure RTS reception, as it may wake up during the period the sender is waiting for a CTS. Our protocol achieves energy saving with the use of short preamble sampling, reducing energy spent in all nodes of the network.

Nevertheless, a short preamble sampling has its consequence: there is a probability of not receiving CTS even when the sender is trying to establish communication. In order to mitigate the consequence and also reduce the total latency, we decide to design our new protocol with anycast communication: instead a single next relay, there is a group of candidates which brings a desired advance towards destination (FCS). The first FCS member to wake up and hear a RTS is selected as the next relay. This softens the sleep-delay, which asynchronous MACs are prone to.

The FCS selection is achieved with a threshold setting, which determines the area of FCS members in each transmission, consequently the average FCS cardinality. For each data packet size, a different FCS cardinality leads to minimum latency.

Simulations were realized to determine the best threshold for each data packet size. Moreover, we compared our new protocol to other state-of-the-art protocols. For the tested parameters, we achieved a global energy consumption at least 50% smaller than all other protocols, while maintaining a competitive latency, slightly higher than AGA-MAC protocol. This is a small price to pay since the energy savings achieved can prolong significantly WSN life time. Our saving is achieved even in stand-by situations, when no traffic is going on at the network.

Acknowledgments. The present work was realized with support of Fundação de Amparo à Pesquisa do Estado de Minas Gerais (FAPEMIG), grant number APQ-03095-16.

References

1. IEEE standard for low-rate wireless networks. IEEE Std 802.15.4-2020 (Revision of IEEE Std 802.15.4-2015), pp. 1–800 (2020)
2. Almusaylim, Z.A., Alhumam, A., Jhanjhi, N.: Proposing a secure RPL based internet of things routing protocol: a review. Ad Hoc Netw. **101**, 102,096 (2020)
3. Bachir, A., Dohler, M., Watteyne, T., Leung, K.: MAC essentials for wireless sensor networks. IEEE Commun. Surv. Tutor. **12**(2), 222–248 (2010)
4. Buettner, M., Yee, G.V., Anderson, E., Han, R.: X-MAC: a short preamble mac protocol for duty-cycled wireless sensor networks. In: Proceedings of the 4th International Conference on Embedded Networked Sensor Systems, SenSys 2006, pp. 307–320. ACM, New York (2006)
5. Davoli, L., Belli, L., Cilfone, A., Ferrari, G.: From micro to macro IoT: challenges and solutions in the integration of IEEE 802.15.4/802.11 and sub-GHz technologies. IEEE Internet Things J. **5**(2), 784–793 (2018)
6. Ghadimi, E., Landsiedel, O., Soldati, P., Duquennoy, S., Johansson, M.: Opportunistic routing in low duty-cycle wireless sensor networks. ACM Trans. Sen. Netw. **10**(4), 1–39 (2014)
7. Gomez, C., Minaburo, A., Toutain, L., Barthel, D., Zuniga, J.C.: IPv6 over LPWANs: connecting low power wide area networks to the internet (of things). IEEE Wirel. Commun. **27**(1), 206–213 (2020)
8. Halder, S., Ghosal, A., Conti, M.: LiMCA: an optimal clustering algorithm for lifetime maximization of internet of things. Wireless Netw. **25**(8), 4459–4477 (2019)

9. Hao, J., Zhang, B., Mouftah, H.T.: Routing protocols for duty cycled wireless sensor networks: a survey. IEEE Commun. Mag. **50**(12), 116–123 (2012)
10. Heimfarth, T., Giacomin, J., De Araujo, J.: AGA-MAC: adaptive geographic anycast mac protocol for wireless sensor networks. In: 2015 IEEE 29th International Conference on Advanced Information Networking and Applications (AINA), pp. 373–381 (2015)
11. Hong, C., Xiong, Z., Zhang, Y.: A hybrid beaconless geographic routing for different packets in WSN. Wireless Netw. **22**, 1107–1120 (2016)
12. Huang, P., Xiao, L., Soltani, S., Mutka, M.W., Xi, N.: The evolution of MAC protocols in wireless sensor networks: a survey. IEEE Comm. Surv. Tutor. **15**, 101–120 (2013)
13. Karp, B., Kung, H.T.: GPSR: greedy perimeter stateless routing for wireless networks. In: Proceedings of the 6th Annual International Conference on Mobile Computing and Networking, MobiCom 2000, pp. 243–254. ACM, New York (2000)
14. Kumar, A., Zhao, M., Wong, K., Guan, Y.L., Chong, P.H.J.: A comprehensive study of IoT and WSN MAC protocols: research issues, challenges and opportunities. IEEE Access **6**, 76228–76262 (2018)
15. Lessmann, J., Heimfarth, T., Janacik, P.: ShoX: an easy to use simulation platform for wireless networks. In: Tenth International Conference on Computer Modeling and Simulation, UKSIM 2008, pp. 410–415 (2008)
16. Liu, S., Fan, K.W., Sinha, P.: CMAC: an energy efficient mac layer protocol using convergent packet forwarding for wireless sensor networks. In: 4th Annual IEEE Communications Society Conference on Sensor, Mesh and Ad Hoc Communications and Networks, SECON 2007, pp. 11–20 (2007)
17. Polastre, J., Hill, J., Culler, D.: Versatile low power media access for wireless sensor networks. In: Proceedings of the 2nd international conference on Embedded networked sensor systems, SenSys 2004, pp. 95–107. ACM, New York (2004)
18. Texas Instruments Inc.: CC2420 2.4 GHz IEEE 802.15.4/zigbee-ready RF transceiver (2013). https://www.ti.com/product/CC2420
19. Zorzi, M., Rao, R.: Geographic random forwarding (GeRaF) for ad hoc and sensor networks: multihop performance. IEEE Trans. Mob. Comput. **2**(4), 337–348 (2003)

Towards an IoT Architecture to Pervasive Environments Through Design Science

Mateus Gonçalo do Nascimento[1]([✉]), Regina M. M. Braga[1], José Maria N. David[1], Mario Antonio Ribeiro Dantas[1], and Fernando A. B. Colugnati[2]

[1] Departament of Computer Science (DCC),
Federal University of Juiz de Fora (UFJF), Juiz de Fora, Brazil
{mateus.goncalo,mario.dantas}@ice.ufjf.br, {regina.braga,
jose.david}@ufjf.edu.br
[2] Medical Intership Departament, Federal University of Juiz de Fora (UFJF), Juiz de Fora, Brazil
fernando.colugnati@medicina.ufjf.br

Abstract. From a computational perspective, the Covid-19 pandemic experienced around the world shows different lessons for society. Among the lessons, collecting data efficiently and effectively in pervasive environments proved to be one of the greatest challenges. Citizens and government agencies need data that portray the reality of the situation experienced in different environments, such as university campuses, neighborhoods and cities. This article presents as a contribution a computational architecture for the development of applications for pervasive environments using IoT and based on simulation. This research was carried out based on design science research through our proposed architecture. Experimental results showed interesting scenarios that could be differentiated for government agencies, considering IoT sensors, during a pandemic such as Covid-19. This work presents a scenario based on the contamination of Covid-19 in a pervasive and intelligent environment. In addition, the architectural approach provides support to scenarios visualization, which allows interesting parameters to decisions.

1 Introduction

The actual Covid-19 has the same challenges in two hundred and thirteen countries [1], especially from the computational perspective. As a result, society urges an essential and more extensive discussion on how several computational technologies (e.g. big data and digital transformation) could be more useful and effective in the future to mitigate this kind of pandemic. In other words, how do these technologies may provide tools for health enhancement and protection for individuals? It is essential to observe that the main advice to all populations worldwide, not different from health specialists in the influenza pandemic in 1918, is to stay at home. This scenario is a strong indication that the big data and digital transformation approaches did not reach an expected level to support the health field, as it could be.

Reinforcing the requirement of knowledge about health digital data, and very tightly coupled to this problem, is the challenge to provide economics advice. The IMF (International Monetary Fund) states that [2] to help lay the foundations for a strong recovery

L. Barolli et al. (Eds.): AINA 2021, LNNS 226, pp. 28–39, 2021.
https://doi.org/10.1007/978-3-030-75075-6_3

after and during the Covid-19 Pandemic, our policy advice will need to adapt to evolving realities. It is essential to understand better the specific challenges, risks, and trade-offs facing every country as they gradually restart their economies. How can IMF advise without any cities and digital health data for a proper data analytics search?

Big data has several definitions and views. As mentioned in [3], to understand big data, it is important to have some historical background. Gartner [3] reported circa 2001 (which is still the go-to definition): Big data contains greater data variety arriving in increasing volumes and ever-higher velocity. Big data is also extensive, deals with more complex data sets, especially from new data sources. These data sets are so voluminous that traditional data processing software cannot manage them. Nevertheless, these massive volumes of data can be used to address business problems you would not have been able to tackle before.

Digital transformation is one of the main goals of today's health field. As reported in [4], studies have observed the quality of evaluation research. Besides, these researchers observed that the persistent lack of progress had led researchers to ask deeper questions about what is occurring when teams evaluate the digital transformation benefits. On the other hand, research works may provide an interesting enhancement in the health area. Luo et al. [5] present an example, with a proposal representing an energy-efficient and highly accurate toothbrushing monitoring system that exploits IMU-based wrist-worn gesture sensing using unmodified toothbrushes. Therefore, an approach which practitioners and researchers could improve together with the work in the health field.

Given the complexity of predicting how Covid-19 expansion will behave in different affected countries, professionals need tools to aid in decision-making. However, the Covid-19 is not the only problem that cities are facing. Agglomeration of people, public transport, basic sanitation, and safety are examples of other challenges. In this context, the development of computational solutions is a key issue to mitigate those problems.

In this paper, we present as a contribution a computational architecture for the development of applications for pervasive environments using IoT and based on simulation. Through simulations, we built scenarios on pervasive environments to visualize the Covid-19 spread and contamination. Our experimental results demonstrated interesting scenarios that can support decision-makers during a pandemic, such as the Covid-19 and the development of an application to visualize the environment.

The paper is organized as follows. Section 2 presents some aspects of digital transformation, big data and related works. Section 3 describes the proposed architecture to support the application development. Section 4 shows evaluations of the present research work. Section 5 presents conclusions and future work.

2 Concepts and Related Work

This section presents some concepts related to digital transformation focusing on e-health and big data. Because of the large variety of areas and environments in e-health, we describe some segments which could be attractive to the present challenge of the Covid-19 pandemic.

2.1 Digital Transformation

As discussed in [6], healthcare is facing the challenge of affordability in a growing and aging population. The authors argue that the progress in data-enabled precision medicine is beginning to transform traditional linear models into an environment of multi-sided market variants. In this scenario, healthcare providers (examples are hospitals, pharmaceutical companies, doctors), on the one side, and healthcare payers (governments, insurance companies, patients) on the other require having a balance between the best possible health provision and cost. The researchers also observe that the future healthcare affordability, patient experience, treatment efficacy, healthcare capacity, and system efficiency will depend on health information exchange platforms' success and leveraging electronic health records.

The work presented in [7] provides an interesting view of how to govern health services digital transformation. The authors mention that such new technologies could materialize and potential benefits. It is also highlighted that may also be accompanied by unintended and/or negative (side) effects in the short or long term. As a result, the authors observed that the introduction, implementation, utilization, and funding of digital health technologies should be carefully evaluated and monitored.

2.2 Big Data

The sensing of the environmental signal may consider, for example, both the home and the people, as is shown in Fig. 1 from research work [8]. This figure provides an example that we developed for an e-health proposal which results in a large amount of useful data from home and mainly from the people inside this environment.

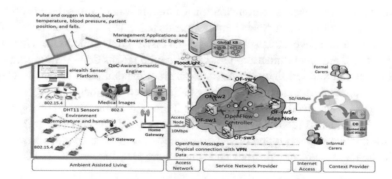

Fig. 1. Ambient assisted living example [8].

Audio/video uploads and file transfers are used to simulate the teleconsulting service and sending medical images. In the application that we developed previously [9], called QoCManApp, the sensed data are collected, quantified, and evaluated, ensuring that only qualified context objects are distributed. Whenever a new data entry is detected in the KB, the inference engine interprets the rules to check for QoE (Quality of Experience) degradation (in this step, the QoC [10] – Quality of Context – parameters associated with the context are analyzed). If there is degradation, the sensed data are discarded. Semantic processing ensures that only accurate, current, valid, complete, and important data be sent to the remote service center.

This research work example presented in [8] was chosen because it is an impressive objective effort that generates e-health big data, without exposing people and health professionals. In [10], an interesting e-health sensor environment could help those conceiving to build an environment to collect that in similar a scenario. On the other hand, an approach for analysis of QoS requirements for e-health services, and mapping to evolved packet system QoS classes, is presented in [11].

2.3 Related Work

Architectural proposals to support the development of IoT applications are not new in the literature. For example, Santana et al. [12] present a study on software platforms for smart cities. The authors propose concepts, define requirements, explain the challenges of a unified reference architecture for the development, testing and validation of software for the area. The architectures and software developed using the IoT paradigm are used in conjunction with cloud computing and/or big data. The research also highlights some relevant application domains (contexts and pervasive environments), such as city sensing, traffic control, city management, health care, public safety, energy management and discarded waste management.

The research carried out by [12] defines non-functional and functional requirements to allow the functioning of a multilevel architecture based on large proportions of data traffic. The authors argue that the architecture can process data on a large scale and enables the development of applications that transform processed data into information. However, the study does not establish how simulations of intelligent environments that use IoT can be adapted to the proposed architecture.

Regarding the use of simulations to represent intelligent environments, Gomes et al. [13] propose constructing an assisted home environment. This work aims to present a computational structure based on a healthcare environment that provides doctors and nurses results in decision-making. The authors present the benefits of having intelligent environment management using IoT sensors. Simulations are used to obtain a person's data generation in a specific environment. However, the research does not address how this environment can be replicated and what is the appropriate computational architecture to serve different simulations.

Through the study of literature, the development of an architecture that contemplates the simulation support will allow visualizing the structure of an intelligent environment more quickly. As a result, applications' construction could be supported for different challenging scenarios, mainly in the IoT context.

3 Proposed Architecture for the Development of Pervasive Applications

The previous section lets us infer that to tackle e-health digital transformation, adopting new technologies must be carefully employed. his section represents the first step in this research contribution. It highlights how we conceived an architecture for the development of pervasive applications using simulations, which is shown in Fig. 2.

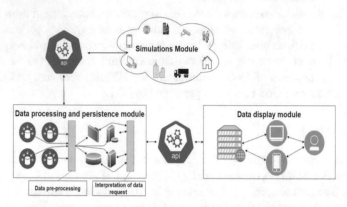

Fig. 2. Simulation-based architecture overview.

The proposed architecture considers the absence of software architectures based on available, flexible and adaptable simulations to different pervasive environments. In this research, we are interested in the challenges imposed by Covid-19 to validate our work. Through the approach explained above, each simulation constructed should represent an intelligent environment and the context in which it is inserted. In this way, contexts such as contamination by Covid-19 can be represented, being responsible for the generation of data on the simulation's events. The proposed architecture should allow research in the intelligent environments' application domain to benefit from constructing a diversity of scenarios through simulations. Those scenarios, on a large scale, use IoT sensors. As a result, it allows validating hypotheses about the application context without the need for a real physical environment.

Each application developed for an intelligent and pervasive environment needs to prioritize its quality attributes. For example, application domains that deal with public safety and another that deals with patients' health probably need different quality attributes. In the above application domains, attributes such as portability, scalability, and interoperability have priority over others, as they deal with the evolution of applications and how to use them in heterogeneous environments. The architecture aims to support the development of IoT applications that use simulations. These simulations-built scenarios allow greater knowledge about the environment it reproduces, assisting in decision-making activities.

The architecture design encompasses three modules that can be communicated to each other through API. Each module includes a set of methods and routines that receive or make data available to interested parties. These modules were designed aimed to enhance previous quality attributes. As mentioned before, Fig. 2 illustrates the proposed architecture.

The **simulation module** allows different contexts – and scenarios – to be built through simulations representing reality as close as possible to real-world contexts. A simulated environment's construction allows complexities, such as access to physical sensors, data from people present in cities and monitoring centers, to be abstracted and easily reproduced. It can also simulate environments that are difficult to reproduce in real cities and pervasive environments or where access to data proves to be a challenge. These advantages could reduce the difficulty of validating and building applications. The main objective is to generate data.

Besides, the communication between objects, like sensors, cameras, and phones, which are part of the constructed scenario, can also be reproduced. Finally, another advantage relates to the costs for running a simulation when compared to have a real scenario or environment. This access to the different objects mentioned above allows us to observe the IoT paradigm and the data they generate.

For the data generated by the simulation to reach the subsequent module, it must be sent to an API, which must have a set of methods capable of handling different data generated by the simulations. Each sensor of the simulation should generate the data and send it. The complexity in dealing with data is abstracted by the API, which should create computational routines that must deal with different data.

The **Data processing and persistence module** predicts that all data generated from the simulations will be processed and persisted, if necessary. A pre-processing data layer allows only important data that is of interest to applications to be stored. After pre-processing, the data will be stored. For example, files can be stored on file servers, data in relational or non-relational databases. This approach is taken to ensure that the correct persistence is carried out. When there is a need to access the data that is persisted, the proposed architecture determines a layer responsible for performing the data request interpretation.

The API that orchestrates the communication between the **data display module** and the **data processing and persistence module** ensures that the necessary data's request always reaches the request interpreter. The request interpreter is responsible for accessing the data, ensuring that they are available and ready to be used whenever requested, in addition to ensuring privacy and integrity. The **data processing and persistence module** is vital for the architecture to function due to storing data.

The **Data display module** is responsible for ensuring that the stored data is interpreted and has some meaning for the application users. These users can be public agencies, private companies, people who use the application, or the decision-makers. This module encompasses three layers. In the first, a server receives all deployed client applications. Through these applications, end-users access to see what happened in the simulations and to support their decision-making. The server sends a request to an API method every time it needs data to generate client application views. As soon as the data request is received, the API communicates with the data processing and persistence module and receives the requested data as a response.

The second layer supports applications that must use the web server that requests and delivers data to be transformed into information. These client applications are used by different end-users, such as citizens and companies, which characterizes the third layer. The third module should address performance, ensuring usability and easy maintenance and security for applications. The coordination between the modules is essential. Different simulated intelligent environments can be validated and tested and allow applications to work for end users and guarantee the desired quality.

4 Evaluation

We conducted this work based on the Design Science Research [14]. This methodology assumes that the solutions (**artifact**) are designed for practical problems. The use of these solutions generates results that address the problem and seek the improvement of theories. In the context of supporting the development of IoT applications for pervasive environments, our research proposes an architecture for the development of pervasive applications using simulations (artifact). For the execution of the DSR, some steps were followed, such as problem definition, literature review and existing solutions discussions, artifact development, evaluation, and discussion of results.

For this purpose, we established the following conjecture. Suppose we offer an architecture for the development of pervasive scenarios. In that case, we can enhance IoT application development to acquire knowledge about the context in which these applications would be used later. As a theoretical basis, this conjecture derives from the knowledge acquired in research already carried out, identified by literature review.

Based on our conjectures, the architecture was designed based on its functional and non-functional requirements. For evaluating the solution [15], an IoT application for Covid-19 was developed and evaluated by a specialist in the area to verify if it is satisfactory to address the research problem.

Additionally, practical and theoretical knowledge can be built concerning the application's use, which implements the proposed architecture, in different contexts and pervasive environments. As contributions to scientific advances, it is expected that this knowledge can support decision-making in the context of the pandemic Covid-19, especially concerning the models used in real contexts of use of the proposed architecture. Through the appearance of the new disease, public and private entities, in addition to civil society, want to know what can happen to the environment in which they operate if isolation measures are taken or if they are not implemented. However, the reproduction of these hypotheses in the real world can lead to deaths, and even social chaos, due to health agencies' saturation.

Figure 3 offers a view of the implemented solution, using the three modules previously proposed. The first module uses simulations to generate data and build the scenario through APIs, the second stores the data in files (.csv) and the PostgreSQL relational database [16]. Finally, the data visualization module displays the data through dashboards.

For experimental purposes and to make it possible to evaluate the proposed architecture based on simulations, a simulation was developed for an intelligent environment and pervasive environment. The simulation was executed on NEC SX-Aurora Tsubasa [17]. This pervasive environment uses sensors, represents everyday life, and deals with the emergence of Covid-19, an infectious disease caused by the new coronavirus.

The created simulation was designed to make it possible to visualize scenarios of social distancing. For the development of the simulation, Siafu [18], an open-source simulator, was used. It allows creating of several intelligent environments, the generation of data, and a graphical interface for the interaction between the simulation persona. Siafu was chosen due to the open distribution of its code, allowing changes to be made according to the project's needs.

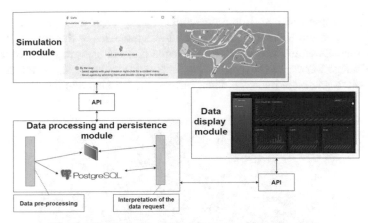

Fig. 3. Simulation-based architecture with applications.

For this assessment, the built and simulated pervasive environment is the Federal University of Juiz de Fora (UFJF) campus, in which standard school days are reproduced. During the school year, students attend classes, have their meals at the university restaurant. Besides, other individuals walk freely through the streets. University employees also perform their daily duties. For this simulation, 23 locations were defined on the Campus. Besides, sensors were distributed, such as levels of cloudiness, rainfall and temperature. Those sensors use IoT paradigm to generate data.

For a faithful portrayal of the campus reality, all 23 defined locations were mapped and IoT sensors were installed. Devices, such as cameras and presence sensors, to map the capacity of an environment, movement and behavior of people using the campus have been allocated. In addition, room and individual temperature sensors, as well as traffic mapping of Campus characters were used. The use of these IoT sensors has proven to be essential for the understanding complex intelligent environment.

The whole construction of the scenario involved understanding which environments are the most frequented at the university and which IoT devices would be important for the mapping. The simulation development process was carried out through the choices made, encompassing designing the environment map, defining the locations, allocating the IoT devices, programming the individual and group behaviors of the characters and, finally, programming how contamination would occur in the environment. The simulation also allows defining interactions between different people and between them and the environment. Thus, the scenario's construction allowed us to predict how people can infect others over the days and how the disease transmission would occur at UFJF.

During the simulated period, contact between infected characters causes the disease to spread. This spread becomes frequent when they are in shared spaces, causing the Covid-19 curve to increase. The spread velocity is represented by the naught reproducibility rate (R0), the average number of new infections someone infected may infect. According to their infection capacity each infected person spreads the disease, the R0, to the others. The characters take 14 days to heal, some of whom may need an ICU, others of a nursery and, finally, some may die. It is possible to establish how many actors will be asymptomatic as well. The proportion of those who might need to be hospitalized is given by a simple constant. It is represented in percentage relative to the number of infected, based on the observed data of Covid-19 epidemics in the city, as the fatality rate does. Impacts on transmission given a social distancing strategy are based on the Imperial College Report 12 [19], providing 3 scenarios to the no action taken. Figure 4 illustrates Siafu graphical interface when running the built simulation.

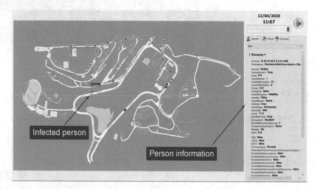

Fig. 4. Siafu execution of UFJF Covid-19 simulation

We illustrated in Fig. 4 the arrangement of persons on the map and information for each person. It is possible to observe a person who is infected, where he/she is traveling, how many days are left for his/her cure and other information about him/her. A video illustrating the simulation of persons moving around the University Campus, the information obtained through the simulation, presented in the data visualization dashboard can be seen at https://drive.google.com/file/d/1A1uXwagb3F41RAVU7dkE 3gPlc6OPBzuN/view.

The simulator was adapted to send data at run time to an API. Each contamination of the simulation agent sends the data of the new contaminant to the API. At the end of a simulation day, there is a survey of the number of infected, cured, and dead persons. Then day's data is sent to an API. After receiving the data, the API sends it to the data processing and persistence module.

Storing the data in this way allows those relevant data to be stored for future studies, such as the development of applications to have data closer to the reality of an intelligent environment. It is also important to store the data generated because the graphical visualization shown in Fig. 4, displayed by Siafu, does not guarantee a perfect understanding of how the Covid-19 infectious curve behaves since the simulator displays only the scenario and individual data.

We developed an application to support data analysis and decision-making. The data display module requests data from the processing and persistence module via the API method. A request is made to the module that has the data, and it returns the information. Accessing the visualization module, the user will see a list of all the simulations already executed. The user will be able to see detailed dashboards when accessing each simulation. Figure 5 illustrates how detailed simulation information can be visualized, such as the number of the infected and general overview of this quantity, number of cured and deaths. It also shows the data day by day of the numbers of the infected and cured people.

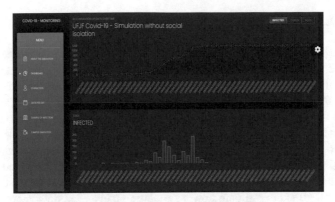

Fig. 5. Dashboard with simulation results

The data visualization module and the dashboard page also have other pages that can be accessed by the menu, such as information about the simulation, character information, data per day, an example of an infection map and a video that shows the entire simulation executed. The simulation information page provides details about the scenario that the simulation was built. For example, the simulation built and executed by the simulation had some social isolation, the number of existing people, and other information. The characters' page informs the name, occupation, current situation, dates of infection, cure and/or death and whether specialized hospital care was needed. Otherwise, the data per day page addresses the number of infected, cured, and dead that the simulation day had. All the data displayed on all pages was generated through the performed simulation. An example of the visualization module can be seen at https://still-savannah-41205-engclient.herokuapp.com/.

We met the scalability attribute through the data persistence module since it had to deal with the execution of several simulation scenarios about Covid-19. Performance and reliability were also met so that no data generated was lost and the persistence module had to increase its performance to handle a large set of data. Interoperability occurred when several applications communicated in an orchestrated manner due to data traffic between applications. The data visualization module was responsible for providing greater usability. It deals directly with the end-user experience when displaying the data, for example, the dashboards that deal with the number of infected, cured, and killed.

As a contribution, through DSR methodology, this work presented an architecture based on data generation through simulations of pervasive environments. The DSR and its steps, as a problem definition; literature review and discussions of existing solutions; artifact development; evaluation; and discussion of results; helped set milestones in the project and allow the delivery of an artifact. The presented architecture aims to support the development of applications for complex scenarios, such as Covid-19, allowing the representation of difficult scenarios to be easily reproduced. As a result, the proposed architecture allows the development of applications and supports strategic decision-making based on end-user data.

5   Conclusion and Future Work

The Covid-19 pandemic scenario reinforces necessary efforts to utilize better of computational technologies, e.g. big data and digital transformation for the healthcare field. These technologies could provide differentiated support in contrast to the present scenario, where the lack of digital data and information worldwide is limiting action in both health and economic areas. In this paper, we presented a contribution in an architecture for the development of pervasive applications with the use of simulations. This architecture was evaluated through the IoT application in the context of pandemic scenarios.

The quality attributes of this architecture aim to ensure that it supports flexibility, scalability, and performance, fulfilling the needs of the developers of the environment. The experiment aimed to evaluate how a simulation can enrich data generation, allowing environments to be reproduced faithfully and resemble reality. The results obtained through simulation of the infection by Covid-19 allow validating the proposed architecture. Besides, data generation can be seen as useful through the visualization of graphs of the applied context.

Through the use of the design science research methodology, it was possible to establish milestones, discuss and evaluate based on the existing architecture, the importance and delivery of the artifact. Furthermore, we presented architecture that uses simulation and pervasive environments focusing on IoT application development and validation. For the health area, the application proposed for the Covid-19 context allows for more assertive studies about how transmission can occur and how to support decision-making by public agencies. Furthermore, different models related to coronavirus isolation could be studied, through simulated scenarios, in-depth.

As future work, each module can be studied individually, and improvements can be proposed guarantee optimization, in the software process. Besides, interoperability in IoT needs to be explored, considering the different levels that need to be addressed by applications, such as, syntactic, semantic and pragmatic.

Acknowledgments. This study was financed in part by the Coordination of Improvement of Higher Education Personnel – Brazil (CAPES) – Finance Code 001, Brazilian National Research Council (CNPq), Petrobras-RED Project and FAPEMIG/Brazil. This work has been supported by UFJF's High-Speed Integrated Research Network (RePesq). https://repesq.ufjf.br/.

References

1. WHO, Coronavirus disease 2019 (COVID-19): situation report, 72. https://apps.who.int/iris/bitstream/handle/10665/331685/nCoVsitrep01Apr2020-eng.pdf. Accessed April 2020
2. IMF. https://www.imf.org/en/Topics/imf-and-covid19. Accessed April 2020
3. Oracle. https://www.oracle.com/big-data/what-is-big-data.html. Accessed April 2020
4. Burton-Jones, A., et al.: Changing the conversation on evaluating digital transformation in healthcare: Insights from an institutional analysis. Inf. Organ. **30**(1), 100255 (2020)

5. Luo, C., et al.: Brush like a dentist: accurate monitoring of toothbrushing via wrist-worn gesture sensing. In: IEEE INFOCOM 2019-IEEE Conference on Computer Communications, pp. 1234–1242. IEEE (2019)
6. Tardieu, H., Daly, D., Esteban-Lauzán, J., Hall, J., Miller, G.: Case study 2: the digital transformation of health care. In: Deliberately Digital, pp. 237–244. Springer, Cham (2020)
7. Ricciardi, W., et al.: How to govern the digital transformation of health services. Eur. J. Public Health 29(Supplement_3), 7–12 (2019)
8. Nazário, D.C., et al.: Quality of context evaluating approach in AAL environment using IoT technology. In: 2017 IEEE 30th International Symposium on Computer-Based Medical Systems (CBMS), pp. 558–563. IEEE (2017)
9. da Silva, M.P., Gonçalves, A.L., Dantas, M.A.R.: A conceptual model for quality of experience management to provide context-aware eHealth services. Future Gener. Comput. Syst. 101, 1041–1061 (2019)
10. Cooking-hacks. e-health sensor platform v20 for arduino and raspberrypi. https://www.coo kinghacks.com/documentation/tutorials/ehealthbiometricsensorplatformarduino-raspberry-pi-medical. Accessed April 2020
11. Skorin-Kapov, L., Matijasevic, M.: Analysis of QoS requirements for e-health services and mapping to evolved packet system QoS classes. Int. J. Telemed. Appl. 2010 (2010). https://doi.org/10.1155/2010/628086. Article ID 628086
12. Santana, E.F.Z., Chaves, A.P., Gerosa, M.A., Kon, F.: Software platforms for smart cities: Concepts, requirements, challenges, and a unified reference architecture. ACM Comput. Surv. (Csur) 50(6), 1–37 (2017)
13. Gomes, E., et al.: An ambient assisted living research approach targeting real-time challenges. In: IECON 2018–44th Annual Conference of the IEEE Industrial Electronics Society, pp. 3079–3083. IEEE (2018)
14. Simon, H.A.: The Sciences of the Artificial. MIT Press, Cambridge (2019)
15. Hevner, A.: A three cycle view of design science research. Scand. J. Inf. Syst. 19(2), 4 (2007)
16. PostgreSQL. https://www.postgresql.org/. Accessed May 2020
17. NEC SX-Aurora TSUBASA, NEC Vector Engine Models. https://www.nec.com/en/global/solutions/hpc/sx/vector_engine.html. Accessed May 2020
18. Siafu Simulator. https://siafusimulator.org/. Accessed May 2020
19. Walker, P.G.T., et al.: The global impact of COVID-19 and strategies for mitigation and suppression. Imperial College London, London (2020)

ASAP - Academic Support Aid Proposal for Student Recommendations

Gabriel Di iorio Silva, Wagno Leão Sergio, Victor Ströele[⊠],
and Mario A. R. Dantas

Institute of Exact Sciences, Federal University of Juiz de Fora (UFJF),
Rua José Lourenço Kelmer, Juiz de Fora, MG, Brazil
{iorio,wagno.leao.sergio,victor.stroele,mario.dantas}@ice.ufjf.br

Abstract. Some research works point out that university students are increasingly presenting depression, anxiety, and even suicidal scenarios during their studies. Numbers indicate that these issues affect both undergraduate and graduate students, dropping their academic performance steadily. However, these disorders can be captured in advance through, for example, heart bit rates and changes in blood pressure. This paper presents the Academic Support Aid Proposal *(ASAP)*, characterized by a model capable of getting data, such as location and body signals, from students and afterward provides some recommendations and advice to them during their period inside the university. The body signals give an emotional context from a specific student and his/her location. These elements help the ASAP in terms of accuracy to the recommendation approach, which is based upon IoT (smart bands) and a machine learning paradigm. The differentiated aspect of the present contribution is based on the use of ubiquitous computing and the proposed architecture.

1 Introduction

The internet has already changed the academic environment in several layers in recent years. With this phenomenon being more and more present in everyday life, it would not take long to start incorporating these technologies in society.

Moreover, with the advent of the Internet of Things (IoT), all the benefits of computational processing now can be carried in non-intrusive ways generating data that can be incorporated intending to help students in their academic lives. That fact can be clearly seen in [1] where they classify into four different groups showing how comprehensive your applications are.

Furthermore, Digital Vortex also addresses this topic in the continuous study over the recent years and how the vortex from time to time. The most recent IMD [2] publication shows us that the educational system still ranks among the center of the vortex, occupying the eighth position. The study intends to illustrate the power, speed, and unpredictability of change linked to digital disruption. Because it is a persistent study that is updated since 2015, it gains more relevance and

L. Barolli et al. (Eds.): AINA 2021, LNNS 226, pp. 40–53, 2021.
https://doi.org/10.1007/978-3-030-75075-6_4

corroborates the idea of the intrinsic relationship between education and the digital scenario.

On the other hand, when we approach a student's academic life, it is also relevant to consider their mental health while at the university. During their graduation period, many students suffer from a lot of psychological pressure. This problem gets more evident while they are doing work presentations or exams. According to the study [3], in a randomly selected sample of students at a single large public university, 4.2% of undergraduates and 3.8% of graduate students screened positive for current panic disorder or generalized anxiety disorder.

The student's academic performance can be directed linked to his anxiety levels [4]. These levels can cause a high heart rate and blood pressure changes, for example, and lead to bad grades or poor performance. With these effects, using wearable sensors like smart bands or smartwatches, we can detect moments like this and help the students with recommendations to normalize their body responses to get the best academic results.

In this work, we present a model able to get the students' location and body signals intending to analyze the collected data and recommend activities and tips during their period at the university dependencies. The location signals are used to see the student's current location and increase recommendations accuracy. Thus, joining both student's location and body signals, the model defines the profile and emotional context using IoT and Machine Learning in the academic environment, aiming to provide students mental health improvement and help them get better academic performances.

This work's differential is based on the use of ubiquitous computing in a new perspective, based on several concepts to solve this so common problem while developing aspects of treatment and maintenance of data and computational architectures.

The work is subdivided as follows: The first section presents the **Introduction** of this paper. The second section offers the **Related Works**. The third section presents the **ASAP** model and all the information about the proposal. In the fourth section, the **Experimental Results and Data Generation** are commented. In the fifth section, the **Conclusion and Future Works** are debated.

2 Related Work

Initially, we observed that monitoring patients' health is frequently addressed, and many studies have tried to solve it. In [5], research was carried out to implant sensors to monitor the elderly to detect their daily activities, thus providing greater independence for them in their homes. In this study, the use of environmental sensors was highlighted, with the justification that wearable sensors offer discomfort to the elderly, increasing the possibility of their rejection of use. The study on ambient assisted living (AAL) carried out in the previous article served as a reference in planning the use of ambient sensors in our research.

On the other hand, in [6], wearable sensors' effectiveness is presented in monitoring students' behavior before and after tests and presentations in an automated way. Also, the work [7] presents a model for detecting and managing emotions through wearable sensors in students to improve their academic performance. The authors pointed out the feasibility of this type of monitoring within the academic environment, reinforcing the proposal for using this type of sensor. The results found in these works showed high performance of the wearable in monitoring the student's well-being in a versatile way. Therefore, in this article, we propose the use of both ambient and wearable sensors for accurate monitoring of the student's condition.

Previous work has shown the ability of machine learning techniques to classify and recognize patterns, which in our case are daily activities and behaviors. In [6] tests were made with algorithms like K-Nearest Neighbor (KNN) and Support Vector Machines (SVM) to classify students' stress levels. [8] uses machine learning algorithms to recognize patterns in daily activities, with the objective of unusual patterns. The conclusions presented led to the choice of these techniques for data classification in the architecture proposed in our article.

As we are in a scenario where a large volume of data will be collected continuously, it is necessary to implement an architecture that can handle such a large processing load. [8] presents a framework that uses a Fog-Cloud paradigm and an architecture that unifies both online and batch processing to manage large volumes of data, based on the Lambda architecture [9]. In [10] the authors show how Cloud and Fog computing can be used to create scalable monitoring and control services from multiple devices in real-time, modular, and capable of handling a high data flow. The previous work highlighted how the Lambda architecture effectively deals with continuous data traffic, both for an immediate response from the system and for a more exhaustive process of aggregating information. This architecture model's choice has as main objectives a high performance in response time and efficiency in processing a high amount of data.

During our research, there are still low efforts in implementing an immediate response from the system when specific patterns in user behavior are detected. Thus, the system presented in this article has as one of its goals to create this additional layer of feedback.

This article aims to recommend actions that can decrease students' stress levels in specific contexts through prior detection of the student's state. Specialized systems such as Recommendation Systems (RS) offer high performance in providing personalized content to users. Besides, as we are addressing a scenario where student behavior is strongly linked to its context, contextual recommendation systems such as those presented in [11] proved to be efficient in achieving the proposed goal. As shown in Fig. 1, several topics of interest in our work have been addressed by the articles, each one focusing on particularity or technique to achieve its goals. We propose to unify the concepts presented to create a new model of real-time monitoring and recommendation for students who suffer from stress.

Research Author	Purpose	Proposes architecture	Model Implementation	Machine learning	Fog-Cloud computing	Lambda architecture	Recommendation system	Ambient sensors	Wearable sensors
Uddin et al. (2018)	A survey on elderly monitoring using ambient sensors.							✓	
Hasanbasic et al. (2019)	Data recognition of wearable sensors.		✓	✓					✓
Gabriel et al. (2019)	Detection and managing emotions on academic environment.	✓							✓
Larcher et al. (2020)	Monitoring daily activities to detect unusual patterns.		✓	✓	✓	✓		✓	
Kiran et al. (2015)	Implementation of lambda architecture for cost-effective processing.	✓	✓			✓			
Myrizakis et al. (2019)	Implementation of a fog-cloud paradigm for a smart home management service.		✓		✓				
Adomavicius et al. (2012)	Presentation of Contextual Recommender Systems.						✓		
ASAP: Academic Support Aid Proposal	Presentation of a monitoring and recommendation system for students through sensors and detection of their conditions.	✓		✓	✓	✓	✓	✓	✓

Fig. 1. Related work comparisons

3   ASAP: Academic Support Aid Proposal

The ASAP model aims to help students at the university dependencies during their everyday routine. With this proposal in mind, we can imagine a scenario where they are all doing activities in different locations, as presented in Fig. 2. Some are doing an exam, running to exercise themselves, or even just walking to their classroom. The students who are anxious while taking an exam may need the recommendations to help lower their stress level and have better grades. All these events are monitored by the ASAP model and help gain more data and increase accuracy while assisting students with their grades and stress levels.

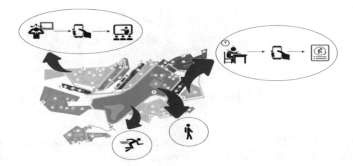

Fig. 2. ASAP inside university environment

Based on a scenario where several students are being monitored by both environment and wearable sensors, using the data generated by them to warn about their stress levels in real-time, we must consider many issues. First of all, we must evaluate how the system responsible for this task will manage the high volume of data generated every second, and, at the same time, this system must also deliver a response to the user in question. This response will consider the student's context and be developed through a considerable amount of processing. This entire process must also be implemented to deliver the information generated to the student in almost real-time. The system's tolerance for infrastructure hardware failures and its scalability to the arrival of new entities should also be evaluated. All these factors are recurrent in the area of Big Data [13]. They have led us to assess a fault-tolerant architecture capable of performing parallel processing with the data flow exposed, handling both the extraction and generation of information and the transmission of the new data to the user to help him/her monitor its status.

Our research done in previous works has led us to choose lambda architecture due to its performance in solving the issues presented. As explained in [17], the lambda architecture aims to be linearly scalable and have a low delay time for both readings and updates, also dealing with hardware and human errors. The lambda architecture implements a distributed processing model among three main layers: the batch layer, the speed layer, and the serving layer, as shown in Fig. 3, each of them responsible for specific functionalities of the architecture.

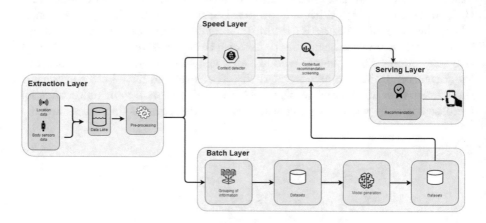

Fig. 3. The ASAP proposal component interactions

The information flow starts at the **Extraction layer** where, initially, all the data is collected using body sensors like smartwatches and the location data using GPS information. The data is then transported to a Data Lake server without processing the information. We adopted the Data Lake approach to enable the storage of a high volume of data with high-velocity [12].

Only after all the data is stored, the pre-processing stage can begin. In this step, all the missing data information that can occur is deleted to be sure of the reliability of the information passed to the construction of the model.

All data entering the system is sent to both the batch and the speed layers and computed in parallel. When the results are obtained, they are indexed by the serving layer to be available for requests.

In the **Batch layer** the management of the primary dataset is made where all incoming data is stored and to make more exhaustive processing of this data so that it is available at the serving layer. The data is divided into clusters using specified Machine Learning algorithms to have the most divergent clusters at the end of this stage. The location information is grouped to see where the student is on the university campus. Each data from each student, when grouped, will give the idea if the student is at the classroom, restaurant, or chemistry department, for example.

In Lambda Architecture, we need to define when the machine learning model has to be rebuilt. We can define it by periods of time or by volume of data ingested in the system. In this work, we chose to reconstruct the model based on the volume of new data because if the amount of data is not that large in some specified period, then the new model will be very similar to the older one, not justifying the processing used in this task.

Moreover, the body sensor data are also grouped but with a difference. To investigate the body sensors data, the history of heart rate frequency, blood pressure, and all measures taken by body sensors must be considered. Thus, the model will be able to identify the student's profile taking into account all the other data provided by the students at the university.

With all the information clustered, the data is transferred to a new dataset, which will provide material to the recommendation algorithm to be trained and lead to the next stage. The model generation stage is responsible for retrieving that data, training and preparing the model to be updated with the new information received, and making the recommendations. We chose the KNN algorithm because its proposal fits the previously seen cluster stage. When the clusters are formed, the most popular recommendation for each cluster is obtained. Then the KNN algorithm is in charge of providing the best recommendation according to the cluster that the student fits. The best recommendation for each cluster is now sent to a new repository and is available to answer every request from the recommendation screening.

While this is done, the **Speed Layer**, unlike the batch layer, has the main function to reduce the delay time in requests dealing only with the most recent data delivered. To ensure that a response is sent in real-time, the speed layer is implemented only with low latency functions and queries, having in some cases [8] the ability to use the data already pre-computed from the batch layer.

At the Speed layer, the context detector is responsible for retrieving the most recent information sent by the body sensors and GPS. With this data, the model can perceive the student's context, like: in the classroom and normal heart rate or at the university square and with high heart rate. Combining those two pieces of information (location and body sensors), ASAP can detect, for example, whether the student is at the university square with a high heart rate, and maybe he is practicing some physical activity. This kind of perception has its origins in the KNN algorithm. If this cluster has its top recommendations related to the practice of physical exercise, he is most likely doing something related. The recommendation screening is in charge of realizing, with the student's context and profile, the best recommendations that were generated previously and detect the most similar cluster to the context.

When the data already computed by the batch layer and detected by the speed layer, it arrives at the **Serving layer** that will be responsible for presenting it to the end-user and making the appropriate updates of information to the user. Here, when the context detection and all the process of recommendation screening are made, the information is displayed on the student's smartphone with some actions to reduce his stress level.

It is worth mentioning that the system also uses a Fog-Cloud computing paradigm to reduce network traffic and server processing load. The lambda architecture components have been separated, so there is a better use of this paradigm. We decided that the Batch and Speed layers would be implemented in the Fog, and the Serving layer implemented in the Cloud node. This was defined because, as the recommendations will already be computed previously, they will be available when the user is in a specific context, thus reducing the client and server's latency. As the Batch layer is responsible for processing recommendations in batch, it's configured in the Fog to facilitate data processing. We also decided that the serving layer should be configured in the Cloud because this way, we facilitate access to the query task and ensure greater security of the respective students' information.

We considered that this architecture fits well for the established purpose through the study. Multiple data sources can be configured in an incremental way so that it does not congest the server and deal with failures of system connections, both in the batch and speed layers. Received data is stored and computed without delays in response, thus updating the generated recommendations.

4   Experimental Results and Data Generation

As the proposal infers a large volume of data, to start viewing the information, there were two important things to develop for the validation of the ASAP: A way to simulate all the body sensors that the students use and a *API (Application programming interface)* to identify the students' location.

4.1   Body Sensors Simulation

The *SIAFU: Open Source Context Simulator* was used to simulate the university environment and generate the body sensors data. We chose the Siafu simulator because it is recognized and used in other projects involving some researchers in this university [14]. To initially understand the generation of data with a lower volume, the simulation was programmed to represent a classroom with some students doing exams during specific periods of the day. The students are involved in four exams during one day with 2 h long each, as shown in the Fig. 4.

All the students get in the classroom and start the exam and, when the test is over, they pass through the door until a new test period begins.

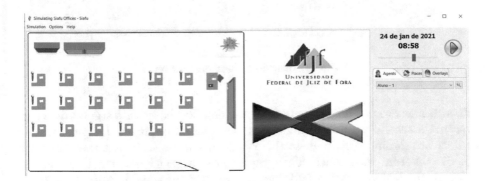

Fig. 4. Body sensors Siafu simulation

With the intention of not biasing the data generated, a randomness system is used. All the students start with the same initial heart rate frequency at the beginning of the simulation. A random number is then generated to indicate whether the heart rate will be increased or decreased if the number is odd or even. Another random number is generated within a limit to specify this variation. The limits of how high or how low the heart rate frequency can variate and the variation were determined using the results of a previous study with real students [7].

All the generated data is exported as a *CSV: comma-separated values* file. We conducted a prior analysis with the *Jupyter Notebook* using *Pandas* library, and it was possible to observe some interesting patterns in certain simulations. The results can be seen in Figs. 5 and 6 where they all cover the same period from 9 AM to 10:55 AM, which represents one of the exams periods.

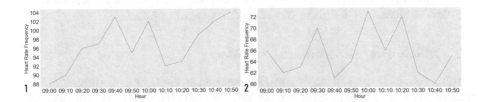

Fig. 5. Heart rate measures graph

In the first graph, it can be verified that this student does not have a well define heart frequency pattern. His heart rate frequency varies a lot and without specific variations. This result is interesting because it represents some expected

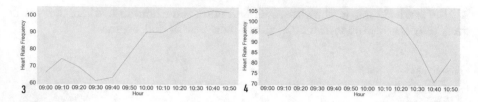

Fig. 6. Heart rate measures graph

difficulties where not all students can have detailed heart rate standards. However, in the second graph, the student has some variation, but it does not increase or decrease significantly, and, in the Y-axis, it can be seen that the values are not distant from each other. This graph can represent a student that remains calm for the test's duration and does not suffer from a significant rise in heart rate. Thus, this case can portray a context where the student does not suffer from stress and anxiety.

Furthermore, in the third graph, there is a case of gradual growth in heart rate, which is related to a classic case when the student gets affected by stress and anxiety and, consequently, increases the heart rate. In these cases, the model must detect this considerable and positive variation. Moreover, there is another different pattern in the fourth graph where the student gets calm as time goes by. His heart rate frequency does not change a lot until the last third of the time, but, in the end, he starts to have frequency getting lower.

Figure 7 shows the boxplot graph of heart rate to help understand the graphics' information. All the students observed in the simulation have specific profiles. Some of them have a heart rate frequency average between 70 and 80, and others between 90 and 100 during the exams. It is also worth mentioning that only three students have values that are considered outliers. Thus, the data generated in the simulation does not have much noise. Moreover, the graph boxes also help visualize how the values of heart rate are distributed to each student and compare their values between them. We can analyze with this distribution of values that the simulation does not bias the data, generating different behaviors for each student that help the model with contrasting profiles.

4.2   Monitoring Environment Solution

Aiming to define the student's context, we use the environment data. Thus, the recommendation system receives data from students monitored by their wearable sensors (in this work, we are using simulated data). Through data extraction and classification, individuals' activities within the defined assisted environment are recognized and used to describe the student's context, which will receive the recommendation system's resources.

In this work, the user's location data is extracted through an application via GPS and Wi-Fi. The proposed system's architecture consists of applications executed locally for data management and another application executed remotely

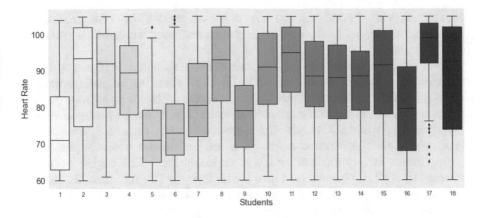

Fig. 7. Boxplot of heart rate

to store users' location data. Therefore, we can identify the environment in which the student finds himself when he presents stress characteristics or other anomalies related to signals such as temperature, blood pressure, blood oxygen, and heartbeat. We can determine whether the student is in a closed or open environment, which allows for richer recommendations by the SR.

The technologies used by the environmental monitoring system are:

- Docker: creation of containers for the services used.
- Docker-compose: installation and configuration of containers.
- Node-RED: creation of automation, access to stored data, and export to dedicated databases.
- FIND3 Internal Positioning: application that scans Wi-Fi and Bluetooth signals and associates them with previously defined environments via Machine Learning.
- MongoDB: dedicated database.
- Mongo Express: the user interface for managing the created databases.

Targeting to collect user data, applications on their phones must have access to a server with the FIND3 application container running. Therefore, for the system to work, it was necessary to grant remote access to students with the server in question both for data extraction and for managing and analyzing them locally.

4.3   Recommendation API

Considering the body sensor data, the data collected from students about their location, and aiming to offer assistance that can help the student manage their stress levels, we have conducted searches for recommendation systems that can be used for experiments and have significant recommendation algorithms presented in the literature. Thus, we decided to use the open-source contextual recommendation engine CARSKit [15]. Besides offering both traditional algorithms and context-aware algorithms, the engine has an architecture that provides fast processing and several parameter setting options, being, therefore, a useful tool for experiments.

In an effort to perform more dynamic experiments with a higher speed, we implemented a Recommendation API [18] that works as an engine wrapper, receiving input from the user, using the information received to interact with the engine, and intercepting the generated data so that it can be post-processed in more dynamic formats. The API was developed in Python and is used through the terminal via the command line. The engine can be used simply, besides opening space for automation and remote executions. The engine results are extracted by API to be converted into JSON format. The API also can store the recommendations and statistics generated in a MongoDB database if the user specifies it, thus ensuring security in performing experiments.

The recommendation API also has some limitations. At the moment, its use is limited only to Linux computers and has a lack of engine regulation parameters. There are also issues of speed and integrations that will be better worked on in later versions. Despite this, the API presented good integration with the engine, and it was efficient in generating results in several use cases.

If it is detected that the student is in a context of stress, anxiety, or nervousness, the **Serving Layer** presents a message to him in order to help him out of this condition. The possible recommendations to be sent can be divided into three groups according to the range of possible recommendations for the student in this work: immediate activities, preventive activities, and indications for quality of sleep. It is important to note that the division into these three groups of activities was made in order to make the study carried out feasible.

Immediate activities are those that can be performed at any time, without the need for an appropriate physical space. Thus, if the student is present in an environment such as an assessment, presentation, or meeting, such measures are effective since they are easy to perform and quick. The immediate activities adopted in this work are:

- *Brain Breaks*: consists of taking a break from the activity that causes you anxiety and doing some pleasurable activity, singing a song you like, or simply interrupting for a few minutes the action that generates stress [20].
- Diaphragmatic Breathing: consists of performing breathing exercises whose main characteristic is exchanging all air in the lungs [16].
- Progressive Muscle Relaxation: consists of a series of simple and easy exercises that help relax the muscles to reduce anxiety levels [16].
- Positive self-talk: consists of talking to yourself with motivating phrases that help to encourage [19].

It is worth mentioning that the immediate activities are not mutually exclusive. That is, students can perform more than one at the same time. For example, if he is starting an assessment, he can take a break (*Brain Breaks*) and combined with diaphragmatic breathing.

Preventive activities are those to prevent the occurrence of more severe cases of heart rate spikes, such as tachycardia or recurrence of large fluctuations in the frequency several times a day. For the student to do these activities, he must be in an appropriate environment. They are:

- Physical Activities: consists of performing aerobic physical activities in order to purposely increase the heart rate for a prolonged period so that endorphins are released into the bloodstream, causing relaxation after the exercise is finished.
- Outdoor relaxation: consists of going to places with direct contact with nature so that the student can relax.
- Seek medical help: in some cases, it is interesting for the user to go to a doctor or therapist as these actions may not have the expected effect.

With the types and recommendations above, we can highlight some examples of recommendations made for each type of student following their profile and context. It is noted through the boxplot graph that student 17, for example, has his heart rate information distributed among high values. Therefore, it is necessary for this student profile to recommend preventive actions since anxiety behavior is recurrent in tests for him. However, as observed in the boxplot of student 5, his frequency is distributed among low values, and he suffers little from anxiety problems during his tests. Thus, it is interesting to recommend immediate activities to correct your specific heart rate increase problems for this profile.

5   Conclusion and Future Works

This article presented the ASAP system, an architecture for monitoring the student's well-being in real-time through sensors and indication of items that can reduce stress and anxiety levels. ASAP aims to help the student manage his emotional condition in the academic environment and school tasks. The research carried out shows viability in the implementation of the system and a great chance of positive feedback on its efficiency operating in a real scenario.

We found some limitations concerning architectural concept tests. First, no data sets were found that could be used to evaluate the system processing flow. Therefore the only data source that can be used is the SIAFU ambient simulator.

Nevertheless, the SIAFU simulator has the necessary resources to perform feasibility tests of the architecture, thus ensuring the exploration of the improvement possibilities for it and defining a basis that can be used in future experiments. Moreover, experts will help us in this project to improve the simulation with a real academic environment. Therefore, this information will be used to train a machine learning model to group the information and recommend actions according to the data. This part is essential because it can explain how our proposal will work with real students.

After the previous effort, a real environment experiment can be performed for data collection. We would evaluate which techniques and parameters will be more efficient in data classification and processing. Once implemented, we will also verify the effect that the recommendation creates on the student's behavior to define which items are the most appropriate for the proposed objective. Besides, we intend to assess security issues and system costs to obtain a model that can be deployed on a large scale and safely.

References

1. Bagheri, M., Movahed, S.H.: The effect of the Internet of Things (IoT) on education business model. In: 2016 12th International Conference on Signal-Image Technology & Internet-Based Systems (SITIS). IEEE (2016). https://doi.org/10.1109/SITIS.2016.74
2. Yokoi, T., et al.: Digital Vortex 2019: continuous and connected change. Lausanne: Global Center for Digital Business Transformation (2019). https://www.imd.org/
3. Eisenberg, D., et al.: Prevalence and correlates of depression, anxiety, and suicidality among university students. Am. J. Orthopsychiatry **77**(4), 534–542 (2007). https://doi.org/10.1037/0002-9432.77.4.534
4. Vitasari, P., et al.: The relationship between study anxiety and academic performance among engineering students. Proc.-Soc. Behav. Sci. **8**, 490–497 (2010). https://doi.org/10.1016/j.sbspro.2010.12.067
5. Uddin, M., Khaksar, W., Torresen, J.: Ambient sensors for elderly care and independent living: a survey. Sensors **18**(7), 2018 (2027). https://doi.org/10.3390/s18072027
6. Hasanbasic, A., et al.: Recognition of stress levels among students with wearable sensors. In: 2019 18th International Symposium INFOTEH-JAHORINA (INFOTEH). IEEE (2019). https://doi.org/10.1109/INFOTEH.2019.8717754
7. Silva, G., et al.: Hold up: modelo de detecção e controle de emoções em ambientes acadêmicos. In: Brazilian Symposium on Computers in Education (Simpósio Brasileiro de Informática na Educação-SBIE), vol. 30, no. 1 (2019). https://doi.org/10.5753/cbie.sbie.2019.139
8. Larcher, L., et al.: Event-driven framework for detecting unusual patterns in AAL environments. In: 2020 IEEE 33rd International Symposium on Computer-Based Medical Systems (CBMS). IEEE (2020). https://doi.org/10.1109/CBMS49503.2020.00065
9. Kiran, M., et al.: Lambda architecture for cost-effective batch and speed big data processing. In: 2015 IEEE International Conference on Big Data (Big Data). IEEE (2015). https://doi.org/10.1109/BigData.2015.7364082
10. Myrizakis, G., Petrakis, E.G.M.: iHome: smart home management as a service in the cloud and the fog. In: International Conference on Advanced Information Networking and Applications. Springer, Cham (2019). https://doi.org/10.1007/978-3-030-15032-7_99
11. Adomavicius, G., Tuzhilin, A.: Context-aware recommender systems. In: Recommender Systems Handbook, pp. 217–253. Springer, Boston (2011). https://doi.org/10.1007/978-0-387-85820-3_7
12. Singh, A., Ahmad, S.: Architecture of data lake. Int. J. Sci. Res. Comput. Sci. Eng. Inf. Technol. **5**(2), 411–414 (2019). https://doi.org/10.32628/CSEIT1952121
13. Chen, M., Mao, S., Liu, Y.: Big data: a survey. Mob. Netw. Appl. **19**(2), 171–209 (2014). https://doi.org/10.1007/s11036-013-0489-0

14. Mendonça, F.M., et al.: EPIDOR: uma abordagem computacional baseada em sistema web e aplicativo móvel para dores crônicas no atual contexto de pandemia do coronavírus. AtoZ: novas práticas em informação e conhecimento **9**(2), 117–128 (2020). https://doi.org/10.5380/atoz.v9i2.74673

15. Yong Z., Mobasher, B., Burke, R.: CARSKit: a java-based context-aware recommendation engine. In: Proceedings of the 15th IEEE International Conference on Data Mining (ICDM) Workshops, Atlantic City, NJ, USA, pp. 1668–1671, November 2015. https://doi.org/10.1109/ICDMW.2015.222

16. de Oliveira, M.A., Duarte, Â.M.M.: Controle de respostas de ansiedade em universitários em situações de exposições orais. Rev. Brasileira Terapia Comport. Cogn. **6**(2), 183–199 (2004). http://pepsic.bvsalud.org/

17. Hausenblas, M., Bijnens, N.: Lambda architecture, vol. 6, p. 2014 (2015). http://lambda-architecture.net/.Luettu

18. Pypi. https://pypi.org/project/carskit-api/. Accessed January 2021

19. Heart. https://www.heart.org/. Accessed January 2021

20. Edutopia. https://www.edutopia.org/. Accessed January 2021

A Method of Expanding Photo-Based User Preference Profile for the Recommendation of Sightseeing Places

Eriko Shibamoto[1] and Kosuke Takano[2(✉)]

[1] Course of Information and Computer Sciences, Graduate School of Engineering,
Kanagawa Institute of Technology, 1030 Shimo-ogino, Atsugi, Kanagawa, Japan
s1985003@cco.kanagawa-it.ac.jp
[2] Department of Information and Computer Sciences, Kanagawa Institute of Technology,
1030 Shimo-ogino, Atsugi, Kanagawa, Japan
takano@ic.kanagawa-it.ac.jp

Abstract. This study proposes a method of expanding photo-based users' preference profiles for the recommendation of sightseeing places. In order to recommend the desired places for the new users, to find the places with unexperienced sceneries and interesting activities. We found some difficulties in conventional studies, involving that the domain of preference analysis is too closed to the personal photo archive. Hence, in this study, we improve to expand the photo-based profiles by predicting users' unexperienced categories of sightseeing places from the photo sets in the photo-based profiles, using a deep neural network, which learning with a sequence pattern of photos taken by different users. Because of the expanded photo-based users' profiles including visual information of unexperienced categories, it allows the recommendation system to suggest the impressive and beautiful spots where the new users who never has the experiences in that places. In the experiment, we evaluate the feasibility of our proposed method that it can expand the photo-based user profiles for guiding the sightseeing places that is expected to influence the user before the user has interests.

1 Introduction

The information recommendation system has become indispensable for finding new knowledge and experience in daily life. For example, many services such as Amazon. com [8] and Spotify [9] recommends items and songs that user is likely to have interests in. In order to recommend appropriate information in those services, it is essential to extract the user's information preference.

Meanwhile, with the wide spread of a mobile phone camera and other types of digital camera, taking photos in every visited places is becoming a daily activity to store and share individual memories and experiences through Social Networking Services (SNS) such as Facebook [10]. Photos have rich visual information more than expressing in words regarding of our impressions, thoughts, intentions, and emotions, so that it seems to be suitable to implicitly extract user's preferences from such a set of photos taken by the user. In addition, the implicit extraction of the user's preference from the photos

L. Barolli et al. (Eds.): AINA 2021, LNNS 226, pp. 54–65, 2021.
https://doi.org/10.1007/978-3-030-75075-6_5

would make it possible to automatically construct the user's preference profile for the information recommendation.

In our previous study, we have proposed a recommendation method of sightseeing places based on the photo-based users' preference profiles [7]. This method utilizes photos taken by the users for the preference analysis, so that it can recommend the sightseeing spots where the users feel impressed when they actually visit. However, since the domain of preference analysis is too closed to the personal photo archive, we found it some difficulties in helping new users to find the places with unexperienced sceneries and interesting activities.

In this study, we propose a method of expanding photo-based users' preference profile for the recommendation of sightseeing places with unexperienced sceneries and interesting activities. Our method constructs a deep neural network to predict the context of photo sets by learning a sequence pattern of photos taken by different users. Then, our method expands the photo-based profiles by predicting users' unexperienced categories of sightseeing places from the photo sets in the photo-based profiles. Since the expanded photo-based users' profiles includes visual information of unexperienced categories, it allows the recommendation system to suggest the impressive and beautiful spots where the new users who never has the experiences in that places before to be interested in and the system can predict that they would be satisfied.

In the experiment, we constructed our prototype using a regression model of photo context that consists of a CNN encoder for the vision processing and a LSTM decoder for the sequence processing. By the several experiments using our prototype, we evaluate the feasibility of the proposed method that it can expand the photo-based user profiles for guiding the sightseeing places that is expected to influence the user before the user has interests.

2 Related Work

There are many researches on recommendation systems for sightseeing spots using image. In [1], Wang et al. propose a new framework of Visual Content Enhanced POI recommendation (VPOI), which incorporates visual contents for POI recommendations. Li et al. develop a method to learn a ranking function by exploiting the textual, visual and user information of photos [2]. Gao et al. studied the content information on LBSNs with respect to POI properties, user interests, and sentiment indications [3].

Besides, some researches focus on the user profile construction for personalized recommendation. Liu propose a personalized recommendation model to capture users' dynamic preference using Gaussian process for implicit feedback data [4]. Meo et al. propose a query expansion and user profile enrichment approach to improve the performance of recommender systems operating on a folksonomy, storing and classifying the tags used to label a set of available resources [5]. Palleti et al. propose a personalized web search system implemented at proxy which adapts to user interests implicitly by constructing user profile with the help of collaborative filtering [6].

In comparison with these conventional recommendation methods, the feature of our method is to expand the photo-based profiles by predicting users' unexperienced categories using the regression and classification capability of deep learning, In our method, a user profile is constructed from the set of photos taken by the user, and the user profile is expanded based on the context of photo sequences so that our method can recommend the sightseeing spots to help the user find the desired ones where they feel impressed.

3   Motivating Example

In the field of a recommendation system, a user preference profile is a personal data that represents the user's interests and preferences, which a recommendation system refers to for providing the personalized information. Therefore, the content of the user profile greatly affects the recommendation quality of the recommendation system. In this study, we use a set of photo data taken by the user as the user preference profile for the personalized recommendation of sightseeing spots, and it is referred to as a photo-based user profile. Since the user is considered to have interests in the photos that the user has taken by him/herself, it would be possible to recommend the sightseeing spots where the user is likely to go based on the photo-based user profile.

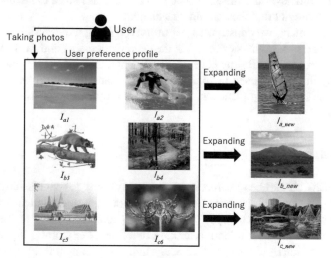

Fig. 1. Expansion of photo-based user preference profile

Meanwhile, it is deemed that we can build a deep neural network, which predicts a set of photos related to the photo-based user profile, based on the temporal relationships and co-occurrence frequency by learning a large number of photos collected from different users. If the photo-based user profile can be expanded by the prediction results of the deep neural network, it would be possible to recommend places where the user has never visited or experienced. For example, suppose that there were many photos of sea and surfing, which a user has taken by him/herself, in the photo-based user profile. In this case, if a windsurfing, which the user has never experienced so far, can be predicted as an activity related to the user's interests based on the photo-based user profile as shown

Fig. 1, then it would be reasonable to recommend places which are the best spots for the windsurfing.

Thus, expanding the photo-based user profile of preference, allow the recommendation system to capture landscapes, activities, gourmet information, and so on, which the user has never visited and photographed, and recommend the impressive and beautiful spots before to be interested in. Furthermore, due to the portability of the expanded user profile, which includes user's information preference, it can be reusable in various applications other than recommendation systems.

4 Proposed Method

In this study, we propose a method of expanding photo-based users' preference profile for the recommendation of sightseeing places by extending our previous recommender system.

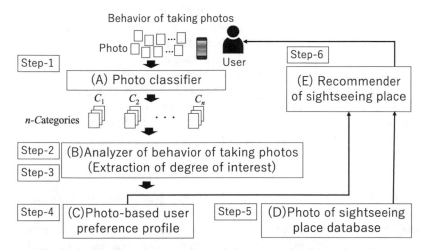

Fig. 2. Architecture of our recommendation system for sightseeing places

In this section, we first take a look at an overview of the proposed recommendation system for sightseeing places in our previous study. Then, we present the detail of our proposed method that expands photo-based users' preference profile for the recommendation of sightseeing places.

4.1 Recommendation System of Sightseeing Places Based on User's Preferences

Figure 2 shows an architecture of our proposed recommendation system for sightseeing places based on the photo-based user preference profile. The feature of this system is to calculate the degree of user's interests in photos actually taken by a user by counting the frequency of photographing and process of editing and clipping, which are regarded as a part of the user's behavior of taking and editing photos of sightseeing places, for the purpose of the recommendation of sightseeing places.

This system is executed in the following steps.

Step-1: Photo classification

A user u takes photos $I_u = \{i_{u1}, i_{u2}, ..., i_{uN}\}$. These photos are classified into n categories C_x $(x = 1, 2, ..., n)$ to characterize a sightseeing spot. In this system, we use the Convolutional Neural Network (CNN) to classify the photo data.

Step-2: Analysis of the behavior of taking and editing photos

First, suppose that a user's behavior of taking and editing photos is $BP = \{B_1, B_2, ..., B_n\}$. Here, each B_i represents an element of taking and editing photos such as B_1: photographing, B_2: exposure adjustment, B_3: contrast adjustment, B_4: white balance adjustment, and B_5: trimming. With respect to counting the frequency of each behavior, they are counted as one for one work regardless of the editing time. In addition, when the user duplicates a photo and edits it, the frequency of edits is counted for each duplicated photo. Finally, the frequency $N_x(B_i)$ is calculated for a category x by summing the various numbers for taking and editing photos.

Step-3: Calculation of the degree of user's preferences for categories

The sum S_x for the frequencies $N_x(B_i)$ in Step-2 is calculated. The categories that the user has a potential interest in can be predicted from S_x.

Step-4: Generation of the user's preference vector

As stated in the above, S_x represents the degrees of the user's preferences for the categories, and this scheme of category C_n can be used to evaluate sightseeing places. Based on this thought, by leveraging the S_x that is calculated in Step-3, our method generates the user's preference vector P_u that represents the user's preferences in each category x $(x = 1, 2, ..., n)$ for the sightseeing places.

$$\mathbf{P}_u = [S_1, S_2, \ldots, S_n] \tag{1}$$

Step-5: Generation of the vector of sightseeing places and recommendation of sightseeing places

Similarly, using the same scheme as the user's preference vector $\mathbf{P}_u$, a feature vector $\mathbf{L}_v = [l_1, l_2, ..., l_n]$ of sightseeing place v is generated. Based on the vectors of the user's preferences and sightseeing places that are generated in Steps 4 and 5, respectively, photos of sightseeing places are recommended to the user. In the recommendation, it seems to be appropriate to suggest sightseeing places whose features are similar to the user's preferences. In this system, sightseeing places are ranked using the cosine measure between the user's preference vector $\mathbf{P}_u$ and the feature vectors of sightseeing place $\mathbf{L}_v$ as follows:

$$sim(\mathbf{P}_u, \mathbf{L}_v) = \mathbf{P}_u \cdot \mathbf{L}_v / |\mathbf{P}_u||\mathbf{L}_v| \tag{2}$$

4.2 Expansion of Photo-Based User Profile

Our proposed method expands a photo-based user profile using a deep neural network after classifying a set of photos into n categories in Step-1 in the previous Sect. 4.1.

The proposed method mainly consists of the following two processes, a learning process and an expansion process.

(1) Learning process: First, sets of photos are collected from photo-based profiles of different users, and association rules regarding patterns of photos appeared in each photo-based user profile are extracted based on the co-occurrence frequency of combinations of photo categories. Then, using a deep neural network, the patterns of photo contexts are learned based on the association rules and a regression model of photo context is created.

(2) Expansion process: The photo-based profile of a target user is expanded by applying the regression model of photo context constructed in (1).

The detail of each process is explained as below.

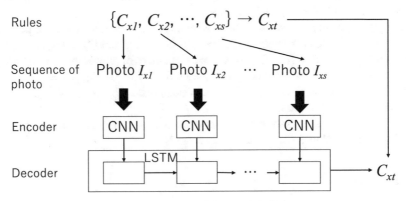

Fig. 3. Architecture of regression model of photo context

[(1) Learning process]
The learning process is executed in the following steps.

Step-1: Different users create their photo-based profiles by taking photos. The photos in the photo-based profile are classified into n-categories C_x ($x = 1, 2, ..., n$).
Step-2: Since each photo set classified for one user in Step-1 can be regarded as one transaction, the apriori algorithm can be applied to photo sets obtained from whole users for extracting a set of association rules, $R = \{r_1, r_2, ..., r_k\}$ according to the specified confidence c and support s. Here, each association rule r_x represents frequent patterns of photo categories and described in the form of $\{C_{x1}, C_{x2}, ..., C_{xs}\} \rightarrow C_{xt}$.
Step-3: Based on the rule set R extracted in Step-2, sequences of photos are used to train a deep neural network, where the photo sequence $I_{x1}, I_{x2}, ..., I_{xs}$, which are corresponding to the left side in r_x, is used as the input and the right side of photo category C_{xt} is used as the answer of a photo context (Fig. 3).

As shown in Fig. 3, the deep neural network consists of an encoder and a decoder, and is referred to as a regression model M of photo context. The encoder is constructed using a convolutional neural network (CNN) for the vision processing, and it receives an input image I_x and encodes it to k-dimension vector. The decoder is constructed using a sequence to sequence model such as LSTM (Long Short-Term Memory) for the sequence processing, and it receives a sequence of k-dimension vectors from the encoder and outputs a photo context label.

[(2) Expansion process]
The expansion process is executed in the following steps.

Step-1: In a photo-based profile of a target user u, a photo set $I_u = \{i_{u1}, i_{u2}, ..., i_{uN}\}$ is classified into n-categories C_x.

Step-2: The regression model M of photo context, which is created in the learning process, receives a sequence of photos as an input and predict a photo category C_{new} as the photo context for the sequence of photos.

For example, suppose that there are 10-photos $i_1, i_2, ..., i_{10}$ in a photo-based profile, and they are classified into two categories, $C_1 = \{i_1, i_2, ..., i_5\}$ and $C_2 = \{i_6, i_7, ..., i_{10}\}$. In this case, if we have a rule $\{C_1, C_2\} \rightarrow C_3$, the sequences of photos that is corresponding to the rule are selected for predicting categories like $s_1 = \{i_1, i_6\}$, $s_2 = \{i_2, i_7\}$, $s_3 = \{i_3, i_8\}$, $s_4 = \{i_4, i_9\}$, $s_5 = \{i_5, i_{10}\}$. Then, new categories $C_{new}^1, C_{new}^2, C_{new}^3, C_{new}^4, C_{new}^5$ are predicted from each sequence.

Step-3: The photo-based profile of the target user u is expanded with the photo I_{new} belonging to the category C_{new} obtained in Step-2.

5 Experiment

We conducted experiments for confirming the feasibility of our method.

5.1 Experiment 1

In Experiment 1, we examine the capability of learning patterns of photo sequence.

5.1.1 Experimental Environment

We prepared two regression models of photo contexts, CIF10 + LSTM and VGG16 + LSTM, as shown in Table 1. We used CNN and LSTM for Encoder and Decoder, respectively, for both models.

For constructing an Encoder in CIF10 + LSTM, we used CIFAR10 dataset [11] for training photos for the 10 category classification and the accuracies of training and validation in 20 epoch were 0.7949 and 0.7150, respectively. In addition, for constructing an Encoder in VGG16 + LSTM, we used Caltech256 dataset [12] with VGG16 for training photos for 20 categories and the accuracies of training and validation in 90 epoch were 0.9145 and 0.8750, respectively.

Besides, we used given rules for the photo sequence learning, where the length of sequence is two, as shown in Table 2 and Table 3. Here, the Encoder in CIF10 + LSTM generates 4,096-dimensinal vector for one photo and the Encoder in VGG16 + LSTM generates 1,024-dimensinal vector, and these vectors are input to each Decoder.

Table 1. Regression models of photo contexts

Model name	Encoder	# of class	Decoder	# of Seq.
CIF10 + LSTM	CNN (CIFAR10)	10	LSTM	2
VGG16 + LSTM	CNN (VGG16 + Caltech256)	20	LSTM	2

Table 2. Rules of photo sequence and context (CIF10 + LSTM)

Rule no.	Category1	Category2	Context of photo
r1	Airplane	Automobile	Bird
r2	Automobile	Bird	Cat
r3	Bird	Cat	Deer
r4	Cat	Deer	Dog
r5	Deer	Dog	Frog
r6	Dog	Frog	Horse
r7	Frog	Horse	Ship
r8	Horse	Ship	Truck
r9	Ship	Truck	Airplane

Table 3. Rules of photo sequence and context (VGG16 + LSTM)

Rule no.	Category1	Category2	Context of photo
s1	Palm-tree	Speed-boat	Waterfall
s2	Iris	Bear	Waterfall
s3	Buddha	Bonsai	Sushi
s4	Butterfly	Cactus	Hibiscus
s5	Dolphin	Iguana	Fern
s6	Waterfall	Penguin	Kayak
s7	Tower-pisa	Spaghetti	Ice-cream-cone
s8	Kayak	Dolphin	Palm-tree
s9	Speed-boat	Waterfall	Palm-tree
s10	Teepee	Tower-pisa	Buddha

5.1.2 Method

We evaluate the accuracies of predicting the photo contexts for the input photo sequences in training and validation.

5.1.3 Results

Figure 4 and 5 shows the results of prediction accuracies of photo context for CIF10 + LSTM and VGG16 + LSTM, respectively. As for CIF10 + LSTM, in 5 epochs, prediction accuracies of both training and validation reached around 1.00 and 0.96. In addition, as for VGG16 + LSTM, in 10 epochs, prediction accuracies of both training and validation reached around 0.90 and 0.86.

These results mean that our prediction models can predict the photo contexts for the specific patterns of photo sequences in two length.

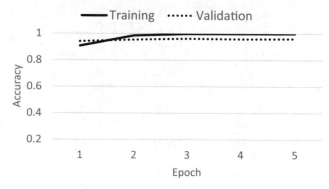

Fig. 4. Prediction accuracy of photo context (CIFAR10 + LSTM)

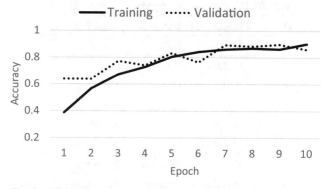

Fig. 5. Prediction accuracy of photo context (VGG16 + LSTM)

5.2 Experiment 2

In Experiment 2, we evaluate the capability of our proposed method to expand photo-based user profiles.

5.2.1 Experimental Environment

In this experiment, we use three pseudo users who have different patterns of photos which are collected from Caltech256 dataset. In photo-based user profiles for each user, User-A has photos of 'palm-tree' and 'speed-boat' categories, User-B has photos of 'palm-tree', 'speed-boat' and 'budda' categories, User-C has photos of 'palm-tree', 'speed-boat', 'buddha' and 'bonsai' categories.

Besides, we use VGG + LSTM for the regression model of photo context which are the same as used in Experiment 1 and the rules as shown in Table 3 are applied for the expansion of photo-based user profiles (Table 4).

Table 4. Photo-based profiles for Pseudo users

User ID	Categories of photos (Number of photos) in photo-based user profile
A	Palm-tree (12), speed-boat (12)
B	Palm-tree (12), speed-boat (12), budda (12)
C	Palm-tree (12), speed-boat (12), budda (12), bonsai (12)

5.2.2 Method

The photo-based profiles for each user is expanded by applying given rules as shown in Table 3, and we discuss about the results of expanding photos in each photo-based profile.

5.2.3 Results

After applying our expansion method, photos are added to each photo-based user profile according to the given rule. Figure 6, 7, and 8 show the example of expanded photos in the photo-based profiles of User-A, User-B, and User-C, respectively.

For User-A, twelve patterns, in which each pattern consists of a sequence of two photos in categories palm-tree and speed-boat, were generated, and the rule s1 were applied to each pattern. As the result, three photos corresponding to three patterns were added into the photo-based profile of User-A. However, predictions for the nine patterns were failed due to the prediction error, and no photos were added to the profile. Here, expanded photos are in the category "waterfall" which is not included in the photo-based profile of User-A. Similarly, one photo were added into the photo-based profile of User-B, and three photos were added into the photo-based profile of User-C, which were not included in the photo-based profiles of User-B and User-C as well.

From these discussions, we could confirm the promising results that our proposed method can expand a photo-based user profile for the further recommendation of sightseeing spots where he/she has never had the experiences in that places before.

Fig. 6. Expanded photos for User-A

Fig. 7. Expanded photos for User-B

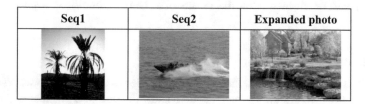

Fig. 8. Expanded photos for User-C

6 Conclusion

In this paper, we have presented a method of expanding photo-based user preference profile for the recommendation of sightseeing places.

In the experiments, we constructed a regression model of photo context that consists of a CNN encoder for the vision processing and a LSTM decoder for the sequence processing. Using this regression model of photo context, we confirmed that this model can learn a pattern of photo sequences from a set of photos and given rules. In addition, we obtained the promising result that our proposed method can expand photo-based user preference profiles, where new photos regarding unexperienced category to users can be added.

However, the number of sequences in the given rules is limited to two this time, and the number of target users is also small. In our future work, we will increase the number of sequences to more than three, and the number of target users for realizing more practical experimental environment. Furthermore, we will evaluate the effectiveness of the proposed method by applying it to the recommendation process of sightseeing places.

References

1. Wang, S., Wang, Y., Tang, J., Shu, K., Ranganath, S., Liu, H.: What your images reveal: exploiting visual contents for point-of-interest recommendation. In: WWW 2017 Proceedings of the 26th International Conference on World Wide Web, pp. 391–400 (2017)
2. Li, X., Pham, T.-A.N., Cong, G., Yuan, Q., Li, X.-L., Krishnaswamy, S.: Where you instagram?: Associating your Instagram photoswith points of interest. In: CIKM, pp. 1231–1240 (2015)
3. Gao, H., Tang, J., Hu, X., Liu, H.: Content-aware point of interest recommendation on location-based social networks. In: AAAI, pp. 1721–1727 (2015)
4. Liu, X.: Modeling users' dynamic preference for personalized recommendation. In: IJCAI 2015: Proceedings of the 24th International Conference on Artificial Intelligence, pp. 1785–1791 (2015)
5. De Meo, P., Quattrone, G., Ursino, D.: A query expansion and user profile enrichment approach to improve the performance of recommender systems operating on a folksonomy. User Model. User-Adap. Inter. **20**, 41–86 (2010)
6. Palleti, P., Karnick, H., Mitra, P.: Personalized web search using probabilistic query expansion. In: WI-IATW 2007, pp. 83–86 (2007)
7. Shibamoto, E., Kittirojrattana, C., Koopipat, C., Hansuebsai, A., Takano, K.: A recommendation system of sightseeing places based on user's behavior of taking and editing photos. In: 2019 IEEE Pacific Rim Conference on Communications, Computers and Signal Processing (PACRIM), pp. 1–6 (2019)
8. Amazon.com. https://www.amazon.com/
9. Spotify. https://www.spotify.com/jp/
10. Facebook. https://ja-jp.facebook.com/
11. The CIFAR-10 dataset. https://www.cs.toronto.edu/~kriz/cifar.html
12. Caltech-256 object category dataset. http://www.vision.caltech.edu/Image_Datasets/Caltech256/

Antlion Optimizer Based Load-Balanced Transaction Scheduling for Maximizing Reliability

Dharmendra Prasad Mahato[1,2]([envelope]) and Van Huy Pham[2]

[1] Department of Computer Science and Engineering,
National Institute of Technology Hamirpur, Hamirpur, Himachal Pradesh, India
dharmendra@tdtu.edu.vn, dpm@nith.ac.in
[2] Faculty of Information Technology, Ton Duc Thang University,
Ho Chi Minh City, Vietnam
phamvanhuy@tdtu.edu.vn

Abstract. This paper presents a load-balanced transaction scheduling algorithm for maximizing reliability. The algorithm is based on the antlion optimizer method. Maximization of reliability in a grid system is an NP-hard problem. The paper compares the proposed algorithm with existing famous meta-heuristic algorithms. The result analysis shows that the proposed algorithm performs better compared to other algorithms.

1 Introduction

Maximizing reliability in grid computing with load-balanced transaction scheduling has not been sufficiently studied. Load-balanced transaction scheduling in grid computing environments becomes a challenging job as both the application and computational resources are heterogeneous [1]. Resource failure makes the problem more complicated. Therefore, some computing nodes become overloaded. Thus a load-balanced scheduling approach is needed as it schedules the job based on the failure possibility of the system constituents. Now, maximizing the reliability in a grid computing system with load-balanced transaction scheduling becomes an NP-hard problem. This paper presents a solution by using the soft computing approach of antlion optimizer (ALO) to optimize system reliability.

ALO [2] works on the concepts of the hunting method of antlions and mimics how they interact with their favorite prey, ants. This method finds the approximate solutions by employing a set of random solutions. This method consists of two populations: a set of ants and a set of antlions. The major contributions in this paper are as follows:

- The paper derives the reliability of grid computing systems considering load-balanced transaction scheduling. The derivation of reliability also considers the deadline-miss fault of the transaction processing.

L. Barolli et al. (Eds.): AINA 2021, LNNS 226, pp. 66–79, 2021.
https://doi.org/10.1007/978-3-030-75075-6_6

- The paper presents a meta-heuristic algorithm, Load Balanced Transaction Scheduling based on Antlion Optimizer (LBTS_ALO) to maximize the reliability in the grid computing systems.

2   Problem Formulation

Let us suppose a set of transactions is represented by T_i as $(i = 1, 2, \ldots, m)$ and are generated by using exponential distribution. It is assumed that each transaction executes within its deadline. If any transaction does not its deadline, it is rolled back or is aborted. Let us assume, N_j represents the set of nodes n as $(j = 1, 2, \ldots, n)$.

2.1   Reliability Model

If a transaction needs to be executed successfully within its deadline, the availability of resources is important to be focused on. The reliability formulation in this paper is influenced by [3–6] in which only node and link faults were considered. But in this paper, we have considered node fault, link fault as well as a deadline-miss fault. The concept of the deadline-miss fault has been influenced by and the concept of steady-state user-perceived availability has been influenced by [7].

As expressed in [3], the reliability of a distributed system is given as

$$R_{k,kb}(X) = \left[\prod_{k=1}^{n} R_k(X) \right] \cdot \left[\prod_{k=1}^{n-1} \prod_{b>k} R_{kb}(X) \right] \qquad (1)$$

Here $R_k(X)$ represents the reliability of a node N_k during a time interval t as $e^{-\gamma_k \sum_{i=1}^{m} x_{ik} e_{ik}}$, where e_{ik} is the expected execution time of transaction i running on node k, γ is the failure rate of node, and total time to execute the transactions at the node N_k is given as $\sum_{i=1}^{m} x_{ik} e_{ik}$. And $R_{kb}(X)$, the communication link reliability at a time interval t is given as $e^{-\sigma_{kb} \sum_{i=1}^{m} \sum_{g \neq i} x_{ik} x_{gb}(cost_{ig}/w_{kb})}$. Here l_{kb} is the communication link from k^{th} to b^{th} node, w_{kb} is the transmission rate of link l_{kb}, and σ_{kb} is the failure rate of communication link l_{kb}, then $\sum_{i=1}^{m} \sum_{g=1}^{m} x_{ik} x_{gb}(cost_{ig}/w_{kb})$ represents the total elapsed time for transmitting the transaction communication via l_{kb}.

The reliability when we consider deadline-miss fault in addition to node and communication link fault is given as

$$R_{k,kb,DM_i}(X) = \left[\prod_{k=1}^{n} R_k(X) \right] \cdot \left[\prod_{k=1}^{n-1} \prod_{b>k} R_{kb}(X) \right] \cdot \left[\prod_{i=1}^{m} R_{DM_i}(X) \right] \qquad (2)$$

Let us suppose $M/M/c$ represents the queue length in the grid computing system. Here first M represents memoryless with λ arrival rate of jobs, second M

represents service rate of jobs with μ and c represents the number of servers. Here the transactions follow the steady-state condition and queuing system model $M/M/c$.

Let us consider $R_{DM_i}(X)$ which represents the probability that there is no deadline-miss fault in the system. The deadline-miss fault rate is represented as ψ_i. Now we can compute $R_{DM_i}(X)$ using the Markov model as

$$R_{DM_i}(X) = e^{-\psi_i \cdot \left[\frac{1}{\mu} + \Pi_0 \cdot \frac{\rho(c\rho)^c}{c!(c\mu - \lambda)(1-\rho)}\right]}, \quad \forall c \in N \tag{3}$$

Here Π_0 is computed as $\Pi_0 = \left[\sum\limits_{i=0}^{c-1} \frac{(c\rho)^i}{i!} + \frac{(c\rho)^c}{c!(1-\rho)}\right]^{-1}$ with $\rho = \frac{\lambda}{c.\mu} < 1$.

Therefore, the reliability by considering the conditional steady-state user-perceived availability of resources is given as

$$R_{k,kb,DM_i,A_\lambda}(X) = \left[\prod_{k=1}^{n} R_k(X).A_\lambda\right] \cdot \left[\prod_{k=1}^{n-1}\prod_{b>k} R_{kb}(X)\right] \cdot \left[\prod_{i=1}^{m} R_{DM_i}(X)\right] \tag{4}$$

Here A_λ gives the steady-state availability of the resources under the load λ [7] as $A_\lambda = \sum_{c=1}^{n} A_{c,\lambda}Q_c$. Here $\forall c = 1, ..., n$ and $A_{c,\lambda}$ is given by $\sum_{i=0}^{K} r_i\Pi_i$ with the steady-state probabilities for the model given as $\Pi_i = \frac{(c\rho)^i}{i!}\Pi_0$, $1 \leq i \leq c - 1$. And therefore, $\Pi_i = \frac{c^c\rho^i}{c!}\Pi_0$, $i \geq c$.

In the above equations, $\rho = \frac{\lambda}{c\mu}$ and Q_c is given as

$$Q_c = \frac{n!}{c!(n-c)!}q^c(1-q)^{n-c}, \tag{5}$$

Here $q = \frac{\eta}{\gamma + \eta}$ is the availability of a single server.

The formulations of some important parameters of the problem are given as follow:

Suppose, $H(t)$ is the response time t, d represents the deadline, then the resource is said to be available at time t if $Pr[H(t) \leq d] \leq \phi$. Then the probability of resource availability $w_d(t)$ at time t is given by $w_d(t) - Pr[D(t) \leq d]$. Now, let us assume the resource availability at time t as $A(t)$ which can be given as $A(t) = Pr[w_d(t) \leq \phi]$.

Let $r_i(t)$ be the function that is the correctness characteristics of the system in the state S_i at time t. Then,

$$r_i(t) = \begin{cases} 1, & \text{if } w_{d|i}(t) \leq \phi \\ 0, & \text{otherwise} \end{cases} \tag{6}$$

We consider the system where the transactions arrive in Poisson fashion with rate λ. It is assumed that the failure, repair, and processing times of transactions are exponentially distributed. The performance of the transaction processing system is affected not only due to failures but also due to transient overloading of the system. Before discussing the reliability model, we introduce the availability model, and based on this we try to model the reliability of the transactions in the grid computing systems. If the system is in state S_i, where $S_i = i$ is the number of transactions in the system, n be the number of servers available at time t and $k \in n$, then $w_{d|k}(t)$ as described in [7] is given as

$$w_{d|k}(t) = \begin{cases} w_{exp}(d), & \forall k < n \\ (w_{erl} \otimes w_{exp})(d), & \forall k \geq n \end{cases} \tag{7}$$

where $w_{exp}(d) = e^{-\mu d}$ and $w_{erl}(d) = \sum_{k=0}^{n-k} \frac{(n\mu d)^k}{k!} e^{-n\mu d}$. Here $\otimes$ represents the convolution operator, and the probability of i transactions in the system supposed to be modeled as $M/M/n$. The $r_i(t)$ is the correctness characteristics function of the system in the state S_i at time t which is given as

$$r_i(t) = \begin{cases} 1, & \text{if resources are available at time } t \\ 0, & \text{otherwise} \end{cases} \tag{8}$$

Let M be the threshold after which all r_i's with $i \geq M$ will be 0. Then, the conditional steady-state availability can be given by $\sum_{i=0}^{M} r_i \Pi_i$

Therefore, under a transaction scheduling $X = \{x_{ik}\}_{1 \leqslant i \leqslant m, 1 \leqslant k \leqslant n}$,

$$A_\lambda(X) = \sum_{c=1}^{n} \sum_{i=0}^{m} r_i \Pi_i \cdot \frac{n!}{c!(n-c)!} q^c (1-q)^{n-c}. \tag{9}$$

2.2   Reliability Model

$$cost(X) = \sum_{k=1}^{c} \sum_{i=1}^{m} \gamma_k x_{ik} e_{ik} r_i \Pi_i . Q_c$$

$$+ \sum_{k=1}^{c-1} \sum_{b>k}^{m} \sum_{i=1}^{} \sum_{i \neq g}^{} \sigma_{kb} x_{ik} x_{gb} \left(\frac{cost_{ig}}{w_{kb}} \right) \tag{10}$$

$$+ \sum_{i=1}^{m} \sum_{k=1}^{n-1} \psi_i x_{ik} \cdot \left[\frac{\lambda}{\mu} + \frac{\Pi_0 \rho (c\rho)^c}{c!(1-\rho)^2} \right]$$

where $\Pi_0 = \left[\sum_{k=0}^{c-1} \frac{(c\rho)^k}{k!} + \frac{(c\rho)^c}{c!(1-\rho)} \right]^{-1}$. The reliability has been expressed regarding $cost(X)$ in [3]. Similarly, the reliability can be expressed as follows:

$$R_{k,kb,DM_i,A_\lambda}(X) = e^{-cost(X)} \tag{11}$$

3   Antlion Optimizer

3.1   Ant Lion Optimizer

The Antlion Optimizer is a recent meta-heuristic approach that models the behavior and interaction of ants and antlions mathematically. This approach has been developed to solve optimization problems considering the steps such as the random walk of ants, building traps, entrapment of ants in traps, catching preys, and implementation of re-building traps.

3.2   Inspiration

Antlions that are also known as doodlebugs come under the Myrmeleontidae family. As demonstrated in Fig. 1, there are generally two phases in their life cycle; larvae (as shown in Fig. 2) and adult. The hunt mechanism of the antlions which attracted the attention of the researchers is interesting. They use this mechanism when they are larvae. They make a small cone-shaped trap to trap the ants (as shown in Fig. 3). In this process, they cleverly sit under the pit and wait for the ants to be trapped. When the ants are trapped, they eat the ants and again amend the pit for the next hunt (as shown in Fig. 4). The main inspiration for the researchers for the ALO algorithm is that the antlions make bigger pit when they are more hungry.

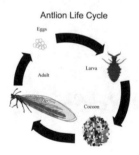

Fig. 1. Ant lion life cycle

Fig. 2. Antlion larva trails (doodles)

3.3   Operators of the ALO Algorithm

There are five main and necessary steps of hunts for larvae. These steps are as; random walk of ants, building traps, entrapment of ants in traps, catching preys, and re-building of traps. The main random walk in this algorithm is as follows (Table 1):

$$\phi(A) = [0, cumsum(2\phi(A_1) - 1), cumsum(2\phi(A_2) - 1), \dots, cumsum(2\phi(A_s) - 1),] \quad (12)$$

Table 1. Definitions

Decision Variables

$$\phi_A = \begin{cases} 1, & \text{if rand} > 0.5 \\ 0, & \text{if rand} \leq 0.5 \end{cases}$$

Fig. 3. Sand pit trap of an antlion

Fig. 4. Sand pit trap with remains of an ant

where *cumsum* calculates the cumulative sum, s is the maximum number of iteration, A shows the step of random walk (iteration in this study), $\phi(A)$ is a stochastic function and *rand* is a random number generated with uniform distribution in the interval of $[0, 1]$.

The location of ants should be stored in the following matrix:

$$M_{Ant} = \begin{bmatrix} A_{1,1} & A_{1,2} & \cdots & A_{1,d} \\ A_{2,1} & A_{2,2} & \cdots & A_{2,n} \\ \vdots & \vdots & \ddots & \vdots \\ A_{m,1} & A_{m,2} & \cdots & A_{m,d} \end{bmatrix} \tag{13}$$

where M_{Ant} is the matrix for saving the position of each ant, $A_{i,j}$ shows the value of the $j-th$ variable (dimension) of $i-th$ ant, m is the number of ants, and d is the number of variables.

The corresponding objective value for each antlion is calculated and stored in the following matrix:

$$M_{OA} = \begin{bmatrix} f(A_{1,1}, A_{1,2}, \cdots, A_{1,d}) \\ f(A_{2,1}, A_{2,2}, \cdots, A_{2,n}) \\ \cdots \\ \cdots \\ f(A_{m,1}, A_{m,2}, \cdots, A_{m,d}) \end{bmatrix} \tag{14}$$

where M_{OA} is the matrix for saving the fitness of each ant, $A_{i,j}$ shows the value of $j-th$ dimension of $i-th$ ant, m is the number of ants, and f is the objective function. In ALO, it is assumed that the antlions also hide somewhere in the search space. To save their positions and fitness values, the following matrices are utilized:

$$
M_{OAL} = \begin{bmatrix} f(AL_{1,1}, AL_{1,2}, \cdots, AL_{1,d}) \\ f(AL_{2,1}, AL_{2,2}, \cdots, AL_{2,n}) \\ \cdots \\ \cdots \\ f(AL_{m,1}, AL_{m,2}, \cdots, AL_{m,d}) \end{bmatrix} \tag{15}
$$

where M_{OAL} is the matrix for saving the fitness of each antlion, $AL_{i,j}$ shows the $j - th$ dimension's value of $i - th$ antlion, m is the number of antlions, and f is the objective function.

3.4   Random Walks of Ants

The random walk discussed above is used in the following equation to update the position of ants:

$$
W_i^A = \frac{(W_i^A - a_i) \times (b_i - c_i^A)}{(d_i^A - a_i)} + c_i \tag{16}
$$

where a_i is the minimum of random walk of $i-th$ variable, b_i is the maximum of random walk in $i - th$ variable, c_i^A is the minimum of $i - th$ variable at $A - th$ iteration, and d_i^A indicates the maximum of $i - th$ variable at $A - th$ iteration.

3.5   Trapping in Antlion's Pits

The impact of antlions on the movement of ants is modeled as follows:

$$
c_i^A = AntLion_j^A + c^A \tag{17}
$$

$$
d_i^A = AntLion_j^A + d^A \tag{18}
$$

where c^A is the minimum of all variables at $A - th$ iteration, d^A indicates the vector including the maximum of all variables at $A - th$ iteration.

3.6   Building Trap

Building traps is done with a roulette wheel is employed. The roulette wheel operator selects antlions based on their fitness during optimization. This mechanism gives high chances to the fitter antlions for catching ants.

3.7   Sliding Ants Towards Antlion

The mathematical model of this step of the hunt is given below. It may be seen in the equation that the radius of ants' random walks hyper-sphere is decreased adaptively.

$$c^A = \frac{c^A}{I} \tag{19}$$

$$d^A = \frac{d^A}{I} \tag{20}$$

where I is a ratio, c^A is the minimum of all variables at $A - th$ iteration, and d^A indicates the vector including the maximum of all variables at $A - th$ iteration.

3.8   Catching Prey and Re-building the Pit

In ALO, catching prey occurs when ants get fitter (dives inside the sand) than its associated antlion. An antlion is then required to update its position to the latest position of the hunted ant to enhance its chance of catching new prey. The following equation simulates this behavior:

$$AntLion_j^A = Ant_i^A, \text{if } f(Ant_i^A) > f(AntLion_j^A) \tag{21}$$

where A shows the current iteration.

3.9   Elitism

Elitism is an important characteristic of evolutionary algorithms that allows them to maintain the best solution(s) obtained at any stage of the optimization process. In this study, the best antlion obtained so far in each iteration is saved and considered as an elite. Since the elite is the fittest antlion, it should be able to affect the movements of all the ants during iterations. Therefore, it is assumed that every ant randomly walks around a selected antlion by the roulette wheel and the elite simultaneously as follows:

$$Ant_i^A = \frac{R_{AL}^A + R_{EL}^A}{2} \tag{22}$$

where R_{AL}^A is the random walk around the antlion selected by the roulette wheel at $A - th$ iteration, R_{EL}^A is the random walk around the elite at $A - th$ iteration, and Ant_i^A indicates the position of $i - th$ ant at $A - th$ iteration.

3.10   Roulette Wheel

Roulette Wheel is a selection strategy used in a swarm algorithm. This strategy is important and affects significantly the convergence rate of the algorithm. The strategy follows the rule as the better fitted an individual, the larger the probability of its survival. The roulette wheel method assumes that the probability of selection is proportional to the fitness of an individual. The method is briefly described as follows [8]. Let us consider, there are N numbers of nodes in a grid system. Each node is selected based on load factor characterized by its fitness $w_j > 0 (j = 1, 2, ..., N)$. Here let us assume that $w_j = \frac{1}{L_j}$. The selection probability of the $j - th$ individual is thus given as

$$p_j = \frac{w_j}{\sum_j^N w_j}, \quad (j = 1, 2, \ldots, N). \tag{23}$$

Let us assume a roulette wheel [8] with sectors of size proportional to w_j for $(j = 1, 2, ..., N)$. The selection of an individual is then equivalent to choosing randomly a point on the wheel and locating the corresponding sector. When a simple search is used, such a location requires $O(N)$ operations while the binary search needs $O(logN)$.

4   Proposed Algorithm

This section proposes the load-balanced transaction scheduling based on ant lion optimizer (LBTS_ALO). In the proposed algorithm, the ants represent nodes and antlions represent transactions in the grid system.

Algorithm 1. LBTS_ALO

1: Initialize the first population of ants and antlions randomly
2: Calculate the fitness of ants and antlions (using Eq. (10))
3: Find the best antlions and assume it as the elite (determined optimum)
4: **while** the end criterion is not satisfied **do**
5:     **for** every ant **do**
6:         Select an antlion using Roulette wheel using Eq. (23)
7:         Update c and d using equations Eqs. (17) and (18)
8:         Create a random walk and normalize it using Eqs. (12) and (16)
9:         **if** $Load_j < \frac{N}{2}$ **then**
10:            Select N_J as the best node
11:         **else**
12:             Go to Step 6.
13:         **end if**
14:         Update the position of ant using Eq. (22)
15:     **end for**
16:     Calculate the fitness of all ants
17:     Replace an antlion with its corresponding ant if it becomes fitter (Eq. (21))
18:     Update elite if an antlion becomes fitter than the elite
19: **end while**
20: **return** elite

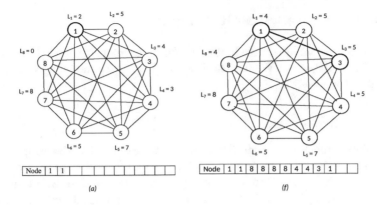

Fig. 5. A working example of the LBTS_ALO when $N = 8$

5   Applying the Algorithm

We applied the proposed algorithm on a complete undirected graph denoted by $G = (N, E)$ where N is the set of nodes; $E = N \times N$ is the set of edges between the nodes gives an illustrative example of how the LBTS_ALO works. Suppose there are m number of transactions $(T_1, T_2, \ldots, T_m)$ which arrive at the system with available nodes (suppose set $(N = j = 1, 2, \ldots N$ has $N = 8)$ as shown in Fig. 5.

Initialization: Line 1 in the proposed algorithm initializes the population of ants and antlions along with the load of each node of the graph as: $L_1 = 2$, $L_2 = 5$, $L_3 = 4$, $L_4 = 3$, $L_5 = 7$, $L_6 = 5$, $L_7 = 8$, and $L_8 = 0$.

Iteration: In this phase, the algorithm selects the feasible solution at each iteration. Initially, at each node, a transaction is randomly placed. Here transaction behaves like antlion and nodes behave like ants. In each iteration, an antlion searches for the ants as its prey and after a successful search, it again updates the fitness values. In each iteration, **while** loop works from lines 4 to 19. Every antlion at each iteration waits for the appropriate nodes with minimum load using Eq. (23). Thus, in one iteration the antlion finds the appropriate node or ant as prey. This best node is a feasible solution. Then each node in the same iteration might have traversed the graph and have searched for the best node. Among all the feasible solutions found in the same iterations, the node which has the highest probability as shown in (Eq. (22) and Eq. (23)) is selected as the optimal solution. Figure 5 applies the proposed algorithm.

Final: Finally the algorithm updates the fitness value of all the nodes. The process continues until the queue for incoming transactions is not empty.

6   Simulation and Result Analysis

The proposed algorithm is evaluated through simulations with Colored Petri Nets (CPNs or CP-nets). We use the Poisson process for modeling various faults occurring in the system.

6.1   Result Evaluation

We modified some known algorithms for comparison with the proposed algorithm. We compared the performances of the proposed algorithm with those algorithms; HBO, ACO, and Randomized algorithm with random selection method [9]. We ran each algorithm 40 times at each time unit value for every problem instance to get the result.

Availability

Availability is an important factor for any system. The maximum availability is needed to maximize the reliability of the system (as shown in Eq. (4) to Eq. (5)). We have the comparative results of availability in grid computing system along with several iterations from 100 to 1000 in 40 runs using the mentioned algorithms as shown in Table 2 and Fig. 6.

Reliability

The reliability of grid computing based transaction processing system considering the conditional steady-state user-perceived availability is computed using Eq. (4) to Eq. (5) and Eq. (11). Since the reliability is based on the availability of the system, we needed the result of the availability to find the reliability of the system. The result is shown in Table 3 and Fig. 7.

Table 2. Availability in grid computing

Time	LBTS_ALO	HBO	ACO	Randomized
100	**0.9981**	0.9977	0.9975	0.9877
200	**0.9933**	0.9935	0.9928	0.9863
300	**0.9953**	0.9958	0.9932	0.9881
400	**0.9929**	0.9939	0.9928	0.9910
500	**0.9899**	0.9881	0.9880	0.9875
600	**0.9915**	0.9896	0.9890	0.9878
700	**0.9945**	0.9931	0.9905	0.9831
800	**0.9875**	0.9887	0.9880	0.9846
900	**0.9880**	0.9890	0.9884	0.9839
1000	**0.9888**	0.9895	0.9879	0.9871

Table 3. Reliability in grid computing

Time	LBTS_ALO	HBO	ACO	Randomized
100	**0.5975**	0.6670	0.6600	0.6100
200	**0.8125**	0.8650	0.7700	0.6300
300	**0.9545**	0.9530	0.7750	0.8330
400	**0.9675**	0.9490	0.8950	0.8890
500	**0.9545**	0.9770	0.9150	0.9200
600	**0.9750**	0.9500	0.9270	0.9432
700	**0.9675**	0.9700	0.9350	0.9503
800	**0.9694**	0.9600	0.9400	0.9482
900	**0.9755**	0.9690	0.9500	0.9625
1000	**0.9785**	0.9730	0.9550	0.9702

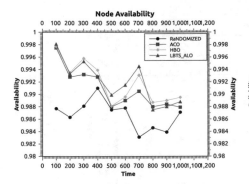

Fig. 6. Availability analysis   **Fig. 7.** Reliability analysis

Miss Ratio

Miss Ratio is one of the important performance parameters for a transaction processing system. For better performance of the system, the miss ratio should be minimum.

Miss Ratio [10] can be calculated as

$$Miss\ Ratio = \frac{T_{miss}}{T_{total}} * 100\% \tag{24}$$

where T_{miss} is the number of transactions missing the deadlines and T_{total} is the total number of handled transactions. The result of miss ratio of transaction in grid computing system is shown in Table 4 and Fig. 8.

Table 4. Miss ratio (%) in grid computing

Time	LBTS_ALO	HBO	ACO	Randomized
100	**32.7500**	33.330	34.000	50.00
200	**16.4510**	13.510	23.075	39.00
300	**03.7955**	04.650	22.454	39.50
400	**02.7205**	05.140	10.484	33.75
500	**03.5261**	06.323	08.475	32.00
600	**01.9430**	04.987	07.317	22.40
700	**02.2488**	03.000	06.472	19.60
800	**02.0450**	04.104	06.000	22.40
900	**01.5853**	03.050	04.914	22.40
1000	**01.4245**	02.650	04.414	11.60

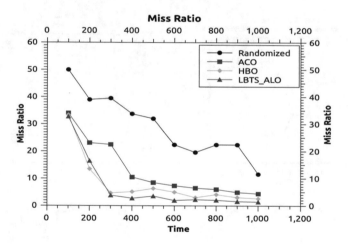

Fig. 8. Miss ratio analysis

7   Conclusion and Future Directions

This paper presents the antlion optimizer-based load-balanced transaction scheduling algorithms. The nature of transaction arrival at the nodes is assumed as exponential distributions. But, the nature of the transactions may follow the Gaussian mixture model. Therefore, in the future, the reliability analysis of load-balanced transaction scheduling would be interesting and useful research.

References

1. Shatz, S.M., Wang, J.P.: Models and algorithms for reliability-oriented task-allocation in redundant distributed-computer systems. IEEE Trans. Reliab. **38**(1), 16–27 (1989)

2. Mirjalili, S.M., Jangir, P., Saremi, S.: Multi-objective ant lion optimizer: a multi-objective optimization algorithm for solving engineering problems. Appl. Intell. **46**, 79–95 (2016)
3. Shatz, S.M., Wang, J.P., Goto, M.: Task allocation for maximizing reliability of distributed computer systems. IEEE Trans. Comput. **41**(9), 1156–1168 (1992)
4. Mahato, D.P., Singh, R.S.: On maximizing reliability of grid transaction processing system considering balanced task allocation using social spider optimization. Swarm Evol. Comput. **38**, 202–217 (2018)
5. Mahato, D.P.: CPNs based reliability modeling for on-demand computing based transaction processing. In: Proceedings of the 47th International Conference on Parallel Processing Companion, pp. 1–4 (2018)
6. Mahato, D.P., Singh, R.S.: Reliability modeling and analysis for deadline-constrained grid service. In: 2018 32nd International Conference on Advanced Information Networking and Applications Workshops (WAINA), pp. 75–81. IEEE (2018)
7. Mainkar, V.: Availability analysis of transaction processing systems based on user-perceived performance. In: 1997 IEEE the Sixteenth Symposium on Reliable Distributed Systems: Proceedings, pp. 10–17 (1997)
8. Lipowski, A., Lipowska, D.: Roulette-wheel selection via stochastic acceptance. Phys. A: Stat. Mech. Appl. **391**(6), 2193–2196 (2012). http://www.sciencedirect.com/science/article/pii/S0378437111009010
9. Chang, R.-S., Chang, J.-S., Lin, P.-S.: An ant algorithm for balanced job scheduling in grids. Future Gener. Comput. Syst. **25**(1), 20–27 (2009). http://www.sciencedirect.com/science/article/pii/S0167739X08000848
10. Tang, F., Li, M., Huang, J.Z.: Real-time transaction processing for autonomic grid applications. Eng. Appl. Artif. Intell. **17**(7), 799–807 (2004). Autonomic Computing Systems. http://www.sciencedirect.com/science/article/pii/S0952197604001228

An Adaptive Large-Scale Trajectory Index for Cloud-Based Moving Object Applications

Omar Alqahtani[⊠] and Tom Altman

Department of Computer Science and Engineering, University of Colorado Denver,
Denver, CO, USA
{omar.alqahtani,tom.altman}@ucdenver.edu

Abstract. The cutting-edge cloud-based computing platforms are a
typical solution for the tremendous volumes of the moving object trajectories and the vast trajectory-driven applications. However, many challenges have been raised by the adopted distributed platforms, the nature
of the trajectories, the diversity in query types, the enormous options of
computing resources, etc. We propose a Dynamic Moving Object Index
that is able to adapt to the changes in a dynamic environment while
maximizing the benefits out of the available resources without any fine-
tuning. It balances the index structure between the spatial and object
localities in order to control the parallelism capacity, the communication overhead, and the computation distribution. The proposed index
has innovative global and local indexes that implement several optimization approaches in order to contain the impact of balancing the locality
pivot in a dynamic environment. We also conduct extensive experiments
on two datasets and various queries including space-based, time-based,
and object-based query types. The experiment study shows a significant
performance improvement compared to existing indexing schemes.

1   Introduction

The availability of location-aware devices leads to skyrocketing volumes of moving object trajectories, which catalyze a wide range of analytic and real-time
applications. Cloud computing is an ideal solution to cope with large-scale data
and various application demands. Indeed, the idea of having computational
resources without the overhead necessary to physically maintain, update, or scale
encourages many organizations, firms, agencies, and businesses of different scales
to move forward and adopt cloud systems. Cloud systems, e.g. Amazon AWS,
Google Cloud Engine, IBM Cloud, Microsoft Azure, Cloudera, etc., are enriched
with computing platforms and software. Spark [3] has been adopted by many
cloud systems, as it offers an in-memory distributed computation platform.

However, processing large-scale moving object trajectories on cloud systems
creates numerous challenges and obstacles because of the distributed platforms,
nature of data, and query types. Generally, cloud systems offer vast options

© The Author(s), under exclusive license to Springer Nature Switzerland AG 2021
L. Barolli et al. (Eds.): AINA 2021, LNNS 226, pp. 80–94, 2021.
https://doi.org/10.1007/978-3-030-75075-6_7

for computing, storing, and communicating resources, which require continuous fine-tuning to maximize the benefits out of the available resources. Also, trajectories are mostly skewed and might cause a computational skewness within the computing nodes (cluster). On the other hand, trajectories have different degrees of compactness which comes as a result of the application's spatial-scope such as city, metropolitan, country, or continent. Multistage and aggregation queries on compact trajectories increase the network traffic, which might cause a performance bottleneck. However, spatial selective queries often do not utilize the parallelism capacity on sparse trajectories.

To overcome the aforementioned challenges, we need to consider the three fundamental adaptation factors: the availability of resources, the nature of trajectories, and the diversity in queries. In this paper, we propose a Dynamic Moving Object Index (*DMOI*) that is capable of adapting in various environments to satisfy a wide range of analytic and real-time applications. *DMOI* leverages the Compulsory Prediction Model (CPModel) [2] to balance both spatial and object localities in a dynamic environment. CPModel is a machine learning model that predicts a Locality Pivot (L_{pivot}) depending on the adaptation factors. To eliminate the impact of the continuance changes in L_{pivot}, *DMOI* provides innovative global and local indexes that implement numerous optimizations including lazy partitioning, virtual trees, traversing on-the-fly, local-optimal locality pivot, tree partitioning, etc.

The main contributions of this work are as follows:

- We propose *DMOI* as an adaptive index for analytic and real-time applications.
- We leverage CPModel to control both spatial and object localities in a dynamic environment.
- We develop innovative global and local indexes to eliminate the effects of the locality pivot fluctuation.
- We evaluate our work by conducting extensive performance experiments comparing various indexing schemes on spatial, temporal, spatio-temporal, continuous, aggregation, and retrieval queries.

The rest of the paper is organized as follows. We discuss the related work in Sect. 2. The index structure, construction, and maintenance are introduced in Sect. 3. We present and analyze the experimental study in Sect. 4. Our concluding remarks are discussed in Sect. 5.

2   Related Work

In this section, we review the related work conducted on moving object trajectories on top of MapReduce platforms. SpatialHadoop [5] is an extension of Hadoop designed to support spatial data by including global and local spatial indexes to speed up the processing of space-based queries. MD-HBase [9] is an extension of HBase which runs on top of Hadoop as a non-relational Key-Value database. MD-HBase linearizes any multi-dimensional index into a single-dimensional index via the Z-order space-filling technique. Hadoop-GIS [1] extends Hive [16], a warehouse Hadoop-based database, by using a grid-based global index and an on-demand local index. PRADASE [8] concentrates on processing trajectories, but it only covers range and trajectory-retrieve queries. It partitions space and time by using a multilevel-grid index as a global index. Another index is used to hash all segments on all the partitions belonging to a single trajectory to speed up the object retrieving query.

GeoSpark [20] is a Spark version of SpatialHadoop in terms of indexing and querying. LoctionSpark [15] distributes the data based on a grid or quadtree global index. Also, it reduces the impacts of query skewness and network communication overhead by tracking query frequencies to reveal cluster hotspots and re-distribute them. SpatialLocation [19] is proposed to process the spatial join through the Spark broadcasting technique and grid index.

DTR-tree [17] distributes trajectories based on 1-dimension only, where the global and local indexes are R-trees built based on the partitioning dimension. Also, a Voronoi-based index is used in [12] to process the top-k similarity queries, a trajectory-based query. Each cell is statically indexed on the temporal dimension. Any trajectory that crosses a partition boundary is split. An auxiliary index is used to trace all the segments that belong to a trajectory. SharkDB [18] indexes trajectories based only on time frames in a column-oriented architecture to process range query and k-NN.

Generally, most of the prior work has focused on spatial or temporal distribution. The resulting distribution partially preserves the spatial and the object localities, in most cases. However, they lack a mechanism to control both localities, which is an essential key to improve the performance and enhance the adaptation capability. The nature of a trajectory, i.e. consecutive timestamped spatial points, creates contradictory domains, which can be seen in the spatial locality and the object locality. As a result, some of the previous contributions optimize their systems to contain this contradiction by focusing on spatial locality and spatial queries, e.g. range query, k-NN, etc., with an object-based auxiliary index, or by only focusing on a particular operator by building an ad-hoc index. To the best of our knowledge, no work has been conducted on a distributed computation that would simultaneously balance an index structure between the spatial and the object localities, while targeting vast query types in a dynamic environment.

In our previous work [2], we proposed a Resilient Moving Object Index (RMOI) for static datasets, where it only targets analytic applications. RMOI adjusts both spatial and object localities by using machine learning models. Those models depend on proportional features to predict L_{pivot}, which is interpreted in the required number of the space-based partitions (α) and the required number of the object-based partitions (β) per α. RMOI starts with the space-based partitioning. Then, each spatial partition is partitioned further based on the object identifier.

In a dynamic environment, the prediction features are continuously changing. For example, when new sub-trajectories arrive after distributing the data and building the global and local indexes, the data size will increase. Consequently, the prediction features, Computation Power Ratio (ComR) and Memory Usage Ratio (MemR), will change, since they depend on the data size. The changes in the features will most likely lead to a change in the previous predicted L_{pivot}. When α or β changes, then all the previous distributed trajectories, including the global and local indexes, need to be adjusted to the new α. Adjusting at this scale will usually be expensive and not practical for real-time applications.

3   Dynamic Moving Object Index

DMOI consists of three main components: partitioning engine, global index, and local index. The Partitioning Engine (PE) is responsible for preparing the partitioning plan and reviewing it periodically. It considers the current time, the near future, and the far future of *DMOI*.

The global index of *DMOI* consists of a spatial tree and object sub-trees. The spatial tree is a binary tree that represents the space-based partitioning, similar to k-d-B-tree [13]. Each spatial node of the spatial tree contains an MBR, node identifier (nid_s), and its level. A spatial node serves as a regular, a leaf, or a lazy-leaf node. A leaf node is used by *DMOI* for the current distribution, where the total number of leaf nodes is α. A lazy-leaf node is used for future distribution, where the total of the lazy-leaf nodes is $L\alpha$. A spatial node could be a leaf and a lazy-leaf at the same time, where $\alpha \leq L\alpha$. The node naming procedure uses the binary digits $\{0,1\}$, however, the name is stored as an integer in nid_s. The children of a spatial node use their parent nid_s and append 0 or 1 to the end of their names. Since they are stored as an integer, the first child inherits its parent nid_s directly. The second child takes its parent nid_s and adds 2^{l_p}, where l_p is the parent level. The root has $nid_s = 0$ and $l = 0$. As a result, the $nids_s$ of a full level l_i are $[0, 1, 2, \ldots, 2^{l_i})$, for $i \geq 0$. The spatial tree expands based on the $nids_s$ order of the last full level, as illustrated in Fig. 1a, to ensure strict order for both the leaf and the lazy-leaf nodes.

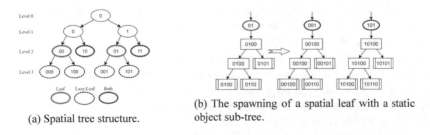

(a) Spatial tree structure.

(b) The spawning of a spatial leaf with a static object sub-tree.

Fig. 1. A spatial tree with a concatenated static object sub-trees.

The object tree is a binary tree that is used during object-based partitioning. Each spatial leaf node is concatenated with an object sub-tree. The object tree is static or dynamic depending on the preferred locality indicator (Γ). The main difference between the static and the dynamic object trees is in the naming procedure. The dynamic object tree is similar to the spatial tree in the naming procedure and will continue from where the spatial tree stopped. The final leaf identifier (nid_g) of the global tree is as follows: $nid_g = (nid_o \cdot nid_s)$, where nid_o is the object node identifier, and the operator $\cdot$ is a string concatenating operator. As a result, the global leaf $nids_g$ are in strict order $[0, \ldots, tpn)$, where tpn is the target partitions number.

On the other hand, the static object nodes depend on a prefix naming technique. The object root offsets the spatial leaf's name by the bit numbers of β. The rest of the object nodes will use this room for their names starting from 0. As a result, $nid_g = (nid_s \cdot nid_o)$. The main benefit of the prefix naming is to reduce the impact of spawning on a spatial node, as illustrated in Fig. 1b. The difference between two object sub-trees, i.e. sibling sub-trees, is only the leftmost bit. So, the migrated segments of any Spark partition related to an object leaf will be moved to only one partition.

DMOI provides a custom-built Spark partitioner that uses a hash map to connect the leaf $nids_g$ of the global index with Spark partitions. Each Spark partition carries nid_g as its identifier, which will be used throughout *DMOI*'s operations. The key of the partitioner is nid_g, while the value is 2-tuple contains the leaf levels and the actual partition identifier used by Spark. *DMOI* leverages the usage of the partitioner to dispense with the dynamic object tree. it computes the $nids_o$ of the dynamic object tree on-the-fly with the help of the partitioner.

The local index contains a grid R-tree (GR-tree) and a binary interval tree (BinTree). Each level of GR-tree is associated with a virtual grid (G). Each node of the level l_i has a unique integer identifier (cid_i) corresponding to its virtual cell at the grid G_i. The cid is composed of two signed shorts x and y corresponding to the space dimensions. The size of the local index is limited to 128 MB, or, roughly, one million segments. So, the two signed shorts of cid are more than enough. The active cells of the leaf grid are hashed in a grid map (GM), which is used to access the leaf nodes directly. The local index allows a top-down traversing starting from the root. Interestingly, each leaf node, or its active grid cell, can compute its ancestor cids to the root, which dramatically reduces the constant time of the traversing complexity.

The GR-tree can expand upward but not the other way. The leaf grid is final where the cell size is computed based on a heuristic method, similar to [7,14]. It ensures that a cell volume can not exceed a certain threshold on a normal distributed data. Although trajectories are skewed, the size of the resulting cells, with a very low threshold, is sufficient for spatial pruning. Also, the segments of the trajectories mostly come in compact packets due to the highways and the main routes, and it is not effective to break them further for more spatial pruning.

Index Construction. The construction of *DMOI* starts with PE to prepare the partitioning plan. First, PE computes α_i, β_i, and tpn_i based on the predicted

$L_{\text{pivot}.i}$, for $1 \leq i \leq 3$. The prediction model CPModel[1] predicts L_{pivot} based on the following features: Computation Power Ratio (ComR), the Memory Usage Ratio (MemR), and the Trajectory Overlapping Indicator (TOver[2]). All the required parameters to compute ComR and MemR are static except the data size. So, to predict $L_{\text{pivot}.1}$, PE uses the available data size. For $L_{\text{pivot}.2}$, PE doubles tpn_1, and then doubles tpn_2 for $L_{\text{pivot}.3}$.

After that, PE finds the preferred locality on the long run (Γ) by doubling the last tpn until MemR ≥ 0.65, which represents the maximum capacity. After predicting $L_{\text{pivot}.\Gamma}$, PE checks if α_Γ is higher than β_Γ, then Γ will indicate that DMOI prefers spatial locality (Γ_{sl}). Otherwise, DMOI prefers object locality (Γ_{ol}).

PE finishes the partitioning plan by finalizing α_{ct}, $L\alpha_{ct}$, and β_{ct}. The idea is to smooth the fluctuations between α_i and β_i for the current and near-future time. If the preferred locality is Γ_{sl}, PE fixes β and allows the global index to only grow on α, and vice versa. As a result, with Γ_{sl}, $\beta_{ct} = \lfloor (\beta_1 + \beta_2 + \beta_3) \div 3 \rfloor$, $L\alpha_{ct} = \lceil tpn_3 \div \beta_{ct} \rceil$, and $\alpha_{ct} = \lceil tpn_1 \div \beta_{ct} \rceil$. With Γ_{ol}, $\alpha_{ct} = L\alpha_{ct} = \alpha_3$ and $\beta_{ct} = \lceil tpn_1 \div \alpha_{ct} \rceil$. When $tpn_{ct} \geq tpn_3$, PE will revisit the partitioning plan.

Now, DMOI is ready to build the global index based on Γ, α_{ct}, $L\alpha_{ct}$, and β_{ct}. It depends on a sample set (ST) from the available trajectories to build the spatial tree similar to building a k-d-B-tree, as illustrated in Algorithm 1. The construction is done level by level following the order of the $nids_s$. If $\Gamma = \Gamma_{sl}$ and $\beta \geq 2$, DMOI builds static object sub-trees concatenated to each spatial leaf. The construction of a static object tree is similar to Algorithm 1, except for the naming procedure as we have explained before. When $\beta = 1$, there is no need for an object tree. If $\Gamma = \Gamma_{ol}$, DMOI depends on a virtual dynamic object tree and no need to construct it.

To build the custom partitioner, DMOI computes the required $nids_g$ on-the-fly without any traversing. With Γ_{ol}, DMOI picks the level ($\lfloor \log_2(tpn - 1) \rfloor$) which is before the last level, since it is guaranteed to be a full level. Next, it checks every nid_g of that level by testing the second child $nid_{g.2}$. If $nid_{g.2} < tpn$, it means this child and its sibling exist in the global tree. Their identifiers and levels will be added to the partitioner. Otherwise, nid_g only will be added with its level. With Γ_{sl}, the spatial tree alone and the static object sub-tree alone are in strict order. So, DMOI finds the spatial leaves, then it finds the object leaves and concatenates them.

Next, DMOI broadcasts the spatial tree nodes to the executor nodes. Then, it launches a Spark job, where each executor helps in tagging the segments ($Segs$). A segment tag ($Seg.tag$) is an integer that matches an nid_g. With Γ_{sl}, each executor inserts Seg into the broadcast spatial tree to find its correspondent spatial leaf nid_s. After that, the executor computes $Seg.tag = (nid_s \ll \beta_{\text{BitNum}}) + (Seg.Tid \text{ MOD } \beta)$, where β_{BitNum} is used to make room

---

[1] CPModel is a light-weight polynomial regression model that is already trained. For more details about CPModel and its features, please refer to our previous work [2].

[2] TOver is used to reveal the ratio of the MBRs of the trajectories with respect to the global scope.

Algorithm 1: Spatial Tree Construction

---

1 level ← 0
2 root ← CreateNode(mbr, level = 0, nid = 0)
3 $N_1[\,] \leftarrow \{\text{ root }\}$
4 $N_2[\,] \leftarrow$ new Array $[\,1 \ll (\,++\text{level}\,)\,]$
5 leafNum ← 1
6 **while** *true* **do**
7   **for** i ← 0 **to** $(N_1.\text{length}) - 1$ **do**
8   leafNum ← BuildLevel($N_1[i]$, N_2, ST, α, leafNum)
9   **if** leafNum $\geq L\alpha$ **then Break** while loop
10    $N_1 \leftarrow N_2$
11    $N_2[\,] \leftarrow$ new Array $[\,1 \ll (\,++\text{level}\,)\,]$

12 **return** root
13 **Function** BuildLevel(node, N_2, ST, α, leafNum):
14   PrepareChildrenMBRs(node.mbr, mbr_1, mbr_2, ST)
15    $node_1 \leftarrow$ CreateNode(mbr_1, level = node.level+1, nid = node.nid)
16    $node_2 \leftarrow$ CreateNode(mbr_2, level = node.level+1, nid = (1≪node.level) + node.nid)
17   node.SetChildrenReferences($node_1$, $node_2$)
18    $N_2[\,node_1.\text{nid}\,] \leftarrow node_1$
19    $N_2[\,node_2.\text{nid}\,] \leftarrow node_2$
20   node.SetLazyLeaf(*false*) // `Default leaf and lazy-leaf: true`
21   **if** $(++\text{leafNum}) \leq \alpha$ **then** node.SetLeaf(*false*)
22   **else** $node_1$.SetLeaf(*false*); $node_2$.SetLeaf(*false*)
23   **return** leafNum

---

for the object leaf nid_o, and $Seg.Tid$ is the segment's trajectory identifier (Tid). Then, the insertion will continue until reaching a spatial lazy-leaf nid_z, where $Seg.tag_z = (nid_z \ll \beta_{\text{BitNum}}) + (Seg.Tid \text{ MOD } \beta)$. The segments will carry $Seg.tag_z$ for future distribution.

With Γ_{ol}, the tagging computation starts with $Seg.tag_z$ as follows: $Seg.tag_z = (Seg.Tid \ll n_{\text{level}}) + nid_s$, where n_{level} is the correspondent spatial leaf's level. Then, $Seg.tag = tpn_{\text{bitmask}} \wedge Seg.tag_z$, where $\wedge$ is a bitwise AND. The bitmask is calculated based on the bit number of tpn. It helps to find nid_g at the last level of the global tree. If nid_g does not exist, the bitmask (tpn_{bitmask}) is temporarily shifted back one bit, and the calculation of $Seg.tag$ will be repeated. The final step in the partitioning phase is to group $Segs$ across the cluster by their $Seg.Tag$.

The last phase in the index construction is building the local index, GR-tree and BinTree. *DMOI* builds a local index for each Spark partition through a Spark job, which will be carried out by the executor nodes. To build a GR-tree, *DMOI* starts by scanning $Segs$ to find $X_{\max}$, $X_{\min}$, $Y_{\max}$, and $Y_{\min}$. Then, it computes the grid range (G_{range}) as follows:

$$G_{\text{range}} = \sqrt{\frac{(X_{\max} - X_{\min}) \times (Y_{\max} - Y_{\min}))}{n}} \times C_{\text{size}} \,,$$

where C_{size} is the estimated cell size $(= 5)$, and n is the maximum number of *Segs* per Spark partition. After that, it finds the correspondent leaf cell identifier (cid_{leaf}) based on the segment's middle point (X_{mid} , Y_{mid}) as follows:

$$X_{cid} = \left\lfloor \frac{(X_{mid} - X_{min})}{G_{range}} \right\rfloor , \quad Y_{cid} = \left\lfloor \frac{(Y_{mid} - Y_{min})}{G_{range}} \right\rfloor$$

$$cid_{leaf} = (Y_{cid} \ll G_{offset}) + (G_{bitmask} \wedge X_{cid}),$$

where G_{offset} is set to 16 (short size), and $G_{bitmask}$ is $2^{16} - 1$. Once a cid_{leaf} is found, *Seg* is inserted in the corresponding leaf node. The GR-tree leaf is retrieved using the grid map (GM) if existed. Otherwise, GR-tree creates a new leaf node, and GM adds a new entry consists of cid_{leaf} as a key and the leaf's array index as a value.

After inserting all the segments and creating the needed leaf nodes, GR-tree completes building the remaining levels recursively. From any node of GR-tree, we can compute its parent cid_i, and its three sibling's identifiers, as follows:

$$cid_i = \left(\left\lfloor \frac{Y_{cid_{i+1}}}{2} \right\rfloor \ll G_{offset} \right) + \left(G_{bitmask} \wedge \left\lfloor \frac{X_{cid_{i+1}}}{2} \right\rfloor \right).$$

As a result, to build an above level l_i, all the GR-tree nodes of level l_{i+1} are stored in a temporary hash map (GM_{i+1}), where each entry has cid_{i+1} as the key and the node as the value. GR-tree also needs another temporary GM_i to store the new parents' nodes. It starts by iterating over the nodes in GM_{i+1} and creates their parents with the proper cid_i. After finishing the iteration, GM_{i+1} is no longer needed. GR-tree continues until there is only one node, i.e. the root. Although the construction requires more space, the time complexity is reduced to linear on average[3].

BinTree is a binary interval tree that is developed by JTS Topology Suite [6] to process temporal queries. GR-tree leaf nodes are stored in an array list, where each leaf maintains its interval during insertion. BinTree is built on the intervals of GR-tree leaves, where the objects are the leaves' indices.

Index Maintenance. The maintenance of *DMOI* involves three procedures: minor, major, and full-scale updates. The first procedure is an insertion process to the local indexes without any global tree spawning. When *DMOI* receives new data, it checks if any Spark partitions exceed the maximum number of *Segs* by using a Spark global accumulator variable (Acc). Next, the driver node tags the new *Segs*, similar to the tagging process during the construction. Then, the tagged *Segs* are broadcast to the cluster nodes as a hash map. After that, through a Spark job, each partition retrieves the tagged *Segs* that carried its nid_g.

Each partition needs to insert the new *Segs* to its GR-tree and update Bin-Tree accordingly. After computing cid_{leaf}, the leaf node can be retrieved via GM. If the leaf node does not exist, GR-tree computes the lineage of the new

---

[3] It is subject to a good hashing method. We use Java 1.8 HashMap.

leaf node, as a list of cid_i, based on the level numbers. If the last cid matches the root, then GR-tree traverses the tree based on the lineage list and creates nonexistent nodes. Otherwise, GR-tree expands upward to reflect the new levels brought by the new leaf. To reduce the path, an optimization could be applied by folding any node that has only one child.

The second procedure, major update, takes place when at least one partition exceeds the maximum size. Since a partitioning process is an expensive Spark job, *DMOI* checks Acc again and considers any partition that exceeds two-third of the maximum size for a partitioning. After that, *DMOI* prepares an updating map, which contains the old and the new $nids_g$.

With Γ_{ol}, *DMOI* computes the new children $nid_{g.1}$ and $nid_{g.2}$ for each $nid_g \in Acc$ that needs to be partitioned. First child inherits its parent nid_g. Second child gets $(1 \ll l_{\mathrm{nid_g}}) + nid_g$. The updating map will have the old nid_g as a key and the new $nid_{g.2}$ and its equivalent level $(l_{\mathrm{nid_{g.2}}})$ as a value. Next, *DMOI* launches a Spark job after broadcasting the update map. Each partition in the update map needs to partition its own local index into two. First, *DMOI* traverses GR-tree and clones the nodes to GR-tree$_2$. When reaching the leaf nodes, *DMOI* tests the segments based on their lazy tags $(Seg.tag_z)$ as follows: $Seg.tag = ((1 \ll l_{nid_{g.2}}) - 1) \wedge Seg.tag_z$. The result is either $nid_{g.1}$ or $nid_{g.2}$ because the object distribution depends on the binary digits of Tid. The first operand of the bitwise $\wedge$ is a bitmask similar to the one we use during the construction with only one extra digit. So, the result of this extra digit, i.e. the output of $\wedge$ on the last digit, is either 0 or 1. With 0, the resulting tag will not change, and it is equivalent to the old nid_g. With 1, the tag is going to be equal to $nid_{g.2}$, since this is how it is created in the first place. Finally, *DMOI* traverses both GR-trees back again to the roots to fix the parents' nodes or delete them if they are empty.

With Γ_{sl}, *DMOI* expands the global tree on α instead of β as discussed previously. First, it finds the spatial leaf that has an object node that exceeds two-third of the maximum size. Then, as illustrated in Fig. 1, *DMOI* pushes down the leaf flag toward the lazy-leaf nodes, if possible. Next, it concatenates the static object sub-tree to the first new leaf and clones it to the second new leaf after fixing the nodes' names. After that, it fills the update map for each object leaf as before with a piece of additional information. It includes the MBR of the second spatial leaf to speed up the partitioning process of GR-tree. The rest is similar to what we have seen with Γ_{ol}.

If the spatial leaf is a lazy-leaf, then it is impossible to spawn that node. *DMOI* will skip spawning that node by not including its object leaves in the update map for one round only. However, it will get its new data. The second time, for this node or any other spatial node, *DMOI* will switch to a full-scale update procedure. The reason for such a practice is to give time for other leaf nodes to spawn. Our observation indicates that the first node hits the tree limits; it most probably has more than two-third, but does not exceed the maximum size. In other words, this node has spawned before at least twice, while other nodes are catching up. So, what triggers the major update procedure, i.e. at least one node exceeds the maximum size, is most likely not this node.

The full-scale update is triggered by the previous situation or when the current $tpn \geq tpn_3$. It is the time for PE to review its partitioning plan and to adjust the index accordingly. However, the preferred locality Γ is the same. PE prepares the new partitioning plan similar to the one during index construction except that the current tpn, α, and β represents tpn_1, α_1, and β_1 for the new round. What is of concern is α when the preferred locality is Γ_{ol}, and β when the preferred locality is Γ_{sl}. So, α will not change if it is within an acceptable range (± 2) when the preferred locality is Γ_{ol}. Again, the next review is set to when tpn reaches the new tpn_3.

The next step is to adjust the global index if one of the variables in question has changed. Otherwise, DMOI will continue as before until the next review bookmark. With Γ_{sl}, the global index needs to expand to push the old lazy-leaf flags down to its new $L\alpha$. If β has changed, DMOI needs to address that as well, similar to the construction phase. With Γ_{ol}, the global index is more likely going to shrink only if α changed. In this case, DMOI will traverse the spatial tree to mark the new leaves and delete the unnecessary spatial nodes. In both cases, DMOI needs to update the partitioner according to the changes.

Finally, the partitions, including their local indexes, may also require an adjustment. With Γ_{ol}, if α changed by ± 2, all the partitions need to update their segments' tags and lazy-tags to reflect the new $nids_g$. After grouping the segments on their new tags, DMOI constructs new local indexes as before. With Γ_{sl}, DMOI also performs the previous steps only if β changes. Otherwise, it only updates $Seg.tag_z$ for all the segments.

Query Processing. We adopt the query processing methodology from [2]. However, the partitions pruning is different. To find the involved partitions in a spatial query, DMOI traverses the spatial tree based on the given query. Then, with Γ_{sl}, it computes $nids_o$ on-the-fly and concatenates them with the found $nids_s$. With Γ_{ol}, DMOI computes $nids_o$ with the help of the partitioner, since it keeps tracking the maximum level of the global tree. For each involved nid_s, DMOI computes all the possible $nids_o$, of the second child only, on the maximum level and checks if they exist. For nonexistent nodes, it recursively checks the upper levels. To find the engaged partitions in a lookup query, with Γ_{sl}, DMOI traverses the spatial tree and returns nid_g that has $nid_o = Tid$ MOD β for each spatial leaf. With Γ_{ol}, DMOI generates all the possible $nids_s$. Next, it concatenates the given Tid to every nid_s and trims them with the bitmask of the maximum level. If the resulted nid_g does not exists in the partitioner, then the bitmask is shifted back one bit. There must be nid_g for each nid_s.

4   Experimental Study

In this section, we present the evaluation of our dynamic index approach. We adopt the indexing schemes of DTR-tree (DTR) [17] and LocationSpark (LSpark) [15], which are designed for Spark platform. DTR depends on one dimension partitioning technique, while LSpark uses a quadtree for partitioning.

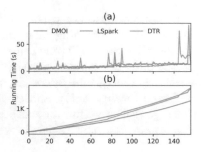
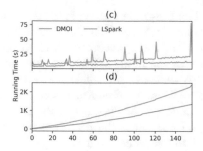

Fig. 2. Index updating: (a) and (c) are the running time for each batch on SF and GT, respectively. (b) and (d) are the accumulative running time on SF and GT.

Experiment Settings. The first one is the German Trajectory dataset (GT) [11]. GT dataset is extracted from Planet GPX of OpenStreetMap database [10]. It contains real spatio-temporal moving object trajectories, covering Germany and parts of its neighboring countries. It covers 636408 km^2 and contains 132 million GPS points for 163 thousand trajectories. The trajectories come from different moving object classes such as vehicles, humans, airplanes, and trains. The second dataset is SF, which is a generated dataset over a real road-network of the San Francisco Bay Area. We used the network-based moving object generator [4] with 20 moving object classes. SF covers 10721 km^2 and contains 119 million points for 140 thousand trajectories.

We conducted the experiments on a six-node cluster using YARN as a resource manager and HDFS as a file system. The nodes were Dell OptiPlex 7040 desktop computers, with quad-core i7 (3.4 GHz), 32 GB of RAM, a 7200 rpm 1 TB HD, and 8M L3 cache, running CentOS 7. We used Apache Spark 2.2.0 with Java 1.8 for our implementation. We adopted Java ParallelOldGC as a garbage collector and Kryo for serialization. The driver node used one node with 8 threads and 24 GB of memory. The worker nodes (executors) used 4 nodes with 24 threads and 64 GB of main memory.

Performance Analysis of Index Maintenance. We sat the batch size of the arriving data to ≈0.55% from the total size. The first 10% is used for index construction. Figure 2a shows the running times of the updating procedure on SF. Although the difference is not clear when comparing a batch by a batch, *DMOI* outperforms DTR and LSpark in the long run, as shown in Fig. 2b. Also, *DMOI* outperforms LSpark on GT dataset, as shown in Figs. 2c and 2d. DTR could not handle GT because of the huge spatial-scope of the dataset.

Performance Analysis of Query Processing. Starting with SF dataset, Figs. 3a and 3b show the average execution time on various selectivity ratios. All the methods have almost the same performance and show an increase with larger selectivity. However, *DMOI* shows a significant speedup on all the other queries. Figure 3c shows that *DMOI* outperforms DTR and LSpark on lookup

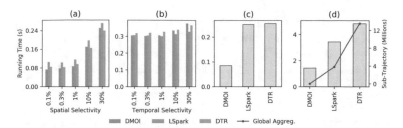

Fig. 3. Average execution time on SF dataset: (**a**) Range Query, (**b**) Interval Query, (**c**) Lookup Query, and (**d**) Longest Query.

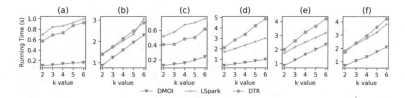

Fig. 4. Average execution time on SF: (**a**) continuous range, (**b**) continuous interval, and (**c**) continuous spatio-temporal queries with small selectivity. Also, (**d**) continuous range, (**e**) continuous interval, and (**f**) continuous spatio-temporal queries with large selectivity.

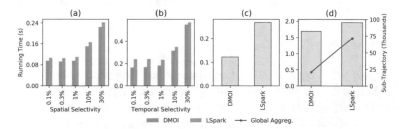

Fig. 5. Average execution time on GT dataset: (**a**) Range Query, (**b**) Interval Query, (**c**) Lookup Query, and (**d**) Longest Query.

query by a factor of 2.99x, on the average. On the longest query, *DMOI* gains a speedup over DTR and LSpark by factors of 3.39x and 2.41x, respectively. The amount of sub-trajectories that are processed globally, i.e. processed through a global aggregation, is also shown in Fig. 3d. Figure 4 shows the execution time of the continuous queries with k = 2, 3, 4, 5, and 6 on small selectivity and large selectivity. In both cases, *DMOI* shows outstanding performance compared to DTR and LSpark.

The results are consistent with GT dataset compared to LSpark only, since DTR could not handle this dataset. Figures 5a and 5b show the average execution time for the range and interval queries, where *DMOI* does slightly better. However, it outperforms LSpark on lookup query and longest query by factors

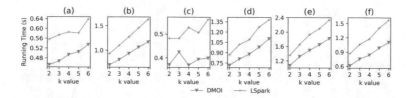

Fig. 6. Average execution time on GT: **(a)** continuous range, **(b)** continuous interval, and **(c)** continuous spatio-temporal queries with small selectivity. Also, **(d)** continuous range, **(e)** continuous interval, and **(f)** continuous spatio-temporal queries with large selectivity.

of 2.2x and 1.16x, respectively, as shown in Figs. 5c and 5d. Also, *DMOI* shows an outstanding performance on the continuous queries, as shown in Fig. 6.

It is worth mentioning that the main goal of the prediction model CPModel is to find a suitable L_{pivot} that would serve all the query types. However, this is not feasible under some circumstances, especially in contradictory situations. As a result, CPModel lowers the bar for few queries without causing a significant performance drop, which helps other queries to perform well. This is the main reason behind the performance of *DMOI* on the range and interval queries, as shown in Fig. 3. However, it kept the performance of those queries at a competitive level and did not sacrifice their performance for the sake of other queries.

5   Conclusion

In this paper, we proposed *DMOI* as a dynamic index for in-memory computation of large-scale trajectories. It is designed to adapt to various cloud environments to satisfy a wide range of real-time and analytic applications. The adaptation capability is achieved by balancing the index structure between the spatial and the object localities, which allows *DMOI* to control the parallelism capacity, the communication traffic, the data loading, and the computation distribution. *DMOI* leverages the prediction model CPModel [2] to be used in a dynamic environment taking into account the computing resources, memory availability, trajectory compactness, spatial-scope, and query types.

Maintaining the index structure over time while controlling the localities is a huge challenge. However, *DMOI* depends on a partitioning plan that considers the current time, the near future, and the far future of the locality pivot. Also, *DMOI* has innovative global and local indexes, which implement many optimizations such as lazy partitioning, virtual trees, traversing on-the-fly, local-optimal locality pivot, tree partitioning, etc. As a result, *DMOI* is able to contain the impact of the updates in the internal structure. We also conduct extensive experiments to validate our indexing method. The results show the adaptation capability of *DMOI*. Also, *DMOI* shows a significant performance improvement on space-based, time-based, and object-based queries.

References

1. Aji, A., Wang, F., Vo, H., Lee, R., Liu, Q., Zhang, X., Saltz, J.: Hadoop GIS: a high performance spatial data warehousing system over MapReduce. Proc. VLDB Endow. **6**(11), 1009–1020 (2013). https://doi.org/10.14778/2536222.2536227

2. Alqahtani, O., Altman, T.: A resilient large-scale trajectory index for cloud-based moving object applications. Appl. Sci. **10**(20) (2020). https://doi.org/10.3390/app10207220

3. Apache Spark. https://spark.apache.org/. Accessed 31 Jan 2021

4. Brinkhoff, T.: A framework for generating network-based moving objects. GeoInformatica **6**(2), 153–180 (2002). https://doi.org/10.1023/A:1015231126594

5. Eldawy, A., Mokbel, M.F.: SpatialHadoop: a mapreduce framework for spatial data. In: Proceedings of the 2015 IEEE 31st International Conference on Data Engineering, pp. 1352–1363 (2015). https://doi.org/10.1109/ICDE.2015.7113382

6. JTS Topology Suite. https://locationtech.github.io/jts/. Accessed 31 Jan 2021

7. Keng Liao, W., Liu, Y., Choudhary, A.: A grid-based clustering algorithm using adaptive mesh refinement. In: Proceedings of the 7th Workshop on Mining Scientific and Engineering Data Sets (2005)

8. Ma, Q., Yang, B., Qian, W., Zhou, A.: Query processing of massive trajectory data based on MapReduce. In: Proceedings of the First International Workshop on Cloud Data Management, CloudDB 2009, pp. 9–16. ACM (2009). https://doi.org/10.1145/1651263.1651266

9. Nishimura, S., Das, S., Agrawal, D., Abbadi, A.E.: MD-HBase: a scalable multi-dimensional data infrastructure for location aware services. In: Proceedings of the 2011 IEEE 12th International Conference on Mobile Data Management, pp. 7–16 (2011)

10. OpenStreetMap contributors: Planet dump. https://planet.osm.org. https://www.openstreetmap.org. Accessed 31 Jan 2021

11. OpenStreetMap contributors: Planet GPX. https://planet.openstreetmap.org/gps/. Accessed 31 Jan 2021

12. Peixoto, D.A., Hung, N.Q.V.: Scalable and fast top-k most similar trajectories search using MapReduce in-memory. In: Proceedings of the Databases Theory and Applications, pp. 228–241. Springer (2016)

13. Robinson, J.T.: The K-D-B-tree: a search structure for large multidimensional dynamic indexes. In: Proceedings of the 1981 ACM SIGMOD International Conference on Management of Data, SIGMOD 1981, pp. 10–18. ACM (1981). https://doi.org/10.1145/582318.582321

14. Schikuta, E.: Grid-clustering: an efficient hierarchical clustering method for very large data sets. In: Proceedings of 13th International Conference on Pattern Recognition, vol. 2, pp. 101–105 (1996). https://doi.org/10.1109/ICPR.1996.546732

15. Tang, M., Yu, Y., Malluhi, Q.M., Ouzzani, M., Aref, W.G.: LocationSpark: a distributed in-memory data management system for big spatial data. Proc. VLDB Endow. **9**(13), 1565–1568 (2016). https://doi.org/10.14778/3007263.3007310

16. Thusoo, A., Sarma, J.S., Jain, N., Shao, Z., Chakka, P., Anthony, S., Liu, H., Wyckoff, P., Murthy, R.: Hive: a warehousing solution over a map-reduce framework. Proc. VLDB Endow. **2**(2), 1626–1629 (2009). https://doi.org/10.14778/1687553.1687609

17. Wang, H., Belhassena, A.: Parallel trajectory search based on distributed index. Inf. Sci. **388–389**, 62–83 (2017)

18. Wang, H., Zheng, K., Xu, J., Zheng, B., Zhou, X., Sadiq, S.: SharkDB: an in-memory column-oriented trajectory storage. In: Proceedings of the 23rd ACM International Conference on Information and Knowledge Management, CIKM 2014, pp. 1409–1418. ACM (2014)
19. You, S., Zhang, J., Gruenwald, L.: Large-scale spatial join query processing in cloud. In: Proceedings of the 31st IEEE International Conference on Data Engineering Workshops (ICDEW), pp. 34–41 (2015). https://doi.org/10.1109/ICDEW.2015.7129541
20. Yu, J., Wu, J., Sarwat, M.: GeoSpark: a cluster computing framework for processing large-scale spatial data. In: Proceedings of the 23rd SIGSPATIAL International Conference on Advances in Geographic Information Systems, SIGSPATIAL 2015, pp. 70:1–70:4. ACM (2015)

High-Level Approach
for the Reconfiguration of Distributed
Algorithms in Wireless Sensor Networks

Emna Taktak[1]([✉]), Mohamed Tounsi[1,2], Mohamed Mosbah[3],
and Ahmed Hadj Kacem[1]

[1] ReDCAD Laboratory, University of Sfax, Sfax, Tunisia
emna.taktak@redcad.org, ahmed.hadjkacem@fsegs.rnu.tn
[2] Common First Year Deanship, Umm Al-Qura University, Mecca, Saudi Arabia
mohamed.tounsi@fsegs.rnu.tn
[3] Univ. Bordeaux, CNRS, Bordeaux INP, LaBRI, UMR 5800, 33400 Talence, France
mohamed.mosbah@u-bordeaux.fr

Abstract. After the deployment of a Wireless Sensor Network (WSN),
we may need to update or even change the algorithm running in the
sensors. WSNs are often deployed on a large scale which makes the manual update of the running algorithms impractical. Moreover, for some
WSNs, sensors are physically inaccessible like sensors deployed in harsh
environments making the manual update impossible. For these situations, wireless reprogramming is a suitable solution. In this paper, we
deal with wireless reprogramming for WSNs using a high-level reconfiguration approach. Our approach is based on a theoretical model which is
the Graph Relabelling System (GRS). Thereby, any algorithm designed
with GRS models will be reconfigurable. We illustrate the proposed approach through a reconfiguration example changing the running algorithm
from fire detection to way detection algorithm. We present the simulations realised with the CupCarbon simulator to validate the reconfiguration approach.

Keywords: Wireless Sensor Network · Graph Relabelling System ·
Reconfiguration algorithm · High-level labelling · Distributed
algorithms

1 Introduction

A Wireless Sensor Network (WSN) incorporates many sensors that communicate together in a distributed way to supervise the environment. The reconfig-
uration/reprogramming of a WSN brings great flexibility and adaptability to
this network. As mentioned in [5], in some cases we just need to adjust some
parameters for the sensors while in other cases the algorithms executed on the
sensors need to be reprogrammed. In this paper, we are interested in the case of
changing the executed algorithms. Thereby, we use the term reconfiguration to

© The Author(s), under exclusive license to Springer Nature Switzerland AG 2021
L. Barolli et al. (Eds.): AINA 2021, LNNS 226, pp. 95–106, 2021.
https://doi.org/10.1007/978-3-030-75075-6_8

refer to the capacity of changing the running algorithms dynamically to WSNs that have been deployed in the field.

Unfortunately, the existing wireless reconfiguration solutions are mainly reprogramming protocols [6,7,11] that do not refer to a theoretical model. Consequently, the proofs for these protocols will be a non-trivial challenge. Indeed, WSN's applications are often deployed for critical systems. Thereby, ensuring the correctness of these applications is important, because any design error can be harmful to human life.

For these reasons in this paper, we propose a reconfiguration approach that refers to the Graph Relabelling System (GRS) as a theoretical model. The GRS is widely used for modelling distributed algorithms for WSNs [1,3,10,12]. It is based on a set of relabelling rules that are closely related to mathematical and logical formulas. Consequently, the correctness of distributed algorithms can be deduced from these rules.

Our approach consists of designing the distributed algorithms with the GRS differently to be easily reconfigured later. Besides, we propose a new reconfiguration algorithm that should be implemented from the beginning with the distributed algorithm to be executed. The reconfiguration algorithm is executed in a distributed manner to spread the changes to be made to change the running algorithm. We illustrate our approach with a fire detection algorithm that we execute to detect the existence of a fire in a forest. We program a planned reconfiguration that should be done at a certain time after executing the fire detection algorithm to change it by an algorithm called the way detection algorithm. The simulations of this reconfiguration example are realised with the CupCarbon simulator to validate the feasibility of our approach.

The paper is organised as follows: Sect. 2 presents some preliminaries. In Sect. 3, we describe how we reconfigure distributed algorithms for WSNs. Section 4 illustrates the reconfiguration approach through an example and presents its simulation results. Section 5 concludes this paper and provides insights for future work.

2   Preliminaries

2.1   The Graph Relabelling System (GRS) for Distributed Algorithms

A GRS expresses how a distributed system operates in an abstract manner. In this model, a WSN is represented by a graph $G = (U, E)$ whose nodes U represent the set of sensors and edges E represent the bidirectional links between nodes [9]. Two nodes are related by an edge if the corresponding sensors have a direct communication link. At every time, each node and each edge is in some particular state, and this state will be encoded by one or many labels. For example, supposing that a node is labelled with a set of labels {label-1, label-2,..., label-l}. The set of labels' values form the **state** of this node.

According to its state and the states of its neighbours, each node may decide to realise an elementary computation step. After this step, the state of this node, its neighbours, and the corresponding edges may have changed according to some specific computation rules. These computation rules are called relabelling rules.

2.2 The Local Computations Model for WSN (LC-WSN)

The local computations model for WSN (LC-WSN) presented in [12] is based on the GRS. In this model, each relabelling rule depends on and modifies only the labels of at most two adjacent nodes and the edge linking them. When using this model, we differentiate between internal relabelling rules and external relabelling rules. The internal relabelling rules are executed by a sensor when it captures new information as presented in Fig. 1. However, the external relabelling rules are executed by a sensor when it communicates with another sensor as presented in Fig. 2.

Fig. 1. An internal relabelling rule

Fig. 2. An external relabelling rule

Definition 1. *A graph relabelling system of a distributed algorithm* [12] *is $S = (L, R, I)$ where:*

- $L = \{$label-1, label-2, ..., label-l$\}$ is a set of labels with l the number of labels,
- R is a finite set of relabelling rules. $R = \{$Rule-1, Rule-2, ..., Rule-k$\}$ with k the number of rules. We have, $R = R_i \cup R_e$ with R_i and R_e are respectively the set of internal and external relabelling rules.
- I is a subset of L called the set of initial labels.

3 Reconfiguring Distributed Algorithms for WSNs

In this section, we explain how we can change or update the algorithm running on a WSN without the need for recompilation or shutting the system down. Our solution does not work on propagating the new code image. Instead, it consists of designing the distributed algorithms in an abstract manner to be easily reconfigured later using a reconfiguration algorithm that we propose. Thus, we propose high-level labels instead of normal labels for the LC-WSN model or any other local computations model. The proposed reconfiguration algorithm to be implemented changes the running algorithm by changing the high-level labels that encode the running algorithm. With the proposed approach, the reconfiguration can be planned from the beginning as it can be also improvised.

Supposing that $S_{current} = \{L_{current}, R_{current}, I_{current}\}$ encodes the initial algorithm to be executed according to Definition 1. This initial algorithm is designed to be used by the reconfiguration algorithm and not to be directly implemented. Indeed, the reconfiguration algorithm is encoded with a GRS with priorities [8] (PGRS) named S_{reconf} and defined as follows:

Definition 2. *The PGRS of the reconfiguration algorithm is* $S_{reconf} = \{L_{reconf}, R_{reconf}, I_{reconf}, >\}$

- $L_{reconf} = \{labelsL, rulesL, initL, v, sw\}$ with:
  - $labelsL$ is the labels' list of the executed algorithm,
  - $rulesL$ is the list of relabelling rules of the executed algorithm with $rulesL = \{$rule-1, rule-2, ..., rule-num$\}$; num is the number of rules. We have, $rulesL = intRules \cup extRules$, with $intRules$ is a set of internal relabelling rules and $extRules$ is a set of external relabelling rules,
  - $initL$ is the list of initial values for the labels of the executed algorithm,
  - v is the version of the algorithm being executed,
  - sw contains the information of the new algorithm to be executed if it exists.
- $R_{reconf} = \{Rchange, Switch, IntRule, ExtRule\}$. These rules are presented respectively in Fig. 4, Fig. 5, Fig. 6, and Fig. 7,
- $I_{reconf} = \{I_{current}, R_{current}, I_{current}, 1, \emptyset\}$, with $R_{current} = R_{iCurrent} \cup R_{eCurrent}$, $intRules = R_{iCurrent}$, and $extRules = R_{eCurrent}$. The label sw can be initialised to be empty or to contain the information of the new algorithm to be executed if the reconfiguration is planned.
- $>$ is defined to be the order of the execution of the rules that should be respected by sensors giving that IntRule $>$ Switch $>$ Rchange $>$ ExtRule.

Thus, to change a running algorithm, we use the high-level labels defined by L_{reconf} that contains the information about the running algorithm as presented in Fig. 3 with {label-1, label-2, ..., label-l} are the ordinary labels encoding the running algorithm.

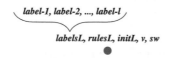

Fig. 3. High-level labels for a node

The number and the definition of the high-level labels do not depend on the executed algorithm because they are defined to be five high-labels for all the sensors. Only the value of these labels depends on the executed algorithm.

As specified in Definition 2, the high-level labels L_{reconf} are modified by the relabelling rules of R_{reconf} containing the rule **Rchange** and the rule **Swicth**. The rule **Rchange** is presented in Fig. 4 with $labelName^s$ encodes the value of the label $labelName$ for a sensor s. Thus, if a sensor s has an old version of the executed algorithm, then it will change it by replacing its current labels noted $labelsL^s$, $rulesL^s$, $initL^s$ and v^s by the values of $s1$ that encode the new algorithm to be executed ($initL^{s1}$, $rulesL^{s1}$ and v^{s1}) because $s1$ has a newer version of the executed algorithm.

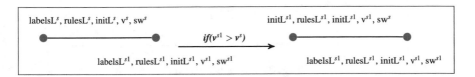

Fig. 4. The rule Rchange

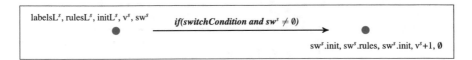

Fig. 5. The rule Switch

The rule **Switch** allows planning the reconfiguration to be automatically initiated by one or many sensors. Thus, as specified in Fig. 5, the chosen sensors execute the **Switch** rule when they verify a defined condition dubbed $switchCondition$ used to trigger the reconfiguration. The reconfiguration is made by changing the labels $labelsL^s$, $rulesL^s$, and $initL^s$ by the information of the new algorithm that the chosen sensor s has in sw^s, incrementing the label v^s and re-initialising the label sw^s.

The difference between **Rchange** and **Switch** is that if a sensor executes the rule **Switch**, then it decides to introduce a new version of the algorithm by itself (change initiative). However, if a sensor executes the rule **Rchange**, then it will change its version of the algorithm following its neighbour sensor (spread of change).

Thereby, if the decision of introducing a new algorithm is programmed from the beginning, the reconfiguration will be done autonomously by the chosen sensor(s) using the **Switch** rule. Then, the spread of the reconfiguration will be realised automatically by the sensors in a distributed manner using the **Rchange** rule.

However, if the decision of introducing a new version of the algorithm is improvised, the system user could initiate the reconfiguration remotely by adding a new sensor to the existing WSN that has only the new algorithm. When this latter communicates with sensors in the existing WSN, the rule **Rchange** will be executed by the existing sensors in the WSN to update their algorithms. This adding procedure can be definitive, i.e., the new sensor will be part of the WSN. The system user can also add a new temporary sensor. In this case, the reconfiguration will be done automatically by the sensors in a distributed manner using only the **Rchange** rule.

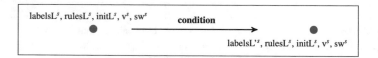

Fig. 6. The rule IntRule

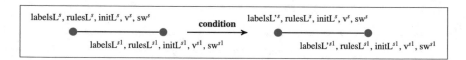

Fig. 7. The rule ExtRule

In addition to the rule **Rchange** and the rule **Swicth**, R_{reconf} contains the rules **IntRule** and **ExtRule** designing how a running algorithm will be executed in an abstract manner. Thus, Fig. 6 presents the rule **IntRule**, this rule presents how an internal rule of the running algorithm will change only the label labelsL^s of a sensor s. The condition **condition** and how labelsL^s will be changed to labelsL'^{s} depends on the executed internal rule from rulesL^s.intRules. For the rule **ExtRule** presented in Fig. 7, it modifies both labelsL^s and labelsL^{s1} representing the labels' list of the two communicating nodes. This change and the exact condition depends on the executed external relabelling rule in rulesL^s.extRules. These two rules should be refined according to the algorithm to be executed.

The resulting algorithm to be implemented on all the sensors allowing their reconfiguration is presented in Algorithm 1. When using the LC-WSN model [12], a sensor s starts by executing the internal relabelling rules of the current algorithm specified as rulesL^s.intRules (line 1 and line 2). In this phase, s updates its label $labelsL$ to $labelsL'$ according to the newly captured values as laid out in Fig. 6. If there are no internal relabelling rules to execute or after executing them, s should check if the switch condition is verified (line 4). If the switch condition is verified and the label sw^s contains the information of the new algorithm, then s realises a reconfiguration by executing the rule **Switch** in Fig. 5 (line 5). If the defined condition is not verified or after executing the **Switch**

Algorithm 1. The reconfiguration algorithm

1: **for each** IntRule $\in$ rulesL^s.intRules **do**
2:     s executes (IntRule)
3: **end for**
4: **if** (switchCondition and sw^s $\neq \emptyset$) **then**
5:     s executes (*Switch*)
6: **end if**
7: **if** $\exists(handshake(s, s1))$ **then**
8:     s sends its state to $s1$
9:     s receives the state of $s1$
10:    **if** $(v^{s1} > v^s)$ **then**
11:        s executes (*Rchange*)
12:    **end if**
13:    **for each** ExtRule $\in$ rulesL^s.extRules **do**
14:        s executes (ExtRule)
15:    **end for**
16: **end if**
17: return to line 1

rule, s does a handshake attempt to communicate with a neighbour $s1$ (line 7). Here, we suppose that we use a handshake algorithm to establish exclusive communications between the pair of sensors [4]. Thereby, if the handshake is not realised, the sensor will return to execute the internal relabelling rules (line 17) checking if there are new captured values. However, if the handshake is realised, then the sensors exchange their states (line 8 and line 9).

After that, to decide whether to execute the reconfiguration or not, s starts by checking its version (line 10). If s and $s1$ have the same version of the algorithm then s will execute its external relabelling rules of this algorithm (line 13 to line 15) as laid out in Fig. 7. Yet, if $s1$ has a newer version then s will execute the reconfiguration by executing the rule **Rchange** laid out in Fig. 4 (line 11).

4   Simulation of a Reconfiguration Example

To simulate the reconfiguration algorithm, we need first to implement the high-level labels in the used simulator which is the CupCarbon simulator. The CupCarbon is a smart city and Internet Of Things WSN Simulator. With CupCarbon [2], a network can be designed and prototyped by an ergonomic and easy interface.

At the first step, we modified the source code of cupCarbon written with Java language in order to treat the label *rulesL* proposed as one of the high-level labels as a set of conditions and instructions executable by the cupCarbon simulator.

In the second step, we implemented the reconfiguration algorithm. Then, we choose to simulate it with an example changing the running algorithm from a fire detection algorithm to a way detection algorithm. In this example, if the fire is detected, the WSN will use the reconfiguration algorithm to replace the fire detection algorithm by the way detection algorithm automatically. In this simulation, we used 50 sensors, with 6 sensors detecting the fire and 1 sensor that will change the executed algorithm from fire detection to way detection.

The fire detection algorithm allows detecting the presence of a fire in a forest. If one or many sensors detect a huge elevation in the temperature value of the environment, they will be on alert state. When a sensor is not informed by the alert state communicates with another sensor informed by the alert state, this latter will be also informed. Normally, this algorithm is specified with a classic GRS named $S_{current}$ with $S_{current} = \{L_{current}, R_{current}, I_{current}\}$ where:

- $L_{current} = \{$Temp, State$\}$ with State $\in \{$normal, alert$\}$ and $Temp$ is the detected temperature of the environment. The label $State$ contains the value $normal$ when there is no detected fire (elevated temperature) in the WSN. However, this label contains $alert$ there is a detected fire in the WSN.
- $R_{current} = \{$FireRule-1, FireRule-2$\}$, these rules are presented in Fig. 8 with $thresh$ is the defined temperature threshold. We have $R_{iCurrent} = \{$FireRule-1$\}$ and $R_{eCurrent} = \{$FireRule-2$\}$,
- $I_{current} = \{0,$ normal$\}$.

In order to be able to reconfigure the fire detection algorithm, it should be designed and implemented differently using the high-level labels. Consequently, instead of $S_{current}$, we design S_{reconf} representing the PGRS of the reconfiguration algorithm of the fire detection algorithm. We have, $S_{reconf} = \{L_{reconf}, R_{reconf}, I_{reconf}, >\}$

- $L_{reconf} = \{labelsL, rulesL, initL, v, sw\}$,
- $R_{reconf} = \{Rchange, Switch, IntRule1, ExtRule1\}$. The rule **RChange** is the same rule presented in Fig. 4. The rule **Switch** is the same rule presented in Fig. 5 with $switchCondition = (alert \in labelsL^s)$. The rule **IntRule1** refines the abstract rule **IntRule** to model an internal relabelling rule of the fire detection algorithm as presented in Fig. 9. The rule **ExtRule1** refines the abstract rule **ExtRule** to model an external relabelling rule of the fire detection algorithm as presented in Fig. 9.
- $I_{reconf} = \{I_{current}, R_{current}, I_{current}, 1, \emptyset\}$, with $R_{current} = R_{iCurrent} \cup R_{eCurrent}$, $intRules = R_{iCurrent}$, and $extRules = R_{eCurrent}$. The label sw is initialised to be empty for all the sensors in the WSN except one sensor named (sc) because the reconfiguration is planned in this example. We have $sw^{sc} = \{init, rules\}$ with init $= I_{new}$ and rules $= R_{new}$. I_{new} and R_{new} encodes the information of the way detection algorithm that we detail in the following paragraph.

Giving that the reconfiguration is planned from the beginning, we put the information of the way detection algorithm on a chosen sensor named (sc). The way detection algorithm proceeds as follows: first, the sensors on fire will be labelled with $noWay$. Second, the sensors that are not on fire will be labelled with way. And finally, when a sensor s labelled $noWay$ communicates with

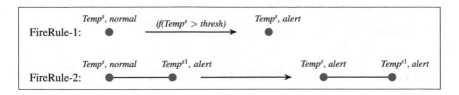

Fig. 8. Rules of the fire detection algorithm

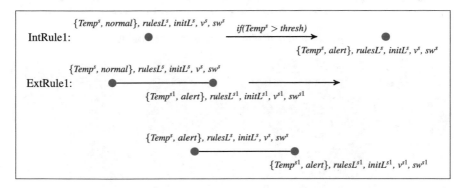

Fig. 9. Rules of the reconfiguration of the fire detection algorithm

another sensor $s1$ labelled *way*, $s1$ will be labelled *danger* instead of *way*. Thus, *danger* indicates that the sensor is near to the fire, *noWay* means that the sensor is in the fire and *way* means that the sensor is in a safe place. This new algorithm is specified with a classic GRS named S_{new} with $S_{new} = \{L_{new}, R_{new}, I_{new}\}$ where:

- $L_{new} = \{\text{Temp}, \text{Path}\}$ with $\text{Path} \in \{\text{way}, \text{noWay}, \text{danger}\}$ and $Temp$ is the detected temperature of the environment,
- $R_{new} = \{\text{WayRule-1}, \text{WayRule-2}, \text{WayRule-3}\}$, these rules are presented in Fig. 10 with *thresh* is the defined temperature threshold. We have $R_{iNew} = \{\text{WayRule-1}, \text{WayRule-2}\}$ and $R_{eNew} = \{\text{WayRule-3}\}$,
- $I_{current} = \{0, \text{way}\}$.

The specification of S_{new} is presented just to show how the way detection algorithm works. In our implementation, we just use I_{new} and R_{new}. Indeed, $sw^{sc} = \{init, rules\}$ with $init = I_{new}$ and $rules = R_{new}$. In addition, we have $switchCondition = (alert \in labelsL^s)$. Consequently, when informed by the existence of a fire guaranteed by $switchCondition$, the sensor (sc) which is the only sensor having the label $sw^{sc} \neq \emptyset$, will execute the rule **Switch** laid out in Fig. 5. The spread of the new algorithm will be realised using the rule **Rchange**.

Figure 11(a) presents a screenshot for the execution of the fire detection algorithm with cupCarbon (the execution of **IntRule1** and ExtRule1). In this figure, the sensors with the fire icon are the sensors detecting the fire in their region.

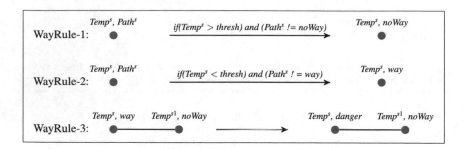

Fig. 10. Rules of the way detection algorithm

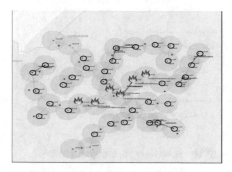

(a) The execution of the fire detection algorithm

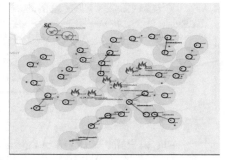

(b) The execution of the reconfiguration and the way detection algorithm

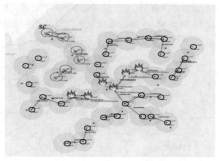

(c) The execution of the reconfiguration, the way and the fire detection algorithm

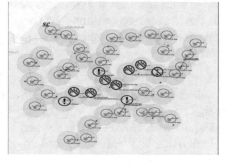

(d) The result of the execution of the way detection algorithm

Fig. 11. A simulation of the reconfiguration example

When a sensor is informed by the alert (its label $labelL$ contains the value $alert$), it will be highlighted with a circle.

In Fig. 11(b), when the sensor (sc) is informed by the alert, it executes the reconfiguration (the **Switch** rule) and thus the way detection algorithm. In this figure, the two sensors with the green checked icons already changed their versions and executed the way detection algorithm. Both of these two sensors have the checked icon because their label $labelL$ contains the value way.

Figure 11(c) shows the spread of the execution of the reconfiguration (the **Rchange** rule) with the way detection algorithm alongside the termination of the execution of the fire detection algorithm in the bottom side of the graph. After the execution of the way detection algorithm, we obtain the result displayed in Fig. 11(d). In this figure, sensors with the checked icon show a safe way. The sensors with the exclamation mark icon represent sensors in danger zones. The sensors in the fire zone are presented with the forbidden icon.

The way detection algorithm is used by firefighters to track the fire and to calm it down without being injured by following the safe path. In our example, if we will never have a fire, the way detection algorithm will never be executed. For this reason, it is not necessary to put its code on all sensors from the beginning. However, when we need to execute it, we give the different sensors the needed rules to be executed (rules of the way detection algorithm).

5 Conclusion and Future Work

In this paper, we presented how we use the high-level labels that we propose to make an algorithm designed with GRS models reconfigurable. Reconfiguration, also known as reprogrammability, is used in this paper to refer to the capacity of being able to change applications dynamically to WSNs that has been deployed in the field. We illustrated our contribution by an example changing the running algorithm from fire detection to way detection. We implemented our reconfiguration algorithm and the example using the cupCarbon simulator.

In future work, we aim at proving the preservation of consistency of the WSN when changing the running algorithm. Besides, we aim to adapt our approach to support the partial reconfiguration of the running algorithm, i.e., changing only some parts of the algorithm for bugs fixing for example.

References

1. Angluin, D., Aspnes, J., Eisenstat, D., Ruppert, E.: The computational power of population protocols. Distrib. Comput. **20**(4), 279–304 (2007)
2. Bounceur, A., Clavier, L., Combeau, P., Marc, O., Vauzelle, R., Masserann, A., Soler, J., Euler, R., Alwajeeh, T., Devendra, V., Noreen, U., Soret, E., Lounis, M.: CupCarbon: a new platform for the design, simulation and 2D/3D visualization of radio propagation and interferences in IoT networks. In: 2018 15th IEEE Annual Consumer Communications Networking Conference (CCNC), pp. 1–4 (2018). https://doi.org/10.1109/CCNC.2018.8319179
3. Chatzigiannakis, I., Michail, O., Nikolaou, S., Pavlogiannis, A., Spirakis, P.G.: Passively mobile communicating machines that use restricted space. Theor. Comput. Sci. **412**(46), 6469–6483 (2011). https://doi.org/10.1016/j.tcs.2011.07.001
4. Fontaine, A., Mosbah, M., Tounsi, M., Zemmari, A.: A fault-tolerant handshake algorithm for local computations. In: 30th International Conference on Advanced Information Networking and Applications Workshops, AINA 2016 Workshops, Crans-Montana, Switzerland, 23–25 March 2016, pp. 475–480 (2016). https://doi.org/10.1109/WAINA.2016.78

5. Helen, L., Redondo, L., Zahariadis, T., Retamosa, D., Panagiotis, K., Papaefstathiou, I., Voliotis, S.: Reconfiguration in wireless sensor networks. In: 2010 Developments in E-systems Engineering, pp. 59–63 (2010). https://doi.org/10.1109/DeSE.2010.17
6. Huang, L., Setia, S.: CORD: energy-efficient reliable bulk data dissemination in sensor networks, pp. 574–582 (2008). https://doi.org/10.1109/INFOCOM.2008.106
7. Hui, J.W., Culler, D.: The dynamic behavior of a data dissemination protocol for network programming at scale. In: Proceedings of the 2nd International Conference on Embedded Networked Sensor Systems, SenSys 2004, pp. 81–94. ACM, New York (2004). https://doi.org/10.1145/1031495.1031506
8. Litovsky, I., Métivier, Y.: Computing with graph relabelling systems with priorities. In: Ehrig, H., Kreowski, H.J., Rozenberg, G. (eds.) Graph Grammars and Their Application to Computer Science, pp. 549–563. Springer, Heidelberg (1991)
9. Litovsky, I., Sopena, E.: Graph relabelling systems and distributed algorithms. In: Handbook of Graph Grammars and Computing by Graph Transformation, pp. 1–56. World Scientific (2001)
10. Michail, O., Chatzigiannakis, I., Spirakis, P.G.: Mediated population protocols. Theor. Comput. Sci. **412**(22), 2434–2450 (2011). https://doi.org/10.1016/j.tcs.2011.02.003
11. Starobinski, D., Trachtenberg, A., Hagedorn, A.: Rateless deluge: over-the-air programming of wireless sensor networks using random linear codes. In: 2008 International Conference on Information Processing in Sensor Networks (IPSN 2008), pp. 457–466. IEEE Computer Society, Los Alamitos (2008). https://doi.org/10.1109/IPSN.2008.9
12. Taktak, E., Tounsi, M., Mosbah, M., Hadj Kacem, A.: Distributed computations in wireless sensor networks by local interactions. In: Ad-hoc, Mobile, and Wireless Networks - 17th International Conference on Ad Hoc Networks and Wireless, ADHOC-NOW 2018, Saint-Malo, France, 5–7 September 2018, Proceedings, pp. 293–304 (2018). https://doi.org/10.1007/978-3-030-00247-3_26

Fast and Scalable Triangle Counting in Graph Streams: The Hybrid Approach

Paramvir Singh[✉], Venkatesh Srinivasan, and Alex Thomo

University of Victoria, Victoria, BC, Canada
{paramvirsingh,srinivas,thomo}@uvic.ca

Abstract. Triangle counting is a major graph problem with several applications in social network analysis, anomaly detection, etc. One of the most popular triangle computational models considered is Edge Streaming in which the edges arrive in the form of a graph stream. We categorize the existing literature into two categories: Fixed Memory (FM) approach, and Fixed Probability (FP) approach. As the size of the graphs grows, several challenges arise such as memory space limitations, and prohibitively long running time. Therefore, both FM and FP categories exhibit some limitations. FP algorithms fail to scale for massive graphs. We identified a limitation of FM category *i.e.* FM algorithms have higher computational time than their FP variants.

In this work, we present a new category called the Hybrid approach that overcomes the limitations of both FM and FP approaches. We present two new algorithms that belong to the hybrid category: Neighbourhood Hybrid Multisampling (NHMS) and Triest/ThinkD Hybrid Sampling (THS) for estimating the number of global triangles in graphs. These algorithms are highly scalable and have better running time than FM and FP variants. We experimentally show that both NHMS and THS outperform state-of-the-art algorithms in space-efficient environments.

Keywords: Triangle counting · Graph streams · Edge sampling

1 Introduction

Counting triangles forms a basis for many network analysis, such as social network analysis, anomaly detection, recommendation system, etc. The triangle count is critical for frequently used triangle connectivity, transitivity coefficient, and clustering coefficient, in the analysis. This task is especially challenging when the network is massive with millions of nodes and edges. Several methods had been proposed that are classified into two categories: exact counting and approximate counting. The exact counting, for triangles, is done through enumeration/listing which touches the triangles one by one [7]. The approximate counting is done by sampling the graph and using probabilistic formulae to estimate the total number of triangles in the whole graph based on the number of triangles found in the sample.

L. Barolli et al. (Eds.): AINA 2021, LNNS 226, pp. 107–119, 2021.
https://doi.org/10.1007/978-3-030-75075-6_9

Edge streaming is a model where a series of edges arrives in order, one at a time. There are many works published on triangle estimation using this model which we classified into two categories. The first category includes algorithms that have a *Fixed Memory (FM)* Budget. These algorithms sample the edges within the fixed memory budget and require the user to input the available amount of memory. Once the available memory is full with edges, the next sampled edge randomly replaces an edge in the memory. This is how the subgraph is maintained within a fixed memory budget.

The second category includes the algorithms that require the user to specify an edge sampling probability p that is fixed for the entire stream, we call them *Fixed Probability (FP)* approach. These algorithms maintain a memory reservoir that doesn't have any size limit. Therefore, if the graph stream arriving is massive, the algorithm often runs out of memory. But, FP algorithms have some benefits over FM algorithms. We analyzed and compared the time complexities of both categories, and show that the FP algorithms are faster than FM algorithms, later in Sect. 3.

However, both categories have their own limitations detailed in Sect. 3. The sample size always grows in FP algorithms due to which they often run out of memory. On the other hand, FM algorithms have a higher running time than FP algorithms.

The idea of using fixed memory and fixed probability together remains unexplored. We explore this idea and the key intention is to overcome the limitations of FM and FP approaches. We form a new category called the Hybrid category. As the name suggests, the Hybrid category utilizes the power of FM algorithms, that make it more scalable, and the power of FP algorithms, that make it the fastest among all available algorithms.

1.1 Contributions

Our contributions are as follows:

- We prove that FP algorithms are faster than FM algorithms. We provide the running time analysis to prove this claim in Sect. 3.
- We propose an algorithm called *Neighborhood Hybrid Multisampling (NHMS)*. Not only is our algorithm highly scalable compared to its FP variant *NMS*, but also is significantly faster. The experimental results validate our claims when we compare our algorithm with others on several real-life datasets.
- We propose an algorithm called *Trièst - ThinkD Hybrid Sampling (THS)* that is extremely efficient. We prove that THS is the fastest algorithm available for Triangle Counting in Graph Streams. Our experimental results show that THS is at least 5 times faster than its FM variant.
- We conduct extensive experimentation and prove that our algorithm could execute graphs with billions of edges with in 16 GB RAM and is thus scalable to large graph datasets, whereas many other algorithms fail to execute on the same machine.

2   Prior Work

The extensive literature is available on the approximation of triangle counting in graph streams. Since in this research we propose that the literature presented until now falls in either the *Fixed Memory (FM)* or *Fixed Probability (FP)* category, we will classify the previous work into these categories in this section.

Pavan et al. [6] presented a Neighbourhood Sampling algorithm that needs to execute multiple copies called estimators to approximate triangles. Neighborhood Sampling is an FM algorithm.

Jha et al. [4] applied Birthday Paradox to get the estimate of triangles in graph stream. Similar to Neighborhood Sampling, Birthday Paradox is also FM algorithm.

Stefani et al. [10] presented a suite of algorithms called *Triest*. *Triest* uses reservoir sampling to sample multiple edges in a fixed memory reservoir. Shin et al. [8] proposed two different algorithms named *ThinkD* (Think before you discard). The first algorithm is $ThinkD_{Fast}$, which falls in an FP category. The other algorithm is $ThinkD_{Acc}$ which handles a dynamic stream with edges insertion and deletion. As we are considering insertion-only streams, $ThinkD_{Acc}$ is no different from *Triest*. Hence, both *Triest* and $ThinkD_{Acc}$ belong to the FM category. For brevity, we call this algorithm *TS*.

Kavassery et al. [5] proposed two algorithms in the paper, Edge-Vertex Multisampling (EVMS), and Neighbourhood Multisampling (NMS). EVMS is an extension to an algorithm by Buriol *et al.* [3] It is categorized as FP algorithm. Kavassery *et al.* [5] presented another algorithm named NMS by modifying the Neighbourhood Sampling [6] using the multisampling approach, similar to EVMS. NMS is also an FP algorithm, it instead uses sampling of edges twice.

None of these works have explored the hybrid approach. We present a new hybrid approach that is both scalable and faster than FM and algorithms. In addition, we present two new hybrid category algorithms (*NHMS* and *THS*, more details in Sect. 4), that perform better than all the works mentioned. *NHMS* is hybrid variant of *NMS*, and *THS* is hybrid variant of *TS*.

3   Fixed Probability vs Fixed Memory Algorithms

3.1   Fixed Probability Algorithms

FP algorithms work as follows: Given a fixed probability p, the algorithm samples the edge from the stream with the probability p and add it to a reservoir, after which the triangle count step is executed for each of the edge in the stream. If a triangle is found, a variable Y is incremented that stores the triangle count in the sample. This continues until the edges in the stream arrive. Whenever the triangle estimates are required, Y is scaled up by some scaling factor and is returned as the final estimate.

Stefani et al. [10] discussed the drawbacks of the algorithms employing FP approach. The limitations of FP algorithms are as follows.

- The input parameter p to be fixed requires in-depth analysis to get the desired approximation quality.
- The reservoir size $|R|$ always grows as there is no limit or restriction to store the number of sampled edges.
- If p is chosen to be large, the algorithm may run out of memory.
- If p is chosen to be small, the algorithms will provide us suboptimal estimates.

3.2   Fixed Memory Algorithms

FM algorithms work as follows: Given a fixed memory budget K, the algorithm samples an edge with probability K/t where t is the time of arrival of an edge, after which the triangle count step is executed for each edge in the stream. Unlike FP algorithms, FM algorithms scale up the count of triangles found for each edge by some scaling factor η and then add it to Y which is the triangle estimation. This continues until the edges in the stream arrive. Whenever the triangle estimates are required, Y is returned directly as the final estimate. We analyzed and found that there are also some drawbacks for FM algorithms as listed below.

- The selection of input parameter, memory budget K, is not clear (even when there is plenty of memory available), and this is similar to choosing input parameter p for FP algorithms. Namely, If K is small, the triangle estimates we get have low accuracy, and if K is large, the algorithm takes longer to run.
- We further observed that FP algorithms are significantly faster than FM algorithms.

3.3   Analysis

FM algorithms first directly fill the reservoir with the edges streamed. Once the reservoir is full, the edge is sampled by probability K/t and is replaced with a random edge in the reservoir. Hence, the reservoir remains almost full always.

In FP algorithms, there is no such reservoir with fixed size. The fixed probability p specified by the user plays an important role here to store the sampled edges in the memory. Let's say there are $|E|$ edges in graph stream Σ, the number of edges sampled by the algorithm will be $p|E|$.

Time Complexity Comparison. Stefani et al. [10] provides a time bound for each edge. For each edge $e = (u, v)$ to compute triangles, TS algorithm requires $O(d(u) + d(v))$ steps where $d(u)$ is the degree of vertex u.

Kavassery et al. [5] does not provide the theoretical time bound proofs for NMS algorithm. We analyzed the NMS algorithm to measure the time complexity that is identified as $(p + q) \sum_{u,v \in R}(d(u) + d(v) + c)$ [9].

While comparing the time complexity of NMS and TS, it is clear that they both depend upon the sum of degree of the vertices of an edge arriving in the graph stream i.e. $O(d(u) + d(v))$. We know that the FM algorithms store more number of edges in the reservoir all the time, whereas, the FP algorithms gradually increases the stored edges in the memory. Therefore, the computation of the shared neighbourhood will be more in FM algorithms. This explains the higher computational times in case of FM algorithms.

4   The Hybrid Approach and Algorithms

The key idea behind the hybrid approach is to utilize the benefits of both FM and FP approaches and overcome their limitations. The hybrid approach algorithms have the combination of the fixed memory budget K and the fixed sampling probability p. We present two new algorithms in this category.

Before moving further directly to the new algorithms, let us consider why the hybrid approach is better than FP and FM algorithms. As we have already discussed the limitations of FP and FM approaches in the previous chapter, the questions below would answer how hybrid algorithms overcome them.

Will Hybrid Algorithms Ever Run Out of Memory Like FP Algorithms? The hybrid approach limits the size of an edge reservoir by K, similar to FM algorithms, whereas in FP algorithms, there is no such limit and the edge reservoir size always grows. Hence, we overpower FP algorithm's biggest limitation.

Are Hybrid Algorithms Faster Than Both FM and FP Algorithms? We have already detailed how FP algorithms are faster than FM algorithms in Sect. 3.3. Hybrid algorithms follow the same trend as FP algorithms until the first K edges in the memory are stored. After the first K edges in FP algorithms, the number of edges keep on growing in the memory and will have more number of neighbours stored, whereas in the hybrid approach, the limit on memory reservoir is set, due to which the traversal time becomes almost constant and does not grow much. Hence, hybrid algorithms are faster than both FP and FM algorithms.

How Should We Select the Value of the Input p and K in Hybrid Algorithms Together? Considering that the size of K in FM algorithms is chosen to be 1% of the graph stream Σ, it is equivalent to assigning p to be 0.01 for FP algorithms. Similarly, $K = 10\%$ of Σ is equivalent to $p = 0.1$.

Hybrid algorithms expect the input parameters similarly. Considering a user wants to store 10% of the edges in memory, the value of K should be 10% of the total number of edges and p should be 0.1. In case some other values are provided, hybrid algorithms will have two possibilities as detailed below:

- If k is small or p is large, the algorithm will not go beyond the k number of the edges in the memory, similar to FM algorithms.
- If k is large or p is small, the algorithm has enough memory to finish its work.

These characteristics of input parameters for Hybrid algorithms prove to be better than FM and FP approaches, as the incorrect input values for both FM or FP algorithms results in sub-optimal estimations. Considering the advantages of the hybrid approach over FM and FP approaches, We developed the two new algorithms that belong to the hybrid category, and are detailed in the subsequent sections.

4.1   Neighborhood Hybrid Multisampling (NHMS)

Neighborhood Hybrid Multisampling (NHMS) is the hybrid variant of NMS algorithm discussed in Sect. 2. The intuition behind this algorithm is similar to NMS algorithm, the only difference is the way edges are sampled using a hybrid approach. To understand this better, let's consider an example where we have two triangles, $t_1 = \{a, b, c\}$ and $t_2 = \{a, b, d\}$ that share an edge ab. Assume the order of the edges arriving in the stream is bc, ab, ca, ad, bd. NHMS will sample t_1 if the first edge bc is sampled in L_1, and the edge ab is a neighbour of a sampled edge from L_1 and is sampled in L_2. When the edge ca arrives, the algorithm checks if there's an edge from L_1 (*i.e.* bc) and L_2 (*i.e.* ab) respectively, that forms a triangle. Similarly, t_2 will be counted on arrival of bd, if ad is sampled in L_1, and we already have ab sampled earlier in L_2, therefore, $\langle ab, ad, bd \rangle$ forms a triangle.

The algorithm requires probabilities p and q just like NMS, along with a memory budget K. We maintain two edge reservoirs L_1 and L_2. In our hybrid approach, we limit the size of both the reservoirs to $K/2$. The pseudo-code can be found in Algorithm 1.

Algorithm 1. NHMS

Input: A graph edge stream Σ, memory budget K, probabilities p and q.
 1: $L_1 \leftarrow \emptyset, L_2 \leftarrow \emptyset, Y \leftarrow \emptyset$
 2: **for each** $e_i = (u, v) \in \Sigma$ **do**
 3:     **if** $\text{coin}(p) = $ "head" **then**
 4:         **if** $|L_1| < K/2$ **then**
 5:             Add e_i to L_1
 6:         **else**
 7:             Replace a random edge in L_1 with c_i
 8:     **if** $e_i \in N(L_1)$ **then**
 9:         **if** $\text{coin}(q) = $ "head" **then**
10:             **if** $|L_2| < K/2$ **then**
11:                 Add e_i to L_2
12:             **else**
13:                 Replace a random edge in L_2 with e_i
14:     **for** every (e_j, e_k) where $e_j \in L_1$, $e_k \in L_2$ such that $time(e_j) < time(e_k)$ and (e_i, e_j, e_k) form a triangle **do**
15:         Add the triangle (e_i, e_j, e_k) to Y
16: Return $|Y|/pq$

Theorem 1 (Time Complexity for NHMS). *Let p and q be the fixed probabilities to sample the edges. The time complexity to process an edge $e = (u, v)$ arriving in the stream by Algorithm 1 is $(p + q) \cdot O(d(u) + d(v))$, where $d(u)$ is the degree of vertex u in the reservoir.*

Proof. Suppose that R is the number of edges processed by Algorithm 1, p is the probability to sample L_1 edges, q is the probability to sample L_2 edges, K is the memory budget, and $d(u)$ is the degree of the vertex u in the reservoir.

The running time of NHMS algorithm is dependent on the degree of the vertices of the edges sampled in L_1 and L_2. The triangle count is calculated by traversing the edges from both L_1 and L_2 set, which are stored in the memory. When an edge (u, v) arrives, the common neighbours are searched for each vertex of that edge in L_1 and L_2 set which takes $pd(u) + qd(v)$ and $pd(v) + qd(u)$ steps. This can be further reduced to $(p + q) \cdot O(d(u) + d(v))$.

Theorem 2 (Space Complexity for NHMS). *Let K be the number of edges to be stored on the memory. The space complexity of Algorithm 1 is $O(K)$.*

Proof. Let K be the number of edges to be stored in the memory. The algorithm maintains two edge reservoirs L_1 and L_2 with the maximum size limit of $K/2$. Therefore, the total memory consumed by algorithm will be $O(K)$.

4.2   Trièst/ThinkD Hybrid Sampling (THS)

Trièst/ThinkD Hybrid Sampling (THS) is the hybrid variant of the TS algorithm discussed in Sect. 2. The algorithm requires a memory budget K just like TS, along with fixed probability p. We maintain an edge reservoir S which would have a maximum size limit K, the triangle counter variable Y.

The intuition behind this algorithm is somewhat similar to NHMS algorithm, the only difference is that we just have one reservoir in this case. Consider an example with triangles $t_1 = \{a, b, c\}$ and $t_2 = \{a, b, d\}$, and the edge stream arrive in an order ab, bc, ca, ad, bd. TS will count t_1 if the edges ab and bc will be stored in S considering it's not full, and ca has neighbouring edges in $S \langle ab, bc \rangle$, that forms a triangle. Now let's consider $S = \{ab, bc, ca\}$ and the reservoir is full, when the new edge arrives, it has to replace an existing edge from S randomly. Therefore, the triangle t_2 will be counted on arrival of bd, if both ad and ab still exists in S. The pseudo-code can be found at Algorithm 2.

Theorem 3 (Time Complexity for THS). *Let p be the fixed probability to sample the edges. The time complexity to process an edge $e = (u, v)$ arriving in the stream by Algorithm 2 is*

$$p \cdot O(d(u) + d(v))$$

where $d(u)$ is the degree of vertex u in the reservoir.

Algorithm 2. THS

Input: A graph stream Σ, sampling probability p
1: $S \leftarrow \emptyset, Y \leftarrow 0$
2: **for each** pair $e = (u, v)$ in Σ **do**
3:     Update(u, v)
4:     Insert(u, v)
5: Estimate(Y)
6: **Function** UPDATE(u, v)
7:     **for each** $w \in N(u) \cap N(v)$ **do**
8:         $Y \leftarrow Y + 1$
9: **Function** INSERT(u, v)
10:     **if** coin$(p) = $ "head" **then**
11:         **if** $|S| < K$ **then**
12:             $S \leftarrow S \cup \{(u, v)\}$
13:         **else**
14:             Replace a random edge in S with (u, v)
15: **Function** ESTIMATE(Y)
16:     **return** Y/p^2

Proof. Suppose R is the number of edges processed by p is the probability to sample S edges, K is memory budget, and $d(u)$ is the degree of vertex u in the reservoir.

The running time of this algorithm is dependent on the degree of the vertices of the edges sampled in memory budget K. The triangle count is calculated by traversing the common neighbours of the vertices of an edge. When an edge (u, v) arrives, the common neighbours are searched for each vertex of that edge in S set which takes $p \cdot O(d(u) + d(v))$ steps.

If $|R| > K$, the algorithm for the intersection of common neighbours requires $O(d(u) + d(v))$ time, where the degrees are w.r.t. the graph formed by a stream so far.

Theorem 4 (Space Complexity for THS). *Let K be the number of edges to be stored on the memory. The space complexity of Algorithm 2 is $O(K)$.*

Proof. Let K be the number of edges to be stored in the memory. The algorithm maintains an edge reservoir S with the maximum size limit of K. Hence, the total memory consumed by the algorithm will be $O(K)$.

It can be formally verified that both NHMS and THS produce unbiased estimates when the reservoirs used are not full. When the reservoirs are full, we still do not observe any bias in the estimations produced by our algorithms, and in practice, our estimations are equal or better than those produced by FM and FP algorithms. While we are not able to show the unbiasedness of our algorithms formally, based on our extensive experiments with large datasets and reservoir sizes of only 1% of the datasets, we conjecture that our algorithms are unbiased or exhibit negligible bias.

5   Experimental Evaluation

5.1   Experimental Settings

Table 1. Summary of real-world graphs

Dataset	Nodes	Edges	Triangles
enron	69,244	254,449	1,067,993
cnr	325,557	2,738,969	20,977,629
dblp	986,324	3,353,618	7,005,235
dewiki	1,532,354	33,093,029	88,611,129
ljournal	5,363,260	49,514,271	411,155,444
arabic	22,744,080	553,903,073	36,895,360,842
twitter	41,652,230	1,202,513,046	34,824,916,864

We implemented and executed all algorithms in Java 8. We also used Webgraph framework [2] because of its great compression ratio in saving or loading graphs. Even though the Xeon server had 64 GB RAM, we did not change the default memory settings of JVM. Server JVM heap configuration is 1/4 of the total System memory available. Hence, the allocated memory that can be used by the whole implementation of algorithm will be the maximum of 16 GB. We had to tweak the JVM heap size for running the original NMS implementation provided by authors on Large graphs like Arabic and Twitter, as it creates a lot of objects and does not scale up in 16 GB RAM.

We perform the evaluation of these algorithms on six real word datasets of varied sizes. All these datasets have been downloaded from the Laboratory of Web Algorithms which provides the compressed form of large datasets using WebGraph [1]. These graphs are then symmetrized and any self-loop is deleted to get the corresponding simple undirected graphs. Table 1 shows the summary of real world graphs used for the experiments.

The comparison of all the algorithms is done considering the total number of edges sampled by the respective algorithms. We followed this criteria to get a fair comparison of all the algorithms (irrespective of FP, FM, or Hybrid approach) as to figure out how the performance of the algorithms looks like when they store the same number of edges. Based on this, our results are discussed in the next section.

5.2   Accuracy

We ran an experiment by sampling 1%, 5%, 10% 15% and 20% of the graph size respectively on THS, NHMS, TS and NMS algorithms and compared their accuracy. The results are plotted in Fig. 1 for all the graph datasets we used.

Clearly, both THS and NHMS have better accuracy than their counterparts. It is important to note that the difference in accuracy is not bigger in large graphs as the error rate is below 0.5% for all the test cases. Even if TS and NMS algorithms are highly accurate, NHMS and THS still have even better accuracy.

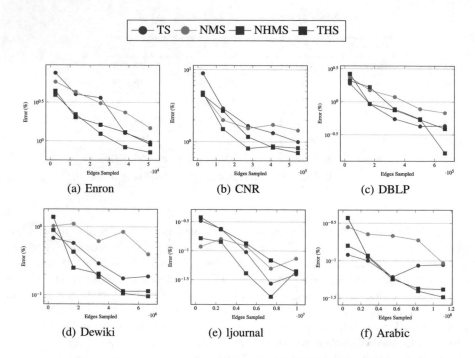

Fig. 1. Error rate for algorithms TS, NMS, NHMS and THS

5.3   Running Time

We ran an experiment in similar setting to accuracy *i.e.* by sampling 1%, 5%, 10% 15% and 20% of the graph size respectively on THS, NHMS, TS and NMS algorithms and compared their running time. We ignored the edge arrival time from the input graph stream to accurately measure the run-time performance of the algorithms. The results are plotted in Fig. 2 for all the graph datasets that we used.

THS versus TS: THS proves to be **5 - 10X faster** than its counterpart TS. Hence, these results validate our claim that Hybrid algorithms are faster than FM algorithms.

NHMS versus NMS: There is only a slight difference between NHMS and NMS. The reason behind this is that NMS is already an FP algorithm and Hybrid algorithms follow the same execution trend as FP algorithms until the

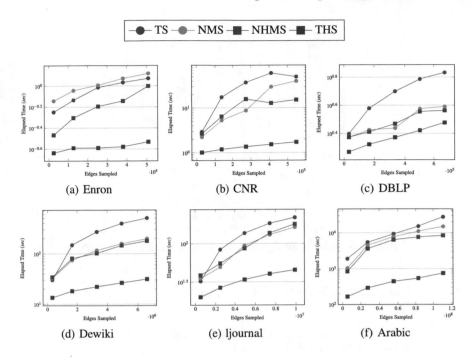

Fig. 2. Run-time for algorithms TS, NMS, NHMS and THS

first k edges are processed, after which the increase in time becomes constant. This would be more clear in the next section on scalability. Hence, NHMS is faster than NMS algorithms after the first k edges are processed.

THS versus NHMS: Among our algorithms, THS outperformed NHMS considering the running time, but there's consistently a trade-off among both between run-time and accuracy. NHMS is more accurate whereas THS is much faster than NHMS. **Overall**, both our hybrid algorithms have better running time in comparison to both FM and FP algorithms. THS outperformed all algorithms including NHMS.

5.4 Scalability

To measure scalability, we ran an experiment on twitter that has more than 1.2 billion edges. Data points including time and the number of edges streamed are collected after every 1 million edges in all the algorithms. The results are shown in Fig. 3.

Scalability of FP Algorithms: FP algorithms are not scalable due to their characteristics. As the edges stored always grows in FP algorithms, elapsed time grows exponentially as the number of edges increases on massive graphs. We can see the results of NMS graph and the algorithm is unable to process whole graph

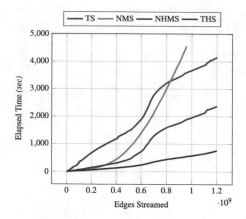

Fig. 3. Scalability analysis of algorithms on Twitter graph

stream and halted even before the billion edges were processed. Hence, NMS is not scalable.

NHMS can Scale NMS: As NHMS is the hybrid variant of NMS, it is important to compare both. NMS does not scale on the large graphs, whereas we see NHMS is able to process the whole graph stream. As we have already discussed, NHMS follows the same trend as FP algorithms until the first k edges. We observe that behavior in Fig. 3 around 600 million edges, where the curve started to flatten, after a slight exponential rise (similar to NMS).

Scalability of THS and Comparison with TS: As seen in Fig. 3, THS scales linearly. In comparison to TS that takes around 4200 s to execute on Twitter graph, THS just takes 760 s that is **5.5X faster**.

   Overall, Both hybrid algorithms scale really well for large graphs and are faster in executing the graph stream than both FM and FP algorithms.

6   Conclusion and Future Work

In this study, We prove that FP algorithms are faster than FM algorithms, due to their characteristics that we discussed in Sect. 3.3. In addition, we present two new algorithms for triangle estimations that belong to the Hybrid category *i.e.* NHMS and THS. In our evaluation, we observed that NHMS and THS perform a lot better than their counterparts, *i.e.* NMS and TS, in both running time and accuracy. THS proves to be at least **5 times faster** than TS in all the cases with varied graph sizes and different experimental settings. NHMS could not only scale NMS in 16 GB memory, but also is approximately **2 times faster** in most cases.

   This research opens up a completely new dimension as we present a new hybrid approach where the community contributions will be proven useful. The theoretical proofs for the accuracy and variance of the estimates of both Hybrid algorithms will provide additional validation of the algorithms. Also, both NHMS and THS can be extended to count sub-graph structures other than triangle.

References

1. Boldi, P., Rosa, M., Santini, M., Vigna, S.: Layered label propagation: a multiresolution coordinate-free ordering for compressing social networks. In: Srinivasan, S., Ramamritham, K., Kumar, A., Ravindra, M.P., Bertino, E., Kumar, R. (eds.) Proceedings of the 20th International Conference on World Wide Web, pp. 587–596. ACM Press (2011)
2. Boldi, P., Vigna, S.: The WebGraph framework I: compression techniques. In: Proceedings of the Thirteenth International World Wide Web Conference (WWW 2004), pp. 595–601. ACM Press, Manhattan (2004)
3. Buriol, L.S., Frahling, G., Leonardi, S., Marchetti-Spaccamela, A., Sohler, C.: Counting triangles in data streams. In: Proceedings of the Twenty-Fifth ACM SIGMOD-SIGACT-SIGART Symposium on Principles of Database Systems, pp. 253–262. ACM (2006)
4. Jha, M., Seshadhri, C., Pinar, A.: A space efficient streaming algorithm for triangle counting using the birthday paradox. In: Proceedings of the 19th ACM SIGKDD International Conference on Knowledge Discovery and Data Mining. pp. 589–597. ACM (2013)
5. Kavassery-Parakkat, N., Hanjani, K.M., Pavan, A.: Improved triangle counting in graph streams: power of multi-sampling. In: 2018 IEEE/ACM International Conference on Advances in Social Networks Analysis and Mining (ASONAM), pp. 33–40. IEEE (2018)
6. Pavan, A., Tangwongsan, K., Tirthapura, S., Wu, K.L.: Counting and sampling triangles from a graph stream. Proc. VLDB Endow. $6(14)$, 1870–1881 (2013). https:// doi.org/10.14778/2556549.2556569. http://dx.doi.org/10.14778/2556549.2556569
7. Santoso, Y., Thomo, A., Srinivasan, V., Chester, S.: Triad enumeration at trillion-scale using a single commodity machine. In: Advances in Database Technology-EDBT 2019, 22nd International Conference on Extending Database Technology, Lisboa, Portugal, 26–29 March 2019, Proceedings. OpenProceedings.org (2019)
8. Shin, K., Kim, J., Hooi, B., Faloutsos, C.: Think before you discard: accurate triangle counting in graph streams with deletions. In: Joint European Conference on Machine Learning and Knowledge Discovery in Databases, pp. 141–157. Springer (2018)
9. Singh, P.: Fast and scalable triangle counting in graph streams: the hybrid approach. Master's thesis, University of Victoria (2020)
10. Stefani, L.D., Epasto, A., Riondato, M., Upfal, E.: TRIEST: counting local and global triangles in fully dynamic streams with fixed memory size. ACM Trans. Knowl. Discov. Data (TKDD) $11(4)$, 43 (2017)

An Empirical Study on Predictability of Software Code Smell Using Deep Learning Models

Himanshu Gupta[1]([✉]), Tanmay Girish Kulkarni[1], Lov Kumar[1],
Lalita Bhanu Murthy Neti[1], and Aneesh Krishna[2]

[1] BITS Pilani, Hyderabad Campus, Hyderabad, India
{f20150339h,f20150647h}@alumni.bits-pilani.ac.in,
{lovkumar,bhanu}@hyderabad.bits-pilani.ac.in
[2] Curtin University, Bentley, Australia
a.krishna@curtin.edu.au

Abstract. Code Smell, similar to a bad smell, is a surface indication of something tainted but in terms of software writing practices. This metric is an indication of a deeper problem lies within the code and is associated with an issue which is prominent to experienced software developers with acceptable coding practices. Recent studies have often observed that codes having code smells are often prone to a higher probability of change in the software development cycle. In this paper, we developed code smell prediction models with the help of features extracted from source code to predict eight types of code smell. Our work also presents the application of data sampling techniques to handle class imbalance problem and feature selection techniques to find relevant feature sets. Previous studies had made use of techniques such as Naive Bayes and Random forest but had not explored deep learning methods to predict code smell. A total of 576 distinct Deep Learning models were trained using the features and datasets mentioned above. The study concluded that the deep learning models which used data from Synthetic Minority Oversampling Technique gave better results in terms of accuracy, AUC with the accuracy of some models improving from 88.47 to 96.84.

Keywords: Code smell · Data sampling techniques · Software metrics · Feature selection

1 Introduction

A code smell is a quantifiable metric which indicates severe problems in complex software development life cycles due to poor programming practices. A code smell by itself may not reflect a programmatic error [1,2] within the software.

H. Gupta and T. G. Kulkarni—The work was done when authors were students in BITS Pilani, Hyderabad Campus.

Instead, it is a harbinger of potential problems in the future during maintenance or when additional functionality is built into the software [3]. A code smell is generally detected by inspecting the source code and searching for sections of the code can be restructured to improve the quality of code. This method is inefficient, especially if developers have to crawl through potentially thousands of lines of code, which can consume a significant amount of time and money to the organization. Based on the internal organization and anatomy of the software, a robust model can be created, which can make this excruciating process a lot simpler. In this work, we have used a set of metrics extracted from the source code of the software as an input to develop multiple models for predicting code smell present in the source code of the software. Time conserving ability and capability to maintain software will be improved if these hidden problems become apparent to developers.

The above-computed metrics are used as an input of the code smell prediction models, so the predictive ability of the models depends on the selection of relevant metrics. In our research, different methods were applied to find the relevant metrics and also methods that help to find these metrics. We observed that the considered datasets to validate this proposed work were highly imbalanced in terms of the number of samples. In order to balance them, we used different smoothing techniques. The primary focus of our work was to evaluate how different smoothing methods and different metric selection method affect the performance of a code smell prediction models.

After successful computation of the above steps, we have used different varieties of deep learning techniques to train the code smell, prediction models. We apply this model on the following code smells- Blob Class (BLOB), Complex Class (CC), Internal Getter/Setter (IGS), Leaking Inner Class (LIC), Long Method (LM), No Low Memory Resolver (NLMR), Member Ignoring Method (MIM), and Swiss Army Knife (SAK) [4]. In this paper, we attempt to answer the following Research Questions (RQ):

- **RQ1: Discuss the ability of selected features over the original features towards detecting Code Smell.** As there were many metrics to select from which the models were to be developed; it was inevitable that some of them were found to be correlated. We aim to determine the features which were related to predicted code smell. We also made sure that the selected features were unrelated to each other. The correlation between the features and resultant code smells were obtained using the Cross-Correlation Analysis and Wilcoxon Sign Rank test [5,6].
- **RQ2: Discuss the ability of Data Sampling Techniques to detect Code Smell.** As the number of code smell and the metrics associated with them varies significantly with source code of the software, There was a need to sample them for creation of an unbiased dataset. This practice ensured that the models which were developed had balanced data to ensure proper training. ADASYN (Adaptive Synthetic Sampling Method) [7] and SMOTE (Synthetic Minority Over-Sampling Method) [8] were used to balance the data.

- **RQ3: Discuss the ability of different deep learning architectures to detect Code Smell.** Eight different deep learning architectures were developed in order to encounter the prediction of code smell. These neural networks were created by varying the number of hidden layers. The performance of these architectures was compared with accuracy, F-Measure and area under the curve.

Organization: The paper is organized as follows: Sect. 2 summarizes the related work. Section 3 gives a detailed overview of all the components used in the experiment. Section 4 describes the research framework pipeline and how different components described in Sect. 3 work together. Section 5 gives the experimental results and Sect. 6 answers the questions asked in Introduction. Finally, we conclude in Sect. 7.

2   Related Work

Several studies were done for code smell detection and how they affect the performance of the system. Yamashita et al. [2] discuses an empirical method which recognises which code smells were recognised as significant for maintainability. The outcome is based on an analysis of an industrial case study done by the author. Khomh et al. [3] explores the connection between code smells and changes proneness in classes. The author has done intensive research to identify whether classes with code smells are more prone to change as compared to other classes and vice versa. The conclusion indicates that specific code smells adversely impact the classes. Coleman et al. [9] demonstrate how software-related decision making can benefit from automated software maintainability analysis. Defect prediction using the Naïve Bayes method has been demonstrated by Wang et al. [10]. The author also analyses the construction of prediction models. Turhan et al. [11] attempt to develop an approach that allows the use of software metrics to assist in defect prediction. The proposed methodology yields statistically better results. Francesca Arcelli Fontana et al. [12] compared 16 different machine learning algorithms on four code smells. He found out that Random Forest performed the best while SVM performed the worst.

There have been several studies for feature selection methods in an unbalanced dataset. Lida Abidi et al. [13] presented an oversampling technique which generates synthetic samples of data using Mahalanobis distance which preserves the covariance of the minority distance.

Our primary contribution is to predict code smells utilizes deep learning in addition to data sampling techniques such as SMOTE and ADASYN to balance the dataset. We further make use of feature selection techniques to improve the accuracy of the models.

3   Research Background

Our research is formulated using the following steps:

- In order to encounter the data imbalance problem, we created two different datasets apart from the original.
- We then select the relevant features from each dataset via cross co-relation analysis and Wilcoxon Sign Rank test.
- After normalizing the datasets, we train eight deep learning models on them and validate them using 5-fold cross-validation.

Table 1. Statistical measures on AUC: sets of metrics.

Code smell type	No smell number	No smell percentage	Smelly number	Smelly percent
BLOB	236	37.59%	393	62.48%
LM	156	24.80%	475	75.20%
SAK	463	73.60%	166	26.40%
CC	188	29.88%	441	70.12%
IGS	277	44.03%	352	55.97%
MIM	261	41.49%	368	58.51%
NLMR	158	25.11%	471	74.89%
LIC	227	36.08%	402	63.92%

3.1   Experimental Dataset

In this paper, we have analyzed the code-bases of 629 open source projects which were scraped from GitHub. This dataset consists of a list of packages along with the corresponding code smells. The Anti-Patterns observed over the repositories are given in Table 1. From the table, it is evident that each code smell varies greatly. Some were very common as to be present in 75.2% of the entire dataset. At the same time, some of the smells were as scarce as being present in just 26.4% of the dataset. Furthermore, we observe that the least commonly observed code smell is the Swiss Army Knife (SAK), and most commonly observed code smell is the Long Method (LM).

3.2   Software Metrics

Software metrics are quantitative measures of different aspects of code features of a software. The different aspects include features like the modularity of a class, size of the class and many other similar characteristics. Although these metrics are used in a variety of tasks related to software engineering like software performance, measuring productivity and other software engineering tasks, we utilized these metrics as features in code smell detection of the software. There were four primary types of software metrics used for this study:

- **Dimensional Metrics:** These metrics are aimed towards providing a quantitative metric for understanding code sizes and modularity. One might believe that more code leads to more features, but it makes the code difficult to handle in the long run [14].
- **Complexity Metrics:** These metrics help us gauge the complexity of applications, which are associated with the fact that an increase in complexity makes it challenging to comprehend and thus, difficult to test and maintain due to a larger number of paths of execution [12].
- **Object Oriented Metrics:** These metrics are used to find the complexity, cohesion and coupling between the software modules [15].
- **Android Oriented Metrics:** These metrics show how Android specific dependencies and operations affect execution speed and User Experience.

3.3   Data Sampling Techniques to Handle Imbalanced Data

As the distribution of the target classes, as shown in Table 1 is skewed within the dataset. Thus, we make use of the following standard data sampling techniques to offset the probability of each class in the dataset:

- ADASYN [7] generates additional data points by using the predefined density distribution of the dataset.
- SMOTE [8] creates additional data points by assuming a uniform distribution of the dataset.

3.4   Feature Selection Techniques

In order to find the relevant code smell metrics, we make use of the Wilcoxon Sign Rank Test [5] and Cross-Correlation analysis [6]. Using the methods mentioned above was important our research as we wanted to use only the statistically significant (highly unrelated to each other and related to output variable) metrics for creation of the models. After we find the significant features with the help of these techniques, we apply cross-correlation analysis to find not related features. For the above premise to work efficiently, the following null hypothesis has been proposed: Feature metric is incapable of finding out a particular code smell. In this study, we have made use of a p-value of 0.05; that is, we reject the hypothesis if the probability of the null hypothesis is below 0.05.

3.5   Deep Learning Model for Classification

In this study, eight deep learning models were used to train the models for predicting different types of code smell. Figures 1.1 and 1.2 show the architecture of the considered deep learning model1 and model2 (DL1 and DL2). Similarly, we are increasing the number of hidden layers for six more deep learning models (DL3: 3 hidden layers, DL4: 4 hidden layers, DL5: 5 hidden layers, DL6: 6 hidden layers, DL7: 7 hidden layers, DL8: 8 hidden layers). The following are the details of the neural networks shown in Figs. 1.1 and 1.2:

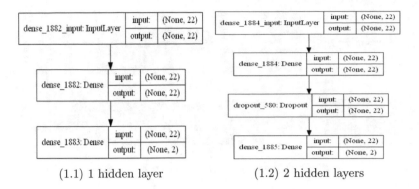

(1.1) 1 hidden layer   (1.2) 2 hidden layers

Fig. 1. Deep learning model framework **Note:** *None* in the figure refers to the Batch size which is flexible.

- The model has an input layer with 22 nodes corresponding to 22 features given as an input to the model.
- It is followed by a hidden layer with 22 nodes and ReLU [16] function is used as an activation function.
- The output layer has 2 nodes with softmax [17] as the activation function.
- For the 2nd model we have added a dropout layer to prevent over-fitting with a probability of 0.2 after the hidden layer.
- Rest of the 6 models follow a similar pattern by adding dense layer and a dropout serially one after another with ReLU being the activation function.
- All the models are trained with a batch size of 32 and 100 epochs.

4   Research Framework

As shown in Fig. 2, SMOTE and ADASYN sampling methods are applied to the dataset to tackle the class imbalance problem. Then, we have made use of statistical significance tests (Cross Co-relation and Wilcoxon Sign rank test) to identify the relevant metrics. These relevant metrics were used to train eight deep learning models and are compared using accuracy and AUC. We have also considered a significance test to validate the considered hypothesis: There is no significant improvement in the predictive ability of the models after applying data sampling techniques and feature selection techniques.

5   Experiment Results

In this work, eight neural networks were used for developing a model to classify different types of code smell by considering extracted features from source code as an input whose details were presented in Subsect. 3.2. SMOTE and ADASYN were used to tackle the class imbalance problem to get three datasets (Original, SMOTE and ADASYN datasets). Further, feature Selection methods

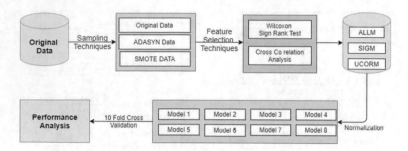

Fig. 2. Flowchart of the research framework

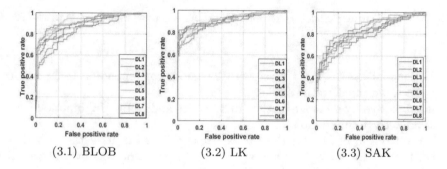

(3.1) BLOB   (3.2) LK   (3.3) SAK

Fig. 3. ROC curve: BLOB, LM. and SAK prediction models

were applied on these datasets to give three feature sets which were All Metrics (ALLM), significant features (SIGM), and uncorrelated significant features (UCORM). For each dataset, eight smells are identified, and eight different models are used, which gives us a total of $(3 \times 3 \times 8 \times 8)$ 576 Distinct Models. The predictive ability of these code-smell prediction models is compared using Accuracy, AUC, and F-Measure values. Figure 3 displays the ROC (Receiver operating characteristic) curve for three code smells (BLOB, LK and SAK). All eight models, which used significant features and trained using SMOTE sampled data, are shown in the figure. The high true positive rate in Fig. 3 suggested that the trained models can predict code smell using different extracted features. Table 2 displays the value of AUC and Accuracy for 192 out of 576 models on SMOTE Data-set. The following observations are extracted from the information present in Table 2:

- The accuracy of the models varies greatly, having a range from 74.52 to 96.84. However, AUC does not vary a lot with a range of .83 to .98.
- The mean model accuracy for SMOTE data is 80.30, and the 75th percentile is 88.14. The mean AUC value was found to be .87 and .94 being the 75th percentile. These values overall indicate high efficiency of the models developed.

Table 2. Accuracy, and AUC values for Deep learning prediction models trained on SMOTE data

	Accuracy								AUC							
	ALLM															
	DL1	DL2	DL3	DL4	DL5	DL6	DL7	DL8	DL1	DL2	DL3	DL4	DL5	DL6	DL7	DL8
BLOB	74.52	75.80	77.71	78.34	75.16	80.25	82.80	83.44	0.83	0.86	0.87	0.88	0.83	0.86	0.87	0.88
LM	90.00	87.89	94.74	93.16	93.16	96.32	94.21	96.32	0.97	0.98	0.99	0.99	0.97	0.98	0.99	0.99
SAK	76.34	77.96	84.95	79.03	82.26	84.95	79.57	84.95	0.85	0.90	0.91	0.91	0.85	0.90	0.91	0.91
CC	82.49	80.79	87.01	86.44	83.62	88.14	88.14	89.27	0.89	0.92	0.92	0.92	0.89	0.92	0.92	0.92
IGS	70.21	73.05	75.18	72.34	72.34	76.60	73.76	78.01	0.79	0.81	0.83	0.84	0.79	0.81	0.83	0.84
MIM	74.32	71.62	78.38	81.08	75.00	82.43	73.65	80.41	0.83	0.89	0.87	0.86	0.83	0.89	0.87	0.86
NLMR	89.95	89.42	94.71	91.53	91.01	92.59	93.12	94.18	0.96	0.98	0.98	0.98	0.96	0.98	0.98	0.98
LIC	81.99	78.26	86.96	85.71	84.47	83.85	83.23	89.44	0.90	0.92	0.92	0.93	0.90	0.92	0.92	0.93
	SIGM															
BLOB	74.52	75.16	73.25	76.43	77.07	74.52	78.98	80.25	0.82	0.84	0.84	0.87	0.82	0.84	0.84	0.87
LM	89.47	88.95	93.68	93.68	94.21	96.84	93.68	95.26	0.96	0.98	0.98	0.98	0.96	0.98	0.98	0.98
SAK	63.98	66.67	65.05	60.22	62.37	64.52	62.90	61.83	0.67	0.71	0.74	0.64	0.67	0.71	0.74	0.64
CC	80.79	82.49	85.31	83.05	84.18	88.7	87.01	87.01	0.88	0.90	0.92	0.93	0.88	0.90	0.92	0.93
IGS	68.09	64.54	73.76	69.50	70.92	73.05	72.34	73.76	0.74	0.79	0.79	0.81	0.74	0.79	0.79	0.81
MIM	71.62	73.65	76.35	70.95	70.95	79.73	72.30	78.38	0.80	0.83	0.83	0.85	0.80	0.83	0.83	0.85
NLMR	89.42	88.89	92.59	88.89	93.65	94.18	92.59	94.71	0.96	0.97	0.98	0.98	0.96	0.97	0.98	0.98
LIC	77.02	81.37	83.23	81.99	85.09	83.23	82.61	85.09	0.90	0.91	0.92	0.92	0.90	0.91	0.92	0.92
	UCORM															
BLOB	80.89	77.71	73.89	77.07	77.07	78.34	79.62	75.16	0.84	0.84	0.84	0.83	0.84	0.84	0.84	0.83
LM	87.89	87.89	89.47	89.47	89.47	90.00	89.47	91.05	0.96	0.96	0.96	0.97	0.96	0.96	0.96	0.97
SAK	64.52	63.98	63.44	63.98	65.05	66.13	65.59	70.97	0.70	0.71	0.73	0.69	0.70	0.71	0.73	0.69
CC	81.36	80.23	84.18	81.36	82.49	86.44	79.66	83.62	0.89	0.89	0.9	0.9	0.89	0.89	0.9	0.9
IGS	65.96	63.83	68.79	68.79	67.38	71.63	66.67	69.50	0.73	0.78	0.79	0.81	0.73	0.78	0.79	0.81
MIM	70.95	70.95	70.95	71.62	70.95	73.65	70.95	71.62	0.81	0.84	0.84	0.82	0.81	0.84	0.84	0.82
NLMR	88.89	89.42	88.36	88.36	90.48	88.36	89.95	87.83	0.95	0.95	0.96	0.96	0.95	0.95	0.96	0.96
LIC	78.26	78.26	78.88	80.12	81.37	83.85	85.09	83.23	0.87	0.88	0.90	0.92	0.87	0.88	0.90	0.92

- The models developed using Significant features (SIGM) perform better as compared to models using Uncorrelated features (UCORM) with mean accuracy being 78.52 for SIGM and 76.52 for UCORM.
- The deep learning models with eight hidden layers have the best performance as compared to models with lesser hidden layers.

6   Comparison

6.1   RQ1: Discuss the Ability of Selected Features over the Original Features Towards Detecting Code Smell

In this section, we analyze the performance of the features used within the model, obtained with the help of the Wilcoxon Sign Rank test and cross-correlation analysis. The box plots give us an easy way to visualize the predictive ability of the trained models. The models trained using different sets of features are further validated by the rank-sum test, which is used to be used for validating the considered hypothesis. In this direction, the following hypotheses are designed and tested using the rank-sum test:

- **Null-Hypothesis:** There is no significant improvement in the performance of the models trained using selected sets of features.
- **Alternate hypothesis** There is a significant improvement in the performance of the models trained using selected sets of features.

Comparison of Different Combinations of Features Using the Box-Plot Diagram: Figure 4 shows the box-plot diagram and descriptive statistics for the performance of the trained models using three different combinations of features, namely: All Metrics (ALLM), significant metrics obtained from the Wilcoxon Sign Rank Test (SIGM), and features obtained from the Cross-Correlation Analysis (UNCORM) in terms of AUC, F-Measure, and Accuracy. The upper and lower edges of the box plot refer to the first and the third quartile value. Also, the top and the bottom line refer to the maximum and minimum value, respectively. The line inside the boxes refers to the average value of the data. In a nutshell, this diagram tells us about distribution for maximum, minimum, percentiles, and dispersion of data. It is evident from Fig. 4 that the models trained by considering all features as an input have better ability to predict code-smell as compared to other sets of features. Figure 4 also suggests that the models trained using significant sets of features have a better ability to predict code-smell as compared to uncorrelated sets of features.

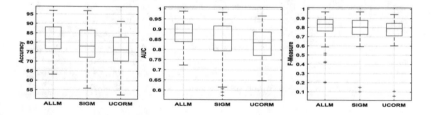

Fig. 4. Results based on significant metrics

Comparison of Different Combinations of Features Using Ranksum Test: This hypothesis has been validated at a confidence level of 95% on the AUC value of the trained models. Hence, the considered null hypothesis is rejected if the p-value is less than 0.05, and the alternate hypothesis is rejected if the p-value is more than 0.05. Table 3c shows the results after applying the rank-sum test on the performance of the models trained using different sets of features. The results in Table 3c suggested that the developed models using different sets of features are significantly different, i.e., the calculated p-value is smaller than 0.05 (alternate hypothesis is accepted).

6.2 RQ2: Discuss the Ability of Data Sampling Techniques to Detect Code Smell

In this question, we analyze the difference in performance for the models trained using datasets generated by the class imbalance techniques. As in the previous

Table 3. Ranksum test

(a) Ranksum Test: Different Model similarity

	DL1	DL2	DL3	DL4	DL5	DL6	DL7	DL8
DL1	1.00	0.04	0.02	0.01	1.00	0.04	0.02	0.01
DL2	0.04	1.00	0.80	0.49	0.04	0.99	0.80	0.49
DL3	0.02	0.80	1.00	0.64	0.02	0.80	0.99	0.64
DL4	0.01	0.49	0.64	1.00	0.01	0.48	0.64	1.00
DL5	1.00	0.04	0.02	0.01	1.00	0.04	0.02	0.01
DL6	0.04	0.99	0.80	0.48	0.04	1.00	0.80	0.49
DL7	0.02	0.80	0.99	0.64	0.02	0.80	1.00	0.64
DL8	0.01	0.49	0.64	1.00	0.01	0.49	0.64	1.00

(b) Different sampling methods

	ORGD	SMOTE	ADASYN
ORGD	1.00	0.00	0.00
SMOTE	0.00	1.00	0.00
ADASYN	0.00	0.00	1.00

(c) Feature Combinations

	ALLM	SIGM	UCORM
ALLM	1.00	0.00	0.00
SIGM	0.00	1.00	0.04
UCORM	0.00	0.04	1.00

section, we make use of the same tools: box-plots, descriptive statistics, and rank-sum test of performance parameters to compare the predictive ability of the trained models using sampled data.

Comparison of Different Samples Using the Box-Plot Diagram: Figure 5 shows the box-plot diagram and descriptive statistics for the performance of the trained models on original data (ORGD), SMOTE sampled data, and ADASYN sampled data in terms of AUC, F-Measure, and Accuracy. From Fig. 5, it is observed that the models trained using SMOTE techniques have better ability to predict code-smell as compared to the original data. The figure also suggested that the 25% of the trained models on sample data have more than 0.94 AUC value.

Comparison of Different Sampling Techniques using Ranksum Test: The rank-sum test further validates the predictive ability models trained on different datasets. The following hypotheses are designed and tested using the rank-sum test:

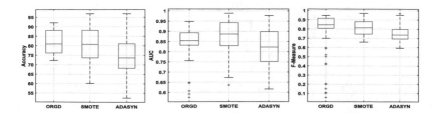

Fig. 5. Result variations on data sampling techniques

- **Null-Hypothesis:** There is no significant improvement in the performance of the models trained on balanced data.
- **Alternate hypothesis:** There is a significant improvement in the performance of the models trained on balanced data.

Table 3b shows the results after applying Ranksum test on the performance of the models trained on different datasets. The p-value smaller than 0.05 of ranksum test present in Table 3b suggested that the models trained on balanced data have significant improvement in predicting different code smells.

6.3   RQ3: Discuss the Ability of Different Deep Learning Models to Detect Code Smell

In this research question, we compare the performance of the code smell prediction models trained using eight different neural networks. We have used graphical analysis with the help of box-plots on Accuracy and Area Under Curve Metrics for each classifier to find the most accurate deep learning model. We have also applied the rank-sum test to validated the following hypotheses:

- **Null-Hypothesis:** There is no significant improvement in the performance of the models while increasing the number of hidden layers.
- **Alternate hypothesis:** There is a significant improvement in the performance of the models while increasing the number of hidden layers.

Comparison of Different Classifiers Using Descriptive Statistics and the Box-Plot Diagram: Figure 6 shows the box-plot diagram for the performance of the trained models using eight deep learning models in terms of AUC, F-Measure, and Accuracy. We see that, as the number of hidden layer increases, a corresponding increase in performance in the prediction of the models. However, from Fig. 6 we can a slight dip in performance in architecture with seven hidden layers (Mean Values: 79.39 accuracy, .86 AUC and .81 F-Measure respectively) as compared to six and eight hidden layers (Mean Values: 81.71 and 81.16 accuracy, .86 and .86 AUC and .81 and .82 F-Measure respectively).

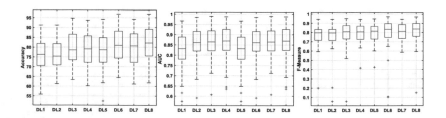

Fig. 6. Results for different deep learning models

Comparison of Different Classifying Techniques using Ranksum Test: The above-considered hypothesis has been validated at a confidence level of 95%

on the AUC value of the trained models. Hence, the null hypothesis is rejected if the p-value is less than 0.05, and the alternate hypothesis is rejected if the p-value is more than 0.05. Table 3a shows the results after applying the rank-sum test on the performance of the models. The results in Table 3a suggested that there is no significant improvement in the performance of the models while increasing the number of hidden layers, i.e., the calculated p-value for most of the pairs are more than 0.05 (alternate hypothesis is rejected).

7   Conclusion

This paper presents the empirical analysis on code smell prediction models developed using data sampling, features selection and neural networks. These models are validated using 5-fold cross-validation, and the prediction ability of these models are compared using three different performance parameters which were AUC, Accuracy, and F-Measure.

Our primary conclusion is that an increase in the number of hidden layer did not lead to a monotonic increase in performance contrary to prior expectation. Furthermore, we see diminishing returns in the performance increase with the addition of a hidden layer. We also observe that the models with eight hidden layers performs the best (higher accuracy, AUC and F-Measure) as compared to other models. The AUC, Accuracy, and F-Measure values of the trained models also suggest that the models trained on balanced data perform better than models developed on original data. The rank-sum test also indicates that the models trained using balanced data have a significant improvement in code smell prediction ability. The results of the rank-sum test also indicate that there is no significant improvement in the performance of the models while increasing the number of hidden layers.

References

1. Azeem, M.I., Palomba, F., Shi, L., Wang, Q.: Machine learning techniques for code smell detection: a systematic literature review and meta-analysis. Inf. Softw. Technol. **108**, 115–138 (2019)
2. Yamashita, A., Moonen, L.: Do code smells reflect important maintainability aspects? In: 2012 28th IEEE International Conference on Software Maintenance (ICSM), pp. 306–315. IEEE (2012)
3. Khomh, F., Di Penta, M., Gueheneuc, Y.-G.: An exploratory study of the impact of code smells on software change-proneness. In: 2009 16th Working Conference on Reverse Engineering, pp. 75–84. IEEE (2009)
4. Hecht, G., Moha, N., Rouvoy, R.: An empirical study of the performance impacts of android code smells. In: Proceedings of the International Conference on Mobile Software Engineering and Systems, pp. 59–69 (2016)
5. Wilcoxon, F., Katti, S.K., Wilcox, R.A.: Critical values and probability levels for the Wilcoxon rank sum test and the Wilcoxon signed rank test. Sel. Tables Math. Stat. **1**, 171–259 (1970)

6. Podobnik, B., Stanley, H.E.: Detrended cross-correlation analysis: a new method for analyzing two nonstationary time series. Phys. Rev. Lett. **100**(8), 084102 (2008)
7. He, H., Bai, Y., Garcia, E.A., Li, S.: ADASYN: adaptive synthetic sampling approach for imbalanced learning. In: 2008 IEEE International Joint Conference on Neural Networks (IEEE World Congress on Computational Intelligence), pp. 1322–1328. IEEE (2008)
8. Bowyer, K.W., Kegelmeyer, W.P., Chawla, N.V., Hall, L.O.: SMOTE : synthetic minority over-sampling technique. Technical report (2011)
9. Coleman, D., Ash, D., Lowther, B., Oman, P.: Using metrics to evaluate software system maintainability. Computer **27**(8), 44–49 (1994)
10. Wang, T., Li, W.: Naive bayes software defect prediction model. In: 2010 International Conference on Computational Intelligence and Software Engineering, pp. 1–4. IEEE (2010)
11. Turhan, B., Bener, A.B.: Software defect prediction: heuristics for weighted naïve bayes. In: ICSOFT (SE), pp. 244–249 (2007)
12. Fontana, F.A., Mäntylä, M.V., Zanoni, M., Marino, A.: Comparing and experimenting machine learning techniques for code smell detection. Empir. Softw. Eng. **21**(3), 1143–1191 (2016)
13. Abdi, L., Hashemi, S.: To combat multi-class imbalanced problems by means of over-sampling techniques. IEEE Trans. Knowl. Data Eng. **28**(1), 238–251 (2015)
14. Abd-El-Hafiz, S.K.: A metrics-based data mining approach for software clone detection. In: 2012 IEEE 36th Annual Computer Software and Applications Conference, pp. 35–41. IEEE (2012)
15. Harrison, R., Counsell, S.J., Nithi, R.V.: An evaluation of the mood set of object-oriented software metrics. IEEE Trans. Softw. Eng. **24**(6), 491–496 (1998)
16. Agarap, A.F.: Deep learning using rectified linear units (ReLU). arXiv preprint arXiv: 1803.08375 (2018)
17. Wang, M., Lu, S., Zhu, D., Lin, J., Wang, Z.: A high-speed and low-complexity architecture for softmax function in deep learning. In: 2018 IEEE Asia Pacific Conference on Circuits and Systems (APCCAS), pp. 223–226. IEEE (2018)

A Big Data Science Solution for Analytics on Moving Objects

Isabelle M. Anderson-Grégoire[1], Kaitlyn A. Horner[1], Carson K. Leung[1(✉)] ⒾⒹ,
Delica S. Leboe-McGowan[1], Anifat M. Olawoyin[1], Beni Reydman[1],
and Alfredo Cuzzocrea[2]

[1] University of Manitoba, Winnipeg, MB, Canada
`kleung@cs.umanitoba.ca`
[2] University of Calabria, Rende, CS, Italy

Abstract. Biodiversity data (e.g., for aquatic organisms, marine creatures and terrestrial animals) and environmental data (e.g., air pollution statistics, water supply and sanitation information, soil contamination data) are examples of big data. Embedded in these big data are implicit, previously unknown and potentially useful information and knowledge that could help improve the ecosystem. As such, data science solutions for big data analytics and mining are in demand. In this paper, we present a data science solution for biodiversity informatics, environmental analytics and sustainability analysis. Specifically, our solution analyzes and mines both biodiversity data and environmental data to examine the impacts of pollution to moving objects. The convex-hull-based method in our solution estimates the pollution exposure to these objects. For evaluation, we conducted case studies on analyzing, mining and visualizing both marine biodiversity data and plastic exposure data to examine the impacts of the plastic exposure to marine creatures. Knowledge discovered by our solution help decision and policy makers to take appropriate actions in building and maintaining a sustainable environment.

Keywords: Data mining · Big data analytics · Biodiversity informatics ·
Pollution · Plastic exposure · Aquatic organism · Marine creature ·
Environmental analytics · Sustainability analysis · Visualization

1 Introduction

Nowadays, big data [1–5] can be found almost everywhere in numerous real-life applications and services. To elaborate, huge amounts of a wide variety of valuable data—in which some may be of different levels of veracity (e.g., precise data, imprecise and uncertain data)—are generated and collected at a very rapid rate from a wide range of rich data sources. Examples include the following:

- biodiversity data [6, 7] (e.g., aquatic organism and marine creature assessments, terrestrial animal stock assessments);
- communication networks [8, 9] and social networks [10–12];

- disease reports [13], as well as epidemiological data and statistics [14–18];
- financial time series [19–21];
- omic data (e.g., genomic data) [22–26];
- urban and transportation data [27–30]; and
- environmental data [31–33] (e.g., air and water temperatures, air pollution statistics, bathymetric and hydrographic data, geo-hazard and debris flows, sea bottom characteristics, sea water samples, sedimentary bed forms, snow samples, soil and salinity samples, volcanic activities, wind speed).

Embedded in these big data are implicit, previously unknown and potentially useful information and valuable knowledge. This explains why the big data are sometimes considered as the "new oil". Hence, data science solutions [34–36]—which aim to discover knowledge and information from big data via data mining algorithms [37–39], machine learning tools [40–43], mathematical and statistical models [44], informatics [45, 46], and visualization [47–52]—for big data analytics and mining are in demand. For instance, analyzing and mining big biodiversity and environmental data could lead to some insights about our environments and ecosystem, and thus could help policy and decision makers take appropriate actions to further enhance the environments and ecosystem.

Over the past decade, the concept of *open data* [53] has become popular. There has been a trend that more big data collections have been made publicly available as open data by scientists, researchers, and non-profit organizations. Moreover, many governments have also led or followed the trend to contribute to the idea of *open government* [54] by making government more accessible to everyone. For example, many member countries (e.g., Canada, France, Ireland, Japan, South Korea) in the Organisation for Economic Co-operation and Development (OECD)[1] have put efforts to (a) make their public sector data available and accessible, as well as to (b) support data re-use. In Canada, many cities and provinces have joined the initiative towards providing open data thereby acknowledging the gained value in doing so. Consequently, as of March 2021, there were 12 open initiatives, 62 open municipalities (e.g., cities of Montréal, Toronto, Vancouver, Winnipeg), and 12 "open provinces" (e.g., Government of Manitoba) across Canada—joined the Canada Open Government Working Group (COGWG)[2]—in providing open data. By doing so, the governments provide (a) an added value of transparency and allow citizens to monitor government activities, and (b) innovation opportunities to citizens in developing new applications to the open data. Citizens also gain some insights about services available in the city from these open data.

As for open data related to biodiversity and environmental data, the following are examples that capture these biodiversity and environmental data:

- Canadian Consortium for Arctic Data Interoperability (CCADI)[3]
- Canadian Watershed Information Network (CanWIN)[4]

---

[1] https://www.oecd.org/gov/digital-government/open-government-data.htm.
[2] https://open.canada.ca/en/maps/open-data-canada.
[3] https://ccadi.ca/.
[4] https://lwbin.cc.umanitoba.ca/.

- Global Biodiversity Information Facility (GBIF)[5]
- Ocean Biodiversity Information System (OBIS)[6]
- Polar Data Catalogue (PDC)[7]

Note that environmental conditions may affect living conditions of species (including human beings). For instance, pollution can be an environmental issue and pose health risks. As a concrete example, air pollution caused by coal burning and/or vehicle emissions may result in acid rain, which may have severe negative impacts on forest agricultural productivity, forest growth, and/or waterways. As another concrete example, contamination of the drinking water supply in a Canadian town of Walkerton caused the E. coli outbreak in year 2000. Over the past few decades, several actions have taken place to help improve the situations. Examples include:

- Arctic Waters Pollution Prevention Act (1970), which aimed to prevent pollution of Canadian Arctic waters
- Canada-United States Air Quality Agreement (1991), which aimed to improve air quality by reducing nitrogen oxide (NO_x) and sulfur dioxide (SO_2) emissions in both countries
- Kyoto Protocol (1997), which aimed to reduce greenhouse gas emissions

In addition to affecting human, pollution (e.g., plastic wastes) also negatively affects other living beings (e.g., aquatic and marine organisms). To elaborate, after gaining widespread popularity during the World War II, mass-produced synthetic plastics quickly became a staple of modern consumerism [55]. This high demand caused global plastic production to increase dramatically, rising from 1.5M tons per year in 1950 to 332M tons per year in 2015 [56]. Beyond just the environmental costs associated with synthesizing commercial plastics from fossil fuels, we should all be concerned about the large quantities of waste generated by the short lifespans of many plastic products. Using data collected in 2010, Jambek et al. [57] estimated that coastal countries are dumping about eight million tons of discarded plastic into the oceans every year. The same qualities that have made us so reliant on plastics—such as their durability, water-resistance, and inability to react with other materials—also explain why they tend to have such deleterious long-term effects on marine ecosystems.

When thinking about the ecological impact of marine litter, people is usually primed to imagine animals caught in plastic bags or six pack rings. Though these sorts of entanglements certainly pose a real and significant threat, they alone do not provide a full picture of how plastic litter negatively affects the health of marine animals. Over time, plastic litter breaks down into smaller pieces of debris and classifies as:

- macroplastics—which are plastic pieces with at least 5 mm in length and thus large enough to be seen with the naked eye; and

---

[5] https://www.gbif.org/.

[6] https://obis.org/.

[7] https://www.polardata.ca/.

- *microplastics*—which are plastic pieces with less than 5 mm in length and thus not easily seen with the naked eye. Consequently, they can be easily ingested by a wide range of marine animals and thus causing harm to these ocean and aquatic lives. In addition to the risk of tissue damage as sharp microplastics travel through an animal's digestive tract, various studies have shown that microplastics can leave behind several forms of toxic chemicals, such as persistent organic pollutants (POPs) [58].

Tiger sharks are just one example of the many species that are likely to suffer from rising microplastic levels in their natural habitats. Prompted by a 71% drop in east coast tiger shark populations, the Australian Marine Conservation Society launched a campaign to have tiger sharks listed as an endangered species[8]. Their arguments focused on how commercial fishing and excessive shark control programs have reduced tiger shark numbers.

Inspired by the aforementioned campaign, we examine the impact of pollution to biodiversity (e.g., marine biodiversity). Specifically, in our current paper, we design and develop a data science solution to analyze and mine both biodiversity data and environmental data for big data analytics and mining—in particular, for biodiversity informatics, environmental analytics and sustainability analysis. To evaluate and to demonstrate the usefulness of our solution, we conduct case studies on analyzing and mining both marine biodiversity data and plastic exposure data to examine the impacts of the plastic exposure to marine creatures. More specifically, we mine shark occurrence data to determine the effects of microplastic concentrations in the ocean on sea animal (e.g., sharks) physiology. If sharks (who have been exposed to fewer microplastics) are significantly healthier than those who travel in regions with higher quantities of marine litter, conservation activists would gain concrete evidence to argue that microplastics are yet another serious threat against the survival of diminishing tiger shark populations.

Key contributions of this paper include our data science solution for biodiversity informatics, environmental analytics and sustainability analysis. It is designed in such a way that it handles the common 7V's that characterize these big data:

- In term of volume, open data can be huge. For example, as of March 2021, there are 56,853 datasets with a total of 1,659,170,565 occurrence records on GBIF.
- In term of velocity, open data keep growing.
- In term of veracity, many of datasets contain (latitudinal, longitudinal)-locations collected by the Global Positioning System (GPS) or the Global Navigation Satellite System (GNSS). These GPS/GNSS locations may be uncertain due to uncertainty inherited from the systems.
- In term of variety, data can be of different formats—including those in comma-separated values (CSV) or tab-separated values (TSV), JavaScript object notation (JSON) and its extension GeoJSON for representing geographical features, resource description framework (RDF), and extensible markup language (XML). Moreover, data can also be obtained from a wide range of rich data sources. For instance, as a

---

[8] https://www.marineconservation.org.au/calls-for-endangered-listing-of-tiger-sharks-as-new-study-reports-a-71-decline-in-three-decades-along-east-coast-australia/.

preview, our solution will analyze and mine two kinds of data—namely, biodiversity data and environmental data (e.g., plastic exposure data).

- In term of value, our solution aims to discover useful information and knowledge from big data.
- In term of validity, our solution aims to enable users (e.g., analysts, researchers) to interpret big data and their related knowledge.
- In term of visibility, our solution aims to provide users (e.g., analysts, researchers) with visualization of big data and their related knowledge.

Our solution constructs convex hulls around locations visited by marine creatures for the determination of pollution data points to be considered in the estimation of exposure to microplastics. Moreover, visualizations of pollution data points superimposed on top of these convex hulls help interpret and validate the knowledge discovered by our solution, and make them visible to decision and policy makers for taking appropriate actions in building and maintaining a sustainable environment. Although we demonstrate our solution in a few case studies (e.g., sharks), it is expected to be applicable to many other living creatures.

The remainder of this paper is organized as follows. The next section presented related works. Section 3 describes our data science solution. Evaluation and conclusions are shown in Sects. 4 and 5, respectively.

2 Related Works

Rather than using tracking data to estimate the amount of pollution that individual animals have likely encountered, marine biologists tend to search for evidence of microplastics in the tissue samples they obtain from aquatic species. Instead of simply counting the number of microplastics caught inside a marine animal's stomach at the time of capture, Fossi et al. [59] employed a less invasive method that infers past microplastic exposure from the concentration of mono-(2-ethylhexyl) phthalates (MEHP) found in tissue samples collected from basking sharks and fin whales. This particular, study relied on the shark tissue samples as a measure of how many microplastics are likely in their environment, whereas our goal is to predict how many microplastics will turn up in a shark's tissues based on existing knowledge of pollution levels in their habitats.

When studies do estimate pollution levels by counting the number of microplastics found directly in oceanic waters, they tend to look at general regions of interest rather than following the exact paths taken by individual animals. For example, Gemanov et al. [60] looked at microplastic abundances across particular manta ray and whale shark feeding areas around Indonesia, but did not attempt to track the specific movements of individuals from these groups. Using the abundance of plastic pollutants in their feeding grounds to calculate an estimated value for mean plastic ingestion, the researchers were able to accurately predict the quantities of microplastics they later found in egested material collected from the studied manta ray population.

Related works that incorporate tracking data for individual marine animals did not use this information to attempt any cross-analysis against pollution levels in the traversed waters. Instead, Heuter et al. [61] used tracking data to argue that various shark species return to home areas more often than previously thought. To provide quantitative measures of the extent to which fishing operations threaten oceanic sharks, Queiroz et al. [62] calculated the spatial overlap between tracked shark movements and fishing vessel routes over the same period of time. In a satellite tracking study that looked specifically at tiger shark movements, Heithaus et al. [63] noted that one of their tagged sharks travelled at least 8,000 km in just under 100 days. With tiger sharks being capable of moving across such large distances, collecting litter data for a single location will often be insufficient.

3 Our Data Science Solution

In this section, we present our advanced intelligent and data science solution for biodiversity informatics, environmental analytics and sustainability analysis. Specifically, our solution analyzes and mines both biodiversity data and environmental data for big data analytics and mining. For example, knowing the flying patterns of moving objects (e.g., birds like Canada goose) and/or their observed occurrences during bird watching helps estimate their active air space. Mapping their air space with those affected by air pollution (e.g., car emissions, chemicals from factories, volcanic gases) then helps determine the negative impacts of air pollution to these flying animals. As another example, knowing the walking patterns, trails or footsteps of wild terrestrial animals (e.g., polar bears) and/or their observed occurrences during animal tracking helps estimate their active areas. Mapping their areas with those affected by land pollution (e.g., solid or liquid waste, groundwater and soil contamination) then helps again determine the negative impacts of land pollution to these moving animals. As a third example, knowing the swimming patterns, trajectories of aquatic and marine animals (e.g., whales, sharks, salmons, lobsters) and/or their observed occurrences during marine wildlife watching helps estimate their active range of territories. Mapping their territorial ranges with those affected by marine pollution (e.g., microplastics) helps determine the negative impacts of marine pollution to these moving marine animals. To illustrate of our idea (and to evaluate our solution), we mine and estimate microplastic exposure among any group of observed marine animals travelling across any range of territories. Our solution mines the available data about the target species and the region of interest, which allows marine biologists to estimate the amount of microplastic exposure associated with animal tracked movements. In general, key steps in our solution include:

1. Gather data concerning pollution (e.g., microplastic) concentrations at many locations within the territories that the animals covered.
2. Visualize the pollution distribution found in Step 1 by clustering the pollution data points into groups (e.g., high, medium, and low concentrations).
3. Identify a group of observed animals to study within the chosen range of habitats.
4. Obtain a list of locations for collections (or each) of the selected animals.
5. Construct the convex hull that covers the positions of the observed data points associated with the targeted animals of interest.
6. Determine pollution data points falling within each of these convex hulls.

7. Compute average pollution concentrations—from the data points included in the convex hull—as estimate of the animal's exposure to pollution over their observed movements.

3.1 Plastic Data and Visual Analytics

Our solution makes good use of a data analytics technique of k-means clustering to sort the pollution (e.g., microplastic) data points into three categories—namely, with low, medium, or high-intensity—that maximize intra-class similarity. With respect to visual analytics of plastic data, our solution helps users visualize these clustered categories by using different colors to represent different levels of population concentrations:

- green to represent lowest concentration clusters,
- yellow to represent medium concentration clusters, and
- red to represent highest concentration clusters.

Plotting these intensity-coded data points on a map allows users to easily visualize the geographic distribution of pollution.

3.2 Estimation of Pollution Exposure to Moving Objects with Convex Hulls

Usually, available data record locations of moving objects—i.e., animals—when they are observed (e.g., observed locations of marine animals at the moments when they surface). To determine their movements between these data points, our solution estimates reasonable boundaries for the territory that marine animals (e.g., sharks) most likely covered during their history of observed occurrences. More specifically, our solution constructs an enclosed area that at least captures all the direct paths between consecutive observed locations. The resulting *convex hulls* serve as estimates of the territory that a marine animal traversed over a particular period of time.

4 Evaluation

To illustrate key ideas of our data science solution and to demonstrate its practicality for biodiversity informatics, environmental analytics and sustainability analysis, we conducted case studies on microplastic exposure among sharks by integrating and examining some biodiversity and environmental data (e.g., GBIF (see footnote 5), AWI-LITTERBASE[9]). According to GBIF, there have been more than 900,000 observed occurrences of sharks, of which about 83% were observed in Australia, 12% in the USA, 2% in the UK, and remainder in various locations around the world (e.g., Bahamas, New Zealand, Netherlands, South Africa). One of its datasets [64] records acoustic tracking—by the Integrated Marine Observing System's Animal Tracking Facility (IMOS ATF) and the Australian Ocean Data Network (AODN)[10]—of movements of tagged

---

[9] https://litterbase.awi.de/.

[10] https://portal.aodn.org.au/.

marine animals in Australian coastal waters for quality controlled detections during 2007–2017. According to AWI-LITTERBASE, plastic composes 62% of global marine litter. Microplastics account for 78% of *micro-litter* (with less than 5 mm in length) from the sea surface, 70% in the water column, and 68% on the seafloor. When combined with other plastic litter (e.g., plastic fibers, plastic pellets, plastic film), they compose more than 90% of mico-litter on beaches, sea surface, water column, and seafloor. As such, more than 3,500 species are negatively affected by litter (e.g., micro-litter). Top-3 types of aquatic life affected by litter include fishes (e.g., sharks), seabirds, and crustaceans (e.g., lobsters, crabs, shrimps). These negative effects are mostly caused by ingestion of litter (which accounts for 38% of the problems), colonization (which accounts for 36%), and entanglement (which accounts for 21%).

After integrating and processing both biodiversity and environmental data, our solution applies k-means clustering to categorize micoplastics according to their concentrations. For easy visualization of the environmental data (micoplastics), those with low concentration (e.g., <3500 items per km^2) are represented by green *dots*, medium concentration (e.g., between 3500 and 10000 items per km^2) in yellow, and high concentration (e.g., >10000 items per km^2) in red in Fig. 1.

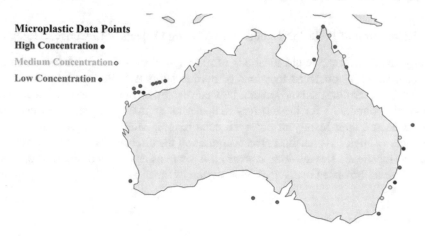

Fig. 1. Distribution of three categories of microplastics along the Australian coastline

For completeness, our solution also clusters data points for (a) macroplastic pollution (with at least 5 mm in length) and (b) their combination for the (micro+macro)-plastic pollution. See Fig. 2, which shows the distribution of macroplastic represented by three types of colored *squares* (green, yellow and red for low, medium and high macroplastic concentrations, respectively). Similarly, the figure also shows the distribution of the combined microplastic & macroplastic represented by three types of colored *squares containing dots of the same colors.*

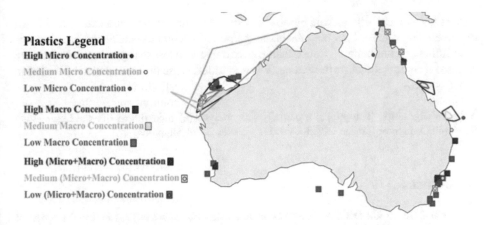

Fig. 2. Pollution distribution with convex hulls enclosing the moving objects of interest

Afterwards, our solution constructs convex hulls covering the positions of the observed data points associated with the moving objects of interest (e.g., sharks) and overlaps the convex hulls with pollution distribution. With their overlaps (e.g., as shown in Fig. 2), our solution then estimates the plastic debris concentration enclosed by the convex hulls. For example, Table 1 shows the average values of the plastic concentration data captured by three sample convex hulls.

Table 1. Plastic debris concentration enclosed by each convex hull (excluding those that did not contain any pollution data points)

Convex hull	Sample #1	Sample #2	Sample #3
#enclosed mirco data points	10	6	1
Avg micro exposure (items/km^2)	2,651	3,878	4,290
#enclosed marco data points	8	5	1
Avg macro exposure (items/km^2)	1,086	1,346	1,790
#enclosed (mico+marco) data points	8	5	1
Avg (mico+marco) exposure (items/km^2)	4,368	5,972	6,080

5 Conclusions

We presented in this paper a data science solution for biodiversity informatics, environmental analytics and sustainability analysis. Our solution analyzes, mines and visualizes both biodiversity data (e.g., observed locations of sharks) and environmental data (e.g., areas of microplastics exposure) to examine the negative impacts of pollution to moving objects (e.g., marine animals like sharks). The convex-hull-based method in our solution estimates the pollution exposure to these objects. For evaluation, we conducted case studies on real-life marine biodiversity data and plastic exposure data. Knowledge discovered by our solution help decision and policy makers to take appropriate actions in building and maintaining a sustainable environment. As ongoing and future work, we explore techniques that further enhance our solution (e.g., by using OLAP [65–70] to process data).

Acknowledgments. This project is partially supported by (a) Natural Sciences and Engineering Research Council of Canada (NSERC) and (b) University of Manitoba.

References

1. Ahmad, A., et al.: Defining human behaviors using big data analytics in social Internet of Things. In: IEEE AINA 2016, pp. 1101–1107 (2016)
2. Barbieru, C., Pop, F.: Soft real-time Hadoop scheduler for big data processing in smart cities. In: IEEE AINA 2016, pp. 863–870 (2016)
3. Jiang, F., Leung, C.K.: A data analytic algorithm for managing, querying, and processing uncertain big data in cloud environments. Algorithms **8**(4), 1175–1194 (2015)
4. Leung, C.K.: Big data analysis and mining. In: Encyclopedia of Information Science and Technology, 4e, pp. 338–348 (2018)
5. Susanto, H., et al.: Revealing storage and speed transmission emerging technology of big data. In: AINA 2019. AISC, vol. 926, pp. 571–583 (2019)
6. Gadelha, L.M.R., et al.: A survey of biodiversity informatics: concepts, practices, and challenges. WIREs DMKD **11**(1), e1394:1-e1394:41 (2021)
7. Ibraheam, M., et al.: Animal species recognition using deep learning. In: AINA 2020. AISC, vol. 1151, pp. 523–532 (2020)
8. Ali, S., et al.: A blockchain-based secure data storage and trading model for wireless sensor networks. In: AINA 2020. AISC, vol. 1151, pp. 499–511 (2020)
9. Kobusinska, A., et al.: Emerging trends, issues and challenges in Internet of Things, big data and cloud computing. FGCS **87**, 416–419 (2018)
10. Fariha, A., et al.: Mining frequent patterns from human interactions in meetings using directed acyclic graphs. In: PAKDD 2013, Part I. LNCS (LNAI), vol. 7818, pp. 38–49 (2013)
11. Jiang, F., et al.: Finding popular friends in social networks. In: CGC 2012, pp. 501–508 (2012)
12. Leung, C.K., Jiang, F.: Big data analytics of social networks for the discovery of "following" patterns. In: DaWaK 2015. LNCS, vol. 9263, pp. 123–135 (2015)
13. Souza, J., et al.: An innovative big data predictive analytics framework over hybrid big data sources with an application for disease analytics. In: AINA 2020. AISC, vol. 1151, pp. 669–680 (2020)
14. Chen, Y., et al.: Temporal data analytics on COVID-19 data with ubiquitous computing. In: IEEE ISPA-BDCloud-SocialCom-SustainCom 2020, pp. 958–965 (2020). https://doi.org/10.1109/ISPA-BDCloud-SocialCom-SustainCom51426.2020.00146
15. Gupta, P., et al.: Vertical data mining from relational data and its application to COVID-19 data. In: Big Data Analyses, Services, and Smart Data. AISC, vol. 899, pp. 106–116 (2021)
16. Leung, C.K., et al.: Big data science on COVID-19 data. In: IEEE BigDataSE 2020, pp. 14–21 (2020). https://doi.org/10.1109/BigDataSE50710.2020.00010
17. Liu, Q., et al.: A two-dimensional sparse matrix profile DenseNet for COVID-19 diagnosis using chest CT images. IEEE Access **8**, 213718–213728 (2020)
18. Shang, S., et al.: Spatial data science of COVID-19 data. In: IEEE HPCC-SmartCity-DSS 2020 (2020)

19. Camara, R.C., et al.: Fuzzy logic-based data analytics on predicting the effect of hurricanes on the stock market. In: FUZZ-IEEE 2018, pp. 576–583 (2018)
20. Chanda, A.K., et al.: A new framework for mining weighted periodic patterns in time series databases. ESWA **79**, 207–224 (2017)
21. Leung, C.K., et al.: A machine learning approach for stock price prediction. In: IDEAS 2014, pp. 274–277 (2014)
22. De Guia, J., et al.: DeepGx: deep learning using gene expression for cancer classification. In: IEEE/ACM ASONAM 2019, pp. 913–920 (2019)
23. Leung, C.K., et al.: Predictive analytics on genomic data with high-performance computing. In: IEEE BIBM 2020, pp. 2187–2194 (2020). https://doi.org/10.1109/BIBM49941.2020.931 2982
24. Pawliszak, T., et al.: Operon-based approach for the inference of rRNA and tRNA evolutionary histories in bacteria. BMC Genomics **21**(Suppl. 2), 252:1–252:14 (2020)
25. Sarumi, O.A., et al.: Spark-based data analytics of sequence motifs in large omics data. Procedia Comput. Sci. **126**, 596–605 (2018)
26. Sarumi, O.A., Leung, C.K.: Adaptive machine learning algorithm and analytics of big genomic data for gene prediction. In: Tracking and Preventing Diseases with Artificial Intelligence (2021)
27. Balbin, P.P.F., et al.: Predictive analytics on open big data for supporting smart transportation services. Procedia Comput. Sci. **176**, 3009–3018 (2020)
28. Chowdhury, N.K., Leung, C.K.: Improved travel time prediction algorithms for intelligent transportation systems. In: KES 2011, Part II. LNCS (LNAI), vol. 6882, pp. 355–365 (2011)
29. Leung, C.K., et al.: Conceptual modeling and smart computing for big transportation data. In: IEEE BigComp 2021, pp. 260–267 (2021). https://doi.org/10.1109/BigComp51126.2021. 00055
30. Leung, C.K., et al.: Urban analytics of big transportation data for supporting smart cities. In: DaWaK 2019. LNCS, vol. 11708, pp. 24–33 (2019)
31. Cox, T.S., et al.: An accurate model for hurricane trajectory prediction. In: IEEE COMPSAC 2018, vol. 2, pp. 534–539 (2018)
32. Mateo, M.A.F., Leung, C.K.: CHARIOT: a comprehensive data integration and quality assurance model for agro-meteorological data. In: Data Quality and High-Dimensional Data Analysis, pp. 21–41 (2009)
33. Sassi, M.S.H., Fourati, L.C.: Architecture for visualizing indoor air quality data with augmented reality based cognitive Internet of Things. In: AINA 2020. AISC, vol. 1151, pp. 405–418 (2020)
34. Cao, L.: Data science: a comprehensive overview. ACM CSUR **50**(3), 43:1–43:42 (2017)
35. Dierckens, K.E., et al.: A data science and engineering solution for fast k-means clustering of big data. In: IEEE TrustCom-BigDataSE-ICESS 2017, pp. 925–932 (2017)
36. Leung, C.K., Jiang, F.: A data science solution for mining interesting patterns from uncertain big data. In: IEEE BDCloud 2014, pp. 235–242 (2014)
37. Chen, Y., et al.: Mining opinion leaders in big social network. In: IEEE AINA 2017, pp. 1012–1018 (2017)
38. Leung, C.K.: Uncertain frequent pattern mining. In: Frequent Pattern Mining, pp. 417–453 (2014)
39. Leung, C.K., et al.: Distributed uncertain data mining for frequent patterns satisfying anti-monotonic constraints. In: IEEE AINA Workshops 2014, pp. 1–6 (2014)
40. Casagrande, L.C., et al.: DeepScheduling: grid computing job scheduler based on deep reinforcement learning. In: AINA 2020. AISC, vol. 1151, pp. 1032–1044 (2020)
41. Leung, C.K., et al.: Explainable machine learning and mining of influential patterns from sparse web. In: IEEE/WIC/ACM WI-IAT 2020 (2020)

42. Leung, C.K., et al.: Machine learning and OLAP on big COVID-19 data. In: IEEE BigData 2020, pp. 5118–5127 (2020)
43. Min, B., et al.: Image classification for agricultural products using transfer learning. In: BigDAS 2020, pp. 48–52 (2020)
44. Leung, C.K.: Mathematical model for propagation of influence in a social network. In: Encyclopedia of Social Network Analysis and Mining, 2e, pp. 1261–1269 (2018)
45. Lee, W., et al.: Reducing noises for recall-oriented patent retrieval. In: IEEE BDCloud 2014, pp. 579–586 (2014)
46. Leung, C.K., et al.: Information technology-based patent retrieval model. In: Springer Handbook of Science and Technology Indicators, pp. 859–874 (2019)
47. Barkwell, K.E., et al.: Big data visualisation and visual analytics for music data mining. In: IV 2018, pp. 235–240 (2018)
48. Braun, P., et al.: Game data mining: clustering and visualization of online game data in cyber-physical worlds. Procedia Comput. Sci. 112, 2259–2268 (2017)
49. Carmichael, C.L., et al.: Visually contrast two collections of frequent patterns. In: IEEE ICDM Workshops 2011, pp. 1128–1135 (2011)
50. Dubois, P.M.J., et al.: An interactive circular visual analytic tool for visualization of web data. In: IEEE/WIC/ACM WI 2016, pp. 709–712 (2016)
51. Leung, C.K., Carmichael, C.L.: FpVAT: a visual analytic tool for supporting frequent pattern mining. ACM SIGKDD Explor. 11(2), 39–48 (2009)
52. Munzner, T., et al.: Visual mining of power sets with large alphabets. Technical report TR-2005-25, Computer Science, UBC, Canada (2005). https://www.cs.ubc.ca/tr/2005/tr-2005-25
53. Audu, A.A., et al.: An intelligent predictive analytics system for transportation analytics on open data towards the development of a smart city. In: CISIS 2019. AISC, vol. 993, pp. 224–236 (2019)
54. Perovich, L.J., et al.: Chemicals in the Creek: designing a situated data physicalization of open government data with the community. IEEE TVCG 27(2), 913–923 (2021)
55. Freinkel, S.: Plastic: A Toxic Love Story (2011)
56. Beckman, E.: The world's plastic problem in numbers. World Economic Forum (2018). https://www.weforum.org/agenda/2018/08/the-world-of-plastics-in-numbers
57. Jambek, J., et al.: Plastic waste inputs from land into the ocean. Science 347, 768–771 (2015)
58. Wright, S.L., et al.: The physical impacts of microplastics on marine organisms: a review. Environ. Poll. 178, 483–492 (2013)
59. Fossi, M.C., et al.: Large filter feeding marine organisms as indicators of microplastic in the pelagic environment: the case studies of the Mediterranean basking shark (*Cetorhinus maximus*) and fin whale (*Balaenoptera physalus*). Mar. Environ. Res. 100, 17–24 (2014)
60. Germanov, E., et al.: Microplastics on the menu: plastics pollute Indonesian Manta ray and whale shark feeding grounds. Front. Mar. Sci. 6, 679:1–679:21 (2019)
61. Hueter, R.E., et al.: Evidence of philopatry in sharks and implications for the management of shark fisheries. J. Northwest Atlantic Fish. Sci. 35, 239–247 (2005)
62. Queiroz, N., et al.: Ocean-wide tracking of pelagic sharks reveals extent of overlap with longline fishing hotspots. PNAS 113(6), 1582–1587 (2016)
63. Heithaus, M.R., et al.: Long-term movements of tiger sharks satellite-tagged in Shark Bay, Western Australia. Mar. Biol. 151, 1455–1461 (2007)
64. Hoenner, X., et al.: Australia's continental-scale acoustic tracking database and its automated quality control process. Sci. Data 5, 170206:1–170206:10 (2018). https://doi.org/10.1038/sdata.2017.206
65. Cuzzocrea, A.: Improving range-sum query evaluation on data cubes via polynomial approximation. DKE 56(2), 85–121 (2006)

66. Cuzzocrea, A., et al.: A hierarchy-driven compression technique for advanced OLAP visualization of multidimensional data cubes. In: DaWaK 2006. LNCS, vol. 4081, pp. 106–119 (2006)
67. Cuzzocrea, A., et al.: OLAP*: effectively and efficiently supporting parallel OLAP over big data. In: MEDI 2013. LNCS, vol. 8216, pp. 38–49 (2013)
68. Cuzzocrea, A., Leung, C.K.: Efficiently compressing OLAP data cubes via R-tree based recursive partitions. In: ISMIS 2012. LNCS (LNAI), vol. 7661, pp. 455–465 (2012)
69. Cuzzocrea, A., Matrangolo, U.: Analytical synopses for approximate query answering in OLAP environments. In: DEXA 2004. LNCS, vol. 3180, pp. 359–370 (2004)
70. Cuzzocrea, A., Serafino, P.: LCS-Hist: taming massive high-dimensional data cube compression. In: EDBT 2009, pp. 768–779 (2009)

Mining Method Based on Semantic Trajectory Frequent Pattern

Chun Liu$^{(\boxtimes)}$ and Xin Li

Hubei University of Technology, Wuhan 40068, China

Abstract. The current trajectory analysis mainly uses clustering method to mine public stay points from multi-user trajectories, calculate user similarity to find hotspots, extract common attributes of approximate crowds, while it's almost no commercial value to calculate any similarity of single-user trajectories. In this paper, a method GPSTOSTFP for mining frequent patterns of individual users based on location semantics is proposed. And an improved Aporior algorithm RGA based on inverse geocoding is proposed. This kind of semantic trajectory frequent pattern mining can effectively identify and mine potential carpooling demands, provide higher precision for location-based intelligent recommendation such as sharing carpooling and HOV lane commute.

1 Introduction

Mobile user trajectory data information can be excavated the space-time characteristics of the user's travel rules, vehicle movement characteristics such as all kinds of information, and the user's network attribute conjoint analysis can also infer the behavior characteristics, interests, social habits, and other important information, to understand the user behavior mode, improving the accuracy of intelligent recommendation and related to intelligence are extremely valuable [1]. Trajectory data are easily affected by weather conditions and sensor noise during the collection process. The difference of data sampling period can easily lead to data redundancy and error. Moreover, trajectory data includes longitude and latitude coordinate values, time stamps, speed and other data types, which are directly calculated and analyzed based on Euclidean distance. It is more convenient, but the results are difficult to understand and directly use.

By transforming the spatiotemporal information in the original trajectory into text data, a trajectory with semantic tags is obtained, which is called semantic trajectory. Semantic trajectory analysis is proposed by Alvares et al. [2], which regarded the semantic trajectory as the position sequence marked with semantic tags. The tag data represented the behaviors that appeared on the trajectory. The trajectory data consistent with the semantic label sequence could be considered as the same trajectory sequence. Semantic trajectory is easy to read and understand, but because the object of operation is a string, its disadvantage is inconvenience in operation, compression and decompression. In [3], a method for compression, and calculating semantic trajectories is proposed. The trajectory sampling points are mapped to the road network to form a semantic trajectory

L. Barolli et al. (Eds.): AINA 2021, LNNS 226, pp. 146–159, 2021.
https://doi.org/10.1007/978-3-030-75075-6_12

sequence by combining the road network data or the POI data set with the road net-work. Road network matching plays an important role in assessing traffic flow, guiding vehicle navigation, predicting vehicle travel routes, finding frequent travel routes from departure point to destination, etc., but matching calculations, multiple compressions and decompressions between multi-layer networks both are cumbersome and difficult to implement in batches.

In the research of frequent sequence mining of trajectory, classical algorithms include Apriori [4], AprioriAll, GSP, SPADE, Prefix Span and so on. In 2013, Luo et al. [5] stud-ied a Time period-based most frequent path (TPMFP) query algorithm, using frequent pattern mining to find the most frequent path in a specific period in a large amount of historical data. In 2015, paper [6] proposed a frequent path sequence mining algo-rithm PMWTI (Path Mining With Topology Information), which fully considered the topology information of logistics network. In 2014, Zhang et al. raised the problem of fine-grained sequence pattern mining for space-time trajectories. They considered that the location points in continuous space were not suitable for sequential pattern mining, but the items consisting of the same semantic location points were more suitable for this problem [7]. Rajkumar Saini et al.[8] (2019) proposed a graph based trajectory classifi-cation algorithm, and this method outperforms the state of the art techniques. Chen et al. [9] (2019) proposed A trajectory frequent pattern mining algorithm based on vague grid sequence (VGS-PrefixSpan), which reduces the time of constructing projected databases and improves the efficiency of the algorithm.

These methods are all clustering mining common residence points, discovering fre-quent patterns and analyzing utilization from the trajectories of multiple users. These methods are useful for extracting common attributes of approximate populations, and can be used to generate position-related recommendation items in collaborative recom-mendation, but because of the group-based analysis, there in a lack of semantic mining analysis of individual behavior patterns, so it is not possible to identify the semantic behavior characteristics of individual users. In order to analyze the frequent patterns of individual users, and make it easier for people to understand the meaning of the frequent patterns and reduce the amount of calculation, this paper designs a semantic trajectory compression (STC) [10] and clustering method based on inverse geocoding [11].

The contributions of this paper are summarized as follows:

- In this paper, a method of inverse geocoding and frequent sequence mining is pro-posed. This method does not need to depend on group characteristics or user manual annotation. It only needs individual user's own historical trajectory data to mine and extract user's calendar behavior patterns, and carry out location matching at the level of geographical labels. It should be suitable for future unmanned and automatic driv-ing platforms such as carpooling recommendation and shared travel. It can provide theoretical and methodological basis.
- We demonstrate our frequent pattern mining algorithm effectively discovers the frequent pattern by using real data sets.

This paper is organized as follows: Sect. 1 introduces the problem and summa-rizes the contribution. Section 2 presents our framework GPSTOSTFP. Section 3 shows experimental results. Section 4 gives the conclusion and prospects the future work.

2   Proposed Work

2.1 Framework

Figure 1 illustrates our framework for mining semantic trajectory frequent patterns from raw GPS dataset (Data format reference Definition 1). We call this framework GPSTOSTFP, which means "GPS Trajectories to Semantic Trajectory Frequent Patterns". We will explain the specific steps of GPSTOSTFP in the following subsections.

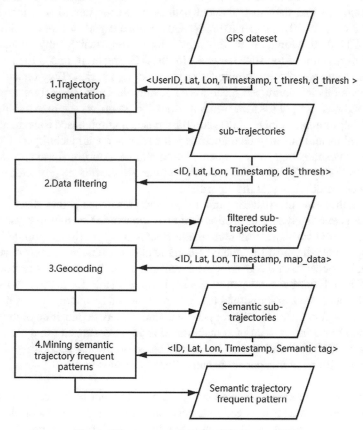

Fig. 1. The process of GPSTOSTFP algorithm

2.2 Related Definitions

Some definitions used in this paper are described as follows.

Definition 1: The position record g at a certain time can be represented by a triples: $g = (u, l, t)$, where u is the user ID, l contains the longitude x and the latitude y, and t is the timestamp.

Definition 2: Given a user u, whose trajectory T_u is a sequence of triples, $(u_1, l_1, t_1), \ldots, (u_i, l_i, t_i), \ldots, (u_n, l_n, t_n)$, where n represents the length of the sequence, and the triples are arranged in timestamp order. The user's path can be sort as $T_u = g_1 \rightarrow g_2 \rightarrow \cdots g_i \rightarrow \cdots g_n$.

Definition 3: The stop point $P = (g, \Delta t)$ is used to calculate the time the user stays at the location through the trajectory's timestamp. If the user stays at a certain location exceeds the time threshold, it means that such location information may hide more interesting events.

Definition 4: Trajectory semantic tag $g' = (l, \varphi)$, where φ represents the semantic tag information. In this paper, the inverse geocoded address is used as φ.

Definition 5: Semantic trajectory sequence $T_s = g_1' \rightarrow g_2' \rightarrow \cdots g_i' \rightarrow \cdots g_n'$ is a series of trajectory semantic tag arranged according to the timestamp, where $g_i' = (l_i, \varphi_i)$ contains the track point l_i in time i and the semantic tag information φ_i. Semantic trajectory sequence is a chronological (*Lat&Lon, Address String*) table.

Definition 6: Frequent trajectory sequences. Supposed that T_f is a non-empty set composed of one or more items, if $T_f \subseteq D, T_f \neq \emptyset$, called item sets T_f in set D. If and only if the frequency of T_f appearance no less than $\theta |D|$ in D, then called T_f is frequent, where $\theta (0 < \theta < 1)$ is the minimum support given by the user, represented by *miniSup*, $|D|$ is the number of transactions contained in data set D, and $\theta |D|$ is the minimum support count, represented by *miniCount*.

According to the above series of definitions. We can know that semantic trajectory frequent pattern mining is actually the process of finding all frequent subsequences T_f satisfying minimum support *minSup* in semantic trajectory database $\{T_{s1}, T_{s2}, \ldots, T_{sn}\}$.

2.3 Trajectory Segmentation

In this study, we named the recorded entity sequences consisting of user ID, GPS positions and timestamps as the raw GPS dataset, which is called GD for short. The format of each entry in GD is defined by Definition 1, where (u, l, t) represents (*user ID, Lat&Lon, Timestamp*). GD records the geographic information of multiple trajectories of entities in different time periods.

GPSTOSTFP's first step is to divide the entity's historical trajectories into multiple sub-trajectories $T_u^1, \ldots, T_u^n$, and each sub-trajectory represents a movement behavior [12]. The key to trajectory segmentation is to find entries that represent the starting and ending points of a single movement behavior, which define the geographic boundaries of a single movement behavior. In this work, we refer to the starting and ending points of a single trajectory as separation points.

The method we use is as follows:

The first step is to convert the timestamps of all entries belonging to the same entity in the GD to the date format (e.g. 2020-01-01) and group these entries by date. The entries in each group represent the movement behavior of the day, which may contain multiple trajectories, such as going to work, going home, and so on.

The second step is to divide each group of entries into multiple sub-trajectories $T_u^1, \ldots, T_u^n$, each of which represents a movement behavior of the entity. As shown in Eq. 1 we find the separation points of a single trajectory through the double constraints of space and time. g_i represents a single entry in GD, which has three labels, u, l and t (see Definition 1). $g_i.t$ represents the timestamp of the entry and $g_i.l$ represents the latitude and longitude of the entry. We convert timestamps into time format (e.g. 13:45:05) and calculate the interval time between two adjacent entries. DIST (See the code in the attachment) is a function that uses the latitude and longitude of two points on the earth to calculate the distance between them. In Eq. 1, the time constraint indicates that the interval between two entries is greater than a time threshold t_{thresh}, and the space constraint means that the distance between the two entries is greater than a space threshold d_{thresh}. The value of d_{thresh} is equal to 2 times URE, where URE is the global average user range error of GPS (data from https://www.gps.gov/systems/gps/perfor mance/accuracy/).

$$g_i, g_{i+1} \in T_u \begin{cases} |g_{i+1}.t - g_i.t| > t_{thresh} \\ DIST(g_i.l, g_{i+1}.l) < d_{thresh} \end{cases} \tag{1}$$

2.4 Data Filtering and Geocoding

GPSTOSTFP's second step is to remove the noise entries in the sub-trajectories $T_u^1, \ldots, T_u^n$, and reverse address encode the latitude and longitude of the entries in the sub-trajectories to enrich each entry with semantic information.

When collecting geographic information, GPS devices are susceptible to weather, electromagnetics, shelter, and other interference, noises may occur in a continuous trajectory. Due to the inherent inaccuracy of GPS, the trajectory data need to be filtered. The commonly used filtering algorithms include Kalman filtering (A typical example of Kalman filtering is to predict the coordinate position and velocity of an object from a limited set of observations containing noise.) [13], particle filtering [14, 15], sliding average filtering [16], median filtering [17] and so on. In this paper, the median filtering method to smooth the sudden noise in the track data is adopted. It should be noted that the sub-trajectories must be segmented before filtering, otherwise the filtering performance may be seriously degraded when the interval time exceeds the threshold. Our operation is to calculate the distance between two adjacent entries for each sub-track. When the distance exceeds dis_{thresh}, we think that there is a noise entry, and then replace the longitude and latitude of the noise point with the average longitude and latitude of the two entries adjacent to the noise entry. The calculation formula of dis_{thresh} is shown in Eq. 2. Where v_{max} represents the fastest speed of the entity moving in the city. We set the specific value of v_{max} as 60km/h.

$$dis_{thresh}(g_i, g_j) = |g_i.t - g_j.t| * v_{max} + d_{thresh} \tag{2}$$

So far, we have completed the grouping of the historical movement trajectories of individual entities by day from GD, and accurately divided the movement trajectories of each group, that is, each day, into several single movement behaviors. However, the longitude and latitude values of each entry in the GPS data we use are accurate to 13 digits after the decimal point, the same geographic may response to many different coordinates and cause redundancy, and GPS trajectories are not intrinsically human readable or semantically meaningful, we can't mine the hidden meaning behind the movement behavior of the entity from GPS data only. Therefore, it is easier to study and apply to transform the latitude and longitude coordinate sequence in GPS data into a location label sequence containing actual semantics. In literature [18], Limited by the technical conditions at that time, reverse geocoding conversion from coordinate values to semantic labels was done through Telephone Yellow Pages which greatly limited the conversion efficiency and prevented it from being popularized and applied. With the development of geographic location service platform technology in recent years, API services based on cloud platform have been quite mature. For example, AMAP's Web API interface cloud services (https://www.amap.com) can provide batch reverse geocoding conversion services, and return the reverse geocoding address structure in the form of JSON or XML as province + city + district/county + development zone + town + village + street (business circle or interest point) + house number. After converting to a geo-tag, coordinate data is replaced by address string. Some continuous coordinate points are usually replaced by the same street or the same place name.

In order to enrich the semantic information of $T_u^1, \ldots, T_u^n$, we obtain the database map_{data} containing the correspondence between latitude and longitude and actual place names through AMP's web API. Then use map_{data} to match the latitude and longitude values of all entries in $T_u^1, \ldots, T_u^n$ with real-world locations. And then we got Semantic trajectory sequences T_s.

2.5 Mining Semantic Trajectory Frequent Patterns

It is a high probability event that a frequent address sequence exceeds 10 different addresses in a semantic trajectory, and the computation for an overlong sequence often exceeds the computing capacity. Therefore, this paper proposes an RGA (Reverse Geocode-based Apriori mining) algorithm based on inverse geocoding to mine frequent patterns of a single user's multi-semantic trajectory. This method is more efficient than the traditional Apriori [19], FP-Growth [20] and PrefixSpan [21] algorithms in frequent pattern mining of semantic trajectory sequence data. The pseudo code of the RGA algorithm is shown in Table 1.

2.6 Framework Summarize

Our proposed framework can be summarized as the following algorithm (Table 2):

Above is the pseudo code and code description of the algorithm. See the attachment for the complete code.

Table 1. The expression of RGA algorithm

Algorithm: RGA agorithm
Parameter:T_s, λ_{time}, $k1$, $k2$, $minSup1$, $minSup2$

//D_{seq} is used to store the results of data compression

1. $D_{seq} = (g_1')$
2.   **for all** $g_i' \in T_s$ **do**
         //choose ROI
3.   **if** $(g'_{i+1}.t - g_i'.t > \lambda_{time})$ **then**
4. $T_{i+1} = (g'_{i+1},)$
5.   Add T_{i+1} to D_{seq}
6.   **elseif** $(g'_{i+1}.\varphi == g_i'.\varphi)$ **then**
7.   Del g'_{i+1} from D_{seq}
8.   **else**
9.   Add g'_{i+1} to T_i
10.   **end for**
11.   **for all** $T_i \in D_{seq}$ **do**
         //seach Top-k1
12.   sub1_prefixs=generate_1_size_freq_item (D_{seq}, minSup1)
         //search and temporary save frequent trajectory sets which consist of sub1_prefixs items
13.   by using prefixspan algorithm
14.   new_projs = generate_proj (D_{seq}, sub1_prefixs)
15.   **if** (sizeof (sub1_prefixs) > k_1) **then break**
16.   **end for**
17.   **for all** $T_i \in$ sub1_prefixs **do**
18.   **if** $(T_i.g'_{i+1}.\text{subaddr}==T_i.g_i'.\text{subaddr})$ **then**
19.   Del g_i' from T_i
20.   **end for**
21.   **for all** sub1_prefixs[i] in D_{seq} **do**
         //search Top-k2
22.   sub2_prefixs=generate_2_size_freq_item (D_{seq},$miniSup2$)
23.   final_projs=generate_proj (new_projs, sub2_prefixs)
24.   **if** (sizeof (sub2_prefixs)>k_2) **then break**;
25.   **end for**
26.   **for all** projection$_i$ in final_projs **do**
27. $T_{si} =$ projection$_i$.$T_{i+1} \cap$ projection$_i$.T_i
28.   **end for,**

Table 2. The expression of GPSTOSTFP algorithm

Algorithm: GPSTOSTFP algorithm

Parameter: GD, t_{thresh}, d_{thresh}, dis_{thresh}, map_{data}, λ_{time}, k, $miniSup$

GPSTOSTFP Input:

(1) GD: a dataset that records GPS data in the format (u, l, t);

Trajectory segmentation Input:

(2) t_{thresh}: time interval threshold of time constraint in trajectory segmentation;

(3) d_{thresh}: spatially constrained spatial distance threshold in trajectory segmentation;

Data filtering Input:

(4) dis_{thresh}: the actual distance threshold of two consecutive entries in GD;

• Note that dis_{thresh} is not a constant, it is a variable related to the time interval of entries participating in the calculation;

Geocoding Input:

(5) map_{data}: a database that records the correspondence between real addresses and latitude and longitude values;

Mining semantic trajectory frequent patterns Input:

(6) λ_{time}: time threshold;

(7) k: The maximum length of a frequent itemset;

(8) $miniSup$: minimum support;

//1. **Trajectory segmentation**

// An empty list is used to store the trajectories of entity u grouped by day.

1. $D'_u = (\ )$

// An empty list is used to store all of the sub-trajectories of entity u.

2. $D_u = (\ )$

3. **for all** $g_i \in GD$ **do:**

// Grouping GPS data of entity u by day

4. $T'_{ui} = SEG\_BY\_DATE(GD)$

5.   Add T'_{ui} to D'_u

6. **end for**

7. **for all** $T'_{ui} \in D'_u$ **do:**

8.   **for all** $g_j \in T'_i$ do:

// Each group of GPS data of entity u is divided into sub-tracks

9. $T_{uk} = SEG\_BY\_TS(t_{thresh}, d_{thresh})$

10.   Add T_{uk} to D_u

11.   **end for**

12. **end for**

// 2. **Data filtering**

//An empty list is used to store the filtered sub-tracks

13. $D^f_u = (\ )$

14. **for all** $T_{uk} \in D_u$ **do:**

//Filter noise entries in sub-tracks

15. $T^f_{uk} = FILTER(T_{uk}, dis_{thresh})$

(continued)

Table 2. (*continued*)

16.	Add T_{uk}^f to D_u^f
17.	**end for**

//3. Geocoding
// An empty list is used to store the semantic sub-tracks

18.	$D_u^s = ()$
19.	**for all** $T_{uk}^f \in D_u^f$ **do:**
20.	**for all** $g_i \in T_{uk}^f$ **do:**

//Use map_{data} to reverse encode entries in filtered sub-tracks

21.	$S_{Li} = GEO(g_i, map_{data})$
22.	Add S_{Li} to T_{si}
23.	**end for**
24.	Add T_{si} to D_u^s
25.	**end for**

//4. Mining semantic trajectory frequent patterns
// An improved aprior method based on inverse address coding is used to extract semantic frequent patterns from semantic trajectory sequences.

26.	$T_f = RGA(T_S, \lambda_{time}, k, miniSup,)$

3 Experiments and Results

3.1 Database

The GPS dataset comes from a car networking project carried out by our laboratory in 2014. This project collected three months of vehicle movement trajectory data for private cars in Wuhan, Hubei province, China. In the process of data acquisition, the data of GPS is collected every 1s. In this experiment, the user data whose ID is 354071018570591 is selected as an example to show the method of frequent trajectory extraction. The basic data types in the original GPS dataset include the vehicle ID, latitude and longitude, timestamps, etc. The data format after privacy clearance is shown in Table 3. Each entry is represented by a quaternion: (u, x, y, t), u means the ID number of the vehicle, t means the time stamp, x, y mean the latitude and longitude respectively. The latitude and longitude data of each entry are merged into one item and represented by l, and then the data format is the same as the g in Definition 1.

During the interception period, the total amount of data generated by the user 354071018570591 is 7095. Draw these GPS data to get the original trajectory map, as shown in Fig. 2.

3.2 Experimental Results

Figure 2 shows the raw data of user 354071018570591, and a large number of GPS entries form multiple messy paths on the map, so it is impossible to directly know the trajectory of the entity and explore the meaning behind the motion behavior of the entity.

Table 3. Data sample

ID	Latitude	Longitude	CreatTime
354071018570591	30.6227221945301	114.316071588546	2013/11/13 13:45:25
354071018570591	30.6227225298062	114.316071420908	2013/11/13 13:45:47
354071018570591	30.622695707716	114.316108385101	2013/11/13 13:46:08
354071018570591	30.6227042991667	114.316118778661	2013/11/13 13:46:21
...			

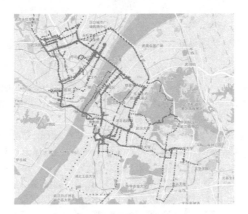

Fig. 2. Original trajectory

The four steps of the GPSTOSTFP framework have been described in detail in Sect. 3, and then the framework will be used to process the raw data.

The first step is to divide the trajectory of the raw data, which can be divided into two sub-processes in total. A sub-process is to convert the timestamp of each entry in the raw data into a date format, and then group the raw trajectories by date.

The second step is to filter the entries in the divided sub-tracks, and use the median method to smooth filter the noise entries.

The third step is to use AMAP's API to get the map_{data}, and then add semantic tags for each entry based on the mapping of the latitude and longitude recorded in the map_{data} with the actual location.

The final step is to use the RGA algorithm to mine the semantic track frequent pattern, which is an improved Apriori algorithm based on GPS inverse coding.

In Fig. 3 we get the longest frequent sequence to visualize them on the map as dark path, the sub-frequent sequence as light path.

3.3 Our Method vs. FP-Growth, Apriori

In this section, we first use the FP-Growth, Apriori algorithm and our method to mine frequent trajectory from the one-monthlong trajectory data of user ID 70591 with different minimum support and compare the results. As shown in Fig. 4, all algorithms spend

Fig. 3. Frequent path graph obtained by mining

more time mining frequent sequences because of the decrease of minimum support. With the increase of the minimum support, the efficiency of each algorithm is getting close to each other, and there is almost no difference.

Our method contains four parts (Trajectory segmentation, Data filtering, Geocoding, Mining semantic trajectory frequent patterns), and our original data can only be used for frequent sequences mining after the first three parts. The two algorithms in the comparison experiment use preprocessed data, and there is no way to mine on the original data. Therefore, the total running time of our method is not dominant in Fig. 4, because the running time here includes two parts: preprocessing and mining frequent sequences.

Therefore, in order to ignore the impact of data preprocessing on algorithm time, we use two no preprocessing required data sets kosarakt and movieItem, to mine frequent sequences, and analyze the advantages and disadvantages of our method and comparison method.

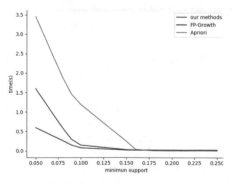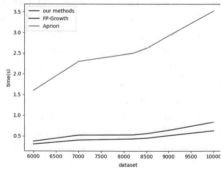

Fig. 4. Minimum support test   **Fig. 5.** Kosarakt minimum support 1/50

Figure 5 shows the experimental results of using the Kosarakt data set to mine frequent sequences. The experiment uses data of different scales to verify the performance of the algorithm. It can be seen from the results that when the minimum support is fixed,

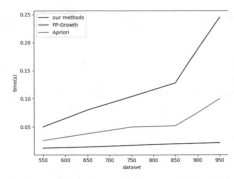 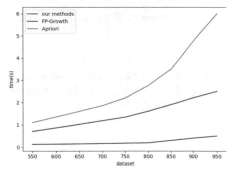

Fig. 6. MovieItem minimum support 1/4   **Fig. 7.** MovieItem minimum support 1/8

the running time of all algorithms will increase with the increase of data size, and the performance of our method is between Apriori and FP-Growth. Since the Apriori algorithm needs to scan the data set multiple times, when the amount of data is larger, the consumption of scanning the data set will increase. The FP-Growth algorithm adopts a divide-and-conquer strategy and only needs to scan the data set twice, so it can effectively reduce the overhead of searching the data set. Our method also reduces the number of times to search data sets and effectively reduces the overhead of searching data sets.

Figures 6 and 7 show the experimental results of frequent sequences mining using movieitem dataset. The feature of this data set is that the item set of a single entry is relatively large, and a large number of entries contain 500 items. The experimental results show that for the long entry dataset, when the minimum support is small, the Apriori algorithm will generate a large number of candidates, the number of scanning data sets increases significantly, and the performance of Apriori algorithm decreases significantly; when the minimum support is large, due to the depth of the generated FP Tree, there are many sub problems to be solved, and the performance of FP growth algorithm decreases significantly. Our method can get a relatively stable result when dealing with long entry data sets.

In general, the method we proposed has the following characteristics:

1. It can convert the original GPS data set into a data set that can be used to mine frequent sequences, and the final result is called semantic frequent trajectories.
2. Our method has better performance when mining frequent sequences on data sets with long entries.

4   Future Work

In this paper the frequency and periodicity in historical trajectory data have been fully identified and utilize, but the infrequent value trajectory or burst information contained in the trajectory are all ignored. In the future we will improve our algorithms to deal with these ignored parts and match the surrounding POI location to help identify the potential meaning and information.

Acknowledgements. Support by the National Key R&D Program in China. Project No. 2017YFC1405403.

Attachment. Data and code download address:
https://github.com/TimothyLx/Mining-method-based-on-semantic-trajectory-frequent-pat
tern-and-carpooling-application.

References

1. Gao, Q., Zhang, F.-L., Wang, R., Zhou, F.: Trajetory big data: a review of key technologies in data processing. J. Softw. **28**(04), 959–992 (2017)
2. Alvares, L.O., Bogorny, V., Kuijpers, B., et al.: A model for enriching trajectories with semantic geographical information. In: Proceedings of the 15th Annual ACM International Symposium on Advances in Geographic Information Systems, p. 22. ACM (2007)
3. Zuo, K., Tao, J., Zeng, H., et al.: Algorithm for trajectory movement pattern mining based on semantic space anonymity. Netinfo Secur. **18**(8), 34–42 (2018)
4. Giannotti, F., Nanni, M., Pinelli, F., et al.: Trajectory pattern mining. In: Proceedings of the 13th ACM SIGKDD International Conference on Knowledge Discovery and Data Mining, pp. 330–339. ACM (2007)
5. Luo, W., Tan, H., Chen, L., et al.: Finding time period-based most frequent path in big trajectory data. In: Proceedings of the 2013 ACM SIGMOD International Conference on Management of Data, pp. 713–724. ACM (2013)
6. Yang, J., Meng, Z., Liang, J.: Logistics frequent path sequence mining algorithm based on topological information. Comput. Sci. **42**(4), 258–262 (2015)
7. Zhang, L., Jiangqin, W., Gao, W.: Hand gesture recognition based on hausdorff distance. J. Image Graph. **7**(11), 1144–1150 (2002)
8. Saini, R., Kumar, P., Roy, P.P., Pal, U: Modeling local and global behavior for trajectory classification using graph based algorithm. Pattern Recogn. Lett. (2019)
9. Chen, Y., Yuan, P., Qiu, M., Pi, D.: An indoor trajectory frequent pattern mining algorithm based on vague grid sequence. Expert Syst. Appl, **118**, 614–624 (2019)
10. Jiang, J., Wang, X.: Review on trajectory data compression. J. East China Normal Univ. (Nat. Sci.) **05**, 61–76 (2015)
11. Peng, X., Gao, S., Chu, X., He, Y., Lu, C.: Clustering method of ship's navigation trajectory set based on spark. Navig. China **40**(03): 49–53+68 (2017)
12. Ji, G., Zhao, B.: Research progress in pattern mining for big spatio-temporal trajectories. J. Data Acquisition Process **30**(01), 47–58 (2015)
13. Xu, Z., Hu, C., Hu, J.: Kalman filtering-based supervisory run-to-run control method for semiconductor diffusion processes. Sci. China (Inf. Sci.), 1–3 (2019). https://kns.cnki.net/kcms/detail/11.5847.TP.20190121.1604.004.html
14. Hightower, J., Borriello, G.: Particle filters for location estimation in ubiquitous computing: a case study. In: Proceedings of the 6th International Conference on Ubiquitous Computing (UbiComp 2004), Nottingham, pp. 88–106 (2004)
15. Gustafsson, F.: Particle filter theory and practice with positioning applications. IEEE Aerosp. Electron. Syst. Mag. **25**(7), 53–82 (2010)
16. Lee, W.C., Krumm, J.: Trajectory preprocessing. In: Proceedings of the Computing with Spatial Trajectories, pp. 3–33. Springer, New York (2011)
17. Zheng, Y.: Trajectory data mining: an overview. ACM Trans. Intell. Syst. Technol. **6**(3), 1–41 (2015)

18. Cao, X., Cong, G., Jensen, C.S.: Mining significant semantic locations from GPS data. Proc. VLDB Endow. **3**(1–2), 1009–1020 (2010)
19. Wang, Y., Lim, E.P., Hwang, S.Y.: Effective group pattern mining using data summarization. Advances in Knowledge Discovery and Data Mining, pp. 713–718. Springer, Heidelberg (2005)
20. Han, J., Pei, J., Yin, Y.: Mining frequent patterns without candidate generation. In: Proceedings of ACM-SIGMOD International Conference on Management of Data, pp. 1–12. ACM Press, New York (2000)
21. Pei, J., Han, J., Mortazavi-Asl, B., Pinto, H.: PrefixSpan: mining sequential patterns efficiently by prefix-projected pattern growth. In: Proceedings of the International Conference on Data Engineering. IEEE Computer Society (2001)

Triangle Enumeration for Billion-Scale Graphs in RDBMS

Aly Ahmed[✉], Keanelek Enns, and Alex Thomo

University of Victoria, Victoria, BC, Canada
{alyahmed,keanelekenns,thomo}@uvic.ca

Abstract. Triangle enumeration is considered a fundamental graph analytics problem with many applications including detecting fake accounts, spam detection, and community searches. Real world graph data sets are growing to unprecedented levels and many of the existing algorithms fail to process them or take a very long time to produce results. Many organizations invest in new hardware and new services in order to be able keep up with the data growth and often neglect the well established and widely used relational database management systems (RDBMSs). In this paper we present a carefully engineered RDBMS solution to the problem of triangle enumeration for very large graphs. We show that RDBMSs are suitable tools for enumerating billions of triangles in billion-scale networks on a consumer grade server. Also, we compare our RDBMS solution's performance to a native graph database and show that our RDBMS solution outperforms by order of magnitude.

1 Introduction

The problem of triangle listing or triangle enumeration can be stated as follows:

Given an undirected graph $G = (V, E)$ with no parallel edges or self loops, output all tuples (a, b, c) such that nodes $a, b, c \in V$ are pairwise connected in G (i.e., they form a triangle). Some algorithms only need to touch triangles or count them, but in this paper, we require that algorithms store them explicitly (hence the term triangle listing).

Triangle enumeration is considered a fundamental graph analytics problem that constitutes a large portion of the computational work required for problems such as calculating a graph's global clustering coefficient (in which the number of triangles each node is involved in is required) [24], finding k-truss decomposition, and calculating the transitivity ratio of a graph. Some of the indirect applications of triangle enumeration include identifying social networks, determining a community's age [25], performing community searches [18, 19], and detecting fake accounts, malicious pages, or instances of web spamming [17, 22].

Triangle enumeration is a nontrivial problem. The optimal worst case for any algorithm's time complexity is $O(|E|^{3/2})$ [22]. Triangle enumeration does not scale very well with large datasets as any node can participate in a triangle with any other node in a graph so long as they have an edge between them, making it difficult to break the problem into smaller pieces. The sizes of datasets continue to grow, and massive datasets

L. Barolli et al. (Eds.): AINA 2021, LNNS 226, pp. 160–173, 2021.
https://doi.org/10.1007/978-3-030-75075-6_13

are becoming more common in business and research. In particular, data in real-world networks is growing to unprecedented levels [10]. For example, Walmart is estimated to create 2.5 petabytes of consumer data every hour [9], Facebook processes tens of billions of likes and messages every day, Google receives 1.2 trillion search requests every year, and Internet of Things (IoT) data is expected to exceed 175 zettabytes by 2025 [1]. Many algorithms either enumerate triangles for graphs that can fit in memory [c.f. [6,7]], are I.O. intensive [c.f. [3,4]] or use distributed systems such as MapReduce [c.f [2,24]].

Relational Database Management Systems (RDBMSs) have been vital tools in storing and manipulating data for many decades [12]. They are commonplace in most businesses, and they are familiar to technical and non-technical users alike. This paper aims to show that they are also useful for analyzing large graphs, specifically in the area of triangle enumeration. The paper also shows how a single machine can use simple partitioning techniques to enumerate billions of triangles efficiently.

In recent times, Graph Databases (GDBs) have grown in popularity. Many businesses may be considering moving to GDBs if they often analyze large graphs. Using dedicated graph databases for graph processing is presumed to provide better performance and scalability over relational databases (c.f. [16]); however, graph databases still have a long way to reach the level of maturity of RDMBSs. Using an RDBMS to implement graph algorithms is, in many situations, more efficient. However, computing graph algorithms using SQL queries is challenging and requires novel thinking. As such, there is active research on the use of novel methods to compute graph analytics on RDBMSs (c.f. [11,13,14,23]). These works have shown that RDBMSs often provide higher efficiency over graph databases for specific analytics tasks.

The contributions of this paper are as follows:

1. We engineer triangle enumeration algorithms in SQL using partitioning and coloring
2. We suggest a modification of the PTE CD [22] algorithm, which we name Source Node Partitioning, which allows us to scale efficiently
3. We compare the performance of a popular open source GDB and a commonly used RDBMS.
4. We give a comparison of the performance of each algorithm on several graphs including billion-scale graph.

2 Triangle Enumeration in RDBMS

The following section introduces state of the art triangle enumeration algorithms and techniques. The basic method used for triangle enumeration, known as the Compact Forward algorithm [20], is explained first. Adaptations created to handle larger graphs that cannot fit into a single machine's memory are then discussed. These adaptations are known as Triangle Type Partitioning [21], and Pre-partitioned Triangle Enumeration [22]. Finally, a further adaptation contributed by this paper that aims to reduce the complexity and the number of queries generated, named Source Node Partitioning, is explained. The SQL implementation of each algorithm developed is illustrated so that readers may recreate these results in an RDBMS of their choice.

2.1 Compact Forward

The compact forward algorithm constitutes the main portion of work done to enumerate triangles in all of the algorithms that follow. Its pseudo code is shown as Algorithm 1. It consists of two main parts: an edge iterator and an orientation technique.

Algorithm 1: Compact Forward

Input   : Undirected graph G = (V, E)
Output: All triangles of G
//Orientation of G
for $(u,v) \in E$ **do**
    if $deg(u) > deg(v)$ **or** $(deg(u) = deg(v)$ **and** $u > v)$ **then**
        | Replace (u,v) with (v,u) in E;
//List triangles of G
for $(u,v) \in E$ **do**
    for $w \in N(u) \cap N(v)$ **do**
        | Output triangle (u,v,w);

Edge Iterator. For any edge $(u,v) \in E$, we can find the triangles associated with it by considering the intersection of the neighbors of u and v (denoted $N(u)$ and $N(v)$ respectively). That is, if (u,v) exists, we can check to see if both (u,w) and (v,w) exist for all such nodes $w \in V$. By performing this operation on all edges, we are guaranteed to enumerate all triangles in the graph. Unfortunately, we might count many duplicates depending on how the undirected graph G is represented.

Orientation Technique. In order to eliminate duplicates, we direct the graph G with a total ordering. Define the degree (or number of neighbouring nodes) of a node v to be $deg(v)$ and assume nodes are represented by unique integer identifiers (i.e. $V \subset \mathbb{N}$). Define a total ordering of the nodes in V, denoted $\rightarrow$, as follows: For all nodes $u,v \in V$, we say $u \rightarrow v$ if and only if $deg(u) < deg(v)$ OR $(deg(u) = deg(v)$ AND $u < v)$. We then arrange all the edges in the graph according to this total order. The resulting graph is a Directed Acyclic Graph or DAG. But how does this help?

Consider a triangle (a,b,c), such that $a \rightarrow b, b \rightarrow c$, and $a \rightarrow c$. When iterating over edge (a,b), we discover (b,c) and (a,c) in the neighbours of a and b. However, when we iterate over edge (b,c), we will not find (c,a) due to the total ordering properties (similarly for edge (a,c), we will not find (c,b)). Thus each triangle is counted exactly once. From now on, when we refer to triangle (a,b,c), we assume the total ordering applies from left to right.

Notice that a total ordering could have been defined on the node identifiers alone assuming they are unique. The reason for involving node degrees in the total ordering is to prevent any given node from having a large list of outgoing neighbours to search through. In the given total ordering, nodes of high degree will have fewer outgoing neighbours and nodes of low degree, though their neighbourhood primarily consists of outgoing neighbours, have few neighbours by definition.

The SQL queries for orienting G are lengthy, yet simple to implement, so we omit them. Algorithm 2 shows the edge iteration component written in SQL and assumes the edge list E has already been oriented.

Algorithm 2: Compact Forward Edge Iteration in SQL

SELECT g1.fromNode AS A, g1.toNode AS B, g2.toNode as C
FROM E g1, E g2
WHERE g1.toNode = g2.fromNode
AND EXISTS (
 SELECT 1 FROM E
 WHERE fromNode = g1.fromNode AND toNode = g2.toNode);

2.2   Triangle Type Partitioning

Suppose G is a large graph that does not fit in a single machine's memory, then the edge list E must be partitioned into smaller lists in order to fit. Even though an RDBMS is designed to handle data that does not fit in main memory, when it has to deal with smaller chunks of the data at a time, the RDBMS can use more efficient join algorithms, such as one-pass joins. However, a problem arises when trying to list triangles in the edge partitions. Consider triangles that have edges in more than one partition, they are certainly missed. As such, we cannot expect the RDBMS query optimizer to be able to automatically find ways to break up the data into chunks to facilitate better query evaluation algorithms. Therefore, we focus here in ways to intelligently partition the data and create independent subtasks.

The triangle type partitioning (TTP) algorithm [22] was designed to resolve this issue. Originally it was designed as a Map Reduce algorithm to be run on distributed systems, but in this paper, it is used on a single machine.

The first step is to colour the nodes using a function $f : V \rightarrow \{0, 1, ..., \rho - 1\}$. This allows every triangle to be classified into the following three types:

- Type 1: All three nodes have the same colour.
- Type 2: Exactly two nodes have the same colour.
- Type 3: Each node has a distinct colour.

A visualization of the triangle types can be seen in Fig. 1.

Let E_{ij} be the set of edges that have endpoints coloured i or j. There are $\binom{\rho}{2}$ such sets. Let E_{ijk} be the set of edges that have endpoints coloured i, j, or k. There are $\binom{\rho}{3}$ such sets as seen in Fig. 2.

No edge in a type 3 triangle has endpoints with the same colour. Knowing this, we can transform the set E_{ijk} into $E'_{ijk} = E_{ijk} - \{(u,v)|f(u) = f(v), (u,v) \in E_{ijk}\}$ where we remove all such edges. This vastly improves performance when enumerating type 3 triangles by reducing the size of the edge lists. On the other hand, each set E_{ij} can contain type 1 and type 2 triangles.

In order to enumerate all triangles, algorithm 2 is run on all $\binom{\rho}{2} + \binom{\rho}{3}$ edge sets (replace E with E_{ij} or E'_{ijk} for all i,j, and k).

Fig. 1. An example of each type of triangle in a node coloured graph

This can be seen in Algorithm 3, which shows the structure for partitioning and generating the SQL queries.

Algorithm 3: Triangle Type Partitioning

---

for $i = 0$ **to** $\rho - 2$ **do**
$\quad$ **for** $j = i+1$ **to** $\rho - 1$ **do**
$\quad\quad$ Generate E_{ij} from E
$\quad\quad$ Run algorithm 2 on E_{ij}
for $i = 0$ **to** $\rho - 3$ **do**
$\quad$ **for** $j = i+1$ **to** $\rho - 2$ **do**
$\quad\quad$ **for** $k = j+1$ **to** $\rho - 1$ **do**
$\quad\quad\quad$ Generate E'_{ijk} from E
$\quad\quad\quad$ Run algorithm 2 on E'_{ijk}
Eliminate duplicates from generated triangles

Consider the sets E_{02} and E_{12} from Fig. 2. The type 1 triangle of colour 2 is counted twice. In fact, this is true of all type 1 triangles.

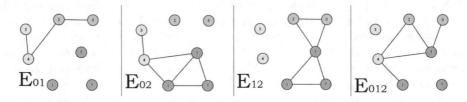

Fig. 2. A visualization of the edge sets from the graph in figure 2 when $\rho = 3$: yellow = 0, blue = 1, pink = 2

Thus we count every type 1 triangle $\rho - 1$ times, and we count type 2 and type 3 triangles exactly once. The duplicate triangles can then be eliminated.

2.3   Prepartitioned Triangle Enumeration - Colour Direction

Triangle type partitioning [22] is an effective method for enumerating triangles in massive graphs that do not fit into memory, yet it is also possible to ensure each triangle is counted exactly once as well as greatly reduce the size of the edge sets to be searched through in the CF algorithm. This is the goal of the Prepartitioned Triangle Enumeration - Colour Direction (PTE CD) method, which expands on the TTP algorithm. Again, this algorithm was proposed in [22] for a Map Reduce setting. Here we adapt it for an RDBMS. Consider how the TTP algorithm pays no attention to the direction of edges when partitioning the edge set, indeed figures 2 and 3 do not display the direction of the edges because that information is irrelevant to the algorithm. PTE CD, however, takes advantage of the direction of each edge in the oriented graph.

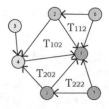

Fig. 3. The graph of Fig. 1 with edge directions shown. The labels indicate which triangle set each triangle belongs to.

Let T_{ijk} be the set of triangles (a,b,c) where $f(a) = i, f(b) = j$, and $f(c) = k$ (i.e. ijk is a permutation of the colours with replacement). For example, $T_{001} \neq T_{010}$ due to the total ordering on the edges in G. This is made more clear in Fig. 3. There are ρ^3 such triangle sets.

Suppose we are trying to find a triangle $(a,b,c) \in T_{ijk}$ and we reach edge (a,b). We know edge (b,c) is in E_{jk} and edge (a,c) is in E_{ik}, which may or may not all be the same edge set from our previous definitions, but more specifically, we know (b,c) goes from colour j to k, and (a,c) goes from colour i to k.

Let E_{ij} be redefined in the following way:

$$E_{ij} = \{(u,v)|f(u) = i, f(v) = j, (u,v) \in E\}$$

There are ρ^2 such edge sets. This new definition allows the algorithm to more precisely choose which edge sets to search through and thereby reduces the total work.

The pseudo code in Algorithm 4 shows the implementation of the PTE CD algorithm. It counts triangles of all types exactly once and improves the performance greatly.

One issue is that it increases the number of edge sets from $\binom{\rho}{2}$ to ρ^2 and the number of enumeration tasks from $\binom{\rho}{2} + \binom{\rho}{3}$ to ρ^3. However, it reduces the number of colours needed to be effective compared to TTP, and since ρ is often relatively small, this does not create a major issue in performance.

Algorithm 4: PTE CD

---

for $i = 0$ **to** $\rho - 1$ **do**
    for $j = 0$ **to** $\rho - 1$ **do**
        | Generate E_{ij} from E
for $i = 0$ **to** $\rho - 1$ **do**
    for $j = 0$ **to** $\rho - 1$ **do**
        for $k = 0$ **to** $\rho - 1$ **do**
            | Run algorithm 2 with E_{ij} as g1, E_{jk} as g2, and E_{ik} in the EXISTS
            clause.

2.4 Source Node Partitioning (SNP)

In this section, we propose another algorithm, called Source Node Partitioning (SNP), that partitions the graph into ρ partitions and generates ρ^2 enumeration tasks. SNP is conceptually simpler than PTE CD. Similar to PTE CD, SNP enumerates each triangle only once and exhibits comparable performance to PTE CD for medium datasets and even better for larger datasets.

The main idea is to partition an oriented input graph G into ρ parts where all the edges (u,v) whose source nodes u are landed in the same partition i based on partitioning function $f(u)$. Each partition i will have a tuple (u, v, j) where j equal the partition number for node v.

$$E_i = \{(u, v, j) | f(u) = i, f(v) = j, (u, v) \in E\}, \ 0 \le i, j \le \rho - 1$$

Now, in order to find a triangle (u, v, k) we only need to check if there is a shared node k between the neighbours of nodes u and v. This is achieved by joining in SQL the tables for E_i and E_j (assuming $f(u) = i$ and $f(v) = j$).

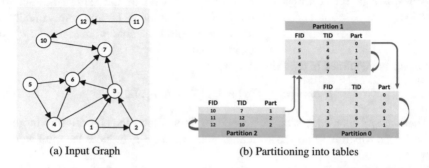

(a) Input Graph   (b) Partitioning into tables

Fig. 4. SNP partitioning

For example, given the graph in Fig. 4 (a), and $\rho = 3$, we partition the edges of the graph into the partition tables given in Fig. 4 (b). Along with the FID and TID attributes, which give the source and target of each edge, we also have a third attribute, called Part, which stores the partition number, of the target node.

To check for instance, if edge $(3, 6)$ is part of a triangle, we need to find the intersection of the neighbors of node 3 with the neighbors of node 6. These neighbors can be extracted from the set of edges $(6, \_)$, which exist in Partition 1 (as indicated by column Part) and the set of edges $(3, \_)$, which exist in the current partition, Partition 0.

Partition oriented graph G based on the source node could be done by several ways for instance, we could use interval partitioning where each partition i will have a unique range of nodes ID or use module function to distribute the source nodes over partitions as shown in Algorithm 5.

In terms of pseudo code we create the partition tables using the queries given in Algorithm 5. Then, we enumerate triangles using Algorithm 6. In a nutshell, it joins two tables, for partitions E_i and E_j, respectively, and limits the scope of the join in the table for E_i to only those nodes that have $Part = j$.

Algorithm 5: SNP: Graph Partitioning

Input: Oriented Graph $G = (V, E)$, number of partitions ρ
Output: A set of ρ partition tables.
for $i = 0$ **to** $\rho - 1$ **do**
  INSERT INTO E_i
    SELECT FID, TID, TID % ρ AS Part
    FROM E
    WHERE FID % $\rho = i$

Algorithm 6: SNP (enumerate triangles)

Input: A set of ρ partition tables E_i.
for $i = 0$ **to** $\rho - 1$ **do**
  for $j = 0$ **to** $\rho - 1$ **do**
    SELECT G1.fid AS A, G1.tid AS B, G2.tid as C
    FROM E_i AS G1, E_j AS G2
    WHERE G1.tid = G2.fid AND G1.part = j AND
      EXISTS
        (SELECT 1 FROM E_i AS G3
        WHERE G3.fid = G1.fid AND G3.tid =G2.tid)

Theorem 1. *SNP Partitioned graph can only distribute a triangle (u, v, k) over a maximum of 2 partitions.*

Proof. To prove it by contradiction let us assume there is a triangle (u, v, k) whose edges exist in three partitions which means neighbors of u and v in an edge (u, v) must exist in three partitions. However, SNP will not separate the edges with the same source node across different partitions, therefore neighbors of u and v will only exist in a maximum of 2 partitions. □

3 Experimental Results

3.1 Setup Configuration

We executed the experiments on a cloud based virtual server running Windows Server 2019 with 4 vCores and 16 GB of RAM.

As RDBMSs we used the latest versions of a commercial database (which we anonymously call CD) As graph database, we used the latest version of a graph database (which we anonymously call GD). We refrain from using the real names of these databases for obvious reasons.

3.2 Datasets

We used four datasets from Stanford's Data collection and four datasets from The Laboratory for Web Algorithmics including a one-billion-edge graph.

The datasets are Web-Google, Pokec, Live-Journal and Orkut (from http://snap.stanford.edu), and Hollywood 2009, Hollywood 2011, UK 2005 and IT 2004 (from http://law.di.unimi.it/webdata). Table 1 shows statistics about the datasets used.

Table 1. Graph datasets

Dataset	# Nodes	# Edges	# Triangles
Web-Google	875,713	5,105,039	13,391,903
Pokec	1,632,803	30,622,564	32,557,458
Live Journal	4,847,571	68,993,773	177,820,130
Orkut	3,072,441	117,185,083	627,584,181
Hollywood 2009	1,139,905	113,891,327	4,916,374,555
Hollywood 2011	2,180,759	228,985,632	7,073,951,555
UK 2005	39,459,921	936,364,282	21,779,347,099
IT 2004	41,291,594	1,150,725,436	47,249,138,589

3.3 Results

Our experiments began with a comparison of the performance of a popular open source graph database and a well known, commonly used RDBMS. The initial idea was to compare the time it took for each database system to compute the clustering coefficient of each graph. However, after running various experiments, it was determined that the graph database likely used approximation algorithms. This theory was then supported when attempting to use the graph database's triangle listing algorithm which took substantially longer than the clustering coefficient calculation, which does not make logical sense if no approximation algorithms are involved because triangle listing is a subset of the computations required for calculating the clustering coefficient. Moreover, the triangle listing algorithm seemed to have a bug at the time of experimentation, and only listed around a tenth of the triangles in the graph.

In order to ensure the measurements for the graph database were as fair as possible, we caused the output to be written to a file instead of writing to standard out (which can often take longer than a computation itself), and we scaled the triangle listing time linearly to match the approximate amount of time it would take to enumerate all of the triangles rather than a fraction of them.

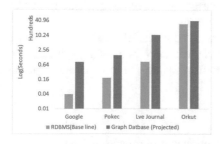

Fig. 5. A logarithmic runtime comparison between a graph database and RDBMS when enumerating triangles on the four smallest datasets: google, pokec, livejournal, and orkut

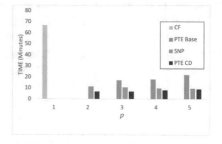

Fig. 6. A comparison of runtimes of the four algorithms discussed in Sect. 3 on the orkut dataset. p is the number of node partitions of the graph, not the number of edge sets created by the algorithms.

Figure 5 shows a comparison between the projected runtime of the graph database and the actual runtime of the RDBMS baseline, which is the compact forward algorithm run on the entire edge set at once. Note that the time scale is logarithmic and we can see that the RDBMS greatly outperforms the graph database when the graph database is required to list all triangles explicitly.

After seeing the drastic performance difference between the two databases in just the baseline case, we did not see value in pursuing any further comparisons with other methods or larger datasets.

Figure 6 compares the performance of all the algorithms discussed in Sect. 2 on the orkut dataset. Smaller datasets did not see much improvement when partitioned, as their edge sets were already quite small. No entry has been given for the PTE BASE method (which is another name for TTP) when $\rho = 2$. This is because the algorithm no longer benefits from the partitioning in this case, and the results would be the same as the baseline CF results.

Note that the yellow bar is the same datapoint that was used in Fig. 5 for the RDBMS baseline on the orkut dataset. Clearly the partitioning algorithms made great improvements over the single table compact forward algorithm. From this we conclude that partitioning can be helpful even when the dataset is able to fit into a machine's memory (as is the case with the orkut dataset). However, as noted before, the benefits become less notable the smaller the dataset becomes and in some cases the performance decrease slightly as the cost of partitioning tables and query planning overcome the savings. As we moved to graphs with billions of triangles, PTE Base reached its performance bottleneck; eliminating duplicate triangles takes a considerable amount of time and resources to perform, therefore, we decided not to test it with other datasets.

We turn our attention to the hollywood datasets for 2009 and 2011 which have over 8 and 11 times more triangles than orkut respectively. Figure 7 shows that PTE CD slightly outperforms the SNP method. Notice, however, that the performance gap begins to close with more node partitions or colours. At first inspection, this seems to be due to the growth rates in the number of queries generated by each method. SNP generates ρ^2 queries compared to PTE CD's ρ^3 queries, which conceivably increases compiling and execution time as well as clutters the SQL script files.

However, the difference is not as clear cut as it may seem. SNP only creates ρ edge sets (or edge partitions), whereas PTE CD creates ρ^2 edge sets. If the desired edge set size is the same relative to the size of a machine's memory (e.g. we want edge sets to be a third the size of a machine's available memory), then PTE CD can use a smaller value for ρ than SNP can. Therefore PTE CD may actually use less queries in practice.

For example, suppose a graph has an edge list with 1.2 million entries, and we desire to use a machine that can fit 300,000 edges, then our edge sets should have about 100,000 edges in them (if we want the database to perform an all-in-memory join). SNP will use $\rho = 12$ to meet this requirement, whereas PTE CD will use $\rho = 4$. Notice that $4^3 = 64 < 12^2 = 144$ in this instance, and PTE CD actually has fewer queries for similar-sized edge sets. Nevertheless, in a very large graph, fewer queries might not be as desired. SQL might need in worst-case scenario memory size of $|E_i| * |E_j| * |E_k|$ to find triangles; hence more scoped queries would be more efficient. Also, the partitioning function used in SNP impacts the performance; Fig. 9 shows a noticeable difference in performance when we used the modulo function, and this could be explained as it is most likely to distribute high degree nodes across all partitions relatively than distribute them using interval partitioning.

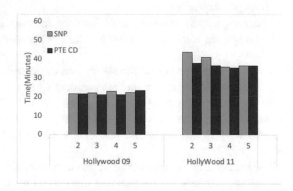

Fig. 7. Runtimes of SNP (blue) and PTE CD (red) algorithms on the hollywood-2009 dataset (~5 billion triangles) and hollywood-2011 dataset (~7 billion triangles). The x axis shows the value of ρ or the number of node partitions.

It is also worth mentioning that the distance between squares increases quickly. In the previous example, PTE CD creates 16 edge sets rather than the ideal 12. Suppose we need to divide the edge list of the graph into 37 edge sets. Then PTE CD must use $\rho = 7$ and create 49 edge sets, which may cause performance issues because the edge sets are too small compared to the optimal value and the time to shuffle the data will increase. Figure 8 shows a decrease in the performance of PTE CD as the number of triangles increase.

It would seem that using the same value of ρ for both methods is in inappropriate comparison, and this would explain the reduction in the performance gap as ρ increases, since SNP performs better for higher values of ρ.

In each case, it may take some testing and experimentation to determining the ideal value for ρ for a given dataset.

3.4   Experiments on Billion-Scale Networks

We show our experiments on two vast datasets, namely UK-2005 and IT-2004, the latter with more than a billion edges. They represent the web network of UK and Italy in 2005 and 2004, respectively. The precise number of nodes and edges is given in Table 1. We ran both the SNP algorithm with $\rho = 20$ and PTE CD algorithm with $\rho = 8$. Both algorithms ran in a single machine and enumerated all triangles in a very reasonable time regarding the graph's size and the number of triangles found in each data set; IT 2004 and UK 2005 contain 47.2 and 21.7 billion triangles, respectively.

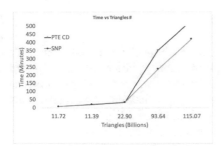

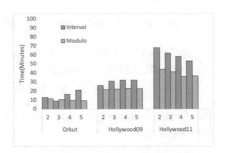

Fig. 8. A comparison of running time between SNP and PTE CD as the number of triangles increased. PTE CD takes more time as partitioning takes longer than SNP.

Fig. 9. SNP: a noticeable difference in running time when partitioning the graph using Interval Partitioning vs using Modulo function.

We compare the running time of the two algorithms in Fig. 10. The result indicates that the SNP algorithm shows better performance than PTE CD. The reason being that SNP scales linearly with ρ in partitioning the tables and quadratically in the number of queries when enumeratimg triangles, however PTE CD scales quadratic to ρ for partitioning tables and cubic to the number of queries performed.

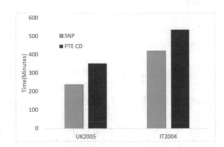

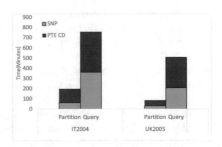

Fig. 10. Results of Triangles Enumeration on very large data sets, IT 2004: 1.15 billion edge graph and UK 2005:0.93 billion edge, Using SNP and PTE CD.

Fig. 11. The time for SNP to build the partitions and run queries is significantly less when datasets get significantly larger.

Figure 11 shows the running time differences to partition the datasets and to enumerate triangles between SNP and PTE CD algorithms. PTE CD takes significantly more time to create ρ^2 tables and execute ρ^3 queries.

4   Conclusion and Future Work

We implemented Triangle Enumeration algorithms, such as CF, PTE Base, and PTE CD, using RDBMSs. We proposed the SNP algorithm that partitions the graph into ρ partitions and generates ρ^2 enumeration tasks. SNP exhibits comparable performance to PTE CD for medium datasets and even better for larger datasets. We experimented with billion scale graphs and enumerated tens of billions of triangles, showing that RDBMSs can perform better than GBDs by multiple orders of magnitude and process massive datasets in a consumer-grade server. One possible direction for future research is to improve performance on RDBMSs by compressing the edge list using variable byte encoding (VBE).

References

1. Mahanthappa, S., Avarkar, B.: Data Formats and Its Research Challenges in IoT: A Survey. Springer (2020)
2. Park, H., Silvestri, F., Kang, U., Pagh, R.: Mapreduce triangle enumeration with guarantees (2014)
3. Chu, S., Cheng, J.: Triangle listing in massive networks. ACM Tkdd **6**, 1–32 (2012)
4. Hu, X., Tao, Y., Chung, C.: Massive graph triangulation (2013)
5. Latapy, M.: Main-memory triangle computations for very large (sparse (power-law)) graphs. Tcs **407**, 458–473 (200)
6. Zhang, Y., Parthasarathy, S.: Extracting analyzing and visualizing triangle k-core motifs within networks (2012)
7. Schank, T.: Algorithmic aspects of triangle-based network analysis (2007)
8. Kelly, R.: Internet of things data to top 1.6 zettabytes by 2022. Campus Technol. **9**, 1536–1233 (201)
9. Mcafee, A., Brynjolfsson, E., Davenport, T., Patil, D., Barton, D.: Big data: the management revolution. Harv. Bus. Rev. **90**, 60–68 (2012)
10. Jagadish, H., Gehrke, J., Labrinidis, A., Papakonstantinou, Y., Patel, J., Ramakrishnan, R., Shahabi, C.: Big data and its technical challenges. CACM **57**, 86–94 (2014)
11. Ahmed, A., Thomo, A.: Pagerank for billion-scale networks in RDBMS (2020)
12. Codd, E.: A relational model of data for large shared data banks. Springer (2002)
13. Agrawal, R., Imielinski, T., Swami, A.: Database mining: a performance perspective. IEEE Tkde **5**, 914–925 (1993)
14. Ordonez, C., Omiecinski, E.: Efficient disk-based K-means clustering for relational databases. IEEE Tkde **16**, 909–921 (2004)
15. Gao, J., Zhou, J., Yu, J., Wang, T.: Shortest path computing in relational DBMSs. IEEE Tkde **26**, 997–1011 (2013)
16. Angles, R., Gutierrez, C.: Survey of graph database models. ACM Comput. Surv. **40**, 1–39 (200)
17. Arifuzzaman, S., Khan, M., Marathe, M.: PATRIC: a parallel algorithm for counting triangles in massive networks (2013)
18. Radicchi, F., Castellano, C., Cecconi, F., Loreto, V., Parisi, D.: Defining and identifying communities in networks. Proc. Natl. Acad. Sci. **101**, 2658–2663 (2004)
19. Berry, J., Hendrickson, B., Laviolette, R., Phillips, C.: Tolerating the community detection resolution limit with edge weighting. Phys. Rev. E **83**, 056119 (2011)
20. Yu, M., Qin, L., Zhang, Y., Zhang, W., Lin, X.: Aot: pushing the efficiency boundary of main-memory triangle listing (2020)

21. Park, H., Chung, C.: An efficient MapReduce algorithm for counting triangles in a very large graph (2013)
22. Park, H., Myaeng, S., Kang, U.: PTE: enumerating trillion triangles on distributed systems (2016)
23. Ahmed, A., Thomo, A.: Computing source-to-target shortest paths for complex networks in RDBMS. J. Comput. Syst. Sci. **89**, 114–129 (2017)
24. Suri, S., Vassilvitskii, S.: Counting triangles and the curse of the last reducer (2011)
25. Leskovec, J., Rajaraman, A., Ullman, J.D.: Mining of Massive Datasets. Cambridge University Press, New York (2020)

An IoT Application Business-Model on Top of Cloud and Fog Nodes

Zakaria Maamar[1(✉)], Mohammed Al-Khafajiy[2], and Murtada Dohan[3,4]

[1] Zayed University, Dubai, UAE
zakaria.maamar@zu.ac.ae
[2] University of Reading, Reading, UK
[3] University of Northampton, Northampton, UK
[4] University of Babylon, Hillah, Iraq

Abstract. This paper discusses the design of a business model dedicated for IoT applications that would be deployed on top of cloud and fog resources. This business model features 2 constructs, flow (specialized into data and collaboration) and placement (specialized into processing and storage). On the one hand, the flow construct is about who sends what and to whom, who collaborates with whom, and what restrictions exist on what to send, to whom to send, and with whom to collaborate. On the other hand, the placement construct is about what and how to fragment, where to store, and what restrictions exist on what and how to fragment, and where to store. The paper also discusses the development of a system built-upon a deep learning model that recommends how the different flows and placements should be formed. These recommendations consider the technical capabilities of cloud and fog resources as well as the networking topology connecting these resources to things.

1 Introduction

On top of services to provision and products to sell, organizations are known for their cultures and business models. A typical example of a successful organization's culture would be *Google* where openness, innovation, excellence that comes with smartness, hands-on approach, and small-company-family rapport are promoted (tinyurl.com/y6ld8wn9). And, a typical example of a successful organization's business model would be *Amazon.com* where millions of customer-enacted e-commerce transactions are successfully performed (www.garyfox.co/amazon-business-model).

To remain competitive and sustain growth, organizations also embrace ICTs with focus lately on the Internet-of-Things (IoT). IoT is about making things like sensors and actuators act over the cyber-physical surrounding so, that, contextualized, smart services are provisioned to users and organizations. According to Gartner (www.gartner.com/newsroom/id/3165317), 6.4 billion connected things were in use in 2016, up 3% from 2015, and will reach 20.8 billion by 2020. Another recent trend in the ICT landscape is coupling IoT with cloud computing and fog computing [15] (fog is *aka* edge). The massive volume of data that things generate, needs to be captured, processed, analyzed, shared, and protected. Data

L. Barolli et al. (Eds.): AINA 2021, LNNS 226, pp. 174–186, 2021.
https://doi.org/10.1007/978-3-030-75075-6_14

range from vegetables' freshness in warehouses to traffic flows on roads and pollution levels in cities. For many years, cloud has been the technology of choice for exposing resources (traditionally software, platform, and infrastructure) as services (*aaS) to organizations that have to "wrestle" with this massive volume of data. According to Gartner too (goo.gl/m9MQXc), *"by 2021, more than half of global enterprises already using cloud today will adopt an all-in cloud strategy"*. However, despite cloud's benefits there exist some concerns about cloud's appropriateness for certain applications (e.g., medical and financial) that have strict non-functional requirements to satisfy in terms of minimizing data latency and protecting sensitive data. Transferring data over public networks to (distant) clouds could take time because of high latency, could be subject to interceptions, alterations, and misuses, and could depend on network availability and reliability. To address data-latency and data-sensitivity concerns, ICT practitioners advocate for fog computing where processing and/or storage facilities are expected to exist "next" (or nearby) to where data is collected minimizing its transfer and avoiding its exposure to unnecessary risks.

In this paper, we examine the appropriateness of developing a business model for IoT applications that would run on top of a cloud/fog infrastructure. This business model should (i) identify the data flows between IoT devices, clouds, and fogs, (ii) expose how collaboration arises between IoT devices, clouds, and fogs, and (iii) track where data processing and storage happen because of the distributed nature of IoT devices, clouds, and fogs. To the best of our knowledge, this is the first step towards defining such a business model in the double context of cloud and fog. Mahmud et al. report the lack of business model for fog environments, only, [10]. The rest of this paper is organized as follows. Section 2 is a brief overview of cloud, fog, and business model. Section 3 details the way we design a business model for IoT applications. Technical details of this design are included in this section, as well. Section 4 concludes the paper and presents some future work.

2   Background

Cloud and Fog in Brief. Cloud is a popular ICT topic that promotes Anything-as-a-Service (*aaS) operation model. This model adopts pay-per-use pricing and consolidates hardware and software resources into data centers. However, despite cloud's popularity, it does not, unfortunately, suit all applications. Cloud is not recommended for latency-critical and data-sensitive applications due to reasons such as high latency added by network connections to data centers and multi-hops/nodes between end-users and data centers that increase the probability of interceptions.

Fog was first introduced by Satyanarayanan et al. in 2009 [12] and generalized by Cisco Systems in 2014 [3] as a new ICT operation-model. The objective is to make processing, storage, and networking facilities "close" to where data is captured and/or stored. The extension from cloud to fog is not trivial due to their subtle similarities and differences. However, their suitability for certain applications remains an open debate [11]. Real-time applications that require almost immediate action and high data protection, would discard cloud in favor of fog. Varghese et al. mention that by 2020, existing electronic devices will generate 43 trillion gigabytes of data that need to be processed in cloud data-centers [14]. However, this way of operating cannot be sustained for a long time due to frequency and latency of communication between these devices and cloud data-centers. Fog would process data closer to its source so, that, network traffic is reduced and both quality-of-service and quality-of-experience are improved.

Business Model in Brief. Despite the critical role of business model in any organization's operation, there is not a common definition of what a business model is. According to Geissdoerfer et al., it is *"a simplified representation of the elements of an organisation and the interaction between these elements for the purpose of its systemic analysis, planning, and communication in face of organizational complexity"* [5]. In a 2015 Business Harvard Review report (hbr.org/2015/01/what-is-a-business-model), it is mentioned that a business model consists of 2 parts: *"Part one includes all the activities associated with making something: designing it, purchasing raw materials, manufacturing, and so on. Part two includes all the activities associated with selling something: finding and reaching customers, transacting a sale, distributing the product, or delivering the service"*. Last but not least, Li sheds light on a business model's key constructs, namely, value proposition (product offerings of the firm, its market segments and its model of revenue generation), value architecture (how a firm senses, creates, distributes, and captures values), and functional architecture (core activities of a firm, namely, product innovation and commercialization, infrastructure for production and delivery, and customer relations management) [8].

3   Business Model for IoT Applications

After an overview of the constructs of our proposed IoT application business-model, we details the types of flows and types of placements and then, present the implementation of a system recommending the definition of these flows and placements.

3.1   Overview

Figure 1 illustrates the constructs we propose to define an IoT application business-model in the context of cloud/fog. The 2 key constructs are flow, specialized into data and collaboration, and placement, specialized into processing and storage.

On the one hand, the flow construct is about who sends what and to whom, who collaborates with whom, and what (time- and location-related) restrictions exist on what to send, to whom to send, and with whom to collaborate. First, the Data Flow (DF) sheds light on how data is transferred from senders to receivers

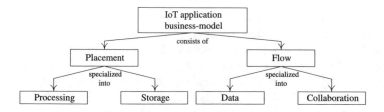

Fig. 1. IoT application business-model's proposed constructs

where senders could be things and fogs, and receivers could be fogs and clouds. Second, the **Collaboration Flow (CF)** sheds light on how, when, and why thing-2-thing, fog-2-fog, and cloud-2-cloud interactions arise. On the other hand, the **placement** construct is about what and how to fragment, where to store, and what (time- and location-related) restrictions exist on what and how to fragment, and where to store. First, the **Processing Placement (PP)** sheds light on how an application's business logic is decomposed into fragments and where the fragments are executed over things *versus* fogs *versus* clouds. Second, the **Storage Placement (SP)** sheds light on where an application's data are spread over things *versus* fogs *versus* clouds.

3.2   Types of Flows

Figure 2 illustrates both the **data flows** and the **collaboration flows** that could be part of an IoT application business-model. First, we rely on our previous work on cloud-fog coordination to recommend how **data flows** should be formed [9,16]. Second, we partially rely on our previous work on fog-2-fog collaboration to recommend how **collaboration flows** should be formed [1].

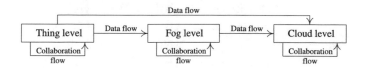

Fig. 2. Data/Collaboration flows in an IoT application business-model

Data flows connect thing and fog together ($DF_{T \rightarrow F}$), thing and cloud together ($DF_{T \rightarrow C}$), and thing and fog and cloud together ($DF_{T \rightarrow F \rightarrow C}$[1]). Compared to the work of Thekkummal et al. who identify 2 **data flows**, from things to clouds and from fogs to clouds where fogs collect data from things [13], we consider a third **data flow** that is from things to fogs since some data do not need to be sent to clouds. It is worth mentioning that although our

---

[1] Pre-processing data at fogs prior to sending the pre-processed data to clouds.

3 specialized data flows could simultaneously exist, we came up in [9, 16] with 6 criteria whose use would permit to either highly-recommend (HR), recommend (R), or not-recommend (NR) which data flow should be formed for a particular IoT application. These criteria are *frequency* (rate of data transfer from things to fogs/clouds; the frequency could be regular, e.g., every 2 h, or continuous), *sensitivity* (nature of data exchanged between things and fogs/clouds; highly-sensitive data should not be exposed longer on networks during transfer), *freshness* (how important data exchanged between things and fogs/clouds should be up-to-date, i.e., recent), *time* (delay that results from withholding/processing data at the thing level until its transfer to fogs/clouds), *volume* (amount of data that things produce and send to fogs/clouds), and *criticality* (demands that fogs/clouds express with regard to data of things; low demands could lead to ignoring certain data). In support of these criteria, we made the assumption that, distance-wise, clouds are **far** from things and fogs are **close** to things. In Table 1, we summarize how the aforementioned criteria, taken independently from each other, assist with recommending the formation of specific data flows: T → C, T → F, and T → F → C. More details about these recommendations are presented in [9].

Table 1. Recommendations to form data flows when criteria are separated [9]

Criterion	Features	T → C	T → F	T → F → C
Frequency	Continuous stream	NR	HR	R
	Regular stream			
	Short gaps	NR	HR	HR
	Long gaps	R	R	R
Sensitivity	High	NR	HR	HR
	Low	R	R	R
Freshness	Highly important	NR	HR	R
	Lowly important	R	R	R
Time	Real-time	NR	HR	HR
	Near real-time	R	HR	HR
	Batch-processing	HR	NR	NR
Volume	High amount	HR	NR	NR
	Low amount	NR	HR	R
Criticality	Highly important	HR	HR	R
	Lowly important	NR	HR	HR

- *Frequency* criterion is dependent on the data stream between things and clouds/fogs. If the stream is continuous (non-stop), then it is highly recommended to involve fogs in all interactions so, that, direct data-transfer to clouds is avoided (i.e., low-latency and low-delay jitter). If the data stream is regular, recommendations will depend on how short *versus* long the gaps are during data transfer.

- *Sensitivity* criterion is about the protection measures to be put in place during data exchange between things and clouds/fogs. If the data is highly sensitive, then it is highly recommended to involve fogs in all interactions so, that, protection is ensured. Otherwise, data could be sent to clouds and fogs.
- *Freshness* criterion is about the data quality to maintain during the exchange between things and clouds/fogs. If the data needs to be highly fresh, then it is highly recommended to involve fogs in all interactions (i.e., subject to be aware of the location of fogs and support to real-time interactions is provided). Otherwise, data could be sent to clouds and/or fogs.
- *Time* criterion is about how soon data is made available for processing. If it is real-time processing, then it is highly recommended to send data to fogs. If it is near real-time (i.e., minutes are acceptable) then it can be sent to clouds and/or fogs. Otherwise, cloud is ideal for data batch-processing.
- *Volume* criterion is about the space constraint over the amount of data collected/produced by things. If this amount is big, it is highly recommended to send data directly to clouds. Otherwise, data could be sent to fogs. In case of a big data amount and data is divisible, then data could be sent over to multiple fogs (i.e., distributed geo-distribution).
- *Criticality* criterion is about ensuring data availability according to fog/cloud demands. If fog/cloud demands are highly important, then it is highly recommended that data should be sent to fog/cloud regardless of the hop number (i.e., geo-distribution).

Contrarily to what we did in [9] where *frequency, sensitivity, freshness, time, volume*, and *criticality* criteria were taken independently from each other, we combined them all using a fuzzy logic-based multi-criteria decision making approach [16]. This approach was demonstrated using a healthcare-driven IoT application along with an in-house testbed that featured real sensors (temperature and humidity DHT11) and fog (rPi2) and cloud (Ubidots) platforms. During the experiments, we modified the *frequency* of streaming data (every 3 s, 5 s, 7 s, and randomly) for each of the 3 data flows, $T \rightarrow C$, $T \rightarrow F$, and $T \rightarrow F \rightarrow C$, and the *volume* (around low and high amount) and *criticality* (around low and high important) of the transmitted data. Upon data messages receipt at an end-point whether fog or cloud, we timestamped these messages prior to storing them. Table 2 includes 2 out of 4 scenarios that summarize the experiments we conducted with focus on the recommendations of forming specific data flows: $T \rightarrow C$, $T \rightarrow F$, and $T \rightarrow F \rightarrow C$. More details about these recommendations are presented in [16].

Collaboration flows connect things together (CF_{T2T}), fogs together (CF_{F2F}), and clouds together (CF_{C2C}) using offloading mechanism that would allow to maintain "acceptable" loads over things, fogs, and clouds. It is not mandatory that the 3 specialized collaboration flows simultaneously exist since this would depend on satisfying under-execution IoT-applications' functional and non-functional requirements. For illustration, we demonstrate how a collaboration flow between fogs could be formed based on our previous work on improving fog performance [1]. We expect adopting the same strategy when developing both collaboration flows between things and collaboration flows between clouds.

Table 2. Recommendations to form data flows when criteria are combined [16]

Scenario #	Criteria	Linguistic values	Recommendations
Scenario 1	Frequency	Regular stream (around short and long gaps)*	T → C is NR; T → F is R; T → F → C is R
	Sensitivity	Around low and high*	
	Freshness	Highly important	
	Time	Real time	
	Volume	High amount	
	Criticality	Lowly important	
Scenario 4	Frequency	Regular stream long gaps	T → C is R; T → F is R; T → F → C is R
	Sensitivity	Low	
	Freshness	Lowly important	
	Time	Near-real time	
	Volume	High amount	
	Criticality	Around lowly and highly important*	

*: Around Val_1 and Val_2: both Val_1 and Val_2 meet the scenario's requirements.

The ICT community already agrees that fog is not a substitute to cloud but a complement; both should work hand-in-hand [2,9]. As per Sect. 2, fog can support, serve, and facilitate services that cloud does not cater well for their needs and requirements. These services are known for being latency sensitive, geo-distributed (e.g., water-pipe monitoring), mobile with high-speed connectivity (e.g., connected vehicles), and largely distributed (e.g., smart energy distribution). However, despite the benefits of fog computing, it happens that fogs working in silos cannot accommodate these services' processing, storage, and communication requirements. Compared to clouds, fogs mean *less* resources, *less* reliability, and *less* latency [7]. Promoting offloading among fogs could be an option, leading to the formation of collaboration flows between fogs according to their ongoing loads and availabilities of their processing, storage, and communication capabilities. This offloading has been the focus of our work in [1].

In compliance with Fig. 2, things periodically collect and generate data from the cyber-physical surroundings and send them to fogs ($DF_{T \rightarrow F}$) and/or clouds ($DF_{T \rightarrow C}$) for processing/storage as deemed necessary. A fog can serve a certain number of requests instantly or offload some to other fogs in the same domain if this fog is congested, which could delay processing these requests. It is worth noting that the importance of fogs being located between things and clouds, makes fogs more accessible/reachable to both things and clouds. Therefore, fog can be used horizontally (i.e., CF_{F2F}) and vertically (i.e., $DF_{T \rightarrow F \rightarrow C}$) in the network to provide the desired services.

By analogy with CF_{F2F}, CF_{T2T} and CF_{C2C} could be formed allowing to develop federations of things and federations of clouds, respectively. Communications in all federations could be either **planned** where links among nodes, i.e., things, fogs, and clouds, are known at design-time and according to a specific business model or **ad-hoc** where links among nodes are formed on the fly and sometimes opportunistically. CF_{T2T} and CF_{C2C} would require a coordination model at the thing level and cloud level, respectively, to ensure that a better load balancing among things and among clouds would be achieved. For instance, the decision of a cloud to collaborate with other clouds to support the processing of a received service's request would depend on the response time. Generally, the response time of a task on a cloud will be computed based on the time required to wait in the queue (in case of loaded cloud), the time to process the received task on the cloud, and the response travel-time that includes both transmission and propagation delays. Meantime, since it is a distributed model, the cloud sends requests for collaboration to all neighboring nodes within its domain to examine the possibility of providing a quicker response time, especially for time-sensitive services. It is worth noting that request-and-response times are considered part of service latency.

3.3   Types of Placements

Figure 3 illustrates both the **processing placement** and the **storage placement** that could be part of an IoT application business-model. In this figure, appropriate IT facilities run on top of clouds, fogs, and things to support the deployment of these 2 types of placements.

To work out and illustrate the details about **processing placement** and **storage placement**, we assume an IoT application that runs on top of a Business Process Management System (BPMS) coupled to a DataBase Management System (DBMS). Therefore, this application's business logic is designed as a set of Business Processes (BPs) that manipulate a set of DataBases (DBs). For the **processing placement**, we present some initiatives that examine **BP fragmentation** where each process fragment would be processed on top of either things (though unlikely), fogs, or clouds. For the **storage placement**, we present some initiatives that examine **data fragmentation** where each data fragment would be stored in things (though less likely), fogs, or clouds. Due to lack of space, only the **processing placement** is discussed.

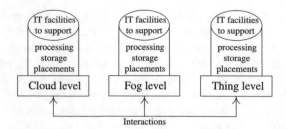

Fig. 3. Processing/Storage placements in an IoT application business-model

Processing placements identify physical locations where an IoT application's BP fragments would be deployed for processing. BP fragmentation has been the subject of many studies shedding light on its rationales, techniques, challenges, benefits, and consequences. Hereafter, we briefly present the works of Cheikhrouhou et al. [4] and Hou et al. [6]. The first two carry out BP fragmentation over clouds/fogs and clouds, respectively, and the last one carries out BP fragmentation in the context of IoT.

Saoussen et al. report that with the continuous advances in ICT and organizations' changing needs, cloud computing has shown some signs of "fatigue" when for instance, real-time applications call for almost zero time-latency. Transferring data to distant clouds is a potential source of delays and opens doors to unwanted interceptions. Luckily, fog computing addresses some clouds' concerns like latency and security. Building upon a previous approach to formally specify and verify cloud resources allocation to BPs using Timed Petri-Net (TPN), Saoussen et al. extended this approach by fragmenting and deploying free-of-violations time-constrained BPs in a mono-cloud and multi-fog context. They resorted to cloud-fog collaboration by verifying at both design-time and run-time where tasks should run (cloud, fog, or either) and where data should be placed (cloud, fog, or either). During BP fragmentation, Saoussen et al. defined $DD(data_i, t_i, t_k, h_i^s, h_k^s)$ that is data dependency between task t_i that produces $data_i$ and task t_k that consumes $data_i$ when t_i is executed in the host h_i^s and t_k is executed in another host h_k^s where a host could be either cloud or fog with respect to Fig. 3.

Hou et al. explore BP fragmentation for distribution purposes with emphasis on IoT-aware BPs. These BPs have to cope with the high volume and velocity of data that IoT generates and hence, need to be transmitted and processed timely. The authors propose a location-based fragmentation approach to partition a BP before applying Kuhn-Munkres algorithm so, that, an optimal deployment of BP fragments is achieved. Fragment collaboration takes place through a topic-based publish/subscribe infrastructure, which allows to reduce network traffic and save process execution-time.

3.4 System Implementation

This section demonstrates a Recommender System ($\mathcal{RS}$) that we developed to provide recommendations for **data flows** and **processing placement** as part of our

IoT application business-model in the context of cloud/fog. The $\mathcal{RS}$ is associated with a Deep Leaning (DL) model that takes as inputs a network topology's constituents (i.e., number of things, number of fog nodes, and number of cloud nodes) and criteria weights (Table 1), and produces as outputs data flows and processing placement recommendations.

Deep Learning Model. The core of our $\mathcal{RS}$ is a DL model that is built in MAT-LAB on Intel core i7-6700HQ, CPU 2.60 GHz, GPU GTX 1070, and RAM 32 GB. This model is a multi-layer classifier that extracts features from the weighted input criteria in order to produce suggestions labeled as highly-recommended.

During the implementation, a sequential development approach (also called *feed-forward* nn model) is adopted allowing each layer's output to be transferred to the next layer and so on. The DL model consists of 5 layers distributed over 1 layer for input, 1 layer for output, and 3 hidden layers for transformation having 20, 20, and 15 neurones, respectively. The hidden layers are used to avoid or help in preventing overfitting. It is worth noting that we experimented more/less number of layers and different number of neurons to get the best accuracy and faster processing time when providing recommendations.

The data used for training was generated by using the 6 criteria related to data flow namely, *frequency, sensitivity, freshness, time, volume,* and *critically.* Figure 4a shows the correlation matrix for the trained data based on these criteria as well as the possible data flows. Hence, the data flow classes/labels are Flow 1, Flow 2, Flow 3, and Flow 4 referring to T→C, T→F, T→C and T→F, and finally T→F and T→F→C, consecutively. From Fig. 4a, we note the data independent and not-related to each other, hence there will be no direct/common indications or features between input data and possible output recommendations. Before the training process, a sequence of operations have been applied to input data, such as manipulate the missing data by mean substitution and normalize the data by standard deviation.

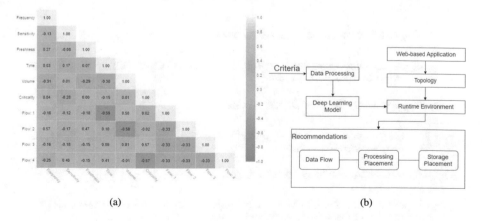

(a)   (b)

Fig. 4. Correlation matrix and recommendation generation process

Recommendation Creation. The process of generating a recommendation is presented in Fig. 4b. A Web-based system has been developed to allow a user to add or design a desired network topology required for an IoT application. Moreover, the IoT application's specifications (based on the 6 criteria presented in Table 1) can also be added to the topology and encapsulated in a JSON message. To this end, the system forwards JSON messages (includes both criteria and topology) to a pre-trained DL model to find the best fitting for data flows and processing placement. It is worth noting that the DL model combines topology and application's criteria (e.g., *frequency*, and *sensitivity*) in one request to recommend the best data flow to find the best routes to.

Results and Evaluation. After passing the topology and criteria to the DL model, the RS proceeds with predicting the best routes for data flows (i.e., T→F, T→C, etc.) and selecting the best potential neighbor node to receive the data. For instance, if there are 2 fog nodes that can be recommended, the RS will recommend either Fog_a or Fog_b based on distance (handled by the domains in the network topology) and send rate (e.g., frequency).

The RS output consists of a predication value for each of the classes representing our data flows, namely Flow 1, Flow 2, Flow 3, and Flow 4. The highest value of a class represents the most recommended class. For instance, for a topology that consists of 1 Cloud node to cover Fog and Thing nodes distributed over 3 *domains* with $domain_1$ having 1 Fog node and 2 Thing nodes, $domain_2$ having 2 Fog nodes and 1 Thing node, and $domain_3$ having 1 Fog node and 1 Thing node, the output recommendation for this topology is presented in Fig. 5. It is

* refer to prediction value of each class

OutputArg = *0.2850 *0.0210 *0.2080 *0.4860

T=>F & T=F=>C

Thing[thing_domain: 1 -- ID: 1 -- Frequency: 39 -- Sensitivity: 75 -- Freshness: 46 -- Time: 50 -- Volume: 98 -- Criticality: 36]
This Flow set to Fog Fog[fog_domain: 2 -- fog_id: 2] OR to Fog[fog_domain: 2 -- fog_id: 2]
Then to Cloud Cloud[cloud_domain: 1 -- cloud_id: 1]

OutputArg = *1.0268 *-0.0118 *-0.1129 *0.0979

T ==>C

Thing[thing_domain: 1 -- ID: 2 -- Frequency: 76 -- Sensitivity: 98 -- Freshness: 54 -- Time: 11 -- Volume: 91 -- Criticality: 27]
This Flow set to Cloud Cloud[cloud_domain: 1 -- cloud_id: 1]

OutputArg = *0.0482 *0.8700 *0.0452 *0.0366

T ==>F

Thing[thing_domain: 2 -- ID: 3 -- Frequency: 88 -- Sensitivity: 77 -- Freshness: 65 -- Time: 32 -- Volume: 31 -- Criticality: 38]
This Flow set to Fog Fog[fog_domain: 2 -- fog_id: 3]

OutputArg = *-0.0043 *-0.0035 *0.9974 *0.0105

T==>C & T==>F

Thing[thing_domain: 3 -- ID: 4 -- Frequency: 1 -- Sensitivity: 46 -- Freshness: 59 -- Time: 69 -- Volume: 78 -- Criticality: 74]
This Flow set to Fog Fog[fog_domain: 3 -- fog_id: 4] OR to Cloud Cloud[cloud_domain: 1 -- cloud_id: 1]

Fig. 5. Recommendation output

noted that each Thing node has it is own criteria that lead to different predictions/recommendatiosn for the data flow and processing placement. In this figure, the processing placement node is the first node in the recommended flow; for instance, in T→F→C flow, the Fog node (i.e., F) is for the processing placement.

4   Conclusion

This paper presented the design of a business model for IoT applications that would be deployed on top of cloud and fog resources. This business model features 2 constructs, flow (specialized into data and collaboration) and placement (specialized into processing and storage). The paper also presented the development of a system built-upon a deep learning model that recommends how the different flows and placements should be formed. These recommendations consider the technical capabilities of cloud and fog resources as well as the networking topology connecting these resources to things. In term of future work, we would like to complete the design of the current system by including the collaboration flow and storage placement and to examine the impact of other criteria like privacy on the collaboration flow.

References

1. Al-Khafajiy, M., Baker, T., Al-Libawy, H., Maamar, Z., Aloqaily, M., Jararweh, Y.: Improving fog computing performance via fog-2-fog collaboration. Future Gener. Comput. Syst. **100** (2019)
2. Al-khafajiy, M., Webster, L., Baker, T., Waraich, A.: Towards fog driven IoT healthcare: challenges and framework of fog computing in healthcare. In: Proceedings of ICFNDS 2018. Amman, Jordan (2018)
3. Bonomi, F., Milito, R., Natarajan, P., Zhu, J.: Fog computing: a platform for internet of things and analytics. In: Big Data and Internet of Things: A Roadmap for Smart Environments, Studies in Computational Intelligence. Cisco, Springer International Publishing (2014)
4. Cheikhrouhou, S., Kallel, S., Guidara, I., Maamar, Z.: Business Process specification, verification, and deployment in a mono-cloud, multi-edge context. Comput. Sci. Inf. Syst. **17**(1), 41 (2020)
5. Geissdoerfer, M., Savaget, P., Evans, S.: The cambridge business model innovation process. In: Proceedings of GCSM 2017, Stellenbosch, South Africa (2017)
6. Hou, S., Zhao, S., Cheng, B., Cheng, Y., Chen, J.: Fragmentation and optimal deployment for IoT-aware business process. In: Proceedings of SCC 2016, San Francisco, USA (2016)
7. Khebbeb, K., Hameurlain, N., Belalab, F.: A Maude-based rewriting approach to model and verify cloud/fog self-adaptation and orchestration. J. Syst. Archit. **110**, 101821 (2020)
8. Li, F.: The digital transformation of business models in the creative industries: a holistic framework and emerging trends. Technovation **92-93** (2020)
9. Maamar, Z., Baker, T., Faci, N., Ugljanin, E., Al-Khafajiy, M., Burégio, V.: Towards a seamless coordination of cloud and fog: illustration through the Internet-of-Things. In: Proceedings of SAC 2019, Limassol, Cyprus (2019)

10. Mahmud, R., Ramamohanarao, K., Buyya, R.: Application management in fog computing environments: a taxonomy, review and future directions. ACM Comput. Surv. **53**(4) (2020)
11. Nieves, E., Hernández, G., Gil González, A., Rodríguez-González, S., Corchado, J.: Fog computing architecture for personalized recommendation of banking products. Expert Syst. Appl. **140** (2020)
12. Satyanarayanan, M., Bahl, P., Cáceres, R., Davies, N.: The case for VM-based cloudlets in mobile computing. IEEE Pervasive Comput. **8**(4) (2009)
13. Thekkummal, N., Jha, D., Puthal, D., James, P., Ranjan, R.: Coordinated data flow control in IoT networks. In: Proceedings of ALGOCLOUD 2019, Munich, Germany (2019)
14. Varghese, B., Wang, N., Nikolopoulos, D., Buyya, R.: Feasibility of fog computing. arXiv preprint arXiv:1701.05451 (2017)
15. Wang, T., Zhang, G., Alam Bhuiyan, M., Liu, A., Jia, W., Xie, M.: A novel trust mechanism based on fog computing in sensor-cloud system. Future Gener. Comput. Syst. **109** (2020)
16. Yahya, F., Maamar, Z., Boukadi, K.: A multi-criteria decision making approach for cloud-fog coordination. In: Proceedings of AINA 2020, Caserta, Italy (2020)

Network Architecture for Agent Communication in Cyber Physical System

Masafumi Katoh[1][(✉)], Tomonori Kubota[1], Akiko Yamada[1], Yuji Nomura[1],
Yuji Kojima[2], and Yuuichi Yamagishi[2]

[1] Fujitsu Limited, Kamikodanaka 4-1-1, Nakahara-ku, Kawasaki 211-8588, Japan
{katou.masafumi,kubota.tomonori,akikoo,nomura.yuji}@fujitsu.com
[2] Fujitsu Limited, Kokura 1-1, Saiwai-ku, Kawasaki 211-0031, Japan
{kojima.yuuji,yamagishi.yuuic}@fujitsu.com

Abstract. Cyber-Physical System (CPS) will gradually be positioned as a social platform due to the progress of Internet of Things (IoT), in which various entities can logically be connected. The base of CPS is the characterization of a physical entity as digital data so that a computer can analyze. In this paper, data for characterizing a person or a machine are categorized into the profile and the context of the subjective entity. Then, we propose to introduce agents figuring out the profile and the context. This paper discusses the network architecture for agent communication. First, we compare topological models that agents exchange static profile to specify agents to be associated by sharing events. Next, we discuss how to isolate network bandwidth which carry mixture of the static profile and the dynamic context for a human-type or a machine-type terminal. Finally, we discuss the physical processing entity for agent program.

1 Introduction

Combining the Web with machine to machine (M2M) [1] or Internet of Things (IoT) [2], sensing data on the physical world can be applied for various context-aware applications [3, 4]. Such trend suggests a Cyber-Physical System (CPS) will gradually be positioned as a social platform, in which various entities can instantly be connected without any physical constraints such as the distance or the weight.

The key of CPS is characterizing a person and a machine as structured data such as profile and context so that Information Technology (IT) can treat. We have proposed before to introduce the personal agent to find out a valuable service in place of the client by knowing the profile and context of the objective person [5]. Now, we challenge to extend the personal agent to the general agent, i.e., for a machine as well as a person. Thereby, every entity can widely overview physical world, find out partners or friends and instantly associate entities by sharing event data.

In Sect. 2 of the paper, we show the characterization of a person and a machine as the profile and the context. Section 3 reviews the trend on applications of context awareness and agent as relevant studies. Section 4 describes our premises called "two-stage filtering" that a transmitting entity knows the profiles of receiving entity by matching both profiles. Section 5 discusses the network architecture for the two-stage filtering by agent

L. Barolli et al. (Eds.): AINA 2021, LNNS 226, pp. 187–199, 2021.
https://doi.org/10.1007/978-3-030-75075-6_15

communication. Section 6 discusses how to isolate the bandwidth to carry static profile and dynamic event for human-type and machine-type devices. Section 7 we discuss the physical processing entity to perform agent program.

2 Characterizing Physical Entity by Profile and Context

We categorize subjective entities on physical world into human and machine. Similarly, we categorize terminal devices for communication into human (H)-type and machine (M)-type [6]. H-type device such as a PC or a mobile phone has interface for the user to input or output data. Reversely, M-type device such as sensor or actuator automatically works by itself without human intervention.

Our prime view is that the entities on the physical world are characterized by static profile data and dynamic context data [5]. The profile is defined as the abstract data to characterize an objective person or a machine. The context is defined as the condition or background, which could impact on the behavior of the objective person or machine. Furthermore, we define an event as the change of context.

Table 1 shows the examples of the profile and the context of person, machine and service. Furthermore, Table 2 shows the examples of profile data and event data to model their traffic aspect. The profile is relatively static, but the context is more dynamic. That is, the occurrence of event is more often than that of profile. For example, the career might be obtained regularly for more than 1 year. In the contrast, the context always changes such as the weather and the location of a mobile object.

Table 1. Example of characterizing person and machine

	Characterization of entity (Profile)	Characterization of environment (Context)
Person	Gender, Age, Family, Nationality, Address, Career, Concern, Mindset	Dialogue, Accompanier, Reading, Smell, Mission, Deadline, Time, Place, Weather
Car	Manufacture, Production date, Owner, Engine, Type, Size, Color	Driver, Passenger, Goal, Condition of traffic and road, Time, Place, Weather
PC	Manufacture, Owner, Type, Processor, OS, Storage, Application	User, Utilization of IT resource, Running tasks, Network condition, Time, Place
Content	Producer, Data volume, Copyright	Demand, ICT environment, Audience

On the other hand, current communication systems are positioned to be designed for a transmitter who already recognizes the existence and the access identifier (ID) of the targeted entity. If people are not sure about the existence or don't know the access ID, they normally use a search engine to get the Uniform Resource Identifier (URI) of the target. However, it is too laborious for people to find out a valuable data source in the huge cyber world due to the constraints of human perception capability [7]. Rather, we'd better build a networking framework, by which we can reach the target without knowing the existence and the access ID. In other words, it is required that every entity can link with others just by specifying characteristics, i.e., profile of the target, for example, by the means of Information/Attribute Centric Network [8].

Table 2. Traffic aspect of profile data and event data

	Example	Occurrence rate	Data length
Profile traffic	Gender, Nationality	(Constant)	Hundreds word in total
	Age, Career, Expertise	Every year	
	Concern, Interested topic	Every month	
	Research item	Every week	
Event traffic	Meeting subject	Every hour	Tens word
	Sensing data on nature	Every minute	
	Place of mobile object, Log	Every second	

3 Related Works

The trend of applications on context and agent are reviewed as the related works.

Context awareness is positioned as an important feature of IT since people unintentionally move under various circumstances [2, 3]. The typical application of context awareness is dynamic resource allocation by referring to users' status and network environment [9] or the service customization depending on personal condition such as user's place [10]. Since the significance of context depends on an individual person, the applications by combining the dynamic context with the static profile have been proposed. For example, a recommendation system [11] or search engines [12, 13] can be customized by referring to the user's profile. Furthermore, [14] has verified that understanding of profile and context is key to activate human communication.

On the other hand, a lot of studies on the agent have already been performed from various viewpoints [15]. The typical application is personal assistance [16]. A personal agent, which plays roles to cyclically transform between tacit and explicit knowledge, has been proposed [17, 18]. The usage of software agent expands still now such as "digital twin", which bridges between physical and cyber worlds [19]. Chat bots based on Artificial Intelligence are implemented in the various devices such as a speaker and a mobile phone [20, 21]. The functional architecture of management for multi-agent system has been shown [22, 23]. Another typical application of agent is for system management. Most of them aim to overcome the weak point of a central management based on Simple Network Management Protocol by mobile agents [24–26]. They have stood on the feature that the mobile agent can move to the entity having the raw data, i.e., vast data must not be moved [27]. However, network architecture for agent communication have not been studied enough.

So, our paper focuses on the network architecture for agent communications.

4 Premise: Two-Stage Filtering

We stand on the premise that the transmitting entities of events should know the profiles of receiving entities to specify which entities could be interested in the event. Reversely, the receiving entities should know the profiles of transmitting entities to specify which could send interesting events. Nevertheless, the transmitting entity might send the event to unspecified entities. If so, events are overflown in the network, and they might not be interesting to most receiving entities. It is clearly not efficient.

Reminding that the occurrence rate of events is much larger than that of profiles, it is allowable that profiles are exchanged among all entities to reduce event traffic by specifying entities to share the event. This suggests two-stage filtering is effective.

On the other hand, data processing capabilities of both person and machines are limited [7]. That is, it is not realistic for them to check all profiles of others or to detect all events. So, we have proposed to introduce a software agent in a network [5, 23]. Figure 1 shows the proposed message flow called "two-stage filtering". First, the transmitting agent focuses narrowly on receiving agents to be associated by matching both profiles. The agent always monitors the context around its entity, detects the change as the event and sends the detected event to only associated agents. Second, the receiving agent filters out the received event by referring the receiver's context. It is noted that Fig. 1 doesn't imply the computing entity on physical world.

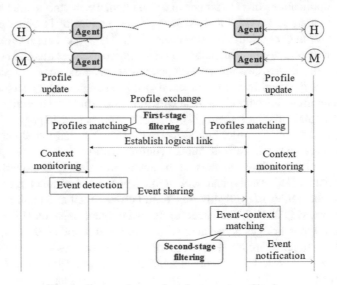

Fig. 1. Proposed procedure for two-stage filtering.

5 Networking Architecture for Two-Stage Filtering by Agents

5.1 Models of Agent Communication for Sharing Profile and Context

We classified the networking of agents into 4 models as shown in Fig. 2. Model (a) is shown as a reference to compare others with two-stage filtering. In model (a), a transmitter's agent sends an event to all unspecified agents without checking the receivers' profile. So, the informed events might not be interesting to most receivers.

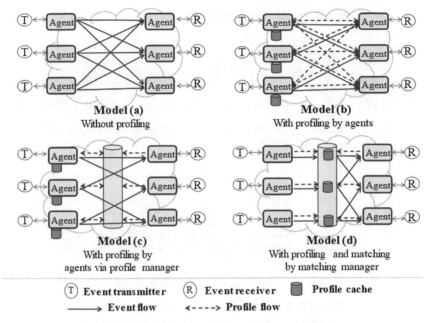

Fig. 2. Models of networking for two-stage filtering.

In models (b), (c) and (d), agents perform two-stage filtering through their communication. That is, an event transmitter's agent sends an event to agents of specified receivers who want to find the event. So, in these models, the transmitter's agent must check receivers' profiles. The difference of (b), (c) and (d) is how to exchange profile and event. Agents in model (b) directly exchange profiles each other, reversely agents in model (c) communicate via single entity named "profile manager". In model (d), there is a central entity named "matching manager" which not only keeps profiles but also matches all pairs of transmitters' and receivers' profiles.

5.2 Evaluation of Event Transfer Delay for 4 Models

We compare event transfer delays of 4 models as a performance evaluation. In every model, delay for event transfer could happen at both the transmitting point and the receiving point. The transmitting agent or the matching manager must send an event to a lot of receivers, so, it takes time to finish sending the event to all receiving agents. On the other hand, the receiving agent must simultaneously receive events from many transmitting agents, so, it could cause queuing delay.

Figure 3 shows the quantitative tendency of event transfer delay of 4 models. Let the number of receivers or transmitters be R or T, respectively. Furthermore, let the ratio of associated receivers or transmitters to total be α or β, respectively.

An average delay at the transmitting agent is proportion to the number of receivers independent of the status of the receivers. That is, it is $R/2$ in model (a), $\alpha R/2$ in model (b) or (c), $\alpha R \beta T/2$ in model (d), respectively. So, the transmitting delay of (b) or (c) is better than (a) and (d).

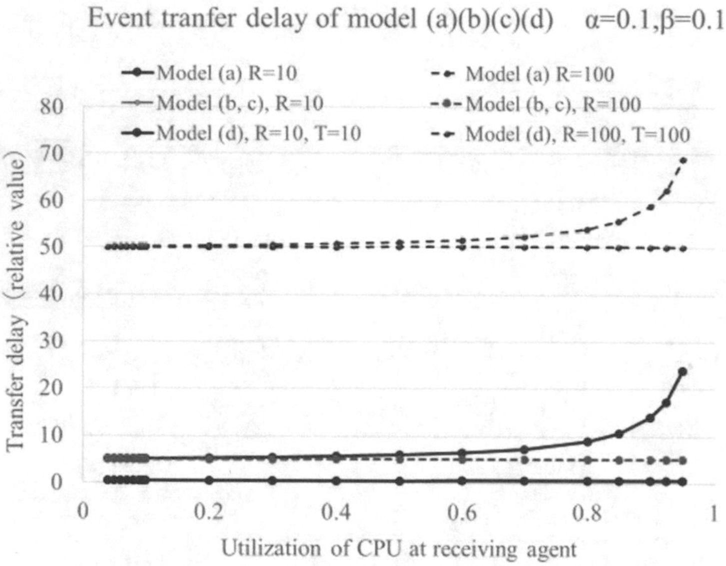

Event tranfer delay of model (a)(b)(c)(d)   α=0.1,β=0.1

Fig. 3. Comparison of event transfer delay for 4 models.

On the other hand, a queuing delay at the receiving agent depends on the utilization of the receiving entity. Here, we assume that the queuing behavior is followed by the classical queuing model M/M/1[28]. The numbers of transmitting agents, which could send events simultaneously to the receiving agent, are T in model (a) and βT in model (b), (c) or (d). Therefore, queueing delay of model (a) increases steeply if the utilization of receiving agent approaches to 1.0. Reversely, the queuing delay of model (b), (c), (d) is negligibly small.

As a result, the models (b) and (c) adopting two-stage filtering is advantageous from the viewpoint of real time event transfer.

5.3   Evaluation of Delay for Profiling in Models (b), (c), (d)

We compare the delays for profiling in models (b), (c) and (d). In model (b) or (c), replied profiles from R receiving agents could concentrate at the transmitting agent. In the contrast, in model (d), the matching manager must match all combinations of profiles for T transmitters and R receivers. That is, the matching load of (d) at the matching manager is T times of that at a transmitting agent of model (b) or (c). Therefore, the profiling delay of model (d) is larger than others unless the matching manager adopts a processor with high performance.

Finally, let us compare the characteristics of profiling delay in models (b) and (c). A transmitting agent gets profiles for R times from R agents in model (b). In the contrast, the agent gets all profiles at one time from the profile manager in model (c). So, the variance of response time to the transmitting agent is different, even though the total volume of profile traffic for a long period is the same. For the evaluation of delay characteristics, the probability distribution of waiting time M(t) is assumed as $M(t) = a e^{-(1-a)t/h}$ based on M/M/1 model. Where, "*a*" is the utilization of the agent, and "*h*" is the mean service time.

Figure 4 shows 90%-tile queuing delay for profiling in models (b) and (c). In model (b), it is calculated such that M(t) is less than a threshold for consecutive *R* times. If the number of receivers *R* becomes large, the delay in model (b) becomes much larger than that of model (c). As a result, we conclude that model (c) is the most reasonable architecture for sharing profiles and events, i.e., two-stage filtering.

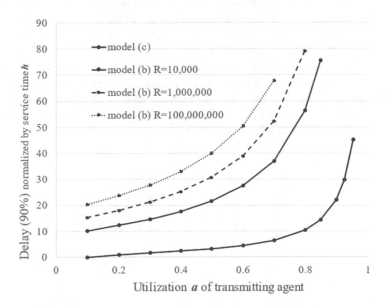

Fig. 4. Comparison of profiling delay in models (b) and (c).

6 Network Design for Agent Communication

In agent communications for CPS, there are two-types of traffic such as profile and event, and two-types of requirements such as for M-type and for H-type device. This section generally discusses how to isolate the network resource for mixture of different traffic. Concretely, we compare two strategies on the usage of resource.

- Hard-isolation: The bandwidth is divided for different types of traffic or requirement.
- Without isolation: A bandwidth is shared for different types of traffic or requirement. Priority queuing is included in this category as a soft-isolation.

6.1 How to Carry Mixture of Static Event Traffic and Dynamic Profile Traffic?

We compare queuing delay for two strategies to carry static profile and dynamic event. Queuing delay of traffic whose interarrival time is modelled as an exponential distribution with a unique mean value can be analyzed by M/M/1. So, in the case of the hard-isolation, each queuing delay of profile or event is calculated by M/M/1. However, if mixture of traffic with different occurrence rates are conveyed in a common resource, the variance of interarrival time is larger than that of exponential distribution. So, we calculate queuing delay by $H_2/M/1$ model [27] which assumes a balanced hyper-exponential as interarrival in the common resource (see Appendix).

Figure 5 shows the typical tendency on queuing delay. Here, the bandwidth without isolation is assumed to be equal to the sum of two bandwidths in the case of hard-isolation. Furthermore, in the case of hard-isolation, the ratio of bandwidth is equal to the ratio γ of profile occurrence rate to event occurrence rate such as 0.1 or 0.01. The gap on occurrence rates corresponds the gap of the coefficient of variance (CoV) such as 2.25 or 7.07. It is clear that queuing delay in the case of bandwidth sharing is larger than that in the case of hard-isolation. The difference becomes large when the gap of interarrival times is large. Furthermore, the networking topology of profile and event in model (c) is different as shown in Fig. 2. As a result, we conclude that the profile and the event should be isolated by different network slices.

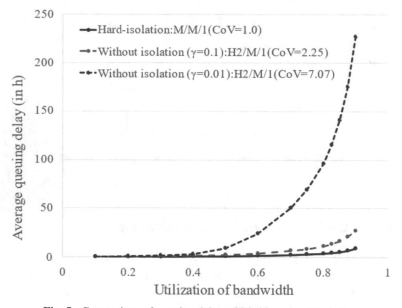

Fig. 5. Comparison of queuing delay with/without hard-isolation.

6.2 How to Carry Event to Human-Type Device or to Machine-Type Device?

The granularity on time recognition by a machine is quite different from by a person. People ordinally recognize second order. On the other hand, a computer can recognize nano- or micro-second due to the CPU whose clock frequency is above GHz, and might not be patient to wait for 1 mill-second even though people don't feel to be waited. So, events for M-type devices should be carried with high priority [24].

If data length or occurrence rate widely diverges, the bandwidth should be isolated [9] as discussed in Sect. 6.1. Rather, the priority control in the common bandwidth seems being more efficient than the hard-isolation since the event traffic itself is identical even though the requirement is different.

Figure 6 shows the quantitative tendency on queuing delay in soft and hard isolation, where the total event load is assumed to be 90% of the bandwidth.

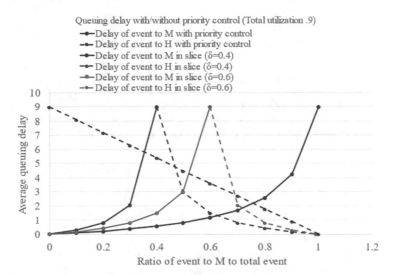

Fig. 6. Comparison of queuing delay with or without priority control.

The horizontal axis represents the ratio of event for M-type terminal to the total. Every straight or dashed line represents the queuing delay for events for M-type or for H-type, respectively. When the ratio of events for M-type increases, the delay of events for M-type increases, but the delay for H-type decreases. Both delays in the case of the soft-isolation are varied within the range which the system is stable. In the contrast, if 40% of the bandwidth is allocated for M-type as shown in red line, but the actual ratio is over 40%, the queuing delay becomes infinitively large and the system is unstable. So, the soft-isolation is more robust and preferable for the mixture of requirements such as to delay-sensitive M-type and delay-tolerant H-type device.

7 Processing Entity for Software Agent

Physical processing entity doesn't matter in the previous discussions. However, agent must actually be performed on a processor. So, the mapping between logical agent and physical processing entity is an important issue. The alternatives of the processing entity are 1) Terminal Device, 2) Gateway between wireless and wired network, 3) Gateway between mobile network and the Internet, 4) Network nodes in the Internet including a Data Center as shown in Fig. 7.

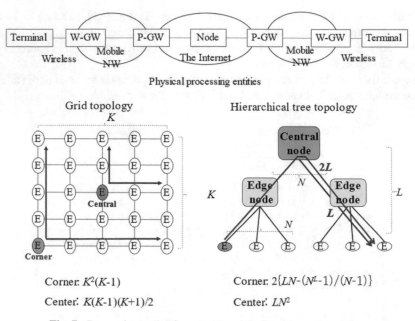

Corner: $K^2(K-1)$ Corner: $2\{LN-(N^L-1)/(N-1)\}$

Center: $K(K-1)(K+1)/2$ Center: LN^2

Fig. 7. Processing entity for agent to minimize propagation delay.

Since the performance depends on the topological position of the processing entity, we discuss the suitable position from the view point of traffic. There are two patterns of traffic flow, that is, between a terminal and an agent and between agents. Since the former is related to the individual entity such as the personal profile, the traffic volume is relatively small. Reversely, since the traffic between agents is for sharing among agents such as the profiles of all entities, the volume is vast. Therefore, we discuss traffic between agents depending on the position of processing entity.

Figure 7 shows models on network topology and formula on the total number of hops for data sharing from the corner with from the central position. Basically, the total number of hops from central position to others is about half of that from the corner.

In the model (c), profiles are exchanged among all agents via the profile manager, and events are directly exchanged among associated agents. Then, the profile manager should be deployed at the central position among all agents to minimize the total propagation delay. Similarly, event delivering function should be deployed at the central position among receiving entities. Therefore, when the agent tends to deliver an event to multiple associated agents, it should move to the central position among them, or ask another processing entity at the central position to deliver the event.

8   Conclusion

Cyber-Physical System (CPS) will gradually be positioned as a social platform due to the penetration of Web and IoT. The key of CPS is the characterizing a person and a machine on physical world as structured data for computing. Then, we categorize characterizing data into the profile of the subjective entity and a context around the entity.

The transmitting or receiving entity should specify receiving or transmitting entities respectively so that the event impacts on receiving entities. Then, we have proposed to have agents perform two-stage filtering to share events among entities to be associated. That is, the transmitting agent focuses narrowly on receiving agents by matching their profiles, and the receiving agent filter informed events based on the receiver's context. Then, we have modelled networking topology for two-stage filtering by agent communication. The suitable architecture is such that all agents exchange profiles via a profile manager and directly send events to associated agents.

Next, the usage of network bandwidth for two-stage filtering was discussed. We have verified that (1) the hard-isolation of bandwidth for mixture of static profile and dynamic event is required, and (2) a priority queuing in soft-isolation of the bandwidth is effective to instantly carry events to delay sensitive machine-type terminals.

Finally, the mapping between a logical agent and physical processing entity was discussed. We have clarified that the agent should move to the central position among associated agents to minimize propagation delay when it shares the event to them.

Acknowledgments. We'd like to extend a special thank you to Dr. Toshitaka Tsuda and Mr. Makoto Murakami for exciting discussion and kind encouragement.

Appendix: Deviation of Queuing Delay on $H_2/M/1$ Model

This is the appendix on calculation of queuing delay in tarrfic mix with different interarrival time in Sect. 6.1. We assume that data interarrival times are independent. Therefore, the queuing behavior can be analyzed by using GI/M/1 model. Let $h(= \mu^{-1})$ and M(t) be the service time and the probability distribution of waiting time, respectively. The probability that waiting time is larger than t is given by the following [28].

$$M(t) = \omega e^{-(1-\omega)\mu t}. \tag{1}$$

Where, ω satisfies the following equation

$$\omega = \delta[(1 - \omega)\mu], \ 0 < \omega < 1. \tag{2}$$

$\delta(x)$ is a Laplace-Stieltjes transform of the probability distribution of data interarrival times $D(x)$.

If the mixture of traffic with different occurrence rates such as profile and event in Table 2 are conveyed in a common bandwidth, the variance of interarrival time is larger than that of exponential distribution. So, we assume that $D(x)$ is modelled as the hyperexponential of order 2, which two exponential distributions are stochastically swiched as the following.

$$D(x) = p_1\left(1 - e^{-\lambda_1 x}\right) + p_2\left(1 - e^{-\lambda_2 x}\right), \lambda_1 > 0, \lambda_2 > 0. \tag{3}$$

$$\delta(S) = p_1\lambda_1/(\lambda_1 + S) + p_1\lambda_2/(\lambda_2 + S), \ p_1 + p_2 = 1, \ p_1 > 0, p_2 > 0. \tag{4}$$

Let us think about a balanced hyper-exponential distribution to simplify the discussion. That is,

$$p_1 / \lambda_1 = p_2 / \lambda_2 \tag{5}$$

The coeffecent of variation (CoV) of interarrival time is as follows.

$$\text{CoV} = \left\{ \left(2p_1(1 - p_1)\right)^{-1} - 1 \right\}^{0.5}. \tag{6}$$

In order to evaluate queuing behavior by using Eq. (1) of GI/M/1, it is necessary to get ω. In the case of interarrival times described by the balanced hyper-exponential, i.e., H_2/M/1 model, the following equation of second degree is derived by substituting Eqs. (4) and (5) into Eq. (2).

$$\omega^2 - (2a + 1)\omega + 4p_1(1 + p_1)a^2 - 2\left\{p_1^2 + (1 - p_1)^2\right\}a = 0. \tag{7}$$

$$a = (p_1 / \lambda_1 + p_2 / \lambda_2)^{-1} \cdot h = (2p_1 / \lambda_1) \cdot h. \tag{8}$$

Then, ω is got by resolving ω $(0 < \omega < 1)$ of Eqs. (7) and (8). Thereby, the queuing delay can be evaluated by substituting the ω into Eq. (1) as shown in Fig. 5.

Incidentally, when $p_1 = p_2 = 0.5$, the interarrival time is described by just an exponential distribution. That is, ω is equal to a by Eqs. (7) and (8). So, the probability function of waiting time M(t) is that of M/M/1 model.

References

1. oneM2M TS 0001 "oneM2M Functional Architecture Baseline Draft" 30 June 2014
2. IEEE communication society "COMSOC 2020 Report," pp. 59–62, December 2011
3. Wiser, M.: "The Computer for the 21st Century" Mobile Computing and Communications Review, vol. 3, no. 3 (1991)
4. Yamaoka, H., et al.: Dracena: a real-time IoT service platform based on flexible composition of data streams. In: 2019 IEEE/SICE International Symposium on System Integrations (SII 2019) (2019)
5. Katoh, M., et al.: Proposal of a personal agent for human centric information networking: collaboration of human profile and human context. In: 2018 International Conference on Information Networking (ICOIN) (2018)
6. Katoh, M., Sato, I., Fukuda, K.: Applicability of a layered architecture adopting distributed data processing to M2M communications. In: 2015 Eighth International Conference on Mobile Computing and Ubiquitous Networking (ICMU) (2015)
7. Card, S.K., Moran, T.P.: The Psychology of Human-Computer Interaction (1983)
8. Kurita, T., Sato, I., Fukuda, K., Tsuda, T.: An extension of information-centric networking for IoT applications. In: 2017 International Conference on Computing, Networking and Communications (ICNC) (2017)
9. Katoh, M., Tajima, Y., Senoo, H., Kimura, D.: Context aware cell selection in heterogenious radio access environment. In: 2017 31st International Conference on Advanced Information Networking and Applications Workshops (WAINA) (2017)
10. Iida, I., Morita, T.: Overview of human-centric computing. FUJITSU Sci. Tech. J. **48**(2), 124–128 (2012)

11. Cantador, I., Bellogin, A., Castells, P.: Ontology-based personalizen and context-aware recommendation of news item. In: 2008 IEEE/WTC/ACM International Conference on Web Intelligence (2008)
12. Gajendragadkar, U., Joshi, S.: User intended context sensitive mining algorithm for search string composition. In: 2015 IEEE/WTC/ACM International Conference on Web Intelligence (2015)
13. Sieg, A., Mobasher, B., Brule, R.: Ontological user profiles for representing context in web search. In: 2007 IEEE/WTC/ACM International Conference on Web Intelligence (2007)
14. Katoh, M., Suga, J., Kojima, Y., Kawai, M.: Application of human profiling by agent for activating human communication. In: The 15th International Conference on Networking and Services (ICNS 2019) (2019)
15. Kurihara, S.: Artificial intelligence. In: IEICE 100th Annalistic publication Section 2, vol. 6, pp. 388–389 (2017) (in Japanese)
16. Huang, T.-C., Yang, C.-S., Bai, S.-W., Wang, S.-H.: An agent and profile management system for mobile users and service providers. In: AINA 2003 (2003)
17. Ismail, S., Ahmad, M.S., Hassan, Z.: Regression analysis on agent roles in personal knowledge management processes. In: 2013 8th International Conference on Information Technology in Asia (CITA) (2013)
18. Zhang, C., Tang, D., Liu, Y., You, J.: A multi-agent architecture for knowledge management system. In: Fifth International Conference on Fuzzy Systems and Knowledge Discovery (2008)
19. Assawaarayakul, C., Srisawat, W., Ayuthaya, S.D.N., Wattanasirichaigoon, S.: Integrate digital twin to exist production system for industry 4.0. In: 2019 4th Technology Innovation Management and Engineering Science International Conference (2019)
20. https://www.nttdocomo.co.jp/service/mydaiz/function/ Accessed April 2019. (in Japanese)
21. Hill, J., Ford, W.R., Farreras, I.G.: Real conversations with artificial intelligence: A comparison between human–human online conversations and human–chatbot conversations. Comput. Hum. Behav. **49**, 245–250 (2015)
22. Abeck, S., Koppel, A., Seitz, J.: A management architecture for multi-agent systems. In: Proceedings of the IEEE Third International Workshop on Systems Management, Newport, RI, USA (1998)
23. Katoh, M., Kubota, T., Yoshida, E., Kojima, Y., Yamagishi, Y.: Proposal of profile and event sharing by agent communication. In: 2020 23rd Conference on Innovation in Clouds, Internet and Networks and Workshops (ICIN) (2020)
24. Katoh, M., Kubota, T., Yasuie, T., Watanabe, Y., Nomura, Y.: Proposal of management agent for delay sensitive IoT communication. In: 2021 International Conference on Information Networking (ICOIN), Jeju Island, Korea (South) (2021)
25. Kazi, R., Morreale, P.: Mobile agents for active network management. In: MILCOM 1999. IEEE Military Communications. Conference Proceedings, Atlantic City, NJ, USA (1999)
26. Guo, F., Zeng, B., Cui, L.: A distributed network management framework based on mobile agents. In: 2009 Third International Conference on Multimedia and Ubiquitous Engineering, Qingdao (2009)
27. Cao, J., Zheng, G., Wang, L.: Research on network fault diagnosis based on mobile agent. In: 2010 International Conference on Networking and Digital Society, Wenzhou (2010)
28. Cooper, R.B.: Introduction to Queuing Theory, New York (1972)

Fi-SWoT: Secure Federated Service Oriented Architecture for the Semantic Internet of Things

Euripides G. M. Petrakis[✉], Ilias Kontochristos, and Karina Rastsinskagia

School of Electrical and Computer Engineering, Technical University of Crete (TUC),
Chania, Greece
petrakis@intelligence.tuc.gr, {ikontochristos,krastsinskagia}@isc.tuc.gr

Abstract. Fi-SWoT is a Federated Service Oriented Architecture (SOA) for the Semantic Web of Things (SWoT). The federation comprises independent and equipotent IoT systems that interact with each other. Fi-SWoT brings together ideas from Semantic Web, context information management, SOA and IoT systems design and, aspires to become a Future Internet (FI) architecture that combines the advantages of all: machine understandable and interoperable devices and services, interconnection and transparency of services and systems in the network. Devices and services of any member of the federation can be discovered and used by users or services over the Web. Fi-SWoT implements advanced system security by means of HTTPS, state-of-the-art user authentication and authorization based on user roles and access policies and, Single Sign-On (SSO) functionality. Fi-SWoT has been implemented in the Google Cloud Platform and evaluated by stressing the system with many thousands of requests.

1 Introduction

The need for federated IoT architectures design has been highlighted in recent reports [7]. The challenge is to provide unified access to devices and services from diverse IoT systems covering (altogether) much wider areas of interest than a single system alone. The Web of Things (WoT) [6] initiative aims at unifying the world of interconnected devices (*Things*) over the Internet. Each device should be published on the Web, be discovered by Web search engines and be reused in applications. The WoT model of W3C [18] defines the requirements for implementing Things as Web objects accessible on the Web. A Thing is described by a JSON object representing its identity (i.e. a URI), properties, purpose (i.e. functionality) and state information and exposes a REST API interface to the Web. Ideally, this requires that Things receive (each one) an IPv6 address and have a Web server installed. However, this is not always possible especially for resource constraint devices. To enable Web functionality for Things, a solution is to deploy a Web Proxy, that runs on a server (or on a gateway) and keeps the virtual images of all Things (i.e. their JSON representation). The Semantic Web

L. Barolli et al. (Eds.): AINA 2021, LNNS 226, pp. 200–214, 2021.
https://doi.org/10.1007/978-3-030-75075-6_16

of Things (SWoT) [5, 12] is the semantic extension of WoT that allows devices and services to become machine understandable and interoperable and be reused in new applications.

The WoT model defines the means for identifying and connecting with Things on the Web (via Web proxies) but, it lacks a mechanism for making them machine understandable. A solution to this problem is to represent Things as Semantic Web Things by converting their JSON descriptions to JSON-LD [8]. Thing descriptions can be de-referenced by associating them with external ontology concepts. Then, the Semantic Web Things can be manipulated using Semantic Web tools such as the SPARQL query language and ontology reasoners. A problem inherent to this solution is that ontologies cannot meet the Quality of Service (QoS) demands of time critical applications [1, 4]; Another is that ontologies are not yet well received by the industry due to their complexity and poor runtime performance (especially in the IoT domain where the amount of data becomes big). In particular, completeness of the representation relates with ontology size and complexity in such a way that, completeness and runtime performance are traded-off.

For referencing to common concepts in the IoT domain, Semantic Sensor Network ontology (SSN) [15] provides definitions for Things (e.g. sensors) and observations (i.e. measurements). Depending on application needs, more general purpose or application specific ontologies can be used as well (e.g. the GEO ontology[1] of W3C for spatially located information). A recent contribution [3], apart from a comprehensive survey of the SWoT ontology domain, proposes ontology patterns for representing dynamic interactions between devices (e.g. their evolution in time). Solutions to the problem of scalability and performance resort to lightweight ontologies such as IoT-Lite [4] or, the Sensor Domain Ontology (SDO) [1]. Both are defined as subsets of SSN and promise sensor data annotation and instantiation in real time. iSWoT [5] suggests decomposing the ontology into parts thereby limiting the size of the ontology graph on which high complexity operations (e.g. reasoning) are applied.

Fi-SWoT combines the advantages of federated and SWoT architectures and aspires to become the first federated IoT architecture incorporating the following desirable characteristics: conformance with state-of-the-art technologies (e.g. SOA design, services virtualization) and W3C standards and, security by design. Fi-SWoT provides (a) advanced protection of the communication over public networks using HTTPS and, (b) protection of all services from un-authorized access and Single Sign-On[2] (SSO) functionality. Fi-SWoT is designed for organizations willing to share information with other organizations for a common objective (e.g. weather stations ally together to create weather forecasts for the whole country, small businesses share content in order to gain deeper insights on business data or design improved business policies). Information on each organization remains under the control of its owner.

---

[1] https://www.w3.org/2003/01/geo/.

[2] https://auth0.com/docs/sso.

Referring to previous work by the authors, iXen [10] introduces the concept of three layered IoT architecture design based on stakeholders interests and the most recent standards of context information management; it is neither federated nor WoT nor SWoT architecture. iSWoT [5] extends the functionality of iXen with WoT and SWoT characteristics. iZen [19] is a federated master-less p2p eco-system comprising equipotent IoT systems (nodes) but, has no WoT or SWoT characteristics. iZen, the same as iSWoT, is a standalone architecture.

The Fi-SWoT prototype is deployed on Google Cloud Platform (GPC) and is tested in a smart city scenario. The scenario is a smart application for cities and households in order to provide useful information to citizens for planning their daily activities and to city authorities for taking decisions (e.g. dealing with city conditions or for improving the quality of life of the citizens). Fi-SWoT is stressed with high data streams and many simultaneous requests addressing data, services and devices in local and foreign nodes. Fi-SWoT is capable of responding in real-time under heavy work loads (i.e. many users issuing multiple requests concurrently).

Issues related to Fi-SWoT design and implementation are discussed in Sect. 2 and in Sect. 3 respectively. Reinforcing overall system security, as well as user authentication, authorization and Single Sign-On (SSO) functionality are discussed in Sect. 4 and Sect. 5. Evaluation results are presented in Sect. 6 followed by conclusions and issues for future research in Sect. 7.

2   Fi-SWoT Federation

Fi-SWoT features an elaborate 3-tier architecture design model. Each layer implements functionality addressing the needs of different user types (stakeholders) namely, administrators, infrastructure owners (i.e. device owners), application owners (i.e. applications developers) and customers (i.e. end-users) who subscribe to applications. The same user may have more than one role. Infrastructure owners are entitled to install and make devices available to application owners who in turn, subscribe to devices in order to create applications; customers subscribe to applications but have no authority to make changes to the system (e.g. to add or remove devices or create applications). Finally, administrators are responsible for installing services, creating user accounts, assigning access rights to users based on roles (i.e. user types) and for monitoring all system operations. Except their competence in controlling the system, they are entitled to perform operations as any other user type (e.g. install or remove devices, applications etc.). Leveraging a 3-layered design, Fi-SWoT is ready to incorporate a business logic (e.g. billing policies) for different types of users and become self-sustainable. All users may benefit from their participation in Fi-SWoT based on their offerings or be charged based on a pay-per-use cost model (left as future work).

Figure 1 illustrates a Fi-SWoT 3-tier system structure comprising three independent systems (nodes) and their interaction. The original iSWoT architecture [5] is remodeled in order to be transformed to a federated one. Each Fi-SWoT

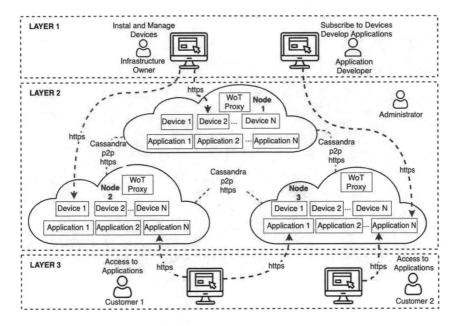

Fig. 1. Fi-SWoT layered architecture.

node publishes services to the Web by means of a Web Proxy. The Web Proxy implements the WoT model approach of W3C and supplies each Fi-SWoT node with a repository of RESTful services for publishing, subscribing, searching and updating information about physical devices (Things). Each Fi-SWoT component features a number of devices and applications which are available for subscription to infrastructure owners and customers respectively. An infrastructure owner in Layer 1 has installed one device in the first Fi-SWoT node and one device in the second node. An application developer in Layer 1 has created an application in the third Fi-SWoT node. Two customers in Layer 3 subscribe to applications in Layer 2 (the first customer subscribes to an application in the second node and one application in the third node; the second customer subscribes to an application in the third node).

Fi-SWoT nodes share devices and services with other nodes but without affecting each other. There is no single point of failure (e.g. upon failure of a node, the remaining nodes will still operate normally). Each node is responsible for the protection (e.g. unauthorized access) of its own devices and services. There is no single master node in charge of login and each node enables federated Single Sign-On (SSO) functionality for its users.

Fi-SWoT aims at enhancing the security of the federated architecture. At first, HTTPS protocol is enabled for services exposing their interface to the Web (i.e. services in charge of the communication between Fi-SWoT nodes using the public network and services enabling users access). This is not necessary for services within the same node (i.e. these services use an internal private network).

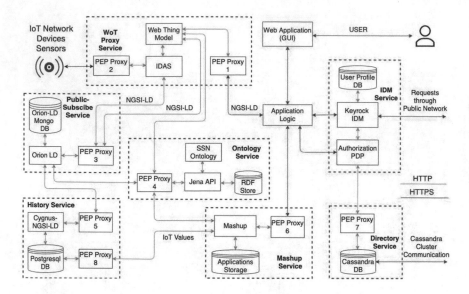

Fig. 2. Fi-SWoT node architecture.

The emphasis is also on protecting all services from unauthorized access. Access to services is granted only to authorized users (or other services) based on user roles and access policies.

3   Fi-SWoT Architecture

Fi-SWoT is a federation of independent nodes. Building upon principles of SOA design, each node is implemented as a composition of RESTful micro-services communicating (a) over a public network with services on the Web or with other nodes and, (b) over a private network with services within the same node. Fi-SWoT nodes are deployed as Virtual Machines (VM) in Google Cloud Platform (GCP). Each VM contains a Docker that encloses services in individual containers. A node can be reached from the Web or by other nodes by means of a public IP address. Users and services can send requests to a service, located in a remote node, via this public IP address followed by the container's published port. Containers located in the same node, utilize private addresses alongside the exposed ports to communicate inside the established private network of Docker.

Figure 2 illustrates the architecture of a Fi-SWoT node (i.e. supported services and their interconnection). IoT devices connect to WoT proxy and IDAS[3] device management service of the node. IDAS provides an interface for physical devices by means of IoT Agents whose purpose is to translate IoT-specific protocols into NGSI-LD. NGSI-LD[4] is an ETSI JSON-LD compliant format for

---

[3] https://fimac.m-iti.org/5a.php.

[4] https://www.etsi.org/deliver/etsi_gs/CIM/001_099/009/01.02.02_60/gs_CIM009v01 0202p.pdf.

management of context information. It describes information being exchanged and entities involved (i.e. sensors that publish measurements and users or services that subscribe to this information). It is the only service which is affected by the property of devices to transmit data using an IoT specific protocol (e.g. CoAP, MQTT, LoRa, Zigbee etc.). Following IDAS, IoT data is transmitted to WoT model service and from there to other services using NGSI-LD. The WoT model service implements all services foreseen in the WoT model of W3C and exposes an API interface for accessing IoT devices and information on the node.

3.1   Back-End Services

Back-end services implement the core functionalities of Fi-SWoT system dedicated to IoT data management. They communicate with each other using REST over HTTP within the private network of the same node.

Devices registered to WoT Proxy publish information to Publish-Subscribe service. ORION-LD[5] implements this service. It is an NGSI-LD server that manages context information processing and distribution. Clients can query, update, and register context information or, they can subscribe to receive notifications upon designated context change. Orion-LD receives an HTTP request when a sensor joins the IoT network through an IoT-Agent of IDAS. It creates a sensor entity with the requested characteristics and stores it in MongoDB. Orion-LD receives notifications when the context is altered, and proceeds to inform all subscribers of the particular device about the relevant change. If a user receives authorization and subscribes to a sensor installed in a remote cloud, Orion-LD proceeds to subscribe to that cloud's Orion-LD instance, in order to receive context changes of the sensor and to inform its subscribers.

History (past) measurements are forwarded to History service. Cygnus-NGSI-LD[6] is a specialized agent compatible with NGSI-LD and is in charge of persisting and creating a historical view of the context. It accepts NGSI-LD data flows from Orion-NGSI-LD (which is the NGSI-LD source of data) and stores them into a PostgreSQL[7] database (for the time being it is not compatible with non-SQL databases like MongoDB). The agent subscribes to each of Orion-LD's sensors, so that it gets notified when a measurement change occurs. Upon notification, the agent receives and stores the data in raw and aggregated (i.e. average values per day, week, month) form in its designated database.

The Ontology service is the knowledge base component that handles IoT general purpose and application specific knowledge. The term Thing stands either for physical entities (e.g. devices) or for virtual entities (i.e. device templates providing definitions for device categories). Physical entities are related (i.e. are instances) of virtual entities in SSN ontology [15]. Virtual entities do not always have physical entities as instances. Virtual entities (i.e. sensor types, observation concepts and actuators) are represented as classes in the SSN ontology. Sensor,

---

[5] https://hub.docker.com/r/fiware/orion-ld.

[6] https://github.com/telefonicaid/fiware-cygnus/tree/master/cygnus-ngsi-ld.

[7] https://www.postgresql.org.

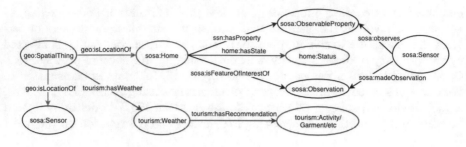

Fig. 3. A fragment of Fi-SWoT ontology.

Observation, Sample, and Actuator (SOSA) ontology [14] lays in the SSN core and represents the most common (elementary) entities, relations and activities involved in sensing, sampling, and actuation.

Fi-SWoT supports seven sensors types (i.e. atmospheric pressure, temperature, humidity, luminosity, precipitation, human presence at home and wind speed). All are defined as subclasses of SOSA class *Sensor* which, in turn, is subclass of SSN class *System*. In particular, classes *ObservableProperty* and *Observation* represent the attribute of any of the above sensor types to measure the corresponding property. SSN and SOSA do not provide definitions for application specific concepts such as geographic location, weather conditions etc. For such domain specific concepts additional ontologies are incorporated into the ontology, such as the Time ontology [17] for temporal properties, the Home and Tourism[8] ontologies for living conditions at homes, and for expressing the correct meaning of Weather concepts and measurements. Figure 3 is a fragment of the ontology illustrating how classes *geo:SpatialThing* of GEO and *sosa:home* are integrated into the ontology.

Fi-SWoT ontology integrates all general purpose and application specific ontologies referred to above. Each time an update operation is executed, a reasoner is invoked (e.g. Pellet [13] or Chronos [2]). The ontology is checked for consistency and, relations which can be inferred from existing ones are added to the ontology as well (e.g. all instances of a class become also instances of its subclasses). Ontology service employees Virtuoso[9] for permanent storage of the ontology graph. The Ontology service interacts with other services using Apache Jena[10] in the following cases: (a) WoT Proxy issues requests for observations, or for physical or virtual entities (e.g. virtual or physical entities are read, created, deleted or updated). A common interaction happens when new observations (e.g. measurements) are received from sensors in which case, the corresponding entities in the ontology are updated; (b) publication and subscription information is updated (e.g. an application subscribes to a physical entity).

---

[8] http://sensormeasurement.appspot.com.

[9] https://virtuoso.openlinksw.com.

[10] https://jena.apache.org.

```
{
    "appname": "AthensApp",
    "appscope": "weather",
    "info": [ {
        "attributes": [              sosa:PrecipitationObservation(?observation) ^
            "temperature",          sosa:hasSimpleResult(?observation, ?result) ^
            "humidity" ],            swrlb:greaterThanOrEqual(?result, 20) ^
        "city": "Athens",                swrlb:lessThan(?result, 50) ->
        "domains": [             sosa:deducesPrecipitation(?observation, ex:HeavyRain)
            "WaterSport",
            "Weather",
            "Garment" ],
        "func": LIVE
    } ]
}
```

Fig. 4. Declaring an application for a city (left) with a rule (right).

The Application Mashup service facilitates development of new applications by reusing information in the History database and the ontology. The service is realized with the aid of Node-Red[11], an open-source flow-based programming tool for the IoT. An application is defined as a composition of methods (i.e. functions) receiving inputs from devices [5]. Figure 4 (left) declares an application for monitoring weather conditions in a city and specifies (in field *appscope*) the name of a rule to be executed (*weather* in this example). Value *func:LIVE* signifies a semantic mashup, that is one that executes SWRL rules [16]. The rule is executed by invoking the reasoner. The rule will examine whether *precipitation* > 20, in which case, the value (i.e. observation) of *sosa:deducesPrecipitation* is *ex:HeavyRain*. If so, the following ontology relations suggest two actions: (a) in regards to domain *Activity*, hiking is not recommended: *ex:HeavyRain tourism: hasRecommendation ex:NoHiking* and, (b) in regards to domain *Garment*, the recommendation is *take umbrella*: *ex:HeavyRain tourism:hasRecommendation ex:TakeUmbrella*. The resulting actions are retrieved from the ontology by issuing the SPARQL query: *SELECT ?s SELECT ?s Rainy tourism:hasRecommendation ?s. ?s rdf:type Garment*.

A key component of Fi-SWoT architecture is the Directory service. Data stored in the directory are physically replicated to multiple nodes and can be discovered by the users of other nodes. The directory is realized by means of a Cassandra database[12] communicating with all other nodes on the network (i.e. Cassandra is distributed by itself and Cassandra nodes form a p2p network). The directory stores information that each node publishes to the network (i.e. disclosed information that can be accessed by the users of each node). Each Fi-SWoT node gets informed about the existence of other nodes in the network (i.e. their identity, locations, contact information, Web access points), as well as about devices and services available for subscription. The role of the Directory

---

[11] https://nodered.org.

[12] http://cassandra.apache.org.

is informative and does not allow access to physical devices and applications by itself.

3.2   Front-End Services

Front-end services provide graphical interfaces to facilitate interaction of users with a node. These services are accessible via the node's public IP and their assigned published port using HTTPS/1.1. The functionalities of each front-end services are briefly mentioned hereafter.

The Web Application comprises the gateway to the Fi-SWoT eco-system, allowing user access over public networks. It is the login endpoint of each node. At login, users are redirected to the graphical user interface of the Identification and Authorization service in order to fill in their credentials. A user database holds user profiles, roles, session information and session history. After successful identification, the users are directed back to the Web Application service where they can choose to access either the customers portal or, the infrastructure owners portal corresponding to their role (Sect. 4). When users select a portal, a role check is performed in order to verify that the user is authorized to access it. Upon successful authentication, all requests are forwarded to Application Logic service. Application Logic orchestrates, controls and executes services in order. When a request is received (from a user or service), it is dispatched to the appropriate service.

The Web Application service provides graphical user interface for interacting with the WoT model service (WoT proxy) for accessing data and services in a node. Application logic is then responsible for dispatching the requests to the appropriate service. These are requests for (a) filtering devices by properties combined with operators (i.e. disjunction, conjunction, negation, location and time operators) [10]; relevant information is retrieved from Publish-Subscribe service, (b) discovering a Fi-SWoT node and for requesting subscription; a request for subscription is sent to the node's administrator for approval; (c) retrieval of historical measurements from sensors; users can specify a time frame or, choose a specific metric of measurements (e.g. maximum or average temperature during a month or average humidity of the week); relevant information is retrieved from History Database; (d) installing new devices, modifying the properties of devices or, for removing devices; the Web Application forwards the requests to the IoT-Agent of IDAS which is turn updates related information in the Publish-Subscribe service.

4   Authorized Access and Single Sign on

The authorization code grant of OAuth2.0[13] and OpenID Connect[14] are incorporated into the APIs of the system. The authorization code grant provides API

---

[13] https://oauth.net/2/.
[14] https://openid.net/connect/.

security and reduces the chance of exposing user credentials by utilizing scoped access tokens. OpenID Connect extends the OAuth 2.0 Protocol and provides enhanced user authentication, ID token validation, and SSO (Single Sign-On) functionality among the nodes.

Identity managers (IDMs) are frameworks consisting of technologies and policies responsible for assigning digital identities to users or services. The Identification and Authorization service is implemented based on a Keyrock IDM[15] image. It orchestrates system authentication and authorization in services based on the OAuth 2.0 protocol. Keyrock IDM is the core of system's security and alongside, Policy Enforcement Points (PEP) and Authorization Policy Decision Point (PDP) services, guarantee that only authenticated and authorized individuals can access protected resources if certain conditions are met. In order to incorporate authorization grant type to services, these must be registered with the IDM.

Policy Enforcement Points (PEP) are implemented based on a PEP-PROXY Wilma[16] image. PEP proxies are application endpoints placed in front of individual services or resources in order to protect them from unauthorized access. PEP proxies intercept client requests to the service they protect and perform authentication before permitting or denying access. For more sophisticated access control, PEP proxies are combined with Policy Decision Points (PDP). After intercepting a request to the protected service, PEP sends the client's attributes to a PDP service which takes the decision to permit or deny access based on relevant registered access policy rule sets. Afterwards, PDP returns its decision to PEP which, in turn, may permit or deny access to the service.

Authorization Policy Decision Point (PDP) AuthZforce[17] implements PDP service. It is responsible to assess requests from users to access protected services. The respective user access rights are encoded by means of XACML[18]. PDP receives a REST request whenever an administrator creates a policy rule for a PEP proxy and stores it in a different XACML domain for each PEP. Afterwards, whenever a PEP forwards a request, PDP checks the rule sets registered in PEP's corresponding domain in order to permit or deny access.

There are two operating scenarios of PEP proxies: the first scenario occurs when a PEP is an intermediate in the communication of two services. The requesting service must include a master key in the header of the request, in order to be authorized by PEP to access the service. The master key is a secret that was defined by the administrator during PEP's creation (each PEP has a different master key). In the second scenario, PEP proxy intercepts the request of a user to the service. In this case, the collaboration of PEP with the Identity Manager and the PDP is necessary to ensure user authentication and authorization. Initially, PEP proxy receives a request for access from a user that includes an OAuth2 token in the header. An OAuth2 token is created during the user's

---

[15] https://fiware-idm.readthedocs.io/en/latest/.
[16] https://fiware-pep-proxy.readthedocs.io/en/latest/.
[17] https://authzforce-ce-fiware.readthedocs.io/en/release-5.1.2/.
[18] https://fiware-tutorials.readthedocs.io/en/latest/administrating-xacml/index.html.

login in the system by Keyrock IDM and represents the identity of the user and that the user is authenticated. PEP exports the token and sends it to Keyrock IDM for validation. Keyrock IDM verifies token validity using its database and responds with the user's role in the organization. PEP forwards the user's role alongside the desired action and the path of the protected resource to PDP to evaluate them. PDP checks the rule sets inside the domain corresponding to the PEP in order to make a decision to permit or deny the request. Finally, PDP returns its decision to PEP which enforces it.

Aiming to minimize the risk of exposing user credentials and to enhance user authorization and authentication, OpenID Connect and the authorization code grant of OAuth 2.0 protocol are implemented on each node. Upon receiving a login request, the Web Application redirects users to its graphical interface, which prompts them to input their e-mail and password. After a successful login, Keyrock IDM responds to the Web Application with the generated authorization code. The Web Application makes a second request to the IDM in order to exchange the received authorization code with an access token and an ID token. IDM returns an access token to the Web Application, alongside a scoped ID token in the form of JSON Web token which includes only the necessary user information. A foreign (remote) node performs signature validation when it receives the ID token in order to verify the authenticity of the sender, and to ensure that the user belongs in a trusted node. Additionally, it extracts the necessary user information from the ID token and binds them to the corresponding user session. This procedure is essential, especially when Single Sign-On (SSO) functionality is used for login to a remote node. In the aforementioned scenario, the Web application will communicate, for authorization and authentication of the user, with the IDM of the node where the user is registered.

5 Communication Security

Communication of Fi-SWoT components is based on the HTTPS/1.1 protocol. It is essential to recognize and enable HTTPS only for components whose communication occurs over public networks, in order to ensure security without heavily affecting performance. Figure 2 presents the mapping of node services. Red-colored lines indicate communication occurring over public networks where HTTPS protocol must be enabled, while green-colored ones indicate communication through the private network that occurs over HTTP. An additional communication channel between devices of the IoT network and IDAS of the WoT service is shown as well. Device communication with Fi-SWoT is typically realized by means of a gateway (i.e. WoT Proxy service may run on a gateway equally well) and low level communication protocols (e.g. CoAP, MQTT, LoRa) most of which apply their own security measures to ensure communication integrity[19] [11]. Securing the IoT network is outside the scope of this paper.

---

[19] https://www.avsystem.com/blog/iot-protocols-and-standards/.

PEP proxies 1 and 6 of Fig. 2, are stationed in the front-end. The graphical user interface of the Web Application and of the User Identification and Authorization services are reachable through the public network as well. Although Cassandra is a back-end service, the inter-node communication occurs over the public network. The modules responsible for HTTPS communication in these services must be enabled. The necessary modifications to enable communication over SSL/TLS are implemented in the Dockerfile that is used to create the services image and in the "docker-compose.yaml" file that creates the containers (i.e. node services) in Docker [9].

Each exposed service creates a private key, encoded by a strong encryption algorithm, and a certificate that includes information about the service owner (e.g. country, state, corporation, email and domain name). The certificate is sent to a trusted Certificate Authority (CA) which tests and verifies the information and proceeds to sign it. Each server service has to own a certificate in order to be able to communicate over the TLS/SSL protocol. The same certificate can be used by multiple services. It contains the server's domain name and the certificate authority that vouches for its authenticity. OpenSSL[20] is utilized for the creation of Certificate Authorities (CA), cryptographic keys, certificate sign requests, and finally of TLS/SSL certificates.

At the beginning of the communication between two services (referred to as client and server), a handshake is performed. During the handshake, the client and server services will agree upon an encryption algorithm to be used (TLS version and cipher suite to be used). The server presents its certificate in order to validate its identity. Finally, the server provides a key for generation of the session keys, that will be used for encryption and decryption, before the actual message transmission begins.

Directory database Cassandra is a back-end component accessible by the cloud's services through the Docker's private network. Although it is a back-end service, it is necessary to communicate over a public network for data exchange with the Cassandra node, placed in each node; thus, incorporation of TLS/SSL is essential. Cassandra database uses SSL/TLS certificates to offer node-to-node encryption.

6   Evaluation

Each Fi-SWoT node is a Virtual Machine (VM) in the Google Cloud Platform. Individual services are deployed within a Docker Engine as containers except Cassandra which is deployed as a service within the VM (it cannot run in a container). Each Docker runs 23 containers (i.e. services). The Fi-SWoT prototype system comprises four nodes (four VMs) and is tested in a smart city scenario with four cities with 1,050 homes each. Each home installs three sensors providing per hour (each one) 100 observations of temperature, humidity, motion and luminosity. Recent measurements are stored in the ontology and in Publish-Subscribe service; history (past) measurements are stored in the History

---

[20] https://www.openssl.org.

Table 1. Average response time (ms) of service calls on local node for varying concurrency.

No.	Service request	Concurrency = 1	Concurrency = 100	Concurrency = 200
1	*Get sensor properties*	17,66	9,121	9,300
2	*Get all sensors*	40,74	29,04	36,21
3	*Update A sensor*	365,0	49,23	49,25
4	*Insert a new sensor*	286,9	72,67	73,00
6	*Get all applications*	96,90	68,11	73,05

database. The History database of each city stores more than 25 Million measurements. The ontology holds recent sensor observations and model relations expressing ontology facts. Overall, the ontology contains 63,063 RDF triples which are kept in main memory.

The requests are issued on the same node (i.e. where the user is connected) but they are distinguished based on whether the results are produced from the same or from foreign (remote) nodes. In the later case, the local node dispatches the requests to the foreign nodes. For 1,000 requests and two nodes (i.e. the local and one foreign node), each node will serve 500 requests; for three nodes, each node will serve 333 requests and so on. The purpose of this experiment is to study the performance of the system under high concurrency, by allowing up to 200 simultaneous requests per node (simulating up to 200 users issuing requests concurrently). To simulate the effects of high workload Apache Benchmark (AB)[21] is installed in each VM. It accepts as input, number of service requests and number of concurrent requests and computes average response time per request. The Web Things Model Service, the Mashup Service, and the Directory Service (Cassandra database) are the main system access points receiving the requests from the Web Application.

Table 1 summarizes the performance of common service requests issued on a local node. The first request addresses the ontology service where all devices (i.e. sensors) are stored. The same result might also be obtained from the MongoDB of Publish-Subscribe service. However, the ontology resides in the main memory and responds much faster. Similar holds for the second query. The third request attempts to update the properties of an existing device. This request will spawn similar update requests to the ontology in the main memory and also to the MongoDB of Publish-Subscribe service where the devices are stored. Similarly, the fourth request will spawn two requests in order to insert a new device in both places. Finally, the sixth request will retrieve all applications from the MondoDB of Mashup service. In all cases, response times improve with the simultaneous execution of requests (i.e. Apache Bench switches to multitasking). Subscription and insertion requests (i.e. update requests in general) are always much slower than simple read (i.e. Get) requests which simply read property values from a service (e.g. a database).

---

[21] https://httpd.apache.org/docs/2.4/programs/ab.html.

Table 2. Average response time (ms) of service calls on local and remote nodes.

No.	Service request	1 node	2 nodes	3 nodes	4 nodes
1	*Get all sensors*	30.20	38.18	40.10	40.40
2	*Subscribe to sensor*	14.57	46.64	52.57	60.73
3	*Get sensor value*	11.75	24,62	26,56	28.75
4	*Subscribe to application*	30.58	50.95	53.62	54.25

Table 2 summarizes the performance of common service requests addressing the local and up to all four nodes (*concurrency* = 100). The first request, queries the ontology service on all four nodes for existing devices. This information is retrieved from the MongoDb database of Publish-Subscribe service on each node. The second request is a subscription to a sensor. This is an update request addressing the Publish-Subscribe service of each individual node separately. The third request retrieves the current value of the specified sensor from the ontology service of each node. The last (fourth) updates the Publish-Subscribe service on each node involved with a subscription to an application. The increased response time for requests addressing foreign nodes account for the time for encryption and decryption of HTTPs and to the additional authorization step of OpenID connect. When HTTPS protocol is enabled, the Certificate installed in the Web Application contains a 2048-bit public key that was generated using the RSA encryption algorithm. Experimental results [9] demonstrate that the HTTPS protocol introduces significant delays (i.e. the time to execute a request doubles in most cases).

7   Conclusions and Future Work

Fi-SWoT succeeds in reinforcing overall system security, user authentication and authorization and supports Single Sign-On functionality among the federation of nodes. There are still areas for improvement and further research, which have not been covered in this study. Upgrading HTTPS/1.1 protocol to HTTPS/2.0 protocol is a challenging task, as HTTPS/2.0 utilizes multiplexing of the requests and of the responses. It can greatly improve system performance, while guaranteeing security of the communication. Securing the communication between the system and the IoT sensors (in cases where the IoT protocols are not providing sufficient security on their own) is a requirement for handing risks due to malicious behavior of IoT devices. Fi-SWoT is currently being extended to support billing policies and functionality for dealing with complex events. Incorporating scalability features for dealing with increased workloads is also an important direction for future work.

References

1. Al-Osta, M., Ahmed, B., Gherbi, A.: A lightweight semantic web-based approach for data annotation on IoT gateways. In: International Conference on Emerging Ubiquitous Systems and Pervasive Networks (EUSPN 2017), pp. 186–193, Lund, Sweden (2017)
2. Anagnostopoulos, E., Batsakis, S., Petrakis, E.G.M.: CHRONOS: a reasoning engine for qualitative temporal information in OWL. In: International Conference in Knowledge Based and Intelligent Information and Engineering Systems (KES 2013), pp. 70–77, Kitakyushu, Japan (2013)
3. Antoniazzi, F., Viola, F.: Building the semantic web of things through a dynamic ontology. IEEE Internet Things J. **6**(6), 10560–10579 (2019)
4. Bermudez-Edo, M., Elsaleh, T., Barnaghi, P., Taylor, K.: IoT-lite: a lightweight semantic model for the internet of things and its use with dynamic semantics. Pers. Ubiquit. Comput. **21**(3), 475—487 (2017)
5. Botonakis, S., Tzavaras, A., Petrakis, E.G.M.: iSWoT: service oriented architecture in the cloud for the semantic web of things. In: Advanced Information Networking and Applications (AINA 2020), pp. 1201–1214, Cham, March 2020
6. Guinard, D.D., Trifa, V.M.: Building the Web of Things. Manning Publications Co., Greenwich (2016)
7. Javeri, P.: Rethinking IoT architecture — the need for distributed systems architecture for the Internet of Things. Medium, white paper, January 2019
8. JSON-LD 1.1: A JSON-based Serialization for Linked Data, W3C Working Draft (2020)
9. Kontochristos, I.: Authorized user access in federated service oriented architectures for the Internet of Things in the cloud. Diploma thesis, Technical University of Crete (TUC), Chania, Crete, Greece (2020)
10. Koundourakis, X., Petrakis, E.G.M.: iXen: content-driven service oriented architecture for the Internet of Things in the cloud. In: Ambient Systems, Networks and Technologies (ANT 2020), pp. 145–152, Warsaw, Poland (2020)
11. Petrakis, E.G.M., Sotiriadis, S., Soultanopoulos, T., Tsiachri Renta, P., Buyya, R., Bessis, N.: Internet of Things as a service (iTaaS): challenges and solutions for management of sensor data on the cloud and the fog. Internet Things **3**–4(9), 156–174 (2018)
12. Rhayem, A., Ben Ahmed Mhiri, M., Gargouri, F.: Semantic web technologies for the Internet of Things: systematic literature review. Internet Things **11**, 1–22 (2020)
13. Sirin, E., Parsia, B., Cuenca Grau, B., Kalyanpur, A., Katz, Y.: Pellet: a practical OWL-DL reasoner. J. Web Semant. **5**(2), 51–53 (2007). 9th Intern. Conference on Ambient Systems, Networks and Technologies (ANT 2018)
14. Semantic sensor network ontology, W3C Recommendation (2017)
15. Semantic sensor network ontology, W3C Recommendation, October 2017
16. SWRL: A semantic web rule language combining OWL and RuleML, W3C Member Submission, May 2004
17. Time ontology in OWL, W3C Candidate Recommendation (2020)
18. Web thing model, W3C member submission, August 2015
19. Zacharia, N., Petrakis, E.G.M.: iZen: secure federated service oriented architecture for the Internet of Things in the cloud. In: Advanced Information Networking and Applications (AINA 2020), pp. 1189–1200, Cham, March 2020

A Systematic Approach for IoT Cyber-Attacks Detection in Smart Cities Using Machine Learning Techniques

Mehdi Houichi[1(✉)], Faouzi Jaidi[1,2], and Adel Bouhoula[3]

[1] Higher School of Communication of Tunis (Sup'Com), LR18TIC01 Digital Security Research Lab, University of Carthage, Tunis, Tunisia
{mehdi.houichi,faouzi.jaidi}@supcom.tn
[2] National School of Engineers of Carthage, University of Carthage, Tunis, Tunisia
[3] Department of Next-Generation Computing, College of Graduate Studies, Arabian Gulf University, Manama, Kingdom of Bahrain
a.bouhoula@agu.edu.bh

Abstract. In these last years, the widespread adoption of the Internet of Things (IoT) concept led to the invention of intelligent cities. Smart cities operate in real time to promote lightness and life quality to citizens in urban cities. Smart city network traffic through IoT systems is growing exponentially though it presents new cyber-security threats. To deal with cyber-security in smart cities, developers need to improve new methods and approaches for detecting infected IoT devices and cyber-attacks. In this paper, we address IoT cyber security challenges, threats and solutions in intelligent cities. We propose an approach for anomaly detection in smart cities applications, networks and systems. Our solution relies on intelligent anomalies as vulnerabilities and threats detection based on different methods and machine learning algorithms. The proposed solution helps in effectively detecting and localizing infected IoT devices as well as generating alerts and reports. To experiment our solution, we used the dataset NSL-KDD to evaluate the accuracy of the model. Obtained results show that our model achieved a high classification accuracy of 99.31% with a low false positive rate.

Keywords: Internet of Things (IoT) · Smart city · Fog layer · Cyber-security · Intrusion detection

1 Introduction

Recently, the spread of the Internet of Things (IoT) has grown significantly in societies around the world. The number of connected IoT devices has already reached 27 billion in 2017. The number of connected devices will grow exponentially. The potential is expected to attain around 125 billion by 2030 and global data transmissions are expected to increase from 20 to 50 percent per year, on average, in the next 15 years [3]. Various smart city applications connect huge number of IoT devices with real-world objects, which actually have very important benefits for city life [4]. However, the large number of IoT devices across different types of services, technologies, devices, and protocols (e.g. wireless, wired, satellite, cellular, Bluetooth, etc.) makes a complexity for managing IoT networks [5, 6]. Therefore, the use of heterogeneous emergent technologies with resource constrained connected devices creates serious cyber security vulnerabilities threatening the daily activities of citizens life. Cyber threats can gain unauthorized access to IoT devices without the knowledge of the qualified user or administrator (for example, the Mirai botnet) [7].

There are two main security challenges in smart city applications. The first challenge is how to detect zero-day attacks. These attacks occur from a variety of IoT device protocols in the cloud data center of a smart city, considering the major attacks are hidden in the IoT devices? The second is how to find a way to detect cyber attacks [7] (e.g. IoT malware attacks, etc.) intelligently from IoT networks before damaging a smart city? Currently, the traditional Intrusion Detection System (IDS) [8] is not designed for IoT networking devices, as these devices have limited resources and functionalities (e.g. smart watches, smart lamps, smart locks, etc.).

Fog Computing was recently designed between cloud and IoT layers to reduce power consumption, storage and latency. In addition, it aims to move the computing process close to the sensors to respond quickly to IoT [9, 10]. Recently, the authors of [11] propose to improve the detection of IoT cyber attacks in the distributed fog layer. IoT attacks detection would be more important for automatically alerting Internet Service Providers (ISPs) or administrators quickly and efficiently.

In the current paper, we propose a system that relies on an intelligent detection method based on machine learning algorithms. It allows detecting attacks with the aim to reduce the false positive rate. Our system is also designed to monitor all IoT traffic in a fog layer and alert the administrator or the service provider of a smart city. Therefore, the proposed system is based on a training model in fog layer networks. This allows how to intelligently learn from training in the distribution of small-scale network traffic near the IoT sensors and distinguish normal behavior from abnormal behavior.

The main contributions of this paper are as follows:

We study main concepts of smart cities and design the architecture of our detection system. It allows detecting IoT cyber-attacks or unusual activities in IoT network traffic from fog layer across. Similarly, our detection system generating comprehensive reports and alerts to notify active agents (e.g. administrators, industry management, ISPs, etc.).

Unlike several related works that detect IoT malicious behavior using signature-based IDS in the cloud data center to detect only known attacks, we propose to detect threats in the fog layer by using machine learning techniques. We identify attacks by extracting statistical metrics from the datasets, including data functionalities and modern attacks [15] that may exist in the traffic of the IoT botnet [16].

We evaluate the performance of the proposed system by using machine learning classification algorithms with the goal to predict normal or malicious behaviors.

The rest of this paper is organized as follows: Sect. 2 introduces the related work. Section 3 provides background concepts relevant to smart city, such as benefits, architecture and detection system. Section 4 discusses the threat model in a smart city and research challenges. We technically detail the architecture of our system in Sect. 5. In

Sect. 6, we evaluate and analyze the performance of the proposed system. Section 7 discusses the results and presents the future plan for this work. Finally, Sect. 8 is reserved for the conclusions and perspectives of our study.

2 Related Works

In this section we review and discuss related works. We mainly focus on research efforts dealing with intrusion detection of IoT cyber-security threats, anomalies and abnormal behaviors within smart cities, particularly IoT-based networks, focusing on machine learning algorithms.

Previous studies discussed traditional IDS methods which have shown that cyber-attacks can take place in different ways: in hosts, networks or both. Proposals concern Host-based IDS (HIDS) [17, 18], Network-based IDS (NIDS) or hybrid solutions [5, 8, 19]. More used detection techniques are based on the analysis of signatures, rules or behaviors. Therefore, the current paper relies on the use of an anomaly-based NIDS. Some anomaly based detection methods are proposed and discussed in [9, 11, 21–23]. There are several techniques that rely on signature-based method. However, this method consumes power and does not detect new attacks. It only detects attacks if they match the stored database [24]. This database is irrelevant for low-capacity IoT devices, since signed methods cannot detect unknown attacks in network traffic.

Most of the existent studies, as well as current works, focusing on network traffic analysis, have not considered the fact that IDS for the IoT network is different from traditional IDS. Therefore, a recent study [8] showed that some IoT networks encounter difficulties with traditional IDS methods due to the limited resources of IoT devices and networks, the use of specific protocols, the power consumption, etc. IDS methods need to be improved for the security of IoT services to protect their qualified users in a large infrastructure like a smart city [12].

Several studies have been conducted to examine the IDS for IoT networks. A recent study [8] focused on intrusion detection within an IoT context, but it has not mentioned the analysis of anomalies based on machine learning algorithms specifically the random forest (RF) [31]. Recently, the authors, in [11], used a deep learning approach to detect cyber-attacks, but they did not use the deviation analysis method.

To our knowledge, there are not a lot of previous studies that addressed our approach. Most of the current IDS methods, mentioned above, do not deal with the specific case of a smart city. In addition, this model learns from normal traffic by training machine learning algorithms to detect harmful behaviors in the smart city.

3 Background

This section contains basic information relevant to this study. We treat the concept of a smart city, its architecture and benefits as well as intrusion detection techniques.

3.1 Smart City Overview

A smart city is a concept that includes and uses IoT technology, smart systems, big data analysis, and information and communication technologies to improve the quality and performance of the various services in a city. It aims to enhance the health and safety of city dwellers, control pollution and reduce resource waste, resource consumption and overall costs. The idea of a smart city is to improve the quality of life for city dwellers through smart technology [4]. Generally, it is the government's responsibility to take care of the smart city by making sure that systems are not vulnerable to cyber-attacks and by ensuring the privacy of residents.

3.2 Smart City Based on Fog Architecture

Intelligent urban architecture is based on the advantages of cloud computing to reduce the latency between cloud and IoT sensor, as shown in Fig. 1. It consists of three layers to display high visual levels in smart urban infrastructure [7, 9–11], including fog layer and IoT sensor layer. The fog layer is an important part of the smart urban architecture that provides data processing and aggregation [14].

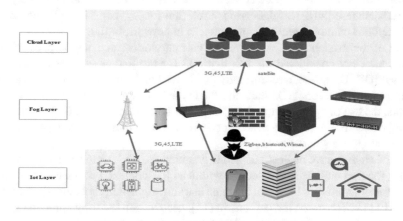

Fig. 1. Smart city based on fog architecture.

1) Cloud Layer: it contains servers to store and manage huge amount of data on top management.
2) Fog layer: it stays between IoT Sensing layer and cloud layer and it make the calculation and management at the edges of the network (e.g. security gateway).
3) IoT Sensing layer: it contains a set of sensors installed in the city to enable the data collection.

3.3 Intrusion Detection Systems

Traditional IDSs are designed to monitor and detect intrusion activity on individual computers or in all network traffic. There are two types of IDS: the first is Host Based

IDS (HIDS) and the second is Network Based IDS (NIDS). HIDS installs software (e.g. antivirus) on each computer to monitor and detect any malicious activity of traffic (intrusion behavior) based only on the computer system connected to the traditional local area network [17, 18]. HIDS can analyze or scan software installed on the computer (for example, application logs, system calls, file systems, etc.). Therefore, HIDS method does not make sense with some IoT devices such as smart lights, clocks, lock doors, etc. These devices have limited functions and resources (e.g. power consumption, delay, computer, low memory, etc.).

The NIDS monitors all network traffic. Therefore, it can detect known or unknown attacks based on a hybrid method that uses both signature-based techniques and variances [6, 21, 27]. NIDS anomaly-based methods are more important to monitor network traffic and detect new attacks [25]. Even the anomaly-based method has high false positive rates; our proposed system promises to reduce false positive rates. However, the signature based method will be cost computation to store attacks in the database and moreover not detect new attacks in future network traffic [8].

Therefore, our detection approach promises to detect attacks using the NIDS method for deviation based on machine learning methods.

4 Problem Statement

We deal in this section with IoT cyber-attacks, threat models and main security challenges. The number of connected devices has grown exponentially with more heterogeneous devices. This leads to an increase of IoT vulnerabilities (such as zero-day attacks, botnets, etc.) to operate attack vectors that can target IoT victims and use them to steal personal information or launch attacks like DDoS over the Internet [3, 7, 28]. For example, IoT botnets can make it easier to compromise vulnerable IoT devices by scanning the Internet and finding the IP address of victims IoT networks. The Mirai botnet phenomenon occurred in late 2016 compromised a large number of IoT devices by launching a DDoS attack on Dyn, which is a DNS provider [7].

The main challenges in protecting IoT devices can be IoT botnets without prior notice from Internet service providers and qualified users. We believe that there are increasing security problems in most smart home networks, as they were designed with vulnerable security techniques, as some companies need profit or have little experience in security [30]. The main security issues of massive IoT devices are embedded in heterogeneous types of devices and protocols that are very complex to protect in a smart city [13]. Therefore, most IoT networks are not adapted with traditional IDS systems that will not accurately detect IoT attack networks [8].

4.1 Adversary Model

We consider, in this model, that the attacker scan the internet to compromise the vulnerable IoT devices connected to various routers in the smart city, such as smart homes, hotels, restaurants, shopping malls, airports, etc. In this way, an attacker, by compromising these IoT devices, can get sensitive data like credit card information, stream videos, send spam, etc.

In the current paper, we suggest a system to identify infected IoT devices in distributing fog networks instead of centralized cloud computing.

4.2 Research Challenges and Assumptions

1) Limited resources: by using the traditional IDS method, we can detect compromised computers, laptops or smartphones on the traditional local network. However, its direct application on IoT devices face challenges as these IoT devices have low functionalities. These IoT devices are very difficult because their resources are limited (e.g. small memory, batteries).
2) Heterogeneous: IoT devices are connected via different protocols. Thus, the increase in IoT devices in the future in smart cities with large infrastructure will lead to serious damage and latency detection with massive IoT data in the cloud centers.
3) High false positive/negative rates: anomaly detection methods can easily generate high false positive and negative rates.

As an assumption, we consider that a lot of heterogeneous IoT devices are connected in the big smart city. Our system promises to detect zero-day attacks and the IoT botnet intelligently by distributing the detection in the fog layer.

5   The Detection Approach

In this section, we propose a smart approach for detecting cyber attacks on fog nodes in a smart city, as illustrated by Fig. 2. The proposed approach defines a framework based on several machine learning algorithms to detect attacks and malicious behavior in future urban IoT networks. We assume this method works to monitor the network traffic passing through each fog node as the fog nodes are closer to the IoT sensors, rather than the large amount of cloud storage in the city to identify normal and abnormal behavior. After detecting fog-level attacks, the framework should automatically notify the cloud security services to inform them to update their system.

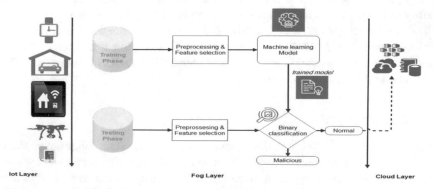

Fig. 2. Proposed approach for detection system model

A-System Design
The system design model consists of multiple components involving a massive amount of IoT devices connected to nebulously distributed networks either privately (e.g. Smart Home Gateway) or publicly in a smart city. By implementing our model on each fog node, it should detect new attacks to alert the security architect and the security administrator as described below:

1) IoT devices: a large number of IoT devices are connected to an entrance in a layer of fog.
2) Gateway: we assume that every private facility that has its own gateway (e.g. smart homes, buildings, shopping centers, schools, etc.) is connected to the master security gateway in the fog layer to process multiple inputs.
3) Detection system model: it can be placed on a main fog node, which can intelligently control the communication and analyze the traffic. This system is based on machine learning methods that improve the performance of algorithms in that system model. From these methods we choose bagging techniques such as the Random Forest (RF) algorithm. we use the NSL-KDD [15] dataset to evaluate our model.

B-Anomaly Detection Based on Machine Learning
As shown before, cyber-attacks can find vulnerable IoT devices in either private or public networks in smart city life. The NIDS can use machine learning algorithms (e.g. Decision Tree, K-Neighbor, Random Forest, etc.) to classify and detect malicious behavior in IoT networks. This can be done using the NIDS with an anomaly analysis method based on machine learning algorithms. This method uses statistical analysis to clean and prepare data for our model predictions. This model can analyze and distinguish the normal traffic from abnormal attacks with reduced False Positive Rates (FPR) [20].

Thus, this detection system model can improve performance for detecting attacks in foggy nodes in smart urban structures than detecting in the cloud center. Thus, these IoT fog networks would ensure light efficiency, less latency and less consumption than the cloud center, which has enormous data in a large smart city infrastructure.

6 Evaluation and Experimental Results

This section presents the analysis and evaluation of our proposed approach, on various parameters based on the NSL-KDD dataset [15]. Therefore, our solution relies on an anomaly detection based approach by using machine learning techniques. In this study, we evaluate the performance of our solution by using the Random Forest algorithm to identify normal network traffic from malicious one.

Thus, this approach can improve performance to effectively detect attacks in the level of fog layer in smart city infrastructure before attending the third layer (cloud layer). As such, these IoT fog networks would ensure lightweight functionality, less latency, and lower power consumption than the cloud computing layer, which has huge amounts of data in a large smart city.

A. NSL-KDD Dataset

Due to the limitations of the original KDD99 dataset [22], we used the NSL-KDD dataset for our experiments. The NSL-KDD dataset solves some inherent problems of the KDD99 dataset. The NSL-KDD dataset is widely used in the development of intrusion detection systems [26]. Although this dataset still has some flaws, it can be used as an effective benchmark dataset to help researchers compare different intrusion detection methods. This study uses the NSL-KDD dataset for intrusion detection simulation. Although the data from NSL-KDD is not specifically used for IoT environment testing, the protocols and attacks involved in this dataset are very valuable for IoT environment intrusion detection research. The types of attacks in this dataset also occur in real-world IoT networks. Protocols such as TCP are becoming more and more important because traditional application layer and cloud platform connections are both important parts of the IoT environment. Therefore, the NSL-KDD dataset is used as a dataset for the Internet of Things [34].

The NSL-KDD dataset has been shown to be added over the KDD99 dataset, such as removing unnecessary entries in the training set and copying entries in the test set [29]. The number of data points in NSL-KDD (KDDTrain+) and test exercises (KDDTest +) is 125,973 examples of the training data and 22,544 examples of the test data, respectively shown in Table 1.

Table 1. The number of instances in the NSL-KDD training and testing dataset

Class	KDDTrain+	KDDTest+
Normal	67343	9711
Anomaly	58630	12833
Total	125973	22544

B. Data Transforming and Normalization

The used dataset contains symbols, continuous and binary values. For example, the feature "protocol type" in the NSL-KDD dataset includes symbolic values such as "TCP", "UDP" and "ICMP". Since many classifiers only accept numeric values, a conversion process is considered. This conversion is very important and has a significant impact on the accuracy of the IDS. In this paper, we replace each single value with an integer to deal with the symbol feature. Therefore, the normalization is a necessary transformation that maps features to the normalized range. A simple and fast method called "one hot encoding" is used in our experiments which permits to transfer the values of "protocol type", "service", and "flag" features to a binary values.

The experiments in the current paper focused on the building classifier to classify into two categories: normal and anomalous (attack). There are 41 features in NSL-KDD dataset [22], including both nominal and numeric features which are shown in Table 2.

Table 2. The features of the NSL-KDD dataset

N°	Feature	Type	N°	Feature	Type feature
1	Duration	Numeric	22	Is_guest_login	Nominal
2	Protocol_type	Nominal	23	Count	Numeric
3	Service	Nominal	24	Srv_count	Numeric
4	Flag	Nominal	25	Serror_rate	Numeric
5	Src_bytes	Numeric	26	Srv_serror_rate	Numeric
6	Dst_bytes	Numeric	27	Rerror_rate	Numeric
7	Land	Nominal	28	Srv_rerror_rate	Numeric
8	Wrong_fragment	Numeric	29	Same_srv_rate	Numeric
9	Urgent	Numeric	30	Diff_srv_rate	Numeric
10	Hot	Numeric	31	Srv_diff_host_rate	Numeric
11	Num_failed_logins	Numeric	32	Dst_host_count	Numeric
12	Logged_in	Nominal	33	Dst_host_srv_count	Numeric
13	Num_compromised	Numeric	34	Dst_host_same_srv	Numeric
14	Root_shell	Numeric	35	Dst_host_diff_srv_rate	Numeric
15	Su_attempted	Numeric	36	Dst_host_same_src_	Numeric
16	Num_root	Numeric	37	Dst_host_srv_diff_host_	Numeric
17	Num_file_creations	Numeric	38	Dst_host_serror_rate	Numeric
18	Num_shells	Numeric	39	Dst_host_srv_serror	Numeric
19	Num_access_files	Numeric	40	Dst_host_rerror_rate	Numeric
20	Num_outbound_cm	Numeric	41	Dst_host_srv_rerror_	Numeric
21	Is_host_login	Nominal			

C. Evaluation Metrics

In this work, we evaluated only the binary classification which classifies the traffic as normal or anomalous. We used known evaluation metrics [32] to assess the performance of the attacks detection in the binary classifications, as shown in Eq. (1).

$$Accuracy = \frac{TP + TN}{(TP + TN + FP + FN)} \tag{1}$$

The *Accuracy* shows the percentage of whole normal and attack data that can be correctly classified with respect to True Positive (TP), True Negative (TN), False Positive (FP), and False Negative (FN) rates. To obtain the percentage of correctly detected attacks; we used the Detection Rate (DR) as shown in Eq. (2). However, to illustrate the percentage of the anomalous detection of attack behaviors, we utilized the False Positive Rate (FPR) as shown in Eq. (3). Thus, the *Accuracy* obtained for our model was 99.31%.

Obtained FPR was 0. 81% and DR was 99%.

$$DR = \frac{TP}{(TP + FN)} \tag{2}$$

$$FPR = \frac{FP}{(TN + FP)} \tag{3}$$

We used more parameters like confusion matrix and other metrics (e.g. precision, recall, F1 score) to evaluate our system. First, we used the confusion matrix to evaluate and count correctly and incorrectly detected instances in normal records as shown in Table 3.

Table 3. Confusion matrix.

Actual	Predicted normal	Predicted attack
Normal	TN	FP
Attack	FN	TP

Figure 3 presented the results obtained by the confusion matrix for our model using the method Random Forest.

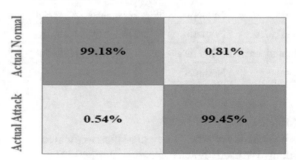

Fig. 3. Confusion matrix for RF test.

Second, we used the *"precision"*, *"recall"* and *"f1-score"* metrics to evaluate our model. The precision metric illustrates how many of the detected abnormal behaviors are correct as shown in Eq. (4), and the recall metric shows how many of the malicious attacks the model detects as shown in Eq. (5). Furthermore, f1-score metrics can gather both precision and recall metrics to obtain the average, as shown in Table 4.

$$Precision = \frac{TP}{(TP + FP)} \tag{4}$$

Table 4. Performance of binary classification

Predicted	Precision	Recall	F1 Score
Normal	99%	99%	99%
Attack	99.06%	99.45%	99%

$$Recall = \frac{TP}{(TP + FN)} \tag{5}$$

However, we obtain the precision with a higher detection of attacks equals to 99.06% and a recall with a higher rate of 99.45%, as shown in Table 4.

Our approach was implemented on windows OS with an Intel core i7 processor and 8 GB of RAM. In addition, we used python programming language and several libraries (e.g. Pandas, Numpy, sklearn, etc.).

D. Comparisons with Other Approaches

To compare our proposed models with the methods from other papers, it should be noted that all the selected methods were trained and tested with NSL-KDD training and testing datasets. We obtained a good result (Accuracy = 99.31%) with our model using random forest compared to the other works, as shown in Table 5.

Table 5. Comparison of our approach with other approaches

Author	Year	Dataset	Classification	Method	Evaluation metric
Gaikwad et al. [33]	2015	NSL-KDD	Binary	Ensemble Naive Bayes	ACC = 81.29%
Pham et al.[2]	2018	NSL-KDD	Binary	Ensemble	ACC = 84.25%
Yuang et al.[1]	2020	NSL-KDD	Binary	RF	ACC = 99.1%
Proposed Approach	2021	NSL-KDD	Binary	RF	ACC = 99.3%

7 Discussion and Perspectives

This paper presented an approach for an anomaly detection system to detect IoT cyber attacks in a smart city. We used a machine learning method specifically the random forest to evaluate our model. We tested our model only with a binary classification and we obtained a good classification result of 99.31% compared to the other works previously mentioned. As shown in Table 5, by using the Random Forest (RF) algorithm, both (Yuang et al. [1] and our proposed) approaches obtained high accuracy rates. While the authors in [1] used the "*max-min*" technique for the normalization of the data, we used the "*one hot encoding*" technique for that purpose.

The work plan of our approach is to train the model among fog nodes and centralize the intrusion detection in a master fog node to detect, identify normal or attack traffic activities and localize it in smart cities. Therefore, we have another goal which is to use n-fold cross validation to evaluate our model, increase the detection rates and reduce false positives rates. We plan also to measure the evaluation and performance metrics by using a multi-classification in the future work with more machine learning algorithms such as Conventional Neural Network (CNN), etc.

8 Conclusion

In this paper, we studied the vision of a smart city security detection system enhancing a traditional IDS for smart city's IoT applications. We presented an approach to detect various IoT attacks in a fog layer instead of a cloud layer. The proposed approach can detect malicious behaviors based on an anomaly-based method by using machine learning techniques. To highlight the efficiency of our proposal, we referred to a case study to highlight the accuracy of our detection approach by using different machine learning techniques and the dataset NSL-KDD. In our ongoing work, we will mainly address the definition of a complete solution for trucking and locating intrusion sources.

References

1. Zhoua, Y., Chenga, G., Jianga, S., Daia, M.: Building an efficient intrusion detection system based on feature selection and ensemble classifier. Comput. Netw. **174**, (2020)
2. Pham, N.T., Foo, E., Suriadi, S., Jeffrey, H., Lahza, H.F.M.: Improving performance of intrusion detection system using ensemble methods and feature selection. In: Proceedings of the Australasian Computer Science Week Multiconference, Brisbane, QLD, Australia, pp. 1–6 (2018)
3. Alqazzaz, A., Alrashdi, I., Aloufi, E., Zohdy, M., Ming, H.: A secure and privacy-preserving framework for smart parking systems. J. Inf. Secur. **9**, 299–314 (2018)
4. Rathore, M.M., Paul, A., Ahmad, A., Chilamkurti, N., Hong, W.-H., Seo, H.: Real-time secure communication for smart city in high-speed big data environment. Future Gener. Comput. Syst. **83**, 638–652 (2018)
5. Garg, S., Kaur, K., Kumar, N., Batra, S., Obaidat, M.S.: Hy-brid classification model for anomaly detection in cloud environment. In: IEEE International Conference on Communications (ICC), pp. 1–7. IEEE (2018)
6. Rathore, M.M., Paul, A., Ahmad, A., Chilamkurti, N., Hong, W.-H., Seo, H.: Real-time secure communication for smart city in high-speed big data environment. Future Gener. Comput. Syst. **83**, 638–652 (2018)
7. Habizadeh, H., Soyata, T., Kantarci, B., Boukerche, A., Kaptan, C.: Sensing, communication and security planes: a new challenge for a smart city system design. Comput. Netw. **144**, 163–200 (2018)
8. Howell, J.: Number of connected IoT devices will surge to 125 billion by 2030, ihs markit says - ihs technology. https://technology.ihs.com/596542/

9. Borgia, E.: The Internet of Things vision: key features, applications and open issues. Comput. Commun. **54**, 1–31 (2014)
10. Restuccia, F., D'Oro, S., Melodia, T.: Securing the Internet of Things: new perspectives and research challenges. IEEE Internet Things J. **1**, 1–14 (2018)
11. Stankovic, J.A.: Research directions for the Internet of Things. IEEE Internet Things J. **1**, 3–9 (2014)
12. Antonakakis, M., April, T., Bailey, M., Bernhard, M., Bursztein, E., Cochran, J., Durumeric, Z., Halderman, J.A., Invernizzi, L., Kallitsis, M., et al.: Understanding the mirai botnet. In: USENIX Security Symposium, pp. 1092–1110 (2017)
13. Zarpelão, B.B., Miani, R.S., Kawakani, C.T., de Alvarenga, S.C.: A survey of intrusion detection in Internet of Things. J. Netw. Comput. Appl. **84**, 25–37 (2017)
14. Santos, J., Leroux, P., Wauters, T., Volckaert, B., Turck, F.D.: Anomaly detection for smart city applications over 5 g low power wide area networks. In: NOMS 2018 – 2018 IEEE/IFIP Network Operations and Management Symposium, pp. 1–9 (2018)
15. Yousefpour, A., Ishigaki, G., Jue, J.P.: Fog computing: towards minimizing delay in the Internet of Things. In: 2017 IEEE International Conference on Edge Computing (EDGE), pp. 17–24. IEEE (2017)
16. Abeshu, A., Chilamkurti, N.: Deep learning: the frontier for distributed attack detection in fog-to-things computing. IEEE Commun. Mag. **56**(2), 169–175 (2018)
17. Roman, R., Zhou, J., Lopez, J.: On the features and challenges of security and privacy in distributed Internet of Things. Comput. Netw. **57**(10), 2266–2279 (2013)
18. Hossain, M.M., Fotouhi, M., Hasan, R.: Towards an analysis of security issues, challenges, and open problems in the Internet of Things. In: 2015 IEEE World Congress on Services (SERVICES), pp. 21–28. IEEE (2015)
19. Habizadeh, H., Soyata, T., Kantarci, B., Boukerche, A., Kaptan, C.: Sensing, communication and security planes: A new challenge for a smart city system design. Comput. Netw. **144**, 163–200 (2018)
20. Moustafa, N., Slay, J.: Unsw-nb15: a comprehensive data set for network intrusion detection systems (unsw-nb15 network data set). In: Military Communications and Information Systems Conference (Mil- CIS), pp. 1–6. IEEE (2015)
21. Koroniotis, N., Moustafa, N., Sitnikova, E., Slay, J.: Towards developing network forensic mechanism for botnet activities in the IoT based on machine learning techniques. In: Mobile Networks and Management: 9th International Conference, MONAMI 2017, Melbourne, Australia, vol. 235, pp. 30–44 (2017)
22. Nobakht, M., Sivaraman, V., Boreli, R.: A host-based intrusion detection and mitigation framework for smart home IoT using openflow. In: 2016 11th International Conference on Availability Reliability and Security (ARES), pp. 147–156. IEEE (2016)
23. Summerville, D.H., Zach, K.M., Chen, Y.: Ultra-lightweight deep packet anomaly detection for Internet of Things devices. In: IEEE 34th International Performance on Computing and Communications Conference (IPCCC), pp. 1–8. IEEE (2015)
24. Raza, S., Wallgren, L., Voigt, T.: Svelte: real-time intrusion detection in the Internet of Things. Ad Hoc Netw. **11**(8), 2661–2674 (2013)
25. Tavallaee, M., Stakhanova, N., Ghorbani, A.A.: Toward credible evaluation of anomaly-based intrusion-detection methods. IEEE Trans. Syst. Man Cybern. Part C (Appl. Rev.) **40**(5), 516–524 (2010)
26. Ahmed, M., Mahmood, A.N., Hu, J.: A survey of network anomaly detection techniques. J. Netw. Comput. Appl. **60**, 19–31 (2016)
27. Prabavathy, S., Sundarakantham, K., Shalinie, S.M.: Design of cognitive fog computing for intrusion detection in Internet of Things. J. Commun. Netw. **20**(3), 291–298 (2018)
28. Oh, D., Kim, D., Ro, W.W.: A malicious pattern detection engine for embedded security systems in the internet of things. Sensors **14**(12), 24 188–24 211 (2014)

29. Moustafa, N., Slay, J.: The evaluation of network anomaly detection systems: Statistical analysis of the unsw-nb15 data set and the comparison with the kdd99 data set. Inf. Secur. J. Global Perspect. **25**(1–3), 18–31 (2016)
30. Rathore, M.M., Paul, A., Ahmad, A., Chilamkurti, N., Hong, W.-H., Seo, H.: Real-time secure communication for smart city in high-speed big data environment. Future Gener. Comput. Syst. **83**, 638–652 (2018)
31. Zhang, J., Zulkernine, M., Haque, A.: Random-forests-based network intrusion detection systems. IEEE Trans. Syst. Man Cybern. Part C (Appl. Rev.) **38**(5), 649–659 (2008)
32. Angrishi, K.: Turning Internet of Things (IoT) into internet of vulnera- bilities (iov): Iot botnets, arXiv preprint arXiv (2017)
33. Schneierl, B.: Security econmics of the internet of things. https://bit.ly/2OBuxBE
34. Liang, C., Shanmugam, B., Azam, S., Karim, A., Islam, A., Zamani, M., Kavianpour, S., Idris, N.B.: Intrusion detection system for the Internet of Things based on blockchain and multi-agent systems. Electronics **9**(7), 1120 (2020)

Integration of Localization and Wireless Power Transfer Using Microwave

Kentaro Hayashi[1]([envelope]), Hikaru Hamase[1], Yuki Tanaka[2], Takuya Fujihashi[1], Shunsuke Saruwatari[1], and Takashi Watanabe[1]

[1] Graduate School of Information Science and Technology, Osaka University, Suita, Japan
hayashi.kentaro@ist.osaka-u.ac.jp
[2] Panasonic Corporation, Osaka, Japan

Abstract. This paper proposes a system to realize simultaneous power supply and localization for indoor Internet of Things (IoT) devices. The power supply to IoT devices is realized via microwave power transfer using numerous distributed access points (APs). The proposed system realizes efficient power transmission by controlling the phase of microwaves from each AP such that the microwaves transmitted from each AP cause constructive interference at an IoT device. The proposed system determines the optimal phase of a signal from each AP based on the RSSI feedback from an IoT device. Furthermore, the proposed system extracts the location of the IoT device from the optimal phases of the power transfer. The mechanism exploits the fact that the phases, which are optimized to make constructive interference at the IoT device, reflects the positional relation between each AP and the IoT device. We evaluated the proposed system using computer simulation, and the evaluation results show that the power supply is improved by about 3.3 times by optimizing the phase. At the same time, decimeter-level localization accuracy was achieved.

1 Introduction

With the development of Internet of Things (IoT) technology, various items around us such as watches, glasses, toothbrushes, and scales are connected to the internet [1]. At the same time, completely new services that utilize the data acquired by these things are appearing: inventory management systems, smart houses, smart factories, health monitoring, and so on.

However, to connect all things to the network in the real meaning, there are two remaining problems. The first problem is how to supply power to IoT devices. The IoT devices currently available on the market consume a lot of power and require frequent battery replacement or charge. Typical attempts to solve this problem include research on energy harvesting [2,3], and power saving [4,5]. Energy harvesting research aims to drive IoT devices autonomously using ambient energy such as sunlight or vibration. However, obtained power depends on the weather, time of day, and location. Power saving research aims to reduce the frequency of battery replacement by reducing the frequency of data collection and transmission. However, to provide high-quality sensing services, it is necessary to acquire more physical space information.

L. Barolli et al. (Eds.): AINA 2021, LNNS 226, pp. 229–239, 2021.
https://doi.org/10.1007/978-3-030-75075-6_18

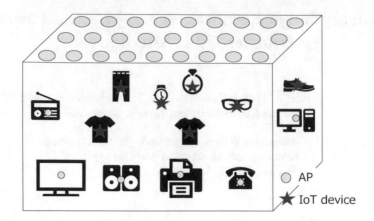

Fig. 1. Assumed environment of the proposed system

The second problem is the localization of the IoT devices. For example, if an IoT device that measures temperature is located near a window or a heater, the meaning of the measured data might differ drastically. If we can add geographical information to the acquired sensor data, higher quality service will be provided. As indoor localization techniques, channel state information (CSI) based methods have been attracting attention. For indoor localization using CSI, two methods have been proposed: the fingerprint-based [6,7] and the Time-of-Arrival (ToA)/Angle-of-Arrival (AoA) based [8–10]. Although both methods achieve high localization accuracy, the device must have a Wi-Fi chip to acquire CSI. Installing Wi-Fi chips in each IoT device is unrealistic in terms of deployment cost.

To this end, this paper proposes a method for integrating wireless power transfer and localization using microwaves. Figure 1 shows an assumed environment of the proposed system. Each device connected to the domestic power source, such as lights, air conditioners, and televisions, acts as an access point (AP). Each AP cooperates to supply power to the IoT devices and locate them. The wireless power transfer is realized by microwave power transfer by each AP. Although the distance attenuation is significant and supplied energy is small, microwave power transfer has two merits: stable power supply regardless of time and environment, and simultaneous data communication using microwaves [11–15]. Each AP distributed in a space has a role as a TX antenna for microwave power transfer. The cooperation of APs generates constructive interference at the location of IoT devices with appropriate phase control of the microwaves transmitted from each AP. The phase of each AP is optimized based on the RSSI feedback from the IoT device. Additionally, the proposed system locates the IoT device based on the optimized phase for power transfer. Our proposed system uses the fact that the phase optimized for the received power of the IoT device reflects the spatial information between the AP and the IoT device. We evaluated the proposed system by simulation. The results show that the proposed method can simultaneously achieve efficient power supply and decimeter-level localization.

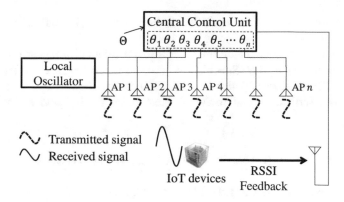

Fig. 2. Network model

The remainder of this paper is organized as follows: Sect. 2 introduces the overview of the proposed system. Section 3 describes the phase optimization algorithm based-on RSSI feedback. Section 4 describes the localization algorithm based-on the optimized phase. Section 5 discusses the evaluation of the proposed method by computer simulations. Finally, Sect. 6 concludes the paper.

2 Proposed System

The proposed system comprised the following two components:

1. Distributed cooperative microwave power transfer (Sect. 3)
2. Locating an IoT device using the optimized phase (Sect. 4)

Figure 2 shows the network model of the proposed system. The system uses many distributed APs to supply power to a wide area with high efficiency. In Fig. 2, there are n number of APs. All APs were connected to a central control unit (CCU) and transmitted the same frequency microwaves while sharing a local oscillator.

The CCU controls the phase of the microwaves from each AP using a phase controller. The set of phases at all APs is called a phase set $\Theta = (\theta_1, \ldots, \theta_n)$, where n represents the number of APs. The IoT device can measure the RSSI and feedback the measured RSSI to the CCU. The CCU optimizes the phase set used for power transmission based on the RSSI feedback from the IoT device. The phase set that can provide the maximum power to the IoT device is called a optimal phase set $\Theta_{\text{opt}} = (\hat{\theta}_1, \ldots, \hat{\theta}_n)$, and the search for the optimal phase set is called phase optimization. The phase optimization procedure is described in detail in Sect. 3. Because IoT devices are mobile, we assumed that the phase optimization is performed periodically.

When there are multiple IoT devices, the phase-controlled cooperative power transfer (PC-CPT) proposed in [16] can be used. In PC-CPT, the optimal phase set is calculated for each IoT device independently. By transmitting microwaves while switching the optimum phase set for each IoT device in a time-sharing manner, it is possible to satisfy the power demand of each IoT device. For simplicity, in this study, we assumed only one IoT device.

In the proposed system, the optimal phase set calculated for power transfer is also used to locate the IoT device. Specifically, we considered the fact that the optimized phase reflects the positional relation between the APs and the IoT device. The localization method is described in detail in Sect. 4. Hereinafter, the IoT device is referred to as a receiver.

3   Distributed Cooperative Microwave Power Transfer

3.1   Calculation of RSSI and Effect of Phase Control

In this section, we calculate the RSSI at the receiver when n APs transmit microwaves. Let P_i denote the power of the signal transmitted from AP i on the receiver, and ϕ_i denote the phase shift on the propagation path between AP i and the receiver. The complex channel gain between AP i and the receiver is expressed by Eq. (1).

$$h_i = \sqrt{P_i} e^{j\phi_i} \tag{1}$$

The received signal is the result of superimposing the signals received from each AP. Therefore, the RSSI at the receiver is expressed by Eq. (2).

$$\text{RSSI} = \left| \sum_{i=1}^{n} h_i \right| = \left| \sum_{i=1}^{n} \sqrt{P_i} e^{j\phi_i} \right| \tag{2}$$

Let $\text{RSSI}(\Theta)$ denote the RSSI at the receiver when controlling phase using the phase set $\Theta = (\theta_1, \ldots, \theta_n)$. $\text{RSSI}(\Theta)$ is expressed by Eq. (3).

$$\text{RSSI}(\Theta) = \left| \sum_{i=1}^{n} \sqrt{P_i} e^{j(\theta_i - \phi_i)} \right| \tag{3}$$

Equation (3) takes the maximum value $\sum_{i=1}^{n} \sqrt{P_i}$ when Eq. (4) holds.

$$\theta_1 - \phi_1 = \theta_i - \phi_i \qquad \forall i \tag{4}$$

When Eq. (4) holds, the phases of the received signal from each AP is equal at the receiver's position. The phase set that satisfies Eq. (4) is the optimal phase set Θ_{opt}.

3.2   Phase Optimization Algorithm Based on RSSI Feedback

The simplest way to perform phase optimization is to receive RSSI feedback for all phase sets and adopt the phase set with the maximum RSSI feedback as the optimal phase set Θ_{opt}. However, when the controllable phase resolution of each AP is P, we need P^n RSSI feedback packets to determine the optimal phase set from all the searches. In this case, the number of RSSI feedback packets required for phase optimization increases exponentially with the number of APs.

To minimize the number of RSSI feedback packets, we use the following procedure:

1. Only AP 1 transmits microwaves under a phase of $\hat{\theta}_1 = 0$.
2. AP 2 starts to transmit microwaves. Based on the RSSI feedback from the IoT device, AP 2 optimizes the phase of the transmitted signal so that the amplitude of the combined signals of APs 1 and 2 at the IoT device is maximized. Thereafter, AP 2 transmits microwaves under a phase of $\hat{\theta}_2$.
3. AP 3 begins to transmit microwaves. Based on the RSSI feedback from the IoT device, AP 3 optimizes the phase of the transmitted signal so that the amplitude of the combined signals of APs 1, 2, and 3 at the IoT device is maximized. Thereafter, AP 3 transmits microwaves under a phase of $\hat{\theta}_3$.
4. The same procedure is repeated until AP n obtains its optimal phase $\hat{\theta}_n$.

When phase optimization is performed according to this algorithm, the number of RSSI feedback packets required is $(n-1)P$. Note that phase optimization is not performed for AP1, so the number of APs for which the phase is optimized is $n-1$.

4   Locating the IoT Device Using the Optimized Phase

4.1   Formulation

The proposed system estimates the receiver's position based on the following two aspects:

- The optimal phase set Θ_{opt} depends on the phase shift on the propagation path between each AP and the receiver (Eq. (4)).
- The phase shift between the two points depends on the distance between them.

More specifically, for every position in space, we computed the optimal phase set that would have been obtained if the receiver had been present at that position. The receiver's estimated position is the coordinate that yields the optimal phase set closest to the actual optimal phase set $\Theta_{\text{opt}} = (\hat{\theta}_1, \ldots, \hat{\theta}_n)$. Mathematically, we find the location (x,y,z) that minimizes the following objective function J.

$$J(x,y,z) = \sum_{i=1}^{n} \left((\hat{\theta}_i - \hat{\theta}_1) - (\psi_i - \psi_1) \right)^2 \tag{5}$$

where ψ_i is the phase delay between AP i and (x,y,z). When no reflected waves are present, ψ_i is expressed as Eq. (6).

$$\psi_i = \frac{2\pi L_i}{\lambda} \bmod 2\pi \tag{6}$$

where λ is the wavelength of the microwaves, and L_i is the distance between AP i and (x,y,z). When the coordinate of AP i is (x_i, y_i, z_i), L_i is expressed as Eq. (7).

$$L_i = \sqrt{(x-x_i)^2 + (y-y_i)^2 + (z-z_i)^2} \tag{7}$$

4.2   Considering Reflected Waves

Next, we consider the effects of reflected waves on Eq. (6). Figure 3 shows the manner in which the microwave transmitted from AP i reaches (x, y, z). In a real environment, in addition to direct waves, reflected waves arrive after being reflected from a floor or wall. Therefore, the relationship between the distance and phase delay shown in Eq. (6) is not applicable. In the proposed system, the reflecting surface's position is assumed to be known, and correction is performed by calculating the combined signal from the direct and reflected waves. For simplicity, the number of reflections was assumed to be one.

To facilitate the treatment of reflected waves, we considered a virtual AP i at a position symmetrical to AP i with respect to the reflecting surface, as shown in Fig. 3. The signal arriving at (x, y, z) after reflecting once from AP i to the reflecting surface can be regarded as a direct wave from the virtual AP to (x, y, z).

First, we considered the contribution of the direct and reflected waves to the phase delay. The path length on the direct wave from AP i to (x, y, z) is L_i, as shown in Eq. (7). Let L_i' denote the path length of the signal transmitted from AP i that reached (x, y, z) after being reflected once by the reflecting surface. L_i' corresponds to the distance from virtual AP i to (x, y, z). According to the Friis transmission formula, when directivity and polarization of the antenna is not considered, the intensity of the microwave decays inversely proportional to the square of the distance. Therefore, the relationship between the amplitude of the direct and reflected waves can be expressed as shown in Eq. (8).

$$B_i = A_i \frac{L_i}{L_i'} \tag{8}$$

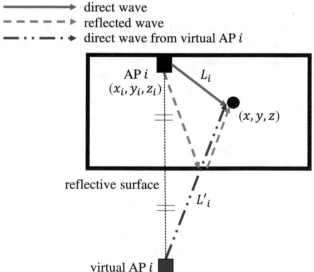

Fig. 3. Considering the virtual AP i

Where A_i denote the amplitude of the direct wave between AP i, and B_i denote the amplitude of the reflected wave. For simplicity, let $A_i = 1$.

The phase shift on the direct wave from AP i to (x,y,z) is ψ_i, as shown in Eq. (6). In addition, let ψ'_i denote the phase shift on the reflected wave from AP i to (x,y,z). The term ψ'_i is expressed using L'_i, as shown in Eq. (9).

$$\psi'_i = \frac{2\pi L'_i}{\lambda} \bmod 2\pi \tag{9}$$

Based on Eqs. (8) and (9), the sum of the signals from the direct wave from AP i and reflected wave from AP i can be expressed as shown in Eq. (10).

$$y_i = e^{\psi_i} + \frac{L_i}{L'_i} e^{\psi'_i} \tag{10}$$

Therefore, when considering the effect of reflected waves, the phase shift from AP i to (x,y,z) is expressed as shown in Eq. (11),

$$\overline{\psi_i} = \arg(y_i) \tag{11}$$

Subsequently, $\overline{\psi_i}$ can be applied as ψ_i to Eq. (5) to calculate the coordinates of the receiver.

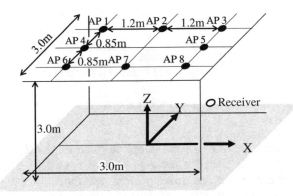

Fig. 4. Antenna placement

5 Evaluation

In this section, we evaluate the phase optimization performance and localization accuracy of the proposed system based on a simulation. In the simulation, the propagation channel between each AP and receiver was calculated via ray launching from the receiver. The frequency and power of the transmitted signal of each AP were set to 920 MHz and 1 W, respectively.

Figure 4 shows the placement of the APs and receiver. The space was a 3 m square cube. The number of APs was eight, and the receivers were randomly placed within the range of $0.25 \leq X \leq 2.75$, $0.25 \leq Y \leq 2.75$, $Z = 1.0$. Each antenna was a dipole antenna.

5.1   Phase Optimization

First, we evaluated the phase optimization algorithm for power transfer proposed in Sect. 3. For the evaluation, we used the following four methods:

1. Optimal: The maximum power that can be supplied under the optimal phase set.
2. Proposed: The phase optimization algorithm described in Sect. 3.2.
3. Random: A method to continue attempting a random phase set.
4. No Control: A method without phase control.

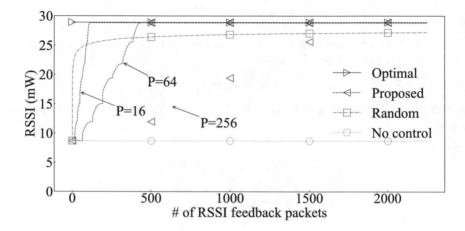

Fig. 5. Transition of RSSI

First, we analyzed the effect of phase optimization on the RSSI. Figure 5 shows the transition of the RSSI for each phase optimization method. The horizontal axis represents the number of RSSI feedback packets, and the vertical axis represents the RSSI at the corresponding step. The resolution of the phase control P was 16, 64, or 256.

As shown in Fig. 5, the phase optimization increased the RSSI. The average RSSI without phase control was approximately 8.2 mW, whereas it was approximately 26.8 mW when the phase of each AP was optimized. By phase optimization, the microwaves from each AP causes constructive interference at the receiver's position. As a result, the received power was improved by about 3.2 times.

Next, we analyzed the convergence speed. In the proposed phase optimization algorithm, the RSSI continued to increase intermittently. This occurred because the phase of each AP was optimized sequentially. For example, if P is 16, then the optimal phase set satisfying Eq. (4) is available in 112 RSSI feedback packets.

As shown in Fig. 5, the effect of P on the RSSI after phase optimization was minimal. A higher phase resolution is expected to yield a higher RSSI after phase optimization. However, the results show that the difference in the RSSI due to the phase resolution was minimal. Because the phase optimization requires a number of RSSI feedback packets proportional to the value of P, this result indicates that the phase optimization can be performed rapidly. Also, the cost of the phase controller can be reduced. The proposed system estimates the receiver's position based on the optimal phase set under a phase resolution of P. If we estimate the receiver's position based on the optimal phase set under a low resolution, then the localization accuracy may be degraded. The effect of P on the localization accuracy is discussed in Sect. 5.2.

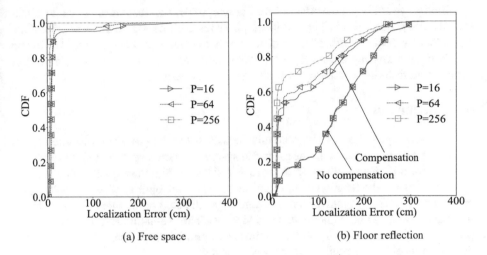

(a) Free space   (b) Floor reflection

Fig. 6. CDF of localization error

5.2   Localization Accuracy

In this section, we evaluate the localization method using the optimized phase proposed in Sect. 4. For the evaluation, we used the following two methods:

1. No Compensation: A method without considering reflected waves, as described in Sect. 4.1
2. Compensation: A method considering reflected waves, as described in Sect. 4.2.

In the no-compensation method, we evaluated the case of free space and the case involving floor reflection, whereas in the compensation method, we evaluated the case involving floor reflection.

   Figure 6 shows the CDF of the localization errors. The horizontal and vertical axes represent the localization error and CDF, respectively. As shown in Fig. 6, a highly accurate estimation was enabled in free space. Meanwhile, when reflected waves were present, the accuracy decreased. This occurred because the relationship between distance and phase difference in Eq. (6) was nullified owing to the effects of the reflected waves.

   Next, we evaluated the effect of correcting the reflected wave described in Sect. 4.2. As shown in Fig. 6(b), the localization accuracy improved by the correction of the reflected wave. It was assumed that the inaccurately estimated position was caused by the non-consideration of the antenna's directivity. When considering the contribution of the direct and reflected waves to the phase delay, Eq. (8) only considers the distance attenuation. In reality, Eq. (8) cannot accurately rectify the phase delay because the signal strength depends not only on the distance, but also on the direction due to the antenna's directivity. A method that considers the antenna's directivity will be considered in future studies.

Finally, we analyzed the effect of the phase optimization resolution on the localization accuracy. As shown in Fig. 6, a lower phase resolution resulted in a lower localization accuracy. More specifically, the difference in average error between the $P = 16$ and $P = 256$ cases was approximately 0.1 m. We believe that this difference is insignificant; hence, high-resolution phase optimization is not necessary for the proposed system.

6 Conclusion

Herein, we proposed a system to simultaneously solve the power and location problems of indoor IoT devices. In the proposed system, numerous distributed APs perform cooperative microwave power transfer to IoT devices. Simultaneously, our proposed system estimates the location of IoT devices based on the optimal phase calculated for power transfer. This system does not require a specific chip or module to be installed in IoT devices and can be implemented at a low cost. The evaluation results show that the power supply is improved by about 3.3 times by optimizing the phase. Simultaneously, decimeter-level localization accuracy was achieved.

Acknowledgments. This work was supported by JSPS KAKENHI Grant Numbers JP19H01101, JP19K11923, and JP18H03231.

References

1. Shah, S.H., Yaqoob, I.: A survey: Internet of Things (IoT) technologies, applications and challenges. In: IEEE Smart Energy Grid Engineering (IEEE SEGE 2016), pp. 381–385, October 2016
2. Lim, C.J., Drieberg, M., Sebastian, P., Hiung, L.H.: A simple solar energy harvester for wireless sensor networks. In: 6th International Conference on Intelligent and Advanced Systems (ICIAS 2016), pp. 1–6, August 2016
3. Kawahara, Y., Bian, X., Shigeta, R., Vyas, R., Tentzeris, M.M., Asami, T.: Power harvesting from microwave oven electromagnetic leakage. In: Proceedings of the 15th International Conference on Ubiquitous Computing (ACM UbiComp 2013), pp. 373–382, September 2013
4. Huang, P., Xiao, L., Soltani, S., Mutka, M.W., Xi, N.: The evolution of MAC protocols in wireless sensor networks: a survey. IEEE Commun. Surv. Tutor. **15**(1), 101–120 (2013)
5. Narusue, Y., Kawahara, Y., Asami, T.: Impedance matching method for any-hop straight wireless power transmission using magnetic resonance. In: IEEE Radio and Wireless Symposium (IEEE RWW 2013), pp. 193–195, January 2013
6. Wang, J., Jiang, H., Xiong, J., Jamieson, K., Chen, X., Fang, D., Xie, B.: LiFS: low human-effort, device-free localization with fine-grained subcarrier information. In: Proceedings of the 22nd Annual International Conference on Mobile Computing and Networking (ACM MobiCom 2016), pp. 495–506, October 2016

7. Wang, X., Gao, L., Mao, S., Pandey, S.: CSI-based fingerprinting for indoor localization: a deep learning approach. IEEE Trans. Veh. Technol. **66**(1), 763–776 (2016)
8. Gong, W., Liu, J.: SiFi: pushing the limit of time-based WiFi localization using a single commodity access point. Proc. ACM Interact. Mob. Wearable Ubiquitous Technol. **2**(1), 1–21 (2018)
9. Kotaru, M., Joshi, K., Bharadia, D., Katti, S.: SpotFi: decimeter level localization using WiFi. In: Proceedings of the 2015 Conference on the ACM Special Interest Group on Data Communication (ACM SIGCOMM 2015), pp. 269–282, August 2015
10. Krishnaveni, V., Kesavamurthy, T., Aparna, B.: Beamforming for direction-of-Arrival (DOA) estimation a survey. Int. J. Comput. Appl. **61**(11), 1–11 (2013)
11. Kassab, H., Louveaux, J.: Simultaneous wireless information and power transfer using rectangular pulse and CP-OFDM. In: IEEE International Conference on Communications (IEEE ICC 2019), pp. 1–6, May 2019
12. Asiedu, D.K.P., Mahama, S., Jeon, S.W., Lee, K.: Joint optimization of multiple-relay amplify-and-forward systems based on simultaneous wireless information and power transfer. In: IEEE International Conference on Communications (IEEE ICC 2018), pp. 1–6, May 2018
13. Perera, T.D.P., Jayakody, D.N.K., Sharma, S.K., Chatzinotas, S., Li, J.: Simultaneous wireless information and power transfer (SWIPT): recent advances and future challenges. IEEE Commun. Surv. Tutor. **20**(1), 264–302 (2017)
14. Fang, Z., Yuan, X., Wang, X.: Distributed energy beamforming for simultaneous wireless information and power transfer in the two-way relay channel. IEEE Signal Process. Lett. **22**(6), 656–660 (2014)
15. Varshney, L.R.: Transporting information and energy simultaneously. In: IEEE International Symposium on Information Theory (IEEE ISIT 2008), pp. 1612–1616, July 2008
16. Kawasaki, J., Hamase, H., Kizaki, K., Saruwatari, S., Watanabe, T.: Phase-controlled cooperative wireless power transfer for backscatter IoT devices. In: IEEE International Conference on Communications (IEEE ICC 2020), pp. 1–6, February 2020

A Usage-Based Insurance Policy Bidding and Support Platform Using Internet of Vehicles Infrastructure and Blockchain Technology

Frank Yeong-Sung Lin[1(✉)], Wen-Yao Lin[1], Kuang-Yen Tai[1], Chiu-Han Hsiao[2], and Hao-Jyun Yang[2]

[1] Department of Information Management, National Taiwan University, Taipei, Taiwan
[2] Research Center for Information Technology Innovation, Academia Sinica, Taipei, Taiwan

Abstract. This paper presents a usage-based insurance (UBI) platform that incorporates Internet of Vehicles (IoV) and blockchain technologies, discussing the potential stakeholders, business models, and interaction modes involved in this platform. Existing UBI products mostly use data on the driver's mileage, driving period, or driving region for more accurate insurance calculations. Automobile UBI encourages customers to continue improving their ability to drive safety and provides a means to smoothly, transparently, and rationally calculate insurance pricing and payout. This paper proposes blockchain architecture to remedy management problems in a UBI environment. A bidding mechanism suitable for the blockchain-based UBI platform was designed to close the information gap between the insurance company and consumer, thus increasing consumer trust in the platform.

Keywords: Usage-based insurance · Internet of Vehicles · Blockchain · Bidding mechanism

1 Introduction

The nature and techniques of monetary transactions are evolving with the development of innovation data processing technologies. Traditionally, a client submitting a claim is expected to sign an agreement with the solicitor. With current advances in computing and mobile communication, however, insurers can pay out compensation automatically to their clients immediately after an accident occurs. Furthermore, the basis for calculating premiums has evolved with the increasing ease of gathering information. For example, the use of usage-based insurance (UBI) premium calculation to obtain a vehicle's status is likely to yield more accurate evaluations of the insured's status. At present, to evaluate a driver's risk level, insurance companies adopt onboard hardware to collect data on how the driver drives [1,2].

Auto UBI products are specifically designed by considering driving behavior. Compared with traditional auto insurance, auto UBI uses data on driving speed, driving period, and braking time to more accurately quantify risk. Auto UBI products can also encourage drivers to continually improve their ability to drive safety and provide a means to smoothly calculate payout and pricing. Auto UBI brings additional benefits, such as providing tracking for stolen cars, roadside assistance, and active emergency notifications. Good driving behavior can make insurance cheaper for the driver [3], which incentivizes good driving habits.

However, more accurate risk evaluation requires more personal driving data to be collected, which encroaches on the driver's privacy. Thus, when establishing a UBI platform, accuracy must be balanced with privacy according to the best practice. At present, privacy is highly valued by many individuals; thus, it must be safeguarded by auto UBI platform designers and users [4].

Intrinsic to blockchain technology is shared database maintenance by peers to trusted parties in an immutable, decentralized, and automated fashion [4,5]. Hence, driving information can be synchronized on a blockchain platform for insurance companies to evaluate drivers' behavior. Furthermore, such a use of blockchain technology potentially enables reliable traffic adjudication, accident prevention, and the preventive maintenance of vehicles [6].

Considering the aforementioned context, the authors of this paper aimed to design a UBI information platform that incorporates blockchain technology, doing so to provide fair prices and accurate information in real time. In particular, this study's contribution is fivefold:

(1) Design a UBI platform and its maintenance infrastructure based on blockchain and onboard equipment.
(2) Plan the driving data synchronization method in the UBI environment to guarantee effective and efficient transmission.
(3) Design a blockchain-based system architecture and analyze its management problems in a UBI environment.
(4) Design a bidding mechanism suitable for the blockchain-based UBI platform, doing so to minimize the information gap between the insurance company and consumer, thus increasing consumer trust in the platform.
(5) Investigate the bidding mechanism, business model, and additional opportunities for development (with respect to technologies or protocols) when using the blockchain-based UBI platform.

2 Related Works

2.1 Current UBI Services

In present-day UBI services, the insurer first provides automobile insurance products based on the driver's driving conduct. The UBI service analyzes the driving data that it gathered (pertaining to, for example, driving speed, braking time, and driving period) to determine the insurance premium for the driver [2]. Distinctive to this study's design is that driving data is saved centrally, making it impossible to know the particular sets of data the company has collected. An analysis of data collected in an auto UBI yields more accurate information of the driving situation. UBI systems typically leverage Internet of Vehicles (IoV) technology, where sensors are installed on a car to record driving behavior. Current UBI systems are also integrated with smartphones through an app to display driving information (on the vehicle's mileage, the driving periods, and even whether the driver brakes frequently). Data from this service can even be combined with the owner's regular maintenance records. These strategies allow insurers to grasp a drivers' driving habits and price their car insurance products more rationally and flexibly [7]. For example, insurers can offer cheaper and more flexible options to a customer with good driving habits (e.g., someone who uses the accelerator in moderation) and who obeys traffic rules because they are less likely to meet an accident or encounter road rage.

Although auto UBI has these advantages, it has its limitations with respect to privacy. Specifically, because auto UBI requires large data sets for big data analysis, it is necessary to obtain the owner's vehicle information and personal information, which deters car owners who value privacy. Addressing this privacy concern will make auto UBI more widely accepted [8].

2.2   Instant Information Transmission Mechanism Through Internet of Vehicles

To ensure that driving data are updated continually, continuously synchronizing the data collected by the IoV infrastructure with the blockchain platform is critical. U.S. patent no. 9 959 764 [9] presents a data synchronizing mechanism. In this mechanism, the onboard unit (OBU) gathers all data obtained from vehicle sensors using an information assortment unit. Subsequently, the onboard framework transmits the data to the insurance agency's system interface [10]. When the insurance agency obtains the data, the data are subject to numerous analyses to identify the vehicle, determine the risk factors, formulate a risk evaluation model, and compute the driver's likelihood of engaging in various behaviors or being in various situations. This study's platform has all gathered data synchronized to a blockchain platform instead, and the system should thus be adapted to the blockchain environment.

2.3   Blockchain-Based Platform Design

(1) DPoS (Delegated Proof of Stake) [11]

The objectives of DPoS are to make blockchain operations more efficient, much faster, and less centralized. The DPoS's consensus mechanism can be easily conceptualized as composed of coin holders needing to vote for delegates, who are responsible for validating transactions and maintaining the blockchain. The adaptation to internal failure in DPoS is improved by the practical Byzantine fault tolerance (BFT). These innovations allow DPoS to ameliorate numerous problems regarding proof of stake (PoS) and proof of work (PoW) [10]. However, DPoS faces a few problems. For example, decentralization can never be attained because although having more validators is better, a network with too many validators is slow.

(2) Tendermint [12]

Tendermint is a consensus mechanism based on the BFT, but it differs from DPoS in its asynchronicity. Tendermint's primary purpose is to improve efficiency, and its asynchronicity improves its fault tolerance. For instance, when confronting a distributed denial-of-service (DDoS) attack, DPoS generates numerous forks, whereas Tendermint does not; its operation stops to ensure consistency [13].

To choose a suitable architecture for this study's platform, operation cost should be the primary consideration because the platform is mainly intended for businesses. In addition, the platform should also be as decentralized as possible to make the mechanism sufficiently fair. Tendermint has several advantages: it is more efficient than PoW methods, more decentralized than proof of authority (PoA) and PoS methods, and quicker at block generation than the DPoS method. A comparison of various blockchain architectures is presented in Table 1. In general, although every blockchain architecture has its benefits, Tendermint is most suited to this study's platform [14].

Table 1. Comparison of blockchain platform architectures.

Items	PoW	PoA	Pure-PoS	DPoS	Tendermint
Consensus basis	Computing power	Authority nodes	Stakes	Stakes	Stakes
Average update speed	20 s–10 min	By assignment	1–10 min	3–40 s	1–3 s
Cost	High	Low	High	Low	Low
Merits	Structure that is widely used	Configuration is very flexible	Low computing power	The system can operate stably under attack	High structural flexibility and update speed
Defects	The efficiency is lower	Low system compatibility	The system has high concentration	Generate more forks when attack happened	The system can operate stably under attack

3   Proposed Business Model and Blockchain-Based System Architecture

3.1   IoV System Infrastructure

Inside an IoV-capable vehicle, an OBU acts as the center that integrates the data collected from all the sensors, including the car computer, dashcam, and automotive radar. The OBU is also responsible for communicating with the blockchain-based UBI platform, uploading corresponding data to it. The structure of such a system inside a vehicle is illustrated in Fig. 1.

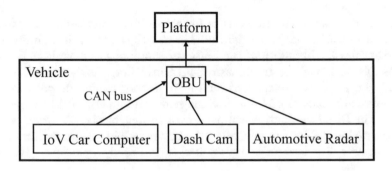

Fig. 1. System structure of an IoV-capable vehicle.

In the IoV system, the driving data are subject to a standardization process and other essential cycles before being uploaded to the blockchain platform (Fig. 2). The platform then begins identifying the vehicle, determining the driver's risk, determining the driver's driving habits, generating a risk judgment model, and eventually, adjusting the premium accordingly. This study modified the system in the aforementioned patent for business use and to ensure that the data are suited to a blockchain environment [15]. The collected driving data are then used to establish a risk evaluation model that yields a more optimal premium, which increases the profitability of the insurance company.

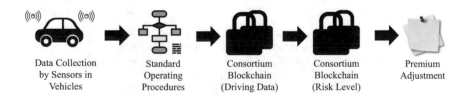

Fig. 2. Dataflow of the UBI platform using blockchain.

3.2   Blockchain-Based UBI Platform

This study designed a system based on Tendermint [16]. Through this architecture, drivers constitute the essential member because their behaviors are recorded in detail on the blockchain. In contrast to the existing auto UBI architectures, this architecture comprises other third-party members, such as car dealers, vehicle manufacturers, vehicle supervisors, vehicle maintenance depots, and banks shown in Fig. 3. Although these members can access the blockchain [17], the data are not public; anyone who requires access to the data must provide a reasonable explanation and comply with the data policy to guarantee that the data are not being misused.

3.3   Bidding Mechanism

Every time a driver drives a car, sensors onboard the car record the driver's behavior. Such behavioral data are then integrated, analyzed, encrypted, and uploaded to the blockchain-based UBI platform. Subsequently, the data traverse a series of validation and confirmation processes from the dealers, manufactures, and maintenance depots. Every insurance agency can fetch the verified data to evaluate the driver's behavior, adjust its premium, and generate a new customized contract that features a range of insurance products for the driver to choose from. This process is illustrated in Fig. 3.

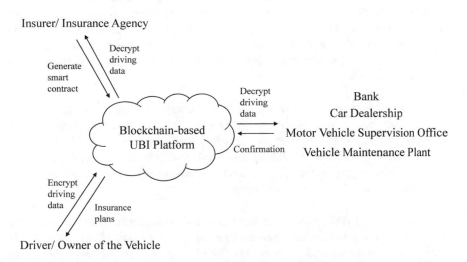

Fig. 3. Designed UBI platform architecture and process of generating customized insurance contracts.

3.4   Advantages of Adopting Blockchain

In contrast to the existing UBI platforms, this study's platform can automatically gather all data from sensors on the vehicle to update the driver's risk assessment model. Furthermore, this study's platform allows data to be immediately uploaded after an accident occurs. Because the platform is synchronized (albeit intermittently) with vehicle maintenance depots, maintenance personnel can guide drivers on how to maintain their vehicle after the accident occurs.

Drivers should be provided information on their risk factors to help them understand the justification behind their insurance prices. This fosters consumer trust in the insurer's pricing decisions. Blockchain technology fosters consumer trust by making data more reliable; this is realized through blockchain's basic principle of decentralizing data storage to make data resistant to modification. Furthermore, because vehicle manufacturers constitute one member of the blockchain, they can endorse the legitimacy of the data collected as being free from manipulation [18].

4   Stakeholders and Fare Adjustment Rules

4.1   Stakeholders on the Blockchain-Based Platform

(1) Driver

The vehicle driver is the most involved in the blockchain platform. Their driving behavior forms the basis of the entire platform. Hence, whether this study's platform is effective depends on whether drivers are willing to participate. This study's platform protects driver data, which appeals to privacy-conscious drivers, and yields potentially cheaper insurance prices as well as tips on safe driving.

(2) Owner of the Vehicle

Because vehicle owners have the right to choose whether to install onboard equipment, the success of this study's service in practice also depends on the willingness of vehicle owners to participate. Through participating, owners better understand the health of their vehicle through the system's provision of driving parameter data, maintenance records, advance maintenance suggestions, and security statements.

(3) Insurer

Insurers constitute the other major actor in this platform. This study's architecture allows insurers to more accurately quantify risk (from data on the driver and vehicle) and improve on their products. To increase their profitability, insurers can also cooperate with vehicle manufacturers and maintenance depots to make driving safer.

4.2   Fare Adjustment Rules

Present-day insurance products consider only the risk posed by the driver and not the vehicle. However, this study's platform allows the history and condition of a vehicle to be detailed through data collected by onboard equipment and from other actors (e.g., maintenance records from depots).

This study's platform enables insurance pricing to be based on two components: driver-related and vehicle-related risks. Insurance plans can thus come in one of three types: first, bill the driver and vehicle owner separately based on their individual risks; second, bill only the vehicle owner based on both driver-related and vehicle-related risks; and third, bill only the driver based on the driver-related risk. Only the third type is used in present-day insurance [19]. Table 2 compares these three plans with existing UBI and non-UBI products.

Table 2. Comparison of insurance plans from this study and existing plans.

Items	Plan 1	Plan 2	Plan 3	Existing UBI	Existing non-UBI
Fare calculating basis	Driver and automobile's risk	Driver and automobile's risk	Only driver's risk	Only driver's risk	Only driver's risk
Responsible for paying insurance	Driver and vehicle owner	Vehicle owner	Vehicle owner or driver	Vehicle owner	Vehicle owner
Fare adjustment methods	By driving behavior and car maintenance history	By driving behavior and car maintenance history	By driving behavior	By driving behavior	By times of hit-and-run

The platform also allows insurers to adjust their pricing depending on how well the driver drives: for example, how frequently the driver speeds, how well the driver brakes, how adequately the driver uses the turn signal, and how regularly the driver maintains the vehicle [20].

5 Discussion

Insurers typically possess expertise that consumers do not have [21], which raises concerns regarding an information asymmetry between the insurer and insured in the traditional insurance model. Because they are put on the blockchain, data on the platform are open and cannot be arbitrarily altered. This study's bidding mechanism ameliorates such an asymmetry, increasing consumer trust in the platform [22]. The mechanism for the adjustment of the insurance premium should be made transparent to drivers. Unlike the present-day insurance model, the bidding mechanism in this study's platform is completely automated for premiums to be constantly updated. Notably, this mechanism accounts for both driver-related and vehicle-related risk; this allows insurance pricing to more accurately reflect actual risk, which, in turn, helps insurance companies be more profitable.

This study's platform also helps drivers better understand their driving habits to help them drive more safely. The data collected on this platform can also be used to adjudicate disputes.

6   Conclusions and Future Work

This study established a blockchain-based UBI platform that leverages IoV infrastructure. This study clarified why blockchain is suitable as the foundation for the platform, how the OBU interacts with the platform, how new insurance plans are generated, and what advantages this service possess. Furthermore, a more refined mechanism for adjusting premiums benefits UBI consumers more; this encourages consumers to use UBI products and share their driving information to insurers for data analysis.

This study's platform offers a win–win situation between drivers and insurance companies: blockchain technology protects the driver's privacy, and automated UBI technology allows insurers to be more profitable. Furthermore, the study's platform allows insurers to provide a reliable and transparent pricing mechanism, which benefit customers. The data gathered on this study's platform also aid automobile manufacturers in refining their products, who otherwise could not access such large quntities of driving data. In general, we believe that this study's platform benefits drivers and insurers and promises a quantum leap in the automobile industry.

Acknowledgement. This work was supported in part by Ministry of Science and Technology (MOST), Taiwan, under Grant Number MOST 109-2221-E-002-144.

References

1. Cohn, A., West, T., Parker, C.: Smart after all: blockchain, smart contracts, parametric insurance, and smart energy grids. Georgetown Law Technol. Rev. **1**(2), 273–304 (2017)
2. Ramamoorthy, R., Gunasekaran, A., Roy, M., Rai, B.K., Senthilkumar, S.A.: Service quality and its impact on customers' behavioural intentions and satisfaction: an empirical study of the Indian life insurance sector. Total Qual. Manage. Bus. Excellence **29**(7–8), 834–847 (2018)
3. Liu, K., Chen, W., Zheng, Z., Li, Z., Liang, W.: A novel debt-credit mechanism for blockchain-based data-trading in internet of vehicles. IEEE Internet Things J. **6**(5), 9098–9111 (2019)
4. Yin, J.L., Chen, B.H.: An advanced driver risk measurement system for usage-based insurance on big driving data. IEEE Trans. Intell. Veh. **3**(4), 585–594 (2018)
5. Viriyasitavat, W., Anuphaptrirong, T., Hoonsopon, D.: When blockchain meets internet of things: characteristics, challenges, and business opportunities. J. Ind. Inf. Integr. **15**, 21–28 (2019)

6. Iredale, G.: 6 Key Blockchain Features You Need to Know Nowx(2020). https://101blockchains.com/introduction-to-blockchain-features/. Accessed on 15 Aug 2021
7. Vettoor, A.S., Paulose, R.: Detection of aggressive driving behavior and driver scoring from smartphone sensors-a study on vehicle telematics and usage-based insurance. J. Comput. Sci. **6**, 125–131 (2019)
8. Zhou, L., Du, S., Zhu, H., Chen, C., Ota, K., Dong, M.: Location privacy in usage-based automotive insurance: attacks and countermeasures. IEEE Trans. Inf. Forensics Secur. **14**(1), 196–211 (2018)
9. Binion, T., Harr, J., Fields, B., Cielocha, S., Balbach, S.J.: Synchronization of Vehicle Sensor Information. U. S. Patent 9 959 764 (2018)
10. Chao, H., Cao, Y., Chen, Y.: Autopilots for small unmanned aerial vehicles: a survey. Int. J. Control Autom. Syst. **8**(1), 36–44 (2010)
11. Yang, F., Zhou, W., Wu, Q., Long, R., Xiong, N.N., Zhou, M.: Delegated proof of stake with downgrade: a secure and efficient blockchain consensus algorithm with downgrade mechanism. IEEE Access **7**, 118,541–118,555 (2019)
12. Assiri, B., Khan, W.Z.: Fair and trustworthy: lock-free enhanced tendermint blockchain algorithm. Telkomnika **18**(4), 2224–2234 (2020)
13. Hebert, C., Di Cerbo, F.: Secure blockchain in the enterprise: a methodology. Pervasive Mob. Comput. **59**, 101,038–101,052 (2019)
14. Lin, W.Y., Lin, F.Y.S., Wu, T.H., Tai, K.Y.: An on-board equipment and blockchain-based automobile insurance and maintenance platform. In: 15th International Conference on Broadband and Wireless Computing, Communication and Applications (BWCCA), pp. 223–232 (2020)
15. Mamoshina, P., Ojomoko, L., Yanovich, Y., Ostrovski, A., Botezatu, A., Prikhodko, P., Zhavoronkov, A.: Converging blockchain and next-generation artificial intelligence technologies to decentralize and accelerate biomedical research and healthcare. Oncotarget **9**(5), 5665–5690 (2018)
16. Lagaillardie, N., Djari, M.A., Gürcan, Ö.: A computational study on fairness of the tendermint blockchain protocol. Information **10**(12), 378–392 (2019)
17. Li, Y., Yang, W., He, P., Chen, C., Wang, X.: Design and management of a distributed hybrid energy system through smart contract and blockchain. Appl. Energy **248**, 390–405 (2019)
18. Wang, J., Han, K., Alexandridis, A., Chen, Z., Zilic, Z., Pang, Y., Piccialli, F.: A blockchain-based ehealthcare system interoperating with Wbans. Futur. Gener. Comput. Syst. **110**, 675–685 (2020)
19. Chen, Y.H., Huang, L.C., Lin, I.C., Hwang, M.S.: Research on blockchain technologies in bidding systems. Int. J. Netw. Secur. **22**(6), 897–904 (2020)
20. Wang, L., Liu, J., Yuan, R., Wu, J., Zhang, D., Zhang, Y., Li, M.: Adaptive bidding strategy for real-time energy management in multi-energy market enhanced by blockchain. Appl. Energy **279**, 115,866–115,882 (2020)
21. Koirala, R.C., Dahal, K., Matalonga, S., Rijal, R.: A supply chain model with blockchain-enabled reverse auction bidding process for transparency and efficiency. In: 2019 13th International Conference on Software, Knowledge, Information Management and Applications (SKIMA), pp. 1–6 (2019)
22. Li, J., Yuan, Y., Wang, F.Y.: A novel GSP auction mechanism for ranking bitcoin transactions in blockchain mining. Dec. Supp. Syst. **124**, 113,094–113,106 (2019)

Shoulder and Trunk Posture Monitoring System Over Time for Seating Persons

Ferdews Tlili[1]([✉]), Rim Haddad[2], Ridha Bouallegue[1], and Raed Shubair[3]

[1] Laboratoire Innov´Com, Ecole Suprieure des Communications de Tunis, Université de Carthage, Tunis, Tunisia
{ferdews.tlili,ridha.bouallegue}@supcom.tn
[2] Université Laval, Quebec City, QC, Canada
rim.haddad@eti.ulaval.ca
[3] New York University, Abu Dhabi, UAE
raed.shubair@nyu.edu

Abstract. The working remotely and online learning has known growth during the last year because of the COVID-19 pandemic spread worldwide. In fact, the remote workers and students remain sitting and slouching on their computers for long hours. Therefore, having a correct sitting posture over time is the greatest way to protect workers from the back pains according to the latest medical researches. In this paper, we present the architecture and design details of the proposed posture monitoring system. The aim of this study is to propose a tracking posture system with complete information about the back posture. The existing posture monitoring systems in literature are limited to trunk flexion monitoring. In this proposal we introduce the shoulder bent monitoring in addition to the trunk flexion monitoring in order to provide complete information about the back posture. The proposed posture monitoring system is a smart belt equipped by inertial sensors to detect the trunk flexion and a shoulder bent to monitor the posture over time. A Smartphone application is developed to notify the person in case of bad posture detection. The proposed system demonstrates encouraging results to monitor the posture over time of seating persons and improves their seating behavior by receiving a real time notification in case of bad posture detection.

Keywords: Posture monitoring · Inertial sensors · Mobile application · Real time system

1 Introduction

Many medical studies improve that the bad posture behavior in working place is the main factor of the back problems. A 3-year prospective cohort study performed on workers of 34 Netherland companies [1] demonstrates that the low back pain risk increases according to the workers trunk flexion over time and the activities. In fact, the study results present that the workers who their trunk flexion is over 60° during 5% of their working time, the workers who their trunk rotation is over 30° during 10% of their working time and the workers who lifted over 25 kg for more than 15 times during working day are the most experienced by the low back pain in different life time [1].

The low back pain has a severe impact on societal and personal patient's quality of life. The low back pain therapy is among the most expensive cares. The total cost of low back pain exceeded $100 billion in United Status [2]. In Brazil, the societal low back pain cost is about US$2.2 billion according to the extracted data from the National database between 2012 and 2016 [3]. The low back pain cost consists of direct and indirect waists. The direct cost is related to the healthcare and the pain treatment. The working day lost and the productivity decreases are the indirect cost of the low back pain that is heavy on the society economy. In [4], the direct cost is about 12% of global cost and the indirect cost is estimated as 88% of the global back pain costs according to Netherland study. In addition, the most of the low back pain patients suffer from a limited quality of life and a partial or total physical disability [5].

In the recent years, the number of the people affected by low back pain is increasing continuously. Many researches are focused to suggest instruments and systems in order to prevent the low back pains. The definition of the body shape is basic information to implement the monitoring posture systems. With the technologies advancements of the interactive video gaming, the cameras with software programs for body recognition are developed to track the body movements. Worawat et al. [6] introduce a neck posture monitoring system based on camera data with sensors information to calculate the neck angle. The proposed system defines and classifies the neck posture in order to detect and notify the user in case of bad posture. The cameras used for posture monitoring systems are based on image processing that can lead to privacy intrusion of the users. In addition, these systems are depended on the set-up implementation. The accuracy and the reliability of these systems are related to the cameras' position.

In the last century, the sensing technologies have known a development and advancement. These technologies provide many information such as position, acceleration, pressure and physiological parameters. A huge amount of data will be treated and processed for healthcare applications. The sensors are characterized by small dimension and portability aspect. In fact, recently, the researchers are focusing on the study of the applicability of the sensing technologies to analyze the spinal posture [7] and the reliability of these technologies to monitor the seated workers [8].

In this paper, we will detail the design and the architecture of the proposed posture monitoring system based on the inertial sensors. In Sect. 2, we review the existing posture monitoring systems. We present the design and the architecture of the system proposal in Sect. 3. In Sect. 4, we introduce the implementation and the first system testing results. We conclude this paper in Sect. 5.

2 Literature Review

In the last decade, the sensing technologies are known advancement in different aspects: connectivity, portability and power management. The sensors are characterized by tiny size. The sensors provide many information such as acceleration, pressure, spatial orientation, Blood pressure, ECG information, etc. These sensors' aspects lead to be wearable devices that can be integrated in different systems [9]. The healthcare applications are known a growth development due to the advancement of the sensing technologies. In [10], Loncar-Turukalo et al. detail in the review the latest advancement of the wearable technologies regarding to the battery efficiency, communication architecture, data analysis and devices miniaturization in the wearable healthcare applications. This review confirms that the wearable technologies attend an acceptable maturity level that can be applied in different healthcare applications. As a part of healthcare applications, the recent researches are focused on the proposal of systems to monitor the posture for rehabilitation and preventing the spine problems caused by the poor posture handling during the day.

Bibbo et al. [11] propose a smart chair for posture monitoring of seated workers. The chair is equipped by textile pressure sensors. The sensors are placed on the backrest and the seat of the chair. The proposed system is able to identify the variation of the posture during the stress tests for seated person. The system looks reliable for the detection of the posture stress level. However, the main drawback of the system is that it is related to a working platform and is not portable for anywhere.

A wearable system is defined in [12] based on the correlation of two sensing technologies for posture monitoring: flex sensors and inertial sensor. The wearable proposed robot provides the posture monitoring, bad posture detection and telerehabilitation assisting via the Smartphone application and cloud server. However, the system presents a complex architecture. And it is limited to define the trunk shape in order to detect bad posture.

In [13], Bootsman et al. propose BackUp which is a smart T-shirt that allow personalized sensor placement for low back posture monitoring. The proposed system contains two inertial measurement units (IMU) and a smart phone application for data analyzing and feedback sending. The experimental results of this system demonstrate positive results to define the low back shape and to detect the bad low back posture. This system is limited to low back posture monitor and the posture shape information. Otherwise, the proposed system does not give precise information about the global spine flexion over time. In addition, the system does not cover the whole posture variation specifically the shoulders movement.

Chung et al. [14] introduce a posture monitoring system composed of three subsystems: notebook computer, smart necklace and Smartphone application. The notebook computer is equipped with a depth camera to collect data, define the structure of skeletal and calculate the reference point for user posture flexion. Then the reference data are sent to the smart necklace to calibrate the inertial sensor. The smart necklace measures the trunk tilt and sends information to the microcontroller for processing. After data analysis, feedback information is sent to the user via an application installed in his Smartphone in order to correct posture in case of bad posture detection. The main drawback of this system is the placement of the three subsystems in close proximity to each other.

In this paper, we propose a posture monitoring system. The proposed system is based on inertial information in order to define the posture information. In addition, a mobile application is developed in order to monitor the posture variation over time and warn the user in case of bad posture detection. We focus in this work to propose a system simple to wear and to use independent of the environment of implementation. A large study of the sensors location is performed in order to have an optimum number of sensors with complete information about the posture flexion. In the next sections, we will detail the design and architecture of the proposed system.

3   System Design and Architecture

3.1   Sensors Positions

In the recent years, the inertial sensors have known evolution thanks to the technology revolution. The small and cheap inertial sensors have enabled wearable and low-cost application of this type of sensor in many healthcare and industrial applications [15, 16]. Compared to the other sensing technologies, the inertial sensors introduce great characteristics to be used for posture monitoring as it is simple to implement on the garment and provide the angle flexion of the body. In this section, we detail the sensors positions study to get complete information about the human posture variations.

The vertebral column is composed of four regions or curves as shown in the Fig. 1. The variation of spinal curves defines the trunk flexion over time.

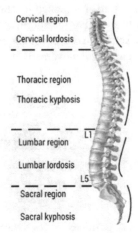

Fig. 1. Spinal regions or curves: Cervical region, Thoracic region, Lumbar region and Sacral region [13]

Clinically, the spinal posture refers to the position of spinal segments with respect to each other and with respect to gravity. In [17], the study quantifies the spinal curves and defines the directions and the angles of the spine curves. In fact, the spinal curves are determined by three main angles according to this study: Thoracic angle, Thoraco-lumbar angle and Lumbar angle as shown in Fig. 2.

The presented angles for spine curves can be measured by the inertial sensors. The tilting angles along axis of inertial sensor modules are able to quantify the spinal curves and monitor the change of the spinal posture. The back posture is not limited to the trunk flexion but also the shoulders tilt. In many working or studying cases the spine posture is fixed, but the bad shoulders bent causes spine stress and back posture disorders.

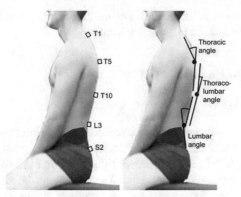

Fig. 2. Angles presentation of the spinal curves: Thoracic angle, Thoraco-lumbar angle and Lumbar angle [17]

The aim of our proposal is to define a simple system for the posture monitoring and bad posture detection. For this case, we propose to place the inertial sensors on the trunk and shoulder as in Fig. 3. The first sensor is placed in Thoracis (T5) position in order to get the information about trunk flexion. For the shoulders bent monitoring, we propose for this study to place the second sensor in one of the shoulders as on the most of time the two shoulders movement has a symmetric movement and provide the same information.

Fig. 3. Belt model: proposed sensors positions (http://www.medical-contact.net/thorax-et-epaule/15-redresse-dos.html)

3.2 Global System Architecture

The proposed posture monitoring system is composed of two main components: A belt equipped with sensors and a mobile application. The smart belt is composed of orthopedic belt easy to wear and comfortable for use and inertial sensors stretched to the belt in the positions as defined in the previous section.

The inertial sensor IMU can be composed of accelerometer, gyroscope and magnetometer or accelerometer and gyroscope. In our study, we choose to use an inertial sensor composed of a 3D accelerometer and 3D gyroscope as the data provided by these components is sufficient for trunk flexion and shoulder bent detection. In addition, the magnetometer is sensitive to the magnetic fields that impact the sensor's measurements. The Three-dimension accelerometer measures the linear acceleration according the 3 axis (x, y and z). Using the acceleration measurement, the tilt angle during accelerometer movement according to 3 axis is defined as detailed in [18] using a basic trigonometric method. The 3D gyroscope measures the rotation velocity according to the 3 axis X, Y and Z. The angle according to an axis is defined by the equation in [19].

The accelerometer angles according to 3 axis (x, y, z) are detailed in the following equations:

$$\theta = \text{arctg}(Ax/\text{sqrt}(Ay^2 + Az^2)) \tag{1}$$

$$\psi = \text{arctg}(Ay/\text{sqrt}(Ax^2 + Az^2)) \tag{2}$$

$$\varphi = \text{arctg}(\text{sqrt}(Ax^2 + Ay^2)/Az) \tag{3}$$

Ax, Ay and Az are the accelerometer measurements according the three axis (x, y and z) [18].

The gyroscope angle is calculated as following [19]:

$$\theta_n = \theta_{n-1} + \omega \times \partial t \tag{4}$$

ω: the angular velocity, θ: angle resulting.

The accelerometer and gyroscope measurements are merged in order to get one information about the tilt angle using the technique complementary filter [19] defined as following:

$$\theta_n = \alpha \times (\theta_{n-1} + \omega \times \partial t) + (1.0 - \alpha) \times \text{Angle\_accel} \tag{5}$$

α is a constant: the commonly used value of α is 0.98 as defined in [19].

Angle_accel: is the accelerometer angle measurement according one of the axis X, Y or Z.

The collected measurement from the IMU is sent to microcontroller for data converting. The sensors are connected to the microcontroller via wire conductors. All collected data from the sensors, are sent to the microcontroller, itself sends data to a cloud server where the data is stored and processed. The using of cloud server to save and treat the collected Data is the best way to store a huge amount of data in real time. The microcontroller is equipped by a Wi-Fi interface that allowed it to send the collected data in real time to the cloud server. The global system architecture is presented in Fig. 4.

A mobile application is designed in order to monitor the posture variation through time and alert the user in case of bad posture detection. The mobile application with its different interfaces will be used for a real time monitoring of the right shoulder, the trunk posture and alert notifications for bad shoulder posture and bad trunk posture. The mobile application flowchart is detailed in Fig. 5.

256 F. Tlili et al.

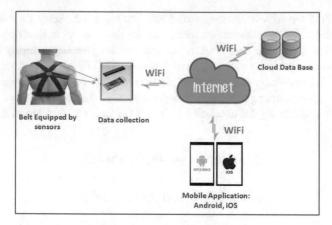

Fig. 4. Global system architecture

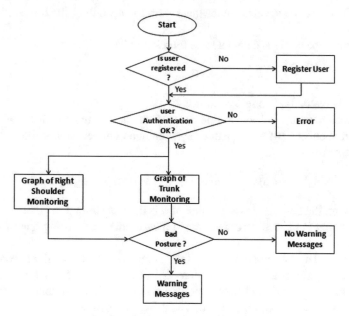

Fig. 5. Mobile application flowchart

4 System Implementations

4.1 Technologies Choice

For system implementation, we made hardware and software technologies choice. We choose a belt that covers the back positions where we planned to stretch the sensors. The chosen belt is easy to wear, comfortable and adjustable according to user body. The belt model is shown in Sect. 3.1 Fig. 3.

The inertial sensors used for the proposed posture monitoring system are the MPU6050. The MPU6050 is composed of 3 axis accelerometer and 3 axis gyroscope [20]. The MPU6050 is small and simple to stick on the Belt as the size of its package 4 mm × 4 mm × 0.9 mm. This sensor model is equipped with I2C interface to communicate with the microcontroller. We use the Raspberry Pi 3 Model B as microcontroller for the system implementation [21]. The Raspberry Pi 3 Model B microcontroller supports a variety of operating systems. This microcontroller model integrates the wire and wireless communication interfaces needed for ensure the communication between different proposed system components.

The database chosen for the posture monitoring system is the Firebase. The Firebase is suitable to the proposed system as it is able to store a huge number of information and integrate many features for real time systems implementation [22].

We choose to develop two mobile applications with different technologies Android and iOS in order to cover the most used operating systems in the smart phones and tablets. Android Studio SDK Version API 28 Android 9.0 (Pie) is the Android environment used for the development of the Android posture monitoring application [23]. Thunkable Drag-and-Drop Builder is used for iOS application development [24]. The Wi-Fi communication technology ensure the communication between different components of proposed system as it is available anywhere especially in working and studying spaces.

4.2 System Results

The proposed posture monitoring system consists of the smart belt and the Mobile application. The inertial sensors are stretched on the belt in the thoracic and shoulder positions. The microcontroller is connected to the sensors via the wired connectors. The collected information from the sensors is converted by the microcontroller. Then the microcontroller processes the collected data from the sensors and sends it to the cloud server. The stored data of the tilt angle degree is for (x, y, z) measurements. The cloud data base is fuelled by the collected data each 5 s. The stored real time data will be processed by the Mobile applications for the user posture monitoring and warning in case of bad posture detection.

The application retrieves this data through a user authentication interface (Login/Register). The mobile application provides a register interface for user registration as shown in the Fig. 6. In addition, an authentication interface is implemented for user access to the application with personnel profile presented in Fig. 7.

The application offers to the user a complete view about the back posture by introducing the trunk flexion monitoring and shoulder bent monitoring. The user can monitor in real time the shoulder bent variation. A real time graph will be displayed containing the variation of tilt angle of the shoulder bent as in Fig. 8. In case of the shoulder bad posture detection, a warning message will be displayed in order to alert the user about the need to adjust his posture as in Fig. 9. In addition, the user can monitor his trunk posture via another interface where a real time graph is displayed containing the trunk variations over time as shown in Fig. 10. When the trunk bad posture detected, a visual message is displayed as shown in Fig. 11. The user receives a visual message in case of

Fig. 6. Registration interface

Fig. 7. Authentication interface

a bad posture detected according to a defined threshold value mentioned in red color in Fig. 9 and Fig. 11.

Fig. 8. Right shoulder monitoring graph

Fig. 9. Warning bad shoulder posture

Fig. 10. Trunk monitoring graph

Fig. 11. Warning bad trunk posture

5 Conclusion

In this paper, we propose the design and architecture of the posture monitoring system. The system consists of smart belt and a mobile application. The smart belt collects the bent variations of the trunk and the shoulder and sends them to the cloud server for data storage. A mobile application installed in the mobile phone of the user provides a personalized interface to the user for the posture tracking variation over time. In addition, the user receives warning messages in real time in case of bad posture detection. The proposed system presents a new architecture by adding the shoulders movement monitoring to the trunk bent tracking in order to have complete information about the posture monitoring for seating person compared to the existing proposed posture monitoring systems. The preliminary results of the proposed posture monitoring system demonstrate encouraged results for improving the posture of seated person in real time.

As a future work, we will test the proposed posture monitoring system on a large scale of working users. Then we will use machine learning algorithms to classify bad postures for men and women separately.

Acknowledgments. We are grateful for the support of the Department of Electrical and Computer Engineering at the New York University of Abu Dhabi (NYU). We would also like to thank SUP'COM' students for their contribution to reaching these results.

References

1. Hoogendoorn, W.E., Bongers, P.M., de Vet, H.C.W., Douwes, M., Koes, B.W., Miedema, M.C., Ariëns, G.A.M., Bouter, L.M.: Flexion and rotation of the trunk and lifting at work are risk factors for low back pain. Spine **25**(23), 3087–3092 (2000)
2. Katz, J.: Lumbar disc disorders and low-back pain: socioeconomic factors and consequences. J. Bone Joint Surg. **88**, 21–24 (2006)

3. Carregaro, R.L., Tottoli, C.R., Rodrigues, D., Bosmans, J.E., da Silva, E.N., van Tulder, M.: Low back pain should be considered a health and research priority in Brazil: lost productivity and healthcare costs between 2012 to 2016. PLoS ONE **15**(4), e0230902 (2020)
4. Lambeek, L., Tulder, M., Swinkels, I., Koppes, L., Anema, J., Mechelen, W.: The trend in total cost of back pain in The Netherlands in the period 2002–2007. Spine **36**(13), 1050–1058 (2010)
5. Geurts, J., Willems, P., Kallewaard, J., Kleef, M., Dirksen, C.: The impact of chronic discogenic low back pain: costs and patients' burden. Pain Res. Manag., 1–8 (2018)
6. Lawanont, W., Inoue, M., Mongkolnam, P., Nukoolkit, C.: Neck posture monitoring system based on image detection and smartphone sensors using the prolonged usage classification concept. IEEJ Trans. Electr. Electron. Eng. **13**(10), 1501–1510 (2018)
7. Simpson, L., Maharaj, M., Mobbs, R.: The role of wearables in spinal posture analysis: a systematic review. BMC Musculoskelet. Disord. **20**(1), 1–14 (2019)
8. Jun, D., Johnston, V., McPhail, S.M., O'Leary, S.: Are measures of postural behavior using motion sensors in seated office workers reliable? Hum. Factors **61**(7), 1141–1161 (2019)
9. Godfrey, A., Hetherington, V., Shum, H., Bonato, P., Lovell, N.H., Stuart, S.: From A to Z: wearable technology explained. Maturitas **113**, 40–47 (2018)
10. Loncar-Turukalo, T., Zdravevski, E., Machado da Silva, J., Chouvarda, I., Trajkovik, V.: Literature on wearable technology for connected health: scoping review of research trends, advances, and barriers. J. Med. Internet Res. **21**(9), (2019)
11. Bibbo, D., Carli, M., Conforto, S., Battisti, F.: A sitting posture monitoring instrument to assess different levels of cognitive engagement. Sensors (Basel, Switzerland) **19**(3), 455 (2019)
12. Zhang, J., Zhang, H., Dong, C., et al.: Architecture and design of a wearable robotic system for body posture monitoring, correction, and rehabilitation assist. Int. J. Soc. Robot. **11**, 423–436 (2019)
13. Bootsman, R., Markopoulos, P., Qi, Q., Wang, Q., Timmermans, A.: Wearable technology for posture monitoring at the workplace. Int. J. Hum. Comput. Stud. **132**, 99–111 (2019)
14. Chung, H.Y., Chung, Y.L., Liang, C.Y.: Design and implementation of a novel system for correcting posture through the use of a wearable necklace sensor. JMIR mHealth uHealth **7**(5), e12293 (2019)
15. El-Sheimy, N., Youssef, A.: Inertial sensors technologies for navigation applications: state of the art and future trends. Satell. Navig. **1**(1), 1–21 (2020)
16. Seel, T., Kok, M., McGinnis, R.: Inertial sensors-applications and challenges in a nutshell. Sensors **20**(21), 6221 (2020)
17. Claus, A., Hides, J., Moseley, L., Hodges, P.: Is 'ideal' sitting posture real? Measurement of spinal curves in four sitting postures. Man. Ther. **14**(4), 404–408 (2008)
18. Fisher, C.J.: Using an accelerometer for inclination sensing. Analog Devices (2010)
19. Christiansen, K.R., Shalamov, A.: Motion Sensors Explainer. W3C Working Group Note (2017)
20. Jian, H.: Design of angle detection system based on MPU6050. In: Proceedings of the 7th International Conference on Education, Management, Information and Computer Science (ICEMC 2017), pp. 6–8 (2016)
21. Ivković, J., Radulovic, B.: The Advantages of Using Raspberry Pi 3 Compared to Raspberry Pi 2 SoC Computers for Sensor System Support, pp. 88–94 (2016)
22. Chatterjee, N., Chakraborty, S., Decosta, A., Nath, A.: Real-time communication application based on android using Google firebase. IJARCSMS (2018)
23. Brandi, P.: Android API Level, backward and forward compatibility. AndroidPub (2019)
24. Ching, A.: A Better Way to Build Apps for iOS. Thunkable (2018)

Agent-Based Modelling Approach
for Decision Making in an IoT Framework

Adil Chekati[1]([✉]), Meriem Riahi[2], and Faouzi Moussa[3]

[1] Faculty of Sciences of Tunis (FST), University of Tunis ElManar,
LIPAH-LR11ES14, 2092 Tunis, Tunisia
adil.chekati@fst.utm.tn
[2] National Higher Engineering School of Tunis, University of Tunis,
1008 Montfleury, Tunisia
meriem.riahi@ensit.rnu.tn
[3] Université de Lorraine, LCOMS, 57070 Metz, France

Abstract. Electronic gadgets and integrated sensors are all around us. Establishing a network of all these objects is therefore what we call the Internet of things (IoT). That allows having digital in every object in our daily lives. IoT applications combine a complex and heterogeneous mixture of constantly evolving hardware and software. What makes decision making and self-adaptation very beneficial for these types of applications. In fact, connected objects required to have the ability to adapt to the needs and state of the user and other objects around them. In this paper, we present a new approach based on a Multi-Agent System (MAS) and taking advantage of its capabilities, to provide the aspect of self-adaptation and decision-making for existing connected objects, as part of a Smart Object framework. Our approach increases the autonomy of decision making by objects, which is significant to make objects smarter and ensure the conversion from connected objects to smart objects.

1 Introduction

We live in a time where connected devices and integrated sensors are all around us, from smartphones and tablets to laptops, and from indoor temperature regulators to home security systems. We live in a new world, a world of smart, where intelligence and connectivity is added to every conceivable object [2]. The emergence of IoT technologies, enables enhancing and interconnecting daily lives objects in an open communication environment over the Internet to perform pervasive actions for helping humans being. In a more comprehensive way, IoT transforms real world objects into smart objects and connect them through Internet [21]. IoT environment is extremely large, complex and contains a significant amount of communications, interactions and data exchanges that should be accomplished correctly. It is a system consisting of heterogeneous physical and virtual connected components, using different languages, platforms and intelligent communication protocols [10].

Pointing out that due to the rapid growth of connected objects and their heterogeneity and also the remarkable evolution of users preferences and activities,

L. Barolli et al. (Eds.): AINA 2021, LNNS 226, pp. 261–272, 2021.
https://doi.org/10.1007/978-3-030-75075-6_21

the simple adoption of this connected objects cannot guarantee user satisfaction nor IoT requirements. To overcome this difficulty, one solution is to incorporate a kind of intelligence into real world objects so that these entities can interact efficiently in real time with changes in their state and environment. In another word, there is a need for a context-aware IoT solutions. Context-awareness refers to a system that uses context to provide relevant information and/or services to the user, where relevancy depends on the user's task [14]. Context aware-ness becomes critical for each object to help it optimize its actions and react intelligently. It is the transition from connected objects to Smart Objects.

The foremost challenge that developers confront when building such IoT sys-tem is the management of a plethora of technologies implemented with various constraints, from different manufacturers, that at the end need to cooperate [16]. Among the challenges associated with existing architectures of IoT is the integra-tion of intelligence aspect into each of the connected objects, which enables us to intelligently manage the heterogeneous data of the dynamic and distributed IoT systems. To deal with these problems, developers can benefit from IoT frame-works through the process of component reuse. Thus, they can set up new IoT systems with the least amount of effort and time. Reuse is the use of previously acquired concepts or objects in a new situation, it involves encoding develop-ment information at different levels of abstraction, storing this representation for future reference, matching of new and old situations [6]. In our case, the reusable artefact is the logical program structures of Multi-Agent System (MAS) such as modules, interfaces and communication language. Furthermore, we believe that using a MAS can add significant advantages over traditional devices that only work as data transmitters. They send data from sensors to a server to gener-ate a situational context and require stimulus from other devices to act on the environment.

Motivating by this, we propose in our research, an agent-based decision mak-ing for self adaptation of smart objects. That enables IoT applications to move forward from the simple collection of IoT data to make autonomous decisions and actions in real time. It consists of benefiting from software reuse methods to repackaging and reapplying muti-agent system's components and modules to require IoT properties. The remainder of this paper is organized as follows: Sect. 2 reviews the most relevant related research work. Section 3 presents the proposed approach based on MAS for Smart Objects framework. Section 4 fol-lows with a case study of the proposed approach. Finally, Sect. 5 concludes and outlines the prospects of this research.

2   Related Work

There are several works, which use agents to deal with some details of IoT such as collaboration, autonomy, and sensing. In this section we discuss the relevant literatures relying on MAS for modelling and developing IoT solutions.

In details, authors in [2] use a combination of complex networks-based and agent-based modeling approaches to elaborate a novel approach to modeling IoT.

Specifically, this approach uses the Cognitive Agent-Based Computing (CABC) framework to simulate complex IoT networks.

Agent-based Cooperative Smart Object (ACOSO) framework, is presented and exploited in [7] and [8] in order to support the Smart Object analysis, design and implementation phases. Assuming that Smart Objects and software agents share multiple features and MAS modeling flexibly helps the conceptualization of autonomous IoT solutions in various contexts, this approach aims to enable the exploitation of the Agent-Based-Computing (ABC) paradigm for modeling IoT systems in terms of MAS. Network-based simulation of this approach confirms its adaptation to tackling intrinsic IoT system complexity.

In [12], an approach for IoT analytics based on learning embodied agents is proposed. In particular, it facilitates the development of Agent-based IoT applications by explicitly capturing their complex and dynamic variabilities and supporting their self-configuration based on a context-aware and machine learning-based approach.

In [15], authors present an agent-based IoT vision. The approach allows the implementation of autonomous and cognitive mechanisms at the thing-level enabling interoperability at the system-level.

The authors in [17] present a Multi-Agent-System based architecture for developing IoT solutions. The idea consists in using MAS features in order to provide an efficient IoT model. The multi-layer architecture takes into accounts the IoT systems requirements and constraints, it controls the connected devices and materials, ensures communication and interactions between different entities, and implements further functionalities that depend on the IoT application domain.

In [13], an architecture for an agent-based ambient intelligence systems is proposed. Which is an architecture for the deployment of devices enhanced with MAS working as Smart Things. It exploits IoT as a middleware for the connectivity and communicability of devices, and uses specialized agents for deploying smart objects.

A framework for IoT systems based on a MAS paradigm is also proposed in [11], the approach is based on Multi-Agent Systems (MAS) and Machine Learning techniques. The authors survey some requirements for developing IoT applications such as: Decentralization, autonomy and actuation. Then propose an Agent-based model to create IoT systems and show how this model meets these requirements.

These relevant contributions which exploit MAS for modelling and developing IoT solution have been compared, according to their provided agent-based features in Table 1. The comparison criterion for the previous approaches in Table 1 indicates: if the solution is a hybrid or pure approach (i.e. fully MAS based or side by side with other paradigms), if it implements mechanisms for autonomicity, cognitivity, virtualization, and cooperativity, and finally, it mentions the simulator or the framework used for implementation.

Table 1. Comparison table between previous approaches

Ref.	Agent model		Agent implementation				Simulator
	Pure	Hybrid	Autonomicity	Virtualization	Cognitivity	Cooperativity	
[2]		X	X		X		AMB Tool
[7]		X	X	X		X	INET
[8]	X		X	X		X	JADE
[12]		X	X				[No]
[13]	X		X	X		X	JADE
[15]	X		X	X	X		OMNet++
[17]	X		X	X	X	X	JADE
[11]		X	X	X	X		JADE

However, the results from the literature do not demonstrate virtualization nor visualization, where heterogeneous objects are interacting in scalable environment. Further, important features mentioned in this paper regarding decision making and self-adaptation are also not covered in the published literature.

3   Multi-Agent System for Smart Objects Framework

Due to the deep heterogeneity of devices and communication protocols in the IoT context, proposing an agent-based approach can be applied to solve these problems. In fact, MAS paradigm enables development of (i) self-configuring, self-protecting and self-optimizing IoT system [1]. (ii) Context-aware and adaptive smart objects IoT systems [20]. Following in this section, we detail the SADM-SmartObjects framework, then a closer look at the decision making approach is illustrated.

3.1   SADM-SmartObject Framework

Framework for self-adaptation and decision making of smart objects called SADM-SmartObjects is introduced for the first time in a previous work [4]. This framework aims to enable the injection of intelligence into IoT devices so that these devices can interact efficiently in real time with changes in its environment and proposes an optimal solution for autonomous decision making. Framework SADM-SmartObjects is based on multi-layer architecture proposed for Self-Adaptation and Decision Making for Smart Objects (Fig. 1).

As explained in [4], this framework is made up of three horizontal layers named respectively: Real-World Object Layer, Virtual Layer and Service Layer.

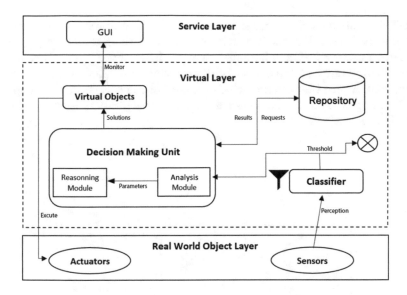

Fig. 1. Architecture of SADM-SmartObjects [4]

3.1.1 Real-World Object Layer
It represents the lower layer, or the physical layer. It monitors, manages and controls the IoT infrastructure. It gathers IoT systems sensors, actuators, tags, and other devices. RWO layer ensures: data perception, communication and connection between objects. It interacts directly with Virtual layer.

3.1.2 Virtual Layer
The middle layer, it ensures the virtualization of every real-world object in the system. It enables achieving the maximum exploitation of data delivered from the RWO layer by analyzing, reasoning and taking the adequate decision, all in real-time. Such functionality can be achieved based on sophisticated modules: Virtual objects, a Classifier is well detailed in [5] and Decision-Making Unit that we are interested in, in this paper:

Decision Making Unit
Represents the intelligent core of Virtual Layer. It represents an IoT solution that enables making sense of all collected data by making decisions and take actions in real-time depending on that data.

- Analysis module: By order, Analysis module is the first to receive a labeled data coming from a classifier. Its role is to check in the Cache storage if any scenario similar to this received data is already existing, if not it forwards this data to next module. This aims to reduce the system response time by saving parameters and scenarios of previous situations.
- Reasoning module: It is responsible of carrying out solution, by enabling objects to act autonomously depending on the state, the measurements and

their environment. It is based on a multi agent system (MAS). Reasoning module in SADM-SmartObjects framework uses the BDI model [9], which is based on a theory of how humans operate: objects have a set of beliefs about the world. They also have a set of desires that they plan to achieve. Finally, they have a set of intentions, which consists of the actions they are currently performing to achieve their desires [18]. Figure 2 presents a close-up look to the Decision Making Unit.

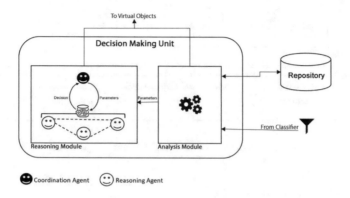

Fig. 2. Decision Making Unit.

3.1.3   Service Layer

The higher layer which can be seen as a front-end one due to its relationship directly with users. It ensures the visualization of the different entities in IoT system, through a Graphic User Interface (GUI). The GUI enables users to monitor, manage, configure and manipulate objects.

3.2   Multi-Agent Based Decision Making Approach

MAS paradigm naturally provides Smart Object autonomy, proactiveness and location, whereas other important features of Smart Object can be specifically defined by other agent-related concepts. Therefore, beyond facilitating resource management and providing communication and coordination, MAS promotes the IoT ecosystem with the following:

- Comprehensive support for the modeling, programming and simulation phases.
- Availability of agent-based methodology for driving the Smart Objects (SO) and IoT systems agentification.
- Natural support to features like cognitivity, context-awareness, autonomicity and proactivity.
- Availability of standards such as FIPA ACL and communication protocols.

SADM-SmartObject's decision making unit distinguishes three kinds of agents: Global Agent, Reasoning Agent and Coordination Agent, as shown in Fig. 2.

3.2.1 Global Agent

In our approach, we define a Global Agent as general abstraction both for living things and for non-living things logically because that includes all things in the system. Most of time it is a reactive agent. Away form the specific physical/technological characteristics of the things their represent "Agents" are defined by means of:

- Identity: An Agent has a unique identity and is addressable.
- Services: Services represents access point for exploiting the capabilities of Agents. That is, depending on the kinds of things and functionalities it abstract: triggering and directing the sensing/computing/actuating capabilities, or accessing some managed resources.
- Goals: refers to the desired state, it can be activated autonomously, or explicitly activated by a user.
- Events: Events represent specific state that can be detected by an Agent.

3.2.2 Reasoning Agent

It is a deliberative agent which represents the core intelligent component of the Decision-Making Unit. Depending on application domains (Crisis Management, SmartHome, eHealth, etc.), many reasoning models can be used such as based logic reasoning, based case reasoning, fuzzy reasoning, etc. Whereas, when we are handling with IoT data that is known by its missing or incomplete sensed data, fuzzy reasoning would be the main choice. The reasoning process is followed up by a decision-making process which is based on the reasoning process results.

3.2.3 Coordination Agent

It is a deliberative agent which aims to manage and control a defined group of Reasoning Agents with the aim to elaborate the adequate decision. Coordination Agent ensures communication, collaboration and even competition between Reasoning Agents using FIPA-ACL-based communications in order to share related data and resources, to accomplish a particular task or to overcome potential conflicts.

This guarantees system coherence and significantly leads to the achievement of framework and agents objectives.

4 Case Study

There has been a recorded rise in the number and intensity of disasters caused by natural phenomena or by human activity around the world in recent years. Such crises adversely impact the countries' Infrastructure, economic growth and sustainable development. Through IoT a vast variety of signals can be acquired and measured during disasters which could be used for meaningful interpretation of events [19]. In this context, one of SADM-SmartObject framework main uses is crisis management. In which the integration of MAS capabilities with the power of IoT that offer more efficient methods to analyze, monitor, predict, and manage crises.

To demonstrate the application of the agent-based decision making approach for SADM-SmartObjects, we consider a case study of a crisis management system, an anti-flood system. In this case:

- Various type of weather sensors (i.e. Temperature, pressure, humidity, wind and rain.) are used to collect weather information along the day, this sensed data is streamed in real-time and many decisions have to be taken instantly to avoid the crisis earlier.
- A local sensor server manages sensors and secure streaming data from sensors to the server application, which is implemented by means of LoRa protocol (Wide Area Network).
- All departments of the ministries and authorities must ensure a total collaboration. Ministries of the Interior, Defense and Health provide information flow continuously about their departments and infrastructures states, in which they communicate, exchange information and generate more consistent data.

The Long Range Network, or LoRa, is both a technology and protocol that allows connected objects to transmit small amounts of data in a low-flow system. It aimed to use LoRa to ensure a reduction in the level of energy consumption beside of its high reliability [3].

Figure 3 of the deployment diagram depictes the deployment of the application architecture.

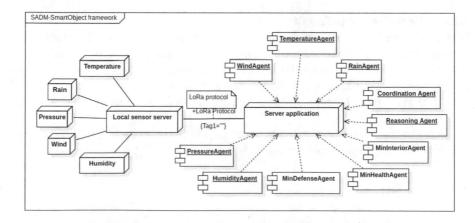

Fig. 3. Deployment diagram of crisis management system

Agents are implemented by means of JADE platform. JADE is selected for the implementation phase mainly for the following reasons:

- Java-based agent middleware.
- Open-source, has a spread community and, over the years, has evolved to run atop novel and heterogeneous computing systems such as Java Micro Edition-enabled and Android-supported devices, as well as on sensor nodes constituting heterogeneous WSN.
- Its middleware provides an effective agent-oriented management/communication infrastructure, that comprises an Agent Management System (AMS), ACL-based message transport system and Directory Facilitator (DF).

The deployment of the application on JADE platform is depicted in Fig. 4. Reasoning Agents and Coordination Agents and are on the same Container (i.e. Container-2), while Global Agents for sensors in another Container (i.e. Container-1).

Although, in this simulation as it is small-scale we consider that the data is already well classified at the server application. Based on sensor nature criterion, a Global Agent is implemented for each real-world sensor, as a virtual representation. We define: AgentHumid, AgentPres, AgentRain, AgentTemp and AgentWind. Also, we distinguish a group Reasoning Agents where a Coordination Agent ensures coherency between them, in order to take the adequate decision.

Figure 4 also shows the interaction between the agents. The measurements from the sensors are transmitted continuously via different sensors agents to Reasoning Agents. While this main processing module is informed about the deviation of weather condition in real time, it provides Coordination Agents with personalized recommendations on how to act to protect the area from any possible threat of floods. It is expected that environmental crisis that may lead to significant welfare losses, and socio-economic consequences are going to be limited by using an effective crisis management system.

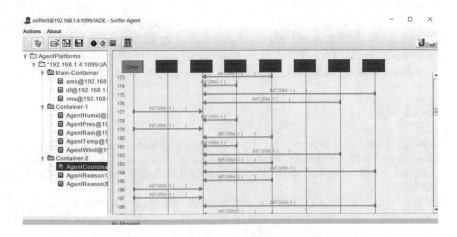

Fig. 4. Interaction between agents

Overall, It illustrates that our agent-based decision making model satisfies its main goals:

- Enable object virtualization, MAS agents are used as virtual entities of the real-world objects. This facilitates managing and controlling those objects.
- Increase the objects autonomy and decision-making capabilities. By taking advantage of MAS potentials, objects are able to cooperate and execute complex behavior without the need of centralized controllers.
- Considering the unlimited communication exchange between agents offered by MAS communication protocol, objects are able to self-adapte depending their environment.
- Develop a flexible architecture model; that could be compatible to be used in different application domains.

From these results, it is clearly exposed the reuse potentials of MAS that are already available and can support the development of platform independent IoT applications for various domains.

5 Conclusion

Internet of things is enabling a new and wide range of decentralized systems from small-scale to large-scale, in which objects are able to actuate autonomously and pro-actively based on reasoning models. In this paper, we have proposed an agent-based architecture for a smart object framework that manages the reasoning and decision-making features and consider IoT requirements, with the aims to benifit of the reusability of agents components. This approach exploits the autonomous characteristics of MAS agents and reuses them to integrate the intelligent aspect to connected objects. Both reactive and deliberative agents are used in a context-aware IoT application for reasoning and decision making because of their ability to act without depending on third-party processing.

As perspectives, we intend to accomplish various case study implementations in several application domains, to ensure covering specific domain properties.

References

1. Ayala, I., Amor, M., Fuentes, L.: The Sol agent platform: enabling group communication and interoperability of self-configuring agents in the Internet of Things. J. Ambient Intell. Smart Environ. **7**, 243–269 (2015). https://doi.org/10.3233/AIS-150304
2. Batool, K., Niazi, M.A.: Modeling the internet of things: a hybrid modeling approach using complex networks and agent-based models. Complex Adapt. Syst. Model. (2017). https://doi.org/10.1186/s40294-017-0043-1

3. Cattani, M., Boano, C.A., Römer, K.: An experimental evaluation of the reliability of lora long-range low-power wireless communication. J. Sens. Actuator Netw. (2017). https://doi.org/10.3390/jsan6020007
4. Chekati, A., Riahi, M., Moussa, F.: Framework for self-adaptation and decision-making of smart objects. Lect. Notes Artif. Intell. **11684**, 297–308 (2019). https://doi.org/10.1007/978-3-030-28374-226
5. Chekati, A., Riahi, M., Moussa, F.: Data classification in internet of things for smart objects framework. In: International Conference on Software Telecommunications and Computer Networks (SoftCOM, pp. 1–6 (2020). https://doi.org/10.23919/SoftCOM50211.2020.9238186.
6. Cybulski, J.L.: Introduction to software reuse. Epartment of Information Systems. The University of Melbourne, Parkville, Australia (1996)
7. Fortino, G., Gravina, R., Russo, W., Savaglio, C.: Modeling and simulating internet-of-things systems: a hybrid agent-oriented approach. Comput. Sci. Eng. (2017). https://doi.org/10.1109/MCSE.2017.3421541
8. Fortino, G., Russo, W., Savaglio, C., et al.: Agent-oriented cooperative smart objects: from IoT system design to implementation. IEEE Trans. Syst. Man Cybern.: Syst. **48**, 1949–1956 (2018). https://doi.org/10.1109/TSMC.2017.2780618
9. Fouad, H., Moskowitz, I.S.: Meta-Agents : Managing Dynamism in the Internet of Things (IoT) with Multi-Agent Networks. The 2018 AAAI Spring Symposium Series (2018)
10. IEEE. Internet of Things (IoT) Ecosystem Study. IEEE (2015). Available via http://iot.ieee.org
11. Nascimento, N.M., Lucena, C.J.P.: FIoT: an agent-based framework for self-adaptive and self-organizing applications based on the Internet of Things. Inf. Sci. (Ny) **378**, 161–176 (2017). https://doi.org/10.1016/j.ins.2016.10.031
12. Nascimento, N., Alencar, P,. Lucena, C., Cowan, D.: An IoT analytics embodied agent model based on context-aware machine learning. In: Proceedings 2018 IEEE International Conference on Big Data (2019)
13. Pantoja, C.E., Viterbo, J., Seghrouchni, A.E.F.: From Thing to Smart Thing: Towards an Architecture for Agent-Based Am I Systems, Smart Innovation, Systems and Technologies, p. 148 (2020)
14. Perera, C., Zaslavsky, A., Christen, P., Georgakopoulos, D.: Context aware computing for the internet of things: aa survey. Communications Surveys Tutorials. IEEE **16**, 414–454 (2014)
15. Savaglio, C., Fortino, G., Zhou, M.: Towards interoperable, cognitive and autonomic IoT systems: an agent-based approach. In: 2016 IEEE 3rd World Forum Internet Things, WF-IoT 2016, pp. 58–63 (2017). https://doi.org/10.1109/WF-IoT.2016.7845459
16. Smiari, P., Bibi, S., Feitosa, D.: Examining the reuse potentials of IoT application frameworks. J. Syst. Software **169**, 110706 (2020). https://doi.org/10.1016/j.jss.2020.110706
17. Sofia, K., Ben, L., Bouaghi, O.: Multi-layer agent based architecture for internet of things systems. J. Inform. Technol. Res. **11**, 32–52 (2018). https://doi.org/10.4018/JITR.2018100103
18. Stone, P.: Learning and multiagent reasoning for autonomous agents. In: Proceedings of the 20th International Joint Conference on Artifical Intelligence IJCAI-07 (2007)
19. Velev, D., Zlateva, P., Zong, X.: Challenges of 5G usability in disaster management. ACM Int. Conf. Proc. Ser. 71–75 (2018). https://doi.org/10.1145/3194452.3194475

20. Wu, Q., Ding, G., Xu, Y., et al.: Cognitive internet of things: a new paradigm beyond connection. IEEE Internet Things J. **1**, 129–143 (2014). https://doi.org/10.1109/JIOT.2014.2311513
21. Xu, X., Bessis, N., Cao, J.: An autonomic agent trust model for IoT systems. Procedia Comput. Sci. **2013**(09), 016 (2013). https://doi.org/10.1016/j.procs

A New Proposal of a Smart Insole for the Monitoring of Elderly Patients

Salma Saidani[1,2(✉)], Rim Haddad[3], Ridha Bouallegue[1], and Raed Shubair[4]

[1] Innov'Com Laboratory, High School of Communication of Tunis
University of Carthage, Tunis, Tunisia
{salma.saidani,ridha.bouallegue}@supcom.tn
[2] National School of Engineers of Tunis (ENIT),
University Tunis EL Manar, Tunis, Tunisia
salma.saidani@enit.rnu.tn
[3] Laval University, Québec, Canada
rim.Haddad@eti.ulaval.ca
[4] New York University (NYU) Abu Dhabi, Abu Dhabi, United Arab Emirates
raed.shubair@nyu.edu

Abstract. The remote control of the physical activities of elderly people during their daily activities to protect them against falls. Physiological parameters monitoring becomes one of the research challenges for the development of real-time survival systems for elderly patients to ensure a healthy life. In this paper, we propose a real-time plantar pressure monitoring system using a smart shoe insole. In this system two applications are developed to monitor the patient, we develop a mobile application and a web application. The patient will use the mobile application for the real-time reception of his plantar pressure where all his gate cycle over time is saved on a database (Firebase) and restored to a Web application used by the doctor for efficient monitoring of the patient, this application authorizes the doctor to control and analyze the medical file patients. The results of the proposed architecture demonstrate the efficiency of the system to control and detect bad plantar pressure.

Keywords: Plantar pressure · Sensors · Microcontroller · Bluetooth · Gait · Mobile-application · Web-application

1 Introduction

Elderly patients need continuous monitoring and control throughout the day, for this, it is necessary to use a very reliable way with a reasonable cost to ensure survival without the need for the presence of an attendant of the patient during his daily activities. In addition, all aging people are threatened with physical and health changes status including changes in memory and brain function [1]. These changes can become a real danger for the elderly if they are not continuously controlled. Falls and chronic disease are both important health issues in older adults [2]. Should be aware that falls can be very dangerous and deadly; and can

L. Barolli et al. (Eds.): AINA 2021, LNNS 226, pp. 273–284, 2021.
https://doi.org/10.1007/978-3-030-75075-6_22

cause serious injuries such as broken bones, head injuries, like wrist, arm, ankle, and hip fractures [3,4]. If the current declining fall death rate is constant, an estimated 43,000 U.S. residents aged 65 and over will die from a fall in 2030 [5].

Falls of elderly persons present a major public health problem in the rate of injury and death increasing every year. The prevention of falls is very important in today's healthcare landscape, where the population is predominantly adult in the world [6]. There are many techniques to control the patient's physical activity, such as determining the number of steps, using smart shoes equipped with sensors.

To measure and control the value of the patient's plantar pressure, we can use an inexpensive and widely used subjective method [7], based on multiple-choice tests, but it may lead to errors in physical activity reports [8,9]. Likewise, we can use objective methods such as pedometers, accelerometers, and heart rate monitors [10]; based on the use of user devices equipped with sensors to detect physical activity (running, walking, jumping), measuring the number of steps, heart rate and temperature of the patient.

The accuracy of these devices depends on the number and types of sensors used in activity monitoring. However, not all of these devices are suitable for daily use in humans disabled [11]. Smart insoles, equipped with sensors appear as a potential solution for discreet control of daily activities [12] since humans wear shoes for several hours a day. In addition, it was suggested that instrumented insoles may be less expensive than available activity monitors [13]. All sensors will be placed on an orthopedic insole to control the gait cycle during different activities of patients. The smart shoe insole system is used to collect the foot pressure value while the subject is inactivity. While we receive the data through Bluetooth communication on a mobile application, if the system detects a high value of gait parameters, an alarm will be sent simultaneously to alert subjects and doctors [1].

In this paper, we present a design of a smart shoe insole system, to evaluate the physical activity of elderly patients. The remainder of the paper is structured as follows: Sect. 2 presents related works to identify the different systems used to measure and analyze the plantar pressure. In Sect. 3 we present the plantar pressure measurement using different equations, in Sect. 4 we describe the proposed model, in Sect. 5 we present the system realization hardware and software implementation, and in Sect. 6 we conclude and outline.

2   Related Works

The plantar pressure measurement systems for monitoring patients have been assessed in a tremendous number of research studies, they generally consist of three main parts: the first part for the detection of parameters, formed by different types of sensors, the second part for the processing and analysis of the information detected by the sensors and the last part for alert feedback to the patient and doctor if the pressure value exceeds a threshold, despite the importance of this part it is rarely defined in existing studies of plantar pressure monitoring systems.

E Klimiec et al. [14], propose a shoe insole system with 8 Polarized polyvinyli-dene fluoride (PVDF)sensors placed in different anatomical zones of the foot, to measure the gait cycle of the patient during walking, many information can be detected by the sensors as; the Gait rhythm, the electric signal, time of area foot contact [14]. This system can be used for the stress tests of people with CVD (cardiovascular Disease).

Shu, Lin et al. [15], present an in-shoe plantar pressure measurement system based on a fabric pressure sensing array with six resistive textile pressure sen-sors, many parameters are calculated using this system as; mean pressure, peak pressure, the center of pressure (COP), and speed of COP [15]. This system is able to compare the shift speed of COP to the threshold value serve to diagnostic musculoskeletal or neurological diseases.

Commercially available systems currently employed by clinicians and researchers to assess the plantar pressures include in-shoe measurement systems; as shown in Table 1 [16]. Those systems were utilized for dynamic and static gait stability analysis [6], gait detection [17], and altered gait characteristics [18]. However, some of these systems use electrical wires to connect in-shoe sensors and data acquisition system around the waist or the ankle which causes a little heavy on the ankle if the patient has diabetes, and uncomfortable for the elderly during their exercise, also for outdoor activities they are not suitable for a long time to provide the measurement, for the data acquisition systems are usually bulky and cannot be configured to connect with different remote receivers such as a smartphone.

Table 1. Commercially systems for plantar pressure [16].

Smart insole	Sensor type	Sensor number	Technology
Arion smart Insoles [19]	Accelerometer Gyroscope GPS	8	Bluetooth
Dynafoot2 [20]	Resistive Accelerometer	58	Bluetooth
Pedar-X Insole [21]	Piezo electric	99	Bluetooth USB Optical fiber

With the purpose of offering a system device able to collect large valid data in real-time during all kinds of activities indoor or outdoor, the plantar pressure measurement system should be able to realize outdoor and indoor measurement in a reliable and correct manner, then the equipment should be comfortable to wear for the patient, also the pressure sensing system must have enough precision and reliability to ensure correct pressure measurement. However, it is rare that plantar pressure measurement systems meet these characteristics, for this reason, we aim to design a smart insole able to afford subjects' needs.

3   Plantar Pressure Measurements

Plantar pressure measurement provides insight into the functioning state of the foot and ankle during the patient's gait cycle, in addition, foot and ankle control is required to maintain weight-bearing flexibility and weight transfer during various daily activities [22,23]. Pressure (P), is defined as the force (F) per unit area (a);

$$P = \frac{F}{a} \tag{1}$$

The measurement of the plantar pressure of the foot by a smart insole equipped with sensors plays a very important role in the detection, analysis, and adjustment of the behavior of patients at risk or suffering from different foot problems or chronic diseases [24]. According to the International System of Units (SI), the unit of measure for force is, the Newton and for pressure is the Pascal [25]and generally in MPa (MeagaPascal) or KPa (kiloPascal) for foot pressure measurement. In addition, $1\,N$ is the force recommended to provide a mass of $1\,kg$, an acceleration of $1\,ms^{-1}$ and $1\,Pa$ produce from $1\,N$ force distributed over an area of $1\,m^2$ [24].

Real-time measurement of plantar pressure force requires that a sensor be placed in the best position in the insole of a shoe for an accurate measurement [26]. The plantar pressure can be calculated in real-time using Eqs. (2–4), the distribution of mean pressure, peak pressure, and center of pressure (COP)[27].

The COP progression is a line path of the center of foot pressure formed by a series of coordinates in the center of pressure as it passes from the rearfoot to the forefoot [27], the COP allows for appropriate calculation of balance control of the patient.

$$Mean = \frac{1}{n} \sum_{i=1}^{n} P_i \tag{2}$$

In addition;

$$Peak = max(P_1, P_2, ...P_n) \tag{3}$$

$$X_{cop} = \frac{\sum_{i=1}^{n} X_i P_i}{\sum_{i=1}^{n} P_i}; Y_{cop} = \frac{\sum_{i=1}^{n} Y_i P_i}{\sum_{i=1}^{n} P_i} \tag{4}$$

For shift speed of COP;

$$speed_{COP} = \frac{u}{\Delta t}(|X_{cop}(t+\Delta t) - X_{cop}(t)|^2 + (|Y_{cop}(t+\Delta t) - Y_{cop}(t)|^2)^{1/2} \tag{5}$$

Where p is the plantar pressure, n indicates the total number of sensors, i represent a certain sensor, X and Y are the coordinates of the total foot area. u is the unit distance between two neighbor coordinate points, and t denotes the time interval. The shift speed is used for clinical research, to determine patient balance walking status, if the COP value exceeds a certain threshold, means the patient is not stable.

4   The Proposed System

The proposed smart insole monitoring system is composed of hardware and software parts to have the best results with good precision. For the hardware part, it consists of an orthopedic insole, resistive sensors, accelerometer, Gyroscope, Micronctroller, electrocardiography (ECG) sensor, and a battery. The software part is composed of the cloud storage database and the development of Web and mobile applications. The database stores all the information collected from the sensors, it is updated in real-time.

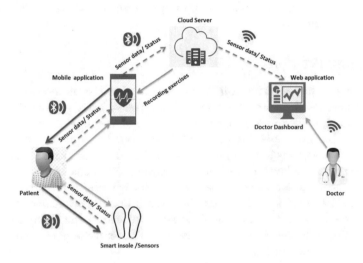

Fig. 1. Global system architecture.

The database is hosted in a cloud server which allows real-time and multiple access at the same time to the stored data. With technological progress and the appearance of telemedicine which is a form of remote medical practice, the use of mobile applications has becomes the best way to monitor a person in real-time. The global architecture of the proposed system is presented as follows in Fig. 1.

The mobile application is intended for the patient to facilitate the follow-up between patients and health professionals. During normal daily patient activity (walking, running, or jumping), sensor signals appear following the presence of pressure in the insole, all the measurements of these signals are collected and converted via the bus interface of the microcontroller then send via Bluetooth to the mobile application.

The web application is intended for the doctor to select, edit, remove, add and, manage the user's profile via Web-based dashboards, also he can access the medical file of each patient to analyze all the data recorded. If a bad plantar pressure value is detected, a visual and audio notification is sent via the mobile application to the doctor to intervene.

4.1 Sensors Type

In the plantar pressure system, the Force Sensing-Resistor (FSR) is selected due to the multiple characteristics and specifics as it offers several advantages too. This sensor is cheap, it is very thin and easy to handle on the insole. The use of the insole equipped with FSR sensors in the sock by the patients offers comfort to them to exercise their daily physical activities. For elderly patients with chronic diseases, they have generally a pain in the foot due to aging or chronic diseases, in case of localized swelling of the patient's foot, the FSR sensors do not develop an additional pressure [28]. FSRs are sensors with a resistance that depends on applied force. The FSR 402 Short is best suited for measuring different specific points in the plantar regions.

4.2 Sensors Positions

The plantar pressure distribution is very sensitive to the method, the tool used for the measurement of pressure and to the physical activity of the patient, for evaluation of the distribution measurements we used the peak pressures [29].

The distribution of plantar pressure in the foot helps us to choose the best positions for the sensors in the foot to ensure good data reception. We have three pressure levels; a low level in the hell, a medium level in the Midfoot, and a high level of precision in the forefoot part of the foot. Using the manual mask, we define the perfect position for the sensors placed in different anatomical zones of the foot. As shown in Fig. 2, the first sensor is placed in the Hallux, the second one in the Medial Forefoot, the third one in the Lateral Forefoot, and the fourth in the Heel, as shown in Fig. 3.

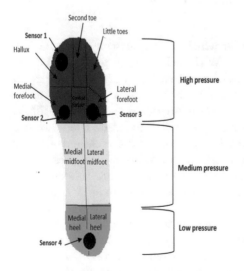

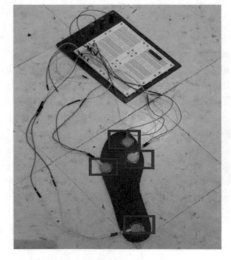

Fig. 2. Plantar pressure distribution   **Fig. 3.** Sensors positions

5 System Implementation

We used for the design of the proposed smart insole as materials; four FSR (Force Sensor Resistor) to measure the physical pressure and weight [30], the ECG to measure the electrical activity of the heart, Gyroscope gy-521 to measure acceleration, orientation, and the inclination of the foot relative to the ground [31], ARDUINO Card to control and retrieve data [32], ATMEGA328P microcontroller [33], Electric wires to connect different components together, Bluetooth HC05 to send data to the laptop, and Secure Digital card (SD card) to save all data's patients.

5.1 Mobile Application

An Android/IOS mobile application is developed for our proposed plantar pressure monitoring system using Thunkable. The Thunkable platform is the best tool to develop an Android application due to its power, simplicity, and free [34]. It allows the development of different applications using smartphone and tablets based on the different integrated libraries. The trustworthiness of mobile applications is related to the technique used to store the data.

In fact, we use cloud storage and a real-time database for our developed mobile application. In addition, we used Firebase as a platform to implement our cloud database, this platform contains many features such as a real-time database, cloud functions, and analysis tools. It provides the structure of the data collected from the various sensors and multi-user access to the database [35].

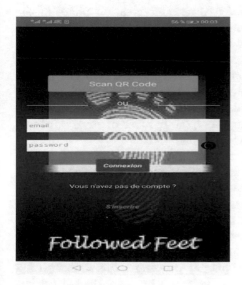

Fig. 4. User interface.   **Fig. 5.** Registration interfaces.

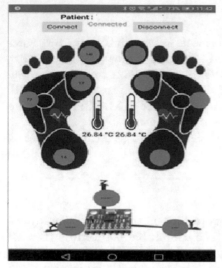

Fig. 6. Patient's data.   **Fig. 7.** Notification.

The patient needs to have complete visibility on his state of health daily in real-time, with the mobile application he can ensure a personalized follow-up while keeping in contact with the care team remotely. In this case, the smart insole is a very recommended solution for all patients, especially the elderly, so that they feel better supported during their activities. A login interface is developed for the mobile application to authenticate the patient, as shown in Fig. 4. As shown in Fig. 5. After the patient's account creation phase in the application, he can now connect to his own account via his email address and password to consult his information and start the exercises, knowing that all authentication information is already saved in a database securely.

Our system for measuring plantar pressure using the smart insole is able to measure the plantar pressure of selected anatomical areas of the foot using the sensors placed on the insole, it can also measure the frequency, heart rate, acceleration, foot angle, body temperature, and the number of steps, all these parameters will be sent via Bluetooth to the mobile phone.

At the same time, the cloud server will send a copy of these physiological parameters to the doctor via Wi-Fi, to a web application, the doctor can consult the medical file of his patients remotely and make the necessary analyzes of the pressure curves of the different sensors based on patient activity. Once the user is connected to the insole via Bluetooth, a connection notification appears and the user can begin to display his parameters including, his angle with respect to the ground, temperature, heart rate, and the position of the sensors in the insole, as shown in Fig. 6. In case of a bad plantar pressure value, a notification as shown in Fig. 7 will send it to the user and doctor, in our case the patient has

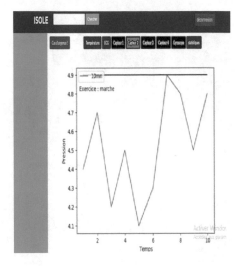

Fig. 8. Create an account.

Fig. 9. Curve of sensor number 2 during walking.

fallen in the right direction, this notification informs that the angle of the foot of the user compared to the normal angle had deviated from the floor.

5.2 Web Application

We aim to monitor elderly patients in their daily lives without disturbing them, the web application was developed for the doctor to supervise and control the health of his patients remotely as shown in Fig. 8.

The doctor must create an account to access the web application, he can then consult the medical file of each patient. In addition, he can intervene in case of bad plantar pressure. All data patients are saved in a Firebase database to ensure high-performance monitoring.

As shown in Fig. 9; in this graph, we control the signal received by sensor number 2 during the walking cycle, we can observe that the plantar pressure in the medial forefoot have a pic after a three pic in 4.7, 4.5 and 4.9 in the different time 2, 4 and 7 s after walking, so we can conclude that's the plantar pressure in the medial foot increase when the patient walking and decrease when he stops. During daily walking, we can observe the variation of signals received by the four sensors during walking, as shown in Fig. 10. The ECG signal is shown in Fig. 11. When the patient starts walking we observe the relation between movement and ECG signal, in fact, the curve of the signal increases and decreases when he stops walking.

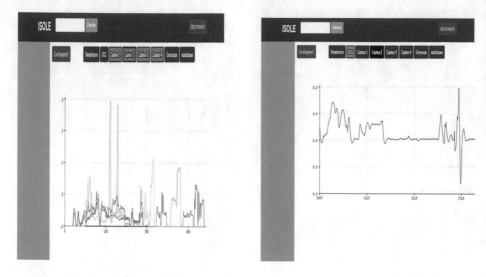

Fig. 10. Curves of four sensors.   **Fig. 11.** ECG curve.

6   Conclusion

In this paper, we detailed the design and the architecture of our proposed smart insole system for plantar pressure monitoring of elderly persons. Our proposed system consists of an Insole equipped with sensors to detect all physiological parameters in real-time, two applications are developed in order to monitor the plantar pressure over time; the mobile application and the web application; we expect through the use of these applications to improve the quality of follow-up with a decrease in the number and/or severity of adverse events that can be happened. An alert in real-time will be sent in case of a bad plantar pressure value detected.

The preliminary tests of our system are performed in Innov'Com Laboratory at Higher School of Communications of Tunis SUP'COM. The first results demonstrate the efficiency of the system to control and detect bad plantar pressure. In the future work, we expect to study and improve the accuracy and reliability of our proposed solution to distinguish men and women from the gait cycle using the different algorithms of data classification.

Acknowledgment. We are grateful for the support of the Department of Electrical and Computer Engineering at the New York University of Abu Dhabi (NYU). We would also like to thank SUP'COM' students for their contribution to reaching these results.

References

1. Belkacem, A.N., et al.: Brain computer interfaces for improving the quality of life of older adults and elderly patients. Front. Neurosci. **14**, 692 (2020)

2. Sibley, K.M., et al.: Chronic disease and falls in community-dwelling Canadians over 65 years old: a population-based study exploring associations with number and pattern of chronic conditions. BMC Geriatr. **14**(1), 22 (2014)
3. Alexander, B.H., Rivara, F.P., Wolf, M.E.: The cost and frequency of hospitalization for fall-related injuries in older adults. Am. J. Public Health **82**(7), 1020–1023 (1992)
4. Sterling, D.A., O'Connor, J.A., Bonadies, J.: Geriatric falls: injury severity is high and disproportionate to mechanism. J. Trauma Inj. Infect. Crit. Care **50**(1), 116–119 (2001)
5. Burns, E., Kakara, R.: Deaths from falls among persons aged 65 years–United States, 2007–2016. Morb. Mortal. Wkly Rep. **67**(18), 509 (2018)
6. Garcés-Gómez, Y.A., et al.: Fall risk in the aging population: fall prevention using smartphones technology and multiscale sample entropy. TELKOMNIKA **18**(6), 3058–3066 (2020)
7. Fini, N.A., Holland, A.E., Keating, J., Simek, J., Bernhardt, J.: How is physical activity monitored in people following stroke? Disabil. Rehabil. **37**(19), 1717–1731 (2015)
8. Freedson, P.S., Miller, K.: Objective monitoring of physical activity using motion sensors and heart rate. Res. Q. Exerc. Sport **71**(sup2), 21–29 (2000)
9. Mudge, S., Stott, N.S.: Timed walking tests correlate with daily step activity in persons with stroke. Arch. Phys. Med. Rehabil. **90**(2), 296–301 (2009)
10. Sirard, J.R., Pate, R.R.: Physical activity assessment in children and adolescents. Sports Med. **31**(6), 439–454 (2001)
11. Hergenroeder, A.L., Gibbs, B.B., Kotlarczyk, M.P., Perera, S., Kowalsky, R.J., Brach, J.S.: Accuracy and acceptability of commercial-grade physical activity monitors in older adults. J. Aging Phys. Act. **27**(2), 222–229 (2019)
12. Hegde, N., Bries, M., Swibas, T., Melanson, E., Sazonov, E.: Automatic recognition of activities of daily living utilizing insole-based and wrist-worn wearable sensors. IEEE J. Biomedical Health Inform. **22**(4), 979–988 (2018)
13. Lopez-Meyer, P., Fulk, G.D., Sazonov, E.S.: Automatic detection of temporal gait parameters in poststroke individuals. IEEE Trans. Inf. Technol. Biomed. **15**(4), 594–601 (2011)
14. Klimiec, E., Jasiewicz, B., Piekarski, J., Zaraska, K., Guzdek, P., Koaszczyski, G: Measuring of foot plantar pressure possible applications in quantitative analysis of human body mobility (2017)
15. Shu, L., Hua, T., Wang, Y., Li, Q., Feng, D.D., Tao, X.: In-shoe plantar pressure measurement and analysis system based on fabric pressure sensing array. IEEE Trans. Inf. Technol. Biomed. **14**(3), 767–775 (2010)
16. Saidani, S., Haddad, R., Mezghani, N., et al.: A survey on smart shoe insole systems. In: 2018 International Conference on Smart Communications and Networking (SmartNets), pp. 1–6. IEEE (2018)
17. Cerda, A.L.: Manejo del transtorno de marcha en adulto mayor. Médica Clínica Las Condes, vol. 25, pp. 265–275 (2014)
18. Shu, L., Hua, T., Wang, Y., Li, Q., Feng, D.D., Tao, X.: In-shoe plantar pressure measurement and analysis system based on fabric pressure sensing array. IEEE Trans. Inf Technol. Biomed. **14**(3), 767–775 (2010)
19. Spender, A., et al.: Wearables and the Internet of Things (2018)
20. Concepts, T., Pitaugier, F.: Dynafoot 2. Techno Concepts (2015)
21. Ramanathan, A.K., et al.: Repeatability of the Pedar-X in-shoe pressure measuring system. Foot Ankle Surg. **16**(2), 70–73 (2010)

22. Soames, R.W.: Foot pressure patterns during gait. J. Biomed. Eng. **7**, 120–126 (1985)

23. Duckworth, T., Betts, R.P., Franks, C.I., Burke, J.: The measurement of pressures under the foot. Foot Ankle **3**, 130–141 (1982)

24. Cavanagh, P.R., Hewitt Jr., F.G., Perry, J.E.: In-shoe plantar pressure measurement: a review. Foot **2**(4), 185–194 (1992)

25. Stefanescu, D.M., Millea, A.: The place of 'Force' in several graphic representations of the international system of units (SI). Sen. Transducers **131**(8), 1 (2011)

26. Deng, C., Tang, W., Liu, L., Chen, B., Li, M., Wang, Z.L.: Self-powered insole plantar pressure mapping system. Adv. Funct. Mater. **28**(29), 1801606 (2018)

27. Zhao, S., Liu, R., Fei, C., Zia, A.W., Jing, L.: Center of pressure progression characteristics under the plantar region for elderly adults plantar pressure force measurement and analysis system. Plos ONE **15**(8), e0237090 (2020)

28. Rana, N.K.: Application of force sensing resistor (FSR) in design of pressure scanning system for plantar pressure measurement. In: 2009 Second International Conference on Computer and Electrical Engineering, vol. 2, pp. 678–685. IEEE, December 2009

29. Chuckpaiwong, B., Nunley, J.A., Mall, N.A., Queen, R.M.: The effect of foot type on in-shoe plantar pressure during walking and running. Gait Posture **28**(3), 405–411 (2008)

30. Badarinath, A.: A truly in-shoe force measurement system. Dissertation, Arizona State University (2018)

31. Sistema Biofoot/IBV, Instituto de Biomecanica de Valencia, Valencia, Spain (2012)

32. Dyer, P.S., Bamberg, S.J.M.: Instrumented insole vs. force plate: a comparison of center of plantar pressure. In: 2011 Annual International Conference of the IEEE Engineering in Medicine and Biology Society, EMBC. IEEE (2011)

33. Sensor Medica. [Catalogue], Italy (2017). 42p

34. Pawlukiewicz, D.: Creating mobile apps with Thunkable. Forever F[r]ame, May 2018

35. Singh, V.: Introduction to FirebaseCoding Gurukul, December 2018

A New Distributed and Probabilistic Approach for Traffic Control in LPWANs

Kawtar Lasri[1(✉)], Yann Ben Maissa[2], Loubna Echabbi[2], Oana Iova[3],
and Fabrice Valois[3]

[1] INPT Rabat, Morocco and Univ Lyon, INSA Lyon, Inria, CITI,
Villeurbanne, France
kawtar.lasri@inria.fr
[2] INPT Rabat, Rabat, Morocco
{benmaissa,echabbi}@inpt.ac.ma
[3] Univ Lyon, INSA Lyon, Inria, CITI, Villeurbanne, France
{oana.iova,fabrice.valois}@insa-lyon.fr

Abstract. Low-Power Wide Area Networks (LPWANs) are wireless
networks with very low power consumption and wide area coverage.
They are capable of supporting the traffic of nearly a thousand nodes
with a duty cycle of less than 1%. However, the gradual densification
of nodes increases the number of collisions and makes it more difficult
to manage the upstream traffic. To mitigate this problem, we propose
a new distributed and probabilistic traffic control algorithm, DiPTC,
which allows nodes to adapt their traffic according to the needs of the
application (e.g., receiving K measurements over a time period) while
being agnostic to the number of nodes and to the network topology. A
control message is broadcast by the gateway to all nodes each period
when the objective is not reached, so that nodes can re-adapt their traffic. We evaluate the proposed solution in simulation and we compare
it with the LoRaWAN protocol. The results show that our algorithm is
able to reach the objective while keeping a low number of collisions, with
a longer network lifetime. Compared to LoRaWAN, our solution shows
a three times increase in the success rate and a decrease by a factor of
10 in the collision rate.

1 Introduction

Data collection is one of the main applications used to deploy sensors in smart
buildings and smart cities. To rapidly provide a data collection infrastructure
at reasonable costs, while covering several kilometers, Low Power Wide Area
Networks (LPWANs) are a very good solution [7].

Despite their attractiveness, these networks suffer from limitations due to the
used frequencies (free ISM band, *e.g.* 868 MHz in Europe) and medium access

This research was partially supported by CAMPUS FRANCE (PHC TOUBKAL 2019,
French-Morocoo bilateral program), Grant Number: 41562UA.

protocol (pure Aloha). In order to achieve long running networks and to support scalable deployments, traffic control is mandatory, especially in two different scenarios. First, in applications where the objective is to get a sampling of a given situation in an area (e.g. environmental monitoring), only a fixed number of measurements per time unit is necessary. On one hand, to receive more information does not bring any added-value and it only leads to network overload, increased collisions and energy consumption. On the other hand, to receive less information leads to a non efficient application. The second scenario is given when an LPWAN is operated by a given telecommunication operator. The operator provides coverage and connectivity for its clients. A service-level agreement (SLA) is defined to specify the capacity provided by the network provider, and the traffic model of the client. Unfortunately, there is no mechanism in LPWAN to manage the traffic sent by a user and the number of nodes connected to the network. Thus, a client can send more data and can connect more end devices, leading to network congestion. We claim that these scenarios supported in LPWANs will be more and more common. Note that mechanisms such as the adaptive data rate (ADR) implemented in LoRaWAN are not able to cope with this challenge. The objective of ADR is to minimize the energy consumption and to adapt the data rate according to the radio link budget and environmental conditions, but it is not the answer to congested networks. ADR is a local mechanism for each end device, without any capabilities to adapt the traffic of a group of nodes.

In this paper, we propose DiPTC, a new distributed, probabilistic and network topology agnostic algorithm for traffic control in LPWANs that allows the network manager and the applications to better steer the data collection. In a nutshell, DiPTC works as follows: (i) A central server (i.e., the network server in LoRaWAN, or core network server in Sigfox) acts as traffic policy enforcer by running a control loop. If the amount of received data does not correspond to the expected value the server schedules a control message to be sent by a gateway to the end devices. (ii) A local node mechanism adapts the traffic intensity taking into account the information received in the control message. It is important to notice that the end devices do not have any neighborhood knowledge and do not use any inter-node communication.

The rest of the paper is organized as follows: after discussing the state of the art for traffic control in wireless sensor networks and LPWANs in Sect. 2, we introduce and explain our proposed solution in Sect. 3. Section 4 describes the simulation setup and the scenarios that are used to validate, discuss and compare our results to the baseline LoRaWAN solution in Sect. 5. We conclude and present our future work in Sect. 6.

2   Related Work

In wireless sensor networks, spatial and temporal data aggregation are presented by many researchers as a solution for traffic reduction, and thus collision reduction. Spatial data aggregation is based on the organisation of nodes in the network and is done by choosing a messenger or a group of messenger device(s) that are responsible to send the gathered data to the gateway [4,6]. To gather the data, a communication between the messenger(s) and the nodes is essential. A temporal data aggregation based on data prediction is described in [5,12], where the authors focus on the temporal aggregation functions using ARIMA and LMS-PCA prediction models. Given an error of exceeded threshold, the nodes adapt their prediction model and send their new model coefficients to the gateway. Those coefficients are used by the gateway to predict data. In order to use data aggregation methods to solve the stated problem, nodes must have a large memory capacity, a neighborhood knowledge and the possibility to communicate between them. Unfortunately these constraints cannot be satisfied in LPWANs, therefore it is not possible to apply temporal aggregation techniques to our problem.

In LoRaWAN, researchers try to reduce the number of collisions and improve scalability by optimizing the allocation of resources, such as spreading factor and transmission power [1,11,13]. Ta *et al.* present LoRa-MAB, a flexible decentralized learning resource allocation algorithm based on the reinforcement learning problem Multi-Armed-Bandit [11]. Their approach far exceeds the results of the classic Adaptive Data Rate (ADR) algorithm implemented in LoRaWAN networks. Besides looking at the spreading factor, Luo *et al.* take into account the periodicity characteristics of traffic usually present in LPWAN applications, to propose a transmission scheduler (S-MAC) [13]. However these approaches only adapt the physical layer parameters of the existing traffic to the environmental conditions. They do not control the amount of traffic sent by the nodes, and they do not ensure that the traffic respects the constraints of the application. To the best of our knowledge no such traffic control mechanism exists today for LPWANs.

3   Towards a Distributed and Probabilistic Approach

We propose to control the uplink traffic through a distributed and probabilistic selection of a set of nodes, denoted as active nodes, which will be responsible to send the required data to the gateway, instead of all the nodes in the network.

3.1   Assumptions

Without loss of generality, we consider the case of a single gateway covering a given area. We also consider that the number of nodes in the network, denoted as N, is unknown. This number can evolve over time depending on successive deployments, hardware failures, or depleted batteries. This approach may target monitoring applications of a physical quantity varying over time, such as pollution [3], where the application needs K measurements over a given area periodically, every ΔT, whatever the number of nodes in the network.

In LPWANs, the downlink is mainly used to acknowledge correctly received packets at the gateway [9]. In this study, we propose to use a broadcast downlink as a source of feedback information, which can indicate, for example, the number of packets actually received during the previous period ΔT as well as the number of the K packets needed by the application. It is possible to consider any other type of message in order to notify if the objective has been reached. Thus, depending on the received information and the local regulations, each node can adapt its traffic to meet the needs of the application. The implementation of the broadcast downlink in an LPWAN is out of the scope of this paper.

3.2   The DiPTC Algorithm

We propose DiTC, an algorithm for controlling the traffic in the network based on the additive increase multiplicative decrease mechanism. When too much data is sent by the nodes, we apply a decreasing policy using the decrease factor x_D. Contrary, when not enough data is sent by the nodes, we apply an increasing policy using the increase factor x_I. This mechanism provides a rapid response by adjusting the traffic intensity with respect to the information transmitted by the gateway at each time period ΔT. To reduce the payload of the downlink message, we consider the feedback information sent by the gateway as a binary value. In other words, the gateway sends 0 when the number of received packets exceeds the needed ones, and 1 when the nodes do not send enough packets. When the objective is reached, no downlink is sent by the gateway, and the nodes continue their current traffic rate. As long as the gateway does not receive the K packets needed by the application per time period ΔT, each node increases its traffic intensity linearly (coefficient x_I). Otherwise, when the gateway receives more than K packets, exceeding the threshold, the nodes reduce exponentially their traffic intensity (coefficient x_D).

The finite state machine in Fig. 1 describes the DiPTC algorithm adopted by each node in the network. At first, each node awaits the reception of the broadcast message (MB) to decide on the number of messages, $m_n(t + \Delta T)$, to transmit in the next period ΔT. The broadcast message MB is sent by the gateway only if the number of received messages is different from the needed one, therefore reducing the downlink traffic. Then, each node decides to increase or to decrease its previous traffic intensity $\alpha_n(t)$ according to the binary value of the broadcast message received from the gateway. In other words, if the value of MB is equal to 1, the number of messages that the node will transmit is increased by x_I, $x_I \in \,]0, 1]$. Otherwise, the node decreases its traffic intensity exponentially using the multiplication factor x_D, $x_D \in \,]0, 1]$. The additive factor x_I and the multiplicative factor x_D are constant over time and equal for all nodes. They represent respectively, the proportion of traffic to increase or to decrease from the previous period. The larger x_I is, the faster is the increase, and the larger x_D is, the slower is the decrease. Next, the node transmits its $m_n(t + \Delta T)$ messages, which are the integer portion of $\alpha_n(t + \Delta T)$. We consider a node active when $m_n(t + \Delta T) > 0$. Finally, in order to avoid the synchronization of traffic intensities for all the nodes we introduced a local variable that works as an adaptation probability. More specifically, each node uses a random variable $V_n(t + \Delta T)$ that follows a Bernoulli distribution of parameter p, to decide if it should take into account ($p = 1$) or not ($p = 0$) the broadcast message.

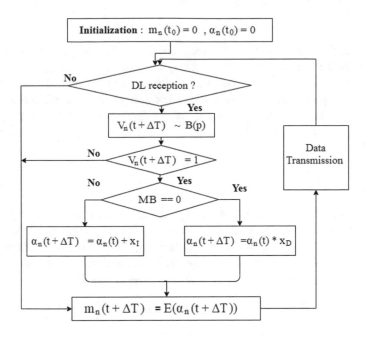

Fig. 1. Finite state machine of the DiPTC algorithm.

4   Simulation Model

While DiPTC can be applied to any LPWAN technology, we chose LoRaWAN as our testing protocol, which we also use as baseline for comparison. We implemented and evaluated our solution in LoRaSim [2], a well know discrete event simulator based on SimPy that allows simulating collisions, the capture effect and interference in LoRaWAN. We enhanced this simulator with a downlink model and a battery depletion model, which we present next.

4.1   Wireless Environment

LoRaSim does not implement downlink communication in LoRaWAN, so we enhanced the simulator by:

1. Adding a downlink model. For the LoRaWAN protocol used as baseline, the downlink is implemented in a unicast mode, to acknowledge the received packets. For DiPTC, the downlink is implemented in a broadcast mode, from the gateway to all the nodes.
2. Modeling the reliability of the downlink through a random variable that follows a Bernoulli distribution of parameter p_{DL} for both solutions.

To model the collisions and the interference on the uplink, we kept the ones used by Roedig et al. [2] for both approaches:

$$P_{rx} = P_{tx} + GL - L_{pl}(d) \tag{1}$$

$$L_{pl}(d) = \bar{L}_{pl}(d_0) + 10\gamma \log(d/d_0) + X_\sigma \qquad (2)$$

where P_{rx} is the power of the received signal, P_{tx} the transmission power, GL the accumulated general gains losses along the communication path, $L_{pl}(d)$ the log-distance path loss model determined by the nature of the communication environment, $L_{pl}(d)$ the path loss in dB at the communication distance d, $\bar{L}_{pl}(d_0)$ the mean path loss at the reference distance d_0, γ the path loss exponent, and $X_\sigma \sim N(0, \sigma^2)$ the normal distribution with zero mean, and σ^2 the variance to account for shadowing.

4.2   Energy Consumption

LoRaSim does not implement any energy consumption model for the nodes. We extended the simulator by adding the energy consumption model described in [8]. The energy consumed E_c after transmitting m messages for each node n, depends on the time on air TOA, the power consumed in the receiver mode $P_{w_{Rx}}$ and the power consumed in the transmitter mode $P_{w_{Tx}}$.

$$E_c = TOA * (P_{w_{Tx}} * m + P_{w_{Rx}}) \text{ for DiPTC} \qquad (3)$$

$$E_c = TOA * m * (P_{w_{Tx}} + P_{w_{Rx}}) \text{ for LoRaWAN} \qquad (4)$$

The values used for the parameters in these models are presented in Table 2.

5   Performance Evaluation

This section presents the evaluation of our proposed algorithm DiPTC. We start by first presenting the setup used in our simulations then we comment on the results.

5.1   Scenarios and Network Parameters

We consider the case of one centered gateway surrounded by N nodes distributed randomly with LoRa parameters chosen randomly. We consider the two simulation scenarios described in Table 1 in order to evaluate the performance of our proposed solution DiPTC, and to compare it against the baseline LoRaWAN. For each scenario, we specify the number of nodes N in the network, the number of measurements K needed by the application, the time period ΔT, the increasing and decreasing parameters (x_I, x_D), the adaptation probability and the simulation time $Simtime$. The adaptation parameters x_I, x_D and p depend on the number of nodes in the network and the traffic intensity, and were chosen accordingly. Given the limited number of pages we will not detail here how they were chosen, their optimization in function of different network conditions is the subject of future work.

In the first scenario, denoted as BASIC, we aim to evaluate DiPTC in the case of a normal traffic, while in the second scenario, denoted as INTENSIVE,

Table 1. Simulation scenarios.   **Table 2.** LoRa and network parameters.

Scenario	BASIC	INTENSIVE
DiPTC		
N	150	150
K	1	10
ΔT	10 min	1 min
x_I	0.5	0.2
x_D	0.5	0.5
p	0.5	0.06
Simtime	1 year	1 year
LoRaWAN		
AVG	1500 min	15 min

Parameters	Values
Payload (PL)	20 bytes
Header (H)	0
Preamble symbols	8
Downlink reception probability p_{DL}	0.99
Transmission power P_{Tx}	14 dBm
Gain and Loss GL	0
Path Loss exponent γ	2.08
Reference distance d_0	40 m
Max. distance to the gateway	300 m
Path Loss at the reference distance $L_{pl}(d_0)$	−127.41 dB
Variance X_σ	N(0,3.57)
Current drawn during the receive mode I_{Rx}	11.2 mA
Current drawn during the Transmission mode I_{Tx}	90 mA
Current drawn during the sleep mode I_{Sleep}	1 μA
Supply voltage	3 V
Battery capacity	30 J

we consider a significantly higher traffic load ($\times 10$) to test the scalability of our proposed solution. While in DiPTC the traffic intensity of each node evolves dynamically according to the feedback information received from the gateway, in baseline LoRaWAN, the traffic intensity is constant. To propose a fair comparison and to respect the application constraints, i.e., receiving K measurements every time period ΔT, in baseline LoRaWAN nodes send their packet following a Poisson distribution with the rate: $AVG = \frac{\Delta T}{K} \times N$.

In baseline LoRaWAN we configured a lost packet to be re-transmitted a maximum of 8 times before being dropped, a value commonly used in different implementations, considering that the standard specifies a maximum number of 15 re-transmissions. In DiPTC there are no re-transmissions, as the nodes do not receive unicast acknowledgments from the gateway.

The values of the propagation model are determined empirically in [2], and those of the energy consumption model can be found in Semtech SX1272 LoRa transceiver Datasheet [10]. Table 2, summarizes these values.

5.2 Performance Evaluation in the BASIC scenario

We first evaluate the performance of DiPTC in the BASIC scenario by looking at the evolution of the number of packets sent and received, the number of collisions, the number of dead nodes, and the downlink and uplink losses throughout the whole lifetime of the network (Fig. 2). This type of detailed analysis allows us to dissect the behavior of DiPTC and better showcase its performance.

The first thing that we can notice in Fig. 2(a) is an alternation of transient and stationary states. Transient states of very short duration (due to the death of a node or a packet loss) are followed by long stationary states. This shows that our DiPTC algorithm is able to converge quickly to the desired number of measurements K, despite environmental challenges, and to remain stable for a significant time. Figure 2(b) shows that the transient states are either caused by the death of a node or an uplink loss or both events. The loss of a downlink does not impact the convergence of our algorithm, as the node continues with the same traffic intensity until it receives a feedback from the gateway telling it otherwise. The frequency of the long duration stationary states is induced by the stability property of our mechanism and by the lack of collisions. The low collision rate is due to a low traffic intensity, consequence of a low number of expected measurements in this scenario ($K = 1$). This low traffic also has a positive effect on the lifetime of a node as it reduces the battery depletion process.

In fact, as shown in the zoom part in Fig. 2(a), the death of node 45 (time: 1794.0) generates a new transient state that lasts for 1 h : 10 min before a new stationary state takes place with a new active node with ID 108. Note that the downlink loss in 1795 h delays the convergence because the feedback information is missing. Thus, we lost $\Delta T = 10$ min but the convergence is not affected. However the uplink losses in 1794.8 h and 1794.3 h extend the transient regime.

If we compare the performance of DiPTC with that of baseline LoRaWAN from Fig. 3(a), the first thing that we notice is that the traffic in LoRaWAN appears to be lower than in DiPTC. Indeed, the maximum number of received

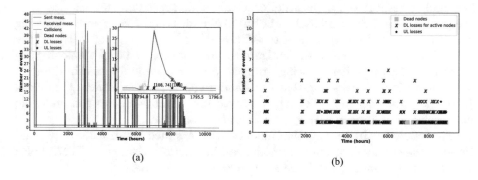

(a)   (b)

Fig. 2. Traffic evolution in the BASIC scenario for DiPTC.

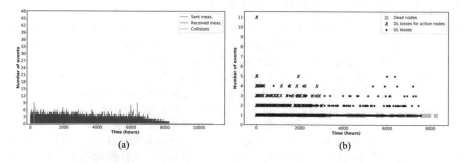

(a)   (b)

Fig. 3. Traffic evolution in the BASIC scenario for LoRaWAN.

messages is 9 messages per period ($\Delta T = 10$ min), while in DiPTC is 43. However, the oscillations in baseline LoRaWAN are more frequent than in DiPTC. This behaviour is due to the fact that the packet generation in baseline LoRaWAN follows a Poisson distribution of rate 0.0006, meaning that, in average, one packet is sent every 1500 min, leading the nodes to traffic less but more often, more *regularly*. In DiPTC, the traffic intensity is controlled by the additive increase multiplicative decrease mechanism that adapts the node activity when the goal of having 1 measure per 10 min is not met, and stabilizes it otherwise.

When the goal of the application is to receive a specific amount of messages per period of time, an important metric that allows us to evaluate the performance of different solutions is the absolute error. Consequently, we plotted in Fig. 4 the application error frequency for both baseline LoRaWAN and DiPTC. We can see that for DiPTC the application error is 0 most of the time, with some cases when it is slightly under. Contrary, LoRaWAN under-performs most of the time.

Finally, Fig. 5 shows the node traffic evolution in the BASIC scenario for baseline LoRaWAN and DiPTC. Note that the active node traffic average corresponds to the traffic of the nodes when $\alpha_n(t) > 0$, and the nodes traffic average correspond to those when $m_n(t) > 0$. These figures illustrate the need to have a data flow control in LoRaWAN. Note that the maximum number of packets that a node can send in DiPTC is 3 times smaller than the baseline LoRaWAN and the network traffic average is smoother than baseline LoRaWAN.

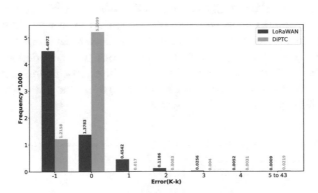

Fig. 4. Error frequency in the BASIC scenario for DiPTC and Baseline LoRaWAN.

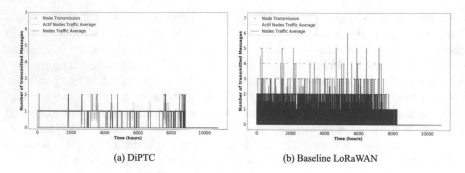

(a) DiPTC   (b) Baseline LoRaWAN

Fig. 5. Nodes traffic evolution in the BASIC scenario.

5.3   Impact of the Traffic Intensity

Fig. 6 illustrates the evolution of the number of collisions, the number of sent and received packets, the number of dead nodes, and the number of downlink and uplink losses, in the case of the INTENSIVE scenario for DiPTC. Unlike the BASIC scenario, the INTENSIVE presents several alternations between long duration transient states and short duration stationary states. These alternations are the result of destabilization produced by several factors: frequent and successive uplink and downlink losses, active node deaths and a high collision probability. The collision probability is higher in this scenario, since we increase the number of needed measurements by the gateway ($K=10$), while reducing the time period ($\Delta T = 1$ min). We observe in Fig. (6(a)) that even with death nodes and downlink losses, the received measurements are close to the objective of 10 measurements per time period. The convergence of our proposal depends mainly on the uplink losses. Moreover, the maximum absolute error in this case is 3, which means that the worst case is to receive 13 messages at the gateway. Considering the network density, this error is acceptable.

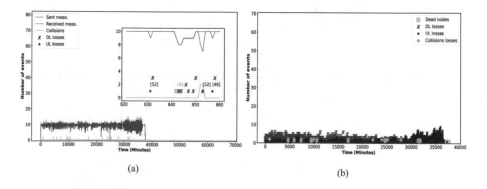

(a)   (b)

Fig. 6. Traffic evolution in the INTENSIVE scenario for DiPTC.

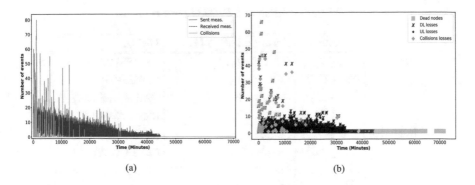

Fig. 7. Traffic evolution in the INTENSIVE scenario for Baseline LoRaWAN

5.4 Overall Performance Evaluation

We finally make and overall performance evaluation of the baseline LoRaWAN and DiPTC using the following metrics:

- **Success rate** μ_c: measures the number of times that the base station receives the K needed measurements per period ΔT.
- **Collisions rate** τ_c: measures the number of packet collisions in the network.
- **Network lifetime** t_l: measures the duration the network is able to support the application requirements, i.e., the nodes are able to send the K needed measurements per period ΔT before the exhaustion of their battery.

We consider the same simulation scenarios and network parameters as above with longer simulation time, $Simtime = 15$ months. Table 3 presents the simulation results of both scenarios. For each metric, we calculated the mean and the standard deviation over 10 simulations.

We note that in the BASIC scenario DiPTC has a success rate of almost 100%, which is 3 times greater than LoRaWAN. This has an obvious positive effect on the network lifetime, which is longer for DiPTC.

In the INTENSIVE scenario DiPTC has a success rate of 29.64%, which is 20 times greater than LoRaWAN. This shows that DiPTC can meet the needs of the application more often than baseline LoRaWAN, even in the case of intensive traffic. Note that the collision rate in DiPTC is 8 time less than in LoRaWAN. Moreover, the network lifetime in DiPTC is considerably longer.

Table 3. Baseline and DiPTC performances.

Sc	Measures	Baseline LoRaWAN	DiPTC
Basic	$(\overline{\mu_c}, \sqrt{\sigma_{\mu_c}})$	(32.45%, 0.12)	(99.18%,0.24)
	$(\overline{\tau_c}, \sqrt{\sigma_{\tau_c}})$	(0%, 0)	(0.0%, 0.0)
	$(\overline{t_l}, \sqrt{\sigma_{t_l}})$	(51823.5 h, 14.84)	(53024 h, 54.29)
	$(\overline{E_c}, \sqrt{\sigma_{E_c}})$	(1044.97 J, 2.82)	(3469.76 J, 140.05)
Intensive	$(\overline{\mu_c}, \sqrt{\sigma_{\mu_c}})$	(1.47%, 1.23)	(29.64%, 0.513)
	$(\overline{\tau_c}, \sqrt{\sigma_{\tau_c}})$	(0.96%, 0.20)	(0.12%,0.02)
	$(\overline{t_l}, \sqrt{\sigma_{t_l}})$	(22563.0 min, 1082.64)	(34102.0 min, 50.00)
	$(\overline{E_c}, \sqrt{\sigma_{E_c}})$	(7081.54 J, 80.9)	(10696.34 J, 24.99)

5.5   Insights

The DiPTC algorithm that we propose shows very good results regarding the convergence time, not only for basic networks, but also for those with an intensive traffic. Thanks to the stability property of the adapting mechanism, our approach manages to converge towards the needed measurements K with a reasonable absolute error. Indeed, DiPTC is able to control the traffic in LPWANs, especially in LoRaWAN, with a low collision probability and only a small overhead. After considering different traffic intensities in our evaluation, we can sum-up the following findings:

1. The reliability of the downlink does not affect the convergence of DiPTC even for low downlink reception probabilities.
2. The death of an active node due to its battery depletion generates a new transient mode of a short duration.
3. The reliability of the uplink has a negative impact on the convergence of DiPTC as it introduces long-term transient states and short-term stationary states.
4. In the BASIC scenario, DiPTC converges exactly towards K measurements per time period. We observe quite long duration for the stationary states compared to the INTENSIVE scenario.
5. In the INTENSIVE scenario, DiPTC converges towards the needed measurements (K) with a reasonable absolute error, but with more frequent and longer transient states.
6. The loss of a transmission during a collision has the same effect as its loss due to the propagation model.
7. Compared to baseline LoRaWAN, DiPTC is able to reach the objective within reasonable deadlines, while keeping a low number of collisions and a low energy cost with a longer network lifetime.

6   Conclusion

In this work, we propose a new distributed and probabilistic algorithm for traffic control in LPWANs. This simple idea, yet powerful, of controlling the traffic using an additive increase multiplicative decrease mechanism and a binary broadcast message sent by the gateway as a feedback information shows very good results. Indeed, DiPTC reduces collision probability and insures convergence to the needs of the application (K measurements per period), within reasonable delays. We also found that compared to the baseline LoRaWAN, DiPTC shows a three times increase in the success rate.

In our future work, we plan to investigate and analyze closely the impact of the traffic adaptation factors x_I and x_D on the equity, the absolute error and the responsiveness of the network, for different traffic and topology scenarios. In fact these parameters affect the duration of the transient states and their oscillation domains, which impact the energy network consumption. Therefore, the values of x_I and x_D introduce a trade-off between the network energy consumption and its responsiveness.

References

1. Azari, A., Cavdar, C.: Self-organized low-power IoT networks: a distributed learning approach. In: IEEE GLOBECOM, Abu Dhabi, UAE (2018)
2. Bor, M.C., Roedig, U., Voigt, T., Alonso, J.M.: Do LoRa low-power wide-area networks scale? In: IEEE MSWiM, Malta (2016)
3. Boubrima, A., Bechkit, W., Rivano, H.: On the deployment of wireless sensor networks for air quality mapping: optimization models and algorithms. IEEE/ACM Trans. Netw. **27**(4), 1629–1642 (2019)
4. Ennajari, H., Maissa, Y.B., Mouline, S.: Energy efficient in-network aggregation algorithms in wireless sensor networks: a survey. In: UNet, Casablanca, Morocco (2016)
5. Liu, C., Wu, K., Tsao, M.: Energy efficient information collection with the ARIMA model in wireless sensor networks. In: IEEE GLOBECOM, St. Louis, MO, USA (2005)
6. Ma, Y., Guo, Y., Tian, X., Ghanem, M.: Distributed clustering-based aggregation algorithm for spatial correlated sensor networks. IEEE Sens. J. **11**(3), 641–648 (2010)
7. Mekki, K., Bajic, F., Chaxel, E., Meyer, F.: A comparative study of LPWAN technologies for large-scale IoT deployment. ICT Express **5**(1), 1–7 (2019)
8. Slabicki, M., Premsankar, G., Di Francesco, M.: Adaptive configuration of LoRa networks for dense IoT deployments. In: IEEE/IFIP NOMS, Taipei, Taiwan (2018)
9. Sornin, N., Eirich, L.M., Kramp, T., Hersent, O.: LoRaWAN specification. LoRa Alliance (2015)
10. SX1272/73. Semtech datasheet - 860 MHz to 1020 MHz Low Power Long Range Transceiver, rev. 4, January 2019
11. Khawam Ta, D.T., et al.: LoRa-MAB: a flexible simulator for decentralized learning resource allocation in IoT networks. In: IEEE Wireless and Mobile Networking Conference, Paris, France, September 2019
12. Tan, L., Wu, M.: Data reduction in wireless sensor networks: a hierarchical LMS prediction approach. IEEE Sens. J. **16**(6), 1708–1715 (2015)
13. Luo, J., Xu, Z.: S-MAC: achieving high scalability via adaptive scheduling in LPWAN. In: IEEE INFOCOM, Virtual Conference, July 2020

A Fog Computing Simulation Approach Adopting the Implementation Science and IoT Wearable Devices to Support Predictions in Healthcare Environments

Thiago G. Thomé[1]([✉]), Victor Ströele[1], Hélady Pinheiro[2],
and Mario A. R. Dantas[1]

[1] Institute of Exact Sciences, Federal University of Juiz de Fora,
Rua José Lourenço Kelmer, Juiz de Fora, MG, Brazil
{thiagogoldoni,victor.stroele,mario.dantas}@ice.ufjf.br
[2] School of Medicine, Federal University of Juiz de Fora,
Rua José Lourenço Kelmer, Juiz de Fora, MG, Brazil

Abstract. The world is facing a grand challenge with the unexpected pandemic caused by the Sars-CoV-2 virus. Several researches are underway to understand more about the virus, its way of spreading in environments and prevention methods. Even in a short period, it was possible to obtain recommendations to assist in the control of contamination, and some parameters of those are the use of masks and social distancing. In this study, we considered the implementation science concept in a simulation effort based on changes in habits and behaviors related based on prevention methods. In addition, in our work we also considered the utilization of wearable IoT devices for monitoring people who live in environments where social isolation is complex. We conceived four scenarios with different prevention approaches and isolation, where the health data of the simulated agents were collected for monitoring and providing predictions. The implementation science approach, together with wearable IoT devices, provided a differentiated view from all environments. Agents that have more preventive habits got contamination rates of 12.11% against the worst scenario, with 77.00%.

1 Introduction

There are pandemics since the beginning of human history and world population has adapted to new realities and habits to overcome the effects and consequences of these unexpected new diseases. In the outbreak of the Covid-19, which is caused by Sars-CoV-2 virus, experts and the World Health Organization recommended increasing social isolation as the main preventive measure [14].

In some environments, such as churches, schools, and universities, there is the possibility of interrupting activities. Nevertheless, there are environments in which isolation and social distance are more complex, leading to other preventive measures as attenuators in contagion. In these environments, such as nursing homes and clinics, social isolation is made more difficult. As the transmission of Covid-19 can occur during the incubation period, intensive monitoring is important in people who live or frequent such environments [6]. Since medical appointments are not recommended, one way to perform people monitoring is due to the use of wearable IoT devices.

The adoption in the use of IoT equipment in recent years, several wearable devices with sensors that allow monitoring human health are easily found on the market. Some of those are smart bands and smartwatches, that are able to capture heart rate, blood pressure, oxygenation level, body temperature, among others vital signs. Which some of these can be important information to assist in the detection of possible contamination of Covid-19.

Ambient Assisted Living (AAL) is based on the use of IoT devices and ways to ensure that elderly people or anyone who needs assistance can have quality of life and independence in their homes. Through a learning engine, this type of light surveillance that uses the collected data can provide to stakeholders warnings about breaking on user habits or standards, which can be related to the health and safety of those being monitored [9].

Due to the huge amount of collected data, filtering and cleaning processes are important to remove useless information and shrink all the collected information. Thus, a fog-cloud cooperative system is proposed to distribute part or all of the tasks that would be focused in just one application, increasing the efficiency of the process [12].

In addition, the Function as a Service (FaaS) concept is proposed to manage the server and to smooth allocating resources and managing servers tasks. In this way, the application can be divided into several functions which are only executed when requested. This type of approach makes the system scalable and has a better cost-benefit in terms of active processing time [4].

Even though it is not always used in health concepts, the main concept of Implementation Science (IS) is focused on the healthcare area [2]. Thereby, the concept of Implementation Science is to reach clinical, community and policy contexts using studies to improve the quality and effectiveness of health services, increasing population's quality of life. From other point of view, Implementation Science seeks to understand how citizens, health professionals, healthcare organizations and politics behaviours to find evidence-based strategies to maximize healthcare investments and interventions.

Through the large amount of data obtained by wearable devices and the use of Implementation Science concepts, it is possible for computer systems to use learning techniques to make predictions related to a person's health, and thereby assist in preventing the contamination of an entire environment.

One of challenges faced in this study was how to adapt a computational proposal that adopts IoT paradigms that allow to understand how the spread of the virus occurs in specific environments. Due to the fact that the health data obtained by IoT devices are from the users themselves and that they tend not to make the data public, there is a great difficulty in the search for databases that meet all the desired characteristics. Therefore, to concatenate Sars-CoV-2 scenario, IoT and environments where social distance is difficult, simulated environments was used.

Using a simulator allows conceiving our IoT paradigm proposal to be adopted and easily demonstrated in indoor Covid-19 contamination, and to reach this objectives, the Siafu simulator was used [10]. Siafu is a simulator that allows to control characteristics of locations, behaviors of agents and the entire context. In addition to having technical features that allow total control of the simulation, the simulator also has a graphical interface and allows the creation and transfer of the generated data.

The approximation of the exposed concepts is represented through the simulation which, with divergent preventive measures, allows the collection of agent's health data and an analysis of the effects of each action that was taken. As a consequence, the attitudes and behaviors of agents can be related to the contamination rates of the environment. Finally, the research question is: *How to conceive a computational proposal adopting the IoT paradigm to easily demonstrate the indoor Covid-19 contamination?*

This paper is organized as follows. Section 2 presents related works, where in Sect. 3 it is introduced materials and methods. The proposed approach is shown in Sect. 4. Experimental environment and results are illustrated in Sect. 5, and finally in Sect. 5 we present conclusions and future work.

2   Related Works

According to study presented in [13], the impossibility of taking care of all infected people can be mitigated through the use of several IoT applications. This case study of look at different IoT efforts in the struggle against the pandemic caused by Sars-CoV-2 virus. From the collection of data in smart cities to the use of mobile applications by the population were analysed. The author emphasizes the importance of securing the information collected, as well as the attention on future works related to storage systems that can manage this large amount of data.

Among the examined hypotheses in [13], several approaches seek the integration between different areas that use IoT, such as locating medical equipment in real time to ease service time for people who are in need. There are other ways that also use another lines of research, such as the use of Artificial Intelligence to monitor and track the spread of the disease in cities and the use of medicine to provide remote medical care through monitoring IoT devices. Due to this great amount of possibilities, the author concludes that the use of Internet of Things can helps controlling the spread of Covid-19's disease.

The research presented in [8] shows an architecture proposal based on simulation for the development of solutions with IoT applications. The work seeks to allow the development of simulations in different contexts of smart cities and the creation of data through the Internet of Things paradigm. The simulations can have several purposes, from the control of public safety to the health of infected patients, and can use specific scenarios, allowing total control of the environments. Therefore, simulations on the dissipation of the Sars-CoV-2 virus can be performed in many different ways.

Simulations were performed in several environments in [8], including the campus of the Federal University of Juiz de Fora (UFJF). Following the simulation process, it was possible to predict how employees, students and other simulation agents can impact in the spread of the virus, allowing the creation of models to understand how new contamination arises. In addition, the collected data was stored for future work, allowing new studies to be developed.

3   Materials and Methods

Due to the unexpected outbreak of Covid-19, the limited knowledge of the virus, and the sharing of fake news, a large part of the world population had no contact with preventive measures guidelines. A need to associate the more respected and advanced studies on contagion arose with so much divergent information present in the media. Thus, simulated tests were considered with results found so far.

In addition to the proposed simulation model that is presented in this paper, the development of the characteristics of the simulated agents, the environment where they reside, the transmission process, and the data collection and transmission are described.

3.1   Simulation Environments

In the research we considered a simulation model as an approach to tackle challenges related to research question. The scope includes the data collection from simulated agents to the computational environment responsible for providing the health predictions. In addition to that, technologies such as fog-cloud cooperation and the use of FaaS (Function as a Service) were also expressed in the main idea. The proposed simulation model is shown in Fig. 1.

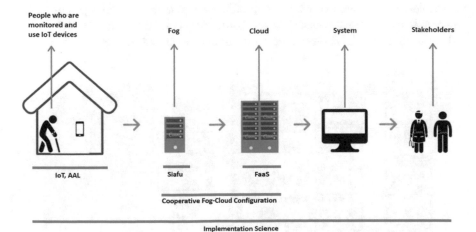

Fig. 1. Simulation model

Following [5], our contribution can be classified in subject areas as an application and referring to the research aspects as a technology approach (i.e. implementation science and IoT wearable devices). In the proposed solution, Siafu Simulator is responsible for simulating the environment where social isolation is complex and the agents, that are the people who need monitoring and use wearable IoT devices in an Ambient Assisted Living. Fogs are devices that correspond to a specific region of agents and their Edges are the mobile smartphones from each agent, that perform computational processes on the agents received information. The cloud uses the Function as a Service concept and is responsible for receiving previously processed information from different Fogs and performing machine learning routines.

In addition, the system represents a web application that has access to the databases located in the cloud and provides predictions, graphs, reports and information that can be used and accessed. The stakeholders in this proposal are the users of the web application who are interested in the generated information, and they can be relatives, doctors, nurses or others health professionals responsible for monitored agents. Finally, the concept of Implementation Science, that assist and support in confronting problems and offering solutions for health interventions and the use of evidence based practices.

Even though they are not present in the image of the proposed model, some important concepts are essential for the consistency of the whole process. Thus, it is important to emphasize the use of security layers to transport the collected information, as they are part of the users' privacy. Moreover, due to the large amount of information, storage systems are also an important part of the model.

For a preliminary appraisal, only the simulation through Siafu, the creation of the data collected by the wearable IoT devices and the mobile application, which works as the Edge of a Fog, were implemented. The environment, the rooms and the characteristics of each agent, just as sleep time, bathroom intervals, and so on, were developed from the beginning. Social distance is complex in this simulated environment, which in this model is a nursing home with 18 agents. No employee or other external agent influences the simulation's progress. The simulated environment consists of six bedrooms, two bathrooms, a living room and a dining room. The described environment is shown in Fig. 2.

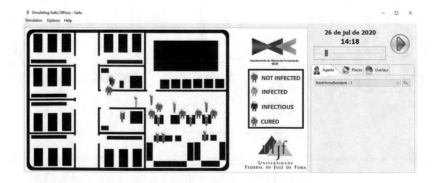

Fig. 2. Simulation environment

An important contribution of this work is the effort related to the entire environment. In other words, since the creation of rooms, the behavior of agents, and health information were developed work from the sketching. Siafu Simulator allowed us with our contribution effort to develop different simulated environments using different techniques. This effort can be translated to a view of a simulation interface.

The situation of the contamination in each agent of the simulation is represented by colors. In blue, agents that have not been contaminated. In yellow, those who have the disease but have not yet begun to spread the virus. In red, the agents that can infect others. Finally, the green agents have passed through the disease's timeline and are now cured.

For the beginning of the contamination simulation, it is considered that a single agent has been contaminated and from that moment the simulation is started. This event is considered to be the beginning of the first day. This situation impacts that the result of the contamination has no influence on the number of employees or external agents of the simulation, which is something very variable. In consequence, the contamination depends exclusively on the residents of the house.

The nodes of a fog can be small devices or even local servers. Representing the fog in the proposed simulation model, the mobile application was developed to collect information from wearable, perform cleaning/filtering routines and send these data to the clouds. The application is being developed and is shown in Fig. 3, where the main screen and an attempt to connect to an IoT device are displayed.

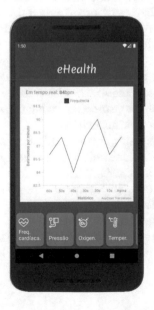

Fig. 3. Ongoing implementation of mobile application

4   Proposed Approach

In the simulated model, we added several personalized features to the agents. Each one has exact schedules to wake up and sleep, intervals to go to the bathroom, and behavior with preventive measures before and after the revelation of the first case in the environment. Due to the exponential advance of cases, it is common for the population to start changing their habits only when someone near them is diagnosed. Therefore, the simulation allows parameterizing all these characteristics. In addition, some health data is collected through wearable IoT devices. Among them, it is important to mention heart rate, blood pressure, body temperature, and blood oxygenation level.

4.1   Test Scenarios

As a result of the recent Covid-19 pandemic, many researchers and scientists around the world have started studies for learning more about the new disease. Due to different ways in which combat is carried out in each region and country, some figures as percentage of asymptomatic individuals are relatively divergent. For the simulation, it was necessary to compile more reputable and recent studies.

Following the studies of [3,6,7], the rate of asymptomatic patients for the simulation was adjusted to 30%. So, for every ten infected people, three will be transmitters and the data captured by wearable will not change, preventing the individual from being identified.

According to [1], in its systematic review and meta-analysis, the use of N95 masks, classified as PFF2 in Brazil, or even other masks can reduce in a significantly way the chance of contagion. Provided that they are used correctly, studies indicate that as PFF2 masks can reduce the chance of contagion by up to 95%, while cotton masks can reach 67%. Another preventive measure analyzed was social isolation, which for a distance of 1 m, reduces the chances of contagion by 82%. In addition to that, for each additional meter, the risk falls by half. Similarly, another form of protection is the use of eye protection, which also has a significant reduction, reaching values of 78% of risk mitigation.

The proposed simulation model allows comparison between different situations. From tests on the effectiveness of prevention methods to how the agents' behavior influences the contamination of an environment. Therefore, four environments were prepared to allow the comparison on the habits and practices of the agents as described in Table 1, Table 2 and Table 3.

Table 1 and Table 2 are responsible for categorizing habits and preventive measures in all environments, grouping them in two. On the other hand, Table 3 presents a different approach than the previous ones, separating the previously created groups.

Table 1. Habits and preventive measures of agents considering the first case revelation in simulated environments 1 and 3

Period of time about the 1st case revelation	Before	After
Use of PFF2/N95 masks	0%	100%
Use of other kind of masks	50%	0%
Use of eye protection	0%	90%
Average increase in interpersonal distance	0 m	1.5 m

In simulated environments 1 and 3, agents changed their habits subtly after the start of the pandemic, with only 50% wearing masks before the first case in the environment. The use of PFF2 masks, the use of eye protection and the habit of increasing the distance were ignored by them.

In contrast, all agents in both simulated environments 2 and 4 used PFF2 or other masks after the beginning of the disease, respectively 90% and 10%. In addition to masks, it is important to note the use of eye protection by 50% of the population and the average increase in distance by one meter.

Furthermore, to increase the precision of the simulation, in environments 1 and 2 when a contaminated agent reaches a clinical picture with symptoms, it is removed from the simulation. In contrast, agents from environments 3 and 4 are not isolated when they experience the first symptoms. Thus, there is no contamination by an agent in which the disease was detected in the first two environments, preventing other agents from continuing their daily routines with an infected agent in which everyone knows its situation.

In all simulated models, the agents' habits and protective measures were similar after the discover of the contamination in the environment. Thus, the diversity and variety between the simulations are found on preventive measures and on the isolation of agents with symptoms. The purpose for these fundamen-

Table 2. Habits and preventive measures of agents considering the first case revelation in simulated environments 2 and 4

Period of time about the 1st case revelation	Before	After
Use of PFF2/N95 masks	90%	100%
Use of other kind of masks	10%	0%
Use of eye protection	50%	90%
Average increase in interpersonal distance	1.0 m	1.5 m

Table 3. Habit of isolating infected agents in simulated environments

Habits	Simulated environments 1 and 2	Simulated environments 3 and 4
Isolation of infected	Yes	No

tals is to understand and demonstrate the consequences of the spread of a virus that transmission starts during the incubation period [6].

5   Experimental Environment and Results

During the test period, 50 executions for each of the simulation environments were done and no unexpected problems occurred. All of them were performed on NEC's SX-Aurora TSUBASA [11] computer, witch use vector processors. In all experiments, the agents behaved according to the defined parameters and the values obtained from the sensors, as heart rate, blood pressure, body temperature and blood oxygenation level, were captured successfully.

Some points of interest in the study are described in Table 4, Table 5, Table 6 and Table 7. Among them, the average values of contaminated agents before and after the disclosure of the beginning of contamination in the environment. Others points are in Fig. 4, as the amount of simultaneously infected agents. It is substantial to highlight that the exposed average values and the effects of the spread of the disease on the environment are the consequences of preventive habits and isolation.

According to the results of the executions, it was detected that around 67.89% of the agents in the simulated environment 1 were contaminated by the end of the simulation. Of the total infected, 65.22% were contaminated before the first case was found. Otherwise, only 12.11% of the agents in the simulated environment 2 were contaminated and only 0.67% of these were contaminated after the identification of the first case in the environment.

On the other hand, when there was no isolation of agents with symptoms, as in simulations 3 and 4, the amounts of contamination were relatively higher when compared to the first two environments. Thus, 77.00% of the agents were contaminated in the simulated environment 3 and 25.89% in the environment 4.

In Fig. 4, as a result of the contamination that occurred, it is possible to analyze a higher concentration of contamination during the first month in the

Table 4. Contaminated agents in simulation environment 1

Situation	Contaminated agents in environment 1
Before first case revelation	65.22%
After first case revelation	2.67%
Total	67.89%

Table 5. Contaminated agents in simulation environment 2

Situation	Contaminated agents in environment 2
Before first case revelation	11.44%
After first case revelation	0.67%
Total	12.11%

Table 6. Contaminated agents in simulation environment 3

Situation	Contaminated agents in environment 3
Before first case revelation	61.56%
After first case revelation	15.44%
Total	77.00%

Table 7. Contaminated agents in simulation environment 4

Situation	Contaminated agents in environment 4
Before first case revelation	11.00%
After first case revelation	14.89%
Total	25.89%

simulated environments 1 and 3. In addition, with a smaller number of contaminated agents, a smoother distribution occurred along the period in environments 2 and 4.

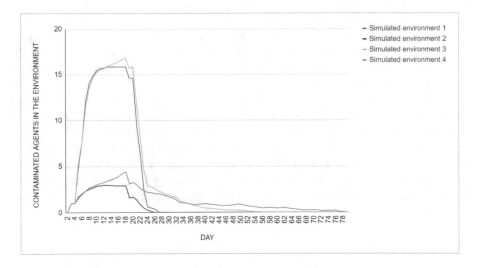

Fig. 4. Contaminated agents in simulation environments

Different analysis can be performed with the obtained results. When observing environments 1 and 2, it shows that even with different prevention habits, the end of contamination was around the twenty-seventh day, which is a significantly better result when compared to environments 3 and 4, that the end were after the fiftieth day. Another way of observing the results is when a comparison between environments 1 and 3 is made with environments 2 and 4. The average amount of contaminated agents at the same time in simulations 1 and 3 reaches values up to five times higher than in the simulated environments 2 and 4. This result highlights the need to isolate contaminated agents immediately after identification.

Finally, a closer look to the results reveals that if a merge is done with the best characteristics of both cases, preventive measures and isolation, the best of both worlds is a environment that was already tested. Through the conferences carried out, the simulated environment 2 with previous use of masks and eye protectors, increased distance and isolation of the infected ones presented the best results when compared to the others. Due to fewer infected agents, it is possible to realize the importance of preventive measures in controlling the spread of Sars-CoV-2 in environments where social isolation is difficult.

6   Conclusions and Future Works

In this work, we adopted the simulation approach, utilizing the Siafu Simulator, where since the creation of rooms, agents' behavior, and health information were developed work from the sketching. Siafu Simulator allowed us with our contribution effort to create diverse simulated environments using different techniques. We considered the implementation science and IoT wearable devices as key elements in our research. It was possible to obtain successful results and, even more importantly, to show how the isolation and preventive habit is essential. The preliminary evaluation of the model, using the simulated environments, showed the relevance of attitudes before knowing the existence of any contaminated agent for a successful prevention.

As future work, we are conceiving to implement features such as cleaning and filtering data in mobile applications. Another important tool is the web system for stakeholders. In parallel, a cooperative fog-cloud system, which could be implemented using the FaaS, is recommended for the best performance and cost management. Finally, it is important to highlight that the use of the implementation science concept will continue in future targeting to provide transparency in developments, which could indicate how to reproduce these efforts from other research groups.

References

1. Chu, D.K., Akl, E.A., Duda, S., et al.: Physical distancing, face masks, and eye protection to prevent person-to-person transmission of SARS-CoV-2 and COVID-19: a systematic review and meta-analysis. The Lancet **395**, 1973–1987 (2020)
2. Decamps, M., Meháut, J.F., Vidal, V., et al.: An implementation science effort in a heterogenous edge computing platform to support a case study of a virtual scenario application. In: International Conference on P2P, Parallel, Grid, Cloud and Internet Computing, vol. 158 (2020)
3. Ferguson, N.M., Laydon, D., Nedjati-Gilani, G., et al.: Impact of non-pharmaceutical interventions (NPIs) to reduce COVID-19 mortality and healthcare demand. Imperial College London (2020)

4. Fox, G.C., Ishakian, V., Muthusamy, V., et al.: Status of serverless computing and Function-as-a-Service (FaaS) in industry and research. ArXiv arxiv:1708.08028, pp. 558–563 (2017)
5. Habibi, P., Farhoudi, M., Kazemian, S., et al.: Fog computing: a comprehensive architectural survey. IEEE Access **8**, 69105–69133 (2020)
6. Lauer, S.A., Grantz, K.H., Bi, Q., et al.: The incubation period of coronavirus disease 2019 (COVID-19) from publicly reported confirmed cases: estimation and application. Ann. Intern. Med. **172**, 577–582 (2020)
7. Liu, Y., Yan, L., Wan, L., et al.: Viral dynamics in mild and severe cases of COVID-19. Lancet Infect Dis. **20**, 656–657 (2020)
8. Nascimento, M.G., Braga, R.R.M., David, J.M.N., et al.: Covid-19: a simulation-based architecture proposal for IoT application development. In: International Conference on High Performance Computing and Simulation (2020, submitted)
9. Nazário, D.C., Campos, P.J., Inacio, E.C., et al.: Quality of context evaluating approach in AAL environment using IoT technology. In: CBMS 2017, pp. 558–563 (2017)
10. NEC (2007) Siafu http://siafusimulator.org/. Accessed July 2020
11. NEC: SX Aurora TSUBASA (2020). https://www.nec.com/en/global/solutions/hpc/sx/vector_engine.html/. Accessed July 2020
12. Salah, F.A., Desprez, F., Lebre, A.: An overview of service placement problem in Fog and Edge computing. Assoc. Comput. Mach. **53** (2020)
13. Singh, R.P., Javaid, M., Haleem, A., et al.: Internet of Things (IoT) applications to fight against COVID-19 pandemic. Diabetes Metab. Syndr. Clin. Res. Rev. **53**, 521–524 (2020)
14. World Health Organization: Coronavirus disease (COVID-19) advice for the public (2020). https://www.who.int/emergencies/diseases/novel-coronavirus-2019/advice-for-public/. Accessed January 2021

Benefits of Using WebRTC Technology for Building of Flying IoT Systems

Robert R. Chodorek[1]([✉]), Agnieszka Chodorek[2], and Krzysztof Wajda[1]

[1] The AGH University of Science and Technology,
Al. Mickiewicza 30, 30-059 Krakow, Poland
chodorek@agh.edu.pl, wajda@kt.agh.edu.pl
[2] Kielce University of Technology, Al. 1000-lecia P.P. 7, 25–314 Kielce, Poland
a.chodorek@tu.kielce.pl

Abstract. The WebRTC technology was intended mainly for audio/ video transmissions with, if necessary, associated data. The use of full-stack WebRTC transmissions, which merges media and non-media streams and flows, enables natural integration of different types of data in one session, common cryptographical protection of the session, and multi-platform applications development. As was shown in our previous papers, these features are essential for building efficient WebRTC-based IoT brokers. This paper is focused on the widely understood adaptability of dual-stack WebRTC-based IoT transmissions. During experiments, conducted in a real environment, the UAV-based IoT system, composed of an air station and a ground station, was transmitted video from a 4K camera and data from sensors through IEEE 802.11ac WLAN. Results show that full-stack WebRTC communication assures good adaptability to network circumstances. The full-stack congestion control is able to do a good job of protecting high priority data coming from sensors, even at the cost of the QoS parameters of the associated video transmission.

1 Introduction

Web Real-Time Communications (WebRTC) [1] is the novel technology that competes with Web-based stand-alone real-time multimedia applications. Although this technology is now at the final stage of standardization, it is widely used for interactive multimedia communication. The role of WebRTC significantly grew up during the current COVID-19 pandemic, due to its multiplatformity.

1.1 WebRTC Technology

WebRTC applications are JavaScript scripts, embedded in Web pages, which are loaded and run in Web browsers together with the Web pages. Web page, written using the HyperText Markup Language (HTML) version 5 (HTML5) and Cascading Style Sheets (CSS), plays the role of the user interface of the WebRTC application.

L. Barolli et al. (Eds.): AINA 2021, LNNS 226, pp. 310–322, 2021.
https://doi.org/10.1007/978-3-030-75075-6_25

The WebRTC protocol stack consists of two half-stacks: one is intended for streaming media (audio and/or video), transmitted in real time, and the other is dedicated to transmitting non-real-time flows. The Real-time Transport Protocol (RTP) is used for transmissions of the media streams. Non-media flows are transmitted using the Stream Control Transmission Protocol (SCTP). The RTP, the RTCP and the SCTP use the User Datagram Protocol (UDP) as an underlying transport protocol. All transmissions are cryptographically protected by the Advanced Encryption Standard (AES) cipher.

SCTP transmissions are congestion controlled with the use of buildin TCP-like congestion control. RTP transmissions are congestion controlled with the use of external mechanisms: the node-side one (simulcasting, which exists in two versions: stream replication simulcast [2] and layered simulcast [3,4]), and the sender-side one (performed by using Google Congestion Control (GCC) [5]). An analysis of the cooperation between the sender-side congestion control and the IEEE 802.11 congestion control was presented in the paper [6].

1.2   WebRTC and Internet of Things

The main area of the applicability of WebRTC is audio- and videoconferencing, and a large number of WebRTC applications in the Internet of Things (IoT) systems were videoconferences associated with non-WebRTC data transmissions. For example, in the paper [7] WebRTC conferencing accompanies the Internet of Things telemetry. A minority of papers are devoted to the usability of WebRTC for data transmission. The method presented in the paper [8] uses the WebRTC audio channel for data transmission. The usage of WebRTC technology for the rapid prototyping of IoT application was presented in [9]. The WebRTC-based extensions to the EURECOM IoT Platform for UAV was proposed in [10].

The combination of the IoT and real-time communication is shown as enabled contextual communication [11]. In the author's previous publications, the WebRTC data channel was used for the building of the IoT broker used for healthcare purposes [12] and for monitoring based on the Unmanned Aerial Vehicle (UAV) [13]. The latter system was also tested for possible co-operation with existing network infrastructure, which may be useful in emergency situations [14].

The main benefits from using WebRTC-based brokers, instead of mixed solutions that combine WebRTC audio/video with IoT made by using dedicated software tools, arise from the dual-stack WebRTC architecture. Besides the contextual communication, such a homogeneous system is characterized by:

- natural integration of different types of data (real-time media streams and non-real-time non-media data flows) in one session,
- common cryptographical protection of the heterogeneous session,
- common run-time environment, the same for audio/video and IoT.

1.3   Motivation, Main Contribution and Organization of This Paper

The full-stack WebRTC offers two half-stack congestion control schemes, related to real-time and non-real-time transmissions. In heterogeneous solutions, separate audio/video and IoT streams compete for bandwidth of a bottleneck link, and final effects are results of these competitions. It is supposed that in the case of homogeneous solution, as the WebRTC is, there is a kind of a trade-of between real-time requirements of audio/video transmissions and stringent requirements of high-priority IoT traffic. However, WebRTC's congestion control strictly depends on implementation, and the final effect of practical co-existence of heterogeneous traffic in shared link still remains unknown.

The aim of this paper is to present results of measurements of heterogeneous traffic, in order to evaluate whether, in practice, full-stack WebRTC is able to protect high-priority IoT traffic at a satisfactory level. Performance evaluation of full-stack WebRTC multimedia transmissions carried out between the air and ground components of the flying IoT system [13] was shown. The main contributions of this paper are:

1. Evaluation of the prototype flying IoT system in terms of the performance of air-to-ground communication carried out in an infrastructure-less 802.11ac WLAN.
2. Evaluation of the prototype flying IoT system in terms of the performance of air-to-ground communication carried out in two infrastructure-based 802.11ac WLANs (a simple and a complex one).
3. Evaluation of the adaptability of heterogeneous WebRTC traffic and, as a result, assessment of the usefulness of WebRTC for multimedia IoT.

The 802.11ac WLAN was selected for transmission of communication traffic because of its high throughput, large enough to transmit 4K video.

The rest of this paper is organized as follows. The next Section presents related work. The third Section describes the flying IoT system boilt using full-stack WebRTC, and the test environment. Performance evaluation of the IoT system is presented in the fourth Section. The fifth Section concludes this paper.

2   Related Work

In the paper [15] transmission function requirements and transmission performance requirements for the HD video transmission from the UAV were analyzed. Various communication networks for the UAV systems (for a single UAV system, multi-UAVs systems, and cooperative multi-UAVs systems) were analyzed in [16].

The analysis of the aerial WiFi network which connects the UAV to the ground station in the test environment was presented in [17]. In the paper [18] the Flying Ad Hoc Network (FANET) WiFi network was presented. The FANET is a modification of the classical MANET network used by the UAVs. The evaluation of the proposed FANET solution was performed using UAV software running in the node.js environment on the Single Board Computer (SBC) installed on the UAV.

The analysis of emergency communication using UAVs was presented in [19]. UAVs can be used to build mobile infrastructure in places where there is a lack of good network infrastructure to give, for example, broadcasting alerts to the ground vehicles in an emergency situation. Another solution to be used in emergency situations was presented in [20]. The UAV works as a flying communication server (a Web server and WebRTC infrastructure) which is used to connect users in the disaster area. Communication between the UAV and users was performed with the use of a WiFi (802.11) network.

The LTE network is often used for transmission between the UAV and its ground station due to the relatively large area covered by such networks [21]. It allows to use of UAV in large areas. The paper [21] presents an analysis of the impact of various conditions in the LTE network transmission channels on the Quality of Experience (QoE) of Full HD 1080p 30 fps video in a flying monitoring system. The analysis was performed using the ns-3 simulator. The analysis shows that LTE is enough to transmit the FHD video only when a good condition channel is selected. The presented results also show that 4K video cannot be sent with satisfactory QoE using this technology.

In [22] a proposal for the use of dynamic adaptive streaming over HTTP (DASH) to energy-efficient adaptive UAV video streaming was presented.

In [23] a proposal of a video streaming system from UAV to many end recipients was presented. Video in the UAV is processed by a specialized Graphics Processing Unit (GPU) and sent via RTP over UDP sockets to ground Media Gateway (Janus-Gateway has been used). Janus-Gateway receives and processes the video stream and then sends it to the final recipients using WebRTC technology.

3   Flying IoT System and Experiments

The flying IoT system, which is the subject of the performance evaluation presented in this paper, is composed of an air station and a ground station. The air station is an IoT system mounted at a UAV. As the air station, the Turnigy SK450 quad copter with Raspberry Pi 3B+ single-board computer (SBC) mounted on board of the quadcopter [13] were used. The SBC worked under control of the Linux Raspbian operating system. The Chromium browser was the run-time environment for WebRTC application.

The ground station was composed of two separate devices: WebRTC multimedia and monitoring station (WMMS), which is an end point of WebRTC transmissions, and the command and control console (CCC), used for piloting the UAV. As the WMMS, a laptop computer working under control the Microsoft Windows operating system was used. The run-time environment for WebRTC application running on the WMMS was the Chrome browser.

What distinguishes this system from other flying IoT systems is the full-stack WebRTC multimedia communication provided between the air station and the ground station. In this solution, both video from a 4K camera and data from sensors are transmitted using WebRTC technology via the IEEE 802.11ac WLAN (the production network, marked in red in Fig. 1, Fig. 2 and Fig. 3; the management and control network, used for pilotage, is marked in yellow). To check the usefulness of this system and to asses its performance, experiments with three different architectures of the WLAN were conducted in a real environment.

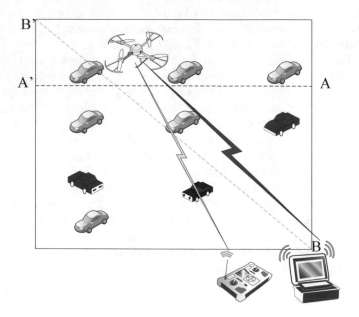

Fig. 1. Infrastructure-less test environment.

Experiments were carried out at the parking lot of AGH University of Science and Technology. The network used for multimedia transmissions was build using IEEE 802.11ac technology. Three network topologies were used (Fig. 1, 2, 3):

- infrastructure-less topology, in which data were transmitted directly between the air station and the ground one, and only stations network adapters were in use,
- simple infrastructure-based topology, in which data were transmitted via single access point,
- complex infrastructure-based topology, in which data were transmitted hop-by-hop, via multiple access points, and wireless connections were accompanied with wired ones.

During experiments, the air station was equipped with an IEEE 802.11ac dual band (2.4 GHz and 5 GHz) network adapter. The ground station was equipped with the IEEE 802.11 Intel® Dual Band Wireless-AC 7260 network adapter. Access points AP1 (Fig. 2 and 3), AP2 and AP3 (Fig. 3) were NETGEAR Nighthawk X4 R7500 AC2350 dual band devices. The key feature, which decided about selection of this type of access points was beamforming able to cooperate with any network adapter, which does not need the active cooperation of the receivers.

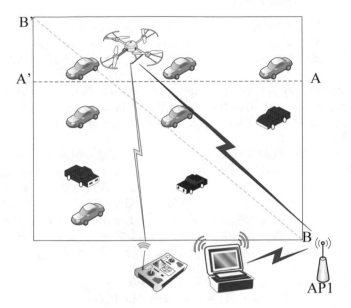

Fig. 2. Simple infrastructure-based test environment.

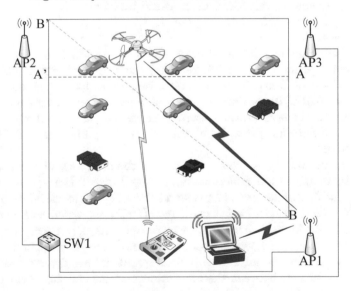

Fig. 3. Complex infrastructure-based test environment.

In the case of infrastructure-less topology, the air station and the ground station forms the Independent Basic Service Set (IBSS) of the IEEE 802.11 network. In the case of simple infrastructure-based one, two Basic Service Sets (BSSs) are created, each composed of a station and the AP1. Stations belonging to different BSSs communicates via common AP1. In the case of complex infrastructure-based topology, the Extended Service Set (ESS) is created, in which BSSs communicates via wired Distribution System (DS). The DS is a Gigabit Ethernet network. The switch SW1 (Fig. 3) was an HP 3500-24G-PoE+ yl Switch.

Experiments were carried at a square parking lot, 70 m width and 70 m long. During experiments, the UAV flies 15 m above the surface of the parking lot. In Fig. 1, Fig. 2 and Fig. 3 two lines have been drawn, which represents trajectories of UAV's flights.

Trajectory A-A' crosses the parking lot, so length of this trajectory is equal to the width of the parking lot (70 m). Points A and A' were situated in the place, in which the strength of WMMS's signal or AP1's signal is small enough to enforce the connection of the air station to AP2 or AP3 (Fig. 3), if they exist, and large enough to allow the air station to keep connected with WMMS or AP1, otherwise (Fig. 1, 2). The point A was located 50 m from the point B. Nearby point A the AP3 access point was placed. The AP2 access point was placed nearby the point A'.

Trajectory B-B' crosses the parking lot diagonally, and the length of this trajectory is 100 m. The point B is the point, in which the air station was placed when experiments with the infrastructure-less topology were performed (Fig. 1). During experiments with infrastructure-based topologies (Fig. 2 and Fig. 3), the AP1 access point was placed at the point A, and the ground station was located about 1 m away.

During experiments, the air station rose to a height of 15 m and then flew to the starting point A or B. After reaching the starting point, it followed trajectories of A-A' or B-B'. To assure measurements at fixed positions, the UAV hovered every 5 m of the horizontal distance between the starting point and the current location of the air station. Aggregated stream, composed of video stream and non-media data flow, received at this phase of experiments, was the subject of measurements.

Experiments were repeated every week, at the same day of the week and at the same time. Results of experiments, averaged over 5 times, are depicted in Fig. 4 and Fig. 5. Results include measured throughput (in megabits per second or in kilobits per second), packet error rate (PER, expressed as a percentage), and jitter (dimensionless). Throughput measurements were carried out for both video stream and data flows. The rest of parameters were read from RTCP reports for video stream only. Note that all quantities are depicted as functions of horizontal distance between the air station and the starting point of a given trajectory.

4   Results

Figure 4 presents throughput of media data (coming from UAV camera) and non-media data (coming from sensors) transmitted between the air station and the ground one. While throughput of media data changes in the function of horizontal distance between the air station and the corresponding node (the

WMMS part of the ground station or corresponding access point), throughput of non-media data stays unchanged at the initial level of 97 kbps. Thus, the entire decline in signal strength, observed when the air station moves away from the corresponding node, was absorbed by the WebRTC video transmission, which has noticeably worse QoS parameters than when the air station and a corresponding node are close each other. The dual-stack WebRTC transmissions caused that low-volume reliable data traffic do not compete for network resources with large-volume 4K video traffic. Moreover, in the case of worsening of a network state it was protected by video stream and, as a result, was able to preserve its QoS parameters.

Figure 4(a) and Fig. 5(a), (b) presents QoS parameters of video traffic obtained when direct transmission between the air station and the ground station was performed. If the air station moves away from the ground station, the throughput of the video transmission decreases. During movement along the A-A' trajectory this decrease was relatively small (about 15% decline between the beginning and the end of trajectory, i.e. at the distance of 70 m), but when the air station moved along the B-B' line, the difference between the beginning and the end of trajectory was about 50% of initial value (with similar, about 15% decline at 70 m). Packet error rates increased then from 0, through about 0.06% (70 m), to 0.2% (100 m), and jitter increased from 30, through about 50 (70 m), to more than 100 (100 m). Decrease of throughput and increase of jitter was controlled (to some extent) by the sender (sending browser congestion control mechanism) through decrease the frame rate.

Replacement of the air station with more effective device, which was the access point AP1 (Fig. 4(c), Fig. 5(c), (d)) have resulted in better performance of the WebRTC video. Although transmissions were carried out via intermediate device, the decline of throughput in distance between stations was not as large, as observed when direct transmissions between the air station and the ground one were conducted. In details, the difference between the beginning and the end of trajectory A-A' was about 10%, and between the beginning and the end of trajectory B-B' was about 20%. Increase of packet error rates between end points of each trajectory was 0.04% (A-A') and 0.8% (B-B'). In the case of A-A' trajectory, replacement of air station with AP1 had little influence on jitter measured at the end point, in practice, but in the case of B-B' trajectory improvement of this QoS parameter was noticeable (61 instead of 102). Generally, after replacement the larger improvement was observed in the case of trajectory B-B'.

The best results give the use of the ESS (Fig. 4(e), Fig. 5(e), (f)). The graph of the video throughput in a function of distance between the air station and the beginning of the trajectory is almost flat. Only at the middle, around the point of attachment to the DS, throughput of video stream decreases a little: from 19.9 Mbps to 1.95 Mbps (at both A-A' and B-B' trajectories). This was caused by non-zero packet error rates, which achieved 0.012% (A-A' trajectory) and 0.011% (B-B' trajectory). The jitter also increased a little: from 30 to 32 (A-A' trajectory) and 33 (B-B' trajectory), and the frame rate was only 2 fps smaller

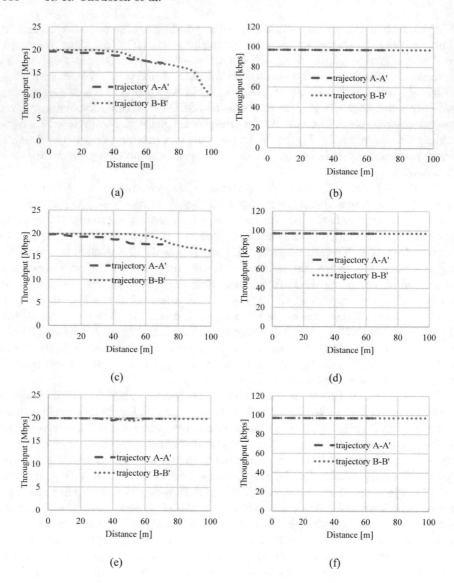

Fig. 4. Throughput of: (a, c, e) video stream, (b, d, f) non-media data in the case of topologies: (a, b) infrastructure-less, (c, d) simple infrastructure-based (e, f) complex infrastructure-based.

than the maximum. Generally, this case shows that in the case of good coverage of monitored area, the monitoring system is able to preserve good quality of service even if the network conditions are bad enough to enforce a change of the point of attachment to the DS.

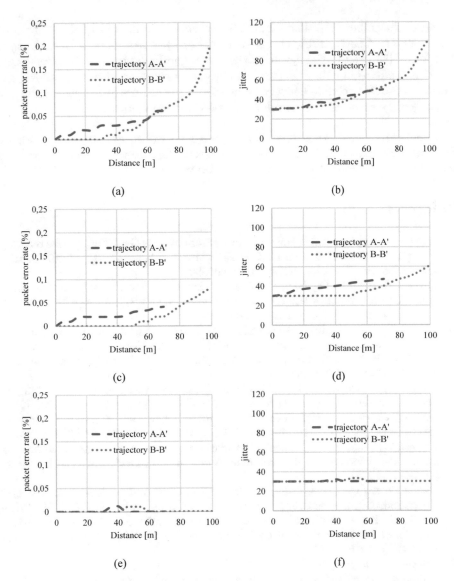

Fig. 5. Packet error rate and jitter: (a, c, e) packet error rate, (b, d, f) jitter in the case of topologies: (a, b) infrastructure-less, (c, d) simple infrastructure-based (e, f) complex infrastructure-based.

5   Conclusions

This paper presents the performance evaluation of full-stack WebRTC multimedia transmissions (both video from a 4K camera and data from sensors) carried out between the air and ground components of the UAV-based IoT system shown in the paper [13]. The flying IoT system was analyzed in three different IEEE 802.11 WLAN architectures in terms of QoS parameters. Results show that WebRTC used in the system allows for a good adaptation of a UAV-based IoT communication subsystem to the current network condition. It allows the system to provide reliable delivery of data from the sensors at the cost of slightly lower QoS parameters of the real-time video transmission.

Acknowledgements. This work was supported by the Polish Ministry of Science and Higher Education with the subvention funds of the Faculty of Computer Science, Electronics and Telecommunications of AGH University.

References

1. Loreto, S., Romano, S.P.: How far are we from WebRTC-1.0? an update on standards and a look at what's next. IEEE Commun. Mag. **55**(7), 200–207 (2017). https://doi.org/10.1109/MCOM.2017.1600283
2. Grozev, B., Politis, G., Ivov, E., Noel, T., Singh, V.: Experimental evaluation of simulcast for WebRTC. IEEE Commun. Stand. Mag. **1**(2), 52–59 (2017). https://doi.org/10.1109/MCOMSTD.2017.1700009
3. Bakar, G., Kirmizioglu, R.A., Tekalp, A.M.: Motion-based adaptive streaming in WebRTC using spatio-temporal scalable VP9 video coding. In: GLOBECOM 2017 - 2017 IEEE Global Communications Conference, pp. 1–6 (2017). https://doi.org/10.1109/GLOCOM.2017.8254127
4. Chodorek, A., Chodorek, R.R., Wajda, K.: Comparison study of the adaptability of layered and stream replication variants of the WebRTC simulcast. In: 2019 International Conference on Software, Telecommunications and Computer Networks (SoftCOM), pp. 1–6 (2019). https://doi.org/10.23919/SOFTCOM.2019.8903887
5. Holmer, S., Lundin, H., Carlucci, G., Cicco, L.D., Mascolo, S.: A google congestion control algorithm for real-time communication. Internet-Draft draft-ietf-rmcat-gcc-02, IETF (2016)
6. Chodorek, A., Chodorek, R.R., Wajda, K.: An analysis of sender-driven WebRTC congestion control coexisting with QoS assurance applied in IEEE 802.11 Wireless LAN. In: 2019 International Conference on Software, Telecommunications and Computer Networks (SoftCOM), pp. 1–5 (2019). https://doi.org/10.23919/SOFTCOM.2019.8903749
7. Gueye, K., Degboe, B.M., Ouya, S., Kag-Teube, N.: Proposition of health care system driven by IoT and KMS for remote monitoring of patients in rural areas: pediatric case. In: 2019 21st International Conference on Advanced Communication Technology (ICACT), pp. 676–680 (2019). https://doi.org/10.23919/ICACT.2019.8702009
8. Shin, D.S.: A study on the tele-medicine robot system with face to face interaction. J. IKEEE **24**(1), 293–301 (2020). https://doi.org/10.7471/ikeee.2020.24.1.293. Institute of Korean Electrical and Electronics Engineers
9. Janak, J., Schulzrinne, H.: Framework for rapid prototyping of distributed IoT applications powered by WebRTC. In: 2016 Principles, Systems and Applications of IP Telecommunications (IPTComm), pp. 1–7 (2016). https://doi.org/10.1109/IPTComm39427.2016.7780249

10. Datta, S.K., Dugelay, J.L., Bonnet, C.: IoT based UAV platform for emergency services. In: IEEE 2018 International Conference on Information and Communication Technology Convergence (ICTC), pp. 144–147 (2018). https://doi.org/10.1109/ICTC.2018.8539671

11. Bubley, D.: Realtime communications. IEEE Internet Things Newslett. (2016). https://iot.ieee.org/newsletter/march-2016/iot-realtimecommunications.html. Accessed 30 Jan 2021

12. Chodorek, R.R., Chodorek, A., Rzym, G., Wajda, K.: A comparison of QoS parameters of WebRTC videoconference with conference bridge placed in private and public cloud. In: 2017 IEEE 26th International Conference on Enabling Technologies Infrastructure for Collaborative Enterprises (WETICE), pp. 86–91 (2017). https://doi.org/10.1109/WETICE.2017.59

13. Chodorek, A., Chodorek, R.R., Wajda, K.: Media and non-media WebRTC communication between a terrestrial station and a drone: the case of a flying IoT system to monitor parking. In 2019 IEEE/ACM 23rd International Symposium on Distributed Simulation and Real Time Applications (DS-RT), pp. 1–4 (2019). https://doi.org/10.1109/DS-RT47707.2019.8958706

14. Chodorek, A., Chodorek, R.R., Wajda, K.: Possibility of using existed WLAN infrastructure as an emergency network for air-to-ground transmissions: the case of WebRTC-based flying IoT system. In: 11th EAI International Conference on Broadband Communications, Networks, and Systems (EAI BROADNETS 2020), pp. 3–21. Springer, Cham (2021). https://doi.org/10.1007/978-3-030-68737-3_1

15. Yang, Q., Yang, J.H.: HD video transmission of multi-rotor Unmanned Aerial Vehicle based on 5G cellular communication network. Comput. Commun. (2020). https://doi.org/10.1016/j.comcom.2020.07.024

16. Hentati, A.I., Fourati, L.C.: Comprehensive survey of UAVs communication networks. Comput. Stand. Interfaces, 103451 (2020). https://doi.org/10.1016/j.csi.2020.103451

17. Guillen-Perez, A., Sanchez-Iborra, R., Cano, M., Sanchez-Aarnoutse, J.C., Garcia-Haro, J.: WiFi networks on drones. In: 2016 ITU Kaleidoscope: ICTs for a Sustainable World (ITU WT), pp. 1-8 (2016). https://doi.org/10.1109/ITU-WT.2016.7805730

18. Bekmezci, I., Sen, I., Erkalkan, E.: Flying ad hoc networks (FANET) test bed implementation. In: 2015 7th International Conference on Recent Advances in Space Technologies (RAST), pp. 665–668 (2015). https://doi.org/10.1109/RAST.2015.7208426

19. Hadiwardoyo, S.A., Hernández-Orallo, E., Calafate, C.T., Cano, J., Manzoni, P.: Evaluating UAV-to-car communications performance: testbed experiments. In: 2018 IEEE 32nd International Conference on Advanced Information Networking and Applications (AINA), pp. 86–92 (2018). https://doi.org/10.1109/AINA.2018.00025

20. Kobayashi, T., Matsuoka, H., Betsumiya, S.: Flying communication server in case of a largescale disaster. In: 2016 IEEE 40th Annual Computer Software and Applications Conference (COMPSAC), pp. 571–576 (2016). https://doi.org/10.1109/COMPSAC.2016.117

21. Naveed, M., Qazi, S., Khawaja, B.A., Mustaqim, M.: Evaluation of video streaming capacity of UAVs with respect to channel variation in 4G-LTE surveillance architecture. In: 2019 8th International Conference on Information and Communication Technologies (ICICT), pp. 149–154 (2019). https://doi.org/10.1109/ICICT47744.2019.9001975

22. Zhan, C., Huang, R.: Energy efficient adaptive video streaming with rotary-wing UAV. IEEE Trans. Veh. Technol. **69**(7), 8040–8044 (2020). https://doi.org/10.1109/TVT.2020.2993303
23. Sacoto-Martins, R., Madeira, J., Matos-Carvalho, J.P., Azevedo, F., Campos, L.M.: Multi-purpose low latency streaming using unmanned aerial vehicles. In: IEEE 12th International Symposium on Communication Systems, Networks and Digital Signal Processing (CSNDSP), pp. 1–6 (2020). https://doi.org/10.1109/CSNDSP49049.2020.9249562

A Neural Network Model for Fuel Consumption Estimation

Chayma Werghui[1,2]([✉]), Amine Kchiche[1], Noureddine Ben Othman[2], and Farouk Kamoun[1]

[1] CRISTAL Laboratory, University of Manouba, Manouba, Tunisia
{Chayma.werghui,Farouk.kamoun}@ensi-uma.tn
[2] ACTIA Engineering Services, Ariana, Tunisia
Noureddine.benothman@actia-engineering.tn

Abstract. There is uncertainty about the methodological assumptions and parameters to be considered in estimating fuel consumption of road vehicles. In fact, recent studies have highlighted how the existing models provide different estimation results accuracy among different scenarios. This study presents an innovative approach based on the neural network algorithm to estimate the fuel consumption of a fleet of gasoline vehicles. The obtained findings show that the proposed neural network model has an accuracy greatly surpassing that of the HBEFA3 model.

1 Introduction

Despite the problems introduced by COVID19, this epidemic has contributed significantly to the reduction of global CO2 emissions during the period of forced confinement. Daily global CO2 emissions were 17% lower in early April 2020 than the average levels recorded in 2019 or just under half due to the changes in people's transportation habits [1]. In fact, the emissions of each country decreased, on average, by -26% [1]. Unfortunately, this reduction was temporary. Because of the transportation fuel consumption, which increased very rapidly before the emergence of covid19 mainly due to the large number of combustion vehicles sold in 2019 (56.52 millions) [2], it becomes necessary to find sustainable solutions to solve problems related to the environment and climate changes.

Intensive work has been so far made by both automotive industrials and scientific researchers to control and reduce fuel consumption in order to minimize emissions and decrease travel costs. The approaches proposed in this context could be classified into three categories: driver's good practices that can save from 10 to 15% of energy [3]; driver's assistance systems whose gain may reach 30% [4]; and manufacturer-related solutions such as new engine technologies and new energies, which can save up to 10% of consumed energy [4]. This classification is illustrated in Fig. 1.

Most of these approaches rely on models that estimate fuel consumption [5] and that are essential for developing and evaluating optimization approaches. According to their granularity and the scope of the considered factors, these models can be classified into 3 categories [5]: black, grey and white box models. The challenge here is to

© The Author(s), under exclusive license to Springer Nature Switzerland AG 2021
L. Barolli et al. (Eds.): AINA 2021, LNNS 226, pp. 323–332, 2021.
https://doi.org/10.1007/978-3-030-75075-6_26

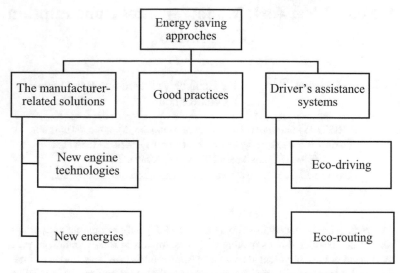

Fig. 1. Classification of energy saving approaches

provide models that take most consumption factors into account to be enough accurate while remaining with acceptable complexity to be applied in real time optimization processes. The problem is that analytical, still simplistic, black box models are less accurate; whereas white box models are very complicated and cannot be integrated in real time optimization applications [5]. In between, grey box models provide a trade-off between complexity and accuracy, but they still remain limited because they overlook several important parameters such as road quality and driving conditions which affect considerably the fuel consumption [6, 7].

As the factors influencing this consumption are very heterogeneous, sometimes not yet well modeled and their correlation with other factors is not yet controlled, we think that machine learning techniques could provide a fairly accurate estimation of fuel consumption.

Our contribution consists in setting up a neural network for estimation. The learning phase of this network is fed by real traces collected from Vehicle Energy Dataset (VED) [8].We also compare its accuracy with that of the most known model and assess its accuracy against real consumption data.

This article is organized as follows. Section 2 represents related work. Section 3 focuses on the use of neural networks; we expose the chosen model and explain the utilized parameters. Section 4 describes the choice of data set. Finally, the obtained results and conclusions are presented in Sects. 5 and 6.

2   Related Work

In the literature, fuel consumption models were classified, according to their granularity and transparency, into three classes.

A white-box fuel consumption model can be constructed based on the physical or chemical processes of an engine, i.e. using mathematical formulae to describe the intake, compression, combustion and exhaust processes of the engine. The main models belonging to this class are the Carbon Balance Method [5] and the Phenomenological Mean Value Model [5]. The basic principle of the Carbon Balance Method is conservation of mass, where, after combustion, the total mass of carbon in the exhaust must be equal to the carbon quality of the fuel before combustion. On the other hand, the Phenomenological Mean Value Model is based on evaluating the internal combustion to predict the amount of torque created and thereby the fuel consumed and exhaust produced. Obviously, the development of white box models requires a thorough understanding of the entire engine system and all its influential sub-processes. The number of parameters to be injected in white-box fuel consumption models is generally high. Furthermore, in some cases, such models will be too complex or even impossible to obtain in a reasonable timeframe due to the complex nature of the fuel system.

In black box fuel consumption models [5], the entire vehicle or, in some cases, only its engine is considered as a black box. Based on the input scale for black box models, there are three different model types, which are engine-based, vehicle-based and modal black-box fuel consumption models. The input variables of engine-based black box fuel consumption models, such as engine speed, engine torque and engine power output, are engine level variables. However, vehicle-level variables, like instantaneous vehicle speed and acceleration, average speed as well as acceleration, are inputs for vehicle-based black box fuel consumption models. Black box fuel consumption models use the basic operating modes (idle, cruise speed, deceleration and acceleration) as inputs. Statistical methods are then applied to provide regression analysis and determine how these variables influence fuel consumption.

In terms of understanding the inner workings of a system, the grey box model is somewhere between the black box and the white box. It involves a partial understanding of the internal system. Among the main models in the grey box fuel consumption model subcategory is Vehicle Transient Emission Simulation Software (VeTESS) [5]. The inputs of VeTESS are vehicle speed and road grade, while its outputs are emissions and fuel consumption rates. Subsequently, fuel consumption is determined using engine map. This model can predict the dynamic fuel consumption of a specific vehicle as long as the speed profile is known.

To synthesize, white box models cannot be used in real time applications because of their complexity, unlike black and grey box models which are simpler but do not provide an accurate estimation of the amount of consumed fuel. Due to this discrepancy between complexity and precision, machine learning could play an important role.

Due to the complexity of and the heterogeneity of the considered factors that could impact the consumption process, new approaches of estimation based on machine learning algorithms have been recently introduced.

Several previous works have indeed used the neural network to estimate fuel consumption such as the study conducted in [9] which used the artificial neural network (ANN) with five inputs namely engine size, speed, fuel type and passengers to predict the cost of fuel consumption. This network has 15 nodes at the hidden layer and estimated fuel consumption as the output node. The works presented in [10] and [11] took into account the slope, radius of curvature, road irregularity and macro-texture of the road surface (measured as sensor depth, measured depth (SMTD). These data were taken from the Highways Agency Pavement Management System (HAPMS), which includes records of articulated trucks traveling at a constant speed. Other data that is included such as total gross vehicle weight, distance traveled, date of travel with time and fuel consumed. These data are provided by Microlise Ltd. The model generated by [10] shows that the variables that have the greatest impact on fuel consumption are gross vehicle weight, road gradient and vehicle speed. In [11] three machine learning techniques were studied: the support vector machine (SVM), random forest (RF) and artificial neural network (ANN). The results showed that all three methods can be used to develop models with good accuracy. The study also shows that the vehicle speed and the gear ratio seem to have a small impact on the fuel consumption of the truck fleet, which is contradictory with the results obtained in [10].

Despite the innovative use of neural networks by these works, we think they still need to be investigated and validated through different datasets and specific features. For example, the work in [10] does consider a constant speed in the dataset, which is in our opinion an unrealistic choice/scenario.

This work aims hence to enhance and complete the efforts to show the accuracy and the pertinence of adopting machine learning approaches in fuel consumption estimations.

In the next section, we explore the machine learning algorithms and go through the choices made to build our model.

3   Prediction Model

Machine learning refers to a set of learning algorithms, obtained from historical data and applied for the resolution of a problem. It permits interpreting data without explicit or deterministic programming. These algorithms are classified into two types: supervised and unsupervised. Supervised algorithm is a machine learning task which consists in learning a prediction function from annotated examples, i.e. they are already associated with a target label or class based on which the algorithm will become able to predict new target on unannotated data. In contrast, unsupervised learning applies training algorithm to reveal similarities and discrepancies between these data and group together those having common characteristics. For supervised learning, we distinguish between regression problems, that predict quantitative problems, and classification problems predicting qualitative variable. The former is the focus of this study.

We chose to use the neural network because it is able to learn from representative examples, by "error back-propagation", and can represent any function. Moreover it is able to integrate heterogeneous factors and to discover hidden correlations between these factors.

In this study, our algorithm (Feedforward neural network) works in 2 steps: a front-propagation phase that goes from input to output and a phase of propagation in the opposite direction. During the first phase, a prediction will be made according to the inputs. Depending on this prediction, an error is calculated and propagated backwards in proportion to its contribution to the error, which represents the back-propagation phase. The more a connection will participate in the error, the more it will be corrected.

The structure of our neural network (Fig. 2) consists of input units, a hidden layer and an output unit. The first component presents our model inputs which will be transmitted to the neurons of the hidden layer. Each neuron of this layer will calculate the weighted sum of the input values. Then, they will be transmitted to the activation function to predict its value. Finally, these received values from the hidden layer are transmitted to the output neuron which will itself apply its activation function to estimate the consumption. For the hidden layer neurons, we used the "RELU" function which is one of the most powerful and useful functions because it has several advantages such as low calculation cost, easy optimization, faster convergence than sigmoid or tanh and it improves the propagation of the gradient. For the output layer neuron, we applied the identity activation function to reduce the calculation time of our model.

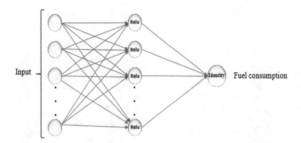

Fig. 2. Feedforward neural network architecture

4 Data Description

For the learning and evaluation phases, we utilized the Vehicle Energy Dataset (VED) [8], a large-scale set of fuel and energy data collected from 383 personal cars in Ann Arbor, Michigan, USA. This open data set captures the vehicles' GPS tracks and their time series data on fuel, energy, speed and auxiliary power consumption. A diverse fleet of 264 gasoline, 92 HEV and 27 PHEV/EV vehicles drove in the real world from November 2017 to November 2018 where the data was collected via on-board OBD-II recorders. The driving scenarios ranged from highways to dense city centers in various driving conditions and seasons. In this work, we focus on the gasoline vehicle.

The dataset is composed of two principal components: Static data and Dynamic data. The static data include vehicle parameters such as Vehicle Engine Configuration, Engine Displacement and Vehicle Weights. The dynamic data involve timestamps naturalist driving records of the fleet such as Timestamp, vehicle speed (Vs), mass air flow (MAF), Engine RPM, Absolute Load, Short Term Fuel Trim Bank (STFT), Long Term Fuel Trim Bank (LTFT), Latitude, Longitude an d Fuel Consumption (FC) (Table 1). Because of the importance of the "instantaneous acceleration" parameter, we calculated it as a function of the instantaneous speed and the timestamp.

In total, the dataset accumulated approximately 100,000 miles. The diverse fleet includes passenger cars (coupes, sedans, convertibles, crossovers and luxury cars) and light trucks (pickup trucks, SUVs, minivans and wagons).

Figure 3 gives an overview of the statistical distribution of the dataset in terms of engine configuration and driving behavior.

Table 1. Acronyms

Acronyms	Definition
Vs	Vehicle Speed
Engine RPM	Engine Revolutions Per Minute
LTFT	Long Term Fuel Trim Bank
STFT	Short Term Fuel Trim Bank
FC	Fuel Consumption
MAF	Mass Air flow

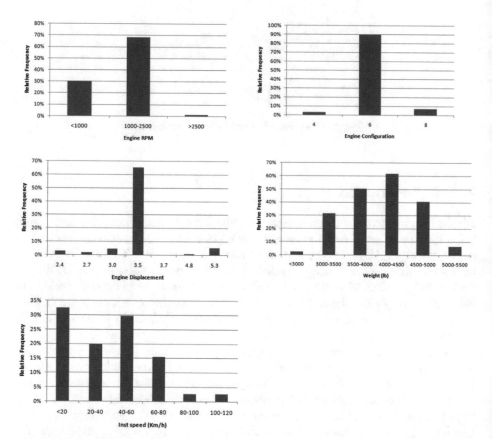

Fig. 3. Probability density of the parameters most influencing fuel consumption

We can clearly observe that, the majority of drivers tend to drive while keeping the engine speed (RPM) in the range between 1000 and 2500, which is in general the recommended driving mode. Some driving stages may still require driving below 1000 RPM or above 2500, which in general increases fuel consumption (Fig. 3).

The majority of the vehicles, as shown in Fig. 3, have a number of cylinders equal to 6, a displacement equal to 3.5 and a mass between 3500 and 5500, which reveals that the most used vehicles in the dataset are LDV (Light duty vehicle).

Instantaneous speed mainly ranges, as expected, between 0 and 60 km/h, while about 20% go beyond this speed, probably in highway parts.

For the training data we trained our model with 120 trips of different durations and for the test data, we took 32 different trips of various durations, as shown in Fig. 4. The time intervals go from 9.51 min to 36.33 min.

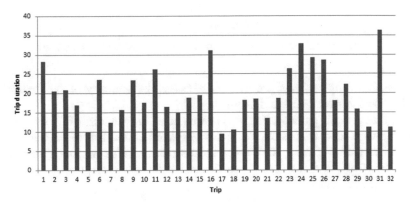

Fig. 4. Different durations of 32 test data trips

To assess the accuracy of our approach, we decided to compare the result of our approach with those of the HBEFA model. This model is developed by INFRAS, a Swiss research institute. It takes into account vehicle's dynamic and engine conditions to estimate emissions and energy consumption for all vehicle categories (PC (passenger cars), LDV (light duty vehicles), HDV (heavy duty vehicles), Buses, and Motor cycles). Each category is divided into different sub-segments (different emission concepts, different fuel types, different piston displacement/weight).

The same trip characteristics have been therefore injected in the HBEFA model to evaluate the fuel consumption estimated by this model and to compare it with the real fuel consumption and the result we obtained with our model.

5   Result and Discussion

As described in the previous sections, we simulated 32 trips with different durations.

Figure 5, shows the mean absolute error or each trip. It clearly shows that our model achieves a good estimation in all trips with a mean absolute error of about only 0.7%. On the other hand, HBEFA model shows unstable accuracy ranging from 2% to 43.9% in some cases. The average mean absolute error is about 14.5% among all trips.

These results attest, in general, of the efficiency of our approach in comparison to the HBEFA model especially with regard to the various scenario considered by the test data (Figs. 3 and 4).

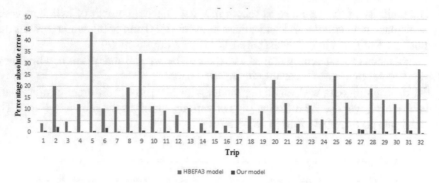

Fig. 5. Percentage absolute error for 32 trips of HBEAF3 and our model

In the following analysis, we will try to assess the accuracy of each model with regard to specific parameters and features.

Classifying the trips according to the weight of the vehicles (Fig. 6), we notice that HBEFA3 is more accurate for vehicles belonging to the class of 4500 lb weighted vehicles. i.e. achieving about 10.6% absolute mean error. This could be explained by the fact that the HBEFA3 model is better calibrated for the heavy category of vehicle giving therefore better results.

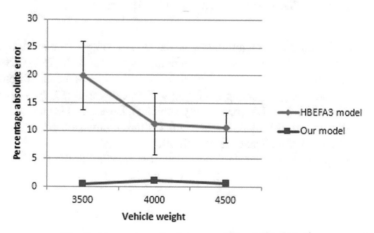

Fig. 6. Percentage absolute error VS vehicle weight

In Fig. 7, we classified the trips according to their durations. The HBEFA3 model seems to be more accurate for longer trips, i.e. in the time interval [25.40] achieving a mean error about 11%. This could be explained by the fact that this model, in contrast to ours, in not able to adapt itself to short time behavior including start-up stages and stop stages. Indeed, for short trip duration, these stages take significant proportion of the total trip duration and hence impacting the overall estimation.

The exact impact of these stages will be more investigated in future works.

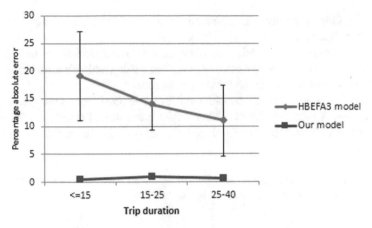

Fig. 7. Percentage absolute error VS trip duration

Lastly, a classification according to the engine displacement (Fig. 8), reveals that the HBEFA3 model is more accurate for 3.6 L, getting a mean absolute error of about 13.7% (while the global one was about 16%). In contrast, our model shows a steady error rate among different engine types, confirming its ability to adapt its model (calibration) according to different parameters.

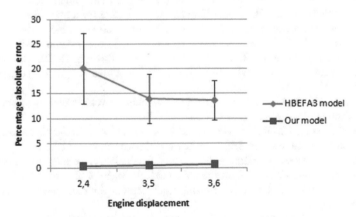

Fig. 8. Percentage absolute error VS engine displacement

As a conclusion, our model achieves a good estimation for different scenarios, while HBEFA3 seems to be customized for some configuration and driving behaviors (meanwhile still far away from our model).

6 Conclusion and Future Works

Applying Neural Network to construct a model for fuel consumption estimation, we obtained very satisfactory results with respect to the widely used HBEFA model with smaller error. Our evaluation was based on experimental data that was quite distributed among different cases.

The application of our model is less complex due to its simple structure that includes a single hidden layer with a low number of neurons, using activation functions having a low calculation cost and the minimum response time.

In future works, we intend to consider and integrate other factors such as road quality and conditions. A step that could further improve our model estimation accuracy and that is impossible or difficult to reproduce for other models, due to the heterogeneity of these factors.

Acknowledgment. «This project is carried out under the MOBIDOC scheme, funded by the EU through the EMORI program and managed by the ANPR.»

References

1. Jackson, R.B., Jones, M.W., Smith, A.J.P., Abernethy, S., Andrew, R.M., De-Gol, A.J., Willis, D.R., Shan, Y., Canadell, J.G., Friedlingstein, P., Creutzig, F., Peters, G.P., Le Quéré, C.: Temporary reduction in daily global CO2 emissions during the COVID-19 forced confinement. Nat. Clim. Change **10**, 647–653 (2020)
2. Holland, M.: Fossil vehicle sales in global freefall. CleanTechnica Report (2020)
3. Eco-conduite, ooreka https://economie-d-energie.ooreka.fr/comprendre/eco-conduite
4. Xie, H., Brown, D., Ma, H.: Eco-driving assistance system for a manual transmission bus based on machine learning. IEEE Trans. Intell. Transp. Syst. **19**, 572–581 (2018)
5. Jin, H., Wang, W., Zhou, M.: A review of vehicle fuel consumption models to evaluate eco-driving and eco-routing. Transp. Res. Part D Transp. Environ. **49**, 203–218 (2016)
6. Rakha, H., Park, S.: Energy and environmental impacts of roadway grades. Transp. Res. Rec. **1987**, 148–160 (2006)
7. Taylor, G.W., Patten, J.D.: Effects of pavement structure on vehicle fuel consumption - phase III. In: National Research Council of Canada (2006)
8. Leblanc, D.J., Peng, H., Oh, G.: Vehicle energy dataset (VED), a large-scale dataset for vehicle energy consumption research. IEEE Trans. Intell. Transp. Syst. 1–11 (2020)
9. Abdalla, A., Noraziah, A., Fauzi, A.A.C., Amer, A.: Prediction of vehicle fuel consumption model based on artificial neural network. Appl. Mech. Mater. **492**, 3–4 (2014)
10. Perrotta, F., Parry, T., Neves, L.C.: Using truck sensors for road pavement performance investigation (2018)
11. Parry, T., Neves, L.C., Perrotta, F.: Application of machine learning for fuel consumption modelling of trucks. In: IEEE International Conference on Big Data, pp. 3810–3815 (2017)

TeleML: Deploying Trained Machine Learning Models in Cross-Platform Applications

Sirojiddin Komolov$^{(\boxtimes)}$, Youssef Youssry Ibrahim, and Manuel Mazzara

Innopolis University, Innopolis, Russia
{s.komolov,m.mazzara}@innopolis.ru, y.ibrahim@innopolis.university

Abstract. Machine Learning (ML) is a fast-growing research and application field which allows developers to utilize different implementation platforms. One of the most common methods for implementing ML consists in deploying the trained model in servers, and evaluating the user data through the communication between a client app and a backend. This paper presents an alternative approach to the problem where the ML model is deployed in the client application, and discusses the pros and cons of this solution. As a case study, a cross-platform application, TeleML, has been developed. TeleML is a Telegram client, allowing users to perform sentiment analysis of posts that appear in chats. The experimental application features only the functionalities that are necessary to perform sentiment analysis on a chat, for example sign-in, iterate over dialogs, and choose a particular discussion, and does not have all the features of the Telegram original client. The findings show that, although the chosen machine learning reaches the state-of-the-art performance and the chosen frameworks allow developers utilize the model in the cross-platform environment, using a single programming language, the approach has limitations regarding user interface and deployment onto the platforms where the performance play an important role, such as mobile environment. The paper discusses technical aspects of the solution as well as limitations and future work.

1 Introduction

Machine learning (ML) has been discussed, researched and utilized mostly in academia and major companies. However, recently, it has become more and more popular and easy-to-use for real world applications, therefore several companies are now adopting ML techniques in their products and services [1]. Even though the approach of deployment of ML models in servers, and accessing them through various services (i.e. micro-service), is becoming a common approach and has its own advantages (for example scalability and maintainability), there are obvious concerns regarding this approach. Regarding deployment of ML models in the client-server architecture the study can identify the following issues:

- security concerns regarding user data sent to servers via various services
- servers can be overloaded by clients' requests, and evaluating user data
- performance decreases from the user viewpoint due to the time spent on additional requests to servers hosting ML models

This work analyses the method of deploying ML models into the client app, with the objective of resolving the aforementioned concerns. The idea of developing all the components of a system in one language is worth attention too since such approach may also simplify some aspects of the project: for example, the limitations related to the requirement of developing skills in another programming language in order to develop a ML model and a client side app. To evaluate the approach, the authors have developed *TeleML*, an application consisting of a Telegram client. The app utilizes a ML model that was trained earlier, and packaged along with the app. The links to the source code of the ML model and TeleML app are presented in the references [23,24]. The ML model performs a process of automatically identifying whether a given text is negative, neutral or positive, which is popular as a *sentiment analysis* task [16].

The following functionalities were implemented in the app: sign-in, sign-out, iterate over dialogs and iterate over messages of a particular chat. The text messages can be classified with the help of the ML model. Google BERT was chosen as the ML algorithm. TeleML is written in Python in Kivy cross-platform development framework. During the development, several problems arose and this paper discusses the affect of the problems onto the final solution.

The current paper discusses the job done in the following sections: Sect. 2 discusses the similar work on the chosen topics; Sect. 3 describes the ML model and the architecture of the app; Sect. 4 presents the achievements of the ML model and the developed app; Sect. 5 contextualizes the findings; Sect. 6 summarizes the job done; and finally the Future Work section discusses the future work based on the findings of the current work.

2   Related Work

At the best of our knowledge, most of the existing work focuses on the ML algorithms and a few papers discuss deployment techniques. For example, [18] is one of the first works applying sentiment analysis on mobile devices, and utilizes the SentiCorr technique for sentiment analysis. Another example, [20] focuses on Android devices and discusses trade-offs between Android constraints (i.e. limited resources and CPU architectures that allow only specific algorithms) and performance of algorithms. This work, however, lacks the implementation of modern state-of-heart algorithms. The research performed in [17] employs 17 sentiment analysis methods in an Android device, and but mainly focuses on the resource utilization, however, does not describe the model deployment techniques. Although [19] describes the utilization of the Natural Language Toolkit (NLTK) to perform sentiment analysis on the twitter tweets in a Rasperry device, it does not provide sufficient explanation about employed sentiment analysis algorithms, aspects of deployment, etc.

The current paper attempts to address the concerns regarding the deployment of a ML model into a software application, specifically on a cross-platform environment, furthermore it discusses the development of both the ML model and the software application.

3 Methodology

The following frameworks and libraries are used in the TeleML app:

- Python - as a programming language for the development of the ML model and client interface
- BERT - as NLP algorithm for the sentiment analysis [2]
- Transformers library by HuggingFace - as a wrapper of the BERT algorithm [3].
- Kivy cross platform framework - for user interface
- *telethon* for interacting with Telegram API in python

The next sub-sections discuss the ML model and the architecture of the app that were built on the above frameworks and libraries.

3.1 ML Model

The aim is to perform sentiment analysis on text, i.e. classifying sentences into negative, neutral and positive. This is a traditional NLP task in the paradigm of Deep Learning. As computers function effectively with numbers and do not have the ability to assimilate text, the first step would be to convert words into numbers, only then, math operations could be done, and vectorize these numbers which is called as tokenization in the NLP tasks [4]. Understanding the meaning of a word involves the awareness of the context and this should be taken into consideration by the NLP technique [5]. Furthermore, achieving a reasonable accuracy requires large amounts of data from which a ML model could learn. The BERT [2] algorithm, which stands for Bidirectional Encoder Representations from Transformers, shows superior results in addressing the above issues comparing to the classical NLP algorithms [6]. One can achieve high performance with the BERT technique, even by training on a relatively small number of training samples which is called fine-tuning, as the BERT comes with pre-trained models trained on a large amount of data. [2] Transformers library from HuggingFace along with python torch library was used to implement Google BERT algorithm [3]. For the tokenization process, BertTokenizer with pre-trained 'bert-base-cased' model was used. When encoding the text using the tokenizer and in the training phase, the recommended parameters presented in the Table 1 were applied.

Table 1. BERT parameters

max_length	160
add_special_tokens	True
return_token_type_ids	False
pad_to_max_length	True
return_attention_mask	True
return_tensors	'pt'
Batch size	16
Learning rate (Adam)	2e-5

Achieving high performance requires higher number of epochs, thus the model was trained in 10 epochs. The Reddit dataset, which has about 37K rows and labelled as negative (−1), neutral (0) and positive (1) [7], was used for fine-tuning. For validation and test, the dataset was split into three: - 80% (29.7K rows) - for training data - 10% (3.7K rows) - for validation data - 10% (3.7K rows) - for test data. The model was fine-tuned, that is trained on a Google Colab machine with 12 GB RAM and 15 GB GPU NVIDIA-SMI 455.45.01. After training the model, it was saved using the torch's save model to be used as pre-trained. It is worth to mention that, although the model was trained with GPU, it was saved with CPU support to allow utilizing the model without access to GPU.

3.2   Architecture of TeleML

The TeleML project follows the MVC principles for separation of concerns [8]. The Fig. 3 demonstrates that the *TeleMLMain* component acts as a controller, while the components of the *pages* module act as a view. The *model* module has two components *classifier* and *tele_utils*, which are responsible for the ML model utilization and working with telegram APIs respectively.

The KV language from the Kivy framework was used to declare the widgets in **.kv* files [9], so that the project is easy to maintain. Because the responsibility for the user interface is shared between **.kv* files (i.e. *views.kv*) and python files (i.e. *views.py*) which utilizes the widgets declared in the **.kv* files. Moreover, the *View* components does not hold any logic, every action is sent to the *controller*, and the *controller* decides what to do. By doing so, it is easier to control the workflow of the project. The communication between views and the controller was implemented through the events with the help of the *pydispatcher* library [10], because it is more clean to apply the code presented in Fig. 1 to notify a receiver about an action rather than the code presented in the Fig. 2 where the widget is trying to find the controller widget from the root window in an error prone way.

As of writing this article, the 1.11.1 version of the Kivy framework has very primitive looking core widgets, thus KivyMD library, which implements the

Google's material design concepts [21], was employed for a better user interface, as material design concepts are believed to improve user experience [22].

```
dispatcher.send(signal=self.signal_on_prev_click,
    sender={})
```

Fig. 1. Code snippet of a sending an event

```
self.get_root_window().children[0]
    .on_dialog_item_press(value_dialog_id,
    value_dialog_title)
```

Fig. 2. Code snippet of a getting the controller widget from the root widget

The *classifier* component, which is responsible for utilizing the ML model, processes the utilization in two phases:

1. Loading pre-trained models
2. Processing sentiment analysis on a given text

The controller initializes the first phase in the *splash-screen*, as it is appropriate time to load resources during the period, when a splash screen is shown. [11] In order to support all the type of machines, the *classifier* is configured to be used with GPU and without GPU.

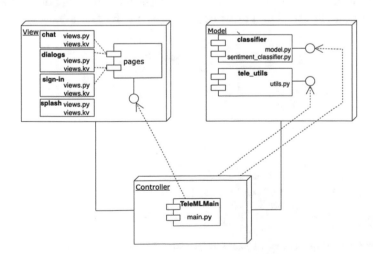

Fig. 3. The architecture of the TeleML application

The next component of the model module is *tele_utils* which uses the *telethon* library to utilize the Telegram APIs [12]. The *telethon* library provides all the necessary features to protect the sensitive data of users, such as encryption and caching an account on the local device, therefore security is provided by saving the *.session* file, which stores the sensitive account information, in user's machine.

As large lists, such as dialogs or messages, can be retrieved from the Telegram's API in chunks, [13] the paging mechanism is implemented in the *tele_utils* component where each page is associated with a subset of items.

4   Results

As a result of the implementation the mentioned architecture and frameworks described in the Methodology section, the TeleML cross-platform app is ready. The following subsections present the obtained results for the ML model, as well as the TeleML app which utilizes the ML model and Telegram APIs into the user interface.

4.1   ML Model

The ML model achieved 94.83% performance. And it took 2 h and 35 min for training and validating the model on the Google Colab machine. The Fig. 4 shows the performance tendency over the epochs while training.

The Table 3 shows the F1-score and performance on each of the predicted classes (i.e. *negative, neutral and positive*). While loading the pre-trained model in the *splash screen* took 4.2 s, and predicting a text *"thank you a lot"* took 0.3 s. The time performance of the model was recorded in the machine, running the TeleML app, with the characteristics presented in the Table 2.

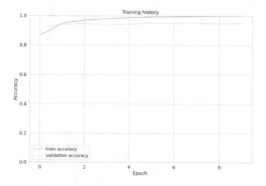

Fig. 4. The training history of the ML model. The graph shows the training and the validation accuracy of the model over the training epochs

Table 2. The performance of the machine running the TeleML

Model	MacBook Pro (15-inch, 2018)
CPU	2,6 GHz 6-Core Intel Core i7
RAM	16 GB 2400 MHz DDR4
Battery charged status	20%
Free RAM status	3,14 GB
CPU load status	6%

Table 3. The classification report of the ML model

–	Precision	Recall	f1-score	Support
Negative	0.93	0.86	0.9	791
Neutral	0.97	0.98	0.98	1333
Positive	0.93	0.96	0.95	1591
–	–	–	–	–
Accuracy			0.95	3715
Macro avg	0.95	0.94	0.94	3715
Weighted avg	0.95	0.95	0.95	3715

4.2   TeleML Application

As the TeleML app was developed on the cross-platform framework *Kivy*, the app is ready to be packaged on the most popular platforms, such as Windows, Linux, OSX, iOS and Android, but currently the app has been run and tested on the OSX platform.

While showing the splash screen, which is represented in the Fig. 5, the two main processes are handled by the *controller*:

- The ML model is loaded onto the memory
- The Telegram account authentication is checked for authentication

The two above processes are done sequentially, that is once the ML model is loaded, the *controller* requests *tele_utils* module for authentication (*telethon* was employed to check the authentication). If an user does not have a valid authentication, then the *controller* shows the *sign-up* screen, otherwise the *dialogs* screen is displayed.

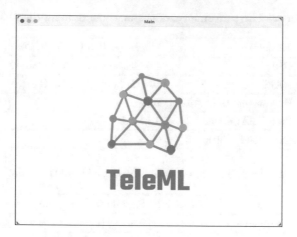

Fig. 5. The user interface of the Splash screen

The *dialogs* screen (see Fig. 6) shows the list of dialogs, and allows users to iterate over dialogs via paging buttons. The list of dialogs are fetched through the *tele_utils* module. Moreover, the screen allows the user to logout of the telegram account by pressing the *sign-out* icon located on the right-top of the screen. If the user clicks on a dialog item, then the pressed dialog's messages are shown in the *chat* screen.

In addition, the *chat* screen (Fig. 7) allows the user iterate over messages via paging buttons. The messages are fetched through *tele_utils* module. After the messages are fetched, the text of the messages (if the text is not empty, of course) are classified into negative, neutral or positive along with confidence score through *classifier* module. Then the parsed messages are displayed as list in the *chat* screen. A user can press the back button (which is located on the top-left corner of the screen) to get back to the *dialogs* screen.

5   Discussions

The methods and achievements were presented in the earlier sections and the following subsections will try to contextualize the ML model and the TeleML app.

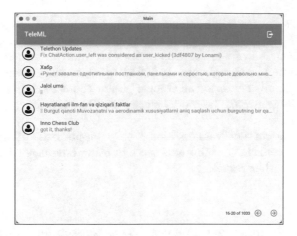

Fig. 6. The user interface of the dialogs screen

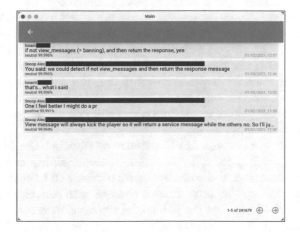

Fig. 7. The user interface of the chat screen

5.1   ML Model

The Fig. 4 demonstrates that the model is performing the effective performance with no significant overfitting. The validation accuracy showed a decent performance which gave us the confidence to utilize the model in the prediction of sentences in the messages.

Talking about the accuracy of the ML model which is 94.83%, the authors have achieved the state-of-art performance. Since the performance of the BERT algorithm is proven to be significantly high compared to other state-of-art algorithms, such as SVM based TF-IDF, GLOVE and FastText, the score that current work achieved is fair.

Although the model is employed for telegram client, it can be applied in the platforms for sentiment analysis which support python environment and meet the specific performance requirements as size of the model is comparatively large, which is 700 MB, especially for platforms with relatively small performance (i.e. mobile platforms). As of writing this article, the limit of an Android APK file to be published on the Google play is 100 MB [14], so the current version of the app can not be published onto the Google play without reducing the model size. Moreover, the large size of the model causes long loading time, which is 4.2 s, in the *splash* screen.

Moreover, the current version model only supports handling English sentences, and if a message contains sentences in other languages, the prediction is not correct and rather random.

5.2   TeleML

As the app is software product, we can discuss the app regarding suitable software quality attributes. A security concern from the user viewpoint is addressed by the app. That is the app stores the *.session* file, which stands for a Telegram account information, locally on the user's machine, and interacts with Telegram API directly with the help of *telethon* library, thus no-third-parties are involved. The next fact is that the classification process is handled locally as the app is shipped along with the pre-trained model. Such an approach is more secure than sending user data to servers and validating the data in servers. However, shipping the pre-trained model along with the app can harm the intellectual property of the developers, because one can easily extract the pre-trained model and use for his own needs.

As the project implements MVC pattern by dividing the logic of the app into modules responsible for different tasks, maintainability can be considered as addressed. This structure allows developers to easily edit the functionalities or add features into the app by only making changes into concrete modules. Moreover, all the modules including the ML model are developed in one programming language - Python gives developers another convenience of programming.

The usability of the app, especially user interface, was addressed by following material design concepts and it allowed to have modern good-looking user interface. However, the Kivy framework does not include enough widgets and mechanisms to achieve a desirable results. For example, the TeleML app has implemented pagination feature in the *dialogs* and *chats* screens on the bottom-right screen of the layouts using buttons on page numbers, which requires a user to spend an extra effort to navigate to the next/previous page. Instead, there are modern approaches, such as fetching and displaying the next page, when a user reached the end of the list view by scrolling, often called as infinite scrolling [15]. But such approach is hard to achieve with the *1.11.1 version* of Kivy framework. In addition, building complex dynamic layouts for an item in the list (i.e. *Recycleview* in the Kivy framework) with the current version of Kivy is hard, hence the user interface is too simple.

The authors have used Python programming language in all the modules of the app, including the user interface, and the interaction with Telegram APIs, therefore for developers of ML, it would not require obtaining knowledge of other programming languages for utilizing the developed ML models. Thus, the approaches taken by the authors could be considered by developers of ML.

6   Conclusions

The TeleML includes several aspects of software development, including user interface development, cross-platform development, working with Telegram APIs, and deployment and utilization of the developed ML model. This combination of software frameworks and libraries gives us a new perspective to the ML development. As the size of the ML model affected the app, i.e. the splash screen took long time to load, the model size is too much for mobile platforms etc., the development phase of the ML model should consider such concerns. Moreover, the ML developers should be responsible for intellectual property of the ML model if the model is going to be shipped along with an app. In addition, when building user interface, the suitable UI framework should be chosen according to the requirements. As the TeleML app shows, the Kivy Framework is not ready yet for complex user interface.

Considering the above issues and approaches, the future work should consider several aspects when deploying the ML model into software systems. First of all, although the BERT algorithm shows significant results, other ML algorithms should be tested as well. More robust and stable frameworks should be considered for the user interface of an application, such as Flutter, React native etc. Moreover, the project should be judged by appropriate metrics which could evaluate the different aspects, such as quality attributes etc. The ways of reducing the size of a pre-trained models, which are to be shipped along with the application, should be considered.

References

1. Ribeiro, M., Grolinger, K., Capretz, M.A.M.: MLaaS: machine learning as a service. In: IEEE 14th International Conference on Machine Learning and Applications (ICMLA), Miami, FL, vol. 2015, pp. 896–902 (2015). https://doi.org/10.1109/ICMLA.2015.152
2. Devlin, J., Chang, M.-W., Lee, K., Toutanova, K.: BERT: pre-training of deep bidirectional transformers for language understanding. arXiv preprint arXiv:1810.04805 (2018)
3. Wolf, T., et al.: HuggingFace's transformers: state-of-the-art natural language processing (2019). arXiv, arXiv:1910.03771
4. Nadkarni, P.M., Lucila, O.-M., Chapman, W.W.: Natural language processing: an introduction," Journal of the American medical informatics association, vol. 18, Issue 5(2011), pp. 544–551 (2011) https://doi.org/10.1136/amiajnl-2011-000464
5. Cambria, E., Poria, S., Gelbukh, A., Thelwall, M.: Sentiment analysis is a big suitcase. IEEE Intell. Syst. **32**(6), 74–80 (2017)
6. González-Carvajal, S., Garrido-Merchán, E.C.: Comparing BERT against traditional machine learning text classification. arXiv preprint arXiv:2005.13012 (2020)

7. Reddit database for sentiment analysis. https://www.kaggle.com/cosmos98/twitter-and-reddit-sentimental-analysis-dataset?select=Reddit_Data.csv. Accessed 6 Jan 2021

8. Goderis, S.: On the Separation of User Interface Concerns: A Programmer's Perspective on the Modularisation of User Interface Code. ASP/VUBPRESS/UPA (2008)

9. KV Language. https://kivy.org/doc/stable/guide/lang.html. Accessed 6 Jan 2021

10. Driscoll, M.: WxPython Recipes: A Problem-Solution Approach. Apress (2017)

11. Tyers, B.: Splash screens and menu. In: Practical GameMaker: Studio, pp. 129–138. Apress, Berkeley, CA (2016)

12. Exo, L.: Telethon: Full-featured Telegram client library for Python 3(1.11. 2)[Python] (2020)

13. Pagination in the Telegram API. https://core.telegram.org/api/offsets. Accessed on 7 Jan 2021

14. Google play APK Expansion Files. https://developer.android.com/google/play/expansion-files. Accessed on 7 Jan 2021

15. Babich, N.: UX: infinite scrolling vs. pagination. UX Planet 1 (2016)

16. Jain, A.P., Dandannavar, P.: Application of machine learning techniques to sentiment analysis. In: 2016 2nd International Conference on Applied and Theoretical Computing and Communication Technology (iCATccT), pp. 628–632. IEEE (2016

17. Messias, J., et al.: Towards sentiment analysis for mobile devices. In: 2016 IEEE/ACM International Conference on Advances in Social Networks Analysis and Mining (ASONAM), San Francisco, CA, pp. 1390–1391 (2016). https://doi.org/10.1109/ASONAM.2016.7752426

18. Chambers, L., Tromp, E., Pechenizkiy, M., Gaber, M.: Mobile Sentiment Analysis (2012). https://doi.org/10.3233/978-1-61499-105-2-470

19. Mogoroase, V.: Applications to Sentiment Analysis Using Raspberry and Different Microcontrollers for Data Streams Classification in the Context of Covid-19. Annals of the University of Craiova, Economic Sciences Series, vol. 1, no. 48 (2020)

20. Tromp, E., Pechenizkiy, M.: SentiCorr: Multilingual Sentiment Analysis of Personal Correspondence. Mathematical Programming, pp. 1247–1250 (2011). https://doi.org/10.1109/ICDMW.2011.152

21. KivyMD: Material Design for Kivy framework. https://github.com/kivymd/KivyMD

22. Ayyal Awwad, A.M., Schindler, C., Luhana, K.K., Ali, Z., Spieler, B.: Improving pocket paint usability via material design compliance and internationalization & localization support on application level. In: Proceedings of the 19th International Conference on Human-Computer Interaction with Mobile Devices and Services (MobileHCI'17). Association for Computing Machinery, New York, NY, USA, Article 99, pp. 1–8 (2017). https://doi.org/10.1145/3098279.3122142

23. Source code of the ML model used in the TeleML app. https://colab.research.google.com/drive/1Ya4uJnmprza3NWIFo4lYMfAt2J5g2yOa?usp=sharing

24. Source code of the TeleML app. https://github.com/Rhtyme/TeleML

PPCSA: Partial Participation-Based Compressed and Secure Aggregation in Federated Learning

Ahmed Moustafa[1], Muhammad Asad[1(✉)], Saima Shaukat[2], and Alexander Norta[3]

[1] Department of Computer Science, Nagoya Institute of Technology, Nagoya 466-8555, Japan
{ahmed,a.muhammad.799}@nitech.ac.jp
[2] Faculty of Computer Science, Lahore Garrison University, Lahore 54792, Pakistan
saimashaukat@lgu.edu.pk
[3] Department of Software Science, Tallinn University of Technology, 19086 Tallinn, Estonia
alexander.norta@taltech.ee

Abstract. Federated Learning (FL) enables users devices (UDs) to collaboratively train a Deep Learning (DL) model on an individual's gathered data, without revealing their privacy sensitive information to the centralized cloud server. Those UDs usually have limited data plans with a slow network connection to a centralized cloud server, which causes limited communication bandwidth between the contributing mobile users. To mitigate this problem, we propose a novel Partial Participation-based Compressed and Secure Aggregation (PPCSA) algorithm. To implement the PPCSA, we use a Sparse Compression Operator (SCO) that reduces the communication bits between the cloud server and the users while maintaining the FL requirements. In particular, PPCSA utilizes a novel compression method and introduces a Local Differential Privacy (LDP) based framework to achieve the communication-efficiency at a new level. Our experiments on a commonly used FL dataset show that PPCSA distinctively outperforms the state-of-the-art schemes in terms of convergence accuracy and communication bits.

1 Introduction

Machine Learning (ML) applications need to process large scale data at the users devices (UDs) of the network such as IoT sensors, or mobile devices [1]. To this end, distributed computing systems received widespread attention, but they failed to protect the user's privacy due to the centralized cloud server involvement for the training of the user's data [2]. To address this challenge, Federated Learning (FL) has emerged as a novel paradigm that trains the user's data based on the learning parameters at the UD while maintaining the user's privacy, as illustrated in Fig. 1. The major objective of FL is to enable the UDs to collaboratively learn a global model by sharing their local parameters without disclosing their private information [3]. To protect the privacy of the user's data, several technology companies have already deployed FL in several applications such as learning sentiment analysis [4], wearable devices [5], and location-based services [6]. However, despite the promise FL brings for such applications, various

L. Barolli et al. (Eds.): AINA 2021, LNNS 226, pp. 345–357, 2021.
https://doi.org/10.1007/978-3-030-75075-6_28

challenges still exists that remain to be resolved [7]. To this end, in this paper, we consider three challenges of FL and propose a solution that simultaneously solves those challenges through the novel PPCSA algorithm:

1. Scalability: In a typical FL network, the whole system is constituted by two major entities; the UDs and the cloud server. Usually, this network requires thousands to millions of UDs to train the learning models. This huge number of UDs sometimes disturbs the learning accuracy due to the limited bandwidth, slow processing, or inactivity during the training process. Therefore, the proposed FL algorithm is designed to operate efficiently with partial participation or random sampling of UDs towards the aim of higher convergence.
2. Communication-Overhead: During the training process, a massive number of UDs are required to upload their locally trained models to a centralized cloud server, which often results in a bottleneck of communication bandwidth. Therefore, it is vital to send such updates infrequent and compressed to design a communication-efficient FL algorithm.
3. Privacy Leakage: Though FL is designed to solve the individuals' privacy issues, however, sharing a small number of system parameters can still cause a considerable privacy concern. Therefore, users need to consider further privacy mechanisms and implement them on local gradients to secure private information.

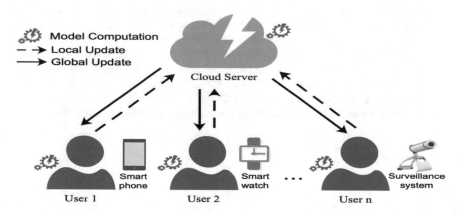

Fig. 1. The general framework of FL where the cloud server sends the shared model to each connected UD that locally trains the shared model using their private data. Afterwards, each UD uploads their local updates to the cloud server, which are then aggregated/averaged to enhance the global shared model.

To resolve the challenges mentioned above, we use the FL setup to find the accurate model among the m distributed UDs, where each UD contains n non-independent and non-identically distributed (*non-I.I.D*) samples from an unknown probability distribution. To keep the originality of these samples, we apply differential privacy on each local UD. The significant contributions in this paper are summarized as follows:

1. The proposed algorithm utilizes stochastic gradient descent (SGD) optimization to reduce the convergence time [8]. In particular, the proposed algorithm allows partial participation of UDs by capturing the constraints of all active UDs. By doing this, only a fraction of total active UDs can contribute to the local models' training.
2. The proposed algorithm allows each UD to send the compressed form of a local update to the cloud server in each epoch of communication. As the cloud server requires large numbers of local models to obtain the global model, the compressed form of local updates results in communication efficiency.
3. The proposed algorithm enables differential privacy at each UD to secure the local gradients before sending the local models to the cloud server. In particular, we enhanced the existing randomized response and perturbed the local gradients through local differential privacy. In this way, the proposed algorithm achieves higher accuracy while maintaining communication efficiency through partial participation.

The rest of the paper is organized as follows: The FL setup and differential privacy are briefly explained in Sect. 2. Section 3 introduces the proposed PPCSA algorithm. Experimental setup and results are evaluated in Sect. 4. An extensive analysis of the proposed algorithm is discussed in Sect. 5. Section 6 concludes the paper and defines future work.

2 Preliminaries

In this section, we briefly explain the considered FL model and then discuss the local differential privacy model, which serves as a foundation of the proposed algorithm.

2.1 Federated Learning

In general, FL is a decentralized learning scheme that enables UDs to learn a shared global model without sharing the individual's private data to the cloud server. This paper uses the generic FL model where the cloud server aims to find an accurate data model among the available UDs those performs significantly concerning the data points. We assume that the data points among the all available UDs are generated through the random probability distribution. In particular, we consider the distributed stochastic learning problem defined as follows;

$$\underbrace{min}_{w} P(w) = \underbrace{min}_{w} \mathcal{R}_\rho [\ell(w, \rho)] \tag{1}$$

where ℓ is calculated by $\ell : \mathcal{F}^q \times \mathcal{F}^y \to \mathcal{F}$ which denotes the stochastic loss function, $w \in \mathcal{F}^q$ is the local model vector, $\rho \in \mathcal{F}^y$ is a random variable with random probability distribution $\wp$ and $P : \mathcal{F}^q \to \mathcal{F}$ denotes the population risk or expected loss. In this FL settings, m distributed UDs generate a collection of n samples that we denote as $\mathcal{S}^j = \rho_1^j, ..., \rho_n^j$ for $j \in [m]$ where, each sample is achieved through the random probability distribution $\wp$.

2.2 Local Differential Privacy

Applying the differential privacy on UDs is a step to collect data from the local devices securely without assuming a trusted data collector. Therefore, our FL settings naturally meet the privacy requirement on this locally differential privacy (LDP). To fulfill this privacy requirement, a set of controlled noises is added to each UD's private data before sending this data to the cloud server. This mechanism should satisfy the definition given below:

Definition 1. $((\varepsilon, \delta)$-LDP) A randomised algorithm $\mathcal{A} : \mathcal{V} \rightarrow \mathcal{R}$ is (ε, δ) UDs-based differentially private if for any $m, m' \in \mathcal{V}$ and $w \in \mathcal{R}$, we have

$$\wp_j[\mathcal{A}(m) = w] \leq n^{\varepsilon} \wp_j[\mathcal{A}(m') = w] + \delta. \tag{2}$$

In the proposed algorithm, we exploit the existing randomized response (RR) technique [9] on the proposed LDP in order to secure the user's privacy while minimizing the communication cost. In particular, the RR allows the cloud server (data collector) to acquire data from the UDs by asking the binary questions from UDs, where the response of UDs should be in a randomized fashion. For instance, each UD flips a coin with the probability of head $1 - \eta$ and sends the updates only when the coin turns head. We exploit a randomized response because it can estimate the unbiased results while satisfying ε-LDP by setting as follows:

$$\eta = \frac{1}{1 + n^{\varepsilon}}. \tag{3}$$

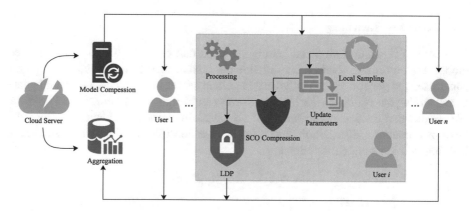

Fig. 2. The execution flow of the partial participation-based compressed and secure aggregation (PPCSA) algorithm. The *User i* shows each step in the local model computation.

3   Proposal of PPCSA Algorithm

This section proposes the novel PPCSA algorithm that simultaneously reduces the communication overhead through partial participation and compression of the gradient updates while providing security to the user's private data through local differential privacy. In particular, here, we briefly explain the partial participation of the UDs, compression, perturbing and then aggregation of the generated local gradients in the following subsections, respectively. In Fig. 2, we show the complete execution flow of the proposed PPCSA.

3.1   Partial Participation of Devices

As we have mention in Sect. 1, the FL network requires thousands to millions of UDs to train the local learning models. The trained models on these UDs are then sent to the centralized cloud server to generate a new global model [10]. The cloud server suffers the following two problems while receiving the local models from the UDs.

1. The cloud server might have limited download bandwidth, which causes the pipeline that results in drastically slow training. In particular, when all the UDs send their updated gradients to the cloud server in each epoch, the cloud server would receive these updates from only a few UDs due to the limited bandwidth. Hence, the pipeline occurs at the cloud server.
2. When all the UDs participate during the whole training process, a considerable communication overhead occurs on the network, which results in high communication costs. Also, generally, all the UDs do not contribute, or participate in each round of the training [11]. As several factors decide whether a UD can participate in training, i.e., a UD should be idle, a UD should be in the cloud server range, a UD should be connected to the wireless network and plugged in during the training, etc.

In the proposed PPCSA algorithm, we assume that from the m UDs, only i numbers of UDs ($i \leq m$) are available for training in each round. Based on point (2) mentioned above, here we also assume that these UDs are randomly distributed in the network [12]. In particular, in each epoch $e = 0, 1, ..., E - 1$ of the training algorithm, the cloud server sends the current global model w_e to all the i participant UDs in subset U_e, which are randomly distributed among the m devices.

3.2   Compression of Local Gradient

In the FL network, the UDs also face the communication bottleneck as they are required to upload the gradient updates, which are trained through multiple iterations of the optimization algorithm (SGD). This gradient updates possibly in gigabytes based on the deep learning (DL) architecture and its millions of parameters. Therefore, the cost of communication from UDs to the cloud server could be slow and expensive because of the limited upload bandwidth. Since minimizing the size of gradient updates from the

UDs is critical, the proposed algorithm PPCSA employs a compression operator on each gradient update to reduce the communication overhead. In particular, each participant UD $i \in U_e$ obtains a local model w_{e,E_l}^i after running E_l local SGD iterations on the latest model w_e received from the cloud server. Once the local model is obtained, each participant UD_i applies a compression operator Q_o on $w_{e,E_l}^i - w_e$. After applying compression on the local models, each participant UD_i perturbs the gradients by adding calibrated noises through the differential privacy and then uploads their perspective models to the cloud server. Once the cloud server receives all the models, it removes the noises through differential privacy and then decodes the compressed updates and combines them to generate a new model w_{e+1}.

3.3 Adding Noise to Local Gradient

In the literature, several authors apply the naive method kRR [13] to perturb the model vector w, which is an LDP mechanism leverage with the Random Response (RR) technique. However, applying kRR directly to FL creates the following drawbacks:

1. Inefficiency in Model Training: in the kRR mechanism, noisy data sampling is set uniformly in the entire dataset. For example, in the image-based dataset, e.g., CIFAR, the kRR samples some rare images due to the uniform sampling, which generates several meaningless tuples that may never appear in the general sampling process before perturbation. This causes a higher level of noise and results in less model accuracy.
2. Limited Privacy Budget: the privacy budget ε represents the quantity of noise applied on the gradients to prevent leakage of data. In kRR mechanism, the ε should satisfy $\varepsilon \geq \ln(|D| - 1) + \ln(1/\eta - 1)$, where $|D|$ is the size of data and η is same as in Eq. (3). In a real-time application, generally $|D|$ is large in size (e.g., over 60,000), and thus a large size of privacy budget would be required.

Considering these drawbacks, we devise a new perturbing mechanism by enhancing the existing RR and name it Enhanced Randomized Response (ERR) that introduces a solution for the aforementioned drawbacks as follows:

1. ERR draws a priority based on the previous samples, and not only non-uniformly but also adaptively samples the noisy terms. Consequently, the drop in model accuracy caused by the noise can be effectively reduced.
2. ERR requires only an essential privacy budget, which has no concern to the size of data $|D|$.

In Algorithm 1, we present the enhanced randomized response (ERR) with the probability of $1 - \eta$.

Algorithm 1. The ERR Mechanism

Input: A tuple from UD (i, ρ_1^j, ρ_2^j)
Output: Perturbed Tuple (i', ρ_1^j, ρ_2^j)
Initialization

1: **if** $x = 0$ **then**
2:      $\rho_1^j = \rho^j$
3: **else**
4:      Sample $i' \sim |m|$
5:      Sample $\rho' \sim \wp$
6:      **if** $\rho' \in \mathscr{P}_i$ **then**
7:          $\rho_1^j = \rho^j$
8:      **else**
9:          $\rho_1^j = \rho'$
10:     **end if**
11:     **return** (i, ρ_1^j, ρ_2^j)
12: **end if**

3.4 Gradient Aggregation

In this subsection, we briefly explain the aggregation process of the proposed PPCSA. The proposed algorithm consists of E epochs and during each epoch every single UD performs E_l local updates, which accumulatively results into total number of $T = E_{E_l}$ iterations. In each epoch $e = 0, ..., E-1$ of the algorithm, the cloud server randomly selects U_e UDs from the total number of available m UDs. Afterwards, the cloud server broadcasts the current model w_e to all the selected UDs U_e, where each selected $UD_i \in U_e$ performs E_l local SGD optimization on its locally available data. In particular, let $w_{e,t}^i$ denotes the model at the UD_i at t-th iteration of the e-th epoch. At each local iteration $t = 0, ..., E_l - 1$, UD_i update its local model by considering the following rule:

$$w_{e,t+1}^i = w_{e,t}^i - \eta(e,t)\nabla G_i(w_{e,t}^i) \tag{4}$$

where the stochastic gradient ∇G_i is computed through the distribution of random samples that are picked from the local dataset S^j. All the selected UDs begin with the same initialization $w_{e,0}^i = w_e$. After E_l local epochs, each UD computes the final update in that particular epoch, which is $w_{e,E_l}^i - w_e$, and compresses the update $Q(w_{e,E_l}^i - w_e)$. In the end, all the selected UDs U_e apply differential privacy to the compressed updates $\eta = \frac{1}{1+e^\varepsilon}(Q(w_{e,E_l}^i - w_e))$ and then upload their gradient updates to the cloud server. The cloud server aggregates these local gradient updates and computes a new global model using the following equation:

$$w_{e+1} = w_e + \frac{1}{i}\sum_{i \in U_e} Q(w_{e,E_l}^i - w_e). \tag{5}$$

This process is repeated for E epochs, or when the desired convergence level is achieved. The complete pseudocode of PPCSA is given in Algorithm 2.

Algorithm 2. Pseudocode of PPCSA

Input: selected participants U_e and the cloud server
Output: new global model
Initialization of PPCSA

1: **for** e = 0, 1, . . . ,E-1 **do**
2:     cloud server randomly selects i UDs U_e
3:     cloud server sends w_e to UDs in U_e
4:     **for** UD $i \in U_e$ **do**
5:         $w^i_{e,0} \leftarrow w_e$
6:         **for** $t = 0, 1, ..., E_l - 1$ **do**
7:             randomly selects a data point $\rho \in S^j$
8:             compute $\nabla G_i(w) = \nabla \ell(w, \rho)$
9:             set $w^i_{e,t+1} \leftarrow w^i_{e,t} - \eta(e,t)\nabla G_i(w^i_{e,t})$
10:            apply $\eta = \frac{1}{1+e^\varepsilon}$
11:        **end for**
12:        send $\frac{1}{1+e^\varepsilon}(Q(w^i_{e,E_l} - w_e))$ to the cloud server
13:    **end for**
14:    server gets $w_{e+1} = w_e + \frac{1}{i}\Sigma_{i\in U_e}Q(w^i_{e,E_l} - w_e)$
15: **end for**

---

4   Evaluations

This section empirically studies the communication/computation tradeoff and evaluates the proposed PPCSA compared to the existing schemes. In particular, we first explain the experimental setup, then we conduct the experiments and compare the proposed PPCSA concerning the different number of participant UDs, compression levels, privacy budgets, and state of the art schemes.

4.1   Experimental Setup

We conduct our experiments on a server with an Intel(R) Core(TM) CPU i7-4980HQ (2.80 GHz) and 16 GB of RAM. The PPCSA algorithm is simulated using TensorFlow[1] in Python. For comparison, we use two benchmarks; Federated Averaging (FedAvg) [14] and Privacy-Preserving Deep Learning (PPDL) [15]. In particular, firstly, we evaluate the proposed PPCSA on three different scenarios of UDs, three different scenarios of compression levels, and three different scenarios of privacy budgets. Then, we set the average number of those parameters and compare the PPCSA with the state-of-the-art schemes. For evaluation, we consider the MNIST[2] dataset where the gradient consists of 60,000 training examples, and each example consists of 28×28 size images. In Table 1, we show the baseline configurations of all the experiments, while the number of UDs, compression levels, and privacy budget varies experiment-wise.

---

[1] https://www.tensorflow.org.
[2] http://yann.lecun.com/exdb/mnist/.

Table 1. Hyper-parameters

Parameter	Value
Network dimensions	$100 \, \mathrm{m}^2 \times 100 \, \mathrm{m}^2$
Global communication rounds $\mathcal{R}$	100
Local epochs e	20
Number of classes	10
Gradient size	32 *bits*
Batch size	20
Learning rate η	0.05
Clients transmission power	200 mW
Local update size	25000 *nats*

4.2 Results

The below plots demonstrate the overall training time for a logistic-regression problem over the MNIST dataset for 100 epochs. We capture the relative cost of communication and computation since our primary focus on communication overhead; therefore, we set *communication/computation* = 100/1 to capture the communication bottleneck. We test the accuracy on *three* different parameters, the number of participant UD_i, the compression levels Q_ℓ, and the number of privacy budgets ε, while all the other hyper-parameters are set as described in Table 1.

4.2.1 Comparison on Different Number of Participant UDs
In Fig. 3, we set three different numbers of participant UDs as $i \in \{25, 50, 100\}$, the compression level is set to 10, and the privacy budget is set to 0.5 to test the accuracy on scalability issue. The plot in Fig. 3 demonstrates the training time verses the training accuracy for each aggregated model to the cloud server in each epoch $e = 1, ..., T/E_\ell$. The experiment shows that the proposed algorithm has a negligible training loss with fewer numbers of UDs, while the higher numbers of UDs produce higher accuracy. This is because a huge part of the gradient is compressed, and differential privacy protects the compressed gradients before aggregation towards the cloud server.

4.2.2 Comparison on Different Compression Levels
Similarly, in Fig. 4, we set three different compression levels as $Q_\ell \in \{5, 10, 20\}$, the participant UDs are set to 50, and the privacy budget is set to 0.5 to test the accuracy w.r.t to communication overhead. The plot in Fig. 4 demonstrates the communication-computation tradeoff. The experiment shows that the higher compression level slows down the convergence time but produces higher accuracy, while the lower level of compression has faster convergence but produces lower accuracy. The reason behind this slow convergence is that each UD has to apply compression on each gradient separately

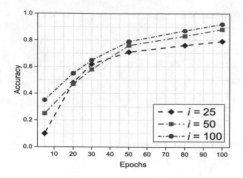
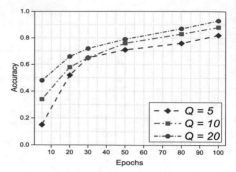

Fig. 3. Accuracy on three different numbers of UDs in 100 communication rounds.

Fig. 4. Accuracy on different levels of compression in 100 communication rounds.

and thus, a higher level of compression results in less growth of compressed-gradient volume.

4.2.3 Comparison on Different Privacy Budgets

In Fig. 5, we set different numbers of privacy budgets as $\varepsilon \in \{0.7, 1.0, 1.5\}$ on local gradients. In contrast, the number of participant UDs are set to 50, and the compression level is set to 10 for testing the accuracy. The graph in Fig. 5, demonstrates that the number of privacy budget has a significant impact on the convergence accuracy. The lower the number of privacy budgets, the lower the accuracy; in contrast, the higher number of privacy budgets produces high accuracy. The fact behind this is that a considerable part of noises is added into the local gradients through differential privacy, which helps secure the gradient data and results in efficient convergence.

4.2.4 Comparison with the Benchmarks

In Fig. 6, we compare the training time of the proposed PPCSA with two benchmarks schemes, FedAvg and PPDL. As demonstrated in Fig. 4, a higher level of compression results into slow convergence, therefore in Fig. 6, we set the optimal number of UDs as $i = 50$, compression level $Q_\ell = 10$ and privacy budget $\varepsilon = 0.5$. The plot in Fig. 6 proves that the proposed algorithm achieves a higher accuracy within a given number of epochs.

Based on all the results demonstrated above, we conclude that the proposed PPCSA solves the three considered challenges in this paper; scalability, communication-overhead, and privacy. Therefore, the proposed PPCSA is not only suitable for small-scale networks but also significant for dense networks.

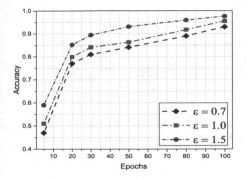

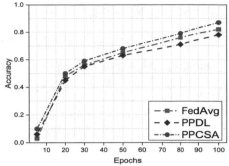

Fig. 5. Accuracy on different numbers of privacy budgets in 100 communication rounds.

Fig. 6. Accuracy comparison with existing approaches in 100 communication rounds.

Table 2. Required bits for download and upload to achieve the targeted accuracy.

Accuracy = 82.1%	Download bits	Upload bits
Baseline	1824 MB	1824 MB
FedAvg	237 MB	256 MB
PPDL	117 MB	284 MB
PPCSA	16.4 MB	164 MB

5 Analysis and Discussion

In this section, we briefly analyze the communication-efficiency and privacy-protection of the proposed algorithm.

5.1 Communication Cost

To further prove the communication efficiency of the proposed algorithm, we briefly explain the required numbers of communicated bits concerning the number of epochs. To achieve meaningful results, we choose the optimal number of UDs, compression level, and privacy budget as $i = 50$, $Q_\ell = 10$, and $\varepsilon = 0.5$, respectively, and compare the results with the benchmarks FedAvg and PPDL. The rest of the parameters are set to the same, as mentioned in Table 1. We train the MNIST dataset until the targeted accuracy is achieved and capture the total number of required bits for uploading and downloading the local-and global model, respectively. The required number of communication bits to achieve the targeted accuracy is given in megabytes (MB) in Table 2.

In Table 2, PPCSA communicates 16.4 MB and 164 MB of data, which is a considerable reduction in communication delay compared to the state-of-art approaches. This is achieved due to the compression of local gradients and the partial participation of UDs.

5.2 Privacy Protection

Here, we briefly analyze the privacy protection of the proposed algorithm.

Theorem 1. *Algorithm 1 satisfies* $(\varepsilon, 2\delta)$*-LDP by setting* $\eta = \frac{1}{2\delta_o n^\varepsilon + 1}$, *where* $\delta_o = \delta - (\delta^{-\frac{1}{\Gamma}} + 1)^{-\Gamma}$, $\Gamma \geq 1$ *is a constant.*

Theorem 1 proves that the untrusted cloud server cannot easily perceive the updates from any two participant UDs; thus, each local gradient's privacy is well protected.

6 Conclusion

In this paper, we address three challenges of federated learning; scalability, communication overhead, and privacy. To this end, we propose the Partial Participation based Compressed and Secure Aggregation (PPCSA) algorithm to solve those challenges. In particular, PPCSA applies three different modules for each gradient update: (1) *partial participation* which captures the availability of user devices and selects the optimal devices for participation, (2) *compression* which compresses each local gradient, (3) *differential privacy* which protects each local gradient during the aggregation towards the cloud server. We further empirically study the communication-computation tradeoff and evaluate the proposed PPCSA in comparison to the existing schemes.

In our future work, we plan to further investigate the proposed PPCSA by considering the adversaries in the network with diverse data learning tasks and high-dimensional datasets.

Acknowledgement. This work has been supported by Grant-in-Aid for Scientific Research [KAKENHI Young Researcher] Grant No. 20K19931.

References

1. Zhu, G., Liu, D., Du, Y., You, C., Zhang, J., Huang, K.: Toward an intelligent edge: wireless communication meets machine learning. IEEE Commun. Mag. **58**(1), 19 (2020)
2. Su, L.: Defending distributed systems against adversarial attacks: consensus, consensus based learning, and statistical learning. ACM SIGMETRICS Perform. Eval. Rev. **47**(3), 24 (2020)
3. Yang, Q., Liu, Y., Chen, T., Tong, Y.: Federated machine learning: concept and applications. ACM Trans. Intell. Syst. Technol. (TIST) **10**(2), 1 (2019)
4. Hu, S., Li, Y., Liu, X., Li, Q., Wu, Z., He, B.: arXiv preprint arXiv:2006.07856 (2020)
5. Xu, J., Glicksberg, B.S., Su, C., Walker, P., Bian, J., Wang, F.: Federated learning for healthcare informatics. J. Healthc. Inform. Res. 1–19 (2020)
6. Samarakoon, S., Bennis, M., Saad, W., Debbah, M.: Distributed federated learning for ultra-reliable low-latency vehicular communications. IEEE Trans. Commun. **68**(2), 1146–1159 (2019)
7. Asad, M., Moustafa, A., Yu, C.: A critical evaluation of privacy and security threats in federated learning. Sensors **20**(24), 7182 (2020)
8. Ketkar, N.: In: Deep Learning with Python, pp. 113–132. Springer (2017)

9. Erlingsson, Ú., Pihur, V., Korolova, A.: In: Proceedings of the 2014 ACM SIGSAC Conference on Computer and Communications Security, pp. 1054–1067 (2014)
10. Hanzely, F., Richtárik, P.: arXiv preprint arXiv:2002.05516 (2020)
11. Asad, M., Moustafa, A., Ito, T., Aslam, M.: arXiv preprint arXiv:2004.02738 (2020)
12. Rahman, S.A., Tout, H., Talhi, C., Mourad, A.: Internet of things intrusion detection: centralized, on-device, or federated learning? IEEE Netw. **34**(6), 310 (2020)
13. Kairouz, P., Bonawitz, K., Ramage, D.: arXiv preprint arXiv:1602.07387 (2016)
14. McMahan, H.B., Moore, E., Ramage, D., Hampson, S., y Arcas, B.A.: arXiv preprint arXiv:1602.05629 (2016)
15. Aono, Y., Hayashi, T., Wang, L., Moriai, S., et al.: Privacy-preserving deep learning via additively homomorphic encryption. IEEE Trans. Inf. Forensics Secur. **13**(5), 1333 (2017)

Granulation of Technological Diagnosis in the Algebra of the n-Pythagorean Fuzzy Sets

Anna Bryniarska[✉]

Institute of Computer Science, Opole University of Technology,
ul. Proszkowska 76, 45-758 Opole, Poland
a.bryniarska@po.edu.pl

Abstract. The paper presents the application of the scale algorithm for computations in two-dimensional spaces. The paper provides a conceptual framework for building algorithms for acquiring diagnostic knowledge about a technical object, for which the symptoms-faults relationship is binary according to the tests. When symptoms and faults are fuzzy, with some scaling, a symptom-fault pairs can be considered as intuitionistic fuzzy sets. In the literature, the deductive theory of n-Pythagorean fuzzy sets (*n-PFS*) is presented. The *n-PFS* objects are a generalization of intuitionistic fuzzy sets (*IFSs*) and Yager *PFSs*. Until now, the values of membership and non-membership functions have been described on a scale of 1:1 and the scale of the quadratic function. Here, the scales of arbitrary power functions are used. The *n-PFS* theory introduces a conceptual apparatus analogous to the classical Zadeh fuzzy sets theory, and it consistently defines the *n-PFSs* algebra. The paper also shows the construction of the fuzzy description language of computations in two-dimensional spaces. The granulation is defined as an interpretation of the descriptive language of technological diagnosis of a gas boiler. It is presented in a certain *n-PFS* algebra defined on the fuzzy pairs of the symptom-fault of this gas boiler.

1 Introduction

Data (information) granulation is an important paradigm for modeling and processing unclear, uncertain and inaccurate knowledge with the usage of information granules as the main mathematical structures in the context of granular calculations [1].

Typically, the range of fuzzy input data is too large to use granular computing algorithms. Scale algorithms are used to use more data. In this paper, the scale algorithm is understood as finding a function that assigns data to values suitable for the application of calculation algorithms. For example, in [2] the use of scale algorithms is proposed. So far, data values have been given on a 1:1 scale and square function scale. Here the scales of arbitrary power functions are used. The paper [2] and the scale algorithmics are used for granular calculations later in this paper.

© The Author(s), under exclusive license to Springer Nature Switzerland AG 2021
L. Barolli et al. (Eds.): AINA 2021, LNNS 226, pp. 358–369, 2021.
https://doi.org/10.1007/978-3-030-75075-6_29

Information granules [1,3], in this paper, relate to the technical objects to be diagnosed. They are mathematical models describing aggregated diagnostic data [4,5]. The following issues are important in technical diagnostics:

1. *Problems of dynamic diagnosis*: Fault detection and isolation in dynamic systems was discussed in many fundamental papers, books, publications and monographs [6–15]. As far as diagnosis of complex industrial plants is concerned, there is a need to decompose the process and the diagnostic system into smaller parts, diagnosed in parallel by separated diagnostic units (Local Fault Diagnosers (LFDs)) [16,17].
2. *Problems of static diagnosis*: methods of signal analysis [18], fault isolability with different forms of faults-symptoms relation [19], generalized reasoning about faults based on the diagnostic matrix [20], diagnostics with application of fuzzy logic [21], and diagnostic test for relation faults-symptoms [22].

Some 2nd type problems are solved in this paper. A conceptual framework is given to build algorithms for acquiring diagnostic knowledge about a technical object, for which the symptoms-faults relationship is binary according to the tests. When symptoms and faults are fuzzy, with some scaling, a symptom-fault pairs can be considered as intuitionistic fuzzy sets. The paper [2] presents, for the first time, the deductive theory of n-Pythagorean fuzzy sets ($n\text{-}PFS$). The $n\text{-}PFS$ objects are a generalization of intuitionistic fuzzy sets (*IFSs*) and Yager *PFSs*. Until now, the values of membership and non-membership functions have been described on a scale of 1:1 and the scale of the quadratic function. Here, the scales of arbitrary power functions are used. The $n\text{-}PFS$ theory introduces a conceptual apparatus analogous to the classical Zadeh fuzzy sets theory, and it consistently defines the $n\text{-}PFSs$ algebra. The granulation is defined as an interpretation of the descriptive language of technological diagnosis of a gas boiler. It is presented in a certain $n\text{-}PFS$ algebra defined on the fuzzy pairs of the symptom-fault of this gas boiler.

This paper is structured as follows. Section 2 introduces the following problem: when symptom-fault matrix is binary, and symptoms and faults are fuzzy with some scaling, then a symptom-fault pairs can be considered as intuitionistic fuzzy sets. Section 3 present the shortened n-Pythagorean fuzzy sets theory conceptual apparatus. Section 4 presents granulation of technological diagnosis of a certain gas boiler in the algebra of n-Pythagorean fuzzy sets.

2   Determination of the Symptom-Fault Diagnostic Matrix

In the process of using the machine, aspects such as machine reliability, efficiency of its use, and the process of its technical operation are very important. If there is a disturbance in this process that affects the above-mentioned aspects, then the machine (assembly, subassembly, element) is damaged. Generally, damage to the machine is an event of the transition of the machine from the undamaged state to the damaged state [23]. The condition of undamaged state is understood as a condition in which the machine fulfills the designated functions and maintains the parameters specified in the technical documentation. On the other hand, the damaged state defines the state of the machine in which it does not meet at least one of the requirements specified in the technical documentation. The main causes of damage are:

1. Structural – resulting from errors in the design and construction of the object,
2. Production (technological) – caused by errors and inaccuracies in technological processes,
3. Operational – resulting from failure to comply with the applicable operating rules or as a result of the impact of external factors unforeseen for the conditions of use of a given facility,
4. Aging – resulting from changes in object parameters during operation.

One of the main goals of technical diagnostics is to indicate faults F of the machine on the basis of diagnostic signals generated by this machine, called symptoms S. Of course, other external influences may also occur, but for now they will be ignored.

A state feature is a physical quantity, related to a machine property, with a measure, pattern and reference level, uniquely describing the value of the instantaneous machine state vector component. The features that characterize an object and its state, that occur only when the object is damaged or not fully undamaged, are called symptoms. Moreover, the diagnostic symptom s defines the measure of the diagnostic signal about the fault f, representing the specific type of fault (signal vector component). On the other hand, the features that determine the condition of an object in the undamaged state are called parameters. The diagnostic parameter is always related to the observable description of the diagnosed object by diagnostic signals (processes). It indirectly determines the values of the features of the object state [13]. The main stage of developing a decision in the diagnostic process is the diagnostic conclusion. The decision, the response of the diagnostic system, reflects the symptom-state diagnosis relationship.

Symptoms and faults (defect, or damage) are conditional and decision attributes, respectively. Let the set $S = \{s_1, s_2, ..., s_l, ..., s_n\}$ be a set of symptoms, and $F = \{f_1, f_2, ..., f_l, ..., f_k\}$ be a set of faults.

The **diagnostic matrix** is r_{ij} for the value defined by the function $r :$ $S \times F \rightarrow V$, where V is the set of data which are the values: $r_{ij} = r(s_i, f_j), i = 1, 2, ..., k, j = 1, 2, ..., n$. This data is data on diagnostic observations, symptom and faults observations, respectively.

The decision table is a special case of a diagnostic matrix. The books [13, 14] present a method of representing diagnostic knowledge in diagnostic matrices. The r_{ij} diagnostic matrix for the symptom values (Table 1) can be defined by the function $r : S \times F \rightarrow V$, where V is the set of data which are the symptom values for the given diagnostic condition: $r_{ij} = r(s_i, f_j)$.

The presence of a symptom value during observation does not mean that there is a relationship between that symptom and fault. If such a relationship

Table 1. Schema of the diagnostic matrix.

$S\backslash F$	$f_1, f_2, ..., f_i, ..., f_k$
s_1	$r_{11}, r_{12}, ..., r_{1i}, ..., r_{1k}$
s_2	$r_{21}, r_{22}, ..., r_{2i}, ..., r_{2k}$
...	...
s_l	$r_{l1}, r_{l2}, ..., r_{li}, ..., r_{lk}$
...	...
s_n	$r_{n1}, r_{n2}, ..., r_{ni}, ..., r_{nk}$

is known, it is the relation $R \subseteq S \times F$, which defines **the binary diagnostic matrix**:

$$r_{ij} = r(s_i, f_j) = \begin{cases} 0, & \text{when } (s_i, f_j) \notin R, \\ 1, & \text{when } (s_i, f_j) = R. \end{cases} \tag{1}$$

Let the set X be the observation space (range) of symptoms s_i and faults f_j. Some **data on these observations are fuzzy** then:

1. r_{ij} are linguistic variables with fuzzy values, and the diagnostic matrix is a matrix of fuzzy sets:

$$r_{ij} = r(s_i, f_j) = \{\langle o, r_{ij}(o)\rangle : o \in X, r_{ij} : X \to [0, 1]\}. \tag{2}$$

2. the relation R after positive tests [22] is binary and some of the symptoms and faults are fuzzy, i.e. the diagnosis (s_i, f_j) corresponds to the pair (μ_i, v_j) of fuzzy sets $\mu_i : X \to [0, 1], v_j : X \to [0, 1]$, and since the observations of symptoms and faults are disjoint, the values of the functions μ_i, v_j, analogically to their frequencies of occurrence, satisfy the condition: $\mu_i(o) + v_j(o) \leq 1$.
   This means that (μ_i, v_j) are the intuitionistic fuzzy sets (*IFSs*) in the sense of [24].
3. Since for the binary relation R, for each diagnosis interpretation, for any symptom is assigned one fault or no fault, then it is correct to assume that $X = S$.

In the further part of this paper, the application meeting points 2. and 3. will be considered.

3   n-Pythagorean Fuzzy Set and Yager Aggregation Operators

This chapter briefly presents the conceptual apparatus introduced in [2].

Definition 1. Further, only triangular, continuous t-norms and s-norms will be considered. Then, the general discussion on the construction of triangular norms using the results of functional equations leads to the theorems given in

[25], according to which: for each continuous t-norm $\bullet_t$ there is a continuous and strictly decreasing function f_t and for each continuous s-norm $\bullet_s$ there is a continuous and strictly increasing function f_s. Functions f_t, f_s are called t-norm and s-norm **generators**, respectively.

Example 1. For the t-norm $x \bullet_t y = \min\{x, y\}$ and any $x, y \in [0, 1]$, the generator is $f_t(x) = 1 - x$.

For the s-norm $x \bullet_s y = \max\{x, y\}$ and any $x, y \in [0, 1]$, the generator is $f_s(x) = x$.

Example 2. For the t-norm $x \bullet_t y = 1 - \min\{1, ((1 - x)^p + (1 - y)^p)^{1/p}\}, p \geq 1$ and any $x, y \in [0, 1]$, the generator is $f_t(x) = 1 - x^p$.

For the s-norm $x \bullet_s y = \min\{1, (x^p + y^p)^{1/p}\}, p \geq 1$ and any $x, y \in [0, 1]$, the generator is $f_s(x) = x^p$.

Definition 2. Operations $\bullet_p, \bullet_l$ specified below will be called respectively the **p-norm** (with properties similar to power) and the **l-norm** (with properties similar to linear functions), and the system $S_{Yager} = \langle [0, 1], \bullet_t, \bullet_s, \bullet_p, \bullet_l, 0, 1 \rangle$ will be called the **Yager system of triangular norms**.

The following notation agreement is accepted when it does not lead to any misunderstandings: $\lambda \bullet_p x = \lambda f_t(x) = x^\lambda$ and $\lambda \bullet_l x = \lambda f_s(x) = \lambda x$.

Definition 3. Let F be the set of all fuzzy sets on a non-empty space X. Any function $p : X \to [0, 1] \times [0, 1]$, defined for any $\mu_p, v_p \in F$ and a natural number $n > 0$, as in the theory fuzzy sets, is: $p = \{\langle x, \langle \mu_p(x), v_p(x) \rangle \rangle : x \in X\}$.

It is called the **n-Pythagorean fuzzy set** (*n-PFS*) if the following condition is satisfied:

$$0 \leq (\mu_p(x))^n + (v_p(x))^n \leq 1, \text{ for any } x \in X. \tag{3}$$

Let *n-PFS* be a set of the n-Pythagorean fuzzy set.

The fuzzy sets μ_p, v_p indicate the membership and non-membership. Zhang and Xu [26] considered $p(x) = \langle \mu_p(x), v_p(x) \rangle$ as the n-Pythagorean fuzzy number (*n-PFN*) represented by $p = \langle \mu_p(x), v_p(x) \rangle$. The notation is used:

$$\textbf{\textit{n-PFN}} =_{df} \{\langle \mu, v \rangle \in [0, 1] \times [0, 1] : 0 \leq \mu^n + v^n \leq 1\}. \tag{4}$$

Fact 3.1. $\textbf{\textit{n-PFN}} = \{p(x) : x \in X, p \in \textit{n-PFS}\}.$

When $n = 1$, then the 1-Pythagorean fuzzy sets are the intuitionistic fuzzy sets (*IFS*), which were studied by Atanassov [24,27,28], and the 2-Pythagorean fuzzy sets are the *PFS* of Yager [29–31].

From simple arithmetic properties of $0 \leq \mu^{n+1} + v^{n+1} \leq \mu^n + v^n \leq 1$ it follows:

Theorem 1. *For any natural number* $n > 1, \textbf{\textit{n-PFN}} \subseteq \textbf{\textit{(n+1)-PFN}} \subseteq [0, 1] \times [0, 1].$

Hence, entering a power scale for membership and non-membership allows to replace $\langle \mu, v \rangle \in [0,1] \times [0,1]$, such that $\mu + v > 1$, with $\langle \mu^n, v^n \rangle \in \boldsymbol{1\text{-}PFN}$, for some n. Thus, the aggregation operations on $\boldsymbol{IFS}$ can be extended to the aggregation operations on $n\text{-}PFS$.

Theorem 2. *In any system* $\boldsymbol{S_{Yager}} = \langle [0,1], \bullet_t, \bullet_s, \bullet_p, \bullet_l, 0, 1 \rangle$, *for any* $\langle \mu_1, v_1 \rangle$, $\langle \mu_2, v_2 \rangle \in \boldsymbol{1\text{-}PFN}$ *and any number* $\lambda \in [0,1]$, *the following conditions are satisfied:*

1. $\langle \mu_1 \bullet_s \mu_2, v_1 \bullet_t v_2 \rangle \in \boldsymbol{1\text{-}PFN}$,
2. $\langle \mu_1 \bullet_t \mu_2, v_1 \bullet_s v_2 \rangle \in \boldsymbol{1\text{-}PFN}$,

Additionally, in some systems $\boldsymbol{S_{Yager}}$, *the following conditions are satisfied:*

3. $\langle \lambda \bullet_l \mu_1, \lambda \bullet_p v_1 \rangle \in \boldsymbol{1\text{-}PFN}$,
4. $\langle \lambda \bullet_p \mu_1, \lambda \bullet_l v_1 \rangle \in \boldsymbol{1\text{-}PFN}$.

Proof. As in paper [2].

Theorem 3. *Let in the system* $\boldsymbol{S_{Yager}} = \langle [0,1], \bullet_t, \bullet_s, \bullet_p, \bullet_l, 0, 1 \rangle$, *conditions 1–4 of Theorem 2 be satisfied. Then, for any natural number* $n \rangle 1$, *for any* $\langle \mu_1, v_1 \rangle, \langle \mu_2, v_2 \rangle \in \boldsymbol{n\text{-}PFN}$ *and any number* $\lambda \in [0,1]$ *the following conditions are fulfilled:*

1. $\langle (\mu_1^n \bullet_s \mu_2^n)^{1/n}, (v_1^n \bullet_t v_2^n)^{1/n} \rangle \in \boldsymbol{n\text{-}PFN}$,
2. $\langle (\mu_1^n \bullet_t \mu_2^n)^{1/n}, (v_1^n \bullet_s v_2^n)^{1/n} \rangle \in \boldsymbol{n\text{-}PFN}$,
3. $\langle (\lambda \bullet_l \mu_1^n)^{1/n}, (\lambda \bullet_p v_1^n)^{1/n} \rangle \in \boldsymbol{n\text{-}PFN}$,
4. $\langle (\lambda \bullet_p \mu_1^n)^{1/n}, (\lambda \bullet_l v_1^n)^{1/n} \rangle \in \boldsymbol{n\text{-}PFN}$.

Proof. $\langle \mu_1, v_1 \rangle, \langle \mu_2, v_2 \rangle \in \boldsymbol{n\text{-}PFN}$ iff $\langle \mu_1^n, v_1^n \rangle, \langle \mu_2^n, v_2^n \rangle \in \boldsymbol{1\text{-}PFN}$.

Then, the conditions of Theorem 2 equivalent to the above conditions 1–4 are satisfied.

Definition 4. In the system $\boldsymbol{S_{Yager}} = \langle [0,1], \bullet_t, \bullet_s, \bullet_p, \bullet_l, 0, 1 \rangle$, the following aggregation operators are defined as **Yager operators on** $\boldsymbol{n\text{-}PFN}$, for any $\langle \mu_1, v_1 \rangle, \langle \mu_2, v_2 \rangle \in \boldsymbol{n\text{-}PFN}$ and any number $\lambda \in [0,1]$.

When conditions 1–4 of Theorem 2 are satisfied, then:

1. $\langle \mu_1, v_1 \rangle \oplus \langle \mu_2, v_2 \rangle = \langle (\mu_1^n \bullet_s \mu_2^n)^{1/n}, (v_1^n \bullet_t v_2^n)^{1/n} \rangle$,
2. $\langle \mu_1, v_1 \rangle \otimes \langle \mu_2, v_2 \rangle = \langle (\mu_1^n \bullet_t \mu_2^n)^{1/n}, (v_1^n \bullet_s v_2^n)^{1/n} \rangle$,
3. $\lambda \langle \mu_1, v_1 \rangle = \langle (\lambda \bullet_l \mu_1^n)^{1/n}, (\lambda \bullet_p v_1^n)^{1/n} \rangle$,
4. $\langle \mu_1, v_1 \rangle^\lambda = \langle (\lambda \bullet_p \mu_1^n)^{1/n}, (\lambda \bullet_l v_1^n)^{1/n} \rangle$.
5. When $\langle (\lambda \bullet_l \mu_1^n)^{1/n}, (\lambda \bullet_p v_1^n)^{1/n} \rangle \notin \boldsymbol{n\text{-}PFN}$, then $\lambda \langle \mu_1, v_1 \rangle = \langle 1, 0 \rangle$,
6. When $\langle (\lambda \bullet_p \mu_1^n)^{1/n}, (\lambda \bullet_l v_1^n)^{1/n} \rangle \notin \boldsymbol{n\text{-}PFN}$, then $\langle \mu_1, v_1 \rangle^\lambda = \langle 0, 1 \rangle$.

Hence, in any system $\boldsymbol{S_{Yager}} = \langle [0,1], \bullet_t, \bullet_s, \bullet_p, \bullet_l, 0, 1 \rangle$, and based on [2], there is:

Theorem 4. *For any* $\langle \mu_1, v_1 \rangle, \langle \mu_2, v_2 \rangle, \langle \mu_3, v_3 \rangle \in \boldsymbol{n\text{-}PFN}$ *and any number* $\lambda \in [0,1]$:

1. $\langle \mu_1, v_1 \rangle \oplus \langle 0, 1 \rangle = \langle \mu_1, v_1 \rangle, \langle \mu_1, v_1 \rangle \oplus \langle 1, 0 \rangle = \langle 1, 0 \rangle,$
2. $\langle \mu_1, v_1 \rangle \oplus \langle \mu_2, v_2 \rangle = \langle \mu_2, v_2 \rangle \oplus \langle \mu_1, v_1 \rangle,$
3. $(\langle \mu_1, v_1 \rangle \oplus \langle \mu_2, v_2 \rangle) \oplus \langle \mu_3, v_3 \rangle = \langle \mu_1, v_1 \rangle \oplus (\langle \mu_2, v_2 \rangle \oplus \langle \mu_3, v_3 \rangle),$
4. $\langle \mu_1, v_1 \rangle \otimes \langle 1, 0 \rangle = \langle \mu_1, v_1 \rangle, \langle \mu_1, v_1 \rangle \otimes \langle 0, 1 \rangle = \langle 0, 1 \rangle,$
5. $\langle \mu_1, v_1 \rangle \otimes \langle \mu_2, v_2 \rangle = \langle \mu_2, v_2 \rangle \otimes \langle \mu_1, v_1 \rangle,$
6. $(\langle \mu_1, v_1 \rangle \otimes \langle \mu_2, v_2 \rangle) \otimes \langle \mu_3, v_3 \rangle = \langle \mu_1, v_1 \rangle \otimes (\langle \mu_2, v_2 \rangle \otimes \langle \mu_3, v_3 \rangle),$
7. $\lambda(\langle \mu_1, v_1 \rangle \oplus \langle \mu_2, v_2 \rangle) = \lambda \langle \mu_1, v_1 \rangle \oplus \lambda \langle \mu_2, v_2 \rangle,$
8. $(\langle \mu_1, v_1 \rangle \otimes \langle \mu_2, v_2 \rangle)^\lambda = \langle \mu_1, v_1 \rangle^\lambda \otimes \langle \mu_2, v_2 \rangle^\lambda,$
9. $(\lambda_1 + \lambda_2)\langle \mu_1, v_1 \rangle = \lambda_1 \langle \mu_1, v_1 \rangle \oplus \lambda_2 \langle \mu_1, v_1 \rangle,$
10. $\langle \mu_1, v_1 \rangle^{\lambda_1 + \lambda_2} = \langle \mu_1, v_1 \rangle^{\lambda_1} \otimes \langle \mu_1, v_1 \rangle^{\lambda_2}.$

Any set $n\text{-}\boldsymbol{PFN}$ can be ordered by the relations $\leq_n$ defined as follows:

Definition 5. For any $\langle \mu_1, v_1 \rangle, \langle \mu_2, v_2 \rangle \in n\text{-}\boldsymbol{PFN}$, $\langle \mu_1, v_1 \rangle \leq_n \langle \mu_2, v_2 \rangle$ iff $\mu_1 \leq \mu_2, v_1 \geq v_2$. The results of the maximum and minimum operation on any $A \subseteq n\text{-}\boldsymbol{PFN}$, determined for the relation $\leq_n$, are denoted by $\max_n A$ and $\min_n A$.

Fact 3.2. *For any* $x, y \in [0, 1]$,

- $\langle 0, 1 \rangle \leq_n \langle 0, x \rangle \leq_n \langle 0, 0 \rangle \leq_n \langle y, 0 \rangle \leq_n \langle 1, 0 \rangle,$
- $\langle 0, 1 \rangle \leq_n \langle x, y \rangle \leq_n \langle 1, 0 \rangle,$ *when* $\langle x, y \rangle \in n\text{-}\boldsymbol{PFN}$.
- $\langle 0, 1 \rangle = \min_n n\text{-}\boldsymbol{PFN}, \langle 1, 0 \rangle = \max_n n\text{-}\boldsymbol{PFN}.$

Fact 3.3. $n\text{-}\boldsymbol{PFN} = \{\langle x, y \rangle \in [0, 1] \times [0, 1] : \langle 0, 1 \rangle \leq_n \langle x, y \rangle \leq_n \langle 1, 0 \rangle\}.$

Summarizing the introduced knowledge about operations and relations in $n\text{-}\boldsymbol{PFN}$, the following operations and inclusion relation can be defined for $n\text{-}PFS$:

Definition 6. For any $p_1, p_2 \in n\text{-}PFS$, and any number $\lambda \in [0, 1]$:

- $p_1 \oplus p_2 = \{\langle x, \langle \mu_{p_1}(x), v_{p_1}(x) \rangle \oplus \langle \mu_{p_2}(x), v_{p_2}(x) \rangle \rangle : x \in X\},$
- $p_1 \otimes p_2 = \{\langle x, \langle \mu_{p_1}(x), v_{p_1}(x) \rangle \otimes \langle \mu_{p_2}(x), v_{p_2}(x) \rangle \rangle : x \in X\},$
- $\lambda_{p_1} = \{\langle x, \lambda \langle \mu_{p_1}(x), v_{p_1}(x) \rangle \rangle : x \in X\},$
- $p_1^\lambda = \{\langle x, \langle \mu_{p_1}(x), v_{p_1}(x) \rangle^\lambda \rangle : x \in X\},$
- $p_1 \subseteq_n p_2$ iff for any $x \in X, \langle \mu_{p_1}(x), v_{p_1}(x) \rangle \leq_n \langle \mu_{p_2}(x), v_{p_2}(x) \rangle.$

The $n\text{-}PFS$ system with the operations and including relation described in Definition 6, will be called an $n\text{-}PFS$ **algebra**.

Let for any $x \in n\text{-}PFS, \mathbf{1}(x) =_{df} \langle 0, 1 \rangle, \mathbf{0}(x) =_{df} \langle 1, 0 \rangle$. From Definition 6 and Theorem 4 it follows that:

Theorem 5. *In the* $n\text{-}PFS$ *algebra, for any* $p_1, p_2, p_3 \in n\text{-}PFN$ *and any number* $\lambda \in [0, 1]$:

1. $p_1 \oplus \mathbf{0} = p_1, p_1 \oplus \mathbf{1} = \mathbf{1},$
2. $p_1 \oplus p_2 = p_2 \oplus p_1,$
3. $(p_1 \oplus p_2) \oplus p_3 = p_1 \oplus (p_2 \oplus p_3),$
4. $p_1 \otimes \mathbf{1} = p_1, p_1 \otimes \mathbf{0} = \mathbf{0},$
5. $p_1 \otimes p_2 = p_2 \otimes p_1,$
6. $(p_1 \otimes p_2) \otimes p_3 = p_1 \otimes (p_2 \otimes p_3),$
7. $\lambda(p_1 \oplus p_2) = \lambda p_1 \oplus \lambda p_2,$
8. $(p_1 \otimes p_2)^\lambda = p_1^\lambda \otimes p_2^\lambda,$
9. $(\lambda_1 + \lambda_2)p_1 = \lambda_1 p_1 \oplus \lambda_2 p_1,$
10. $p_1^{\lambda_1 + \lambda_2} = p_1^{\lambda_1} \otimes p_1^{\lambda_2}.$

4   Technical Diagnosis Granule System

This chapter analyzes diagnostic data regarding an exemplary gas boiler, based on the doctoral dissertation by Bryniarska. The diagnostic conceptual apparatus from Sect. 2 and its interpretation in the n-Pythagorean fuzzy sets theory are used.

It is assumed that the diagnosis: symptom-fault (s_i, f_j) corresponds to the pair (μ_i, v_j) of the fuzzy sets $\mu_i : X \to [0, 1], v_j : X \to [0, 1]$. Since the observations of the symptoms and faults are disjoint, the values of the functions μ_i, v_j on the scale of the n-th power satisfy the condition: $\mu_i(o)^n + v_j(o)^n \leq 1$.

This means that (μ_i, v_j) are the n-Pythagorean fuzzy sets $(PFSs)$ in sense of [2]. Such n-Pythagorean fuzzy set will also be referred to as the **technical diagnosis granule**. Since, for the binary relation R of each interpretation of the diagnosis, for any symptom is assigned one fault or none, it is correct to assume that $X = S$.

The number n of the scale power is searched as follows: for each pair (μ, v) of the function values μ_i, v_j there is the smallest natural number n such that $\mu^{n-1} + v^{n-1} > 1$ and $\mu^n + v^n \leq 1$. Then, $(\mu, v) \in \boldsymbol{n\text{-}PFS}$. From among the numbers n determined in this way, the largest one is chosen.

The largest number n for which all $(\mu, v) \in \boldsymbol{n\text{-}PFS}$ is $n = 4$. This means that the technical symptom-fault diagnosis of the gas boiler can most widely be interpreted using the 4-Pythagorean fuzzy sets. Furthermore, the **granulation of technological diagnosis in the algebra of 4-Pythagorean fuzzy sets** will be considered.

Table 2 shows the binary diagnostic matrix $r_{ij} = r(s_i, f_j)$ of the symptom-fault relationship of the gas boiler, which are gathered in the Table 3.

It is assumed that: if $r_{ij}(s_i, f_j) = 0$, then symptom s_i or fault f_j occurs at the level 0 of their detection during the diagnosis: $(0, 1), (0, 0), (1, 0)$, i.e., there is no symptom and there is a fault, or there is neither symptom nor fault, or there is a symptom and no fault. When $r_{ij}(s_i, f_j) = 1$ and $s_i = \mu, f_j = v$, because $(0, 1), (0, 0), (1, 0), (\mu, v) \in \boldsymbol{4\text{-}PFS}$, then there exists the 4-Pythagorean fuzzy set p such that $p_{ij}(s_k) = (\mu, v)$ for $k = i$, and $p_{ij}(s_k) = (0, 1)$, for $k \notin i$. Such a 4-Pythagorean fuzzy set will be referred to as the **elementary granule of technical diagnosis**. The diagnosis: *you can smell gas with the degree 0.9 so there is a leakage on the boiler with the degree 0.4* is described by the elementary granule $p_{5,10}(s_5) = (0, 9, 0, 4), p_{5,10}(s_k) = (0, 1)$, for $k \notin 5$. It can be shown that any technical diagnosis granule in the *4-PFS* algebra is the sum $\oplus$ of some elementary granules.

Let the symptom descriptions and the descriptions (s_i, f_j) of the diagnosis belong to the L_{diag} language of diagnosis. The description of the diagnosis will be called the **elementary concept**.

The granulation of technological diagnosis in the algebra of n-Pythagorean fuzzy sets is $\boldsymbol{Gr} = \langle 4\text{-}PFS,^{Gr} \rangle$, where $^{Gr} : L_{diag} \to L_{4\text{-}PFS}$ and $L_{4\text{-}PFS}$ is a language of the *4-PFS* algebra. This language L_{diag} also includes concepts which are subsets of $S \times F$. Using the symbols explained in Definition 6, the meaning of L_{diag} expressions is determined by the granularity $\boldsymbol{Gr}$ as follows:

Table 2. Relation of symptoms with faults, degree of their fuzziness and determination of n-PFS.

Symptom	Fault	Number n for n-PFS
(s1) The boiler does not turn on = 0,8	(f1) No power (fuse problem, short circuit) = 0,4	2
	(f2) Boiler protection triggered = 0,8	4
	(f3) Gas meter defective = 0,4	2
	(f4) No gas supply (or low gas pressure) - gas regulator or high power gas valve = 0,4	2
	(f5) No water in the system or boiler (or low pressure) = 0,6	2
(s2) Noises, crackles, strange sounds come from the boiler = 0,5	(f6) Circulation pump defective = 0,4	1
(s3) The boiler turns itself off = 0,8	(f7) Badly made chimney = 0,8	4
	(f8) Air in the radiators = 0,4	2
(s4) The boiler works in the CH mode, but not in the DHW mode = 0,5	(f9) Air in the boiler = 0,9	3
(s5) You smell gas = 0,9	(f10) Leakage on the boiler = 0,4	2
(s6) Hot case of boiler = 0,7	(f11) Defective gas regulator = 0,4	2
(s7) Audible or visible spark, but burner does not ignite = 0,9	(f12) Faulty electronics (electronic module, electronic parts) = 0,4	2
	(f13) Ignition electrode defective = 0,4	2
(s8) The burner lights up and goes out after a few seconds = 0, 8	(f14) Damaged or contaminated water system = 0,6	3
	(f15) Defective ionization electrode = 0,8	4
(s9) The burner goes out after a long time = 0,6	(f16) Boiler overheating = 0,4	1
	(f17) Defective gas unit = 0,4	1
(s10) The burner is working but there is a small flame 0,7	(f18) Clogged filters, nozzles or burner components = 0,4	2
(s11) Poor boiler efficiency (CH) = 0,5	(f19) Heating water temperature sensor defective = 0,6	2
	(f20) Clogged, dirty CH water filter = 0,8	2
(s12) No hot water or too low DHW temperature = 0,4	(f21) Clogged, dirty DHW water filter = 0,8	2
	(f22) Dirty or damaged heater = 0,4	1
	(f23) Dirty secondary heat exchanger = 0,6	1

for any $x \in S$, and any concept names C, D:

- **G0.** *Granulation of the elementary concept:* $(s_i, f_j)^{Gr} = p_{ij}$, when $r_{ij}(s_i, f_j) = 1$; and $(s_i, f_j)^{Gr} = \mathbf{0}$, when $r_{ij}(s_i, f_j) = 0$;
- **G1.** *Granulation of the concept:* $\{(x_1, y_1), (x_2, y_2), ..., (x_k, y_k)\}$: $\{(x_1, y_1), (x_2, y_2), ..., (x_k, y_k)\}^{Gr} = (x_1, y_1)^{Gr} \oplus (x_2, y_2)^{Gr} \oplus ... \oplus (x_k, y_k)^{Gr}$
- **G2.** *Granulation of the full concept:* $\top^{Gr}(x) = \langle 0, 1 \rangle$, where $\top = S \times F$;
- **G3.** *Granulation of the empty concept:* $\bot^{Gr}(x) = \langle 0, 1 \rangle$, where $\bot = \varnothing$;
- **G4.** *Granulation of the concept repeatability in the alternative of concepts with the degree:* $\lambda \in [0, 1] : (\lambda C)^I(x) = (\lambda C^{Gr})(x)$;
- **G5.** *Granulation of the concept repeatability in the conjunction of concepts with the degree:* $\lambda \in [0, 1] : (C^\lambda)^I(x) = (C^{Gr})^\lambda(x)$;
- **G6.** *Granulation of concept conjunctions:* $(C \sqcap D)^{Gr}(x) = (C^{Gr} \otimes D^{Gr})(x)$;

Table 3. Binary symptom-fault relationship of a gas boiler.

r_{ij}	f_1	f_2	f_3	f_4	f_5	f_6	f_7	f_8	f_9	f_{10}	f_{11}	f_{12}	f_{13}	f_{14}	f_{15}	f_{16}	f_{17}	f_{18}	f_{19}	f_{20}	f_{21}	f_{22}	f_{23}
s_1	1	1	1	1	1	0	0	0	0	0	0	0	0	0	0	0	0	0	0	0	0	0	0
s_2	0	0	0	0	0	1	0	0	0	0	0	0	0	0	0	0	0	0	0	0	0	0	0
s_3	0	0	0	0	0	0	1	1	0	0	0	0	0	0	0	0	0	0	0	0	0	0	0
s_4	0	0	0	0	0	0	0	0	0	1	0	0	0	0	0	0	0	0	0	0	0	0	0
s_5	0	0	0	0	0	0	0	0	0	0	1	0	0	0	0	0	0	0	0	0	0	0	0
s_6	0	0	0	0	0	0	0	0	0	0	0	1	0	0	0	0	0	0	0	0	0	0	0
s_7	0	0	0	0	0	0	0	0	0	0	0	0	1	1	0	0	0	0	0	0	0	0	0
s_8	0	0	0	0	0	0	0	0	0	0	0	0	0	1	1	0	0	0	0	0	0	0	0
s_9	0	0	0	0	0	0	0	0	0	0	0	0	0	0	1	1	0	0	0	0	0	0	0
s_{10}	0	0	0	0	0	0	0	0	0	0	0	0	0	0	0	0	1	0	0	0	0	0	0
s_{11}	0	0	0	0	0	0	0	0	0	0	0	0	0	0	0	0	0	1	1	0	0	0	0
s_{12}	0	0	0	0	0	0	0	0	0	0	0	0	0	0	0	0	0	0	0	1	1	1

- **G7.** *Granulation of concept alternatives*: $(C \sqcup D)^{Gr}(x) = (C^{Gr} \oplus D^{Gr})(x)$;
- **G8.** *Granulation of concepts inclusion*: $(C \sqsubseteq D)^{Gr} = "C^{Gr} \leq_n D^{Gr"}$,
- **G9.** *Granulation of concept equality*: $(C = D)^{Gr} = "C^{Gr} = D^{Gr"}$,

The concepts $\lambda C, C^{\lambda}$ are concepts repeated respectively in the alternative and conjunction with parallel or linear repetition of the C concept diagnosis. It is caused by the uncertainty, inaccuracy or imprecision of the results of this diagnosis. The number $\lambda \in (0, 1]$ is the degree of diagnosis that requires this repetition. When $\lambda = 1$, then the diagnosis is certain and does not need to be repeated. If the diagnosis can be repeated a maximum of 10 times, then $\lambda = 1 - i/10$, where i is the number repetitions, i.e. $\lambda = 1, 9/10, 8/10, 7/10, 6/10, 5/10, 4/10, 3/10, 2/10, 1/10, 0$. The concepts inclusion $C \sqsubseteq D$ can be understood intuitively as a relations of the diagnosis preference relations D on C.

When the granulation $\boldsymbol{Gr}$ meets the conditions **G0 – G9**, it is called the **concept fuzzification**. If, as a result of fuzzification, only a few characteristic functions on the space S, belonging to the *n-PFS* algebra, are obtained for all concepts, then such granulation is called **exact**.

5   Conclusion

This paper presents the first application of the n-Pythagorean fuzzy sets theory and the scale algorithm. In order to show this application, the diagnosis language L_{diag} has been proposed with an interpretation in the *n-PFS* algebra. The concept of diagnosis is limited to retrieving knowledge about the symptom-fault system, assuming that the relation r linking symptoms and faults is binary, and the faults are uniquely assigned to symptoms.

In the future, it is important to analyze more examples of the technical diagnosis granulation. There is also a problem to solve concerning determination of the granulation of such a diagnosis in which the relation r will be fuzzy.

The conceptual apparatus can also be used in information technology, for example to formulate the theory of diagnosis of information retrieval on the

Web [32]. Another application may be to develop the domain of computer science referred to here as scale algorithm.

Appendix

Symbols used in paper:

$F(f_1, ..., f_k)$ – faults, $S(s_1, ..., s_n)$ – symptoms;
$V(r_{11}, ..., r_{ij})$ – the diagnostic matrix;
R – the relation between symptom and fault, the binary diagnostic matrix;
X – the observation space;
o – an observation;
μ_i, v_i – fuzzy sets;
$\bullet_t$ – the t-norm; $\bullet_s$ – the s-norm;
$\bullet_p$ – the p-norm (with properties similar to power);
$\bullet_l$ – the l-norm (with properties similar to linear functions);
f_t – the t-norm generator, function; f_s – the s-norm generator, function;
$\lambda \in [0, 1]$ – any number;
$p_1, p_2, ...$ – elements of the n-PFS system;
L_{diag} – the language of diagnosis;
$\boldsymbol{Gr}$ – the granulation of technological diagnosis in the algebra of n-PFS.

References

1. Pedrycz, W.: Allocation of information granularity in optimization and decision-making models: towards building the foundations of granular computing. Eur. J. Oper. Res. **232**(1), 137–145 (2014)
2. Bryniarska, A.: The n-Pythagorean fuzzy sets. Symmetry **12**, 1772 (2020). https://doi.org/10.3390/sym12111772
3. Bryniarska, A.: Certain information granule system as a result of sets approximation by fuzzy context. Int. J. Approx. Reason. **111**, 1–20 (2019). https://doi.org/10.1016/j.ijar.2019.04.012
4. Baumeister J.: Agile Development of Diagnostic Knowledge Systems. infix, Akademische Verlagsgesellschaft Aka GmbH, Berlin (2004)
5. Belard, N., Pencole, Y., Combacau, M.: A theory of meta-diagnosis: reasoning about diagnostic systems. In: Twenty-Second International Joint Conference on Artificial Intelligence (IJCAI 2011), pp. 731–737 (2011)
6. Himmelblau, D.M.: Fault Detection and Diagnosis in Chemical and Petrochemical Processes. Elsevier, Amsterdam (1978)
7. Reiter, R.: A theory of diagnosis from first principles. Artif. Intell. **32**(1), 57–96 (1987)
8. Basseville, M., Nikiforov, I.: Detection of Abrupt Changes - Theory and Application. Prentice-Hall, Englewood Cliffs (1993)
9. Gertler, J.: Fault Detection and Diagnosis in Engineering Systems. CRC Press, Boca Raton (1998)
10. Chen, J., Patton, R.: Robust Model Based Fault Diagnosis for Dynamic Systems. Kluver Academic Publishers, Boston (1999)

11. Blanke, M., Kinnaert, M., Lunze, J., Staroswiecki, M.: Diagnosis and Fault-Tolerant Control. Springer, Berlin (2003)
12. Simani, S., Patton, R., Fantuzzi, C.: Model-Based Fault Diagnosis in Dynamic Systems Using Identification Techniques. Springer-Verlag New York Inc., Secaucus (2003)
13. Korbicz, J., Kościelny, J., Kowalczuk, Z., Cholewa, W.: Fault Diagnosis. Models, Artificial Intelligence, Applications. Springer, Heidelberg (2004)
14. Korbicz, J., Kowal, M. (ed.): Intelligent Systems in Technical and Medical Diagnostics. Springer, Heidelberg (2014)
15. Witczak, M.: Modelling and Estimation Strategies for Fault Diagnosis of Non-Linear Systems. From Analytical to Soft Computing Approaches. Springer, Berlin (2007)
16. Boem, F., Ferrari, R.M.G., Parisini, T., Polycarpou, M.M.: Distributed fault diagnosis for nonlinear systems. Preprints of the 8th IFAC Symposium on Fault Detection, Supervision and Safety of Technical Processes, pp. 1089–1094 (2012)
17. Czichos, H. (ed.): Handbook of Technical Diagnostics Fundamentals and Application to Structures and Systems, Springer, Heidelberg (2013)
18. Cholewa, W., Korbicz, J., Moczulski, W., Timofiejczuk, A.: Methods of signal analysis. In: Korbicz, J., Kowalczuk, Z., Kościelny, J.M., Cholewa, W. (eds.) Fault Diagnosis. Springer, Heidelberg (2004). https://doi.org/10.1007/978-3-642-18615-8_4
19. Kościelny, J.M., Syfert, M., Rostek, K., Sztyber, A.: Fault isolability with different forms of faults-symptoms relation. Int. J. Appl. Math. Comput. Sci. **26**(4), 815–826 (2016)
20. Bartyś, M.: Generalized reasoning about faults based on the diagnostic matrix. Int. J. Appl. Math. Comput. Sci. **23**(2), 407–417 (2013)
21. Kościelny, M.J., Bartys, M., Syfert, M.: Diagnostics of industrial processes in decentralised structures with application of fuzzy logic. In: Proceedings of the 17th World Congress the International Federation of Automatic Control Seoul, Korea, 6–11 July 2008 (2008)
22. Kościelny, J.M., Sztyber, A.: Decomposition of complex diagnostic systems. IFAC-PapersOnLine **51**(24), 755–762 (2018)
23. Moczulski, W.: Technical Diagnostics: Methods of Knowledge Acquisition, [in Polish]. OW Silesian University of Technology, Gliwice (2002)
24. Atanassov, K.T.: Intuitionistic fuzzy sets. Fuzzy Sets Syst. **20**, 87–96 (1986)
25. Aczel, J.: Lectures on Functional Equations and Their Applications. Academic Press, New York (1966)
26. Zhang, X., Xu, Z.: Extension of TOPSIS to multiple criteria decision making with Pythagorean fuzzy sets. Int. J. Intell. Syst. **29**, 1061–1078 (2014)
27. Atanassov, K.T.: On the modal operators defined over the intuitionistic fuzzy sets. Notes Intuit. Fuzzy Sets **10**(1), 7–12 (2004)
28. Atanassov, K.T.: On Intuitionistic Fuzzy Sets Theory. Springer, Heidelberg (2012)
29. Yager, R.R.: Aggregation operators and fuzzy systems modeling. Fuzzy Sets Syst. **67**, 129–145 (1994)
30. Yager, R.R.: Pythagorean fuzzy subsets. In: Proceedings of the 2013 Joint IFSAWorld Congress and NAFIPS Annual Meeting (IFSA/NAFIPS), Edmonton, AB, Canada, 2013, pp. 57–61 (2013)
31. Yager, R.R.: Pythagorean membership grades in multicriteria decision making. IEEE Trans. Fuzzy Syst. **22**, 958–965 (2014)
32. Bryniarska, A.: The auto-diagnosis of granulation of information retrieval on the web. Algorithms **13**, 264 (2020). https://doi.org/10.3390/a13100264

Automatic Kidney Volume Estimation System Using Transfer Learning Techniques

Chiu-Han Hsiao[1(✉)], Ming-Chi Tsai[2], Frank Yeong-Sung Lin[2], Ping-Cherng Lin[1], Feng-Jung Yang[3], Shao-Yu Yang[4], Sung-Yi Wang[2], Pin-Ruei Liu[2], and Yennun Huang[1]

[1] Research Center for Information Technology Innovation, Academia Sinica, Taipei, Taiwan
chiuhanhsiao@citi.sinica.edu.tw
[2] Department of Information Management, National Taiwan University, Taipei, Taiwan
[3] Department of Internal Medicine, National Taiwan University Hospital Yunlin Branch, Yunlin, Taiwan
[4] Department of Internal Medicine, National Taiwan University Hospital, Taipei, Taiwan

Abstract. Deep learning technology is widely used in medicine. The automation of medical image classification and segmentation is essential and inevitable. This study proposes a transfer learning–based kidney segmentation model with an encoder–decoder architecture. Transfer learning was introduced through the utilization of the parameters from other organ segmentation models as the initial input parameters. The results indicated that the transfer learning–based method outperforms the single-organ segmentation model. Experiments with different encoders, such as ResNet-50 and VGG-16, were implemented under the same Unet structure. The proposed method using transfer learning under the ResNet-50 encoder achieved the best Dice score of 0.9689. The proposed model's use of two public data sets from online competitions means that it requires fewer computing resources. The difference in Dice scores between our model and 3D Unet (Isensee) was less than 1%. The average difference between the estimated kidney volume and the ground truth was only 1.4%, reflecting a seven times higher accuracy than that of conventional kidney volume estimation in clinical medicine.

Keywords: Image segmentation · Transfer learning · Total kidney volume

1 Introduction

Total kidney volume (TKV) strongly affects kidney function in patients with autosomal dominant polycystic kidney disease or chronic kidney disease [1,2].

© The Author(s), under exclusive license to Springer Nature Switzerland AG 2021
L. Barolli et al. (Eds.): AINA 2021, LNNS 226, pp. 370–381, 2021.
https://doi.org/10.1007/978-3-030-75075-6_30

Thus, TKV is commonly used as an evaluation index of kidney function. Furthermore, TKV can be used to evaluate the treatment efficacy of potential therapeutic regimens [3]. Accurate TKV calculation is vital in disease staging and prognosis monitoring. With current kidney imaging methods, such as computed tomography (CT) and magnetic resonance imaging (MRI), doctors can calculate TKV manually [4]. However, current TKV calculation methods are time and labor intensive. Some details in medical images are difficult to observe with the naked eye, and calculation accuracy largely depends on a doctor's experience. Because tumorous kidneys have irregular shapes, automatic kidney volume calculation is challenging. Chagot *et al.* [5] proposed a TKV calculation method for ultrasound images based on the use of NEFROVOL software. Deep learning–based organ segmentation algorithms have substantially enhanced kidney segmentation performance [6]. Sharma *et al.* [7] adopted VGG-16 and a fully convolutional neural network (CNN) to compute TKV from CT images. Bazgir *et al.* [8] proposed a modified three-dimensional (3D) Unet model for kidney segmentation in MRI images. With the Kidney Tumor Segmentation 2019 (KiTS19) Challenge data set, Isensee *et al.* [9] used a 3D residual Unet and generated a Dice score of 97.37%, winning the team first place in the challenge. However, most 3D segmentation methods require considerable computing resources and training time. The method proposed in this paper uses an alternative approach to improve current kidney segmentation methods with transfer learning technology. A lightweight two-dimensional encoder–decoder architecture was established for automatic kidney volume calculation.

2   Methods

2.1   Transfer Learning

Transfer learning technology was adopted because the training data for kidney segmentation are limited. Transfer learning can transfer pretrained weights from one task to another; this method can be a practical solution when annotated data are limited. To enhance the segmentation performance of the model, the Liver Tumor Segmentation (LiTS) Challenge data set and a private data set (NTU) were the source data, and the KiTS data set was used as the target data set in this paper [10,11]. The liver data set was used as the source because liver segmentation, which focuses only on one region, is easier than kidney segmentation is. The proposed model can accurately detect organ features.

2.2   Image Processing

In this paper, the kidney tissue area in a CT image was segmented using the artificial neural network (ANN) technique. This technique has been used in several image recognition applications, the most basic of which is image classification. However, the purpose of this study was to make the neural network identify and highlight the kidney region when CT images are used as inputs, which constitutes image segmentation. In image classification, the most commonly used

ANN is a CNN. Instead of using all the pixels of a raw image as inputs, which increases complexity, CNNs use the convolutional layer, which identifies key features that might help to predict the highlighted region [12]. By using the pooling layer, a CNN can compress an image without losing substantial information from the original image, yielding a low-dimensional but more delicate image. Furthermore, in image segmentation, the upsampling layer is required before output to make the output image size match that of the input image. Many CNN models perform well in image segmentation. The main model used herein is the most popular one, namely Unet.

2.3   Data Format

2.3.1   NIfTI Format

Compared with the ANALYZE format, the NIfTI (.nii) format significantly improves image direction [13]. The anatomical coordinate system, also called the patient coordinate system, is the most crucial coordinate system in medical imaging areas. It is a continuous 3D space. The six directions, namely left, right, anterior, posterior, superior, and inferior relative to a standing patient, are measured along positive and negative axial directions in the coordinate system [13]. In Fig. 1, the positive direction of each axis in a coordinate system is usually represented by its first letter, such as "RAS" and "LAS". The NIfTI format uses a right-anterior-superior (RAS) coordinate system to store data. In other words, the right, anterior, and superior directions of a patient are positive. However, the original data might not use RAS coordinates. Thus, an affine transform was implemented using an affine matrix to convert the directions of an image into an RAS coordinate system. Each image file had a unique affine matrix recorded in the header of the NIfTI file; thus, the original directions of the images could be obtained from affine matrices in the headers. In the KiTS data set, the coordination system is left-posterior-inferior (LPI).

2.3.2   DICOM Format

Images in the private National Taiwan University (NTU) data set are in Digital Imaging and Communications in Medicine (DICOM) format (.dcm). DICOM is an international standard medical data format used to communicate, store, access, print, process, and illustrate medical images [14]. It was proposed by the American College of Radiology and National Electrical Manufacturers Association to resolve inconsistencies among manufacturers [13]. Unlike the NIfTI format, the DICOM format uses a left-posterior-superior (LPS) coordinate system. Based on the analysis of the 131 cases in LiTS dataset, cases 0–52 have the LAS orientation, which is needed to be flipped horizontally to make the orientation consistent. Cases 68–82 have RAS orientation, which is needed to be flipped horizontally and vertically to make the orientation consistent. The other cases have the desired orientation of LPS. A more detailed illustration is presented in Fig. 1 [13] and Fig. 2.

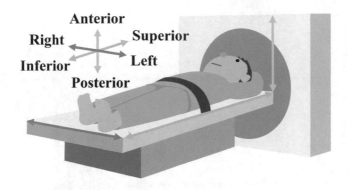

Fig. 1. Demonstration of the image directions.

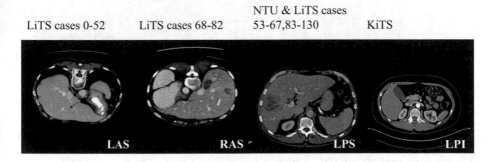

Fig. 2. Demonstration of the image direction of different data sets.

2.4   Unet

Unet has an encoder–decoder structure. In the encoder, an original input image is passed through multiple convolutional layers and pooling layers to extract the key features in the image and reduce its dimensionality. In the decoder, the compressed image passes through multiple upsampling layers, which enlarge it to its original size. However, enlarging an image does not add information to it; therefore, immediately before reaching the upsampling layer, the image concatenates with other images with the same dimensionality from the decoder outputs; this process adds some information from the original image as a reference during enlargement. Figure 3 presents an example of a basic Unet structure.

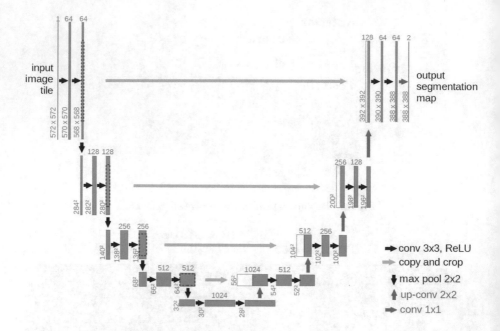

Fig. 3. Unet model structure.

3   Experiments and Results

3.1   Source Data Set: LiTS17 and NTU

The LiTS data set was retrieved from the LiTS Challenge on the Codalab website [11]. The challenge invites competitors to design an algorithm for liver tumor segmentation. The training data set contains data for 130 patients, and the validation data set contains data for 70 patients. Because only the training data set contains liver and liver tumor labels, the proposed model was trained and validated using only the challenge's training data set. The CT and annotation images in the data set have dimensions of 512×512 pixels. Details of the label meanings are listed as follows:

0: non-liver area
1: normal-liver area
2: liver-tumor area

The CT images were in NIfTI (.nii) format. Therefore, before training, the corresponding program was required to access the images or a specific program needed to be written to convert them to PNG format. In addition to the private NTU data set from NTU Hospital, Yunlin Branch (institutional review board number: 201801124RINB), were used in the experiments. The NTU data set

includes the data of 205 patients; most of the CT images have dimensions of
512 × 512 pixels, and the marking method is identical to that for the LiTS and
KiTS data. Although the NTU data set has both arterial and portal venous
phase information, only the portal venous phase was used for the experiment.

3.2   Target Data Set: KiTS19

The KiTS data set originated from the 2019 KiTS Challenge on the Grand Challenge website [10]. The data set consists of data from 300 patients, of which only
210 have labels. Therefore, only 210 cases were used for training and validation.
Most of the CT images in the data set have dimensions of 512 × 512 pixels, and
the marking method was identical to that used for the LiTS data set. The files
downloaded from the website were in Gnu Zip format (.gz); therefore, decompression was required to convert them into NIfTI format. The tumor areas were
regarded as normal regions to train the model (i.e., labels of 2 were modified
to 1). However, in some patients, tumors present outside the normal organ area
(e.g., on the organ surface). In addition, to present the irregularity of a tumor's
shape, a CT image might include multiple kidneys, resulting in failure to predict
a normal organ area. Consequently, normal organs and tumor areas were treated
separately.

3.3   Evaluation Metrics

The Dice score represents the accuracy of region-of-interest segmentation from
an image for performance evaluation, as expressed in (1). It is the ratio of two
times the correctly predicted area to the sum of the ground truth area (X)
and prediction area (Y). The Dice score, which ranges from 0 to 1, indicates
the similarity between the prediction and the ground truth. A Dice score of 1
indicates that the prediction completely matches the ground truth.

$$\text{Dice} = \frac{2 \times |X \cap Y|}{|X| + |Y|} \tag{1}$$

3.4   Implementation Parameters

Windowing using Hounsfield units (HUs) was conducted to filter unnecessary
information from the CT images. For the liver images, the window width was
200, and the window location was 150 [15]. For the kidney images, the window
width was 600, and the window location was 100, which were selected manually
to ensure that the window contained 95% of the kidney region. Binary cross
entropy was the loss function, and Adadelta was the optimizer with a learning
rate of 1.0. The batch size was 2 due to GPU memory limitation. All models
were trained for 100 epochs. A computer equipped Intel Core i5-9400F central
processing unit (CPU) with 64 GB of random access memory (RAM) and an
NVIDIA GeForce RTX 2070 SUPER graphics processing unit (GPU) with 8 GB
of RAM were used in the experiments.

3.5   Results

The proposed transfer learning framework is illustrated in Fig. 4. The model was trained with the LiTS17 and NTU training data set, and a liver segmentation model was established. The pretrained model weights were then set as initial parameters for the training of the kidney segmentation model with the KiTS19 data set. The expected function of the kidney segmentation model is presented in Fig. 5. The liver segmentation results are presented in Table 1 and Fig. 6. ResNet-50 outperformed VGG-16 in liver segmentation. Both VGG-16 and ResNet-50 were selected for accuracy evaluation experiments.

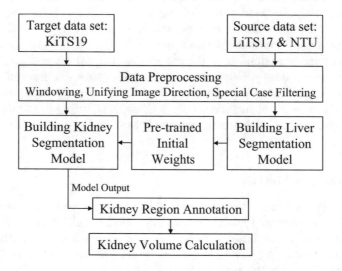

Fig. 4. Proposed experimental procedures.

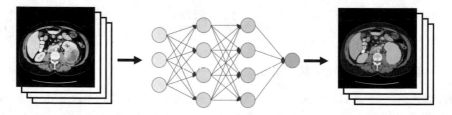

Fig. 5. Demonstration of kidney segmentation.

Table 1. Model performance in liver segmentation.

Encoder	Decoder	Dice per case	Recall per case	Precision per case
VGG-16	Unet	0.9463	0.9389	0.9611
ResNet-50	Unet	**0.9566**	0.9634	0.9502

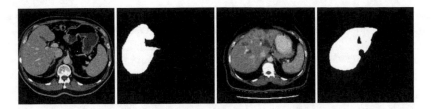

Fig. 6. The liver segmentation results.

The results of kidney segmentation using VGG-16 and ResNet-50 are presented in Table 2. The ResNet-50 encoder outperformed the VGG-16 encoder. The usage of transfer learning from the liver data set could enhance the Dice score by nearly 1% for both model architectures. As shown in Fig. 7, the segmentation model successfully annotated the kidney regions in abdomen CT images. In addition, the 3D demonstration of kidney segmentation indicated that the proposed model has high stability and consistency.

Table 2. Model performance in kidney segmentation.

Encoder	Decoder	Transfer learning	Data set	Dice per case	Recall per case	Precision per case
VGG-16	Unet		KiTS19	0.9271	0.9371	0.9233
VGG-16	Unet	V	Source: LiTS17 & NTU, Target: KiTS19	0.9345	0.9124	0.9664
ResNet-50	Unet		KiTS19	0.9590	0.9688	0.9516
ResNet-50	Unet	V	Source: LiTS17 & NTU, Target: KiTS19	**0.9689**	0.9742	0.9233

TKV is conventionally calculated as presented in (2) [16]. In the calculation, the kidneys are assumed to be approximately ellipsoid. However, the irregular shapes of tumorous kidneys make this equation unsuitable for calculating TKV, resulting in large errors.

$$\text{TKV} = \text{Length} \times \text{Width} \times \text{Depth} \times \frac{\pi}{6} \tag{2}$$

A comparison of TKV calculation methods is summarized in Table 3. The proposed automatic kidney volume calculation method has an average difference of only 1.43% from the ground truth, which is approximately seven times lower than the differences obtained from conventional volume calculations.

Table 3. Comparison of kidney volume calculation methods. |AVG| indicates the average of the absolute values of the volume differences for each case.

Case	Ground Truth (ml)	Calculator (ml)	Difference (%)	Ours (ml)	Difference (%)
200	562.43	491	−12.7	563.64	0.2
201	608.21	563	−7.4	610.35	0.4
202	482.99	572	18.4	490.96	1.7
203	192.49	164	−14.8	195.06	1.3
204	529	433	−18.1	526.61	−0.5
205	331.09	339	2.4	329.46	−0.5
206	466.18	401	−14	446.79	−4.2
207	484.03	456	−5.8	487.96	0.8
208	398.45	402	0.9	379.64	−4.7
209	479.41	430	−10.3	479.52	0.02
|AVG|			10.48		**1.43**

4   Discussion

In the discussion, we find that our model still has some potential improvements for better segmentation performance. Although some modifications may not yield significant enhancement, the strategies to fine-tune the hyper-parameters are listed below. First, the selection of windowing methods, which is the range of HU values, can be slightly different from various data sources. Some adaptive window selection algorithms may filter more non-kidney information and retain regions of interest. Second, the basic encoder, such as VGG-16 and ResNet-50, can segment the kidney region with high dice score. Some state-of-the-art encoders, such as EfficientNet and FixEfficientNet, may further improve the model. Third, for the transfer learning methods, the source data set selection may further improve the segmentation results. Some medical image segmentation data sets with numerous cases may generate more robust pre-trained weights.

The training environments of the 10 leaders in the KiTS 2019 competition are summarized in Table 4. Considering manuscript availability and data integrity, the rankings in the table reflect those during the KiTS 2019 competition; therefore, the ranking information is not up to date. Three groups did not provide their training environments, but their manuscripts mentioned that their original training procedures required substantial GPU resources. The proposed model is cost-effective; it requires fewer computing resources than previous studies. Because the experimental environment in this study is cost-effective, only one GPU with 8 GB RAM was used, and the training time is approximately 3 d, which is similar to that of previous models.

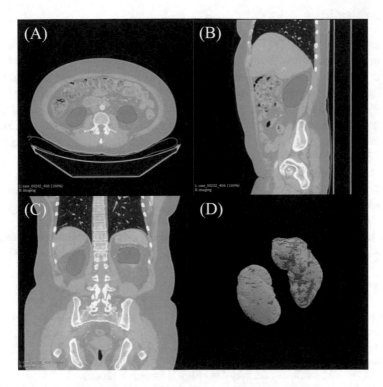

Fig. 7. Kidney segmentation results with the combined ResNet-50 and Unet model; (A) axial plane, (B) sagittal plane, (C) coronal plane, and (D) 3D demonstration. The red region denotes the predication, and the green region denotes the ground truth.

Table 4. Training environments of the 10 leaders in the KiTS 2019 competition.

Rank	Name	Training environment
1	Isensee	Nvidia Titan XP with 12 GB VRAM
2	PingAn Technology Co.	NVIDIA Tesla V100
3	Shanghai United Imaging Intelligence Inc.	Unknown
4	Institute of Computing Technology	NVIDIA Tesla V100 with 32 GB
5	Nanjing University of Science and Technology	4 Nvidia Titan XP (GPU) and 2 Intel Xeon E5-2650V4 (CPU)
6	Department of Electrical Engineering and Automation, Anhui University	Unknown
7	Department of Electronic Engineering, Fudan University	2 NVIDIA Tesla V100 with 32 GB
8	Nanjing University of Science and Technology	Unknown
9	NVIDIA	8 NVIDIA Tesla V100 with 16 GB (DGX-1 server)
10	Ball Chen	NVIDIA Tesla P40

5   Conclusions

To accurately calculate TKV, this study developed a deep learning–based kidney segmentation algorithm. However, most kidney segmentation methods based on deep learning have high computing resource requirements and a long training time. Transfer learning was employed in the present study, with the parameters from other organ segmentation models used as the initial input parameters. The results indicate that the use of transfer learning enhances model performance significantly. The proposed method with transfer learning in the lightweight ResNet-50 encoder achieved a Dice score of 0.9689. By utilizing two popular data sets from online competitions, our study achieved cost efficiency. Moreover, the automatic volume calculator yielded a volume difference of only 1.43% compared with the ground truth.

Acknowledgement. This work was supported in part by Ministry of Science and Technology (MOST), Taiwan, under Grant Number MOST 109-2221-E-002-144.

References

1. Tangri, N., Hougen, I., Alam, A., Perrone, R., McFarlane, P., Pei, Y.: Total kidney volume as a biomarker of disease progression in autosomal dominant polycystic kidney disease. Can. J. Kidney Health Dis. **4**, 1–6 (2017)
2. Levy, M., Feingold, J.: Estimating prevalence in single-gene kidney diseases progressing to renal failure. Kidney Int. **58**(3), 925–943 (2000)
3. Grantham, J., Torres, V.: The importance of total kidney volume in evaluating progression of polycystic kidney disease. Nat. Rev. Nephrol. **12**, 667–678 (2016)
4. Chapman, A.B., Wei, W.: Imaging approaches to patients with polycystic kidney disease. Semin. Nephrol. **31**(3), 237–244 (2011)
5. Chagot, L., et al.: Clinical kidney volume measurement accuracy using NEFRO-VOL. In: 2018 IEEE International Symposium on Medical Measurements and Applications (MeMeA), pp. 1–6 (2018)
6. Kaur, R., Juneja, M.: A survey of kidney segmentation techniques in CT images. Curr. Med. Imaging Rev. **14**(2), 238–250 (2018)
7. Sharma, K.: Automatic segmentation of kidneys using deep learning for total kidney volume quantification in autosomal dominant polycystic kidney disease. Sci. Rep. **7**(2049), 1–10 (2017)
8. Bazgir, O., Barck, K., Carano, R.A.D., Weimer, R.M., Xie, L.: Kidney segmentation using 3D U-Net localized with expectation maximization. In: 2020 IEEE Southwest Symposium on Image Analysis and Interpretation (SSIAI), pp. 22–25 (2020)
9. Isensee, F., Maier-Hein, K.H.: An attempt at beating the 3D U-net. arXiv:1908.02182
10. KiTS19 Challenge Homepage (2019). https://kits19.grand-challenge.org/. Accessed on 14 Jan 2021
11. Liver Tumor Segmentation Challenge (2017). https://competitions.codalab.org/competitions/17094. Accessed 14 Jan 2021
12. Kruthiventi, S.S.S., Ayush, K., Babu, R.V.: DeepFix: a fully convolutional network for predicting human eye fixations. IEEE Trans. Image Process. **26**(9), 4446–4456 (2017)

13. Li, X., Morgan, P., Ashburner, J., Smith, J., Rorden, C.: The first step for neu-roimaging data analysis: DICOM to NifTI conversion. Neurosci. Methods **264**, 47–56 (2016)
14. About DICOM: Overview (2021). https://www.dicomstandard.org/about. Accessed14 Jan 2021
15. Alomari, R.S., Kompalli, S., Chaudhary, V.: Segmentation of the liver from abdom-inal CT using Markov random field model and GVF snakes. In: 2008 International Conference on Complex, Intelligent and Software Intensive Systems (CISIS), pp. 293–298 (2008)
16. Irazabal, M.V., et al.: Imaging classification of autosomal dominant polycystic kidney disease: a simple model for selecting patients for clinical trials. J. Am. Soc. Nephrol. **26**, 161–187 (2015)

Detecting Interaction Activities While Walking Using Smartphone Sensors

Lukas Ehrmann[1]([✉]), Marvin Stolle[1], Eric Klieme[1], Christian Tietz[2], and Christoph Meinel[1]

[1] Hasso Plattner Institute (HPI), University of Potsdam, Potsdam, Germany
{lukas.ehrmann,marvin.stolle}@student.hpi.de,
{eric.klieme,christoph.meinel}@hpi.de
[2] Senacor Technologies AG, Berlin, Germany
christian.tietz@senacor.com

Abstract. An alternative to passwords for the verification of a user's identity is authentication based on behavioural biometrics such as gait-based verification. To have good classification performances, knowledge of the specific activity or location while walking is required. In the related discipline of *Human Activity Recognition*, much work has been done to identify typical activities such as walking vs running using accelerometer and gyroscope with high accuracy. Yet, related work is missing fine-grained evaluations of activities while walking such as *typing a text message* or *recording a voice message*. This work presents a three-stage machine learning approach to classify interaction activities while smartphone users are walking. The first stage classifies the rough location of the phone, the second stage distinguishes between *landscape* and *portrait* orientation, and the third stage classifies the final interaction activity. Using a dataset of 30 persons executing several activities with a smartphone during walking, we compared the performance of different machine learning designs. Finally, a *Random Forest* classifier performs best and achieves an accuracy of 98% for the first stage, 98% for the second stage, and 75% for the combined three-stage approach.

1 Introduction

Smartphones have become constant companions in our everyday life. The Google I/O conference announced in 2019 that there are almost 2.5 billion monthly active Android smartphone devices. Smartphones offer the ability to recognize human interaction activities using a variety of built-in sensors and processing power. Besides, the acceptance of carrying a smartphone for an extended period of time is high.

While gait-based verification approaches already reach high accuracies using data sampled from walking while *carrying* the phone in a fixed location like a pocket, the many ways of *interacting* with the phone while walking pose new challenges. Recent work showed that good verification performance is still possible if the specific activity is known and particular classifiers are created based

L. Barolli et al. (Eds.): AINA 2021, LNNS 226, pp. 382–393, 2021.
https://doi.org/10.1007/978-3-030-75075-6_31

on *activity subsets* [6]. To collect sufficient training data but also to decide about the right classifier for the final verification, the particular activity needs to be recognized reliably. Although many different *Human Activity Recognition* (HAR) approaches exist with high accuracies using machine or deep learning with accelerometer or gyroscope data, they mostly focus on the distinction of very fundamental activities such as *running, walking, sitting*, or *walking stairs*. To the best of our knowledge, a generic HAR approach for the fine-grained distinction of activities during the interaction with smartphones while walking such as *texting* or *calling* using sensor data only is yet missing. While a fusion of information on the currently running application, the recent user interactions gathered from system-side APIs and a general HAR may solve this problem as well, it requires a deep integration into the running operating system and a decent permission system for access. In contrast, sensor access is still possible without additional permissions in the major OS platforms (iOS, Android) and is thus independently usable. This work presents a multistage machine learning approach that is capable of this interaction activity recognition using sensor data only. Our detailed contributions are, as follows:

- We show that the distinction of eight different interactions while walking with a smartphone is possible using only sensor data with an accuracy of up to 75% using Random Forest classifiers (see Sect. 4.1).
- For the distinction, we design and evaluate a three-stage classification approach and prove the validity of the stages using state of the art unsupervised clustering techniques as shown in Sect. 3.3.
- We present that proximity sensor, light sensor, and the accelerometer's z-axis are very important for the determination of the general location of the phone while the x- and y-axis are most important for the orientation determination and frequency features are the most important for the final activity determination using the ANOVA F-test (see Sects. 3.2 and 4.4).

2 Related Work

Related work shows that accelerometer and gyroscope data is already sufficient to recognize human physical activities with a high accuracy. An accelerometer sensor can trace motion while a gyroscope is capable of tracing orientation data. The work of Sweta et al. [11] underlined that these sensors are very suitable even if different machine learning algorithms are used. They classified the activities *walking, jumping, running, sitting* and gathered labelled data using a self-made Android application with a simple interface where four volunteers participated. They collected data with a sampling rate 50 Hz and processed the data by sampling it into 2.56 s sliding windows with a 50% overlap. Features such as *arithmetic mean values, standard deviation of values, median absolute deviation (MAD)*, or the *interquartile range* are extracted from the data. For classification, they compared *J48 (decision tree), Support Vector Machine (SVM)* and *Multinomial Logistic Regression (MLR)*. Overall the SVM outperformed the other classifiers with an accuracy of 99.90%. While the work of Sweta et al. [11]

focused on the machine learning algorithm comparing accuracies, other research put focus on recognizing many different activities and even include exotic activities. One example is the work of L. Atallah et al. [1] where they not only used *eating* in their activity set but also focused on the influence of different positions of wearable accelerometers.

Typically, five aspects of *HAR* are subject to variation in related work and essential for the final evaluation performance that is expressed as *accuracy, precision, recall,* or *F1-score* most of the time.

- *Activity selection*: Activity selection defines which activities the work targets for recognition. Most of the related work used activities like *running, walking, sitting, standing* for their approach but also include special activities like e.g. *jumping* in the work of Sweta et al. [11].
- *Sampling rate*: Through related work, this aspect differs and depends on the sensor device that was used. E.g. the work of C.A. Ronao et al. [9] gathered data of accelerometer and gyroscope with a sampling rate 50 Hz, Z. Chen et al. [3] used a sampling rate 20 Hz of the accelerometer sensor and J. Figueiredo et al. [4] sampled GPS sensor data 1 Hz.
- *Sensor selection*: This attribute describes the hardware sensors that are used to gather data. Most of the related work picked accelerometer and gyroscope as their sensors of choice [2,8,11,12]; some also include GPS [4] or experiment with other sensors [10].
- *Feature extraction*: Feature extraction is used to process raw data that was gathered from sensors before training a classifier. Referring to Lara et al. [7] extracted features can be classified into either *time* or *frequency domain*. Prior to extraction, the recorded time-series data is usually split into windows. Additionally, an overlap is applied to the windows so that the amount of data increases. Table 1 shows different features from related work assigned to their domain.
- *Classifier choice*: Different machine learning algorithms are used to classify data. The work of R.A. Voicu et al. [12] used a multi-layer perceptron (MLP); Dănuţ Ilisei and Dan Mircea Suciu [5] applied Naive Bayes and Decision Tree classifiers;

Table 1. Different extracted features approaches from related work [2–4,7].

Domain	Features
Time Domain	Mean, Median, Minimum, Maximum, Standard Deviation, Variance, Average Absolute Difference, Correlation Coefficient, Autocorrelation, Kurtosis, Skewness Interquartile Range, Zero Crossing Rate
Frequency Domain	Fourier Transform, Discrete Cousin Transform, Median Frequency, Maximum Frequency, Fundamental Frequency, Total Energy

To summarize, related work in *Human Activity Recognition* varies the five different aspects *sensor selection, sampling rate, activity selection, feature extraction,* and *classifier choice* to increase classification accuracy for activity recognition. While diverse ways of HAR were already proposed, the typical set of activities consists of rather basic types such as *running, walking,* or *sitting* missing a further distinction related to many ways of today's interaction with smartphones while walking. In this work we will show how interaction activities while walking with a smartphone could be recognized and extend the amount of activities being recognizable.

3 Approach

In this part, we describe the *dataset* we use for evaluating our classification approach, how we *pre-process our data,* the *three-stage classification model,* which *machine learning* techniques we use, and how we *validate* them.

3.1 Dataset

For our work, we use the dataset from Klieme et al. [6]. In their experiment with 30 participants, the data was collected in a semi-supervised manner where each participant was led through the recording of up to 18 different *carrying* and *interaction* activities with the help of a smartphone app deployed on a Google Pixel. For each activity, the procedure was similar and at least 30 s of the respective activity were recorded.

Table 2. Activities recorded in the experiment of Klieme et al. [6].

Group	Activities
Carrying	Front/Back Right/Left Pocket Trousers, Holding Right/Left Hand, Jacket Outer/Inner Right/Left Pocket, Jacket Breast Pocket
Interaction	Texting Landscape, Texting Portrait, Reading Landscape, Reading Portrait, Listening Voice Message, Recording Voice Message, Telephoning

Please see Table 2 for a summary of all activities that were considered. Due to the dataset containing various carrying activities, we use the three most used carrying activities (*back right pocket, holding right hand, jacket outer pocket*) and cluster them in a new activity called *Carrying*. In addition to accelerometer and gyroscope sensor data we extract light and proximity sensors data of the dataset.

3.2 Data Preprocessing

The dataset contains the following two artifacts: multiple different measurements taken on the same timestamps and measurements unequally spread over time. To fix this issue, the data is processed with a two-stage preprocessing pipeline as shown in Fig. 1.

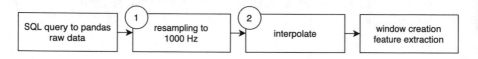

Fig. 1. Preprocessing pipeline used to prepare raw data for input to an ML classifier.

The first stage resamples the data to a frequency 1000 Hz. Thereby it creates a data frame with fixed periods of 1 ms as an index. In the case of multiple measurements on a single millisecond, it calculates the mean and assigns it to the respective timestamp. The second stage interpolates and thereby fills data points with no measurement. A linear interpolation method is used, which treats the values as equally spaced. For feature extraction, data is further preprocessed using the *sliding window* technique.

As stated in Sect. 2, related work proposed different window sizes and overlaps. Based on the various statements, we decide to evaluate a selection of combinations. Therefore we use window sizes from 500 ms to 15000 ms with a step size of 500 ms, and various window overlaps from 0 to 95%. For each sliding window, we extract 110 features shown in Table 3. The Fast Fourier Transform (FFT) algorithm is used for transformation into the frequency domain.

Table 3. Extracted features in each window in the final step of the preprocessing pipeline.

Domain	Sensor(s)	Methods	Count
Time Domain	Light	mean	1
	Proximity	mean	1
	Accelerometer axis: x, y, z Gyroscope axis: x, y, z	Mean, median, median absolute deviation, variance, max, min, kurtosis, skewness, interquantile range	54
Frequency Domain	Accelerometer axis: x, y, z Gyroscope axis: x, y, z	Sum of power in bins: 0.5 Hz to 10.5 Hz, binwidth: 1 Hz	54

3.3 Three-Stage Classification Model

Referring to the varying characteristics of the targeted activities for recognition (see Subsect. 3.1) and [6], a one-stage classification is quite challenging and activity subsets based on the rough position of the phone related to the body can be helpful: For example, if a user is telephoning, or interacting with voice messages he would most likely have the phone around his head while a user reading a text or watching a video would most likely have the phone in front of the body. Afterwards, further clustering might be helpful again e.g. by activity (such as texting, reading/watching) or orientation (landscape vs. portrait activities). In the end, a classifier would then have to decide only between 2–4 activities.

To design the stages, we use an unsupervised machine learning algorithm called *t-SNE (t-distributed Stochastic Neighbor Embedding)*. T-SNE is able to cluster high dimensional data and visualize the result in low dimensional space. We applied the t-SNE algorithm with *one* to *eight* clusters (covering *carrying* and the different interaction activities presented in Table 2), various perplexities, and various labels (orientation, location and interaction activity). We labeled the datapoints by orientation and various number of clusters for location. We achieved the best separation with the location labels and three clusters as shown in Fig. 2.

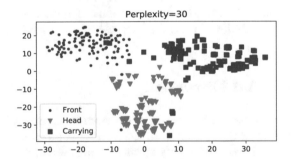

Fig. 2. T-SNE clustering in 2 dimensions identifies a decent clustering of the eight interaction activities into the clusters of *front, head,* and *carrying* activities.

We applied this algorithm as well for the *front, head,* and *carrying* subsets of activities to determine the second and the third stage. Please refer to Fig. 3 for the final stages of our proposed multi-stage classification.

3.4 Classifiers and Validation

Referring to the diverse approaches presented in related work the following classifiers are evaluated for each stage: Support Vector Machines (SVM), K-Nearest-Neighbor classifier (KNN), Random Forest (RF), Multilayer Perceptron (MLP), and AdaBoost.

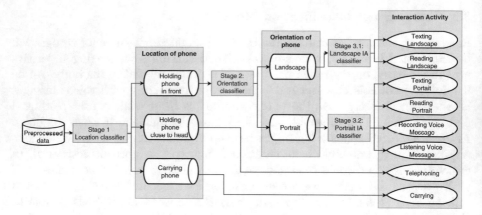

Fig. 3. The final three-stage classification approach first determines the location of the phone, continues with its orientation, and finally decides on the particular interaction activity.

For validation, we design a specific five-fold cross-validation to split the dataset randomly into five folds, each containing six unique users. Since each user generated a different number of samples, each fold gets balanced so that every interaction activity has the same amount of data across all folds. This procedure guarantees that we use data from yet unseen participants in our test dataset.

4   Evaluation and Discussion

The results are evaluated using a top-down approach. Firstly, we assess and discuss the *combined classification* result, secondly the *pipeline design and the classifier choice*, thirdly the *window size and the overlap*, and finally, the *feature engineering*.

4.1   Combined Classification

With our three-stage classification approach, we achieved a combined accuracy for classifying the interaction activities of **74.72%**, a precision score of 76.02%, a recall score of 74.72%, and an F1-Score of 74.16%. The classification of inter-action activities is based on the following three stages and four classifiers (see Fig. 3) with their respective accuracies. The first stage, which classifies the location (1), achieved an accuracy of 98%, 98% for the second stage which classifies the orientation (2), 85% for the landscape activity classifier (3.1) in the third stage, and 72% for the portrait activity classifier (3.2) in the third stage.

Figure 4 shows the confusion matrix of the combined three-stage classification.

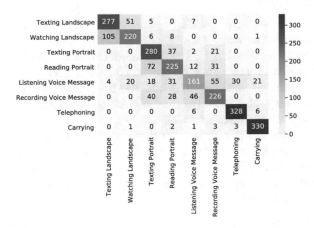

Fig. 4. Confusion matrix for classifying the Interaction Activities making use of the model shown in Fig. 3.

The first stage of the classifier separates *Telephoning, Carrying*, and all other activities. As reflected by the high accuracy of the first classifier, there were only a small number of misclassifications as presented in Fig. 4. There is confusion between the predicted label *Listening Voice Message* and *Telephoning*. We assume that some users listened to the voice message while holding the phone in front, others holding the phone at the head. The different location while listening to a voice message was not labelled in the SMOMBIES dataset.

The second stage of the classifier separates landscape (*Texting Landscape, Watching Landscape*) and portrait activities (*Texting Portrait, Reading Portrait, Listening Voice Message, Recording Voice Message*). This stage has a very high accuracy which reflects the low amount of misclassifications between the portrait and landscape clusters.

The poor classification performance of interaction activities executed while holding the phone in the same position (e.g. in front landscape, in front portrait) is represented by the last stage. We assume that the sensor data differences between interaction activities while holding the phone in the same position are minimal. Referring to this work and the used dataset, we see limited possibilities to improve classification between those activities. The tilt of the phone supports the better classification between *Listening Voice Message, Recording Voice Message* and *Texting Portrait, Reading Portrait*.

It can be identified that all metrics hold nearly the same value ($Precision \approx Recall \longrightarrow FP \approx FN$). This proves that we evaluated the classifier with balanced data.

4.2 Pipeline Design and Classifier Choice

Making use of hyperparameter tuning and various window sizes and overlaps (see Subsect. 4.3) we analyzed the classifiers mentioned in Subsect. 3.4 and the following pipeline designs: (1) A single-stage approach, where a Random Forest classifier reached an accuracy of 65.39% as maximum. (2) A two-stage approach while classifying the location in the first and the interaction activity in the second stage, which reached an accuracy of 70.12%. (3) A two-stage approach, classifying the orientation in the first and the interaction activity in the stage, reaching an accuracy of 62.3%. (4) And a three-stage (shown in Fig. 3), which achieved the highest accuracy of 74.72%.

The RF classifier outperforms the other classifiers (KNN, SVM, AdaBoost, MLP) independent of the window size and the overlap and achieves the highest accuracy in every stage. Figure 5 shows an example for the first stage. Also, it has lower computational resource requirements and an automated feature selection which supports classification.

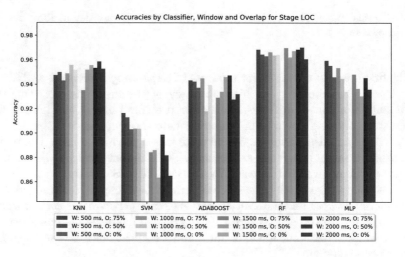

Fig. 5. Random Forest classifier outperforms other evaluated classifiers in the first stage, which classifies the location of the phone.

4.3   Window Size and Overlap

The window sizes and overlaps presented in Subsect. 3.2 were used for this analysis. The analysis showed that larger window sizes perform better than lower window sizes since frequency features are more precise for longer windows. Figure 6 shows the impact of different window sizes and overlaps on the accuracy. The best results can be achieved with an overlap of 15000 ms and 93.75% overlap. A disadvantage of large window sizes is that it is difficult to use in practice. E.g. an activity which requires the user to read a short text message could take less than **15000 ms**. Besides, the analysis showed that a higher overlap increases performance to a certain extend. We achieved the highest accuracies with an overlap of **93.75%**. This brings the advantage of having more data available for training the classifier.

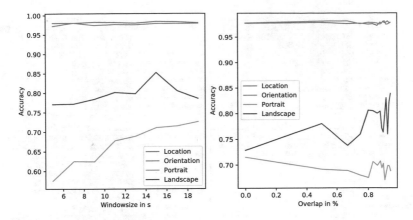

Fig. 6. Different window sizes and overlaps vs accuracies for the RF classifier

4.4 Feature Engineering

As discussed in Subsect. 4.3, we achieved the highest accuracies for Interaction Activities classification with a window size of 15000 ms and a overlap of 93.75%. These values are used to evaluate the features. It showed that for each axis, both gyroscope, and accelerometer the features *median, mean, max, min* amongst each other, the features *variance* and *MAD* amongst each other and the *frequency* features amongst each other are strongly correlated.

Correlation is part of the quality of features but does not indicate the contribution of a feature to classification. To analyze the contribution of a feature to classification, we use an ANOVA F-test. Features with a higher F-Value can contribute more to the respective classification. Figure 7 shows the F-Values of features for each classifier stage (*location, orientation, landscape* and *portrait*).

The *median of the z-axis of the accelerometer* holds the highest F-Value for the location classifier, thus is a very important feature. The second highest F-Value is assigned to the *mean of the z-axis of the accelerometer*. But since the *mean* and *median* of the z-axis accelerometer are highly correlated they will not increase classification accuracy in a way as an independent variable with the same F-Value would do. By including the proximity and light sensor, two uncorrelated features where introduced which contribute significantly to the classification of the location of the phone. Comparing the F-Values between the various classifiers (Fig. 7) we identify that the z-axis of the accelerometer has the highest impact on the classification of the location supported by the proximity sensor. The x- and the y-axis of the accelerometer have the highest impact on classifying the orientation of the phone. Especially frequency features support the classification of the interaction activity at the stages of landscape and portrait classifier. The peak F-Value decreases with each ML stage. For classifiers with lower F-Values, lower accuracies are expected.

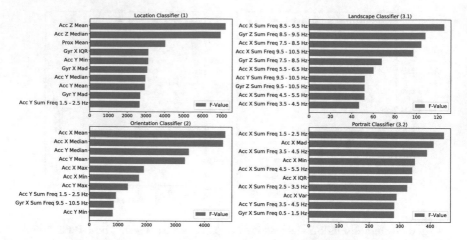

Fig. 7. ANOVA F-Values for 10 highest ranked features for each classifier shown in Fig. 3.

5   Conclusion and Future Work

This paper proved that selected interaction activities can be recognized with a good accuracy using a three-stage classification model. Specifically, interaction activities like *carrying* with 97.06% or *telephoning* with 96.47% are highly distinguishable and therefore suitable for activity subset based authentication. In the end, we achieved an average accuracy of approximately 75% for all eight activities in the combined classifier including activities such as *writing a text message* or *watching a video* while walking with the smartphone. We achieved this using techniques of machine learning and data of accelerometer, gyroscope, proximity, and light sensors. We implemented a three-stage classification approach that was validated using dimensionality reduction techniques to visualize high dimensional data. Furthermore, we designed a balanced k-fold cross-validation to prevent using the data of one user in both train and test folds and each classifier was optimized using a self-implemented parameter tuning technique to find the best window size, window overlap, and classifier and its hyperparameters. Finally, we evaluated and discussed the use of the sensors and the three-stage classification approach. The main driver for classifying interaction activities is the location and the tilt of the phone. For future work, we consider the following aspects as important:

- Our approach was based on interaction data while walking. This can be extended by using another base activity like sitting or laying.
- We considered eight interaction activities while walking but did not cover all common activities. This can be extended with other interaction activities like playing games, taking pictures, selfies, etc.
- Other features might contribute to classification. This could include additional sensors or other statistical summarising methods.

- Further experiments could be conducted to have more fine-grained labels of the *listening* and *recording voice message* activities to know about the location and whether it was done in front of the body or similar to phone calls, i.e. having the phone next to the ear.

References

1. Atallah, L., Lo, B., King, R., Yang, G.: Sensor positioning for activity recognition using wearable accelerometers. IEEE Trans. Biomed. Circuits Syst. **5**(4), 320–329 (2011). https://doi.org/10.1109/TBCAS.2011.2160540
2. Chen, Z., Jiang, C., Xiang, S., Ding, J., Wu, M., Li, X.: smartphone sensor based human activity recognition using feature fusion and maximum full a posteriori. IEEE Trans. Instrum. Meas. 1 (2019). https://doi.org/10.1109/tim.2019.2945467
3. Chen, Z., Zhu, Q., Soh, Y.C., Zhang, L.: Robust human activity recognition using smartphone sensors via CT-PCA and online SVM. IEEE Trans. Industr. Inf. **13**(6), 3070–3080 (2017). https://doi.org/10.1109/TII.2017.2712746
4. Figueiredo, J., et al.: Recognition of human activity based on sparse data collected from smartphone sensors. In: 6th IEEE Portuguese Meeting on Bioengineering, ENBENG 2019 - Proceedings (2019). https://doi.org/10.1109/ENBENG.2019.8692447
5. Ilisei, D., Dan, M.S.: Human-activity recognition with smartphone sensors. Springer (2019). https://doi.org/10.1007/978-3-030-40907-4
6. Klieme, E., Tietz, C., Meinel, C.: Beware of SMOMBIES: verification of users based on activities while walking. In: Proceedings - 17th IEEE International Conference on Trust, Security and Privacy in Computing and Communications and 12th IEEE International Conference on Big Data Science and Engineering, Trustcom/BigDataSE 2018, pp. 651–660 (2018). https://doi.org/10.1109/TrustCom/BigDataSE.2018.00096
7. Lara, Ó.D., Labrador, M.A.: A survey on human activity recognition using wearable sensors. IEEE Commun. Surv. Tuts. **15**(3), 1192–1209 (2013). https://doi.org/10.1109/SURV.2012.110112.00192
8. Lawal, I.A., Bano, S.: Deep human activity recognition using wearable sensors. In: ACM International Conference Proceeding Series, pp. 45–48 (2019). https://doi.org/10.1145/3316782.3321538
9. Ronao, C.A., Cho, S.B.: Human activity recognition with smartphone sensors using deep learning neural networks. Expert Syst. Appl. **59**, 235–244 (2016). https://doi.org/10.1016/j.eswa.2016.04.032
10. Su, X., Tong, H., Ji, P.: Accelerometer-based activity recognition on smartphone. In: CIKM 2014 - Proceedings of the 2014 ACM International Conference on Information and Knowledge Management, vol. 1, pp. 2021–2023 (2014). https://doi.org/10.1145/2661829.2661836
11. Sweta, J., Sadare, A., K. Shreesha, P.: Human activity recognition using smartphone sensor data, vol. 472. Springer, Singapore (2020). https://doi.org/10.1007/978-3-319-39904-1_4
12. Voicu, R.A., Dobre, C., Bajenaru, L., Ciobanu, R.I.: Human physical activity recognition using smartphone sensors. Sensors (Switz.) **19**(3), 1–18 (2019). https://doi.org/10.3390/s19030458

Training Effective Neural Networks on Structured Data with Federated Learning

Anastasia Pustozerova[1]([✉]) [iD], Andreas Rauber[1,2] [iD], and Rudolf Mayer[1,2] [iD]

[1] SBA Research, Vienna, Austria
{apustozerova,rmayer}@sba-research.org
[2] Vienna University of Technology, Vienna, Austria

Abstract. Federated Learning decreases privacy risks when training Machine Learning (ML) models on distributed data, as it removes the need for sharing and centralizing sensitive data. However, this learning paradigm can also influence the effectiveness of the obtained prediction models. In this paper, we specifically study Neural Networks, as a powerful and popular ML model, and contrast the impact of Federated Learning on the effectiveness compared to a centralized approach – when data is aggregated at one place before processing – to assess to what extent Federated Learning is suited as a replacement. We also analyze the effect of non-independent and identically distributed (non-iid) data on effectiveness and convergence speed (efficiency) of Federated Learning. Based on this, we show in which scenarios (depending on the dataset, the number of nodes in the setting and data distribution) Federated Learning can be successfully employed.

Keywords: Federated machine learning · Effectiveness evaluation

1 Introduction

The data used for training Machine Learning (ML) models often contains sensitive information, and thus must be protected from adversarial access and use. Federated Learning (FL) helps to reduce privacy risks by training ML models locally, and thus removing the need of transferring training data. It eliminates the possibility for an adversary to obtain all the training data on a centralized server. In FL, only the parameters of the learned model (e.g. weights or gradients in case of neural networks) are shared, with an aggregator or other participants.

Despite the benefits that FL entails, the effectiveness of the approach is crucial. It is imperative, for any privacy-preserving method, to achieve utility of the learned model close to a fictional, idealized (if privacy was not a concern) centralized setting. In this paper, we therefore study whether FL allows training models of a quality comparable to centralized training, and further compare it to training only on local data without collaboration. We consider different FL settings and scenarios, and formulate recommendations for using FL in different settings. We specifically focus on Neural Networks. While much of the literature focuses on image data, we take a closer look at structured (relational) data, which has many use cases e.g. in medicine, businesses and other industries.

One can distinguish FL by the coordination and aggregation strategy, into sequential and parallel learning. McMahan et al. [1] describe the *parallel* setting, where processing nodes independently and simultaneously train local models, and subsequently send them to an aggregator that computes a global model, e.g. with the *Federated Averaging* algorithm [1]. In *sequential* FL, sometimes also referred to cyclic incremental learning [2], the models are trained and shared incrementally from one node to the following in sequence. This approach does not require a dedicated aggregator, thus completely avoiding a centralized instance. Sequential learning becomes inefficient with a large number of nodes. It can, however, be a viable alternative in settings with a smaller number of nodes, e.g. when several medical institutions want to train a collaborative model, but are not able to share sensitive data.

In this work, we:

- Analyse the behaviour and performance of Federated Learning on datasets not previously considered in the literature
- Investigate how sequential and parallel Federated Learning of Neural Networks perform on structured data, and compare the effectiveness to the (idealized target) baseline results of models trained on centralized data.
- Study parallel and sequential learning with varying numbers of processing nodes in the federation, and analyze the influence on models effectiveness
- Show the impact of different distributions in the data on the model quality, and identify which Federated Learning setting (sequential or parallel) is more beneficial to use in scenarios with (i) equal distribution of data among the nodes, and (ii) non-iid data

This paper is structured as follows. Section 2 discusses related work. In Sect. 3, we describe the FL setup and implementation, datasets, ML models and the choice of architecture and hyper-parameters. In Sect. 4, we present the results of our evaluation. In Sect. 5, we provide conclusions and recommendations for successful federated training and provide an outlook on future work.

2   Related Work

A large share of current research in Federated Learning is dedicated to collaborative medical data processing [2,3] due to strict privacy policies discouraging and legal regulations limiting the sharing of patients data. Federated Learning gathered significant attention as a method allowing to let data distributed on mobile devices reside there, while training effective machine learning models [4].

Federated Learning, however, poses several challenges. Communication costs can be high, especially when the number of processing nodes is large [4]. The heterogeneity of the systems (nodes) involved is a further challenge [5]. Unbalanced

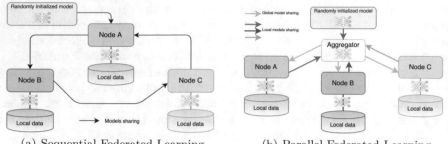

(a) Sequential Federated Learning   (b) Parallel Federated Learning

Fig. 1. Federated Learning architectures

and not independent and identically distributed data (non-iid data) can increase the complexity of the training process, and also increase communication costs [6]. Moreover, security and privacy risks are an issue in Federated Learning [7]. Federated Learning allows to avoid explicit data sharing, however, the models transferred during training can leak sensitive information about local data [8].

With this wide range of challenges to address, the foremost critical remains the effectiveness of the approach – particularly whether Federated Learning results in lower quality models. Federated Learning will be considered a viable alternative only if it can achieve accuracy that is comparable, or even on par, with the ideal baseline of a centralized learning approach, and thus can be used to achieve high-quality predictions that are useful in real-world settings.

Sheller et al. [2] considered parallel and sequential learning and concluded that parallel learning gives more reliable result than cyclic incremental learning, allowing to reach 99% of the accuracy of the model trained on centralized data. In this work, we perform a more structured and broader comparison of parallel and sequential learning, using multiple datasets, and specifically testing on non-iid distributed data. Nilsson et al. [9] evaluate the effectiveness of Federated Learning on image dataset (MNIST), considering three different averaging algorithms in parallel learning. They conclude that *Federated Averaging* performs the best on non-iid data. We, therefore, use the *Federated Averaging* algorithm in the parallel learning setting in our study. McMahan et al. [1] evaluate Federated Learning performance on image and text data, and also show that *Federated Averaging* allows training effective machine learning with non-iid data. We extend this evaluation, by performing an evaluation on various structured datasets in different Federated Learning settings: sequential vs. parallel learning, iid vs. non-iid data, and a varying number of participants in the federation.

3   Study Design

In this section, we describe different Federated Learning setups, datasets and pre-processing steps, the architectures we use for neural networks and the hyperparameters choice, and finally, the settings with non-iid data.

Sequential Federated Learning (see Fig. 1(a)) starts with a randomly initialized model, which is then sent to the first node in the sequence (*Node A*). After receiving the random model, *Node A* trains it on its local data, and then sends the model to the next node in the sequence (*Node B*). The process continues until the last node in the sequence has trained the model, and the second *federated cycle* starts when the last node in the sequence sends the model to the first one. The model training process, once the federated network is set up, can be orchestrated in a peer-to-peer fashion, without the requirement of a central coordinator. This can eliminate potential single points of failure and a central point that might be subject to attacks. Also, variations on the sequence of nodes might be introduced, i.e. that it is different in each cycle.

Parallel Federated Learning (see Fig. 1(b)) starts with the random initialization of a model and sending it to every node. Then, each node trains the model locally on its data and sends the trained model back to the aggregator. The aggregator averages the collected, locally trained models, and thus creates a new *global* model. We can calculate global model's weights W_{global} as an average of corresponding weights from all local models W_k, $k = 1, ..., n$: $W_{global} = \frac{\sum_{k=1}^{n} W_k}{n}$, where n is the number of processing nodes in the federation. After calculating a global model, the coordinator sends it back to every node for training during the subsequent *federated cycle* and so on.

Among different types of Federated Learning architecture, we focus on parallel learning as on the most spread form of Federated Learning and on sequential learning as the basic form of decentralized Federated Learning.

In our study, we first perform a grid search in centralized learning to find optimal hyper-parameters (learning rate, number of epochs, number of iterations and batch size) for the baseline comparison. We use these hyper-parameters for training models in parallel and sequential learning and, if needed, we tune some of the parameters also using grid search. We assume that each node trains their model locally with the same number of epochs in both parallel and sequential Federated Learning. We then compare the accuracy of the models trained in centralized and federated settings. We take the results of centralized training as the baseline, as it represents the ideal scenario and can serve as an upper bound of effectiveness for the final model.

We assume that each node has a fixed, unique local dataset. We follow several approaches to distribute the original (centralized) benchmark datasets utilized in our study among the nodes (see Sect. 3.3).

We implement our sequential and parallel Federated Learning coordination architectures using Python 3.7 and the *Pytorch* framework.

Table 1. Datasets

Dataset	# Samples	# Features	Target	Data types
Purchase-2–100	32,000	600	2,10,20,50,100 classes	Binary
Location	5,280	293	30 classes	Binary
Breast Cancer	683	9	2 classes	Categorical
Adult	48842	14	2 classes	Categorical, numerical

3.1   Datasets

To analyze a broad range of cases, we select datasets with different characteristics (cf. Table 1), to study how Federated Learning deals with these different settings. We consider eight different classification tasks (*Purchase* dataset is used with five targets) and evaluate FL performance on these tasks.

The ***Purchase*** dataset is derived from the "Acquire Valued Shoppers"[1] dataset. As in [10], we apply *k-means clustering* to create 5 different classification tasks, with 2, 10, 20, 50 and 100 classes, respectively, where each class corresponds to a group of individuals with similar purchase behavior. We denote these classification tasks as *Purchase-2*, *Purchase-10*, and so on, in the remainder of this paper. We use 30,000 instances for training and the rest for testing).

The ***Location*** dataset is based on check-in data from the mobile phone app Foursquare, from April 2012 to September 2013[2]. Based on the description in [10], we again use *k-means* clustering to create a classification task with 30 classes, representing similar user groups. The final dataset contains 5,280 records, representing unique users (1,280 are used for testing, the remainder for training). The 293 binary features represent characteristics of the places users have visited.

The ***Breast Cancer*** dataset[3], is a frequently used small-scale benchmark medical dataset. It contains 683 instances and nine categorical attributes, and a target attribute denoting the class of the instance (benign or malignant). We used 400 instances for training and 283 instances as a test set.

The ***Adult*** dataset is derived from the US Census Database[4]. The task is to predict if a person earns more or less than 50K. We use 45,000 instances for training and 3,842 for testing.

The processed and clustered versions of all the datasets are available on Zenodo[5]. We divide all dataset with five random splits into training and test data, i.e. a repeated holdout validation. In the evaluation section, all the plots and tables depict mean results among the five data splits.

---

[1] https://www.kaggle.com/c/acquire-valued-shoppers-challenge/data.

[2] https://sites.google.com/site/yangdingqi/home/foursquare-dataset.

[3] https://archive.ics.uci.edu/ml/datasets/breast+cancer+wisconsin+(original).

[4] http://archive.ics.uci.edu/ml/datasets/Adult.

[5] https://doi.org/10.5281/zenodo.4562403.

3.2   Trained Models

In this paper, we focus on the performance of neural networks in federated learning, and thus employ a Multi-Layer Perceptron (MLP) as a prediction model for all classification tasks. For the *Purchase* and *Location* datasets, we build on the architectures in [10], and compare our results to their baseline. The network has one hidden layer of 128 neurons, *Tanh* activation function, and a *Softmax* layer. Slightly extending the architecture from [10], we further add a *dropout* layer (0.5) for the classification tasks on the *Purchase* dataset with ten or more classes, as the regularization from the dropout layer leads to higher accuracy.

As benchmark machine learning model for *Breast Cancer* dataset we follow [11]. The model we consider is a neural network with one hidden layer of nine nodes, a *ReLU* activation function on the hidden layer, and *Tanh* activation on the last layer. For *Adult* dataset, we use a fully connected neural network with one hidden layer of 64 neurons and *Tanh* activation, and *Sigmoid* activation on the last layer. We use Adam optimizer with 10^{-3} learning rate for all datasets, but *Adult*, where we use a learning rate of 10^{-4}. As a loss function for 2-targets classification tasks we use *Binary Cross Entropy Loss*. For the rest of the tasks, we use the *negative log likelihood loss* frequently applied for multi-classification tasks.

3.3   Data Distribution in Federation

We perform experiments with training data either (i) distributed roughly equally among the processing nodes, and also investigate the influence of (ii) non-iid data. To simulate an equal distribution, we randomly share the data among the nodes, in a way that they have the same number of instances.

To simulate a setting with non-iid data, we follow similar procedure used in [1]. We thus first sort the dataset by target label, and then split the dataset into several shards. The number of these shards is proportional to the number of nodes in the setting and different for datasets (n is the number of nodes in Federated Learning): *Breast Cancer* - $2n$ shards; *Adult, Purchase-2* - $4n$ shards; *Location, Purchase-10 – Purchase-100* - $10n$ shards. We simulate the size of the local training sets such that all the training data is normally distributed among the nodes. Then we assign to each node corresponding number of shards.

We compare the effectiveness of FL on non-iid data not only with the centralized baseline, which provides an upper bound of achievable accuracy, but also with the average accuracy score of the models trained just locally. This represents the alternative of not participating in FL, and provides a lower bound. We can thus investigate in which scenarios and to which extent is it beneficial to participate in federated learning, instead of just training a model on local data.

4   Effectiveness Evaluation

To evaluate the effectiveness of the models, we measure their prediction accuracy on a test set. We perform experiments on five different random splits of the data into training and test set, and report the mean and standard deviation of the accuracy on the test sets.

In the detailed analysis below, we consider two aspects of the effectiveness – the overall, final accuracy reached on centralized versus federated settings, as well as the time respectively number of iterations/cycles of training required to converge to the optimum accuracy. In some settings, this training time requirement can vary considerably between centralized and federated settings, and thus becomes an important aspect of the training parameters.

Table 2. *Location, Adult, Breast Cancer* datasets, mean and standard deviation of accuracy in % on the tests set for different number of nodes in parallel and sequential Federated Learning

Dataset	Centralized	2 nodes		8 nodes		32 nodes	
	Data	Sequential	Parallel	Sequential	Parallel	Sequential	Parallel
Purchase-2	96.4 ± 0.3	96.0 ± 0.4	97.7 ± 0.2	96.4 ± 0.3	96.0 ± 0.4	96.2 ± 0.2	95.8 ± 0.4
Purchase-10	84.0 ± 0.5	83.4 ± 0.5	84.3 ± 0.6	83.7 ± 0.5	84.1 ± 0.5	83.1 ± 0.8	**80.2 ± 1.1**
Purchase-20	79.0 ± 0.8	78.8 ± 0.9	79.5 ± 0.7	77.9 ± 0.7	78.8 ± 0.7	78.6 ± 1.0	**75.9 ± 1.1**
Purchase-50	73.8 ± 1.0	72.2 ± 1.1	74.8 ± 1.0	72.6 ± 0.7	75.5 ± 0.7	71.6 ± 1.6	74.3 ± 1.1
Purchase-100	64.3 ± 0.6	**61.8 ± 1.8**	66.6 ± 1.5	63.1 ± 1.5	66.4 ± 1.2	62.3 ± 1.0	67.3 ± 1.3
Location	80.3 ± 0.8	80.4 ± 1.1	80.6 ± 1.4	79.0 ± 0.7	81.5 ± 1.4	**76.5 ± 1.3**	78.1 ± 0.7
Breast Cancer	97.5 ± 1.2	97.3 ± 1.2	95.6 ± 1.3	97.4 ± 1.1	96.8 ± 0.9	97.5 ± 1.2	96.2 ± 1.4
Adult	86.1 ± 0.7	86.1 ± 0.5	85.9 ± 0.5	86.2 ± 0.7	85.9 ± 0.4	85.6 ± 0.8	85.2 ± 0.5

4.1 Equal Distribution of the Data

To ensure that the centralized baseline is adequate, we compare the results for *Purchase, Location* and *Adult* datasets in centralized setting to the benchmark from [10], for *Breast Cancer* dataset to [11]. For each dataset we train centralized models with accuracy scores close to stated in the benchmark examples.

Table 2 shows the results for the scenario of equal distribution of the data among the nodes. We observe that both parallel and sequential Federated Learning in the majority of the cases allow reaching an accuracy score close to the one achieved by the models trained on centralized data. We highlighted the few cases when the accuracy score was lower to more than 3%, than centralized baseline, e.g. sequential learning with 32 nodes on *Location* dataset. However, the deviation from the baseline accuracy is at most 4% in all considered scenarios.

Sequential learning performs worse than parallel on the classification tasks that exhibit a larger number of classes, i.e. *Location, Purchase-50* and *Purchase-100*. Sequential and parallel Federated Learning performed very similar on classification tasks with two targets (*Purchase-2, Adult, Breast Cancer*), both achieving an accuracy close to the baseline of centralized learning.

From the results on *Breast Cancer* (shown in Figs. 2(a) and (b)) and *Adult* datasets, we notice that sequential Federated Learning manages to reach the

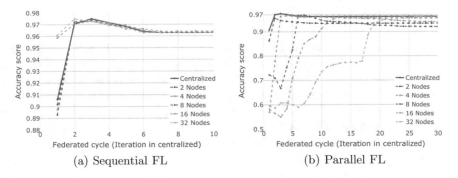

(a) Sequential FL   (b) Parallel FL

Fig. 2. Federated Learning on *Breast Cancer* dataset with equal distribution of the data among the nodes

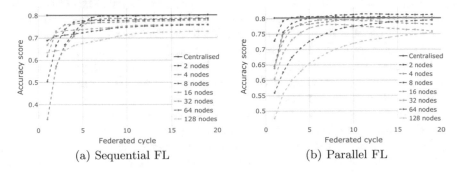

(a) Sequential FL   (b) Parallel FL

Fig. 3. Federated Learning on *Location* dataset with equal distribution of the data among the nodes

baseline accuracy with less federated cycles than parallel. The speed of model convergence in parallel learning drops with an increasing number of nodes in the setting, e.g. with eight nodes it takes seven federated cycles to reach the baseline accuracy, while with 32 nodes, it takes 23 federated cycles (see Fig. 2(b)). However, one should consider that in sequential Federated Learning a node cannot start training until the previous one finished training, and sent a model to the successor. This can make sequential federated learning considerably less efficient than the parallel, especially with a larger number of nodes in the setting.

Sequential learning performs overall worse than parallel on *Location* dataset. The quality of the final model drops up to 5% with 32 and 64 nodes, and up to 8% with 128 nodes (see Fig. 3(a)). The accuracy of the models trained in parallel learning deviates from the baseline accuracy up to 5%. This worst-case occurs with 128 nodes (see Fig. 3(b)). Sequential Federated Learning is more sensitive to the increasing number of nodes than parallel.

4.2   Non-iid Data

Non-iid data reduces the quality of the models trained with both parallel and sequential Federated Learning in the majority of considered settings. Moreover, it increases the number of cycles needed for models to converge. Remember that on the *Adult* dataset, sequential and parallel learning (denoted as Sequential and Parallel in the Fig. 4(a)) allowed reaching similar accuracy results, only up to 1% less than centralized, with data distributed equally among the nodes. With non-iid data, both sequential and parallel reach similar scores to each other, which are, however, up to 6% worse than centralized setting. One can see that the average score of models trained only on local data is up to 5% lower (e.g. on 64 nodes) or similar (e.g. four nodes) to accuracy reached with federated learning. Further, Figs. 4(b) and (c) shows that it takes more federated cycles for models to converge with non-iid data. In the case with eight nodes (see Fig. 4(b)), models trained with sequential learning need around ten cycles to converge, and models trained with parallel learning even more than 40. Both settings only mange to reaching an accuracy 6% lower than the centralized baseline and 2% higher than lower baseline. We notice the same trend for the settings with four and more than eight nodes. In settings with two nodes (see Fig. 4(b)) parallel learning performs better than sequential converging on the tenth cycle versus 20th. The accuracy of models trained with sequential learning is on average 1% less than in parallel and 3% better than the local average.

We notice that sequential Federated Learning performs better than parallel on non-iid data on the *Breast Cancer* dataset, allowing to reach model convergence for less number of federated cycles. This is especially the case with a larger number of nodes. Parallel learning also results in worse accuracy scores, for up to 5% lower accuracy than in centralized settings (sequential learning allows training models with up to 2% lower accuracy). However, both parallel and sequential learning require more federated cycles to train quality models on non-iid data (in different scenarios up to 50 more cycles).

Non-iid data drops federated learning performance on Location dataset even more. Figure 5(a) shows that on equally distributed data both parallel and sequential federated learning allow reaching accuracy score up to 4% lower than centralized baseline. Sequential and parallel learning had similar performance on non-iid data, which was up to 15% worse than centralized learning. We also notice that with a larger number of nodes in the setting the accuracy of the final model decreases. However, one can also see that the average of local training is dramatically decreasing with an increasing number of nodes. This can be explained by an unequal representation of each class in different nodes. In the considered non-iid setting, some of the nodes did not have instances from some classes at all and therefore could not learn any information about these classes. This is why with 32 nodes local average accuracy is only 15% while federated learning allows reaching 65–67% accuracy. Despite sequential and parallel federated learning allow reaching similar accuracy score, Fig. 5(b) shows that the speed of convergence is rather different. While parallel learning converges on 15th federated cycle, sequential allows reaching similar accuracy only around the 50th

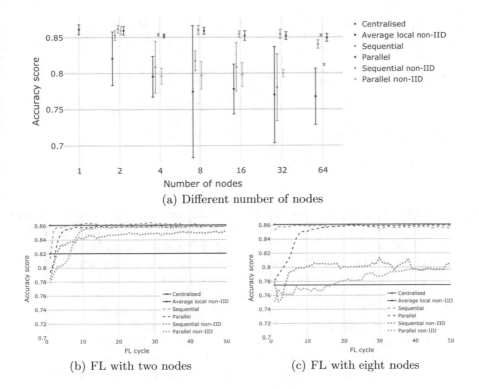

(a) Different number of nodes

(b) FL with two nodes   (c) FL with eight nodes

Fig. 4. Comparison of the effectiveness with differing number of nodes Federated Learning on *Adult* dataset, equally distributed and non-iid data

cycle. Moreover, parallel learning allows training models simultaneously, which can make the overall federated learning process more efficient than sequential.

Both sequential and parallel learning perform well on *Purchase-2* dataset allowing to reach centralized accuracy with up to 1% difference. Non-iid data has less impact on models quality on *Purchase-2* dataset, in comparison to classification tasks with a larger number of classes (*Purchase-10 – Purchase-100*). While sequential learning showed better performance on the datasets with two target classes (*Breast Cancer, Adult, Purchase-2*), the parallel setting allows reaching more stable and efficient results on the classification tasks with 10 and more classes in the setting (*Location, Purchase-10 – Purchase-100*).

4.3   Main Findings and Recommendations

From our experimental evaluation on eight different classification tasks on structured, relational datasets, we can draw the following findings:

– In the settings *with equal distribution of the data among the nodes, Federated Learning allows reaching accuracy close to centralized learning* (at most up to 5% difference in accuracy), with similar hyper-parameters and the same model architecture used in centralized settings.

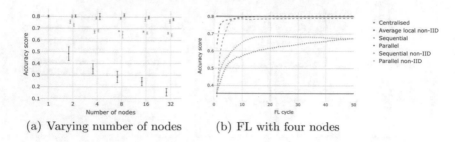

(a) Varying number of nodes   (b) FL with four nodes

Fig. 5. Federated Learning on *Location* dataset with non-iid data

- *Non-iid data demands more federated cycles for a model to converge than equally distributed data.* The speed of convergence decreases with an increasing number of nodes in federated learning. It is more difficult to reach baseline accuracy with non-iid data for both sequential and parallel federated learning, however both give *a higher score comparing to only local training.*
- We note that *classification tasks with two classes are less influenced by non-iid data*, than tasks with a larger number of target classes. The effect is especially pronounced in the settings with 16 nodes and more on all considered datasets. This is likely caused by the fact that with many classes and many nodes, the number of data items per node in each class is becoming very small, and it is thus very difficult for the individual models to train a generalizing model on that class.
- Parallel Federated Learning performs better than sequential on classification tasks with a large number of classes – especially with a larger number of nodes in the setting (more than four nodes). It allows to reach higher accuracy of the models, and takes less federated cycles for models to converge. That holds for both equally distributed and non-iid data. It is thus a *recommendation to use parallel Federated Learning on classification tasks with ten and more classes*
- Sequential learning showed good performance on classification tasks with two target classes, allowing to reach close to baseline accuracy (at most up to 2% lower), with less than ten cycles even with non-iid data. This approach, therefore, is a viable solution allowing to avoid centralized aggregation of the models as in parallel learning. It is thus a *recommendation to use sequential Federated Learning on two-classes classification tasks.*

5   Conclusion

Federated Learning allows to perform privacy-preserving machine learning on sensitive data, and thus offers an alternative to settings where data needs to be centralized and/or anonymised for processing. Federated Learning needs to achieve the effectiveness similar to centralized setting to be considered a viable alternative. In this paper, we thus studied the effectiveness of classification algorithms on multiple datasets.

We showed that Federated Learning allows reaching a baseline accuracy in settings with equally distributed data, comparable to models trained on centralized data (at most with up to 5% drop in accuracy score). The hyper-parameters applied in centralized learning (e.g. the number of epochs, iterations, learning rate) can be used to train effective models in sequential and parallel Federated Learning. With non-iid data, training good quality models with Federated Learning can results in significantly lower accuracy and can entail higher communication costs due to the larger number of federated cycles needed for the model to convergence.

Future work will focus on extending this analysis to additional datasets, different machine learning algorithms and a thorough investigation of privacy threats in Federated Learning.

Acknowledgments. This work was partially funded from the European Union's Horizon 2020 research and innovation programme under grant agreement No 826078 (FeatureCloud). SBA Research (SBA-K1) is a COMET Centre within the COMET – Competence Centers for Excellent Technologies Programme and funded by BMK, BMDW, and the federal state of Vienna. The COMET Programme is managed by FFG.

References

1. McMahan, B., Moore, E., Ramage, D., Hampson, S., y Arcas, B.A.: Communication-efficient learning of deep networks from decentralized data. In: International Conference on Artificial Intelligence and Statistics, Fort Lauderdale, FL, USA, (2017). PMLR
2. Sheller, M.J., Reina, G.A., Edwards, B., Martin, J., Bakas, S.: Multi-institutional deep learning modeling without sharing patient data: a feasibility study on brain tumor segmentation. In: International Workshop on Brain Lesion (BrainLes), in conjunction with MICCAI (2018)
3. Rieke, N., Hancox, J., Li, W., et al.: The future of digital health with federated learning. NPJ Digit. Med. **3**(1) (2020)
4. Konečný, J., McMahan, H.B., Yu, F.X. Richtárik, P., Suresh, A.T., Bacon, D.: Federated learning: strategies for improving communication efficiency. In: NIPS Workshop on Private Multi-Party Machine Learning (2016)
5. Nishio, T., Yonetani, R.: Client selection for federated learning with heterogeneous resources in mobile edge. In: IEEE International Conference on Communications (2019)
6. Sattler, F., Wiedemann, S., Müller, K.-R., Samek, W.: Robust and communication-efficient federated learning from non-i.i.d. data. IEEE Trans. Neural Netw. Learn. Syst. (2019)
7. Lyu, L., Han, Yu., Zhao, J., Yang, Q.: Threats to Federated Learning. Springer, Cham (2020)
8. Truex, S., Liu, L., Gursoy, M., Lei, Yu., Wei, W.: Demystifying membership inference attacks in machine learning as a service. IEEE Trans. Serv. Comput. (2019)

9. Nilsson, A., Smith, S., Ulm, G., Gustavsson, E., Jirstrand, M.: A performance evaluation of federated learning algorithms. In: Workshop on Distributed Infrastructures for Deep Learning. ACM (2018)
10. Shokri, R., Stronati, M., Song, C., Shmatikov, V.: Membership inference attacks against machine learning models. In: IEEE Symposium on Security and Privacy (SP) (2017)
11. Marcano-Cedeño, A., Buendía-Buendía, F.S., Andina, D.: Breast cancer classification applying artificial metaplasticity. In: Bioinspired Applications in Artificial and Natural Computation. Springer, Berlin, Heidelberg (2009)

A Multi-view Active Learning Approach for the Hierarchical Multi-label Classification of Research Papers

Abir Masmoudi[✉], Hatem Bellaaj, and Mohamed Jmaiel

ReDCAD Laboratory, University of Sfax, Sfax, Tunisia
{abir.masmoudi,hatem.bellaaj,mohamed.jmaiel}@redcad.org

Abstract. In this paper, we focus on the hierarchical multi-label classification task of scientific papers, which consists in assigning to a paper the set of relevant classes, which are organized in a hierarchy. The difficulty of manually constructing sufficient labeled datasets renders challenging the automatic classification task of research papers according to hierarchical labels. Multi-view active learning is a widely adopted method to address this issue, by iteratively selecting the most useful unlabeled samples for the multi-view classifiers exploiting disjoint data' views, and querying an oracle on their real labels. However, none of the state of the art studies in this field is proposed for the hierarchical multi-label classification task. In this paper, we propose an effective multi-view active learning framework for the hierarchical multi-label classification task, applied on scientific papers. Our approach adopts a novel selection strategy that relies on both uncertainty and representativeness criteria when selecting the most informative unlabeled samples in each iteration. Experimental results on a real world dataset of ACM research papers show the efficiency of our approach over several baseline methods.

1 Introduction

The growing number of published scientific papers renders necessary their automatic classification, which facilitates their further search and retrieval by researchers, students, and professors. For that, machine-learning methods have been widely used to solve the problem of research papers classification, which allows to automatically associate each paper with its relevant label(s). In many real cases, a paper may be assigned to several labels simultaneously that are organized in a hierarchy structure. For instance, the ACM classification tree and the MESH thesaurus are typical hierarchical classification schemas used to classify research papers. Hence, research papers classification can be considered as a typical case of Hierarchical Multi-label Classification (HMC) problems. In the HMC problem, the assignment of a given label is implicitly automating the assignment of all its parent labels (so-called the hierarchy constraint). Moreover, the assigned labels for a paper can belong to different paths in the label hierarchy.

Several HMC methods are proposed in the literature that can be categorized into two main groups: local and global approaches [1]. The local approach builds a set of local classifiers on the label hierarchy, where each classifier is used for predicting a specific

© The Author(s), under exclusive license to Springer Nature Switzerland AG 2021
L. Barolli et al. (Eds.): AINA 2021, LNNS 226, pp. 407–420, 2021.
https://doi.org/10.1007/978-3-030-75075-6_33

label, parent label or hierarchical level. At the prediction phase, the final classification is obtained by combining the local classifiers' outputs while taking into account the hierarchy constraint. The global approach builds a single model that deals with all labels in the hierarchy at once, and identifies the relevant labels for an unseen example. The success of these HMC methods depends extremely on the size of the labeled training data. However, in real world situations, it is very difficult to manually annotate huge amounts of training samples according to hierarchical labels, which are sufficient for achieving good classification performances.

This challenging issue calls for the use of Active Learning (AL) [2], which facilitates creating high-quality datasets with less human labeling costs by exploiting labeled and unlabeled data. More precisely, AL is an iterative learning process that selects the most informative unlabeled samples for the classifier based on certain criteria. The selected samples are assigned with their ground-truth labels by an oracle (e.g., expert), and added to the initial labeled set. The batch-mode AL, wherein a batch of samples is selected in each iteration, is an efficient method to speed up the learning process. In batch-mode AL, three criteria can be considered which are: uncertainty, representativeness, and diversity [3]. The uncertainty criterion chooses the unlabeled samples whose predicted labels by the classifier are assigned with less certainty. The representativeness criterion selects samples that well represent the underlying distribution of unlabeled data. Thus, samples residing in dense regions of the feature space are the most representative ones, since they have a high similarity towards a large set of unlabeled samples. As for the diversity criterion, it promotes choosing scattered samples in the feature space, which ensures that selected samples are not located in a specific high-density region.

Multi-View Learning (MVL) [4] is another important paradigm, that reduces the required amount of labeled data. In MVL, data are described by multiple complementary views, where each view is sufficient for the learning step, and independent from the other views. Furthermore, learning is made by a committee of classifiers, which use disjoint feature subsets corresponding to different data' views. Hence, by combining the multi-view classifiers' outputs, their individual errors are minimized, which avoids the need for larger training datasets.

MVL has been successfully integrated into several AL studies (e.g., [5, 6]). The first MVAL algorithm, called Co-testing [5], selects contention samples, i.e., samples on which multi-view classifiers provide different labels, among which some samples are chosen as the most uncertain ones. Many MVAL studies (e.g. [6]) have been further suggested in binary and single-label settings. They often focus on quantifying the disagreement degree among multi-view classifiers on the predicted labels of candidate unlabeled samples, and choose the samples with high disagreement levels as the most uncertain ones. However, existing MVAL methods assess the uncertainty of unlabeled samples across the different views, assumed to be assigned with a single-label. Thus, none of the proposed MVAL methods can be directly applied in HMC settings wherein samples are assigned with many labels simultaneously that are hierarchically structured.

Differently from existing studies, in this work, we propose a novel MVAL approach for the HMC task, which is applied on scientific papers while using content and bibliographic coupling information as two distinct papers' views. Our proposed approach applies an effective two-steps selection mechanism for identifying the batch of the most informative unlabeled samples in each MVAL round based on uncertainty and representativeness criteria. Firstly, a set of the most uncertain samples are selected by maximizing the disagreement information on all their contention labels across the disjoint views. Then, from the preselected set of samples with maximum uncertainty, a final batch of the most representative ones are selected based on their similarity degrees towards their K-nearest neighbors. The most informative samples, which are the most uncertain and representative ones, are further assigned with their true labels by an oracle, and added to the labeled training set. Hence, the classifiers are retrained on the updated labeled set during the following iteration. The described process repeats until the stopping condition is satisfied (i.e., the maximum number of iterations is achieved).

Experimental results on a dataset of ACM research papers prove the advantage of the proposed approach compared to some baseline methods.

The remainder of this paper is organized as follows. Section 2 reviews state-of-the art researches on AL and MVAL domains. Section 3 describes the newly proposed approach. Section 4 presents the experimental results of the proposed approach with regards to several baselines. Section 5 concludes the paper, and outlines future research lines.

2 Literature Review

As explained in [3], there are three different criteria that can be used for selecting the batch of the most useful samples in each AL iteration: which are uncertainty, representativeness, and diversity. Several AL studies rely only on the uncertainty criteria for identifying the most informative samples. In uncertainty-based AL approaches, the classifier queries the samples whose predicted labels are assigned with less certainty, i.e., the classifier is less sure about the labels. For single-label classification tasks with a probabilistic classifier, a simple method is to choose the predicted positive samples having their posterior probabilities that are near to 0.5 as the most uncertain ones [2]. Entropy is also a widely adopted uncertainty measure that chooses the sample having the maximum entropy of its estimated label.

In multi-label settings, computing the sample's uncertainty is a more complicated task since it should consider all its predicted labels for deducing the overall score. For instance, in [7], the independent uncertainties over all labels are assessed based on the outputs of independent binary classifiers, and then a weighted combination of the computed uncertainties is performed to select the most uncertain samples. In HMC settings, few uncertainty measures [8, 9] have been proposed. However, the suggested uncertainty measures query instance/label pairs instead of a single sample associated with all its true labels.

The main issue encountered when considering uncertainty criteria only is that it is prone to select outliers, which exhibit a strong ambiguity in labels prediction for the learner, but have no benefit on the classifier's performance. The representativeness criterion aims to solve this problem by evaluating whether an example well represents the underlying distribution of unlabeled data. A straightforward method [10] is to compute the sample's representativeness based on their similarity degree towards all unlabeled data. The main idea is that a representative sample is expected to share a high similarity with many unlabeled samples. Some recent methods [11, 12] characterize the sample's representativeness based on its k-nearest neighbors from the unlabeled set.

Other AL studies (e.g. [13]) suggest selecting uncertain samples, but also the most diverse ones in the feature space, thus reducing the redundancy issue between selected samples in the same batch. They typically rely on clustering techniques that group similar samples into the same clusters. In fact, since samples in the same cluster represent similar information, a single instance is often selected from each cluster to ensure the differences between the final selected examples.

Multi-view Active Learning. Multi-view active learning is a domain resulting from the combination of active learning and multi-view learning fields that seeks adding the most informative unlabeled samples for the multi-view classifiers exploiting disjoint data' views to the labeled set. Co-testing is the first proposed MVAL method that selects contention samples from which a batch of the most informative unlabeled samples are considered as the most uncertain ones for the classifiers. Inspired by co-testing method, most MVAL studies focus on quantifying the disagreement information on labels predictions for discovering the most uncertain unlabeled samples. For instance, in [14, 15], the authors use a disagree-based query selection strategy for single-label settings that quantifies the sample's uncertainty as the number of unique label estimates across the different views. Samples with maximum disagreement levels are picked as the most informative ones. In [16], contention samples are firstly selected based on the disagreement among the multiple views. Then, the overall entropy of each contention sample is estimated by making a sum of its independent entropy score of its classification in each view. Samples with the highest entropy scores are considered as the most informative ones. In [17], the uncertainty is estimated by making a trade-off between the disagreement level among the different views, and the independent uncertainty in each view. The former is computed based on the number of unique predicted labels among classifiers. The latter is estimated as a summation of the view-wise posterior distributions, while using breaking ties [18] as a measure for computing the individual uncertainty of the predicted label in each view.

All the above-mentioned uncertainty measures are designed for binary and single-label classification tasks, wherein a sample is associated with a single label. The measure presented in [19], which is applied in multi-label settings, quantifies the overall uncertainty score of a sample/label pair by summing the independent uncertainty in each view separately, and also by considering the disagreement information across the different views. Their used metric is a label-based uncertainty measure that assesses the uncertainty degree at the level of sample/label pairs, and does not allow computing the unified uncertainty of a sample on its different labels across the different views.

Based on the above analysis, we have deduced that none of the proposed MVAL studies was proposed for the HMC task. Moreover, the suggested instance-based uncertainty measures in these MVAL studies assume that each sample is assigned to only a single label, and thus are not directly applicable in HMC settings, as they should consider the multiple labels assigned to instances.

3 Proposed Approach

In this section, we present our batch-mode MVAL approach for the HMC task, applied on scientific papers. In our approach, two hierarchical multi-label classifiers are firstly learnt using distinct feature vector representations of the initial labeled data corresponding to different data' views. Then, they are iteratively updated with the most informative unlabeled samples (with their true labels) based on *uncertainty* and *representativeness* criteria.

In the following, we firstly define some notations, and problem definition. Secondly, we present the used hierarchical multi-label classifier. Thirdly, we explain how we quantify the sample's uncertainty and representativeness, respectively. Finally, we present the suggested algorithm.

3.1 Notations

In a typical MVAL-HMC framework, we have an input space $X = X^{(1)} \times X^{(2)}$ where $X^{(1)}$ and $X^{(2)}$ refer to two distinct views of a sample. $Y = \{0, 1\}^{lb}$ is the label space, where $lb = |Q|$ is the number of labels, and $Q = \{q_1, q_2, \ldots, q_{lb}\}$ is the set of labels, which are organized in a hierarchy $(Q, \leq_h)$ where $\leq_h$ is a partial order representing parent-child relationships ($\forall q_1, q_2 \in Q : q_1 \leq_h q_2 \Leftrightarrow q_2$ is a child class of q_1).

Each instance i in the input space is represented as $x_i = \left(x_i^{(1)}, x_i^{(2)} \right)$, where $x_i^{(1)}$ corresponds to the feature vector in the first view while $x_i^{(2)}$ refers to the feature vector in the second view. Each sample i may be equally assigned with a set of labels from Q, represented in a labels vector $y_i = \{y_{i1}, y_{ik}, \ldots, y_{i|Q|}\} \in \{0, 1\}^{lb}$, where $y_{ik} = 1$ if the kth label is relevant to x_i, 0 otherwise.

Given a very small set $L = \{(x_i, y_i)\}_{i=0}^{l}$ containing l labeled samples, and a larger set $U = \{(x_i)\}_{i=l}^{l+u}$ of u unlabeled samples, i.e., $u \gg l$. The main goal is to enhance the performance of two multi-view hierarchical multi-label classifiers $C^{(1)}$ and $C^{(2)}$, such as $C^{(j)} : X^{(j)} \to Y$, which exploit both L and U training sets, when predicting the labels of unseen test examples.

3.2 Hierarchical Multi-label Classifier

The hierarchical multi-label classifier, used in this work, adopts a local approach that trains a multi-label classifier at each parent node of the hierarchy, which is responsible in predicting its corresponding children nodes. At the prediction step, it follows the top-down fashion, and thus surfs the label hierarchy from root to leaf nodes for determining the relevant labels. As a multi-label classifier, we use the Binary Relevance (BR) method [20] that builds a separate binary classifier for each label in the hierarchy. Here, we adopt Random Forest, which is a rule-based learner, as a basic binary classifier. The latter is chosen because rule-based classifiers are more effective in dealing with the sampling bias problem with regards to margin-based classifiers.

3.3 Uncertainty

In MVAL, the samples' uncertainties is often quantified based on the disagreement value on their labels predictions across the different views. In HMC setting, hierarchical multi-label classifiers may disagree, for a specific sample, on several labels, which are referred to as contention labels (i.e., labels on which classifiers give different predictions). In this work, the unified uncertainty degree of an unlabelled sample x_i is computed by aggregating the disagreement scores on all its contention labels, and normalizing the obtained score by the number of contention labels[1], as defined by the following equation:

$$Unc(x_i) = \frac{1}{|\hat{Q}_i|} \sum_{q \in \hat{Q}_i} Dis(x_i, q) \tag{1}$$

Where $\hat{Q}_i$ is the set of contention labels for x_i, and $Dis(x_i, q)$ refers to the disagreement level for x_i with respect to contention label q.

The disagreement level for an unlabeled sample with respect to a contention label can be computed based on the classifiers' outputs (i.e., confidence scores). Let $C_q^{(v)}\left(x_i^{(v)}\right)$ the binary prediction value of a contention label q for an unlabeled sample x_i by the hierarchical multi-label classifier $C^{(v)}$, which uses the features of the vth view, $v = 1, 2$. Its absolute real value $|C_q^{(v)}\left(x_i^{(v)}\right)| \in [0, 1]$ is considered as the confidence score.

Hence, the disagreement level $Dis(x_i, q)$ for a contention label $q\left(i.e., C_q^{(1)}\left(x_i^{(1)}\right) \neq C_q^{(2)}\left(x_i^{(2)}\right)\right)$ is formally defined as:

$$Dis(x_i, q) = \left| |C_q^{(1)}\left(x_i^{(1)}\right)| - |C_q^{(2)}\left(x_i^{(2)}\right)| \right| \tag{2}$$

According to Eq. (2), the higher the difference between the confidence scores is, the higher the disagreement level is between classifiers with respect to contention label q. When the disagreement level is large, one of the classifiers has surely provided a confident prediction on label q while the other has completely failed. In such case, we expect that including the unlabeled sample x_i (with its ground-truth labels) would minimize the errors of the mistaken classifier for the contention label q while simultaneously reinforcing the assumptions of the correct classifier. Consequently, samples with the highest disagreement levels on all their contention labels are the most uncertain ones for the classifiers, as defined in (1).

3.4 Representativeness

Representativeness criterion chooses the samples that well represent the underlying distribution of unlabeled data, which avoids selecting outliers. Some AL studies [11, 12] assess the samples' representativeness based on their k-nearest neighbours in the unlabeled set. Their main idea is that an example is considered representative if it has a high

---

[1] Our measure favors selecting samples with high disagreement level on all their contention labels irrespective of the number of contention labels.

similarity towards its k most similar unlabeled samples. Hence, the representativeness degree of an unlabeled example x_i can be written as:

$$Rep(x_i) = \sum_{x_j \in N(x_i, U, k)} sim(x_i, x_j) \tag{3}$$

Where $N(x_i, U, k)$ represents the k most similar examples to x_i from the unlabeled set U, and $sim(x_i, x_j)$ refers to the similarity between samples x_i and x_j based on certain distance metric.

In this work, the similarity between two samples x_i and x_j is calculated based on their Euclidian distance as follows:

$$Sim(x_i, x_j) = \frac{1}{ED(x_i, x_j)} \tag{4}$$

Where $ED(x_i, x_j)$ refers to the Euclidian distance between x_i and x_j.

3.5 Proposed Algorithm

The pseudo code of the proposed batch-mode MVAL algorithm for the HMC of scientific papers is illustrated in Algorithm 1. The proposed algorithm takes as input a labeled set L and an unlabeled set U. N, p, k, and m parameters denote the number of iterations, the number of selected samples based on uncertainty criterion, the number of nearest neighbouring samples, and the final batch size, respectively. At each iteration, two hierarchical multi-label classifiers $C^{(1)}$ and $C^{(2)}$ are initially trained using distinct views of the labeled set L (Lines 3–4). Then, we choose a batch of the most informative samples from the unlabeled set U by applying our selective sampling strategy that relies on our adopted uncertainty and representativeness measures (Lines 5–14). Specifically, the uncertainty level is firstly computed for each unlabeled example in U by considering the disagreement level on its contention labels, as defined by Eq. (1). The p unlabeled examples having the highest uncertainty degrees are subsequently selected, and put in another set $\hat{U}$. Then, we compute the representativeness' scores of the candidate samples in $\hat{U}$ based on their k nearest neighbours, as defined in Eq. (3). We choose a subset of m samples having the highest representativeness scores to be included into the final batch H, which are the most informative ones. Each sample in H is assigned with its ground-truth labels by an oracle, and moved from the unlabeled set U to the labeled set L (line 15–19). Thus, in the following iteration, the classifiers are retrained on the augmented labeled set. Once the iterative training is finished (i.e., the maximum number of iterations N is achieved), we get two final classifiers $C^{(1)}$ and $C^{(2)}$, which are used for determining the labels for each test sample. In this work, we adopt the consensus principle in final labels predictions, i.e., a label is assigned to an example only when it is judged relevant by both classifiers.

Algorithm 1
===

INPUT
L: labeled set
U: unlabeled set
$X^{(1)}$, $X^{(2)}$: two feature sets
$C^{(1)}$, $C^{(2)}$: two hierarchical multi-label classifiers
N : maximum number of iterations
p : number of candidate samples based on their
uncertainty degree
k : number of nearest neighboring samples
m : batch size of the most informative samples
BEGIN
1:       Iterations=0
2:       Repeat
3:       Train hierarchical multi-label classifier $C^{(1)}$
         using L based on $X^{(1)}$
4:       Train   hierarchical   multi-label   classifier
         $C^{(2)}$ using L based on $X^{(2)}$
5:        For each x_i in U do
6:         Compute $Unc(x_i)$ as in Equation (1)
7:        End For
8:        Select the p unlabeled examples having the
9:        highest uncertainty scores and put them in a
          pool $\hat{U}$
10:       For each x_i in $\hat{U}$ do
11:        Compute $Rep(x_i)$ as in Equation (3)
12:       End For
13:       Select the m unlabeled examples having the
14:       highest representativeness scores and put them
          in a pool H
15:       For each x_i in H do
16:        Query an oracle on the true labels $\breve{y}_i$ for x_i
17:        Add the example $(x_i, \breve{y}_i)$ to L ($L=L \cup (x_i, \breve{y}_i)$)
18:        Remove it from U ($U= U- (x_i)$)
19:       End for
20:      iterations= iterations+1
21:      Until (iterations>N)
OUTPUT
Two final classifiers $C^{(1)}$ and $C^{(2)}$, which are trained on
the augmented set L, and their outputs are combined to
predict the labels of testing samples.

4   Evaluation Study

4.1   Dataset

To evaluate the performance of the proposed approach, we use a dataset of ACM research papers, wherein papers are annotated according to the ACM hierarchical taxonomy, that we have constructed in our previous work [21]. Papers were automatically extracted from the ACM website using a Web data extraction tool[2]. The extracted raw set contains papers having all their metadata (i.e., title, abstract, cited references, and their categories). Each paper in the dataset has two feature vectors corresponding to its content and bibliographic coupling information.

Content-Based Feature Vector. We generate a content-based feature vector for each paper in the raw dataset. The first step is to apply some pre-processing tasks for extracting the vocabulary terms from all papers' contents. More precisely, the abstracts and titles of all papers are firstly segmented by applying Part-Of-Speech (POS) tagging to keep only the words that consist of nouns, verbs and adjectives. Then, the retrieved words are converted into their base forms using the Porter Stemmer algorithm, while removing the low-frequency terms. Once choosing the vocabulary terms, a paper-term matrix P is built where each cell $P_{i,j}$ indicates the importance of the jth term feature in the ith paper, which is computed based on TF weighting schema, and scaled to the range [0–1]. Thus, each row P_i refers to the BoW representation of the ith paper.

Bibliographic Coupling-Based Feature Vector. We built a bibliographic coupling-based feature vector for each paper using a bibliometric similarity matrix whose cells contain the pairwise similarities between all papers from the input collection. The latter are computed using DescriptiveBC [22], an improved bibliographic coupling measure. Specifically, DescriptiveBC estimates the similarity between two papers by searching for each reference in one of the papers, the most similar reference in the other paper, and vice versa. In DescriptiveBC, the similarity between two references is computed based on the common terms in their titles. For more details about DescriptiveBC measure, the reader may refer to [22].

Based on the computed similarities between papers using DescriptiveBC measure, we construct a bibliometric similarity matrix M wherein each cell to refers to the similarity score between a pair of papers i and j, and each row M_i refer to the bibliographic coupling-based feature vector of a paper i.

Dataset Description. Table 1 provides a description of the used dataset with a set of statistical properties like the number of samples (S), the number of features (F), the number of labels (Lb), the number of distinct label combinations (Dl), and the depth of the hierarchy (d). In our experiments, our dataset is divided into three parts with equal sizes using stratified sampling: one partition is used as the test set, and two remaining partitions are considered as the training dataset.

As we seek to assess the impact of the size of the initial labelled training set, we have adopted varied ratios of labelled examples number (Ln) to unlabelled samples number (Un) in all conducted experiments, that is (Ln:Un) = (1:6) and (1:7). Note that all experiments are repeated 10 times with 10 different random divisions of the dataset

---

[2] https://www.connotate.com/connotateexpress/.

Table 1. Description of our experimental dataset.

Dataset	Properties				
	S	F	Lb	Dl	d
ACM	3170	7617	16	151	2

(i.e. 10-fold cross validation), and the obtained results are averaged to deduce the final scores.

4.2 Experiments and Results

In this section, we discuss the evaluation results of the proposed approach with regards to several comparative methods on our experimental dataset. To be fair, we compare our proposed approach only with *instance-based* MVAL baselines, which are:

- **Random:** it randomly selects a set of m instances from the unlabeled pool at each iteration.
- **MVAL-U:** an MVAL method that chooses the m most informative samples based solely on the suggested uncertainty measure, presented in Sect. 3.3.
- **MVAL-ContentionSample:** an MVAL method that chooses a subset of m randomly chosen contention samples (i.e., samples on which classifiers provide different predictions for at least one of the labels, similar to the principle adopted in naïve Co-testing method [5]).
- **MVAL-UD:** an MVAL method that identifies the p most uncertain samples based on our proposed uncertainty measure. Then, it applies the K-means clustering technique used in [13] to group the p most uncertain samples into m clusters. Finally, the sample with the highest uncertainty score from each cluster is included into the final batch of the most informative samples, to guarantee the diversity criteria between the selected most uncertain samples.

In all the comparative approaches, we keep using the same hierarchical multi-label classifier, described in Sect. 3.2. All these methods were implemented in the Mulan package, which is an open source Java library for multi-label learning. The different parameters are fixed as follows: $N = 50$, $p = 60$, $k = 20$ and $m = 10$. Moreover, the multi-view classifiers' outputs are combined based on the consensus principle, as in our proposed approach.

In this work, we use the Micro-F1 and Macro-F1 measures, which are two multi-label evaluation metrics that are widely used in HMC contexts. The Micro-F1 measure computes the precision and recall, which are computed over all sample-label pairs, which gives the opportunity for the majority labels to influence the scores. As for the Macro-F1 measure, it computes the precision and recall values for each label separately, and then averages to get the overall score. It gives equal importance for each label, and thus provides more emphasis for the minority labels to influence the Macro-F1 scores.

Figures 1 and 2 show the experimental results of the different methods in terms of Micro-F1 and Macro-F1 scores, respectively, on our experimental dataset, at various labeling rates. From Fig. 1, many conclusions can be made. Firstly, we found that, under various labeling rates, the performance results of MVAL-U method, which relies on our proposed uncertainty measure, do not always reach superior results as compared to the naïve random selection method during the iterative learning process. This can be explained by the fact, that choosing the most uncertain samples is prone to selecting redundant samples in the feature space and unuseful ones (i.e., outliers), especially when the batch size is quite large, which degrades the efficiency of MVAL-U method.

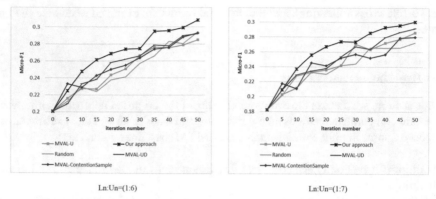

Ln:Un=(1:6)   Ln:Un=(1:7)

Fig. 1. Experimental results on ACM dataset where the horizontal axis is the iteration number, and the vertical axis refers to the Micro-F1 score.

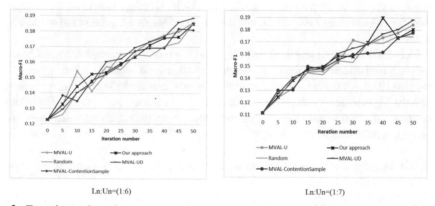

Ln:Un=(1:6)   Ln:Un=(1:7)

Fig. 2. Experimental results on our experimental dataset in terms of Macro-F1 scores at different learning iterations.

Second, we see that MVAL-UD method mostly provide superior results as compared to MVAL-U and Random methods across different learning iterations under varied sizes of the initial labeled training data. Obtained results prove the efficiency of combining both the samples' uncertainty and diversity when choosing the most informative unlabeled samples for achieving higher overall classification results than random selection strategy. Third, we see that our approach is almost outperforming MVAL-U method, which rely on our adopted uncertainty measure respectively. This proves the efficiency of our suggested combination strategy in better selecting the most useful unlabeled samples, which helps achieving superior results across the iterative learning process. Moreover, we found that our proposed approach obtains better results than MVAL-UD method, which demonstrates that, on our experimental dataset, combining uncertainty and representativeness criteria is more useful than uncertainty and diversity criteria. We can conclude that our approach is the best method since it succeeds in achieving superior performance results upon the other comparative baselines (i.e., Random, and MVAL-ContentionSample) in terms of Micro-F1 scores in most cases.

By looking at Fig. 2, we observe that our proposed approach do not reach the highest Macro-F1 results compared to the other baseline methods. As the Macro-F1 scores are dominated by the performances results on the minority labels, we deduce that our selection strategy do not guarantee a more balanced learning of the different labels. These results lead us to conclude that our proposed approach tend to add samples associated with the majority labels as the most informative ones, to the detriment of the minority ones.

5   Conclusion

In this paper, we present an MVAL approach for the HMC task, applied on scientific papers, while using content and bibliographic coupling information as two distinct papers' views. To the best of our knowledge, none of the proposed MVAL studies have been proposed for the HMC task, in which labels may belong to a complex hierarchical structure. Our proposed MVAL approach suggests an effective selection mechanism for choosing a batch of the most informative samples for the multi-view classifiers exploiting disjoint data' views based on uncertainty and representativeness criteria. Experiments validate the efficiency of the suggested approach against some MVAL baselines, and show the usefulness of combining the samples' uncertainty and representativeness for reaching the highest classification results in terms of Micro-F1 scores. As future work, we intend to study the impact of other influential parameters in our approach as the use of other kinds of HMC learners.

References

1. Silla, C.N., Freitas, A.A.: A survey of hierarchical classification across different application domains. Data Min. Knowl. Disc. **22**(1), 31–72 (2011)
2. Settles, B.: Active learning literature survey, computer sciences. Technical Report 1648. University of Wisconsin–Madison (2009)
3. He, T., Zhang, S., Xin, J., Zhao, P., Wu, J., Xian, X., Cui, Z.: An active learning approach with uncertainty, representativeness, and diversity. Sci. World J. (2014)
4. Zhao, J., Xie, X., Xu, X., Sun, S.: Multi-view learning overview: recent progress and new challenges. Inf. Fusion **38**, 43–54 (2017)
5. Muslea, I., Minton, S., Knoblock, C.A.: Active learning with multiple views. J. Artif. Intell. Res. **27**, 203–233 (2006)
6. Hu, J., He, Z., Li, J., He, L., Wang, Y.: 3D-Gabor inspired multiview active learning for spectral-spatial hyperspectral image classification. Remote Sens. **10**(7), 1070 (2018)
7. Brinker, K.: On active learning in multi-label classification. In: From Data and Information Analysis to Knowledge Engineering, pp. 206–213. Springer, Berlin, Heidelberg (2006)
8. Nakano, F.K., Cerri, R., Vens, C.: Active learning for hierarchical multi-label classification. Data Min. Knowl. Disc. **34**(5), 1496–1530 (2020)
9. Yan, Y., Huang, S.J.: Cost-effective active learning for hierarchical multi-label classification. In: IJCAI, pp. 2962–2968 (2018)

10. Settles, B., Craven, M.: An analysis of active learning strategies for sequence labeling tasks. In: Proceedings of the 2008 Conference on Empirical Methods in Natural Language Processing, pp. 1070–1079 (2008)
11. Zhu, J., Wang, H., Tsou, B.K., Ma, M.: Active learning with sampling by uncertainty and density for data annotations. IEEE Trans. Audio, Speech Lang. Process. **18**(6), 1323–1331 (2009)
12. Yang, K., Cai, Y., Cai, Z., Xie, H., Wong, T.L., Chan, W.H.: Top K representative: a method to select representative samples based on K nearest neighbors. Int. J. Mach. Learn. Cybern. 1–11 (2019)
13. Patra, S., Bruzzone, L.: A batch-mode active learning technique based on multiple uncertainty for SVM classifier. IEEE Geosci. Remote Sens. Lett. **9**(3), 497–501 (2011)
14. Di, W., Crawford, M.M.: Multi-view adaptive disagreement based active learning for hyperspectral image classification. In: 2010 IEEE International Geoscience and Remote Sensing Symposium, pp. 1374–1377 (2010)
15. Zhang, Z., Pasolli, E., Crawford, M.M.: An adaptive multiview active learning approach for spectral-spatial classification of hyperspectral images. IEEE Trans. Geosci. Remote Sens. **58**(4), 2557–2570 (2019)
16. Zhang, Y., Lv, D., Zhao, Y.: Multiple-view active learning for environmental sound classification. Int. J. Online Biomed. Eng. (iJOE) **12**(12), 49–54 (2016)
17. Hu, J., He, Z., Li, J., He, L., Wang, Y.: 3D-Gabor inspired multiview active learning for spectral-spatial hyperspectral image classification. Remote Sens. **10**(7), 1070 (2018)
18. Liu, C., He, L., Li, Z., Li, J.: Feature-driven active learning for hyperspectral image classification. IEEE Trans. Geosci. Remote Sens. **56**(1), 341–354 (2017)
19. Zhang, X., Cheng, J., Xu, C., Lu, H., Ma, S.: Multi-view multi-label active learning for image classification. In: 2009 IEEE International Conference on Multimedia and Expo, pp. 258–261 (2009)
20. Zhang, M.L., Li, Y.K., Liu, X.Y., Geng, X.: Binary relevance for multi-label learning: an overview. Front. Comput. Sci. **12**(2), 191–202 (2018)
21. Masmoudi, A., Bellaaj, H., Drira, K., Jmaiel, M.: A co-training-based approach for the hierarchical multi-label classification of research papers. Expert Syst. e12613 (2020)
22. Liu, R.L.: A new bibliographic coupling measure with descriptive capability. Scientometrics **110**(2), 915–935 (2017)

Blue-White Veil Classification of Dermoscopy Images Using Convolutional Neural Networks and Invariant Dataset Augmentation

Piotr Milczarski[1]([✉]) [ID], Michał Beczkowski[1] [ID], and Norbert Borowski[2] [ID]

[1] Faculty of Physics and Applied Informatics, Department of Computer Science, University of Lodz, Pomorska str. 149/153, 90-236 Lodz, Poland
{piotr.milczarski,michal.beczkowski}@uni.lodz.pl

[2] Faculty of Physics and Applied Informatics, Department of Nuclear Physics and Radiation Safety, University of Lodz, Pomorska str. 149/153, 90-236 Lodz, Poland
norbert.borowski@uni.lodz.pl

Abstract. In the dermoscopy, the Three-Point Checklist of Dermoscopy and the Seven-Point Checklist are proved to be sufficient screening methods in the skin lesions assessments checking by dermatology expert. In the methods, there is a criterion of blue-whitish veil appearance within the lesion and it can be classified using CNN classifiers. In the paper, we show the results of CNN application to the problem of the assessment of whether the blue-white veil is present or absent within the lesion. We build the neural network with the help of the available VGG19, Xception and Inception-ResNet-v2 pretrained convolutional neural networks, trained, validated and tested on the prepared images taken from the PH2, using the invariant dataset augmentation. The original authors' approach using the defined invariant dataset augmentation for expanding the test set by seven copies invariantly transformed from original images shows that the classification characteristics like accuracy and true positive rate as well as the F1 and MCC tests can be much higher (5–20%) than using only original images. In the paper, the confusion matrix parameters result in: 98–100% accuracy, 98–100% true positive rate, 0.0–2.3% false positive rate, tests F1 = 0.95 and MCC = 0.95 as well as AUC value close to 1. That general approach can provide higher results while using CNN networks in other disciplines not only in dermatology and dermoscopy.

Keywords: Convolutional neural networks · Dermoscopy · Blue-white veil classification · Invariant dataset augmentation

1 Introduction

1.1 Dermatological Screening Methods of Skin Lesions

In the dermoscopy, the Three-Point Checklist of Dermoscopy (3PCLD) [1–3] and the Seven-Point Checklist (7PCL) [4–6] are defined and proved to be the sufficient screening methods in the skin lesions assessments by the dermatology experts. In the 3PCLD methodology, there are criteria of asymmetry in shape, hue and structure distribution within the lesion defined and it can have value either 0, 1 or 2, presence or absence of the pigmented network and blue-white veil. The 7PCL checklist takes into account pigment network and blue-white veil as the assessment factors, but also streaks pigmentation, regression structures, dots and globules and vascular structures. The next example of the screening method is the ABCD rule, which also defines the symmetry/asymmetry of the lesion [7–9]. In the paper, we show the results of CNN application to the problem of the blue-white veil within the skin lesion in the dermoscopic images.

In the presented research paper, we build the neural network with the help of the chosen available pretrained convolutional neural networks Xception (XN), VGG19 [10] and Inception-ResNet-v2 (IRN2). From the available networks e.g. Xception and Inception-ResNet-v2, the VGG19 requires medium computational power and provides promising results even with a relatively small but well-described PH2 dataset [11]. We compare the best results achieved by the mentioned pretrained network. The results vary from 91 to more than 99%.

Melanoma is a life-threatening disease that is completely cured if removed in the early stages [12, 13]. This statement is also confirmed by the statistics shown by the European Cancer Information System [16] and the American Cancer Society [17]. Therefore, the proper treatment of the lesions that clinically might be recognized as melanoma is warranted while minimizing the excision of benign lesions.

The three or seven criteria were important in distinguishing malignant from pigmented skin lesions. These three criteria are asymmetry atypical pigment network and blue-white structures (a combination of earlier categories of blue-whitish veil and regression structures) a preliminary calculations showed that presences of any of two of these criteria indicates a high likelihood of melanoma.

The blue-white veil as a feature in dermoscopic approaches has been researched using the feature-based and appearance-based machine learning methods. In appearance-based methods, the CNN networks have been applied to the image dataset [4, 22, 23]. In [4] the authors have used Inception V3 pretrained network achieving in blue-white veil classification 87.6% accuracy, 96.6% true positive ratio (sensitivity), 49.3 specificity (50.7% false positive ratio), 89.0% precision and 0.87 area under curve for the ROC. In the feature-based methods [14, 24–27] the problem is how to derive the blue-white veil features. In the paper [25], the authors tested on a set of 179 dermoscopy images and achieved a detection error rate lower than 15% using the C4.5 method. In the paper [28], the authors used the VGG19 network and achieved the average accuracy for the blue-whitish veil 90.5, with the average true positive rate of 96.7% and the average false positive rate of 10.9%.

The paper is organized as follows. In Sect. 2 we present and discuss methods of used datasets to acquire dermoscopy lesion information. Blue-whitish veil clinical description and used methods of verification are thoroughly described. The Blue-whitish veil dermatological search methods are discussed in Sect. 3. Research and discussion of the results are presented in Sect. 4. The final results are given in the following section. Finally, Sect. 6 presents the conclusions.

1.2 Dermatological Datasets

We conduct our research on a certain type of data sets which contains clinical description of skin lesion. The PH2 dataset [11] consists of dermoscopic images which are described as follows. The dermoscopic images were obtained at the Dermatology Service of Hospital Pedro Hispano (Matosinhos, Portugal) under the same conditions through the Tuebinger Mole Analyzer system using a magnification of 20 times. They are 8-bit RGB color images with a resolution of 768 × 560 pixels.

This image database contains a total of 200 dermoscopic images of melanocytic lesions, including 80 common nevi, 80 atypical nevi, and 40 melanomas. The PH2 database includes medical annotation of all the images namely medical segmentation of the lesion, clinical and histological diagnosis and the assessment of several dermoscopic criteria (colors; pigment network; dots/globules; streaks; regression areas; blue-whitish veil) [11, 15, 18].

The ISIC Archive contains the largest publicly available collection of quality controlled dermoscopic images of skin lesions [19]. Presently, the ISIC Archive contains over 24,000 dermoscopic images, which were collected from leading clinical centers internationally and acquired from a variety of devices within each center. The other examples of the dermatological datasets can be found in [20, 21].

2 The Invariant Dataset Augmentation

The Invariant Dataset Augmentation is based on the authors' idea of how to expand the available image dataset using the available geometrical invariant transformations and not to lose fragile features. The blue-white veil can be less visible after a rotation by any angle when a pixel is calculated from the neighboring pixels.

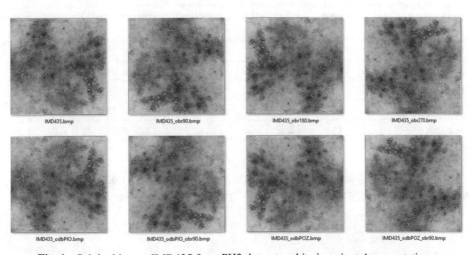

Fig. 1. Original image IMD435 from PH2 dataset and its invariant Augmentations.

These transformations do not change the asymmetry of shape, hue and structure distributions, as well as other features that are in the original copy of the image that are taken into account in 7PCL and 3PCLD. We have chosen seven image transformations: rotation by 90°, 180° and 270°, mirror reflection by a vertical and horizontal axis of the images and their rotations by 90°. The example result of the seven transformations on the image with id IMD435 from the PH2 dataset is presented in Fig. 1, where there is the original image in the upper left corner.

These transformations do not change the pixels, they are pixel invariant, mutually unambiguous and reversible. Seven new copies for each image are achieved. Altogether, we have 1600 images based on the PH2 dataset with each asymmetry feature eight-fold increased. These images can be taken to build CNN networks for the classification of the images and finding their features e.g. the blue-white veil, the type of the skin lesion asymmetry and the pigment network.

3 Pretrained CNN and Their Accuracy

The pretrained Convolutional Neural Networks have different features that should be taken into account when choosing a network to apply to a given problem. The most important characteristics are network accuracy, true positive and negative ratio, speed, and size. While selecting a network these features should be taken into account. Currently, we can choose within several pretrained networks. The chosen five network characteristics are given in Table 1. The network depth is defined as the largest number of sequential convolutional or fully connected layers on a path from the input layer to the output layer. The inputs to all networks are RGB images.

Table 1. Pretrained Convolutional Neural Networks features

Network	Depth	Size [MB]	Parameters [Millions]	Image input size	Everage accuracy [%]
VGG19	19	535	144	224 × 224	70
Xception	71	85	22.9	299 × 299	80<
Inception-ResNet-v2	164	209	55.9	299 × 299	80
NASNet-Large	*	360	88.9	331 × 331	>80
Alexnet	8	227	61	227 × 227	55

The simplest and one of the oldest CNN networks is AlexNet which architecture is briefly shown in Fig. 2 and the dimensions after each main operation.

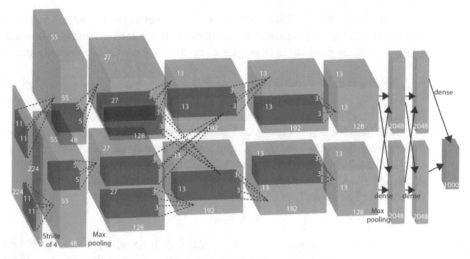

Fig. 2. CNN with its general layers model based on AlexNet.

4 General Method Description

In the research, we used the PH2 dataset as the training and testing sets. The general method can be described using the following steps:

1. The image set preprocessing to meet the CNN networks requirements (see Table 1):

   a. Scaling the images to the input size of the network 224px (or 299px for networks) to their shortest dimension (in our case height) using the Bicubic Sharper algorithm in Photoshop.
   b. Cropping the images as a square image. In the result, we achieved 200 images scaled to 224 × 224 px or 299 × 299 px respectively (see Table 1).
   c. Using the Invariant Dataset Augmentation we have achieved 1600 images based on the PH2 dataset with each asymmetry feature eight-fold increased. The transformations do not change the asymmetry of shape, hue and structure distributions, as well as other features that are in the original copy of the image that are taken into account in 7PCL and 3PCLD.

2. Setting up one of the chosen convolutional neural networks e.g. Xception, Inception-ResNet-v2.
3. Network training on:

   a. The dataset consisting of eight copies of the PH2 images from step 1c of the procedure.
   b. The dataset consisting of the original images of the PH2 images from step 1b of the procedure.

Both training datasets are built out of 75% of the images from the dataset obtained in the point 1b for each class and their corresponding 7 IDA transformations, see Table 2. Resulting networks in steps 3.a and 3.b are saved for future testing.

4. Network testing on:

    a. (for 3.a) the 25% of the original dataset from the step 1b.
    b. (for 3.b)) the 25% of the dataset containing the original images and their copies respectively, different from the training set described in step 3.b.
        The procedures in Steps 3 and 4 are repeated five times for each set of training, validation and test sets resulting in twenty networks and confusion matrices for the test sets. The example of the confusion matrix parameters are given in Tables 3 and 4.

5. Accuracy, true positive rate etc. defined and calculated according to Eq. (1)-(6) and their average values with the variance, minimum and maximum values are calculated for twenty (4rounds x5) networks obtained in steps 3–4.
6. The Accuracy + Error Type I (correct classification plus overestimation) – in the case of malignant melanoma the final diagnosis is usually after histopathological research. That is why it is better if the screening method overestimates the diagnosis than the opposite (underestimation error type II – False Negative). We consider that the best method has the biggest Accuracy + Error Type I.

The confusion matrix parameters are defined according to the following formulas:

$$ACC = (TP + TN)/N \tag{1}$$

$$TPR = TP/(TP + FN) \tag{2}$$

$$FPR = FN/(FP + TN) \tag{3}$$

$$Prec = TP/(TP + FP) \tag{4}$$

$$F1 = 2TP/(2TP + FP + FN) \tag{5}$$

$$MCC = (TP * TN - FP * FN)/\sqrt{(TP + FP)(TP + FN)(TN + FP)(TN + FN)} \tag{6}$$

where:

- N – a number of all cases;
- true positive, TP – number of positive results i.e. correctly classified cases;
- true negative, TN – number of negative results i.e. correctly classified cases;
- false positive, FP – number of negative results i.e. wrongly classified cases as positive ones;

- false negative, FN – number of positive results i.e. wrongly classified cases as negative ones, also called Type II error;
- accuracy, ACC;
- true positive rate, TPR, also called Recall;
- false positive rate, FPR;
- precision, Prec.;
- score test F1;
- the area under curve, AUC for the receiver operating characteristic curve, ROC;
- Matthews correlation coefficient, MCC.

To show the value of using the Invariant Dataset Augmentation we have compared this methodology with training and testing on the same image sets. The first one (marked as BW1) is using the IDA images. In the second one (BW2) we have used only the original 200 images, 150 for the train set and 50 for the train one. That set division and images were corresponding to the first approach. i.e. if IMD002 and its copies were in the training set in BW1, this image was also in the training set for BW2.

Table 2. Number of the images in the original PH2 dataset and augmented one.

Number of images	Original PH2 dataset			Invariant dataset Augmentation		
	Total	Train	Test	Total	Train	Test
Blue-white veil present	36	27	9	288	216	72
Blue-white veil absent	164	123	41	1312	984	328
Total	200	150	50	1600	1200	400

5   Results of the Classification Using Modified Pretrained CNNs

5.1   Hardware Description

The research has been conducted using Matlab 2019b with up-to-date versions of Deep Learning Toolbox™ (v. 12) on two computers independently (configurations are shown below), to cross-check the results. With the Deep Learning Toolbox, you can perform transfer learning with pretrained deep network models shown in Tables 3 and 4. The computer configurations are different but it only affects the time of execution in CNN networks training and at the end training. The calculated average accuracies, as well as their maximum and minimum ones, show results close to each other while running on both machines. The operating system on both computers is Microsoft Windows 10 Pro and the configurations are as follows:

- Set 1. Processor: Intel(R) Core(TM) i7-8700K CPU @ 3.70GHz (12 CPUs), Memory: 64GB RAM, Graphics Card: NVIDIA GTX 1080Ti with 11GB of Graphics RAM.

- Set 2. Processor: Intel(R) Core(TM) i7-6800HK CPU @ 3.60GHz (8 CPUs), Memory: 16GB RAM, Graphics Card: NVIDIA GTX 1070 with 8GB of Graphics RAM.

The second machine specification is used for the test of the procedure and checking whether the classification parameters depend on their hardware.

5.2   Results Using the Same Datasets

The chosen factors derived from the confusion matrix are presented in Tables 3 and 4 for BW1 and BW2. We can see that opposite to the accuracy the true positive rate (TPR) is much higher in the BW2 approach. This shows that the sensitivity of the method is high. In the T18 columns, the test results are shown using each image obtained from the IDA procedure as a separate one.

The worst scenario procedure used in the columns IDA works according to the following procedure. The original image and its DM copies are classified by the CNN networks. Then if in the eight results there is at least one positive case the result of the classification for the original image is positive. Else if all the result values are negative the result of the classification is negative. The procedure advantage is the higher value

Table 3. The classifications results for the BW1 training using IDA transformations with the test results for single original image set (T1), multiplicated dataset with 8 copies of the original file (T8) and using the worst-case scenario (IDA) classification. The chosen confusion matrix factors accuracy (ACC), true positive rate (TPR), false positive rate (FPR), area under curve (AUC) for the receiver operating characteristic curve (ROC) with their average (AVG), variance (VAR), minimum (MIN) and maximum (MAX) values and F1 score and Matthews correlation coefficient (MCC) for the chosen CNN network.

CM factor		VGG19			XN			IRN2		
		T1	T8	IDA	T1	T8	IDA	T1	T8	IDA
ACC [%]	AVG	94.2	94.1	93.8	93.0	93.2	92.8	92.2	92.6	91.5
	VAR	2.8	2.2	2.6	1.3	1.4	1.6	1.9	1.3	1.9
	Min	86.0	89.0	88.0	92.0	91.5	90.0	88.0	89.5	86.0
	Max	98.0	97.8	98.0	96.0	96.5	96.0	96.0	94.5	96.0
TPR [%]	AVG	76.7	77.6	82.7	72.2	69.9	84.4	68.3	67.8	76.1
	VAR	12.1	9.6	10.23	7.5	7.9	6.5	8.1	9.3	8.8
	Min	55.6	62.5	66.7	55.6	58.3	66.7	44.4	47.2	66.7
	Max	88.9	88.9	100	88.9	83.3	88.9	77.8	79.2	88.9
FPR [%]	AVG	2.0	2.3	3.8	2.4	1.7	5.4	2.6	1.9	5.1
	VAR	2.3	1.8	1.6	1.7	1.0	1.7	1.4	1.4	3.2
	Min	0.0	0.3	2.4	0.0	0.3	2.4	0.0	0.3	2.4
	Max	7.3	5.2	7.3	4.9	3.4	7.3	7.3	5.8	12.2
Test	F1	0.94	0.93	0.95	0.84	0.90	0.89	0.88	0.81	0.89
	MCC	0.93	0.91	0.94	0.81	0.88	0.86	0.86	0.78	0.86
AUC	AVG		.966			.969			.957	
	VAR		.027			.009			.012	
	Min		.914			.950			.930	
	Max		.996			.985			.971	

of true positive ratio (TPR) that is correlated with a higher false positive (FPR) which is a drawback of the procedure. Nonetheless, the worst scenario procedure is finding more positive cases than in the traditional classifications for a single file (T1) or treating each file as a separate file.

To achieve a high validation accuracy i.e. 100% or close and a low validation loss i.e. around 0.5 or less we have trained the networks using 30 epochs in the BW1 cases and from 60 (VGG19, IRN2) to 100 (Xception, VGG19) in the BW2 cases. The higher values of the epochs have not changed the validation accuracy and the loss as well as the testing values of the accuracy, true positive rate etc.

Table 4. The classifications result for the BW2 training using only original images as a training set with the test results for a single original image set (T1), multiplicated dataset with 8 copies of the original file (T8) and using the worst-case scenario (IDA) classification. The chosen confusion matrix factors like in Table 3.

CM factor		VGG19			XN			IRN2		
		T1	T8	IDA	T1	T8	IDA	T1	T8	IDA
ACC [%]	AVG	93.2	93.6	94.0	92.1	91.9	92.1	91.0	90.8	90.3
	VAR	2.9	3.0	3.0	1.5	0.7	2.9	3.0	1.2	1.9
	Min	88.0	89.6	90.0	88.0	90.3	86.0	86.0	85.5	88.0
	Max	98.0	98.5	100	94.0	93.0	98.0	96.0	93.3	96.0
TPR [%]	AVG	77.8	79.6	89.4	66.7	61.8	81.7	62.2	58.4	73.9
	VAR	11.7	10.2	8.2	7.9	7.1	9.5	13.3	16.1	10.1
	Min	55.6	66.7	77.8	55.6	51.4	66.7	33.3	23.6	55.6
	Max	100	97.2	100	77.8	73.6	88.9	77.8	76.4	88.9
FPR [%]	AVG	3.4	3.3	5.0	2.3	1.5	5.6	2.7	2.1	6.1
	VAR	2.2	2.0	2.2	1.6	1.1	3.9	1.7	1.3	2.7
	Min	0.0	0.0	0.0	0.0	0.0	0.0	0.0	0.6	2.4
	Max	7.3	7.0	7.3	4.9	3.0	14.6	4.9	5.2	12.2
Test	F1	0.95	0.96	1.0	0.82	0.79	0.94	0.88	0.80	0.89
	MCC	0.94	0.95	1.0	0.79	0.74	0.94	0.86	0.76	0.86
AUC	AVG	.973	.970		.956	.959		.947	.957	
	VAR	.024	.024		.019	.014		.022	.015	
	Min	.916	.923		.908	.927		.911	.931	
	Max	1.00	.999		.981	.976		.984	.972	

The time of training depends on the network and number of the training images and the machine specification. The times for the BW1 approach and machine 1 specification have varied from around 12 min for VGG19, 40 min for Xception and 60 min for IRN2. The times for BW2 and the same machine have varied from around 5 min for VGG19 (60 epochs), 14 min for XN (100 epochs) and 21 min for IRN2 (60 epochs).

In both approaches the confusion matrix factors are used where possible: accuracy (ACC), true positive rate (TPR), false positive rate (FPR), area under curve (AUC) for the receiver operating characteristic curve (ROC) with their average (AVG), variance (VAR), minimum (MIN) and maximum (MAX) values and F1 score and Matthews correlation coefficient (MCC) for the chosen CNN network.

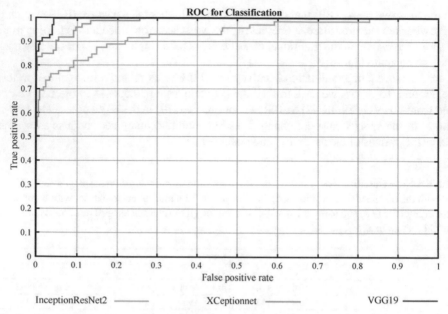

Fig. 3. The examples of the best receiver operating characteristic curve (ROC) with the highest value of the area under curve (AUC) for the three chosen CNNs.

The examples of the best receiver operating characteristic curve (ROC) with the highest value of the area under curve (AUC) for the three chosen CNNs are shown below in Fig. 3.

6 Conclusions

The correct selection of specific features, in particular, the blue-whitish veil has an impact on the analysis of specific disease fragments. One of the research methodologies which is used to conduct proper features segmentation based on specific disease feature of skin lesion is the Three-Point Checklist of Dermatology (3PCLD) as in Seven-Point Checklist (7PCL).

In the research, we have used the three pretrained CNN networks: Xception, VGG19 [10] and Inception-ResNet-v2 as well as PH2 dermoscopic image dataset [11]. The results achieved are quite promising. The average accuracy was above 90% reaching 100%. The networks usually quite well classified the images of the lesions. The images with the present blue-white veil were rarely underestimated.

In the paper, we show the advantages of using in test as well as in train and validation not only the original image but also its seven defined invariant copies. That general approach can provide higher results while using CNN networks in other disciplines not only dermatology.

The original authors' approach using the defined Invariant Dataset Augmentation shows that the classification characteristics like accuracy and true positive rate as well as the F1 and MCC tests can be much higher (5–20%) than using only original images.

In the paper for the reaching 98–100% accuracy, 98–100% true positive rate, 0.0–2.3% false positive rate with tests $F1 = 0.95$, $MCC = 0.95$ as well as AUC reaching the value 1.

References

1. Soyer, H.P., Argenziano, G., Zalaudek, I., et al.: Three-point checklist of dermoscopy. A new screening method for early detection of melanoma. Dermatology **208**(1), 27–31 (2004)
2. Argenziano, G., Soyer, H.P., et al.: Dermoscopy of pigmented skin lesions: results of a consensus meeting via the Internet. J. Am. Acad. Dermatol. **48**(9), 679–693 (2003)
3. Milczarski, P.: Symmetry of Hue Distribution in the Images. LNCS, vol. 10842, pp. 48–61. Springer (2018)
4. Kawahara, J., Daneshvar, S., Argenziano, G., Hamarneh, G.: Seven-point checklist and skin lesion classification using multitask multimodal neural nets. IEEE J. Biomed. Health Inform. **23**(2), 538–546 (2019)
5. Argenziano, G., Fabbrocini, G., et al.: Epiluminescence microscopy for the diagnosis of doubtful melanocytic skin lesions. Comparison of the ABCD rule of dermatoscopy and a new 7-point checklist based on pattern analysis. Arch. Dermatol. **134**, 1563–1570 (1998)
6. Carrera, C., Marchetti, M.A., Dusza, S.W., Argenziano, G., et al.: Validity and reliability of dermoscopic criteria used to differentiate nevi from melanoma: a web-based international dermoscopy society study. JAMA Dermatol. **152**(7), 798–806 (2016)
7. Nachbar, F., Stolz, W., Merkle, T., et al.: The ABCD rule of dermatoscopy. high prospective value in the diagnosis of doubtful melanocytic skin lesions. J. Am. Acad. Dermatol. **30**(4), 551–559 (1994)
8. Milczarski, P., Stawska, Z., Maslanka, P.: Skin lesions dermatological shape asymmetry measures. In: Proceedings of the IEEE 9th International Conference on Intelligent Data Acquisition and Advanced Computing Systems: Technology and Applications, IDAACS, pp. 1056–1062 (2017)
9. Menzies, S.W., Zalaudek, I.: Why perform Dermoscopy? The evidence for its role in the routine management of pigmented skin lesions. Arch Dermatol. **142**, 1211–1222 (2006)
10. Simonyan, K., Zisserman, A.: Very deep convolutional networks for large-scale image recognition. In: Conference Track Proceedings of 3rd International Conference on Learning Representations (ICRL), San Diego, USA (2015)
11. Mendoncca, T., Ferreira, P.M., Marques, J.S., Marcal, A.R.S., Rozeira, J.: PH2 – a dermoscopic image database for research and benchmarking. In: 35th Annual International Conference of the IEEE Engineering in Medicine and Biology Society (EMBC), Osaka, pp. 5437–5440 (2013)
12. Was, L., Milczarski, P., Stawska, Z., Wiak, S., Maslanka, P., Kot, M.: Verification of results in the acquiring knowledge process based on IBL methodology. In: Artificial Intelligence and Soft Computing, ICAISC 2018. LNCS, vol. 10841, pp. 750–760. Springer (2018)
13. Celebi, M.E., Kingravi, H.A., Uddin, B.: A methodological approach to the classification of dermoscopy images. Comput. Med. Imaging Graph. **31**(6), 362–373 (2007)
14. Was, L., Milczarski, P., Stawska, Z., et al.: Analysis of skin diseases using segmentation and color hue in reference to melanocytic lesions. In: Artificial Intelligence and Soft Computing, ICAISC 2017. LNCS, vol. 10245, pp. 677–689. Springer (2017)
15. Milczarski, P., Stawska, Z., Was, L., Wiak, S., Kot, M.: New dermatological asymmetry measure of skin lesions. Int. J. Neural Netw. Adv. Appl. 32–38 (2017)
16. European Cancer Information System (ECIS). https://ecis.jrc.ec.europa.eu. Accessed 21 Feb 2020

17. ACS – American Cancer Society. https://www.cancer.org/research/cancer-facts-statistics.html. Accessed 21 Feb 2020

18. Milczarski, P., Stawska, Z.: Classification of skin lesions shape asymmetry using machine learning methods. In: AINA Workshops Proceedings 2020. Advances in Intelligent Systems and Computing, vol. 1150, pp. 1274–1286 (2020)

19. The International Skin Imaging Collaboration: Melanoma Project. https://isdis.net/isic-project/. Accessed 21 Mar 2020

20. Argenziano, G., Soyer, H.P., De Giorgi, V., et al.: Interactive Atlas of Dermoscopy. Milan, Italy, EDRA Medical Publishing & New Media (2002)

21. Menzies, S.W., Crotty, K.A., Ingwar, C., McCarthy, W.H.: An Atlas of Surface Microscopy of Pigmented Skin Lesions. Dermoscopy. McGraw-Hill, Australia (2003)

22. Esteva, A., Kuprel, B., Novoa, R.A., et al.: Dermatologist-level classification of skin cancer with deep neural networks. Nature **542**, 115–118 (2017)

23. He, K., Zhang, X., Ren, S., Sun, J.: Deep residual learning for image recognition. In: Proceedings of the IEEE Conference on Computer Vision Pattern Recognition, pp. 770–778 (2016)

24. Madooei, A., Drew, M.S., Sadeghi, M., Atkins, M.S.: Automatic detection of blue-white veil by discrete colour matching in dermoscopy images. In: Proceedings of the MICCAI 2013, pp. 453–460 (2013)

25. Jaworek-Korjakowska, J., Kłeczek, P., Grzegorzek, M., Shirahama, K.: Automatic detection of blue-whitish veil as the primary dermoscopic feature. In: Lecture Notes in Computer Science (Including Subseries Lecture Notes in Artificial Intelligence and Lecture Notes in Bioinformatics). LNAI, vol. 10245, pp. 649–657. Springer (2017)

26. Celebi, M.E., et al.: Automatic detection of blue-white veil and related structures in dermoscopy images. CMIG **32**(8), 670–677 (2008)

27. Di Leo, G., Fabbrocini, G., Paolillo, A., Rescigno, O., Sommella, P.: Toward an automatic diagnosis system for skin lesions: estimation of blue-whitish veil and regression structures. In: International Multi-Conference on Systems, Signals & Devices, SSD 2009 (2009)

28. Milczarski, P., Was, L.: Blue-white veil classification in dermoscopy images of the skin lesions using convolutional neural networks. In: Artificial Intelligence and Soft Computing, ICAISC 2020. LNCS, vol. 12415, pp. 636–645. Springer (2020)

Smilax: Statistical Machine Learning Autoscaler Agent for Apache FLINK

Panagiotis Giannakopoulos and Euripides G. M. Petrakis[✉]

School of Electrical and Computer Engineering, Technical University of Crete (TUC),
Chania, Greece
pgiannakopoulos1@isc.tuc.gr, petrakis@intelligence.tuc.gr

Abstract. Smilax is a statistical machine learning autoscaler agent for applications running on Apache Flink. Smilax agent acts proactively by predicting the forthcoming workload in order to adjust the allocation of workers to the actual needs of an application ahead of time. During an online training phase, Smilax builds a model which maps the performance of the application to the minimum number of servers. During the work (optimal) phase, Smilax maintains the performance of the application within acceptable limits (i.e. defined in the form of SLAs) while minimizing the utilization of resources. The effectiveness of Smilax is assessed experimentally by running a data intensive fraud detection application.

1 Introduction

The relationships between service providers and customers are important for achieving high level of satisfaction and trust. In cloud computing, the service provider - customer relationship is not arbitrary but it is shaped by a Service Level Agreement (SLA). The SLA specifies obligations and penalties in case of non-compliance with the agreement. There is a direct relationship between quality of service and amount of computing resources allocated to the client's application. Typically, the operation of the provider is assisted by software agents which monitor changes in the performance and regulate the allocation of computing resources to the application either reactively (as they occur) or, proactively (i.e. ahead of time). Reactive scaling policies are easy to implement but, leave room for both over or under-utilization of resources [1,8]. Proactive scaling policies are capable of predicting possible SLA violations and make optimal resource allocations decisions but, come with a complex implementation and requires a-priori knowledge of the workload for model training. The training of the model can be carried-out either offline or online.

Stream processing enables a variety of brand-new applications characterized by increased data generation and the low latency response. Apache Flink[1] is a distributed processing engine for stateful computations over unbounded and bounded data streams in critical-mission applications such as, fraud detection

---

[1] https://flink.apache.org/.

© The Author(s), under exclusive license to Springer Nature Switzerland AG 2021
L. Barolli et al. (Eds.): AINA 2021, LNNS 226, pp. 433–444, 2021.
https://doi.org/10.1007/978-3-030-75075-6_35

(i.e. detection of suspicious transactions), anomaly detection (i.e. detection rare or suspicions events), rule-based alerting (i.e. identification of data which satisfy one or more rules) and many more. However, if streaming data is generated at different speeds, Apache Flink cannot automatically and optimally adjust the utilization of its computing resources. Existing resource allocation policies for Apach Flink are all reactive or, the resource scaling decisions resort to human operators who monitor the performance of the system.

Smilax is an autonomous agent which monitors and maintains the performance of Apache Flink within acceptable limits (i.e. defined in the form of SLAs) while minimizing the utilization of computing resources. During a training phase, a reactive scaler collects workload and performance information and adjusts (scales-up or down) the number of servers whenever the performance limit (i.e. the SLA) is violated. During the optimal phase, the agent explores the performance and builds a statistical machine learning model which registers the optimal mapping between workload, performance and number of servers. Model fitting takes place at run time from production data. As soon as the model is deemed stable, the agent switches to optimal mode. The model is then used for predicting the performance of the application and for making scaling decisions proactively (i.e. before SLA violations occur). The stability of the model is constantly monitored and as soon as a model change is detected (e.g. the workload becomes unpredictable), the agent switches back to reactive mode to start collecting new data in order to build a new performance model.

Apache Flink and Smilax are deployed on Docker Swarm[2], a low-footprint virtualization platform based on Docker containerization. Smilax is evaluated on a fraud detection application which runs an 6.5-h workload that produces up to four thousands records per second. The experimental results demonstrate that Smilax always makes accurate predictions and results in less SLA violations and better overall utilization of computing resources compared to its reactive counterpart.

Related work on autoscaling is discussed in Sect. 2. An introduction to Flink and how an autoscaler for Flink can be designed is discussed in Sect. 3. Smilax solution is presented in Sect. 4 followed by experimental results in Sect. 5. Conclusions, system extensions and issues for future research are discussed in Sect. 6.

2   Related Work

In regards to proactive scaling for stream processing, the ideas are not quite mature yet and have not been incorporated into commercial real-time analytics platforms. Initial ideas for a statistical machine learning model for the scaling of resources are discussed in [4]. Arabnejad et al. [2] compare two different autoscaling types of Reinforcement Learning (RL), which is SARSA and Q-learning. The autoscaler dynamically resizes Web applications in order to meet the quality of

---

[2] https://docs.docker.com/engine/swarm/.

service requirements. Bibal Benifa and D. Dejey [3] propose the RLPAS algorithm, which applies RL using a neural network in order to reduce the time for convergence to an optimal policy. Rossi, Nardelli and Cardellini [5] propose RL solutions for controlling the horizontal and vertical elasticity of container-based applications in order to cope with varying workloads. These autoscalers do not adapt their scaling model to changes of the application's behavior at run-time.

The following solutions are all reactive: DS2 [7] enables automatic scaling of Apache Flink applications. A controller assesses the running application at operator level in order to detect possible bottlenecks in the data-flow (i.e. operators that slow down the whole application). In contrast to Smilax which monitors and scales applications at job level (i.e. multiple operators or tasks may execute in a job), DS2 is designed to adjust the parallelism of each operator separately in order to maintain high throughput. Autopilot[3] is a proprietary solution for Ververica platform which is designed to drive multiple high throughput, low latency stream processing applications on Apache Flink. There are also solutions which have been incorporated into the real-time analytics platforms of commercial cloud providers: Apache Heron[4] is the stream processing engine of Twitter; Dataflow[5] is a serverless autoscaling solution that supports automatic partitioning and re-balancing of input data streams to servers in the Google Cloud Platform.

3   Smilax Ecosystem

Apache Flink provides an extensive toolbox of operators for implementing transformations on data streams (e.g. filtering, updating state, aggregating). The data-flows or jobs (i.e. operations chained together) form directed graphs (Job Graphs), that start with one or more sources and end at one or more sinks. The Flink cluster consists of a Job Manager and a number Task Managers (workers). The Job Manager controls the operation of the entire cluster: schedules the workers, reacts to finished or failed tasks, load balances the workload among Task Managers, coordinates checkpoints and recovery from failures. The Task Managers are the machines (servers) which execute the tasks of a workflow. A task represents a chain of one or more operators that can be executed in a single thread or server. A task can be executed in parallel (on separate Task Managers). Each parallel instance of a task is a subtask. The number of subtasks running in parallel is the parallelism of that particular task.

The number of allocated Task Managers varies over time and it is regulated by Smilax agent. Smilax agent monitors the operation of all tasks and, depending on workload and performance, decides to change the parallelism of a task (i.e. scale-up or down). Flink is particularly flexible, but making the most out of it can become a challenging task that requires in depth understanding of its underlying architecture (especially in the case of multiple workflows with many

---

[3] https://docs.ververica.com/user_guide/application_operations/autopilot.html.

[4] https://incubator.apache.org/clutch/heron.html.

[5] https://cloud.google.com/dataflow.

operators executing in multiple layers). For simplicity of the discussion, the following assumptions do apply in Smilax: each Task Manager (worker) runs the entire Job Graph (workflow), which means that the number of allocated workers is identical to the parallelism of the job. Rescaling actions (e.g. adding or removing a worker) will modify the parallelism of all operators of a subtask at the same time. Changing the parallelism of individual operators would require that Smilax monitors each operator separately and takes scaling decisions based on the performance of each individual operator, in the example of DS2 [7]. If more than one workflow run on the same Flink cluster, taking optimal scaling decisions for each individual workflow requires monitoring the performance of each workflow separately (i.e. a separate model must be built for each workflow).

An application receives data records (or events) from streaming sources such as Apache Kafka[6]. Kafka, queues data from application sources like databases, sensors, mobile devices, cloud services etc. Kafka reads data streams in topics and in parallel (i.e. events are appended to more than one partitions defined for that topic). The incoming workload is monitored by inspecting the Kafka topics which are the data sources of the running job. The workload represents the number of records per second the system receives. Kafka queues are empty if the system consumes (processes) the received data at a rate higher than the production rate; otherwise, the data remains in the queue (slow records). The average length of Kafka queues is an indicator of whether the system can keep up with the data production rate. Prometheus service[7] is responsible for the monitoring of running applications. Prometheus retrieves the Kafka metrics by querying the HTTP endpoint of JMX[8] (i.e. Prometheus cannot connect to Kafka directly). Apache Zookeeper[9] is a coordination service for the Kafka queues.

The percentile of slow records is computed as *queue-length/workload*. In Smilax, quality of service is represented by an SLA metric which is defined as the percentile of slow records per second that a client (e.g. application owner or user) can accept. In this work, the threshold is 90% (i.e. less than 10% of the number of records can remain in the queue or more than 90% of the records are processed instantly). Smilax, collects information from Kafka queues and adjusts (scales-up or down) the number of workers as soon as the SLA is violated.

Figure 1 is an abstract architecture of Smilax ecosystem. Each shape with dotted line represents a Virtual Machine (VM) with a Docker environment installed. Boxes with solid lines within a Docker represent containers running the specified services. The entire cluster runs on Docker Swarm using Docker images of Apache Flink[10]. Prometheus service runs in a separate container and is responsible for monitoring the system in terms of allocated resources (i.e. number of servers), workload and performance. The containers within this Docker are configured to run either as Job Manager or Task Managers. The cluster is deployed as Flink

---

[6] https://kafka.apache.org.

[7] https://prometheus.io.

[8] https://docs.oracle.com/en/java/javase/15/jmx/.

[9] https://zookeeper.apache.org.

[10] https://flink.apache.org/news/2020/08/20/flink-docker.html.

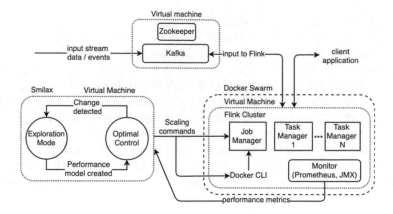

Fig. 1. Flink deployment with Kafka and Smilax autoscaler.

Session Cluster so that the lifecycle of the running job is independent from the lifetime of the cluster. In any other case (i.e. application mode or per-job mode), the cluster would shut down prior to rescaling.

Initially, all services are constrained into one single VM. New servers can be added into the same VM as well. Once the capacity of the VM is exhausted, new servers will be added in a new VM (i.e. a Swarm with two nodes). Deployment of new servers is a three layer process. Figure 2 illustrates the three rescaling layers. Scaling commands address the CLI of Flink layer (top-most layer). However, the option to scale Flink using CLI[11] is temporarily disabled in version 1.11 (or older) but will be soon supported in a future version. The only way to change the parallelism of a job is to first stop the job (and take save-point) and, then re-run the job with the new parallelism. As a result, incoming records remain in Kafka until the job recovers. In the mean-time, Smilax discards all performance metrics and no scaling action takes place.

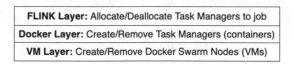

Fig. 2. Resource adjustment layers.

Scaling actions issued on Flink layer (i.e. for allocating new server nodes in containers) address also the Docker layer (Docker CLI). A number of Task Managers can be in hot-standby (e.g. can be nodes de-allocated recently due to scale-down). If the required resources are there, the number of servers in hot-standby is reduced by the number of requested servers and these servers will

---

[11] https://issues.apache.org/jira/browse/FLINK-12312.

join the cluster again. If not enough servers are in hot-standby, additional servers will be allocated by addressing the Docker layer, provided that the existing VM has the capacity to accommodate the new servers. Otherwise, the request is propagated to the VM infrastructure layer (i.e. Openstack in this work) which manages the allocation and de-allocation of VMs [1]. For scaling down, if the job reached the minimum parallelism allowed, the operation is aborted; otherwise the operation is propagated from Flink to Docker layer. The de-allocated containers are put in hot-standby for future use. If their number exceed a pre-defined limit, they are removed. Notice that, Docker has no knowledge whether a container is in use or it is idle. Information on which containers are active (are receiving records) is obtained from Flink. Finally, if the VM is empty it is removed as well.

4 Smilax

The autoscaler implements a controller which is in a constant switch between two states, referred to as training (or exploration) phase and, work (or optimal) phase respectively [6].

4.1 Exploration Phase

Smilax explores the behavior of the application under varying workloads and different parallelism (i.e. number of servers). The controller takes scaling decisions reactively (i.e. as soon as the SLA is violated).

Let W_{max} be the capacity of a single Task Manager (server). The capacity represents the maximum incoming records rate which a single server can process without violating the SLA. If s_t and w_t are the number of servers and workload respectively when SLA is violated for first time, then $W_{max} = w_t/s_t$. In order to use a better approximation, the capacity is updated each time SLA is violated and the average W_{max} value is computed. The reactive scaler checks periodically (i.e. every 15 s) whether a scale-up or a scale-down decision must be taken. If S and W are the number of servers and the workload when the SLA is violated, then the system needs $S' = W/W_{max}$ servers and $S' - S$ servers must be added (assuming that the capacity of each server is the same). Conversely, if the number of servers is more than $W/W_{max}/0.9$ (i.e. there are at least 10% more servers than needed), the cluster is over-utilized and the autoscaler will remove the extra servers. This condition must be satisfied for a number of 10 consecutive samples (taken every 15 s).

The performance model captures the relation between workload, number of servers, and percentage of slow records. The model is described by one dependent and two independent variables: $y = f(X) = f(x_1, x_2)$ where, x_1 is the workload, x_2 is the number of servers and y is the percentage of slow records. The model is trained using non-linear regression. The degree of the polynomial is selected by applying brute-force search (i.e. the solution is searched exhaustively). This is a three step iterative process during which the method will test all degrees from 1 through *maxDegree* by applying Algorithm 1. The computational complexity of the method increases with *maxDegree*. For each degree, a model is created using non-linear regression on the collected measurements. The Route-Mean-Square-Error (RMSE) is computed by comparing the actual against the predicted value of the model and describes how concentrated the dataset is around the regression line. The degree of the model is the one with the minimum RMSE value.

Algorithm 1. Computing the best-fit degree for dataset (X,y).

1: **procedure** FINDDEGREE(X,y,maxDegree)
2:     $minRMSE \leftarrow \infty$
3:     $bestDegree \leftarrow 0$
4:     $degree \leftarrow 0$
5:     **while** $degree < maxDegree$ **do**
6:         $model \leftarrow NonLinearRegression(X, y, degree)$
7:         $y_{actual} \leftarrow y$
8:         $y_{predicted} \leftarrow model.predict(X)$
9:         $RMSE \leftarrow MeasureRMSE(y_{actual}, y_{predicted})$
10:        **if** $RMSE < minRMSE$ **then**
11:            $minRMSE \leftarrow RMSE$
12:            $bestDegree \leftarrow degree$
13:        $degree \leftarrow degree + 1$
       return $bestDegree$

The stability of the model is checked periodically by applying a Bootstrapping technique [10]. This is also a two stage process: in the first phase, the method checks whether the dataset comprises enough data to create an accurate model. If the standard deviation of the predicted values is less than the model stability threshold λ, the method proceeds to phase two whose purpose is to assess that the model is accurate to predict the performance of the application for each parallelism [6]. The stability threshold is the maximum acceptable error of the model and it is user defined (i.e. $\lambda = 0.05$ for both phases). Once this phase is completed (i.e. the model is deemed stable), the controller switches to optimal control.

4.2   Optimal Phase

The autoscaler takes scaling decisions proactively (i.e. the model attempts to predict SLA violations before they occur). Smilax no longer uses instant metrics in order to take decisions about parallelism. Instead, near future predictions of the workload are used to determine both, the performance and the optimal parallelism of the system. This is also a two stage process: (a) in the first phase, future predictions of the workload are derived based on past (i.e. recent) values by applying linear regression. Assuming that the rate will not change in the near future, the workload is predicted based on the slope of this curve representing the rate of change of the workload (i.e. whether it increases, decreases or it is steady). For the next 60 s and for every 5 s, the output is an array with 12 values. (b) For each future workload and according to the model, the performance (i.e. percentile of slow records) of the application can be predicted as well. For each predicted value of the workload, the performance takes a value for each possible parallelism. The optimal parallelism S_{target} is the minimum parallelism which satisfies the SLA (i.e. the percentage of slow records per second is less than 10%). Algorithm 2 illustrates this process.

Algorithm 2. Scaling policy during optimal control

1: **procedure** PROACTIVESCALER
2:     $w_{future} \leftarrow WorkloadPredictor()$
3:     $parallelismSet \leftarrow [1 \ldots n_{max}]$
4:     $S_{target} \leftarrow n_{max}$
5:     **for** $n_i \in parallelismSet$ **do**   ▷ select optimal parallelism
6:         $performancePoints \leftarrow Predict(n_i, w_{future})$
7:         $evaluation \leftarrow CheckViolation(performancePoints)$
8:         **if** $evaluation$ **then**
9:             $S_{target} \leftarrow n_i$
10:            break
11:    $S_{new} \leftarrow Hysterisis(S_{target})$
12:    $scale(S_{new})$
13: **procedure** WORKLOADPREDICTOR
14:    $w_{past} \leftarrow GetPastWorkload(10mins)$
15:    $slope \leftarrow LinearRegression(w_{past})$
16:    $w_{future} \leftarrow slope.predict(1min)$
       return w_{future}
17: **procedure** CHECKVIOLATION(performancePoints)
18:    **for each** $point_i \in performancePoints$ **do**
19:        **if** $point_i \geq SLA$ **then return** $false$
       return $true$
20: **procedure** HYSTERISIS(S_{target})
21:    $S_{old} \leftarrow CurrentNumberOfTaskmanagers$
22:    **if** $S_{target} > S_{old}$ **then**
23:        $S_{new} \leftarrow S_{old} + \alpha \cdot (S_{target} - S_{old})$
24:    **else if** $S_{target} < S_{old}$ **then**
25:        $S_{new} \leftarrow S_{old} + \beta \cdot (S_{target} - S_{old})$
26:    **else**
27:        $S_{new} \leftarrow S_{old}$
       return S_{new}

To prevent rapid oscillations in parallelism values, hysteresis gains α and β are defined in the range [0,1]. The final parallelism S_{new} is computed as:

$$S_{new} = \begin{cases} S_{old} + \alpha(S_{target} - S_{old}), & \text{if } S_{target} > S_{old} \\ S_{old} + \beta(S_{target} - S_{old}), & \text{if } S_{target} < S_{old} \end{cases} \tag{1}$$

Parameter α specifies how quickly the system will make the transition from S_old to S_target. A low value means that new servers will be added gradually but, this could cause SLA violations (since the number of servers is less than needed); a value close to 1 means that servers will be added quickly (or at once if $\alpha = 1$). A value close to 1 should be preferred in order to avoid SLA violations. Parameter β specifies how quickly the system will scale down from S_old to S_target. A value close to 1 will cause servers to be de-allocated quickly. A high or low value will not cause SLA violations but could cause under-utilization of resources (i.e. if more than S_{target} servers are retained for long). Henceforth, $\alpha = 0.9$ and $\beta = 0.4$.

Changes in the environment such as system updates or hardware failures could lead the application to behave in an unpredictable way. Change point detection is applied to detect whether the model is capable of predicting the behavior of the application. Smilax computes the percentile (per second) of the residuals $\|actualPerformance - predictedPerformance\|$. Residual values are collected in periods of 5 s. The residuals of an accurate model must be close or equal to zero. Online change detection [9] captures abrupt changes in the streaming data. The function returns a score (i.e. prediction error) which is constantly compared against a user defined threshold (0.08 in this work). In case of very low threshold, the controller will mark the performance model as inaccurate for very small deviations. Conversely, a high threshold could cause SLA violations. If the model is no longer valid (i.e. it is no longer capable of predicting the performance and take scaling decisions proactively), the controller switches again to exploration mode in order to train a new model.

5   Experiments

The purpose of the following set of experiments is to assess the performance of Smilax autoscaler for adjusting the resources of an Apache Flink cluster running a Click Fraud Detection application. This type of fraud occurs on the Internet in pay-per-click (PPC) online advertising applications. Website owners post advertisements and receive re-numeration based on how many Web users click on the advertisements. Fraud occurs when a person or software imitates a legitimate user by clicking on an advertisement without having an actual interest in it. The application receives records from a Kafka topic with elements User-IP, User-ID, time-stamp and event type (e.g. "click"). The production rate of these records represents the workload of the application. The application attempts to detect fraud by searching for the following patterns every 60 s: (a) Counts of User ID's per unique IP address, (b) Counts of IP addresses per unique User ID and, (c) Click-Through Rate (CTR) per User ID.

Flink runs on a VM (8 CPUs, 16 Gb Ram) on the Openstack infrastructure of TUC and starts with 1 Task Manager in a container (1 CPU with 2 Gb RAM). Prometheus is deployed on a separate container (1 CPU with 1 Gb RAM). Smilax runs on a second VM (4CPUs with 8 Gb RAM). A third VM runs Kafka with Zookeeper and JMX exporter (4 CPUs with 8 Gb RAM).

The duration of the experiment is 6.5 h or 1,834 samples. During exploration mode, samples are taken every 15 s while, during optimal mode, samples are taken every 5 s. The workload is generated using Faban[12] which runs on a fourth VM (2 CPUs with 4 Gb RAM). Faban can be used for the generation of benchmarks based on real workload distributions. The workload follows a Gaussian distribution in two stages. Smilax switched to optimal control after 4 h. This is the duration of the training (exploration) phase during which Stability Check applies every 2 h (threshold $\lambda = 0.05$). Figure 3 illustrates (a) workload distribution at the top most graph, (b) predicted parallelism (i.e. number of servers allocated) in the middle and, (c) points in time where SLA violations occurred (slow records) the third (bottom) graph. All three graphs have the same x-axis showing times in number of samples. The accuracy of the model did not change in the optimal phase and no SLA violations occurred.

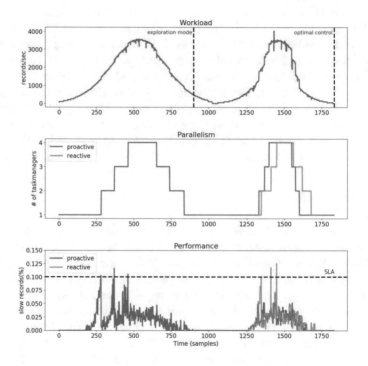

Fig. 3. Workload, predicted values of parallelism and slow records.

The resource allocation policy is not the same during the two phases. During the training phase, the autoscaler allocates as many servers are required (Sect. 4.1). During the optimal phase, allocation of resources is dictated by Eq. 1. Compared to a pure reactive policy the results of this experiment in Fig. 3 reveal three important advantages of proactive over reactive scaling: (a) allocation of resources ahead of time (i.e. before SLA violations occur), (b) the reactive scaler takes greedy decisions ending-up in sub-optimal allocation of resources (i.e. in

---

[12] http://faban.org.

this experiment, the reactive autoscaler utilized 1.85 servers per minute as compared to the proactive autoscaler which utilized 1.80 servers per minute) and, (c) less charges to service providers due to SLA violations. As shown in the last (performance) graph, the reactive policy encountered SLA violations in three cases during the optimal phase causing the respective scaling actions. Instead, the proactive policy triggered all scaling actions ahead of time and avoided all SLA violations.

Figure 4 illustrates the performance model that was built during the training phase and represents performance (i.e. percentage of slow records) as a function of the workload (i.e. number of input records per second) for various cases of parallelism (i.e. number of servers). The model explored the performance from 1 up to 4 servers. As expected, the percentage of slow records (and subsequently) SLA violations decrease with the number of servers. Although selecting the maximum parallelism (i.e. 4 servers) would be a safe choice, this would result in under-utilization of computing resources. The model indicates that, for a given workload, the optimal choice is to select the minimum parallelism that achieves no SLA violations.

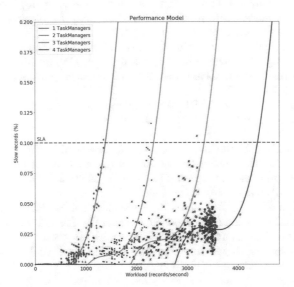

Fig. 4. Performance model of the Fraud Detection application.

6 Conclusions

Smilax is application agnostic and can be used to support optimal scaling decisions proactively. As a proof of concept, the paper shows how Smilax supports optimal scaling in applications running on Apache Flink so that they use as few resources as possible while maintaining their performance within acceptable limits. Hence, clients are not charged for idle or under-utilized resources and providers are not charged for violating their SLAs with their clients. Smilax builds-upon early ideas by Bodik [4]. The original work has been improved on certain methodology aspects including, algorithmic model construction, model validity, incorporation within a state-of-the-art streaming platform (i.e. Apache Flink) and verification in a high impact fraud detection use case.

The assumptions in regards to Apache Flink customization have to be relaxed. A more effective autoscaler would take scaling decisions at a finer granularity level (i.e. operator level). Optimizing change point detection in order to capture gradual changes of the performance (and not only abrupt change), improving the learning method in order to handle unexpected changes of the workload (e.g. outliers), estimating parameters λ, α, β automatically (i.e. using machine learning) to match the peculiarities of different workloads, are also important issues for future research.

References

1. Alexiou, M., Petrakis, E.G.M.: Elixir: an agent for supporting elasticity in Docker Swarm. In Advanced Information Networking and Applications (AINA 2020), Caserta, Italy, vol. 1151, pp. 1114–1125 (2020)
2. Arabnejad, H., Pahl, C., Jamshidi, P., Estrada, G.: A comparison of reinforcement learning techniques for fuzzy cloud auto-scaling. In: 17th IEEE/ACM International Symposium on Cluster, Cloud and Grid Computing (CCGRID 2017), Madrid, Spain, pp. 64–73 (2017)
3. Bibal Benifa, J.V., Dejay, D.: RLPAS: reinforcement learning-based proactive autoscaler for resource provisioning in cloud environment. Mob. Netw. Appl. **24**(4), 1348–1363 (2019)
4. Bodik, P., Griffith, R., Sutton, C.A., Fox, A., Jordan, M.I., Patterson, D.A.: Statistical machine learning makes automatic control practical for internet datacenters. In: Hot Topics in Cloud Computing (HoTCloud 2009), San Diego, California, USA, pp. 195–203. USENIX Association (2009)
5. Rossi, F., Nardelli, M., Cardellini, V.: Horizontal and vertical scaling of container-based applications using reinforcement learning. In: IEEE 12th International Conference on Cloud Computing (CLOUD 2019), Milan, Italy, pp. 329–338 (2019)
6. Giannakopoulos, P.: Supporting elasticity in flink. Technical report, ECE School, Technical Univ. of Crete (TUC), Chania, Greece (2020)
7. Kalavri, V., et al.: Three steps is all you need: fast, accurate, automatic scaling decisions for distributed streaming dataflows. In: 13th USENIX Symposium on Operating Systems Design and Implementation (OSDI 2018), Carlsbad, CA, pp. 783–798 (2018)
8. Sharma, P., Chaufournier, L., Shenoy, P., Tay, Y.C.: Containers and virtual machines at scale: a comparative study. In: 17th International Middleware Conference, pp. 1:1–1:13 (2016)
9. Takeuchi, J., Yamanishi, K.: A unifying framework for detecting outliers and change points from time series. IEEE Trans. Knowl. Data Eng. **18**(4), 482–492 (2006)
10. Yu, H., et al.: Bootstrapping estimates of stability for clusters, observations and model selection. Comput. Stat. **34**(1), 349–372 (2019)

Application Relocation Method for Distributed Cloud Environment Considering E2E Delay and Cost Variation

Tetsu Joh[1,2]([✉]), Takayuki Warabino[1,2], Masaki Suzuki[1,2], Yusuke Suzuki[1,2], and Tomohiro Otani[1,2]

[1] KDDI CORPORATION, 3-10-10, Iidabashi, Chiyoda-Ku, Tokyo, Japan
{te-jyo,warabino,masaki-suzuki,uu-suzuki,otani}@kddi-research.jp
[2] KDDI Research, Inc., 2-1-15 Ohara, Fujimino-Shi, Saitama, Japan

Abstract. Multi-access Edge Computing (MEC) is attracting attention as a way of realizing the diverse services expected in 5G. Since resources of the MEC hosts are limited, optimal Application (App) allocation methods for the Distributed Cloud Environment (DCE) consisting of the MEC hosts and Central Cloud hosts are being actively studied. This paper proposes an App relocation method for the Common Service Infrastructure, which manages diverse Apps across service providers, to enhance the User Equipment accommodation efficiency of the DCE. The proposed method achieves a high success rate and efficient App relocation by selecting a combination of the relocated App and destination host while simultaneously considering the E2E delay variation of each App affected by App relocation and the cost associated with App relocation. Numerical simulations show that the proposed method reduces the number of initial App allocation failures by up to 73.6% compared to the existing method.

1 Introduction

Multi-access Edge Computing (MEC) is considered a key enabler to realize the diverse services expected in 5G [1]. MEC is an edge computing technology where computing capabilities, called MEC hosts, are widely distributed at the edge of a mobile network. Since an Application (App) that processes a service can be allocated in the proximity of the User Equipment (UE), MEC has features such as low response delay and high bandwidth transmission compared to the conventional Centralized Cloud (CC) [2–4]. From this context, MEC is expected to accommodate delay sensitive and/or traffic-hungry services. Similar to Infrastructure as a Service (IaaS) in CC, sharing the resources of MEC hosts deployed by a Mobile Network Operator (MNO) with service providers is considered to be one use case of using MEC [2, 5, 6].

Optimal App allocation methods for the Distributed Cloud Environment (DCE) consisting of multiple MEC hosts and hosts located in the CC (CC hosts) are being actively investigated [3, 7]. The MEC hosts have limited resources due to limited space and power supply compared to the CC hosts. Therefore, the Quality of Service (QoS) is affected when processing load concentrates on a specific MEC host for reasons that include local

© The Author(s), under exclusive license to Springer Nature Switzerland AG 2021
L. Barolli et al. (Eds.): AINA 2021, LNNS 226, pp. 445–457, 2021.
https://doi.org/10.1007/978-3-030-75075-6_36

congestion in the mobile network. In addition, as a distributed architecture, an MEC system is expected to be costly to build and manage, possibly resulting in a higher resource usage cost than that of the CC. Accordingly, methods to save power and reduce the cost of the system while satisfying the QoS by optimizing App allocation in DCE have been studied [8, 11].

We are developing a Common Service Infrastructure (CSI) which enables service providers to rapidly deploy new advanced services without needing to be aware of the detailed mechanism in the DCE. A system overview of the CSI is shown in Fig. 1. Each service provider registers service information such as the App program and required delay for service execution in the CSI. According to a service request from the UE, the CSI selects the most suitable host to allocate the App based on the required delay, the location of the UE, the status of MEC and CC hosts, and the status of links connecting the hosts, and allocates the App to the selected host. The CSI constantly monitors the status of hosts and links even after the App allocation, and performs App relocation if any service does not meet the required delay.

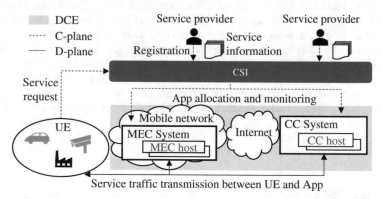

Fig. 1. System overview of the Common Service Infrastructure (CSI)

Two primary benefits can be provided by adopting the CSI. The first is to simplify the barriers for service providers when launching new services by reducing the cost and the time required to launch a service. If the CSI is not utilized, service providers would be required to monitor the status of all hosts and links in the network constantly for optimizing App allocation. Building such a system would take substantial capital investment and time. Utilizing the CSI will lower these building costs and enable rapid service launch. Second is the improved efficiency in resource usage of the entire system by sharing the CSI among multiple service providers. Sharing the CSI enables optimizing App allocation across diverse Apps of multiple service providers (regarded as global optimization). Rather than optimizing App allocation by an individual service provider (regarded as partial optimization), the resource utilization efficiency of the entire DCE can be increased. Additionally, compared to individual monitoring by each service provider, monitoring the status of hosts and links by the CSI will reduce the processing load on the hosts and the traffic in the network required for monitoring.

This paper proposes an App relocation method for the CSI with the aim of improving the UE accommodation efficiency of the DCE while maintaining the QoS. The proposed method achieves a high success rate and efficient App relocation through selecting a combination of the relocated App and its destination host by using a range of information maintained in the CSI. In the proposed method, non-threshold-exceeding Apps, whose E2E delay between the App and UE satisfies the threshold, are considered to be candidates for the relocated App. This greatly contributes to enhancement of the App relocation success rate. Additionally, efficient App relocation is performed through simultaneous consideration of the E2E delay variation of each App affected by the App relocation, and the cost associated with the App relocation (variation in the resource consumption cost, the communication cost of App transfer) in the selection process.

2 Related Work

The European Telecommunications Standards Institute (ETSI) Industry Specification Group (ISG) MEC is leading an international initiative on the standardization of edge computing. The ISG defines a MEC reference architecture [9] consisting of functional elements that enable App allocation, e.g., instantiation, relocation, and termination. [9] shows the factors that should be considered in App allocation, such as service information (required delay, amount of required resource, etc.), the status of the MEC host, and the status of the network in the MEC system. However, the specific selection method for App allocation is not shown.

In some R&D communities, the App allocation method is one of the major topics in MEC [3, 7, 8], and various studies have been conducted. These efforts can be roughly classified into two categories: initial App allocation, in which a new App is allocated to a host, and App relocation, in which an App running on one host is relocated to another.

2.1 Initial App Allocation

Optimization in initial App allocation is performed to enhance the efficiency of the MEC system. Various factors affecting the efficiency of the system, such as QoS, resource consumption cost, and power consumption, depend on the allocation of the App. In order to improve these factors, optimal App allocation methods in initial App allocation are being studied.

M. Berno et al. propose a method for balancing the Apps across the available hosts, controlling the amount of resources shared by one host with another, and minimizing the total energy in [8]. The conflicting goals of energy consumption, App balancing, and active host number are optimized while hard-processing deadlines and heterogeneous requirements of the App (deadlines and required computing resources) are jointly considered.

M. Vondra and Z. Becvar propose a MEC host selection algorithm that increases user satisfaction with experienced delay of data transmission and computing in [10]. Experienced delay associated with offloading an App is optimized based on App parameters, user requirements, and status of the MEC system such as high and low throughput in the backhaul.

Q. Li et al. propose an energy-efficient computation offloading and resource alloca-tion scheme to minimize the system cost in [11]. The system cost of DCE in heteroge-neous networks is minimized by jointly optimizing the computation offloading strategy, transmission power, and computation resource allocation.

2.2 App Relocation

The E2E delay of an App once allocated (allocated App) always varies depending on the UE dynamics in the mobile network and the load status of hosts and links, and in some cases exceeds the required delay of the App. For improving the E2E delay in such a situation, methods for relocating such App from one MEC host to another are studied.

W. Zhang et al. propose a comprehensive edge cloud relocation decision system that makes optimal App relocation decisions in [12]. In the relocation decision (when and where to relocate), E2E delay of the App is expected based on the UE mobility and the variation in link quality and host load. When it exceeds the threshold, App relocation is performed while taking the E2E delay improvement effect and the relocation cost (e.g., service down time) into account.

S. Maheshwari et al. propose a relocation framework for an App virtualized by container technology in the MEC systems in [13]. The App whose E2E delay, expressed as the sum of the network delay between the UE and the MEC host and the processing delay of the container on the MEC host, exceeds the threshold will be relocated to the host with the lowest relocation cost calculated based on the memory consumption of the App, from among the MEC hosts whose current CPU load is lower than the load threshold.

3   System Architecture

The system architecture of the CSI we are developing is shown in Fig. 2. The CSI receives a service request from the UE on behalf of the service provider, performs initial App allocation, E2E delay monitoring of the allocated App, and our proposed App relocation through the processing in each functional block as shown in Fig. 2.

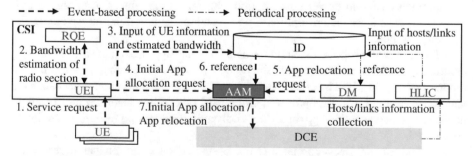

Fig. 2. System architecture of the Common Service Infrastructure (CSI)

The Internal Database (ID) stores information required for initial App allocation and App relocation. The information includes service information (required delay, required resource amount, App program, etc.), network topology, host information (maximum resource amount, resources usage) and link information (maximum bandwidth, bandwidth usage), and UE information (identifier, radio quality information).

The Hosts and Links Information Collector (HLIC) periodically collects the hosts and the links information in the DCE and stores collected information in the ID.

The User Equipment Interconnector (UEI) receives a service request from the UE and requests the App Allocation Manager (AAM) to perform initial App allocation. In receiving the service request, the UE information obtained from the UE, and estimated available bandwidth in the radio section of the UE provided by the Radio Quality Estimator (RQE) are registered in the ID by the UEI.

The RQE estimates the available bandwidth in the radio section of the UE based on radio quality information extracted by the UEI from the UE information. This estimated value, as well as other information stored on the ID, is used by the AAM and Delay Monitor (DM) when estimating the E2E delay.

The DM periodically monitors the E2E delay of the allocated App in the DCE to determine if the threshold is exceeded. When the threshold-exceeding App is detected, the DM requests the AAM to perform App relocation.

The AAM performs initial App allocation and App relocation in response to requests from the UEI and the DM. The host to which the App will be allocated is selected based on consideration of the E2E delay estimated based on the information stored in the ID and cost of resource consumption. The proposed App relocation method is running on the AAM.

4 Methodology

4.1 Motivation

The UE accommodation efficiency of DCE will be enhanced through periodical App relocation by the CSI. The load of the host where the new App is allocated will rise, resulting in E2E delay increments for the allocated App on the host. QoS will not be satisfied if the E2E delay of the allocated App exceeds the required delay, therefore, the new App needs to be allocated to a host where no such allocated App will be present due to the initial App allocation. When the number of such allocated Apps increases with the variation in hosts and links load, the frequency of initial App allocation failure will increase because the number of hosts to which new Apps can be allocated is limited. This frequency can be improved by maintaining the E2E delay of the allocated App under a certain level through App relocation, which enhances the efficiency of UE accommodation in the DCE.

However, existing App relocation methods such as [12, 13], in which only the threshold-exceeding App is targeted for relocation, may face the problem of insufficient relocation when the number of UEs making service requests increases. Since the required delay of the threshold-exceeding Apps on the MEC host is low, these Apps need to be relocated to the MEC host located in close proximity to the UE. Therefore, the number of MEC hosts where the E2E delay of the threshold-exceeding Apps can be improved and the required delay of the allocated Apps can be satisfied is limited, which degrades the success rate of App relocation.

This problem can be improved through the selection of a relocated App on the host where the threshold-exceeding App is located, with consideration given to the non-threshold-exceeding App. When the non-threshold-exceeding App is relocated, the load on the host will be reduced, which can indirectly improve the E2E delay of the threshold-exceeding App. The non-threshold-exceeding App includes an App whose E2E delay is sufficiently lower than the required delay and can be relocated to a host with a higher delay such as a CC host. Therefore, the success rate of the App relocation can be maintained even when the number of UEs making service requests increases by considering the non-threshold-exceeding App in the App relocation.

For such relocated App selection, simultaneous consideration of E2E delay variation affected by the App relocation and the cost associated with the App relocation is important. Since the relocated App potentially suffers from degradation of service quality [12, 14], it is necessary to suppress the number of relocated Apps from a QoS perspective. With App relocation, the E2E delay of allocated Apps on the source host will decrease, whereas the E2E delay of allocated Apps on the destination host will increase. By considering these E2E delay variations can achieve efficient E2E delay improvement in App relocation. As a result, the occurrence of threshold-exceeding Apps will be suppressed and the number of relocated Apps will be reduced. Additionally, the resource consumption cost of the hosts and links varies due to App relocation, and there is a communication cost for transferring the information of the relocated App from the source host to the destination host [14]. Therefore, a relocated App selection method that can simultaneously take these factors into account is required.

4.2   Proposed App Relocation Method

In this paper, Delay-variation Resource-cost-variation Communication-cost Based Selection (DRCBS) is proposed. It is an algorithm for selecting a combination of relocated App and destination host with simultaneous consideration of E2E delay variation indicator Id, resource consumption cost variation indicator Ir, and communication cost indicator Ic. When DRCBS is applied to a host where a threshold-exceeding App has been located, relocation indicator I, which is expressed in formula (1), is computed for each combination of App on the host and its relocation destination host, and the combination with the largest I value is selected.

$$I_{h,h'}^i = W_1 Id_{h,h'}^i + W_2 Ir_{h,h'}^i - W_3 Ic_{h,h'}^i \qquad (1)$$

The i denotes App, the h denotes the host where the threshold-exceeding App is located, the h' denotes the relocation destination host, and the W denotes the weight factor of each indicator. Each indicator in formula (1) is computed by using App array I = $\{1, ..., i, ..., I\}$ and host array H = $\{1, ..., h, ..., h', ..., H, -1\}$ (-1 denotes the CC host). In the following, allocation array $X_i = \{x_i^1, \cdots, x_i^h, \cdots, x_i^{h'}, \cdots x_i^H, x_i^{-1}\}$ of App i, and all App allocation array $X = \{X_1, ..., X_i, ..., X_I\}$ are used. The x_i^h is 1 if App i is allocated at host h, and 0 if it is not. Additionally, each App can be allocated to only one host ($\Sigma_h X_i = 1$).

E2E Delay Variation Indicator Id. The Id indicates variation in the E2E delay of each App on the source and destination host of App relocation. This indicator is denoted by the allocation array X_{before} and X_{after} before/after App relocation, the E2E delay $d^i(X)$ of App i for given allocation array X, and the required delay D^i for the service of App i, as shown in formula (2).

$$Id^i_{h,h'} = \Sigma_i(d^i(X_{before}) - d^i(X_{after}))\Big/ D^i \qquad (2)$$

The larger the $Id^i_{h,h'}$, the more effective the improvement of E2E delay is by relocation. The $d^i(X)$ is denoted by the data delivery delay $d^i_{dlv}(X)$ in the network segment between the UE and App i, and the computation delay $d^i_{cmp}(X)$ on the host, as shown in formula (3).

$$d^i(X) = d^i_{dlv}(X) + d^i_{cmp}(X) \qquad (3)$$

Resource Consumption Cost Variation Indicator *Ir*. The *Ir* indicates the cost variation associated with the change in resource consumption of hosts and links caused by the App relocation. *Ir* is denoted by the resource consumption cost $c^i(X)$ of App i for given allocation array X, and expected cost C^i of App i, as shown in formula (4).

$$Ir^i_{h,h'} = (c^i(X_{before}) - c^i(X_{after}))\Big/ C^i \qquad (4)$$

The larger the $Ir^i_{h,h'}$, the greater the efficiency of the resource consumption cost will be in the App relocation. The $c^i(X)$ is denoted by the host resource consumption cost $C^i_{hst}(X)$ and the link resource consumption cost $C^i_{\ln k}(X)$ as shown in formula (5).

$$c^i(X_i) = c^i_{hst}(X_i) + c^i_{\ln k}(X_i) \qquad (5)$$

Communication Cost Indicator *Ic*. The *Ic* indicates the communication cost incurred by transferring the information of the relocated App maintained on the memory resources of the relocation source host to the relocation destination host. The *Ic* is denoted by the transferred information volume T^{reloc}_{mem}, as shown in formula (6).

$$Ic^k_{h,h'} \propto T^{reloc}_{mem} \qquad (6)$$

5   Evaluation

The proposed method is evaluated through numerical simulation. The DRCBS and the following methods are compared in terms of the success rate of App relocation in the threshold-exceeding App occurrence event, the number of initial App allocation failures, the number of relocated Apps along with the average number of threshold-exceeding Apps during the evaluation period, and the App relocation associated cost.

NR. App relocation is not performed in the over-threshold App occurrence event.

TO. [12, 13] based method. The over-threshold App is relocated to the host identified according to the E2E delay improvement and the communication cost of App relocation.

DBS. Relocated App and destination host are selected in consideration of the *Id*. Cost efficiency of DRCBS is evaluated by comparison with this method.

RCBS. Relocated App and destination host are selected in consideration of the *Ir* and *Ic*. E2E delay improvement of DRCBS is evaluated by comparison with this method.

5.1 Conditions of the Numerical Simulation

We conducted simulations in which the CSI iterates initial App allocation to the DCE in response to the service request of UEs that occur in every period. During the simulation, the CSI performed operations based on each method presented previously for the over-threshold App occurrence event.

We assumed that the DCE consists of three MEC hosts co-located in a different base station and one CC host. The simulation settings regarding DCE are shown in Table 1.

Table 1. Configuration information relating to the distributed cloud environment.

Parameter	Values
Packet transmission delay D_{trn} (RTT) [ms]	20 (between UE and nearest MEC host)
	30 (between UE and non-nearest MEC host)
	70 (between UE and the CC host)
Maximum bandwidth of links [Gbps]	3 (mobile network section)
	10 (Internet section)
Maximum processing power P_m^h [Gcycle/s]	112 (MEC host), 336 (CC host)
Unit price of CPU C_{cpu} [\$/Gcycle]	$8.33*10^{-6}$ (MEC host), 2.78×10^{-6} (CC host)
Unit price of memory C_{mem} [\$/Gbit]	$3.33*10^{-6}$ (MEC host), 1.11×10^{-6} (CC host)
Unit price of data transfer C_{bit} [\$/Gbit]	$167*10^{-6}$ (between UE and non-nearest MEC host, or MEC host and MEC host)
	$583*10^{-6}$ (between UE/MEC host and CC host)

The service requests of UEs were generated according to a scenario in which the service status (initiate, continue, or terminate) of UEs is simulated in each period. The service status of UEs was pre-defined according to the Markov chain model based on the state transition probability of the App with which the UE communicates. Table 2 shows the App specifications of the three assumed services. The services are video distribution services such as VR (App1 in Table 2), video analysis services such as IoT surveillance cameras (App2), and real-time information collection and distribution services such as automated driving support (App3). It was assumed that the radio area where the UE is located and the service requested by the UE are constantly the same.

In the initial App allocation, the App was allocated to a host whose value obtained by Eq. (7) is the lowest among the hosts where the E2E delay of each App can be satisfied with the required delay and the resource consumption of the hosts and links does not exceed the maximum resource capacity.

$$\Sigma_i d^i(X) \Big/ D^i + \Sigma_i c^i(X) \Big/ C^i \tag{7}$$

In this evaluation, Eqs. (3), (5), and (6) were derived from Eqs. (8) to (12).

$$d_{dlv}^i(X) = D_{trn}(X, r, i) + T_U^i \Big/ th_{upl}(X, r, i) + T_D^i \Big/ th_{dwl}(X, r, i) \tag{8}$$

Table 2. App specifications.

Parameter	App1	App2	App3
Required E2E delay D^i [ms]	50	500	100
Processing period (PP) [ms]	30	30	100
Transfered data size from UE to App per PP T_U^i [Gbit]	$0.03*10^{-3}$	$1.2*10^{-3}$	$0.3*10^{-3x}$
Transfered data size from UE to App per PP T_D^i [Gbit]	$1.2*10^{-3}$	$0.03*10^{-3}$	$0.3*10^{-3}$
Required processing power PS P^i [Gcycle]	0.15	0.105	0.1
Required memory size M^i [Gbit]	0.03	0.03	0.01
Expected cost per PP C^i [$]	$1.56*10^{-6}$	$1.18*10^{-6}$	$0.97*10^{-6}$
State transition probability of initiation	1/90	1/60	1/120
State transition probability of termination	1/30	1/20	1/40
Assumed processing	Video distribution	Video analysis	Real-time information
E2E delay threshold	$0.95*D^i$		

$$d_{cmp}^{jn}(X) = d_{cmp}^{jn-1}(X) + \sum_{n}^{J}(J - n + 1)(P^{jn} - P^{jn-1})/P_m^h \qquad (9)$$

$$c_{hst}^i(X_i) = P^i C_{cpu}(X_i) + M^i C_{mem}(X_i) \qquad (10)$$

$$c_{lnk}^i(X_i) = (T_i^U + T_i^D)C_{bit}(r, i) \qquad (11)$$

$$Im_{h,h'}^{\prime k} = M^i C_{bit}(h, h') \qquad (12)$$

The $D_{trn}(X, r, i)$ denotes the packet transmission delay in the network section between the radio area r of the UE and App i for given allocation array X. As shown in Table 1, the packet transmission delay D_{trn} was assumed to be approximated by RTT. The $th_{upl}(X, r, i)$ and the $th_{dwl}(X, r, i)$ denote the throughput of uplink and downlink between the radio area r of the UE and App i for X. By referring to [10], we assumed that the th_{upl} and th_{dwl} are the minimum available bandwidth among the available bandwidths of each link in the network that the packet traverses, and are also assumed to be the available bandwidth in the radio section of the UE. The radio section throughput of UEs in Uplink and Downlink were configured in the range of 80–200 Mbps and 40–400 Mbps, respectively. The T_U^i and the T_D^i denote the data size transferred from UE to App and from App to UE per processing period of App i. The $J(X, h) = \{j1,...,jn,...jJ\}$

denotes the set of Apps on host h in the allocation set X aligned in ascending order based on the required computation power. The P^{jn} denotes the required processing power per processing period of App jn. The P_m^h denotes the maximum processing power of host h. We assumed that host h assigns the CPU power equally to each App on h, with reference to [15]. Note that $d_{cmp}^{j1-1}(X)$ and P^{j1-1} are 0. The M^i denotes the required memory size of App i. The $C_{cpu}(X_i)$ and the $C_{mem}(X_i)$ denote the unit price of CPU and memory in the allocation array X_i of App i. The $C_{bit}(r, i)$ denotes the unit cost of data transmission between radio area r and App i.

5.2 Result

We set 150 periods during which the maximum number of UEs possibly to be accommodated in the DCE (maximum UE) is constant (100, 110, 120, 130 units), and compared the methods in these periods.

Figure 3 shows the results for the success rate of App relocation and the number of initial App allocation failures. In Fig. 3 (a), the success rate of App relocation for each method to select the relocated App (DRCBS, DBS, and RCBS) is higher than that of TO for any maximum UE, and the success rate is maintained even when the maximum UE is increased. On the other hand, in TO, the success rate of App relocation decreased as the maximum UE increased. Figure 3 (b) shows that the number of initial App allocation failures is reduced by App relocation, and the reduction effect is enhanced by the methods to select the relocated App. The number of initial App allocation failures in DRCBS was 73.6% less than that of TO when the maximum UE is 120, and even when the maximum UE is 130, where almost no App relocation took place in TO, it was 60.6% less. Accordingly, DRCBS and other methods to select a relocated App improve the success rate of App relocation and reduce the number of initial App allocation failures.

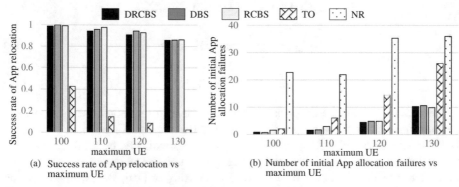

Fig. 3. Success rate of App relocation and the number of initial App allocation failures ($W1 = 1$, $W2 = 1$, $W3 = 0.01$). In Fig. 3 (a), 1 on the y-axis means 100%.

Figure 4 shows the results relating to the efficiency of App relocation. In Fig. 4 (a), the effect based on consideration of the E2E delay variation indicator Id is evaluated through a comparison of DRCBS and RCBS. The number of relocated Apps for RCBS, which does not consider the E2E delay variation indicator Id, was constantly higher than that of DRCBS. On the other hand, there is no significant difference in the average number of threshold-exceeding Apps during the evaluation period. These results show that the number of relocated Apps can be suppressed by considering the E2E delay variation indicator Id. In Fig. 4 (b), the effect when cost indicator Ir and Ic are taken into consideration is evaluated through a comparison of DRCBS and DBS. In terms of the App relocation associated cost, which is part of the overall system cost and represented by $W_2 Ir - W_3 Ic$, the DBS without considering Ir and Ic was constantly lower than that of DRCBS. Thus, the cost efficiency associated with App relocation can be enhanced by taking Ir and Ic into consideration. Results shown in Fig. 4 indicate that DRCBS performs efficient App relocation by simultaneously considering the E2E delay variation of each App affected by the App relocation and the cost associated with the App relocation.

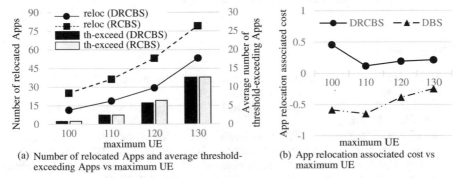

(a) Number of relocated Apps and average threshold-exceeding Apps vs maximum UE

(b) App relocation associated cost vs maximum UE

Fig. 4. Results relating to the efficiency of App relocation ($W1 = 1$, $W2 = 1$, $W3 = 0.01$). In (b), a positive value for the y-axis means that the cost after App relocation is decreased and a negative value means that the cost is increased after App relocation.

6 Conclusion

In this paper, an efficient App relocation method, DRCBS, is proposed for the CSI, which allocates App processing with different requirements of delay and resource amount, to a DCE consisting of the MEC host and the CC host. The DRCBS enables the CSI to improve UE accommodation efficiency of the DCE while maintaining QoS of each App allocated in the DCE. The DRCBS achieves a high success rate and efficient App relocation by selecting a combination of the relocated App including the non-threshold-exceeding App and the destination host. Efficient App relocation is performed through simultaneously considering the E2E delay variation of each App affected by the App relocation and the cost associated with the App relocation in the selection process. The results of numerical simulations show that the proposed method reduces the number of initial App allocation failures in DRCBS by up to 73.6% compared to the existing method. Additionally, DRCBS was more efficient in terms of the number of relocated

Apps and the App relocation associated cost compared to other methods for selecting a relocated App.

Enhancement of App relocation efficiency through optimizing the weight value W is left for future work. The results of the evaluation are expected to be further improved when the weight value W, which was treated as common static values across services in this paper, is configured with an optimal value for each service individually. For example, the App relocation associated cost can be more efficiently optimized by setting a higher ratio of W2 for an App with long operating periods, such as resident services, and a higher ratio of W3 for an App with short operating periods. Therefore, a future research topic is a method to determining the optimal weight value for each service dynamically in response to the type and characteristics of the services treated by the CSI.

Acknowledgments. This work was conducted as part of the project entitled "Research and development for innovative AI network integrated infrastructure technologies (JPMI00316)" supported by the Ministry of Internal Affairs and Communications, Japan.

References

1. ITU-R: IMT Vision – Framework and overall objectives of the future development of IMT for 2020 and beyond. Itu-R M.2083-0, ITU (2015)
2. Abbas, N., Zhang, Y., Taherkordi, A., Skeie, T.: Mobile edge computing: a survey. IEEE IoT J. **5**, 450–465 (2018). IEEE
3. Mao, Y., You, C., Zhang, J., Huang, K., Letaief, K.B.: A survey on mobile edge computing: the communication perspective. IEEE Commun. Surv. Tut. **19**, 2322–2358 (2017) . IEEE
4. Ahmed, A., Ahmed, E.: A survey on mobile edge computing. In: 10th International Conference on Intelligent Systems and Control, ISCO 2016, pp. 1–8. IEEE (2016)
5. Bellavista, P., Berrocal, J., Corradi, A., Das, S.K., Foschini, L., Zanni, A.: A survey on fog computing for the Internet of Things. Pervasive Mob. Comput. **52**, 71–99 (2019). IEEE
6. Luo, Y., Qiu, S.: Optimal resource reservation scheme for maximizing profit of service providers in edge computing federation. In: 2019 IEEE International Conference. IEEE (2019)
7. Porambage, P., Okwuibe, J., Liyanage, M., Ylianttila, M., Taleb, T.: Survey on multi-access edge computing for Internet of Things realization. IEEE Commun. Surv. Tut. **20**, 2961–2991 (2018). IEEE
8. Berno, M., Alcaraz, J.J., Rossi, M.: On the allocation of computing tasks under QoS constraints in hierarchical MEC architectures. In: International Conference on FMEC 2019, pp. 876–879. IEEE (2019)
9. ETSI: MEC 003 - V2.1.1 - Multi-access edge computing (MEC). Framework and Reference Architecture, Vol. 1, pp. 1–21. ETSI (2019)
10. Vondra, M., Becvar, Z.: QoS-ensuring distribution of computation load among cloud-enabled small cells. In: International Conference on CloudNet 2014, pp. 197–203. IEEE (2014)
11. Li, Q., Zhao, J., Gong, Y., Zhang, Q.: Energy-efficient computation offloading and resource allocation in fog computing for Internet of Everything. China Commun. **16**, 32–41. IEEE (2019)
12. Zhang, W., Hu, Y., Zhang, Y., Raychaudhuri, D.: SEGUE: quality of service aware edge cloud service migration. In: International Conference on CloudCom2016, pp. 344–351. IEEE (2016)
13. Maheshwari, S., Choudhury, S., Seskar, I., Raychaudhuri, D.: Traffic-aware dynamic container migration for real-time support in mobile edge clouds. In: International Conference on ANTS2018, pp. 1–6. IEEE (2018)

14. Wang, S., Xu, J., Zhang, N., Liu, Y.: A survey on service migration in mobile edge computing. IEEE Access **6**, 23511–23528 (2018). IEEE
15. The Linux Kernel documentation: https://www.kernel.org/doc/html/latest/scheduler/sched-design-CFS.html. The kernel development community

A Feasibility Study of Log-Based Monitoring for Multi-cloud Storage Systems

Muhammad I. H. Sukmana$^{(\boxtimes)}$, Justus Cöster, Wenzel Puenter,
Kennedy A. Torkura, Feng Cheng, and Christoph Meinel

Hasso Plattner Institute, University of Potsdam, Potsdam, Germany
{muhammad.sukmana,kennedy.torkura,feng.cheng,christoph.meinel}@hpi.de,
{justus.coester,wenzel.puenter}@student.hpi.de

Abstract. With more cloud customers are storing their data in multiple Cloud Service Providers (CSPs), they are responsible for managing the data in the multi-cloud storage environment, including monitoring the events on the cloud. They could monitor various cloud storage services by collecting, processing, and analyzing the cloud storage log files generated by multiple CSPs. In this paper, we investigate the feasibility of log-based monitoring for multi-cloud storage systems. We evaluate the current state of cloud object storage services and their logging functionality by analyzing cloud storage log files generated by a proof-of-concept cloud storage broker system using the three largest public CSPs: Amazon Web Services, Google Cloud Platform, and Microsoft Azure. We discover the logging functionality of cloud storage services could create severe security and reliability issues for cloud customers monitoring the multi-cloud storage systems due to cloud storage log files might not record unauthenticated and unauthorized requests with unpredictable delivery time.

Keywords: Multi-cloud storage system · Cloud object storage services monitoring · Log monitoring · Amazon Web Services · Google Cloud Platform · Microsoft Azure

1 Introduction

An increasing amount of data are now stored in the cloud wherein 2019 48% of corporate data is stored in the cloud - a 13% increase in the last three years [19]. Many cloud customers even utilize cloud storage services from different cloud service providers (CSPs) for their services and applications since it provides better data availability and service reliability than using the services from a single CSP [6,8].

As cloud customers store their data in multiple CSPs, they are responsible for securely managing their data in the cloud due to the shared responsibility model in cloud computing [11]. One of the cloud customer's cloud data management

L. Barolli et al. (Eds.): AINA 2021, LNNS 226, pp. 458–471, 2021.
https://doi.org/10.1007/978-3-030-75075-6_37

tasks is to monitor the activities of cloud storage services and their stored data. They could monitor their multi-cloud storage systems by collecting, processing, and analyzing log files generated by cloud storage services or **cloud storage log files** [17].

In this paper, we investigate the feasibility of monitoring events happening in a multi-cloud storage system using cloud storage log files generated by cloud object storage services from the three most used public CSPs: Amazon Web Services (AWS)[1], Google Cloud Platform (GCP)[2], and Microsoft Azure (Azure)[3].

Our main contributions are as follows:

- We investigate the current state of cloud object storage services and their logging functionality in AWS, GCP, and Azure.
- We compare, correlate, and analyze cloud storage log files generated from our proof-of-concept cloud storage broker (CSB) system [16] by simulating different file activities from various actors to determine if cloud storage log files are suitable for monitoring the events in multi-cloud storage systems.
- We determine the current logging functionality of cloud object storage services from AWS, GCP, and Azure could create **severe security and reliability issues** for cloud customers due to inconsistent cloud storage log delivery and unauthenticated and unauthorized requests might not be recorded in the log files.

The structure of the paper is as follows: Sect. 2 shows several related works in cloud storage monitoring. In Sect. 3, we describe the overview of multi-cloud storage monitoring and its challenges. Section 4 discusses the current state of cloud object storage services and their logging functionality in AWS, GCP, and Azure. Section 5 describes our cloud storage log files gathering process by simulating various file activities on our CSB system [16]. In Sect. 6, we compare, process, and analyze cloud storage log files and CSB system log file generated from our evaluation scenario to evaluate log-based monitoring for the multi-cloud storage system. Finally, Sect. 7 concludes our paper and presents future work.

2   Related Works

Several works have proposed various monitoring systems for cloud storage services.

De Marco et al. [1] utilized the AWS S3 server access log to monitor and detect the violations in the Service Level Agreement (SLA) between AWS with cloud customers. Garion et al. [3] analyzed large amounts of cloud object storage service's log entries using Apache Spark to monitor the service's performance, estimate the potential for archiving the storage services, and detect security threats and anomalies of customer behavior. [2] developed a system to monitor

---

[1] https://aws.amazon.com.

[2] https://cloud.google.com.

[3] https://azure.microsoft.com.

Infrastructure-as-a-Service storage service's usage and analyze the file access patterns based on several files' parameters, e.g., access frequency, size, and replication. Torkura et al. [20] proposed a cloud threat detection and incident response for multi-cloud storage systems called Slingshot by aggregating and analyzing cloud logs from AWS CloudTrail and GCP Logging with the cloud security assessment alerts.

Our work is different from the related works above as we analyze the feasibility of monitoring the events in a multi-cloud storage environment using cloud storage log files of the three most used public CSPs on the market. We collect, correlate, and analyze the cloud storage log files by simulating different request types executed by various actors to access the files stored in multiple CSPs.

3   The Challenges of Log-Based Multi-Cloud Storage Monitoring

The multi-cloud storage approach refers to a technique to store the data in multiple CSPs [8]. It provides better data interoperability and availability than utilizing the services from a single CSP as the data could migrate between CSPs and still be retrieved if one or several CSPs are unreachable [6]. The approach utilizes distributed data storage strategies, e.g., erasure coding, replication, or fragmentation, where multiple data parts are stored in different CSPs [6,8].

As cloud customers migrate their data to the cloud, cloud customers effectively relinquish the physical control of their data to the CSP [22]. Meanwhile, cloud customers must securely manage their data stored on various CSPs as part of the CSP's shared responsibility model to ensure that the data can only be accessed by authorized actors [11]. Cloud customer's data could be accessed or modified by non-authorized actors, such as the CSP or anonymous Internet users, without their knowledge due to misconfigured cloud storage services [21,22].

One of the cloud customer's cloud data management responsibilities is monitoring the latest activities and state of the cloud storage services and the data stored in the cloud [16,18]. Cloud customers could monitor cloud storage services by collecting, processing, and analyzing the generated **cloud storage log files**. The log files provide information on the events happening in the cloud storage services, such as the requester's information, request types, and response information [5,20]. From the cloud customer's perspective, cloud monitoring can be used for multiple purposes, such as cloud resource usage cost and performance monitoring [3], check for SLA violation [1], or discover malicious activities and misconfiguration [3,20].

However, there are several challenges faced by cloud customers to monitor the activities in a multi-cloud storage environment using cloud storage log files.

Each CSP utilizes various hardware and software resources to build cloud environments where it is not required to comply with cloud computing standards while providing cloud services to cloud consumers [16]. It causes each CSP to have custom mechanisms, and implementations of the same cloud storage service, including its API, data model, and logging functionality [7].

Due to the shared responsibility model in cloud computing, cloud customers must collect, process, and analyze cloud storage log files to monitor the storage activities in multiple CSPs while resolving the heterogeneity of cloud storage services [17]. Although cloud customers could utilize cloud log management services, e.g., AWS CloudWatch[4] or GCP Logging[5], these services lack cross-CSP collaboration functionality to manage the cloud storage log files from various CSPs.

Cloud customers could not influence how the cloud storage services would behave, including what information on the cloud storage log files is produced from the events in cloud storage services. For example, there is a lack of response information from the executed requests to cloud storage services and cloud storage log files. The response information might contain null response code and response message of the executed request. It is due to the CSP not returning the request identifier of the executed request in the response information as it may not be propagated correctly in various CSP components [23].

4   The State of Public Cloud Object Storage Services

Cloud object storage services is an Infrastructure-as-a-Service that provides long-term, highly available, and cheap data storage on the cloud. It provides a key-value store interface to store arbitrary objects [4]. Cloud customers could store any amount or size of the data in the cloud as **objects** or **blobs** in the **buckets** or **containers** [16].

Cloud customers are expected to securely configure their buckets and objects to ensure only authorized actors could access the data. Buckets and objects could be configured through policy and access control list (ACL) to specify which entities and their privileges are allowed to access the resources [16].

In general, there are four types of requests made to the cloud storage services based on the configuration of the bucket and their objects and the returned responses:

- **Unauthenticated Request:** The request is executed by actors that are not authenticated to the CSP's domain.
- **Authenticated Request:** The request is executed by the CSP customers, CSP services, or other actors that exist in the CSP's domain.
- **Authorized Request:** The request is executed by actors with the correct privileges to access the services, buckets, or objects and returns a successful response.
- **Unauthorized Request:** The request is executed by actors with incorrect privileges to access the services, buckets, or objects and returns a failed response.

---

[4] https://aws.amazon.com/cloudwatch/.
[5] https://cloud.google.com/logging.

Cloud storage services provide logging functionality that records all events of the requests sent to the services and their responses in **cloud storage log files**. There are two types of cloud storage log provided by the CSP:

- **Storage Access Log:** Storage access log file records the events happening in a selected monitored bucket and delivers the logs to a target bucket. It provides a structured and simple data format, such as CSV. Cloud customers need to enable the logging configuration for each monitored bucket or the cloud storage services.
- **Cloud Activity Log:** Cloud activity log file records the events in the cloud resources and services in cloud customer's CSP environment, including cloud storage services or a specific bucket. It provides semi-structured data commonly in JSON format to accommodate different cloud resource's data models.

Cloud customers could expect several characteristics of the cloud storage log files to monitor the events happening in the cloud storage services:

- **Reliable Log Delivery:** Cloud storage log must be delivered after the event is happening in cloud storage services.
- **Record Request Information:** Cloud storage log must record the information of all request types sent by the requester to cloud storage services.
- **Record Response Information:** Cloud storage log must record the information of the response based on the executed request.
- **Consistent Log Values:** Cloud storage log should have consistent log values for different request types.
- **Consistent Number of Log Fields:** Cloud storage log should have consistent log field numbers for different request types.

We look at the current state of cloud object storage services from three public CSPs: Amazon Web Services, Google Cloud Platform, and Microsoft Azure.

4.1   Amazon Web Services

AWS provides its object storage service through Simple Storage Service (S3)[6] service. It allows for pre-signed URL or signed URL creation with customized parameters with a validity period of up to 7 days[7].

AWS S3 provides the **server access log** [9] by enabling the bucket logging option in the monitored AWS S3 bucket. Events in the monitored bucket will be periodically written as log files and delivered on a best-effort basis within a few hours of the events to the target bucket by AWS S3 Log Delivery group.

AWS CloudTrail[8] can be used to provide the **cloud activity log** for AWS S3 services. It continuously monitors and logs all events in the AWS infrastructure,

---

[6] https://aws.amazon.com/s3/.

[7] https://docs.aws.amazon.com/AmazonS3/latest/dev/ShareObjectPreSignedURL.html.

[8] https://aws.amazon.com/cloudtrail/.

including AWS S3's bucket-level and object-level API calls in the monitored bucket or AWS S3 account [10]. A trail needs to be enabled to deliver the log files around every 5 min to the specified target bucket or AWS CloudWatch.

4.2 Google Cloud Platform

GCP provides its object storage service through the Storage[9] service. It allows for signed URL creation with customized parameters valid of up to 7 days[10].

GCP Storage provides the **usage log** [12] to monitor activities happening in the GCS bucket. After the bucket logging option in the monitored GCP Storage bucket has been enabled, all events happening in the monitored bucket will be periodically collected and written as log files by the GCP Storage Analytics group to the target bucket. The log files are delivered hourly, approximately 15 min after the end of the hour. Duplicate log entry could exist for the log files created in the same hour that could be detected by checking the $s\_request\_id$ field for duplicate values.

GCP Logging generates the **cloud audit log** that records activities in a GCP project, folder, or organization [13]. In this paper, we consider only the **data access audit log** from the cloud audit log to record the events in the monitored GCP Storage bucket. The cloud audit log files are delivered approximately every hour to the GCP Storage target bucket, GCP Pub/Sub[11], or GCP BigQuery[12].

4.3 Microsoft Azure

Azure provides its object storage service through the Storage Blob (Blob)[13] service. The objects, or blobs, are stored in the container, or a bucket, of a storage account. Azure Blob does not allow for shared access signatures (SAS) or signed URL creation with customized parameters[14].

Azure Blob provides the **storage analytics log** [14] that records detailed events happening in the storage account on the best-effort basis. The log files are delivered to a container named $logs up to every hour. Duplicate log entries could be detected by checking the value of *RequestId* and *Operation* fields. We utilize version 2.0 of the storage analytics log specifically for the Blob service for this paper.

We utilize the **storage resource log** generated by Azure Monitor[15] as the cloud activity log to provide detailed diagnostic and auditing information of the events within the Azure infrastructure, including the storage account [15]. The log files are delivered to Azure Event Hubs[16] or the storage account every hour.

---

[9] https://cloud.google.com/storage.
[10] https://cloud.google.com/storage/docs/access-control/signed-urls.
[11] https://cloud.google.com/pubsub/.
[12] https://cloud.google.com/bigquery/.
[13] https://azure.microsoft.com/en-us/services/storage/.
[14] https://docs.microsoft.com/en-us/azure/storage/common/storage-sas-overview.
[15] https://azure.microsoft.com/en-us/services/monitor/.
[16] https://azure.microsoft.com/en-us/services/event-hubs/.

5   Data Gathering

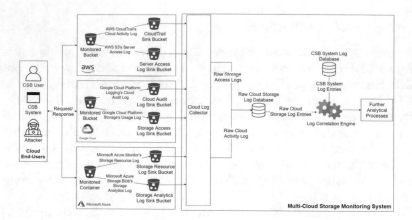

Fig. 1. Overview of multi-cloud storage monitoring system used by the cloud storage broker system for the evaluation

We collect cloud storage log files generated from the activities in a proof-of-concept cloud storage broker (CSB) system to evaluate the logging functionality of cloud object storage services of AWS, GCP, and Azure. CSB is a third-party multi-cloud storage system that provides Storage-as-a-Service that mediates the relationship between CSB users as the cloud end-users and multiple CSPs [16].

We simulated file activities of 10 CSB users with the same public IP address to upload, download, and delete the files stored in the multiple CSPs via the CSB system using the application programming interface (API) and signed URLs.

We also simulated download file activities done by the **attacker** representing anonymous Internet user, malicious CSB users, other cloud customers, and IAM entities in the CSB system, following the threat model in [21] as follows:

- Attacker requests the object storage URLs.
- Attacker accesses the objects using their CSP credential.
- Attacker requests object signed URLs generated using their CSP credential.
- Attacker requests expired object authorized signed URLs.
- Attacker requests modified object authorized signed URLs.

We develop a multi-cloud storage monitoring system for the CSB system to collect, process, and analyze cloud storage log files generated by the activities between the CSB users, attackers, the CSB system, and multiple CSPs, as can be seen in Fig. 1. The CSB system utilizes one credential for each CSP to manage CSB user's files and cloud storage log files. Three separate buckets in each CSP are used to store CSB users' files, storage access log files, and cloud activity log files.

6   Evaluation

We collect, process, correlate, and analyze the cloud storage log files from AWS S3, GCP Storage, and Azure Storage services generated from our evaluation scenario. We then discuss the feasibility of multi-cloud storage environment monitoring using cloud storage log files from various CSPs.

6.1   Cloud Storage Log Comparison and Correlation

We compare and correlate the storage access log and the cloud activity log from multiple CSPs generated from the CSB system's evaluation scenario.

6.1.1   Amazon Web Services S3

AWS CloudTrail's cloud activity log has more fields and more detailed information about the events in the AWS S3 service compared with AWS S3's server access log. However, it does not have a consistent number of log fields as it depends on the request types received by the AWS S3 service. The values on both log types are consistent for almost all fields, except for the request method in the *Operation* field (server access log) and *eventName* field (cloud activity log).

We observed the server access log files are delivered around **20 to 40 minutes** after the event recorded in the CSB system log, while the cloud activity log files are delivered about **10 minutes** after the event. The log files consist of one or multiple log entries where the files could be generated **unsorted** regardless of when the requests happened. The timestamp in the *Time* field (server access log) and *eventTime* field (cloud activity log) are almost identical with **1-second** maximum difference.

We could identify the requester using the *Requester* field (server access log) and the *userIdentity* field (cloud activity log). The request made using API or pre-signed URL could be differentiated using *Authentication Type* field (server access log) and *authenticationMethod* field (cloud activity log). The pre-signed URL's signature only appears in the *Request-URI* field in the server access log anonymized.

The server access log records all of the attacker's unauthenticated and unauthorized requests, while cloud activity log **does not log** object access requests using expired signed URL and modified signed URL. This is because AWS CloudTrail logs failed authorization requests and unauthenticated requests, but it does not log the request with failed authentication [10]. We also discover several attempts accessing the CSB system's bucket by other actors outside of our attacker requests from unauthorized AWS accounts and AWS Config service[17].

We could correlate the server access log with the cloud activity log using the *Request ID* fields in both log types to generate **one-to-one correlation result**. However, several server access log entries of attackers' unauthorized requests might not be correlated since they do not have the corresponding cloud activity log entries.

---

[17] https://aws.amazon.com/config/.

Table 1. Comparison of cloud storage log files from Amazon Web Services, Google Cloud Platform, and Microsoft Azure. $^+$= Signature anonymized, *= Using custom user agent

Category	AWS S3 Server Access Log	AWS Cloud- Trail Cloud Activity Log	GCP Storage Usage Log	GCP Logging Cloud Activity Log	Azure Blob Analyt- ics Log	Azure Monitor Resource Log
Duplicate entry	✗	✗	✓	✓	✓	✓
Signature of signed URL	✓ $^+$	✗	✓	✗	✓ $^+$	✓ $^+$
Object access methods differentiation	✓	✓	✓	✗	✓	✓
Consistent number of log fields	✓	✗	✓	✗	✓	✓
Consistent log values	✓	✓	✗	✓	✓	✓
Authorized API request	✓	✓	✓	✓ *	✓	✓
Authorized signed URL request	✓	✓	✓	✓ *	✓	✓
Unauthenticated object storage URL request	✓	✓	✓	✓ *	✗	✗
Unauthorized API request	✓	✓	✓	✓	✗	✗
Unauthorized signed URL request	✓	✓	✓	✓ *	✗	✗
Expired signed URL request	✓	✗	✓	✗	✗	✗
Modified signed URL request	✓	✗	✓	✗	✗	✗

6.1.2   Google Cloud Platform Storage

GCP Storage's storage usage log provides a consistent number of log fields for different requests. Meanwhile, GCP Logging's cloud audit log contains more fields, although it does not have the *Request ID* and *Request URI* fields with different field numbers depending on the request types. The values on both log types are almost consistent except for the request method in the *cs_method* field (storage usage log) and *methodName* field (cloud audit log).

We observed that the storage usage log files and cloud audit log files are delivered between **1 to 2 hours** after the event recorded in the CSB system log. The log files consist of one or multiple log entries where the files could be generated **unsorted** regardless of when the requests happened.

We could only differentiate the request from the signed URL and API in the storage usage log by checking X-Goog-Signature parameter in the *cs_uri* field or the *cs_method* field. Since the cloud audit log does not differentiate the requests made by API or signed URL, we executed both storage access methods with custom user-agents, which also help to identify the requests by the attacker and CSB system.

The storage usage log records all attacker's unauthenticated and unauthorized requests. However, the cloud audit log **does not record** unauthenticated requests with expired or modified signed URL. We also discover a HEAD HTTP request from an unknown IP address to get the bucket's list of objects or configuration.

We could not directly correlate the storage usage log with the cloud audit log since the cloud audit log does not have the *Request ID* field. Although we could correlate using the fields existing in both log types, e.g., *IP address, User agent,*

Object, Request method, and *Response code*, it could create **one-to-many log correlation**, where one storage usage log entry is correlated with multiple cloud audit log entries. We could resolve this by selecting the correlated log entry with the minimum timestamp difference between *time_micros* field (storage usage log) and *timestamp* field (cloud audit log) where it is observed to be around **4 to 140 milliseconds**.

6.1.3 Microsoft Azure Storage Blob

Azure Blob's storage analytics log provides more comprehensive information about the event happening in the storage account than Azure Monitor's storage resource log. The values and the number of fields on both log types are consistent for all request types. We only focus on the events on our monitored bucket by filtering on the *request-url* field (storage analytics log) and the *uri* field (storage resource log).

We observed the storage analytics log files and storage resource log are delivered around **1 hour** after the event recorded in the CSB system log. Both log files consist of one or multiple log entries where the files could be generated **sorted** based on the request's timestamp. We observed the timestamp difference between the storage analytics log and storage resource log is between **1 millisecond to 24 seconds**, which corresponds to *end-to-end-latency-in-ms* field in the storage analytics log.

We could differentiate successful requests from the shared access signature and API in both log types by checking the *authentication-type* field in the storage analytics log and *type* field in the storage resource log. The signature embedded in the signed URL in the storage resource log and storage analytics log are anonymized in the *request-url* field (storage analytics log) and the *uri* field (storage resource log).

Both storage analytics log and storage resource log **do not record** any requests made by the attackers. This is due to Azure to only record certain anonymous request types, but it does not log failed unauthenticated and unauthorized requests to the storage account [14]. We could not determine if there are unauthorized or unauthenticated requests made to our container and blobs outside our attacker scenario.

We could generate **one-to-one log correlation result** using the *correlationId* field (storage resource log) and the *requestIdHeader* field (storage analytics log).

6.2 CSB System Log and Cloud Storage Log Correlation

We face several challenges as we correlate the CSB system log files with cloud storage log files from multiple CSPs from our evaluation scenario to provide full timeline activities of the CSB user's files stored in the cloud.

Cloud customers first need to retrieve newly created cloud storage log files from the log file sink buckets in various CSPs . The cloud storage log files then need to be checked for duplicate entries, as explained in Sect. 4. Finally, we need to parse certain values in cloud storage log files to be uniform following the CSB system log format before correlating it with the CSB system log file.

Although the responses of the executed CSB's file requests do not return the request identifier (ID), we could still correlate the CSB system log files with cloud storage log files using the similar log fields, such as *Timestamp, IP address, User agent, Bucket, Object, Request method, Response code*, and *Request URI* (if available). This could generate a **one-to-many correlation result** due to the CSB system log entry's timestamp is slightly late compared to the timestamp in cloud storage log entries. We could resolve this issue by selecting the correlated log entry with the smallest time difference between the CSB system log entry and cloud storage log entry. Finally, the correlated log information could then be processed for further analysis, such as discovering suspicious activities in the CSB environment.

Since all requests to multiple CSPs should be authorized by the CSB system before it is executed, we used the CSB system log as the **source of truth** for our evaluation, where cloud storage log entry should have the corresponding CSB system log entry. If there are uncorrelated cloud storage log entries, the CSB system needs to investigate it by identifying the nature of the request and its requester based on the response message and checking the configurations of the bucket and the objects.

6.3 Summary

In general, the cloud activity log provides more detailed event information in the cloud object storage services than the storage access log. Cloud customers could use either or both cloud storage log types to monitor the events in a multi-cloud storage environment. However, cloud customers could face several issues while monitoring the multi-cloud storage environment using cloud storage log files.

The cloud storage log files from multiple CSPs are delivered **inconsistently** and **unpredictably** where the files are available **up to 2 hours** after the actual events happened. This would make cloud storage log files unfeasible for real-time monitoring and analysis. The cloud customers need to wait until new log files are available in the sink buckets before processing and analyzing the log files.

We conclude cloud storage log files might have **different information quality** as they might contain **incomplete and inconsistent** information about the events happening in the cloud object storage services. Also, several CSPs intentionally **do not record unauthorized and unauthenticated requests** by design [10,14]. This could create an **information gap** issue for cloud customers and making it difficult to correlate the information of executed requests from the multi-cloud storage systems with the cloud storage log files due to the absence of the request ID in the cloud storage log entry, where it might create **one-to-many** log correlation result.

The combination of inconsistent and incomplete information in the cloud storage log and unpredictable log delivery time could create **reliability and security issues** for cloud customers. An attacker could launch attacks on a multi-cloud storage environment during the time gap between the actual attacks and the cloud storage log files delivered to cloud customers. Cloud customers will be unable to detect and react to the attacks in real-time as they need to wait until the log files are available where the attacks might not be recorded in the log files.

We recommend several aspects to be improved by the CSPs to allow cloud customers better monitor the events happening in the multi-cloud storage systems:

- Cloud storage log files should record complete and consistent information of all event types happening in the cloud storage services.
- Cloud storage log files should be available as soon as the CSP has processed the requests to allow real-time event monitoring for cloud customers.
- The CSP should use a standardized cloud logging format to help cloud customers process and correlate log information from different CSPs.
- The CSP should improve cross-CSP collaboration functionality for the log management services to monitor the events in various CSPs in one platform.

7   Conclusion and Future Works

In this paper, we investigate the feasibility of monitoring the events on cloud storage services from multiple CSPs in a multi-cloud storage system using cloud storage log files. We collect, process, and analyze cloud storage log files of Amazon Web Services, Google Cloud Platform, and Microsoft Azure generated from a proof-of-concept cloud storage broker system by simulating different file activities from various actors in the system. We conclude that monitoring the events in the multi-cloud storage systems utilizing cloud storage log files could create reliability and security issues due to unpredictable log file delivery and inconsistent and incomplete event information with unrecorded unauthorized and unauthenticated requests.

We would like to combine the log-based monitoring process with other cloud resource management processes to provide a holistic multi-cloud security management process for cloud customers, e.g., periodic resource discovery and resource assessment against cloud security best practices.

Acknowledgment. We would like to thank Bundesdruckerei GmbH for the support for this paper.

References

1. De Marco, L., Ferrucci, F., Kechadi, T.: Slafm: A service level agreements formal model for cloud computing. In: The 5th International Conference on Cloud Computing and Service Science (CLOSER 2015), Lisbon, Portugal, 20–22 May 2015 (2015)
2. Devarajan, A.A., SudalaiMuthu, T.: Cloud storage monitoring system analyzing through file access pattern. In: 2019 International Conference on Computational Intelligence in Data Science (ICCIDS), pp. 1–6. IEEE (2019)

3. Garion, S., Kolodner, H., Adir, A., Aharoni, E., Greenberg, L.: Big data analysis of cloud storage logs using spark. In: Proceedings of the 10th ACM International Systems and Storage Conference, p. 1 (2017)
4. Huang, W., Ganjali, A., Kim, B.H., Oh, S., Lie, D.: The state of public infrastructure-as-a-service cloud security. ACM Comput. Surv. (CSUR) **47**(4), 1–31 (2015)
5. Khan, S., Gani, A., Wahab, A.W.A., Bagiwa, M.A., Shiraz, M., Khan, S.U., Buyya, R., Zomaya, A.Y.: Cloud log forensics: foundations, state of the art, and future directions. ACM Comput. Surv. (CSUR) **49**(1), 1–42 (2016)
6. Nachiappan, R., Javadi, B., Calheiros, R.N., Matawie, K.M.: Cloud storage reliability for big data applications: a state of the art survey. J. Netw. Comput. Appl. **97**, 35–47 (2017)
7. Pichan, A., Lazarescu, M., Soh, S.T.: Cloud forensics: technical challenges, solutions and comparative analysis. Digit. Invest. **13**, 38–57 (2015)
8. Rafique, A., Van Landuyt, D., Reniers, V., Joosen, W.: Towards an adaptive middleware for efficient multi-cloud data storage. In: Proceedings of the 4th Workshop on CrossCloud Infrastructures & Platforms, pp. 1–6 (2017)
9. Amazon Web Services: Amazon s3 server access logging (2020). https://docs.aws.amazon.com/AmazonS3/latest/dev/ServerLogs.html. Accessed 09 Jun 2020
10. Amazon Web Services: logging amazon s3 API calls using aws cloudtrail (2020). https://docs.aws.amazon.com/AmazonS3/latest/dev/cloudtrail-logging.html. Accessed 18 Jun 2020
11. Amazon Web Services: shared responsibility model (2020). https://aws.amazon.com/compliance/shared-responsibility-model/. Accessed 19 Nov 2020
12. Google Cloud Platform: access logs & storage logs (2020). https://cloud.google.com/storage/docs/access-logs. Accessed 05 Jun 2020
13. Google Cloud Platform: cloud audit logs with cloud storage (2020). https://cloud.google.com/storage/docs/audit-logs. Accessed 23 Jun 2020
14. Microsoft Azure: Azure storage analytics logging—microsoft docs (2020). https://docs.microsoft.com/en-us/azure/storage/common/storage-analytics-logging?tabs=dotnet. Accessed 09 Jun 2020
15. Microsoft Azure: create diagnostic settings to send platform logs and metrics to different destinations (2020). https://docs.microsoft.com/en-us/azure/azure-monitor/platform/diagnostic-settings. Accessed 18 Sep 2020
16. Sukmana, M.I., Torkura, K.A., Graupner, H., Cheng, F., Meinel, C.: Unified cloud access control model for cloud storage broker. In: 2019 International Conference on Information Networking (ICOIN), pp. 60–65. IEEE (2019)
17. Sukmana, M.I., Torkura, K.A., Prasetyo, S.D., Cheng, F., Meinel, C.: A brokerage approach for secure multi-cloud storage resource management. In: 16th EAI International Conference on Security and Privacy in Communication Networks (SecureComm) 2020. Springer (2020)
18. Syed, H.J., Gani, A., Ahmad, R.W., Khan, M.K., Ahmed, A.I.A.: Cloud monitoring: a review, taxonomy, and open research issues. J. Netw. Comput. Appl. **98**, 11–26 (2017)
19. Thales: 2019 global cloud security study (2019). https://cpl.thalesgroup.com/cloud-security-research. Accessed 19 Oct 2020
20. Torkura, K.A., Sukmana, M.I.H., Cheng, F., Meinel, C.: Slingshot - automated threat detection and incident response in multi cloud storage systems. In: 2019 IEEE 18th International Symposium on Network Computing and Applications (NCA), pp. 1–5 (2019). https://doi.org/10.1109/NCA.2019.8935040

21. Torkura, K.A., Sukmana, M.I.H., Meinig, M., Kayem, A.V.D.M., Cheng, F., Graupner, H., Meinel, C.: Securing cloud storage brokerage systems through threat models. In: 2018 IEEE 32nd International Conference on Advanced Information Networking and Applications (AINA), pp. 759–768 (2018). https://doi.org/10. 1109/AINA.2018.00114
22. Wang, C., Ren, K., Lou, W., Li, J.: Toward publicly auditable secure cloud data storage services. IEEE Netw. **24**(4), 19–24 (2010)
23. Yu, X., Joshi, P., Xu, J., Jin, G., Zhang, H., Jiang, G.: Cloudseer: workflow monitoring of cloud infrastructures via interleaved logs. ACM SIGARCH Comput. Archit. News **44**(2), 489–502 (2016)

Detecting Performance Degradation in Cloud Systems Using LSTM Autoencoders

Spyridon Chouliaras$^{(\boxtimes)}$ and Stelios Sotiriadis

Birkbeck, University of London, Malet Street, Bloomsbury, London WC1E 7HX, UK
{s.chouliaras,stelios}@dcs.bbk.ac.uk

Abstract. Cloud computing technology is on the rise as it provides an easy to scale environment for Internet users in terms of computational resources. At the same time, cloud providers manage this demand for computational power by offering a pay per use model for virtualized resources. Yet, it is a challenging issue to administer the variety of different cloud applications and ensure high performance by identifying failures and errors on runtime. Distributed applications are error-prone, and creating a platform to support minimum hardware and software failures is a key challenge. In this work, we focus on anomaly detection of data storage systems, and we propose a solution for detecting performance degradation of cloud deployed systems in real time. We use Long Short-Term Memory (LSTM) Autoencoders for learning the normal representations and reconstruct the input sequences. Then, we used the reconstructed errors of the LSTM Autoencoders on unseen time series data to detect abnormal behaviours. We used state-of-the-art benchmarks such as TPCx-IoT and YCSB to evaluate the performance of HBase and MongoDB systems. Our experimental analysis shows the ability of the proposed approach to detect abnormal behaviours in cloud systems.

1 Introduction

Nowadays, cloud computing offers a state-of-the-art technology for organisations that are looking for scalable application development. In a cloud system, applications often deployed as virtual machines (VM) that provide customised technical characteristics, operating systems and software packages. As the demand is on the rise, cloud providers are promising cloud elasticity and scalability on their systems as well as infinite amounts of virtual resources. Cloud scalability is the ability of the system to accommodate larger loads, while cloud elasticity is the ability of the system to scale with loads dynamically [1]. In addition, data are becoming massive in terms of volume, variety and velocity, while different techniques have been implemented in order to deal with this new phenomenon. A NoSQL (meaning 'not only SQL') has come to describe a large class of databases sufficient to deal with massive amounts of data, supporting also non-schema structure and real-time analysis [2]. Such systems can store data from different

L. Barolli et al. (Eds.): AINA 2021, LNNS 226, pp. 472–481, 2021.
https://doi.org/10.1007/978-3-030-75075-6_38

sources without a specific structure, a characteristic that makes them powerful for different applications such as e-commerce, navigation systems, web services and IoT. However, the robustness and the reliability of cloud systems remains a key problem for information technology companies as cloud systems could suffer from hardware, software and network failures. Therefore, efficient ways on detecting abnormalities are more important than ever, since anomalies in the system can cause performance degradation, unreasonable energy consumption as well as false scaling estimation from automatised algorithms. The latter, indicates the importance of finding techniques in order to detect abnormal cases based on real time usage and develop algorithms able to predict future failures.

We present Long Short-Term Memory (LSTM) Autoencoders for detecting performance degradation in cloud systems. LSTM Autoencoders have been proved effective for time series learning and anomaly detection [3]. The model trained beforehand on normal sequence representations and learned to reconstruct the input sequence with the lowest possible error. Then, the reconstructed error used as a score in order to classify future time series as normal or abnormal. New unseen observations that have lower reconstruction error than a given threshold are being classified as normal while observations that have higher reconstruction error are being classified as abnormal. The threshold value of the reconstructed error has been tuned as a hyper-parameter based on its ability to detect abnormal cases of various systems.

2   Related Works

Anomaly detection refers to the process of identifying data points, events or observations that deviate from the normal data distribution. Deep learning based anomaly detection methods have been proposed in many domains [4–7]. Amongst them, authors proposed LSTM networks as powerful sequence learners that promise a substantial improvement over the conventional Artificial Neural Networks. In extension to this, we propose LSTM based Autoencoders to detect outliers that indicate application performance degradation. LSTM Autoencoders combine the ability of the Autoencoder to extract the most representative information with the advantage of the LSTM to effectively process sequential data. For that reason, LSTM Autoencoders have been applied widely for anomaly detection [8–11]. In [8] authors proposed an LSTM based Encoder-Decoder scheme for Anomaly detection as an alternative of standard approaches. In their findings, the LSTM Autoencoder learns to reconstruct normal time-series behaviour and by using the reconstruction error effectively detects anomalies from predictable, unpredictable, periodic, aperiodic and quasi-periodic time series. In [9] authors proposed an LSTM Autoencoder network-based method combined with a one-class support vector machine algorithm used for anomaly detection. The LSTM Autoencoder network trained on normal representations and used to calculate a prediction error vector. Then, the one-class Support Vector Machine algorithm used to separate the abnormal observations from normal samples based on these prediction error vectors.

In [10] authors proposed CANolo, an intrusion detection system based on LSTM Autoencoder to identify anomalies in Controller Area Networks. Their framework automatically trained on controller area networks streams to build a model based on the legitimate data sequences. Then the reconstructed error used to detect anomalies between legitimate sequences and simulated attacks. In [11] authors used mobile traffic data and proposed an LSTM Autoencoder to reconstruct mobile traffic samples and an LSTM traffic predictor to predict the traffic of future time instants. In both cases they analyzed the reconstructed and the predicted error to assess if the mobile traffic presents anomalies or not.

In this work, we propose LSTM Autoencoders to reconstruct normal sequence representations of application metric data streamed from MongoDB and HBase systems. Then unseen observations are being classified as normal or abnormal on the fly based on the reconstruction error produced by the LSTM Autoencoder.

3   Methodology

A variety of metrics captured over a period of time from monitoring systems to support our method. Prometheus[1] has been used as a monitoring system for MongoDB[2] and HBase[3] to collect application and system metrics. In addition, Yahoo! Cloud Serving Benchmark (YCSB) [12] and TPCx-IoT[4] workloads used to execute various intensive tasks. Then, under the assumption that YCSB and TPCx-IoT generate repeatable patterns, we created a dataset to train deep learning models that learn normal application behaviours. For that purpose an LSTM Autoencoder has been trained and used for pattern recognition of long-range dependencies and anomaly detection.

3.1   Long Short-Term Memory Networks

Long Short-Term Memory (LSTM) model is a type of a Recurrent Neural Network that promises a substantial improvement in sequential data with temporal sequences and long-range dependencies [13]. An LSTM model contains special units called memory cells that are organised inside the recurrent hidden layer. Each memory cell contains an input gate, an output gate and a forget gate. The forget gate has been subsequently added in order to enable processing continuous input streams that are not segmented into subsequences [14]. The forget gate gives the ability to LSTM cell to learn to reset itself at appropriate times and prevent the network to break down in situations where time series grow arbitrary through time (e.g. continuous input streams) [15].

An LSTM receives an input sequence $x = (x_1, x_2, x_3, ..., x_T)$ and maps it to an output sequence $y = (y_1, y_2, y_3, ..., y_T)$ by calculating the following equations iteratively from t = 1 to T:

---

[1] https://prometheus.io/.
[2] https://www.mongodb.com/.
[3] https://hbase.apache.org/.
[4] http://www.tpc.org/tpcx-iot/.

$$i_t = \sigma(W_{ix}x_t + W_{im}m_{t-1} + W_{ic}c_{t-1} + b_i) \tag{1}$$

$$f_t = \sigma(W_{fx}x_t + W_{fm}m_{t-1} + W_{fc}c_{t-1} + b_f) \tag{2}$$

$$c_t = f_t \odot c_{t-1} + i_t \odot g(W_{cx}x_t + W_{cm}m_{t-1} + b_c) \tag{3}$$

$$o_t = \sigma(W_{ox}x_t + W_{om}m_{t-1} + W_{oc}c_t + b_o) \tag{4}$$

$$m_t = o_t \odot h(c_t) \tag{5}$$

$$y_t = \phi(W_{ym}m_t + b_y) \tag{6}$$

where the i is the input gate that uses the logistic sigmoid function denoted as σ in order to control the information that flows into the cell, f is the forget gate that uses the logistic sigmoid function to decide what information to keep outside the cell state, o is the output gate that uses the logistic sigmoid function to control the information that flows out of the cell, c is the cell activation vectors, m is the output activation vector where $\odot$ is the element-wise product of the vectors, g is the cell input activation function, h is the cell output activation function which is usually the *tanh* function, ϕ is the output activation function of the output sequence y which is usually the *softmax* function, W terms denote weight matrices and b terms denote bias vectors.

3.2   LSTM Autoencoder for Anomaly Detection

Autoencoder is a fundamental paradigm of unsupervised learning with the ability to reconstruct input features with the least possible amount of distortion [16]. The Autoencoder consists of an encoder and a decoder. The encoder extracts the hidden features from the input data and encodes the information into a learned representation vector. Then, the decoder receives the encoded vector as an input and reconstructs the original sequence with a reconstruction error.

LSTM Autoencoder combines the ability of the Autoencoder to extract the most representative information with the advantage of the LSTM to process sequential data with long-range dependencies. We use normal data to train the LSTM Autoencoder that learns to reconstruct normal univariate time series data while on the contrary produces high reconstruction error on abnormal time series. Thus, the reconstruction error used as a score to classify future data points as normal or abnormal. Figure 1 shows the architecture of the LSTM Autoencoder. We use one LSTM layer for the encoder and one LSTM layer for the decoder. The encoder takes the input sequence $(x_1, x_2, ..., x_n)$ and encodes the information into an encoded feature vector. Then, the encoded feature vector is being used as an input for the decoder layer that tries to reconstruct the original sequence into $(\widehat{x_1}, \widehat{x_2}, ..., \widehat{x_n})$.

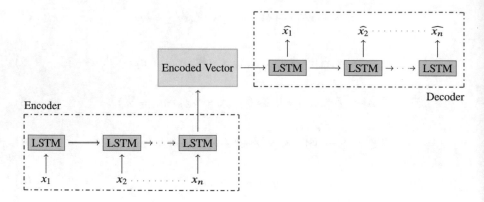

Fig. 1. Univariate time series data encoding based on LSTM autoencoder

The LSTM Autoencoder is trained to minimize the Mean Absolute Error (MAE) that measures the average of the absolute differences between the observed and the reconstructed observation as follows

$$MAE = \frac{\sum_{i=1}^{n} |x_i - \widehat{x_i}|}{n} \tag{7}$$

where x_i and $\widehat{x_i}$ are the actual and the reconstructed observation respectively. Then, the reconstruction error is used as a score to detect future outliers. Since the LSTM Autoencoder trained on normal sequences, it produces low reconstruction error on future normal series while on the contrary generates high reconstruction error on abnormal sequences. Thus, new observations are being classified as normal if the reconstruction error is lower than user's threshold or abnormal if it is not.

4   Detecting Performance Degradation in Cloud Systems

In this section, experimental scenarios are being discussed and visualized to demonstrate the effectiveness of our solution. We verify the performance of our method based on YCSB and TPCx-IoT as real world workload benchmarks that execute in MongoDB and HBase system respectively. It includes, (a) the experimental setup and the benchmark analysis, (b) LSTM Autoencoder for anomaly detection of MongoDB system and (c) LSTM Autoencoder for anomaly detection of HBase system.

4.1   Experimental Setup and Benchmark Analysis

We developed the experiments using two infrastructures to support MongoDB and HBase deployment. The first environment consists of a single Virtual Machine (VM) that is running a linux operating system with 2 CPU cores, 8 GB RAM and 512 GB Hard Disk Drive (HDD). The VM hosts MongoDB as a real world NoSQL application system to be monitored by Prometheus. While YCSB workload executes a mix of 50/50 reads and writes, Prometheus collects and stores applications metrics with equally spaced points in time. The second environment consists a cluster of 2 nodes each configured with 2 CPU cores, 8 GB RAM and 256 GB HDD. Each node consists an HBase server that runs on top of Hadoop Distributed File System (HDFS). The TPCx-IoT workload used as representative of activities typical in IoT gateway systems. Similarly, Prometheus collects and stores application metrics in real time.

4.2   Performance Anomaly Detection of MongoDB System

In this section, we evaluate the performance of LSTM Autoencoder to detect anomalies in MongoDB system based on application throughput. Figure 2 shows application throughput under YCSB workload execution. The first wavelet, illustrates the load phase of YCSB workload where 200,000 records loaded in the database. The second wavelet shows the run stage where YCSB uses a mix of 50% read and 50% write operations. In total, 5 load and 5 run stages have been used alternately to create 10 representative wavelets. Figure 2 suggest that executing YCSB workload with the same configurations generate a repeatable application throughput pattern. Consequently, the LSTM Autoencoder learns to reconstruct normal throughput representations by minimizing the loss function, that is, the mean absolute error. The reconstruction error has been used as a score to detect future abnormalities in the throughput.

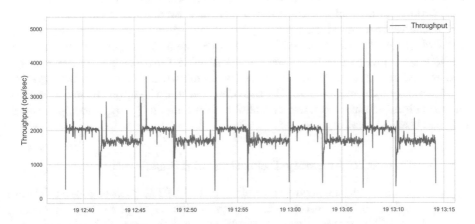

Fig. 2. Throughput of MongoDB on YCSB workload.

To demonstrate the effectiveness of our method, we used Stress linux package to introduce additional intensive tasks that impact application performance. Figure 3 shows a continuity of Fig. 2 with two additional wavelets that represent the load and the run stage of YCSB workload. While YCSB executed the load stage under normal conditions (11th wavelet), in the run stage (12th wavelet) we execute Stress package as an additional intensive task. The latter, impacts the application throughput as it creates abnormal patterns shown in Fig. 3.

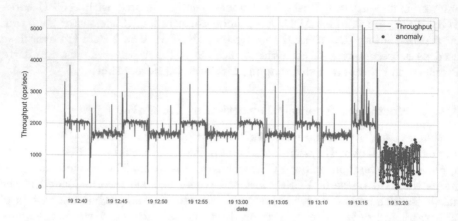

Fig. 3. Throughput of MongoDB on YCSB workload with Stress.

We effectively detect throughput degradation by using LSTM Autoencoder reconstruction error. Since the LSTM Autoencoder trained to reconstruct normal throughput patterns, it produces high reconstruction error on throughput values that deviate from the normal data distribution. Then we classify throughput observations as normal or abnormal based on a predefined application threshold. If the reconstructed error of a given observation surpasses the application threshold, then the observation is tagged as abnormal. The application threshold consists a tuning parameter based on application functionality. If the application is resilient in throughput fluctuations, a high threshold value is suggested to detect only extreme outliers. On the other hand, if the application is sensitive on throughput changes, a lower threshold will be less forgiving and more values will be classified as abnormal.

4.3   Performance Anomaly Detection of HBase System

In this section, we demonstrate the ability of the LSTM Autoencoder to detect abnormal values in IoT sensor data. We deployed an HBase cluster with 2 nodes while TPCx-IoT benchmark used to model sensor data generated by power substations. Figure 4 shows the throughput of the HBase cluster with 2 nodes while we execute TPCx-IoT workload. In total the TPCx-IoT workload generated 8 wavelets that used as a representative normal input sequence for the LSTM Autoencoder. Then, the LSTM Autoencoder learns to reconstruct the input sequence with the lowest possible error rate. We use the reconstruction error as a score in order to classify future throughput values as normal or abnormal.

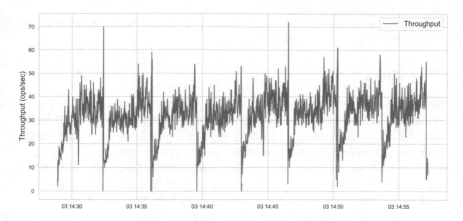

Fig. 4. Throughput of HBase on TPCx-IoT workload.

Figure 5 is a continuity of Fig. 4 that shows the execution of the TPCx-IoT workload while at the same time network delays introduced in the system. We demonstrate the ability of the LSTM Autoencoder to detect abnormal throughput values by artificially injecting two network delays of 200 ms into the network system. The first delay injected between 14:58 and 15:00 while the second delay injected between 15:07 and 15:09. As a result, the throughput of the system plummets to zero between the time window of the injected network delay.

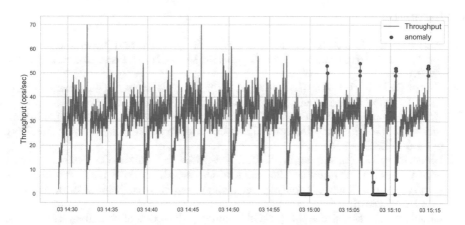

Fig. 5. Throughput of HBase on TPCx-IoT workload with network delay.

As Fig. 5 shows LSTM Autoencoder successfully detects abnormal cases including the throughput values between the network delay time windows. As discussed earlier, we classify throughput observations as normal or abnormal based on a predefined application threshold. Thus, observations with a reconstruction error higher than the threshold were tagged as abnormal. The threshold has been tuned according to application functionality to detect values that significantly vary from the normal distribution.

5   Conclusion

In this study, we presented LSTM Autoencoders for detecting performance degradation in cloud systems. We evaluated the ability of our method by using various NoSQL systems and real world workload benchmarks. YCSB workload used to execute read and write operations in MongoDB that generate repeatable throughput patterns. Additionally, to show the ability of our method to adapt in various applications, we used an HBase cluster and executed TPCx-IoT benchmark that models sensor data. In our findings, LSTM Autoencoders successfully detected performance outliers in both systems under different scenarios such as intensive system tasks and network system delays. The LSTM Autoencoders trained on normal sequence representations to minimize the reconstruction error. Then, a threshold application parameter used as an indicator to classify future values as normal or abnormal.

Our method detects abnormal application behaviours under various tasks. We believe that our accurate and low overhead anomaly detection tool introduces a real time approach by using application metrics, as strong indicators of application overall performance. Detecting degradation of application performance increases the reliability of cloud systems and promises strong monitoring tools to cloud system administrators.

References

1. Sotiriadis, S., Bessis, N., Amza, C., Buyya, R.: Vertical and horizontal elasticity for dynamic virtual machine reconfiguration. IEEE Trans. Serv. Comput. **99**, 1–1 (2016)
2. Gudivada, V.N., Rao, D., Raghavan, V.V.: Nosql systems for big data management. In: 2014 IEEE World Congress on Services, pp. 190–197. IEEE (2014)
3. Ghrib, Z., Jaziri, R., Romdhane, R.: Hybrid approach for anomaly detection in time series data. In: 2020 International Joint Conference on Neural Networks (IJCNN), pp. 1–7. IEEE (2020)
4. Bhattacharyya, A., Jandaghi, S.A.J., Sotiriadis, S., Amza, C.: Semantic aware online detection of resource anomalies on the cloud. In: 2016 IEEE International Conference on Cloud Computing Technology and Science (CloudCom), pp. 134–143. IEEE (2016)
5. Chouliaras, S., Sotiriadis, S.: Real time anomaly detection of NoSQL systems based on resource usage monitoring. IEEE Trans. Indus. Inf. **16**, 6042–6049 (2019)
6. Malaiya, R.K., Kwon, D., Kim, J., Suh, S.C., Kim, H., Kim, I.: An empirical evaluation of deep learning for network anomaly detection. In: 2018 International Conference on Computing, Networking and Communications (ICNC), pp. 893–898. IEEE (2018)
7. Feng, C., Li, T., Chana, D.: Multi-level anomaly detection in industrial control systems via package signatures and LSTM networks. In: 2017 47th Annual IEEE/IFIP International Conference on Dependable Systems and Networks (DSN), pp. 261–272. IEEE (2017)

8. Malhotra, P., Ramakrishnan, A., Anand, G., Vig, L., Agarwal, P., Shroff., G.: LSTM-based encoder-decoder for multi-sensor anomaly detection. *arXiv preprint* arXiv:1607.00148 (2016)

9. Nguyen, H.D., Tran, K.P., Thomassey, S., Hamad, M.: Forecasting and anomaly detection approaches using LSTM and LSTM autoencoder techniques with the applications in supply chain management. Int. J. Inf. Manage. 102282 (2020)

10. Longari, S., Valcarcel, D.H.N., Zago, M., Carminati, M., Zanero, S.: Cannolo: an anomaly detection system based on LSTM autoencoders for controller area network. IEEE Trans. Netw. Serv. Manage. (2020)

11. Trinh, H.D., Zeydan, E., Giupponi, L., Dini, P.: Detecting mobile traffic anomalies through physical control channel fingerprinting: a deep semi-supervised approach. IEEE Access **7**, 152187–152201 (2019)

12. Cooper, B.F., Silberstein, A., Tam, E., Ramakrishnan, R., Sears, R.: Benchmarking cloud serving systems with YCSB. In: Proceedings of the 1st ACM Symposium on Cloud Computing, pp. 143–154 (2010)

13. Hochreiter, S., Schmidhuber, J.: Long short-term memory. Neural Comput. **9**(8), 1735–1780 (1997)

14. Sak, H., Senior, A.W., Beaufays, F.: Long short-term memory recurrent neural network architectures for large scale acoustic modeling (2014)

15. Gers, F.A., Schmidhuber, J., Cummins, F.: Learning to forget: continual prediction with LSTM (1999)

16. Baldi, P.: Autoencoders, unsupervised learning, and deep architectures. In: Proceedings of ICML Workshop on Unsupervised and Transfer Learning, pp. 37–49 (2012)

An Ontology for Composite Cloud Services Description

Wafa Hidri$^{(\boxtimes)}$, Riadh Hadj M'tir, and Narjès Bellamine Ben Saoud

National School of Computer Science, RIADI Laboratory,
Manouba University, Manouba, Tunisia
`wafa.hidri@ensi-uma.tn`, `riadh.hadjmtir@riadi.rnu.tn`,
`narjes.bellamine@ensi.rnu.tn`

Abstract. In last few years, cloud services are described and analyzed as independent atomic services without taking into account their relationships with other services existing in cloud environment. This later is a complex ecosystem of multiple services, delivered by different cloud providers in different formats using their own vocabulary. These heterogenous services are involved as service components in the creation of a value-added cloud composite applications. The composition of these services is challenging and can decrease portability and interoperability of cloud offerings. In this paper, we propose a model for a complete definition and description of cloud services, considering them as composite services. Five service description aspects are considered in this work; the business, technical, operational, structural and semantic aspects. Therefore, we propose to use an ontology representation, extended from the Linked USDL language, which enables an effective description and a rich specification of the domain knowledge of composite cloud services. In addition, we show how the proposed description could be used in a concrete example and how semantic rules are implemented to infer additive knowledge in order to select relevant cloud services for composition.

Keywords: Cloud service description · Structural aspect · Ontology

1 Introduction

Cloud computing has recently emerged as the sum of software and Utility Computing [1] services. Currently, cloud services are provided as a monolithic cloud stack [3] that contains four levels of service offerings: Business Process (BPaaS), Software (SaaS), Platform (PaaS) and Infrastructure (IaaS) as a service [2]. In fact, services are defined at separate levels but they are not behaving independently. In single and multi-cloud environment, there are many services that can have different relationships between similar services involved in the same level and between different services referencing to different levels. For example, cloud SaaS applications are supported by the required PaaS platform to be executed. This later is installed on the required infrastructure. Thus, cloud services are not only defined by its functional features, but also by its dependencies.

Cloud application developers compose SaaS services at the application level and attribute dynamically the corresponding platform and infrastructure services in order to

© The Author(s), under exclusive license to Springer Nature Switzerland AG 2021
L. Barolli et al. (Eds.): AINA 2021, LNNS 226, pp. 482–494, 2021.
https://doi.org/10.1007/978-3-030-75075-6_39

create service-based composite applications. These end-to-end applications are created by forming service compositions or service syndications [4] at each cloud level. Hence, developers move these cloud applications from monolithic approach [3] to syndicated one [4]. Consequently, cloud services are not isolated but they interact and communicate with other services to form a composite one. In addition, the structure of such services, called *composite services*, must be defined and modeled to specify both the functional and the deployment levels and all the dependencies between them.

Diverse existing models, languages and ontologies have been covered cloud service description from only some aspects and have been described cloud services as independent entities without taking into consideration the existing dependencies between services. To overcome the shortcomings of existing service descriptions works, we propose a semantic model that generalizes the description of all cloud service types covering all aspects (technical, business, operational, structural, and semantic). Based on Linked USDL language [9], we propose *LinkedCSOnto*, a novel ontology for cloud service description. First, we capture interactions and dependencies in the context of cloud services. Then, we show how to extend Linked USDL to support these interactions. Finally, we discuss the use of *LinkedCSOnto* to discover and compose cloud services.

The rest of this paper is structured as follows: Sect. 2 introduces the cloud architecture and its components. Section 3 discusses several related works. Section 4 presents the proposed conceptual meta-model that specify the structural aspect of a composite cloud service. Section 5 describes the Linked USDL extension corresponding to the proposed meta-model, the new concepts added and their relationship to existing modules. Finally, Sect. 6 presents the ontology extended from Linked USDL, and Sect. 7 concludes with some future work.

2   Cloud Architecture, Services and Relations

In cloud environment, the cloud service is a composite service because it is able to be integrated with the others by sharing resources or services. Its structure is essentially a dynamic organization that contains a set of cloud service components and a set of relationships. As illustrated by (Fig. 1), within the cloud layered architecture there are different types of relationships and interactions between services and between actors. Three layers are presented; i) The user layer defines actors which can be considered as provider or consumer according to the situation. Each provider offers a set of services and requires a set of other ones from other providers. Existing interactions between service providers are defined by the offered and required services and resources across layers, ii) The service layer defines two main types of services (business and IT services). Business services such as business process, software and platform services. IT services are IaaS resources which are required by the business services to be deployed, and iii) In the interaction layer, each cloud service is linked to a capability of another service that meets its requirements. This type of relationships is called *inter-services dependencies*. For instance, to compose two services from two different levels, each service's need should be associated with the corresponding one or more service's offers. There are four main types of dependencies or relationships in this architecture. The first relationship is between actors (cloud users) and cloud services while the three others relationships are between services across layers (Fig. 1):

1. Access-to relationships: they are connections between end-users and BPaaS services or SaaS services, between developer and PaaS or IaaS services and between administrator and IaaS services.
2. Implemented-by relationships: they are connections between activities of BPaaS services and SaaS services [6].
3. Deployed-on relationships: they are connections between BPaaS or SaaS services and PaaS services, i.e. when service in BPaaS or SaaS layer uses a PaaS service.
4. Hosted-on relationships: between all business services and IaaS services, i.e. when service in BPaaS, SaaS or PaaS layer uses an IaaS service.

All these connections between cloud services are vertical [7] relationships from business to infrastructure services.

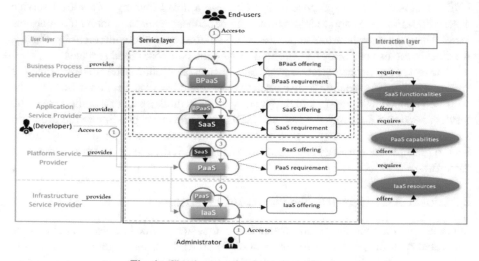

Fig. 1. Cloud computing layered architecture.

Another type of relationships is defined generally across multiple cloud services as follow in Fig. 2; it is called *intra-services dependencies*. For instance, to compose two services from the same level (two SaaS services), the developer needs a cross SaaS composition, based on a functional matching (input/output matching), following the business logic of the required business process and taking into account the deployment constraints of the linked SaaSs. This type of horizontal [7] relationship is close to traditional services composition [8]. It is defined between all cloud services belongs to the same layer in single cloud (PaaS1C2_PaaS2C2) or across multiple clouds as shown in Fig. 2 (BPaaSC1_BPaaS2C2, SaaSC1_SaaS2C2, IaaSC1_IaaS1C2). So, all these interactions must be taken into consideration when describing cloud services to capture its business dependencies and deployment relationships.

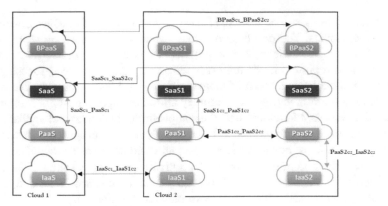

Fig. 2. Cloud services and their relationships in single cloud and across multiple clouds.

3 Related Work

Several approaches have been proposed for specifying and describing cloud services. They generally assume that services are atomic and independent entities covering different aspects such as technical, operational, business and semantic aspect. The majority of cloud service description works are semantic approaches. There are lot of interest in the development of ontologies to describe and specify cloud services in order to simplify the discovery, selection and composition of them and to automate their deployment and management. In fact, Afify et al. [17] proposed a semantic-based system, based on a unified ontology, that facilitates the SaaS publication, discovery and selection processes. The proposed ontology combines services domain knowledge, SaaS characteristics, QoS metrics, and real SaaS offers. Moscato et al. [16] proposed mOSAIC ontology which describes cloud resources, parameters and features, in order to compose cloud services by using composition patterns and orchestration, verify the soundness of orchestrated services and automatically analyze QoS of composed cloud services. Rekik et al. [13] proposed a cloud service ontology that takes into account the functional and non-functional properties, attributes and relations of infrastructure, platform and software services in order to enhance cloud service discovery and selection. Bassiliades et al. [19] provided an ontology for PaaS which is an extension of DOLCE+DnS Ultralite (DUL) ontology that defines the PaaS characteristics and parameters as classes. The proposed PaaSport ontology allows a best semantic matchmaking and ranking of cloud PaaS offering. Alfazi et al. [10] developed a comprehensive ontology based on the NIST [2] cloud computing standard to discover and categorize cloud services into several clusters in real environments. Sun et al. [14] proposed a unified semantic cloud service description model, extended from the basic structure of USDL [9], according to a set of identified cloud-service-specific requirements and attributes. In addition, authors included an additional module (transaction module) to model the rating of cloud services. The authors in [18] have proposed a comprehensive ontology that covers all functional and non-functional features of cloud services and have focused on the semantic interrelationships among these features. Zhang et al. [15] presented an OWL-based ontology that defines and formalizes functional and non-functional concepts, attributes and relations of infrastructure services. The work in [12] has covered technical, business, operational, and semantic aspects. The proposed ontology description is based on USDL [9] and WSMO (Web Service Modeling Ontology). In [11], authors proposed a semantic based representation to represent both cloud services and virtual appliances, focusing on the functional characteristics of both of them and trying to identify the possible relations existing among them. The proposed ontology allows to support customers in choosing the best suitable offer and to automate the composition of services and appliances. The Blueprint Approach is proposed in [5] where authors have proposed an abstract and uniform specification of cloud services across all three layers of the cloud stack. They formalize the semantics of their proposed specification in order to allow cloud service-based applications engineers to more precisely match, compare select and assemble multiple alternative SaaS, PaaS, and IaaS specifications (Table 1).

Table 1. Classification of cloud service semantic description approaches according to several service aspect.

Research work	Technical aspect	Operational aspect	Business aspect	Functional aspect	Non-Functional aspect	Structural aspect
CSDM [14]	+	+	+	+	+	
mOSAIC [16]	+	+		+	+	
CSO [13]			+	+	+	
PaaSport [19]				+	+	
Alfazi et al. [10]				+		
CSD-USDL [12]	+	+	+	+	+	
SaaSOntology [17]			+	+	+	
CoCoOn [15]	+	+		+	+	+
Di Martino et al. [11]				+		+
Blueprint [5]	+	+		+	+	+
CloudFNF [18]				+	+	
LinkedCSOnto	+	+	+	+	+	+

We conclude, from the studied semantic approaches, that several service facets have been defined and presented to describe cloud services. However, these works generally have focused on one side of description which considers cloud service as a simple functional black box and have neglected its structure and internal architecture in the cloud system, that means their components and the dependencies and connections among them. Compared to the studied approaches, our work combines the internal and the external requirements in the same semantic model to describe composite cloud services covering all service aspects.

4 Composite Cloud Service Meta-model

Depending on the cloud stack and service's dependent and deployment connections, a cloud service is considered as composite service, composed of service components (i.e. atomic cloud service) and the connections that can exist between them. Its architecture is defined in terms of the service topology connecting the service components. Various different service topologies can define a composite service. We will define the structural aspect of this composite service using a conceptual model as shown in Fig. 3. While the technical, operational, business aspects are modeled and defined in previous research works, the novelty of our work is to include the internal architecture of cloud composite service. In the following, the definition of a set of elements or concepts of the proposed meta-model is presented:

- *Service*: represents a cloud service provided by a cloud provider that can be composed with another one.
- *CompositeService*: represents a cloud composite service, defined by a set of service topologies.
- *ServiceTopology*: represents a logic structure of a composite service that groups a set of atomic cloud services and their relationships.
- *ServiceComponent*: represents an atomic cloud service, it is a service component of a composite cloud service.
- *Requirement*: represents the conditions of the good functioning of the service. It has different types.
- *Capability*: represents the functionality exposed or offered by a component service.
- *ServiceRelationship*: represents a link or a connection between services. It is defined by a set of dependent and deployment constraints.

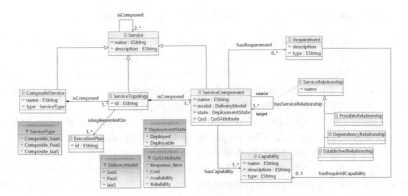

Fig. 3. Composite cloud service meta-model.

5 Linked USDL Extension

To cover the semantic aspect in describing composite cloud service, we need to bring semantics to the previous meta-model. Therefore, to link the structural service aspect, defined by the conceptual previously model, to other existing ones; we extend Linked USDL [9] that covers three aspects: technical, operational, and business.

5.1 Linked USDL Overview

We propose to extend this language because it represents a semantic description of services and it is the most comprehensive because it is divided into five modules [9] which have different purposes. This division allows to cover all service facets, use only those most important and to easily extend the lacking ones. Each module is a set of concepts and properties. There are as follows:

- USDL-core: is the basic module of Linked USDL, it covers the operational aspects of a service.
- USDL-price: covers the concepts that describe the price structure of a service.
- USDL agreement: covers the SLA module which describes the quality of the service provided, such as response time and availability.
- USDL-sec: covers the security properties of a service.
- USDL-ipr: covers the rights and obligations to use a service.

The modelling of services covering the important key aspects that a service can provide and the possibility for novel extensions are considered as the main advantages of Linked USDL. However, this later does not consider all concepts that efficient describe cloud services' connectivity considering them as *linked composite services*. To overcome this limitation, we propose an extension of the operational facet of Linked USDL, so that the composability connections of elementary service are incorporated for better describing composite service structure in cloud environments.

5.2 Linked Composite Cloud Service Description Ontology

In this section, we propose an ontology representation extended from the Linked USDL language. This extension is defined by introducing a set of new concepts that describe the structure of composite cloud service: service components and their different connections. At run-time, this composite service structure is considered as the execution plan that contains service components and the connections among them.

The proposed ontology allows the discovery and the composition of cloud atomic service based on their dependencies captured in the description of composite services. As shown in Fig. 4, the extension is defined by an inheritance relationship between the *Service* concept and its sub-concepts *CompositeService*, *ServiceTemplate* and *Service-Component*. These three new concepts consist of describing the composition dependency between them in order to specify the structure of a composite service. Our goal is to describe the characteristics of service component and its different dependencies which are the main new concepts. A reflexive association *hasServiceRelationship* defines the composability between two service components in terms of two roles; service component source and service component target. The *ServiceComponent* concept has two relationships with the *Requirement* and the *Capability* concepts respectively. These relationships are represented using two Object Properties *hasRequirement* and *hasCapability*. Each service component' requirement has a required capability. This dependency is represented using an Object Property *hasRequiredCapability*. Each service component has its context such as the location, time or deployment State. This relationship, represented by the Object Property *hasContext*, is included to assure a flexible description of services. We describe the consumer needs and requirements by including the *ConsumerRequest* concept.

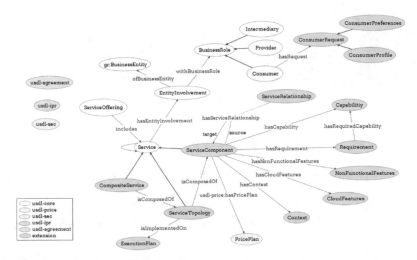

Fig. 4. Linked USDL extension covering the structural service aspect.

6 The Proposed Ontology LinkedCSOnto

The *LinkedCSOnto* is defined in the Web Ontology Language (OWL) and designed using Protégé v4.3 editor as shown in Fig. 5. To evaluate our ontology, we associate real world instances to the proposed concepts and we choose the Fact++ reasoner which is available with Protégé in order to capture inconsistency errors. In this work, we propose a qualitative evaluation to improve the quality of the proposed ontology, which includes consistency detection, duplication and disjunction verification and the errors correction phase. The qualitative evaluation is done by the Fact++ reasoner in order to check the ontology coherence. Some errors are detected after applying the reasoner. So, we must omit a disjoint relation between some classes and define some instances to correct the errors and obtain a valid ontology. To increase the quality of *LinkedCSOnto* in order to be more used, it needs to be compared to existing standards or real corpus. So, a quantitative evaluation will be done as a future work.

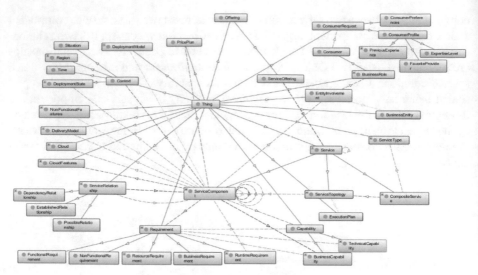

Fig. 5. LinkedCSOnto.

We introduce the main concepts and their definitions:

− *EntiyInvolvment*: defined in the usdl-core module, is an instance involved in the cloud system.
− *BusinessRole*: defined in the usdl-core module, captures the role of participants involved in the cloud system such as provider, consumer or intermediary like creator or broker.
− *BusinessEntity*: belongs to the *GoodRelations* ontology, defined in the usdl-core module and represents the legal agent that look for an offering.
− *CloudFeatures*: the characteristics of a cloud service component such as pricing model, location, category, etc.
− *DeliveryModel*: the type of cloud service component; BPaaS, SaaS, PaaS or IaaS.
− *NonFunctionalFeatures*: the QoS attributes of a service such as response time, cost or reliability.
− *PricePlan*: the price specification of a service.
− *Requirement*: a need required by a service component which ensures its proper functioning and execution.
− *Capability*: the function of a service component, it can be business or technical.
− *ServiceRelationship*: a connection between service components, there are three types of relationships: possible, dependent and established relationships.
− *ServiceType*: the type of a cloud composite service, for example Composite SaaS or Composite IaaS.
− *ExecutionPlan*: the implementation of an instance of service topology, means that an execution plan of a composite service.

6.1   Concrete Example to Instantiate *LinkedCSOnto*

To illustrate the applicability of our ontology, let us consider the following composite cloud service (Inventory Management service) as shown in Fig. 6. In the abstract work-flow below, each task represents an atomic cloud component service. These components can be SaaS, PaaS or IaaS services interconnected and grouped in an instance service template. The implementation of the whole composite service is considered as a concrete workflow. The tasks *Ordering* and *Inventory* are realized by two SaaS services deployed on *Inventory management* PaaS service which needs *Storage* IaaS service to host and store inventory content.

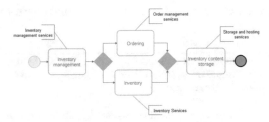

Fig. 6. Concrete example of composite cloud service.

Figure 7 shows how some concepts of our ontology are used to describe the composite service defined previously, and how the object properties and classes were instantiated according to the existing dependencies among services mentioned in the concrete example above.

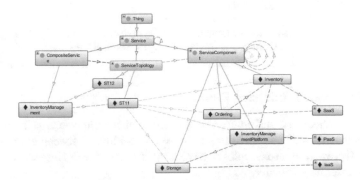

Fig. 7. Inventory Management service ontology representation.

6.2   Querying the Ontology

In the following blocks, we present some examples of SPARQL queries that can be specified to retrieve important information about services. These queries can be used in the composition process in order to discover and select services that meet user preferences and constraints.

- Query 1: How to retrieve the name of service components which have a SaaS service model?

```
PREFIX ns: <http://www.owl-ontologies.com/LinkedCSOnto.owl#>
PREFIX rdf: <http://www.w3.org/1999/02/22-rdf-syntax-ns#>
SELECT ?servicecomponent ?name
WHERE {
        ?servicecomponent ns:hasSaaSModel ns:SaaS .
        ?servicecomponent rdf:type ns:ServiceComponent .
        ?servicecomponent ns:ServiceComponentName ?name
        }
```

- Query 2: How to retrieve the service components which have the PaaS capability (Database)?

```
PREFIX ns: <http://www.owl-ontologies.com/LinkedCSOnto.owl#>
PREFIX rdf: <http://www.w3.org/1999/02/22-rdf-syntax-ns#>
SELECT ?servicecomponent
WHERE {
        ?servicecomponent ns:hasPaaSCapability ns:Database .
        ?servicecomponent rdf:type ns:ServiceComponent
        }
```

- Query 3: How to retrieve the service components which have a Dependency Relationship with the service component Storage?

```
PREFIX ns: <http://www.owl-ontologies.com/LinkedCSOnto.owl#>
PREFIX rdf: <http://www.w3.org/1999/02/22-rdf-syntax-ns#>
SELECT ?servicecomponent
WHERE {
        ?servicecomponent ns:hasDependencyRelationship ?Storage .
        ?servicecomponent rdf:type ns:ServiceComponent .
        }
```

6.3 SWRL Rules for Inferring Knowledge

To compose services in cloud environment to create new composite one, there are some constraints that should be verified to guarantee compatibility between composed cloud services. We present some examples of SWRL rules to describe the important constraints in order de infer new knowledge.

- Constraints on inter-service composition (compatibility across layers): verifies if a PaaS service can be combined with an IaaS service, i.e. if the capabilities of an IaaS service satisfy the PaaS technical requirements.

Rule1:
PaaS(?P) ∧ IaaS(?I) ∧ hasRequirement(?P,?X) ∧ hasCapability(?I,?Y) ∧
hasCPUPower(?X,?CPUPx) ∧ hasCPUPower(?Y,?CPUPy) ∧ hasStorageSize(?X,?STx)
∧ hasStorageSize(?Y,?STy) ∧ hasOSVersion(?X,?Vx) ∧ hasOSVersion(?Y,?Vy) ∧
swrlb:Equal(?CPUPx, ?CPUPy) ∧ swrlb:lessThan(?CPUPx, ?CPUPy) ∧
swrlb:Equal(?STx, ?STy) ∧ swrlb:lessThan(?STx, ?STy) ∧ swrlb:Equal(?Vx, ?Vy)
→Combine(?P,?I)

- Constraints on intra-service composition (compatibility cross SaaS): verifies if a SaaS service can be combined with another SaaS service, i.e. if they have the same features.

Rule2:
SaaS(?S1) ∧ SaaS(?S2) ∧ SaaS(?S3) ∧ hasRequirement(?S1,?X1) ∧
hasRequirement(?S2,?X2) ∧ hasCapability(?S3,?Y3) ∧ hasCategory(?X1,?Catx1) ∧
hasCategory(?X2,?Catx2) ∧ hasCategory(?Y3,?Caty3) ∧ hasLocation(?X1,?Lx1) ∧
hasLocation(?X2,?Lx2) ∧ hasLocation(?Y3,?Lx1) ∧ swrlb:Equal(?Catx1, ?Caty3) ∧
swrlb:Equal(?Catx2, ?Caty3) →Combine(?S1,?S3)

7   Conclusion

In this work, we focus on the description of composite cloud services. The proposed approach is based on a conceptual model proposed to capture the structural aspect of composite cloud service. To obtain a complete and comprehensive representation, we introduce a new extension of Linked USDL ontology. The designed result ontology covers five service aspects; structural, semantic, business, operational and the technical. We defined the features of cloud services and their dependencies in order to introduce an efficient composability model. As future work, we plan to exploit the proposed ontology as the core of an automatic discovery and composition system allowing to query the ontology and extract the relevant cloud services according to the user request.

References

1. Armbrust, M., Fox, A., Griffith, R., Joseph, A.D., Katz, R.H., Konwinski, A., Lee, G., Patterson, D.A., Rabkin, A., Stoica, I., Zaharia, M.: Above the clouds: a Berkeley view of cloud computing. Dept. Electrical Eng. and Comput. Sciences, University of California, Berkeley, Rep. UCB/EECS, vol. 28, no. 13 (2009)
2. Mell, P., Grance, T., et al.: The nist definition of cloud computing (2011)
3. Papazoglou, M.P., van den Heuvel, W.-J.: Blueprinting the cloud. IEEE Internet Comput. **15**(6), 74–79 (2011)
4. Papazoglou, M.P.: Cloud blueprints for integrating and managing cloud federations. In: Software Service and Application Engineering, pp. 102–119. Springer (2012)
5. Nguyen, D.K., Lelli, F., Papazoglou, M.P., van den Heuvel, W.-J.: Issue in automatic combination of cloud services. In: 2012 IEEE 10th International Symposium on Parallel and Distributed Processing with Applications, pp. 487–493. IEEE (2012)

6. Hidri, W., Hadj M'tir, R., Bellamine Ben-Saoud, N., Ghedira-Guegan, C.: A meta-model for context-aware adaptive business process as a service in collaborative cloud environment. Procedia Comput. Sci. **164**, 177–186 (2019)

7. Mietzner, R., Fehling, C., Karastoyanova, D., Leymann, F.: Combining horizontal and vertical composition of services. In: 2010 IEEE International Conference on Service-Oriented Computing and Applications (SOCA), pp. 1–8. IEEE (2010)

8. Benzadri, Z., Hameurlain, N., Belala, F., Bouanaka, C.: A theoretical approach for modelling cloud services composition. In: 2016 International Conference on Advanced Aspects of Software Engineering (ICAASE), pp. 1–8. IEEE (2016)

9. Cardoso, J., Pedrinaci, C.: Evolution and overview of linked USDL. In: International Conference on Exploring Services Science, pp. 50–64. Springer (2015)

10. Alfazi, A., Sheng, Q.Z., Qin, Y., Noor, T.H.: Ontology-based automatic cloud service categorization for enhancing cloud service discovery. In: 2015 IEEE 19th International Enterprise Distributed Object Computing Conference, pp. 151–158. IEEE (2015)

11. Di Martino, B., Cretella, G., Esposito, A.: Towards a unified OWL ontology of cloud vendors' appliances and services at PaaS and SaaS level. In: 2014 Eighth International Conference on Complex, Intelligent and Software Intensive Systems, pp. 570–575. IEEE (2014)

12. Ghazouani, S., Slimani, Y.: Towards a standardized cloud service description based on USDL. J. Syst. Softw. **132**, 1–20 (2017)

13. Rekik, M., Boukadi, K., Ben-Abdallah, H.: Cloud description ontology for service discovery and selection. In: 2015 10th International Joint Conference on Software Technologies (ICSOFT), vol. 1, pp. 1–11. IEEE (2015)

14. Sun, L., Ma, J., Wang, H., Zhang, Y., Yong, J.: Cloud service description model: an extension of USDL for cloud services. IEEE Trans. Serv. Comput. **11**(2), 354–368 (2015)

15. Zhang, M., Ranjan, R., Haller, A., Georgakopoulos, D., Menzel, M., Nepal, S.: An ontology-based system for cloud infrastructure services' discovery. In: 8th International Conference on Collaborative Computing: Networking, Applications and Worksharing (CollaborateCom), pp. 524–530. IEEE (2012)

16. Moscato, F., Aversa, R., Di Martino, B., Fortis, T.-F., Munteanu, V.: An analysis of mosaic ontology for cloud resources annotation. In: 2011 Federated Conference on Computer Science and Information Systems (FedCSIS), pp. 973–980. IEEE (2011)

17. Afify, Y.M., Moawad, I.F., Badr, N., Tolba, M.F.: A semantic-based software as-a-service (SaaS) discovery and selection system. In: 2013 8th International Conference on Computer Engineering & Systems (ICCES), pp. 57–63. IEEE (2013)

18. Al-Sayed, M.M., Hassan, H.A., Omara, F.A.: CloudFNF: an ontology structure for functional and non-functional features of cloud services. J. Parallel Distrib. Comput. **141**, 143–173 (2020)

19. Bassiliades, N., Symeonidis, M., Gouvas, P., Kontopoulos, E., Meditskos, G., Vlahavas, I.: Paasport semantic model: an ontology for a platform-as-a-service semantically interoperable marketplace. Data Knowl. Eng. **113**, 81–115 (2018)

GPU Accelerated Bayesian Inference
for Quasi-Identifier Discovery
in High-Dimensional Data

Nikolai J. Podlesny[(✉)], Anne V. D. M. Kayem, and Christoph Meinel

Hasso-Plattner-Institute, Potsdam, Germany
{Nikolai.Podlesny,Anne.Kayem,Christoph.Meinel}@hpi.de

Abstract. Determining unique attribute combinations as quasi-identifiers is a common starting point for both re-identification attacks and data anonymisation schemes. The efficient discovery of those quasi-identifiers (QIDs) has been a combinatoric nightmare, actually an enumeration problem [1–3] given its W2-complete nature [4–6]. Proper privacy guarantees are required to fulfil highest ethical standards and privacy legislation like CCPA or GDPR, yet also enable the most modern data-driven business model based on monetising corporate data pools. In this work, we offer three main contributions: First, we contribute an algorithm that vectorises the QID search. This QID discovery is based on Bayesian inference detection, which usually suffers a state-space explosion for large-scale datasets. By utilising GPU acceleration to execute the vectorised algorithm, we counter the state-space-explosion issue raised by Bayesian networks. Second, we show its applicability to anonymising high-dimensional data which suffers high information-loss when using standard anonymisation approaches. Third, we offer an empirical model that compares multiple optimisations to discover all QIDs in near real-time, even in large-scale datasets. The latter becomes extremely useful for instances in digital health settings where algorithmic execution time can influence life-and-death triage. Finally, we point out that the same approach can foster de-anonymisation attacks on already published datasets. A demonstration is enclosed to re-identify individuals from Mount Vernon, NY and Southern California in a published Twitter dataset on US Presidential Election 2020.

1 Introduction

The search and transformation of privacy-related information are currently receiving increasing attention. Novel data sources like continuous sensor streams offer a whole new dimension for generating data-driven insights and value, like monetisation efforts. Yet, with increasing observations both in size and type, it remains a delicate balance to protect individuals' privacy. Data privacy has been a strong objective in the Western Hemisphere for a while and has been valued not only as part of the highest ethical standards but through various regulations and extreme penalties when failing to comply. Despite recent efforts, there are still a few areas with extensive research demand including digital health environments with distributed data repositories among others. Particularly for digital health, its data characteristics are shaped by many describing attributes

© The Author(s), under exclusive license to Springer Nature Switzerland AG 2021
L. Barolli et al. (Eds.): AINA 2021, LNNS 226, pp. 495–508, 2021.
https://doi.org/10.1007/978-3-030-75075-6_40

(i.e., genetics, disease and adherence records), a long and unique medical history (i.e., over the lifetime of a patient) which in combination form massive, high dimensional datasets. These data are typically stored in a distributed data repository, which corresponds to a challenging setting by nature. Not only various data transformations and data redundancy are involved, but intuitively one would first try to combine, and then to process raw data sources which often expose high privacy risks through potentially combining personal health information (PHI) and personally identifiable information (PII) records unnecessarily.

Problem Statement. In those high-dimensional, large-scale and distributed health data environments protecting individual privacy is crucial, but extremely challenging. One of the key ingredients for both data anonymisation efforts and re-identification attacks is the discovery of attribute value inferences, also known as quasi-identifiers (unique attribute combinations). These quasi-identifiers (QIDs) are usually the starting point of re-identification initiatives like in the instance of Governor William Weld medical information re-identification [7,8]. The search for QIDs has been proven to be significantly computation-intensive, in fact, both NP-hard and W2-complete. Despite such complexity, the determination of quasi-identifier is absolutely critical as the first step for preserving individuals and their sensitive data.

Contribution. Recent technology developments offer a new perspective, even for analysing large-scale datasets with complex processing procedures. One of these achievements is the utilisation of GPUs with massive parallelisation and unified memory management, wherefore we offer the following contributions in this work:

- offer an approach to locate quasi-identifiers candidates with Bayesian inferences on very large-scale over millions of combinations
- present an algorithm which executes traditional scalar based QID search scheme as vector-based calculation to accelerate compute even for high-dimensional datasets massively
- empirical assess different optimisation schemes including massive-parallelisation and state-space aggregations to counter state-space explosion during inference detection
- show how contributed enhancements can be used to foster re-identification in probabilistic privacy approaches and demonstrate actual re-identification on an already published dataset

Outline. The rest of the paper is structured in the following manner: Affiliated work is summarised in Sect. 2. We introduce Bayesian inferences, their usage serving as an indicator for quasi-identifiers (QIDs) and latest state-space aggregation approaches in Sect. 3. In Sect. 4, we discuss privacy challenges particular to continuous data streams and how traditional data processing can be projected into the data stream context. Algorithmic details on transforming traditional, scalar-based Bayesian inference detection into the context of vectors and tensors for GPU hardware will be offered in Sect. 5. An empirical assessment and comparison of both optimisation approaches are contributed in Sect. 6. Section 7 finally concludes our results and suggests avenues for future work.

2   State-of-the-Art

The research conducted with GPU-accelerated Bayesian inference learning, especially in the privacy community, is relatively light to the best of our knowledge. There has been work done on GPU-acceleration for naive Bayes approaches [9–12] or Bayesian learning for linear models [13].

In the field of data anonymisation, extensive exploration of GPU utilisation is also quite limited so far. There are a few instances of leveraging GPU acceleration for facial recognition and their data processing. Yet, we are not aware of an empirical analysis on GPU based syntactic or semantic data anonymisation.

Especially around the data streams setting, anonymisation methods currently struggle to balance near real-time processing, minimise information-loss, and ensure privacy. Ng et al. [14], and Kalidoss et al. [15] addressed these challenges, for instances, through utilising some Map-Reduce approach. Horizontal scaling, however, quickly reaches its limits for high-dimensional datasets as well, wherein some Digital Health settings hundreds or thousands of describing attributes exist. Solanki et al. [16] offered a heuristic-based approach to apply perturbation and k-anonymity to data streams for eliminating sensitive attributes. While this is the right step, we believe the exact elimination of all QIDs is necessary to ensure privacy holistically. To achieve this successfully, a radical new approach was needed, like the utilisation of GPU compute capacities.

3   Bayesian Inferences as QID Indicators

Earlier, Podlesny et al. made the case of identifying de-anonymisation risks, also in distributed environments, through utilising Bayesian networks and their inherent attribute value inferences [17]. Those inferences are derived from the likelihood of the appearance of two or more associated attribute data values, for instances that gender *female* and prescription *cortisol* appear together (see Table 1). By determining such likelihood of attribute value appearance for all combinations, an extended adjacency matrix can be compiled (see Table 2). Here, the adjacency matrix does not only hold only a binary option whether there is a connection or not, but also the probabilistic likelihood of appearance. Same matrix representation can then be projected on a graph-based representation. Subsequently, these data can be utilised to form cyclic inferences, describing cycles within the graph or rows and columns within the adjacency matrix. Deduction of these cycle inferences then serve as metrics like the *mean* or *sum* of cycle inferences to triage high-risk attribute combinations serving as quasi-identifiers (QID).

Table 1. Sample health dataset

Gender	ZIP	Drug	Disease
M	60617	remdesivir	COVID
F	64123	cortisol	COVID
M	60617	cortisol	COVID
M	10001	remdesivir	COVID

Table 2. Adjacency matrix representation

	m	f	10001	64123	60617	Cortisol	Remdesivir	COVID
M	0	0	0.5	0	0.5	0.6	0.3	1
F	0	0	0.5	0	0.5	0.6	0.1	1
10001	1	0	0	0	0	1	0	1
64123	0	1	0	0	0	0	1	1
60617	1	0	0	0	0	0.5	0.5	1
Cortisol	1	0	0.5	0	0.5	0	0	1
Remdesivir	0	1	0	0.5	0.5	0	0	1
COVID	0.75	0.25	0.25	0.25	0.5	0.5	0.5	0

As the reader might anticipate, this processing is quite computation intensive. Assessing all attribute value combinations for a designated adjacency matrix arises to a combinatorial problem of $O(2^n)$. Previous work evaluated mathematical optimisation schemes with multigrid and manifold solver to counter this heavy enumeration processing to reduce state space explosions [18].

State Aggregation in Bayesian Networks. While building adjacency matrices, considering various attribute combination options with different tuples length quickly result in a combinatorial problem. This is not being reduced by the fact, that not only attribute combinations, but also attribute value combinations needs to be assessed to discover unique patterns potentially re-identifying individuals. Such attribute value combinations can be, for instances, hospitalisation events and demographics like in the case of Governor William Weld re-identification [7].

One way to counter the combinatorial problem is to introduce solver methods like multigrid and manifolds [17]. In a Bayesian network or its adjacency matrix representation, each unique attribute value is a node and edges indicate probabilistic probabilities (inferences) as the likelihood of joint appearances in a designated dataset. Here, multigrid and manifold approaches can aggregate those nodes in all probability or given some threshold t for its likelihood. On a massive scale, combining those states significantly throttles previously outlined state-space-explosion [18]. The *exact approach* when being in all probability may be weakened to an *approximated one* for accelerating computation time but involves higher information loss or potentially missing a quasi-identifier candidate. For the approximation, nodes are aggregated given some threshold for high likelihoods of collective appearances like a conditional probability $>95\%$. The determination of such a threshold should be realised greedily but balances computation efforts over information-loss. The following example briefly illustrates the approach of aggregating nodes:

Example 1. Picking up the previous example from Fig. 1, in case of $f \rightarrow 12160$ being in all probability, f_12160 is derived as a new state and simultaneously conflating the original states and transitions.

Algorithm 1: Calculate probabilistic linkages

---

1 Calculate probabilistic linkages *(table, settings)*;
   Input   : Table *table* containing the dataset
                Object *settings* with setting values
   Output: Array *result* including all attribute value tuples and their probability
2 result = initialize a table with columns ["value1","value2","likelihood"]
3 partial_tables = initialize an empty list
4 // *Prepare the dataset for parallelise its execution*;
5 colcom = create a list of permutations of *table*'s columns with length of 2
6 create a process pool:
7   split *colcom* in chunks and execute build_probability_for_column for each chunk in
         the process pool
8   append the outcome from each process to the partial_tables
9 close the process pool
10 result = concatenate the partial_tables

Parallelised Computation Through GPU Acceleration. A different perspective on addressing combinatorial problems is to parallelise their execution. Until now, such parallelisation has been effectively limited to the number of CPU cores available in the processing node. For n CPU cores, essentially only n computation can be processed at the same time appreciating an up to 30% performance uplift with hyper-threading implementing simultaneous multithreading. Nayahi et al. [19] explored options to scale-out calculations, following the horizontal scaling approach with a MapReduce principle of *divide and conquer* large process tasks among many nodes [20]. Yet, the more nodes are introduced, the higher the network I/O and infrastructural costs limiting its practical scalability.

Observing the latest developments in GPU acceleration, a similar methodology can be applied here as well. Traditionally, graphics processing units (GPU) promise a high compute density, high computations per memory access, with deep pipelines (>100 stages), high throughput and increased latency tolerance. Their memory capacities are still limited, lacking a similar large L1-L3 cache as with CPU architecture. While individual CPU cores are faster and smarter than an individual GPU core instruction set, the sheer number of GPU cores and the massive amount of parallelism that they offer more than make up the single-core clock speed difference and limited instruction sets (see Fig. 1).

As one can not simply execute the same code base on GPU, a few pre-processing steps are required.

- pre-fetch original dataset and convert the designated dataset into Apache Arrow format for unified memory management between CPU L1-L3 and GPU memory
- apply dictionary encoding of attribute values to have a numeric representation
- generate attribute value combinations and compose vectors

Now, having the pre-processed dataset in GPU memory, and its computation, we leverage existing CUDA frameworks to execute the probabilistic appearances' vectorised computation massively. Section 5 will detail out the implementation steps.

Algorithm 2: Building probabilities for attribute value tuples

---

1 build probability for column $(coltuple, table)$;
   Input   : Array $coltuple$ as list of two column names,
                  Table $table$ containing the dataset
   Output: Array $result$ including all attribute value tuples and their probability
2 // *Transform table to adjacency format through GROUP BY method*;
3 group = initialise a table with columns ["value1","value2","likelihood"]
4 group = group $table$ by $coltuple$ and count group appearance
5 column_size = count the total column length
6 // *Calculate their conditional probability*;
7 group['likelihood'] = divide each group size by $column\_size$ and add as column
8 return group

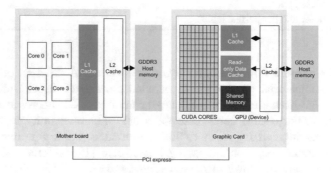

Fig. 1. GPU hardware architecture

This leaves us with two options to practically realise quasi-identifier discovery through the Bayesian network model. Number one is with mathematical solver as greedy methodology. The second is through massive parallelisation. While the former has been introduced in [17, 18], we will address in-depth the latter of massive-parallelisation and offer detailed benchmarking results, runtime and resource allocation in the following.

4   Projection on Data Streams

Over the past years, the demand for data anonymisation approaches in data streams grew. More and more data feeds from various sensors track a variety of metrics. In manufacturing, these can be belt speed, acceleration, or rotations. In healthcare, sensors are used to monitor patients' vitals like their heartbeat or oxygen saturation. Most prominently, fitness sensors as consumer hardware became suitable for everyday use generating massive information flows with often highly sensitive data (movement patterns, health or vital status).

While structured data in traditional settings needed to be processed to provide privacy aware duplicates of a static data repository, new domains like mobile health apps require proper processing on continuous data flow. There is, however, a major difference between static and stream content which also reflects on the anonymisation methodology itself. In a static context, the dataset can be easily observed and analysed as a full picture. For stream processing, only a tiny snapshot of the entire stream is available and processed at once. This snapshot may be an individual data record or (micro-) batch of information. For anonymising this standalone snapshot, the previous data stream events must be considered to avoid the linkage of distributed quasi-identifiers among various separated yet linked events. This causality can be illustrated by extending the previous example from Table 1:

Example 2. Given the data sample from Table 1 with four patient records, their gender, ZIP, disease diagnosis and drug prescription. Receiving now an additional snapshot of two more records like {F, 10001, remdesivir, COVID} and {F, 10002, hydroxychloroquine, COVID} it is hard to determine their risk of private data exposure lacking the previous context. With the knowledge of the existing table, it becomes quickly clear that the second record sticks out given both a new ZIP and drug prescription while the first one blends in (see Table 3).

The transparency of having references is critical to assess privacy risks. We recognise, it is unfeasible to re-evaluate for every snapshot the entire table on large-scale with state-of-the-art. The reader may notice a balance between grouping incoming records into batches and updating the reference table to compare against. The larger a batch, the less incremental and iterative processing on the data stream history must be completed. At the same time, a large batch also implies high information-loss through the absence of granularity. Exactly this balance will become one of the assessment objectives for the empirical analysis in Sect. 6. Significant uplift of the processing speed of common anonymisation approaches is needed to achieve any near real-time anonymisation of a continuous data stream. Solving this compute challenge, also solves the existing bottlenecks of anonymisation in data stream processing.

Table 3. Extended Sample health dataset through stream data snapshot

Gender	ZIP	Drug	Disease
M	60617	remdesivir	COVID
F	64123	cortisol	COVID
M	60617	cortisol	COVID
M	10001	remdesivir	COVID
F	10001	remdesivir	COVID
F	10002	hydroxychloroquine	COVID

In the following, we will outline the algorithmic breakdown and implementation of GPU accelerated Bayesian inference detection. Achieving such a vectorised processing for Bayesian inferences promises magnitudes of computation increase.

5   GPU-Acceleration

Traditional compute implementations of Bayesian inference detection as outlined by Podlesny et al. [17, 18] is currently purely scalar based, as most of the CPU work. Single operations are translating and executed on the CPU instruction set. The GPU instruction set is a bit more limited and optimised, particularly for vector-based computation to fully unleash GPU's massive-parallelisation. Yet, being able to compute large vector calculations at once, and efficiently, it promises a tremendous acceleration.

The challenge comes down to projecting workload from a scalar based to a vector-based calculation to leverage this compute opportunity. For Bayesian inferences, the task at hand is to compute the probability of appearances as inter-dependency between various events A and X, where A occurs knowingly that X already occurred: $P(A|X)$. As described in Sect. 3, these events technically answer to each attribute value in the designated dataset. To generate $P(A|X)$, the inter-dependency, we count how often A and X appear together in a record over the total appearances of A. Thankfully, trivial mathematical operations like counting and grouping have been already implemented as vectorised operation and made available in open source library like cuDF[1] from RAPIDS[2] (NVIDIA initiative). cuDF offers an interface as most of us know from python pandas[3], yet runs already on vectorised CUDA[4]. CUDA is a parallel computing API for Nvidia GPU chips, which offers various features to utilise Nvidia's GPU like scattered reads, shared and unified memory management, and fully supported integer and bitwise operations. Accessing these functionalities, cuDF on top offers a commonly known interface which we are using for Algorithm 3 to calculate Bayesian inferences. Section 10 outlined the necessary (pre-)processing steps, including pre-fetching the original dataset, converting it into Apache Arrow format for unified memory management and generating attribute value combinations to compose the vectors. The same procedure is reflected in the pseudo-code (see Algorithm 3).

---

[1] https://github.com/rapidsai/cudf.

[2] https://rapids.ai/.

[3] https://pandas.pydata.org/.

[4] https://developer.nvidia.com/cuda-zone.

Algorithm 3: Vectorised computation of Bayesian inferences

---

1 Calculate probabilistic linkages $(self, table)$;

 Input   : Object $self$ including the CUDA runtime environment

   Table $table$ as Apache Arrow table containing the dataset

 Output: Array $result$ including all attribute value tuples and their probability

2 // Prepare the data table;

3 colnames = extract attribute names as columns from $table$

4 candidates = generate all permutations of $colnames$ as set

5 tableEncoded = apply dictionary encoding of the $table$ to replace varchar with integer

6 result = initialize result a table with columns ["fromValue","toValue","likelihood"]

7 // Parallelize execution;

8 for $candidate$ in $candidates$:

9   fromAttribute = get first set element of $candidate$

10   toAttribute = get last set element of $candidate$

11   countVector = group $table$ by candidate set and count attribute value pairs

12   weightedCountVector = divide attribute value occurrences in $countVector$ by their individual occurrences

13   append $results$ with $weightedCountVector$ where the $fromAttribute$ value answers to the $fromValue$ and $toAttribute$ value to $toValue$ and its weightedCount as $likelihood$

14 return result

Under the hood, the group by and count statement is dissembled, stored and distributed as device instructions and variables to each GPU core and GPU RAM (see CUDA docs[5]).

```
1 cudaMemcpy(destination, source, bitwidth,
       cudaMemcpyHostToDevice);
```

On the shared memory, each CUDA core computes the algorithm on its allocated data chunk, returning the calculated results back to the CPU and main memory for assembling (see CUDA docs[6]).

```
1 cudaMemcpy(destination, source, bitwidth,
       cudaMemcpyDeviceToHost);
```

In this way, the vector calculation can simultaneously calculate 5120 operations for a single Tesla V100, or 51200 operations for 10x Tesla V100.

---

[5] https://docs.nvidia.com/cuda/cuda-runtime-api/group__CUDART__MEMORY.html.

[6] https://docs.nvidia.com/cuda/cuda-runtime-api/group__CUDART__TYPES.html.

To verify soundness, we compare the expected number of QIDs with its reference implementation on CPUs and compare all CPU based found QIDs against the GPU implementation. As this is not a probabilistic implementation, there is no surprise that both results sets match precisely. In contrast, optimisation to aggregate the state-space through multigrid and manifolds described in Sect. 3 do generate more false positives and may cause a slightly higher information-loss on balance to reducing runtime complexity. Both aspects of time complexity and type I/II errors will be assessed deeper through an empirical model in the following section.

6   Evaluation

The optimisation approaches of mathematical nature and GPU acceleration promise impressive performance gains given the underlying enumeration complexity. In the following empirical study, the surplus through either of those methodologies will be quantitatively assessed. To ensure repeatability, the experiments are conducted on publicly available datasets with state-of-the-art hardware resources.

Hardware. Our examination runs on a GPU-accelerated high-performance compute cluster, housing 160 CPU cores (E5-2698 v4), 760GB RAM, and 10x Tesla V100 with 5120 CUDA cores each and a combined Tensor performance of 1120 TFlops. The execution environment for GPU related experiments will be restricted to one dedicated CPU core and a single, dedicated Tesla V100 GPU. For CPU related experiments, the runtime environment on the compute cluster is restricted to 10x dedicated cores.

Data Sets. A semi-synthetic dataset was created to allow for a reproducible assessment and the disclosure of raw data samples for comparison. Various sources have been concatenated, including official government websites, public statistical data and datasets as part of previous publications to assemble semi-synthetic health dataset. A complete list of all attributes is publicly available on github.com [21]. Naturally, this dataset includes various of quasi-identifiers (QIDs) which amount increases over the number of available describing attributes (see Fig. 2a). The variability of our k-anonymity requirement does not significantly alter the QID amount growth in their high-dimensions.

6.1   Comparing Runtime Complexity

The most noticeable difference between running the native scalar based Bayesian inference implementation, its state aggregation optimisation and vector-based GPU acceleration is the difference in runtime. Figure 2b delineates the execution time in seconds over increasing number of attributes. While multigrid reduces the state-space explosion through aggregations by some factors, the vector-based implementation outperforms both by magnitudes.

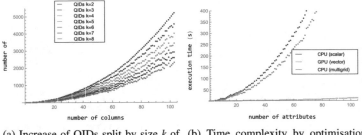

(a) Increase of QIDs split by size k of k-anonymity requirement

(b) Time complexity by optimisation approach

Fig. 2. Runtime comparison

Varying Number of Records. The dimension of records (rows) does not significantly impact the performance. For small datasets, this is different, yet given a reasonable size, additional rows usually do not introduce new attribute values but rather represent different composition of various attribute values where their binomial coefficient remains the same as its time complexity.

Varying Number of Attributes. With increasing columns, so more describing attributes and exponential growth of attribute combinations serving as QID candidates, all methodologies denote rising processing time (see Fig. 2b). The sheer amount of massive-parallelisation in the GPU implementation yet throttles this increase. For very large n currently out-of-scope, however, we do expect a boost.

Varying Cluster Sizes. Recognising the potential of simultaneously conducting calculations on 5120 CUDA cores per hardware component, the question of scaling limitations also emerges with GPU resources. Currently, not all libraries support multi-GPU setups, yet their implementation is still worth it. Figure 3a compares the traditional scalar based implementation on CPU, against vectorised implementation on 1x and 10x GPU nodes. Given the axis size, only a minor difference is visible, yet converting same to a log scale Fig. 3b illustration is quite impressive. There is a slight overhead in the beginning to divide and conquer the workload across multiple nodes by the CPU, plus a potential overhead in bus I/O between main memory and GPU memories. Yet especially the anticipated time growth for n attributes larger than our setup looks extremely promising.

Resource Utilisation. To determine the GPU nodes' utilisation during the experiments, we aggregate second based measurements to five second intervals. Figure 4a visualises the compute utilisation over time, Fig. 4b the memory usage. Both Figures demonstrate plenty of resources left, constituting numerous opportunities to grow data sizes. Given the progress achieved so far, we leave additional optimisation tweaks for future work, yet, clearly highlighting 20% unused compute capacity and 70% free memory space. In

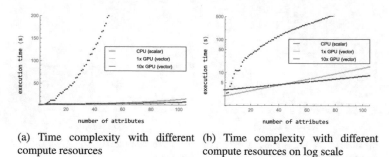

(a) Time complexity with different compute resources

(b) Time complexity with different compute resources on log scale

Fig. 3. Runtime comparison by resources

fact, in all experiments we have conducted, not a single GPU node max out available capacities overall.

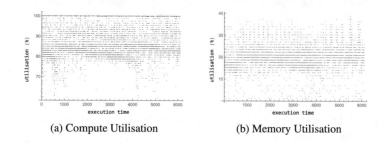

(a) Compute Utilisation

(b) Memory Utilisation

Fig. 4. GPU resource utilisation during experiments

6.2   Accelerated QID Discovery Success Cases

The previous evaluation results indicate a novel opportunity to ensure privacy while minimising information-loss. Yet, the same methodology can be flipped and used to attack existing, published datasets to re-identify individuals if not processed correctly in the first place. Particularly against probabilistic approaches, the sheer computation power can be used to easily brute-force de-anonymisation by try-and-error linkage guessing. Two examples should illustrate the successful utilisation of the accelerated QID discovery as the basis for de-anonymisation.

Figure 3a delineates, that processing 2^{100} attribute combinations for QID discovery becomes available at the fingertip. As a first sample, we explored a publicly available social media dataset about the US Presidential Election 2020 with more than 20 million records [22]. In this dataset with just 11 columns per record, we could re-identify individuals based on unique attribute combinations like their posting daytime, keywords and retweet count. We have removed the full last name but for demonstration reference *Kristine A.* from southern California or *Cameron R.* from Mount Vernon, NY. Cameron completed Boston College as an undergrad in 2019 with a major in Computer Science and a minor in Art History. As part of future work, we will demonstrate these re-identification implications against common heuristic privacy applications in more depth on a larger scale study.

7   Conclusions and Future Work

The complete discovery of all quasi-identifiers on large-scale has always been a challenge given its enumeration problem. Utilising Bayesian inferences to detect QIDs offered promising results yet lacked scalability constraints until now. In our work, we introduced and showcased multiple optimisation efforts and compared those to the state-of-the-art. Namely, we presented an approach to locate quasi-identifiers candidates based on the Bayesian inferences on very large-scale, efficiently and fast for millions of combinations. To accelerate the algorithmic runtime, state-space aggregation and algorithmic optimisation have been applied. For the latter, we introduced an algorithm which executes traditional scalar-based QID search scheme in a vector and tensor-based manner to allow massive-parallelisation in computing. We achieved this by utilising novel GPU acceleration for the underlying enumeration problem. Further, we offered empirical models and compared this massive-parallelisation against classical optimisation approaches to counter state-space explosion during inference computation. Our results promise magnitudes in decreasing execution time and enable near real-time applications around large-scale (stream) data anonymisation. As success cases, we demonstrated that the accelerate QID discovery successfully re-identified individuals in an already published social media dataset.

References

1. Nickolls, J., Dally, W.J.: The GPU computing era. IEEE Micro. **30**(2), 56–69 (2010)
2. Owens, J.D., Houston, M., Luebke, D., Green, S., Stone, J.E., Phillips, J.C.: GPU computing. Proc. IEEE **96**(5), 879–899 (2008)
3. Cook, C., Zhao, H., Sato, T., Hiromoto, M., Tan, S.X.D.: GPU-based ising computing for solving max-cut combinatorial optimization problems. Integration **69**, 335–344 (2019)
4. Podlesny, N.J., Kayem, A.V., Meinel, C.: Attribute compartmentation and greedy UCC discovery for high-dimensional data anonymization. In: Proceedings of the Ninth ACM Conference on Data and Application Security and Privacy, pp. 109–119. ACM (2019)
5. Bläsius, T., Friedrich, T., Lischeid, J., Meeks, K., Schirneck, M.: Efficiently enumerating hitting sets of hypergraphs arising in data profiling. In: Algorithm Engineering and Experiments (ALENEX), pp. 130–143 (2019)
6. Bläsius, T., Friedrich, T., Schirneck, M.: The parameterized complexity of dependency detection in relational databases. In: Guo, J., Hermelin, D. (eds.) International Symposium on Parameterized and Exact Computation (IPEC), Leibniz International Proceedings in Informatics (LIPIcs), Dagstuhl, Germany, vol. 63, pp. 6:1–6:13 (2016). Schloss Dagstuhl–Leibniz-Zentrum fuer Informatik
7. Barth-Jones, D.: The're-identification'of governor William Weld's medical information: a critical re-examination of health data identification risks and privacy protections, then and now. Then and Now (July 2012) (2012)
8. Price, W.N., Cohen, I.G.: Privacy in the age of medical big data. Nature Med. **25**(1), 37–43 (2019)

9. Zhu, L., Jin, H., Zheng, R., Feng, X.: Effective Naive Bayes nearest neighbor based image classification on GPU. J. Supercomput. **68**(2), 820–848 (2014)
10. Viegas, F., Gonçalves, M.A., Martins, W., Rocha, L.: Parallel lazy semi-Naive Bayes strategies for effective and efficient document classification. In: Proceedings of the 24th ACM International on Conference on Information and Knowledge Management, pp. 1071–1080 (2015)
11. Andrade, G., Viegas, F., Ramos, G.S., Almeida, J., Rocha, L., Gonçalves, M., Ferreira, R.: GPU-NB: a fast CUDA-based implementation of Naive Bayes. In: 2013 25th International Symposium on Computer Architecture and High Performance Computing, pp. 168–175. IEEE (2013)
12. Chen, F.C., Jahanshahi, M.R.: NB-CNN: deep learning-based crack detection using convolutional neural network and Naïve Bayes data fusion. IEEE Trans. Ind. Electron. **65**(5), 4392–4400 (2017)
13. Gruber, L., et al.: GPU-accelerated Bayesian learning and forecasting in simultaneous graphical dynamic linear models. Bayesian Anal. **11**(1), 125–149 (2016)
14. Ng, W.S., Kirchberg, M., Bressan, S., Tan, K.L.: Towards a privacy-aware stream data management system for cloud applications. Int. J. Web Grid Serv. **7**(3), 246–267 (2011)
15. Kalidoss, T., Sannasi, G., Lakshmanan, S., Kanagasabai, K., Kannan, A.: Data anonymisation of vertically partitioned data using map reduce techniques on cloud. Int. J. Commun. Netw. Distrib. Syst. **20**(4), 519–531 (2018)
16. Solanki, P., Garg, S., Chhinkaniwala, H.: Heuristic-based hybrid privacy-preserving data stream mining approach using SD-perturbation and multi-iterative k-anonymisation. Int. J. Knowl. Eng. Data Min. **5**(4), 306–332 (2018)
17. Podlesny, N.J., Kayem, A.V., Meinel, C.: Towards identifying de-anonymisation risks in distributed health data silos. In: International Conference on Database and Expert Systems Applications, pp. 33–43. Springer (2019)
18. Podlesny, N.J., Kayem, A.V., Meinel, C.: Identifying data exposure across high-dimensional health data silos through Bayesian networks optimised by multigrid and manifold. In: IEEE 17th International Conference on Dependable, Autonomic and Secure Computing, DASC 2019. IEEE (2019)
19. Nayahi, J.J.V., Kavitha, V.: Privacy and utility preserving data clustering for data anonymization and distribution on hadoop. Future Gener. Comput. Syst. **74**, 393–408 (2017)
20. Dean, J., Ghemawat, S.: MapReduce: simplified data processing on large clusters (2004)
21. Podlesny, N.J.: Synthetic genome data (2021)
22. IBRAHIM SABUNCU. USA Nov.2020 election 20 mil. tweets (with sentiment and party name labels) dataset (2020)

Improvement of the Matrix for Simple Matrix Encryption Scheme

Wenchao Liu, Yingnan Zhang, Xu An Wang$^{(\boxtimes)}$, and Zhong Wang

Engineering University of PAP, Xi'an, China

Abstract. Multivariable public key cryptography is a scheme that can meet the anti-quantum characteristics. The Simple Matrix scheme is constructed using the operations between three matrices. An improved version of the Simple Matrix scheme is proposed in this paper. The public key mapping of the new scheme includes two matrices made up of random cubic polynomials as elements, so that the rank attack is infeasible for the new scheme, and the algebraic attack breaks the system is at least as hard as solving a set of random quadratic equations. One of the private key matrices is improved so that the ratio of the ciphertext to the plaintext of the new scheme is greater than or equal to 2 times, which breaks the fixed ciphertext and plaintext ratio of the Simple Matrix scheme. The new scheme not only has a flexible ratio of the ciphertext to the plaintext to meet different needs, but also increases the redundancy of the ciphertext and further strengthens the security.

Keywords: Multivariable public key cryptography · Simple matrix scheme · The ratio of the ciphertext to the plaintext · Security analysis

1 Introduction

Public key cryptography plays a very important role in the field of computer security today. The traditional public-key encryption scheme is mainly based on the knowledge of classical number theory. Typical examples are RSA, elliptic curve cryptography ECC, Attribute-based public key cryptography, etc. [1]. In 1994, Peter Shor proposed a polynomial time algorithm Shor algorithm that runs on a quantum computer, making the factorization of large integers and solving the discrete logarithm problem on the Abel group no longer difficult in polynomial time [2]. Its time complexity is approximately O (log3N) (N means the number of bits is N), which poses a threat to the traditional public-key cryptography based on the classical number theory. Moreover, with the rapid development of computer hardware technology, the computing power of computers has increased dramatically. After the emergence of the traditional computer, a computer with quantum bits as a computing unit has appeared. They are called quantum computers. After the emergence of large-scale quantum computers, most of the traditional public key cryptography will be easily breached.

© The Author(s), under exclusive license to Springer Nature Switzerland AG 2021
L. Barolli et al. (Eds.): AINA 2021, LNNS 226, pp. 509–521, 2021.
https://doi.org/10.1007/978-3-030-75075-6_41

At present, there are several major anti-quantum cryptosystems, Coding-based cryptosystem, Lattice-based cryptosystem, Hash-based cryptosystem, and Multivariate-based cryptosystem [3]. Why are these cryptosystems able to effectively counter quantum computer attacks? Because they are constructed on the basis of NP difficulties, quantum computers have no significant advantages over ordinary computers in solving such problems. The Multivariate-based cryptosystem in anti-quantum cryptography is not only effective against quantum computation attacks [4], but also has more advantages in running efficiency and consumes less resources than traditional public key cryptography [5]. Although there are many practical multivariate signature schemes, the number of effective and secure multivariate encryption schemes is still limited.

At the PQCrypto2013 conference, Chengdong Tao et al. proposed a new multivariable public-key encryption scheme, the Simple Matrix (abbreviated as ABC) scheme [6], which can resist most attacks against multivariate public-key cryptographic schemes. In the following year, Jintai Ding et al. proposed an improvement to the Simple Matrix scheme. The Cubic Simple Matrix scheme made the original quadratic multivariable polynomials of public keys become cubic polynomials [7]. This improvement can effectively resist the rank attack [8], and makes the difficulty of algebraic attack is similar to the difficulty of solving a random quadratic polynomials system. Both the Simple Matrix scheme and the Cubic Simple Matrix scheme control the ratio between the ciphertext and the plaintext in a double relationship, which makes the application inflexible enough to meet various demands. This paper proposes a new cryptographic scheme based on the Simple Matrix. It guarantees security while adding redundancy compared to the Simple Matrix scheme and provides a flexible ratio of ciphertext to plaintext.

In the Sect. 2 of this paper, the basic Simple Matrix scheme is introduced. The Sect. 3 introduces the improved version of the Simple Matrix scheme. The Sect. 4 analyzes the security of the newly proposed scheme. Experiments and analysis were conducted on the efficiency of the program's resistance to attack. The last section summarized the full text and found insufficient results

2   Multivariable Public Key Cryptosystem and Simple Matrix Scheme

2.1   Bipolar System

Current multivariate cryptosystems can be divided into Bipolar, Mixed, Polynomials (IP) and Multivariate Quadratic (MQ) [9]. Most of the multivariable public key cryptosystems are based on bipolar systems. There are fewer mixed modes [10], and the other two are fewer.

The public key of a bipolar system consists of two parts: one is the finite field K of the system and its domain structure, let $K = GF(q)$ and the other is the m polynomials of the public key map P. The private keys are randomly selected reversible linear affine transformations L_1 and L_2, and the center map F [11].

There is a limited domain $K = GF(q)$. In a bipolar multivariate cryptosystem, the public key is $k_n \to k_m$, the map F: $F(x_1, x_2, \cdots, x_n) = (f_1, f_2, \cdots, f_m)$, where f_1 is an n-ary polynomial in $F(x_1, x_2, \cdots, x_n)$.

Its center map is $\bar{F}$ and the construction process is as follows:

(1) $\bar{F}(\bar{x}_1, \bar{x}_2, \cdots, \bar{x}_n) = (\bar{f}_1, \bar{f}_2, \cdots, \bar{f}_m)$, and $\bar{f}_i \in K[x_1, x_2, \cdots, x_n]$
(2) The calculation of the map $\bar{F}(\bar{x}_1, \bar{x}_2, \cdots, \bar{x}_n) = (\bar{y}_1, \bar{y}_2, \cdots, \bar{y}_m)$ can be easily solved, and correspondingly the only original image can be quickly found.

Encryption process: Known messages that require encryption are $X = (x_1, x_2, \cdots, x_n)$. Substituting X as the plaintext into the public key polynomial, the ciphertext is calculated $Y = F(X) = (y_1, y_2, \cdots, y_m)$.

Decryption process: The ciphertext message is $Y = (y_1, y_2, \cdots, y_m)$ decryption is the solving equation: $Y = F(X)$.

Solve first $L_1^{-1}(Y) = \bar{Y}$, then calculate $\bar{X} = \bar{F}^{-1}(\bar{Y})$, recalculate the plaintext $X = L_2^{-1}(\bar{X})$.

2.2 Hybrid System

The public key of the hybrid multivariable public-key cryptosystem is also a map that is used in a finite field. The map using $k_{n+m} \to k_l$ is denoted as H: $H(x_1, x_2, \cdots, x_n, y_1, y_2, \cdots, y_m) = (h_1, \cdots, h_l)$ where $h_i \in K[x_1, x_2, \cdots, x_n, y_1, y_2, \cdots, y_m]$. To construct such an encryption scheme requires finding the corresponding center map $\bar{H} : K^{n+m} \to K^l$

Center maps to meet for any given $\bar{x}_1$, $\bar{x}_2$, $\cdots$, $\bar{x}_n$ the equations: $\bar{H}(\bar{x}_1, \bar{x}_2, \cdots, \bar{x}_n, \bar{y}_1, \bar{y}_2, \cdots, \bar{y}_m) = (0, \cdots, 0)$ are easy to solve. And in most cases, it is a linear system of equations about the variables $\bar{y}_1$, $\bar{y}_2$, $\cdots$, $\bar{y}_m$. After entering the plaintext, we can easily get the ciphertext. If we give another m variables $\bar{x}_1$, $\bar{x}_2$, $\cdots$, $\bar{x}_n$, you get a quadratic equation set about the variable $\bar{H}(\bar{x}_1, \bar{x}_2, \cdots, \bar{x}_n, \bar{y}_1, \bar{y}_2, \cdots, \bar{y}_m) = (0, \cdots, 0)$. This random quadratic system is a system of equations with a special structure that can be easily solved, that is, get the plaintext. After finding the center map that meets the requirements, H can be expressed as follows:

$$H = L_3 \circ \bar{H} \circ (L_1 \times L_2)$$

The definitions of L_1 and L_2 are the same as the linear maps S and T defined in the bipolar system. L_3 is a linear mapping from F_l to F_l.

The public key of the hybrid system includes: 1) the finite field K where the system is located and its domain structure; 2) the polynomial equations that constitutes the public key map H.

The construction idea of the mixed multivariable public key cryptosystem is similar to that of the bipolar system. The affine transformation is used to hide the center map. The difference is that the hybrid system uses three affine transformations L_1, L_2, and L_3 to hide the equations $H(X, Y) = (0, \cdots, 0)$. At present, there are few multivariable public key cryptosystems based on hybrid systems, of which the famous one is Patranin's Dragon encryption system [8].

2.3   Simple Matrix Scheme

The Simple Matrix algorithm structure is rather special and consists of three matrices. The main parameters are n, m, s. Where n represents the length of the plaintext and m represents the length of the ciphertext. The relationship between them satisfies $m = 2n$, that is, the length of the ciphertext is twice that of the plaintext, and the plaintext variable is used $X = (x_1, x_2, \cdots, x_n)$, $s \in N$ and $s2 = n$.

Then, define the three small matrices A, B, and C as $s \times s$, as shown below:

$$A = \begin{bmatrix} p_1 & \cdots & p_s \\ \vdots & \ddots & \vdots \\ p_{(s-1)s+1} & \cdots & p_n \end{bmatrix}$$

$$B = \begin{bmatrix} b_1 & \cdots & b_s \\ \vdots & \ddots & \vdots \\ b_{(s-1)s+1} & \cdots & b_n \end{bmatrix}$$

$$C = \begin{bmatrix} c_1 & \cdots & c_s \\ \vdots & \ddots & \vdots \\ c_{(s-1)s+1} & \cdots & c_n \end{bmatrix}$$

Among them, the n elements of A are the variables X themselves; each element of matrix B and C is a random linear combination of variables $X = X_1, X_2, \cdot, x_n$.

Then define two matrices E_1, E_2 with dimensions $s \times s$. Among them, $E1 = A \cdot B$, $E2 = A \cdot C$, and in front of the description of the matrix A, B, C, it can be seen that the elements in E_1, E_2 are all quadratic polynomials about the variables. There are m quadratic polynomials for both E_1 and E_2. These m quadratic polynomials are combined into the center map F of the Simple Matrix algorithm: $K^n \rightarrow K^m$

Next, similar to the bipolar mode construction, there are two reversible linear maps $T : K^n \rightarrow K^n$, $S : K^m \rightarrow K^M$, and they are combined with the center map F to obtain the public key. $P = S \circ F \circ T : K^n \rightarrow K^m$. The private key contains two matrices B, C and reversible linear map S, T.

The encryption process of Simple Matrix scheme is as follows:

Encrypt the plaintext $X = (x_1, x_2, \cdots, x_n)$ and use the public key P to operate on it to get the ciphertext $Y = (y_1, y_2, \cdots, y_n)$: $Y = (y_1, y_2, \cdots, y_m) = P(x_1, x_2, \cdots, x_n)$.

The decryption process of Simple Matrix scheme:

Decrypting ciphertext $Y = (y_1, y_2, \cdots, y_m)$ can be divided into the following three steps.

(1) Using a reversible linear map S to compute $Z : Z = (z_1, z_2, \cdots, z_m) = S^{-1}(y_1, y_2, \cdots, y_m)$.

(2) As can be seen from the structure of the Simple Matrix, each polynomial of the center map F is an element in the matrix E_1, E_2, which are all quadratic polynomials composed of variables. The value of each element in E_1, E_2 is $Z = z_1, z_2, \cdots, z_m$:

$$
E_1 = \begin{bmatrix} z_1 & \cdots & z_s \\ \vdots & \ddots & \vdots \\ z_{(s-1)s+1} & \cdots & z_n \end{bmatrix}
$$

$$
E_2 = \begin{bmatrix} z_{n+1} & \cdots & z_{n+s} \\ \vdots & \ddots & \vdots \\ z_{n+(s-1)s+1} & \cdots & z_m \end{bmatrix}
$$

(3) For the center map, find the original image, $F(W) = Z$, $W = (w_1, w_2, \cdots, w_n)$. There are three solutions for different situations depending on the center map.

First, if the matrix E1 is invertible, then W can be obtained by formulating n linear equations for the variable W according to the formula $B \cdot E_1^{-1} \cdot E_2 = C$.

Next, if the matrix E_2 is invertible, the linear equations for the variable W can also be obtained from the equation $C \cdot E_1^{-1} \cdot E_2 = B$. Finally, if E_1, E_2 are all irreversible, but A is invertible. Then consider the element in the A^{-1} matrix as a new variable $(p_1, p_2, \cdots, p_n)$. There are $A^{-1}E_1 = B, A^{-1}E_2 = C$, so there are m linear equations about the variables $(w_1, w_2, \cdots, w_n)$, $(p_1, p_2, \cdots, p_n)$, so that the variable W can be solved. If the matrices A, E_1, E_2 are irreversible, they cannot be decrypted.

(4) Using the reversible linear map T to compute W and get the plaintext X: $X = (x_1, x_2, \cdots, x_n) = T^{-1} (w_1, w_2, \cdots, w_n)$

3   Improved Version of Simple Matrix

3.1   Structural Thought

From the introduction in the previous section, it can be seen that the ratio of the ciphertext to the plaintext in the Simple Matrix scheme is fixed to double, which leads to a solution that is not flexible enough to meet multiple requirements. This section proposes an improved version that makes the ratio of ciphertext to plaintext more flexible, changes its public key from the original quadratic polynomial to a cubic polynomial, providing greater security.

First, there are three small matrices A, B, and C. Among them, the dimension of matrix A is also $s \times s$, $s \in N$, and $s^2 = n$. Unlike the Simple Matrix scheme, the composition of its elements is the quadratic polynomial of the variable $X = (x_1, x_2, \cdots, x_n)$ rather than the variable itself.

$$
A = \begin{bmatrix} a_{11} & \cdots & a_{1s} \\ \vdots & \ddots & \vdots \\ a_{s1} & \cdots & a_{ss} \end{bmatrix}
$$

$$a_{ij} = \sum_{r=1}^{n} \sum_{s=r}^{n} \alpha_{rs}^{(ij)} x_r x_s + \sum_{r=1}^{n} \beta_r^{(ij)} x_r + \varepsilon$$

The dimensions of the other two matrices B and C are $s \times s$, $s \times v$, $s \in N$, $v \in N$ respectively. Unlike the Simple Matrix scheme, the dimensions of the matrix C become flexible and variable. The elements in the matrix are the same as the elements of the B, C matrix in the Simple Matrix are the linear combinations of the variables $X = (x_1, x_2, \cdots, x_n)$.

$$B = \begin{bmatrix} b_{11} & \cdots & b_{1\,s} \\ \vdots & \ddots & \vdots \\ b_{s1} & \cdots & b_{ss} \end{bmatrix}$$

$$C = \begin{bmatrix} c_{11} & c_{12} & \cdots & c_{1\,s} & \cdots & c_{1v} \\ c_{21} & c_2 & \cdots & c_{2\,s} & \cdots & c_{2v} \\ \vdots & \vdots & & \vdots & \vdots \\ c_{s1} & c_{s2} & \cdots & c_{ss} & \cdots & c_{sv} \end{bmatrix}$$

$$b_{ij} = \sum_{r=1}^{n} \alpha_r^{(j)} x_r + \mu, c_{ij} = \sum_{m=1}^{n} \alpha_m^{(j)} x_m + \theta$$

Then, two matrices E_1, E_2 are calculated, in which $E_1 = A \circ B$, E_1 has the dimension $s \times s$, and $E_2 = A \circ C$ has the dimension $s \times v$. From the foregoing, the elements in the matrix E_1, E_2 are all cubic polynomials.

$$E_1 = \begin{bmatrix} e_{11} & \cdots & e_{1\,s} \\ \vdots & \ddots & \vdots \\ e_{s1} & \cdots & e_{ss} \end{bmatrix}$$

$$E_2 = \begin{bmatrix} \bar{e}_{11} & \bar{e}_{12} & \cdots & \bar{e}_{1\,s} & \cdots & \bar{e}_{1v} \\ \bar{e}_{21} & \bar{e}_n & \cdots & \bar{e}_{2\,s} & \cdots & \bar{e}_{2v} \\ \vdots & \vdots & & \vdots & & \vdots \\ \bar{e}_{s1} & \bar{e}_{s2} & \cdots & \bar{e}_s & \cdots & \bar{e}_{sv} \end{bmatrix}$$

$$e_{ij} = \sum_{r=1}^{n} \sum_{s=r}^{n} \sum_{t=s}^{n} \alpha_{ist}^{(ij)} x_r x_s x_t$$
$$+ \sum_{r=1}^{n} \sum_{s=r}^{n} \beta_{rs}^{(ij)} x_r x_S + \sum_{r=1}^{n} \sigma_r^{(ij)} x_r + \delta$$
$$\bar{e}_{ij} = \sum_{a=1}^{n} \sum_{b=a}^{n} \sum_{c=b}^{n} \alpha_{abc}^{(ij)} x_a x_b x_c$$
$$+ \sum_{a=1}^{n} \sum_{b=a}^{n} \beta_{ab}^{(ij)} x_a x_b + \sum_{a=1}^{n} \sigma_a^{(ij)} x_a + \gamma$$

The elements in E_1, E_2 have $m = s^2 + s \times v$. These m elements are all cubic polynomials made up of variables $X = (x_1, x_2, \cdots, x_n)$. The center map is $H : K^n \to K^m$. It also requires two reversible affine transformations $T : K_n \to K_n$, $S : K^m \to K^m$ to hide the center map. The public key of the scheme is $P = S \circ H \circ T : K_n \to K_n$, and the private key is the map S, T and the matrix B, C.

When encrypting, the plaintext can be directly brought into the public key P. When decrypting, first look at whether matrix E_1 is invertible, if invertible, we can obtain $s \times s$ equations for n variables by solving the equations $B \cdot E_1^{-1} \cdot E_2 = C$. If the matrix E_1 is irreversible, it depends on whether the matrix A is invertible. If the matrix A is invertible, the elements in the matrix A^{-1} can be regarded as new elements and multiplied by the matrix E_1, E_2, respectively, resulting in $A^{-1}E_1 = B$, $A^{-1}E_2 = C$, there are $2n(2s^2)$ variables, $m = s^2 + s \times v$ linear equations, $m \geq 2n$, so the plaintext variable X can be solved finally.

3.2   Encryption and Decryption Process

(1) Encryption process: similar to the Simple Matrix scheme, the plaintext $X = (x_1, x_2, \cdots, x_n)$ is brought into the formula of the public key P to obtain the ciphertext $Y = (y_1, y_2, \cdots, y_m)$
(2) Decryption process:
   ① The ciphertext $Y = (y_1, y_2, \cdots, y_m)$ is processed first, and the operation is performed on the information Y using the reversible affine transformation S to obtain the variable W.

$$W = (w_1, w_2, \cdots, w_m) = S^{-1}Y = S^{-1}(y_1, y_2, \cdots, y_m)$$

② The inverse of the center map H, $Z = H^{-1}(W)$.

First of all, if the matrix E_1 is invertible, then W can be obtained by formulating $s \times s$ linear equations for the variable Z according to the formula $B \times E_1^{-1} \times E_2 = C$.

Next, if E_1 is irreversible, but A is invertible. Matrix A consisting of quadratic polynomials of the variable $Z = (z_1, z_2, \cdots, z_n)$. If the matrix A is invertible, Next, if E_1 is irreversible, but A is invertible. Matrix A consisting of quadratic polynomials of the variable $Z = (z_1, z_2, \cdots, z_n)$, If the matrix A is invertible, the element in the matrix A^{-1} is treated as a new variable $\bar{Z} = (\bar{z}_1, \bar{z}_2, \cdots, \bar{z}_n)$ and then calculate $A^{-1}E_1 = B$ and $A^{-1}E_2 = C$.

$$\begin{bmatrix} z_{11} & \cdots & z_{1s} \\ \vdots & \ddots & \vdots \\ - & & - \\ z_{s1} & \cdots & z_{ss} \end{bmatrix} \cdot \begin{bmatrix} w_{11} & \cdots & w_{1s} \\ \vdots & \ddots & \vdots \\ w_{s1} & \cdots & w_{ss} \end{bmatrix}$$

$$= \begin{bmatrix} b_{11} & \cdots & b_{1s} \\ \vdots & \ddots & \vdots \\ b_{s1} & \cdots & b_{ss} \end{bmatrix} = B$$

$$\begin{bmatrix} \bar{z}_{11} & \cdots & \bar{z}_{1s} \\ \vdots & \ddots & \vdots \\ \bar{z}_{s1} & \cdots & \bar{z}_{ss} \end{bmatrix} \cdot$$

$$\begin{bmatrix} W_{ss+1} & W_{ss+2} & \cdots & W_{ss+s} & \cdots & W_{ss+v} \\ W_{ss+v+1} & W_{ss+v+2} & \cdots & W_{ss+v+s} & \cdots & W_{ss+2v} \\ \vdots & \vdots & & \vdots & & \vdots \\ W_{ss+(s-1)v+1} & W_{ss+(s-1)v+2} & \cdots & W_{ss+ss} & \cdots & W_{ss+sv} \end{bmatrix}$$

$$= \begin{bmatrix} c_{11} & c_{12} & \cdots & c_{1s} & \cdots & c_{1v} \\ c_{21} & c_{22} & \cdots & c_{2s} & \cdots & c_{2v} \\ \vdots & \vdots & & \vdots & \vdots & \\ c_{s1} & c_{s2} & \cdots & c_{ss} & \cdots & c_{sv} \end{bmatrix} = C$$

We can get $s^2 + s \times v$ equations about the variables $Z = (z_1, z_2, \cdots, z_n)$ and $\bar{Z} = (\bar{z}_1, \bar{z}_2, \cdots, \bar{z}_n)$. By combining the above $s \times s$ and $s \times v$ equations, a system of equations with the number $m = s^2 + s \times v$ and the number of variables $2n(2s^2)$ can be obtained. Since $m \geq 2n$, the solution can be solved and get variable $Z = (z_1, z_2, \cdots, z_n)$.

Decryption fails if matrix A is irreversible.

③ Using the affine transformation T to perform an inverse operation on the variable $Z = (z_1, z_2, \cdots, z_n)$ to obtain the variable $X = (x_1, x_2, \cdots, x_n) = T^{-1}Z = T^{-1}(z_1, z_2, \cdots, z_n)$.

From the above decryption process, it can be seen that the probability of decryption failure of the improved version of the Simple Matrix scheme is the same as the probability of decryption failure of the original Simple Matrix scheme, that is, if a finite field of order q exists and the matrix is a square matrix and is in the finite field, the probability of the matrix being irreversible is $1/q$, that is, there is a possibility of $1/q$ that decryption fails.

4   Security Analysis

4.1   Formal Safety Certification

The security of the Improved version of Simple Matrix scheme was proved and formalized reduction was used to prove that the scheme satisfies IND-CPA security under the standard model.

Definition 1. If the attacker O cannot break the Improved version of Simple Matrix scheme in the following game in a polynomial time that cannot be ignored, it is said that the scheme has IND-CPA security under the selected plaintext attack.

The attack game between the attacker O and the challenger P based on the Improved version of Simple Matrix scheme is defined as follows:

(1) The challenger P generates a key pair (sk, pk) through a key generation algorithm and an initialization algorithm, and transmits the key pair to the attacker O.

(2) The attacker O can freely access the random encryption oracle Enc.

(3) Challenge phase: Attacker O randomly generates two equal-length plaintexts $m0$, $m1$ and sends them to challenger P. Challenger P throws a fair coin, encrypts plaintext mb and obtains ciphertext, Sent to attacker O. Since then, the attacker O still has free access to the encryption oracle Enc.

(4) Guessing phase: Attacker O outputs $\bar{b}$ as guess.

If $b = \bar{b}$, the attacker O challenges successfully and defines the advantages of its success as follows:

$$\text{Adv}_O^{\text{IND-CPA}} = |2\Pr(b = \bar{b}) - 1|$$

The following is safety regulations.

Theorem 1. For the Improved version of Simple Matrix scheme, if there is an IND-CPA attacker O who can win the above challenge game with the advantage $\text{Adv}_O^{\text{IND}}$ (expressed as δ_{CPA}), then there is an attack algorithm that can solve multivariable public key cryptography in polynomial time. The theoretical basis for the MQ problem.

Proof. Given an MQ problem instance of the Improved version of Simple Matrix scheme $Y = \bar{H}(X)$, this example is attacked using an algorithm α to break the challenge, that is, the plaintext is calculated. The attacker O acts as a subroutine for the algorithm α, and the algorithm plays the challenger O in the attack game.

(1) The algorithm α generates public and private key pairs (sk, pk) in the MQ problem instance and informs the attacker O of the public key pk.

(2) The algorithm α acts as an encryption oracle and responds to the attacker O's encrypted challenge. Attacker O randomly selected messages to send an encrypted request. The algorithm randomly selects the invertible matrix A and calculates the ciphertext $Y = (y_1, y_2, \cdots, y_m) = \bar{H}(X)$ according to the encryption method, and then transmits the encrypted result to the attacker O.

(3) Challenge phase: Attacker O randomly outputs two equal length plaintexts m_0, m_1. Then, the algorithm throws a fair coin $b \in 0, 1$, encrypts the plaintext m_b and obtains the ciphertext θ and returns it to the attacker O. After that, the attacker O can still freely access the encryption oracle.

(4) Guessing phase: Attacker O outputs $\bar{b}$ as guess.

The probability that attacker O challenges to success is, then its advantage of winning is: $\Pr(b = \bar{b})$, then its advantage of winning is:

$$\text{Adv}_O^{\text{IND-CPA}}(O) = |2\Pr(b = \bar{b}) - 1| = \delta_{\text{CPA}}$$

It can be known from the descriptions of the construction ideas and schemes of the core mapping in Sect. 3.1 that since the quadratic polynomials of the plaintext variable in the matrix A is randomly selected before each encryption, the probability of the same result of twice encryption is very low for the same plaintext m. Reference [7] pointed out that when directly attacking the MQ problem, the time complexity μ of the direct attack under $O(2^100)$ security is: $\mu \geq 3 \cdot \partial \cdot \tau^2 \geq 2^{102}$ (τ is the number of highest-order automorphisms monomials, ∂ is the number of non-zero monomials in each polynomial). In addition, Improved version of Simple Matrix increases the number of equations in the core mapping, further improving the redundancy, and further increasing the complexity of the scheme. Therefore, attacker O obtains b in polynomial time by continuously querying m_0's and m_1's encryptions cannot be obtained. From the above, the attacker O guesses the probability of success is $1/2$, so δ_α is small enough to be an insignificant amount that cannot be calculated, and $\delta_\alpha \leq \delta_{CPA}$.

Based on the above proof process and analysis, the difficulty of attacking the Improved version of Simple Matrix scheme can be reduced to the difficulty of solving the MQ problem, and it is proved that it satisfies the IND-CPA security under the standard model.

4.2   Algebraic Attacks

The XL algorithm [15] and the Grobner base algorithm [16] are the main methods for solving nonlinear equations, and they are also an effective method for attacking multivariable public key cryptography. In a multivariable public key cryptosystem, a ciphertext variable is substituted into a public key polynomial equation group to obtain a set of polynomial equations $Y = F(X)$ about the plaintext variable. The equation set is directly solved and the plaintext is obtained. This is the main idea of an algebraic attack or a direct attack. The essence of the XL algorithm is to calculate the Grobner basis, which can be seen as a redundant variant of the F4 algorithm [17]. The most effective way to find the Grobner basis is the F4 and F5 algorithms proposed by Faugere. Using the F5 algorithm, it is possible to break the challenge 1 for the HFE system [18]. The computational complexity of the F5 algorithm to solve the multivariate cryptosystem by computing the Grobner basis is approximately $2^{0.873n}$ [19], where n is the number of variables.

When using algebraic attacks to test the Improved Simple Matrix scheme, the attacker is faced with a system of $m(m \geq 2n)$polynomial cubic polynomial equations consisting of n variables. As previously mentioned, the system is obtained by multiplying a matrix A containing randomly selected polynomial quadratic polynomials with matrices B and C containing multiple linear combinations (ignoring the reversible affine transformations S and T). The solution to this cubic system is at least as difficult as the solution of a multivariate quadratic system with randomly chosen coefficients.

5   Efficiency Analysis

5.1   Analysis of the Ratio of Ciphertext and Plaintext

From the description of the Improved version of Simple Matrix in the previous section, the plaintext length is n, and there are $s \in N$, $n = 2s$ while the ciphertext length is m, m satisfies $m = s^2 + s \times v$, v represents the number of columns of matrix C. If you want to get the plaintext, then the number of linear equations should be greater than or equal to the number of variables. From the description of the decryption process in Sect. 3, we know that $m \geq 2n$, that is, $s^2 + s \times v \geq 2s^2$. Thus, it can be shown that the matrix C is relatively flexible matrices, as long as the number of its rows is equal to s, and the number of the columns satisfy the $v \geq s$. It can be decrypted under the condition that matrix E_1 is invertible or at least matrix A is invertible. The ratio of the ciphertext to the plaintext is expressed as:

$$\gamma = \frac{m}{n} = \frac{s \times s + s \times v}{s^2} = \frac{s+v}{s} = 1 + \frac{v}{s} \geq 2, (v \geq s)$$

From the above inequality, we can see that under the condition of $v \geq s$, Improved version of the Simple Matrix scheme can break the fixed limit of 2 times between the ciphertext and the plaintext, reduce some of the connections between the plaintext and the ciphertext, providing a further improvement. Flexible and secure ratio is able to meet different encryption requirements. For example, when coding-based ciphers are combined, flexible ratios allow more dimension choices when choosing codewords [20].

5.2 Attack Efficiency

The following experiments compare the direct attacks effects of the original Simple Matrix scheme and Improved Simple Matrix scheme, (m, n) represent the ciphertext and the plaintext scale, respectively. To make it easier to compare the ability to resist direct attack, the ratio of ciphertext to plaintext of Improved version of the Simple Matrix are 2 times the same as the original scheme. Both schemes are performed on domain $GF(256)$. The experimental comparison results are shown in Table 1:

Table 1. Direct attack effect comparison table

		Simple Matrix scheme	Improved version of the Simple Matrix scheme
(8,4)	Time(s)	–	0.82
	Memory(MB)	–	5.19
(14,7)	Time(s)	–	2.80
	Memory(MB)	–	9.88
(18,9)	Time(s)	0.03	15.41
	Memory(MB)	3.04	18.21
(28,14)	Time(s)	0.66	–
	Memory(MB)	6.31	–

Through experimental comparison, it can be found that Improved version of the Simple Matrix scheme is more complex than the Simple Matrix scheme, which takes longer and is more secure. With the expansion of the scale of the ciphertext and plaintext, the expended time consumption increases exponentially.

6   Conclusion

The Simple Matrix scheme uses three matrix operations to construct a center map. In this paper, the original Simple Matrix scheme is improved, and the quadratic polynomial in the original public key map is replaced by a cubic polynomial, which can effectively defend against attack methods such as rank attacks and improve the security of the scheme. It breaks the fixed relationship between the ratio of the ciphertext to the plaintext of the original Simple Matrix scheme, which increases system redundancy and reduces some of the connections between the plaintext and the ciphertext, making the scheme more flexible and reliable. However, in the improved scheme, the length of the plaintext whose ciphertext length is greater than or equal to 2 times is further improved. The next step is to expand the value range of the ratio of the ciphertext to the plaintext and provide a more efficient solution for encryption and decryption.

Acknowledgment. This work was supported by National Key R&D Program of China (Grant No. 2017YFB0802000), National Natural Science Foundation of China (Grant Nos. U1636114).

References

1. Rivest, R.L., Shamir, A., Adleman, L.: A method for obtaining digital signatures and public-key cryptosystems. Commun. ACM **21**(2), 120–126 (1978)
2. Shor, P.: Polynomial-time algorithms for prime factorization and discrete logarithms on a quantum computer. SIAM J. Comput. **26**(5), 1484–1509 (1994)
3. Bernstein, D.J., Buchmann, J., Dahmen, E. (eds.): Post Quantum Cryptography. Springer, Heidelberg (2009)
4. Barreto, P.S., et al.: A panorama of post-quantum cryptography. In: Open Problems in Mathematics and Computational Science, pp. 387–439. Springer International Publishing (2014)
5. Bogdanov, A., Eisenbarth, T., Rupp, A., Wolf, C.: Time-area optimized publickey engines: MQ-cryptosystems as replacement for elliptic curves? In: Oswald, E., Rohatgi, P. (eds.) CHES 2008, LNCS, vol. 5154, pp. 45–61. Springer, Heidelberg (2008)
6. Tao, C., Diene, A., Tang, S., Ding, J.: Simple Matrix Scheme for Encryption. In: Gaborit, P. (ed.) PQCrypto 2013, LNCS, vol. 7932, pp. 231–242. Springer, Heidelberg (2013)
7. Ding, J., Petzoldt, A., Wang, L.C.: The cubic simple matrix encryption scheme. In: Post-Quantum Cryptography, pp. 76–87. Springer International Publishing (2014)
8. Kipnis, A., Shamir, A.: Cryptanalysis of the HFE public key cryptosystem. In: Advances in Cryptology – CRYPTO 1999, pp. 19–30. Springer, Heidelberg (1999)

9. Patarin, J.: Asymmetric cryptography with a hidden monomial. In: Advances in Cryptology-CRYPTO 1996, pp. 45–60. Springer, Heidelberg (1996)
10. Ding, J., Gower, J.E., Schmidt, D.S.: Multivariate Public Key Cryptosystems. Springer, Heidelberg (2006)
11. Ding, D.: Analysis and Improvement of TTS Scheme in Multivariable Public Key Cryptosystem. Xidian University (2013)
12. Goubin, L., Courtois, N.: Cryptanalysis of the TTM cryptosystem. In: Okamoto, T. (ed.) ASIACRYPT 2000, LNCS, vol. 1976, pp. 44–57. Springer, Heidelberg (2000)
13. Coppersmith, D., Stern, J., Vaudenay, S.: Attacks on the birational permutation signature schemes. In: Stinson, D.R. (ed.) CRYPTO 1993, LNCS, vol. 773, pp. 435–443. Springer, Heidelberg (1994)
14. Ding, J., Schmidt, D.: Rainbow, a new multivariable polynomial signature scheme. In: Applied Cryptography and Network Security, pp. 164–175. Springer, Heidelberg (2005)
15. Courtois, N.T., Klimov, A., Patarin, J., Shamir, A.: Efficient algorithms for solving overdefined systems of multivariate polynomial equations. In: Preneel, B. (ed.) EUROCRYPT 2000, LNCS, vol. 1807, pp. 392–407. Springer, Heidelberg (2000)
16. Faugere, J.C.: A new efficient algorithm for computing Gröbner bases without reduction to zero (F5). In: Proceedings of the 2002 International Symposium on Symbolic and Algebraic Computation, pp. 75–83. ACM (2002)
17. Faugere, J.C.: A new efficient algorithm for computing Gröbner bases (F4). J. Pure Appl. Algebra **139**, 61–88 (1999)
18. Ding, J., Buchmann, J., Mohamed, M.S.E., Mohamed, W.S.A.E., Weinmann, R.-P.: Mutant XL. Talk at the First International Conference on Symbolic Computation and Cryptography (SCC 2008), Beijing (2008)
19. Mohamed, M.S.E., Cabarcas, D., Ding, J., Buchmann, J., Bulygin, S.: MXL3: an efficient algorithm for computing Gröbner bases of zero-dimensional ideals. In: Lee, D., Hong, S. (eds.) ICISC 2009, LNCS, vol. 5984, pp. 87–100. Springer, Heidelberg (2010)
20. Han, Y., Lan, J., Yang, X., Wang, J.: Multivariable encryption scheme combined with low rank error correction coding. J. Huazhong Univ. Sci. Technol. (Nat. Sci. Ed.) **44**(03), 71–76 (2016)

Lifestyle Authentication Using a Correlation Between Activity and GPS/Wi-Fi Data

Akira Miyazawa[✉], Tran Phuong Thao, and Rie Shigetomi Yamaguchi

Graduate School of Information Science and Technology, The University of Tokyo, 7-3-1 Hongo, Bunkyo, Tokyo 113–8656, Japan
{sgong,tpthao,yamaguchi}@yamagula.ic.i.u-tokyo.ac.jp

Abstract. In recent years, lifestyle authentication, which combines multiple personal behavioral data for authentication, has been proposed as a new authentication method in addition to traditional knowledge-based authentication, possession-based authentication, and biometrics-based authentication. In previous research on lifestyle authentication, authentication scores of each authentication element were often calculated independently and used for the final authentication, ignoring the correlation between each element. It was also often difficult to apply lifestyle authentication methods in the real world because they required a large amount of preliminary data. In this paper, we propose a new method that solves these problems by using the correlation between GPS/Wi-Fi data from smartphones and activity data (activity types that are inferred from the metabolic equivalent of task (MET)) from activity trackers. We applied our method to the data collected in the MITHRA project, which is a proof-of-concept experiment of lifestyle authentication. As a result, we achieved an equal error rate (EER) of 0.087 and 0.130 when ideal data were obtained and not obtained, respectively.

Keywords: Lifestyle authentication · Behavioral authentication · GPS · Wi-Fi · Activity tracker

1 Introduction

Over the past decade, with the spread of digital devices, an increasing number of users are concerned about cybersecurity. In particular, there is a growing interest in authentication technologies that identify legitimate users and protect users from unauthorized access. There are three traditional commonly-used authentication methods to identify individuals: knowledge-based authentication (such as passwords), possession-based authentication (such as physical security keys), and biometric-based authentication (such as fingerprint, face, or eyes). In addition, behavioral authentication and lifestyle authentication, which we define as an authentication method that combines multiple behavioral authentications to use various types of different behavioral data, have been proposed as the fourth authentication method. One characteristic of behavioral authentication and lifestyle authentication is that they do not need users' explicit action

© The Author(s), under exclusive license to Springer Nature Switzerland AG 2021
L. Barolli et al. (Eds.): AINA 2021, LNNS 226, pp. 522–535, 2021.
https://doi.org/10.1007/978-3-030-75075-6_42

to authenticate individuals, unlike the three traditional authentication methods. For example, a lifestyle authentication method that compares a user's past location data with current location data to authenticate individuals does not continuously require explicit actions from the user because they can be obtained without the user's permission or action after the user's initial approval. Thus, these authentication methods are expected to improve convenience compared to the three traditional authentication methods.

Several behavioral/lifestyle authentication methods have been proposed, including [13,14] that use individual differences in smartphone usage patterns, [16,18] that use physical and behavioral characteristics measured by activity trackers, and [5,6] that use GPS and Wi-Fi logs to extract behavioral information. Two main research approaches about behavioral/lifestyle authentication of identifying individuals are: 1) To use the historical data of the target for a period of time to form an authentication template and then match the data in the newest time window (e.g., the most recent 24-hour data) with the authentication template to identify the target [2,5,8]. 2) To prepare both the data of the target and others (other individuals who are joining the system along with the target) and classify them using supervised classification methods such as SVM (Support Vector Machine) by considering the authentication as a classification problem [1,17,18]. These two methods have the problem of requiring a large amount of preliminary data. Also, when combining multiple authentication factors in behavioral/lifestyle authentication, the scores of each factor were often calculated independently and then used for the final authentication, without using the correlation that should exist among individual authentication factors. For example, GPS location data and acceleration data obtained from the same person should have some correlations (e.g., when the location data show a movement of the user, then the acceleration data should also show an indication of the movement), but most of the previous studies use these data independently to generate authentication results.

To address these problems, this paper proposes a new correlation-based lifestyle authentication method that meets the following contributions:

- We introduce a new lifestyle authentication method that uses the correlation between activity history (activity types inferred from MET) and GPS/Wi-Fi history.
- We show that our method can achieve the same level of authentication accuracy compared to existing behavioral/lifestyle authentication by experiments using real-world data. In the experiment, we achieved an equal error rate (EER) of 0.087 and 0.130 when ideal data were obtained and not obtained, respectively.
- Since our method does not require preliminary template generation nor training, it is more applicable in the real-world.

2 Related Work

2.1 Behavioral Authentication

Lifestyle and behavioral authentication have been proposed as the fourth authentication method to be an alternative to the traditional knowledge-based authentication, possession-based authentication, and biometrics-based authentication [19]. Recently, these authentication methods have been actively studied due to the widespread of smart devices [1,9,16]. Previously, it was necessary to wear a unique device to collect personal behavior patterns, which was a significant barrier to real-world applications. However, with the widespread of devices equipped with a wide variety of sensors like smartphones and smartwatches, it has become possible to retrieve behavioral data without using dedicated devices and with little to no burden on the individual. These backgrounds have made real-world applications much easier.

The following section will describe existing behavioral authentication research in three aspects: authentication factors, data acquisition, and authentication result generation.

2.1.1   Authentication Factor

Existing behavioral authentications can be classified into three major types (biometrics, device operation, and location) based on the authentication factors they use.

Biometrics

Traditional biometrics-based authentication relies on dedicated sensors to read individual characteristics (such as fingerprints, irises, face, etc.) and perform the authentication. While these methods can efficiently and accurately authenticate individuals, they also require an enrollment process in advance and sometimes raise privacy concerns.

Unlike traditional biometrics-based authentication, behavioral authentication does not use the features that are generally considered time-invariant but the features that vary over time. This kind of information is not a feature that can identify an individual solely by itself, but it can be utilized to identify an individual if used over a certain period of time (e.g., one day). For example, existing studies focusing on this kind of information include those focusing on gait [2,9] and multiple activity-related data (calorie burn, step count, heart rate, etc.) [18].

Device Operation

Device operation-based authentication focuses on the difference in each individual's habit to use a device such as smartphones. For example, Monrose et al. [8] proposed a method to identify individuals based on individual differences in keystrokes on a computer keyboard. Later, with the spread of smart devices, many researchers began to apply this authentication method to smartphones, such as using individual differences in keystrokes on a software keyboard [12] and touch habits [13,14].

Location

Location-based authentication gathers various information directly or indirectly linked to the user's location, such as GPS and nearby Wi-Fi access points, and utilizes them for authentication. In general, human behavior shows regular patterns, such as visiting a specific place every day. Individual authentication is possible by extracting such behavioral patterns from location information. Such methods are often used as one of the authentication elements of lifestyle authentication because it regularly uses behavioral data over a relatively long time. As an example of location-based authentication, some studies focus on the trend of nearby Wi-Fi access points and use it to extract the user's behavior pattern for authentication [5, 6].

2.1.2   Data Acquisition

Behavioral authentication proposed so far can be roughly classified into two types based on data acquisition methods: (1) using the data acquired from a single device and (2) synthesizing the data acquired from multiple devices.

Data acquisition using a single device has been widely used before the widespread of smart devices and is a standard method in relatively old research, such as the research on gait authentication using a dedicated device attached to the ankle [2] as mentioned above. On the other hand, in the past few years, smartwatches and activity trackers equipped with accelerometers and heart rate sensors have become popular [3]. It is expected that by using sensors installed on these wearable devices, together with the data from sensors on smartphones, we will be able to perform behavioral authentication with better accuracy than with a single device.

Because of these backgrounds, an increasing number of behavioral authentication research focuses on utilizing the data from multiple devices. For example, Lee et al. [7] proposed a method using accelerometers and gyroscopes of both smartphones and smartwatches. In the research, they concluded that using the same type of sensor data from multiple devices for authentication significantly improved the authentication accuracy compared to using the data from a single device. Some methods extend [7] by incorporating machine learning into model generation to achieve authentication for multiple devices [20].

2.1.3   Result Generation

In behavioral authentication, behavioral data of individuals over a certain period of time is used for authentication. There are two major methods for generating authentication results in existing research on behavioral authentication.

The first method is to collect the preliminary data of the authentication target for a certain period of time, generate an authentication template for the individual based on the data, and then compare the template with new input for authentication. The advantage of such a template-based method is that it is relatively easy to apply in the real world because no other person's data is involved in generating the authentication template. However, since human behavior changes over time, there is a problem with updating the template once the template is generated.

The second method is to use supervised classification methods such as SVM and Random Forest to perform authentication by considering the generation of authentication results as a classification problem. This method generally has higher authentication accuracy than the template-based method and has the advantage of not requiring a dedicated template generation algorithm. Still, this classification-based method has several problems: (1) since it requires other people's data to authenticate, it is challenging to implement when it comes to real-world applications, (2) the result may be affected by a biased dataset, and (3) the result is unpredictable if the data of a third party who does not appear in the training dataset are supplied on authentication.

2.2 MITHRA Project

In this research, we use the data collected in the MITHRA (Multi-factor Identification/auTHentication ReseArch) project [15], a proof-of-concept experiment of lifestyle authentication conducted in our laboratory. This experiment was lasted for about three and a half months, from January 11, 2017, to April 26, 2017, and various behavioral data were collected from the experiment participants' smartphones. The data collected included device identifier and information (OS and version), IP addresses, BSSIDs of the nearby Wi-Fi access points, and GPS location collected from the experimental smartphone application (MITHRA application). Some participants also provided usage data from third-party applications in addition to the MITHRA application. The breakdown of participants in the experiment is as follows:

- MITHRA only: 5849 participants
- MITHRA and Manga One (Book App): 7582 participants
- MITHRA and Shofoo! (Flyer App): 2594 participants
- MITHRA, Manga One, and Shofoo!: 2 participants

When collecting information from the MITHRA application, the privacy policy page was displayed on the screen before the start of data collection, and data were collected only when the participant agreed to it. After the start of the experiment, participants were able to change data collection preferences at will, and no personal information other than mentioned above (e.g., height, age, weight, etc.) was collected.

In addition, about 100 participants wore an activity tracker (Omron Healthcare HJA-750C) to collect their daily activities. Most of the activity tracker data were collected in a form linked to the experimental data collected by the MITHRA application. Also, prior to the experiment, we obtained the appropriate permission from the university's ethics committee.

3   Our Proposed Method

In the proposed method, authentication is based on the correlation between different types of data collected from two devices: GPS and nearby Wi-Fi access point data from a smartphone (hereinafter referred to as "location data") and activity data collected from an activity tracker (hereinafter referred to as "activity data"). Specifically, the type of activity (no activity, household activity, walking) inferred from the activity data and the type of movement (stationary, walking, moving by vehicle) inferred from the location data are matched to identify whether the data belong to the authentication target or someone else.

In our method, only one day's worth of data is required for the authentication. This mitigates the problem of preparing preliminary data for a certain period of time, which existed in both the template-based method and classification-based method described in Sect. 2.1.3.

3.1   Location Data

The proposed method first determines the type of movement by calculating the user's speed of movement over a certain period of time from GPS location data (latitude and longitude information) obtained from a smartphone. We define the location data (longitude and latitude) measured at a specific time t_1 as $l_1(\text{lon}_{t_1}, \text{lat}_{t_1})$. The distance d [m] traveled from time t_1 to time t_2 can be calculated as follows (note that r is the equatorial radius $r = 6378137$ m).

$$d = r \arccos(\sin \text{lon}_{t_1} \sin \text{lon}_{t_2} + \cos \text{lon}_{t_1} \cos \text{lon}_{t_2} \cos(\text{lat}_{t_2} - \text{lat}_{t_1})) \quad (1)$$

The typical walking speed of a human is about 75 m per minute [4]. However, since the accuracy of GPS in smartphones often fluctuates, the location information may not be reflected immediately even if a person starts moving. In such a case, the user may appear to be moving at a very high speed or not moving at all, even though the user is actually walking. To suppress this problem, in this method, movements of between 5 and 300 m per minute are treated as walking, below that as stationary, and above that as moving by vehicle.

Also, to reduce the degradation of authentication accuracy even if the user turns off GPS for some reason or the lack of GPS data occurs due to OS limitations, we treat the data as walking if the data satisfies the following Eqs. (2) or (3) when the movement speed is less than 0.001 m and the device is not connected to Wi-Fi.

$$n(W_{t_1}) \neq 0 \wedge n(W_{t_2}) = 0 \quad (2)$$

$$n(W_{t_2}) \neq 0 \wedge \frac{n(W_{t_1} \cap W_{t_2})}{n(W_{t_2})} < 0.2 \quad (3)$$

where W_{t_1} and W_{t_2} denote the set of BSSIDs collected at time t_1 and t_2, respectively, and $n(W)$ denotes the number of elements in the set W. We set the threshold of Eq. (3) to 0.2 because it delivered the best combination of lowering the false rejection rate (FRR) and limiting the increase of the false acceptance rate (FAR) in the experiment.

3.2   Activity Data

In this method, the activity type derived from the acceleration data measured by the activity tracker is used for authentication. There is a method proposed by Ohkawara et al. [11] to estimate the activity type from acceleration data. The activity tracker used in this study (HJA-750C by Omron Healthcare) estimates the activity type by this method [10].

In their method, the activity type is estimated from the acceleration data by the following procedure. First, let ACC_{fil} be the composite acceleration of the three axes after passing through a high-pass filter with a cutoff frequency of $0.7\,Hz$, and ACC_{unfil} be the same composite acceleration without passing through the high-pass filter. Then, based on these composite accelerations, the activity type can be split into three types (no activity, household activity, walking) as follows.

- No activity: $ACC_{fil} < 29.9\,mG$
- Household activity: $ACC_{fil} \geq 29.9\,mG \,\wedge\, ACC_{unfil}/ACC_{fil} \geq 1.16$
- Walking: $ACC_{fil} \geq 29.9\,mG \,\wedge\, ACC_{unfil}/ACC_{fil} < 1.16$

In this study, we used the same method as [11] to identify the activity type.

3.3 Authentication

In the proposed method, the type of movement obtained by processing GPS and Wi-Fi data in Sect. 3.1 is used as input, and the final authentication is performed by comparing and calculating the correlation with the activity type obtained by the activity tracker in Sect. 3.2. In this section, we will describe the details of the concrete procedure.

3.3.1 Extraction of Walking Periods

We extract a series of walking periods from the activity data labeled with three types. First, $a_t \in \{stop, live, walk\}$ is defined as the activity type estimated from the activity tracker log at time t. Note that "stop" denotes no activity, "live" denotes household activity, and "walk" denotes walking. Next, we define $ex(a, t_1, t_2)$ as the maximum time range (between t_1 and t_2) in which activities other than $a \in \{stop, live, walk\}$ appear in succession.

In general, it is considered that there are small stops in one series of walking, such as waiting for a traffic light, so we define a series of walking $wp(t_1, t_2)$ from time t_1 to t_2 as the most extensive range of time that satisfies all of the following conditions.

$$a_{t_1}, a_{t_2} = walk \tag{4}$$

$$t_2 - t_1 \geq 5\,min \tag{5}$$

$$ex(walk, t_1, t_2) < 3\,min \tag{6}$$

In order to avoid unintentional recognition due to malfunction of the measurement device, the data used for authentication is limited to a series that lasts at least $5\,min$, as shown in Eq. (5).

3.3.2   Comparison with Location Data

The authentication is performed by comparing the walking period calculated in Sect. 3.3.1 and the movement type estimated from the GPS and Wi-Fi data (hereinafter referred to as GPS data for simplicity) calculated in Sect. 3.1.

First, we define $\mathrm{gcount}(t_1, t_2)$ as the number of samples of GPS data obtained during time $t_1 \leq t \leq t_2$. Walking period $\mathrm{wp}(t_1, t_2)$ is used for authentication only if the corresponding GPS data satisfies the following conditions shown in the equation below. Note that t_s is the latest GPS measurement time that satisfies $t_s \leq t_1$, and t_e is the oldest GPS measurement time that satisfies $t_e \geq t_2$, and $\mathrm{G_{int}}$ is a regular interval in which the smartphone records its location data.

$$\frac{t_e - t_s}{\mathrm{gcount}(t_s, t_e)} \leq 2\mathrm{G_{int}} \ [\mathrm{min}] \tag{7}$$

In our experiment, $\mathrm{G_{int}}$ is set to 5 min since the MITHRA application logged the location data every five minutes. The reason for such a restriction is to suppress misjudgments caused by excessively long GPS/Wi-Fi measurement intervals due to a malfunction of the experimental application.

The following condition is used to determine whether the inferred movement type by GPS and Wi-Fi at the measurement time $t_s \leq t \leq t_e$ coincides with the walking period $\mathrm{wp}(t_1, t_2)$, where v_{t_n} is the average speed of movement between GPS measurement time t_n and the next measurement time t_{n+1}.

$$\exists v_{t_n}, \ v_{\mathrm{low}} \ [\mathrm{m/min}] < v_{t_n} < v_{\mathrm{high}} \ [\mathrm{m/min}] \ (t_s \leq t_n < t_e) \tag{8}$$

where v_{high} and v_{low} are the upper and lower thresholds (constants) of the GPS movement speed at which the user is considered to be walking. These thresholds will differ according to each smartphone's GPS accuracy used in the experiment. In our experiment, we set $v_{\mathrm{high}} = 300$ and $v_{\mathrm{low}} = 5$, respectively, which produced the best EER result as described in Sect. 3.1.

Furthermore, there may be a case where the activity type calculated from the activity data and the movement type calculated from the location data are inconsistent, such as when someone else's data is given as an input. To lower the authentication score in such cases, the location data is considered inconsistent and penalized when the following conditions are satisfied when the activity type is other than walking. Note that $t_1' \leq t \leq t_2'$ is a period in which the activity type is other than walking, t_s' is the latest GPS measurement time that satisfies $t_s' \leq t_1'$, and t_e' is the oldest GPS measurement time that satisfies $t_e' \geq t_2'$. Besides, let $\mathrm{gcount_{ac}}(t_1, t_2)$ be the number of GPS measurement samples that satisfy Eq. (8) between t_1 and t_2.

$$\mathrm{gcount}(t_s', t_e') > \mathrm{C_{thr}} \ \wedge \ \frac{\mathrm{gcount_{ac}}(t_s', t_e')}{\mathrm{gcount}(t_s', t_e')} \geq \mathrm{W_{thr}} \tag{9}$$

where $\mathrm{C_{thr}}$ restricts the minimum count of continuous non-walking periods used for inconsistency detection, and $\mathrm{W_{thr}}$ is a threshold to determine whether the activity and location data contradict each other. These thresholds will also differ according to the interval of data collection and devices used in the experiment. In our experiment, we used $C_{thr} = 30$ and $W_{thr} = 0.4$ as they provided the best EER result.

In this method, one day's worth of data is used to determine the final authentication result. The purpose of this is to improve the authentication accuracy by using the multiple walking periods included in the entire day since human activities generally show a periodicity in one day.

If we define the total number of independent walking periods in a day as $wcount_{all}$, and the number of walking periods satisfying Eq. (8) minus the number of non-walking periods satisfying Eq. (9) as $wcount_{ac}$, the concordance rate P is calculated by the following equation.

$$P = \frac{wcount_{ac}}{wcount_{all}} \tag{10}$$

When the concordance rate exceeds a certain threshold λ, it is considered to be the same person, and when it falls below, it is considered to be someone else.

4   Experiment

4.1   Experimental Setup

In this experiment, among the behavioral data collected in the MITHRA project [15], we used the GPS/Wi-Fi data collected from the experiment participants' smartphones and the activity history data collected from activity trackers. The GPS/Wi-Fi data were collected at a five-minute interval (in an ideal case), containing the smartphone's latitude and longitude information and BSSIDs of the nearby Wi-Fi access points. The activity tracker's data is collected at a 1-minute interval, containing the amount of activity (METs) and the activity type (see Sect. 3.2). We used Python 3.9.1 and `pandas` library for data processing. The experiment was conducted on a computer with AMD Ryzen 7 1700X CPU (8 cores, 16 threads) and 64 GB of RAM.

4.2   Data Selection and Parameter Setting

From the subjects who participated in the MITHRA project, 64 subjects who participated in the experiment using both GPS/Wi-Fi history from their smartphones and activity trackers were selected. In those 64 subjects, 29 subjects who used Android devices and wore the activity tracker for more than 30 min a day were selected to be used in this experiment. The reason for restricting operating systems to Android is that it is difficult for iOS devices to regularly collect GPS/Wi-Fi information due to OS limitations, which often causes data loss. The limitation of the activity tracker is meant to extract only the data of the participants who wore the activity tracker for a certain period of time in order to evaluate the proposed method effectively. Also, as a reference, we applied the method to 14 subjects who used Android devices and wore the activity tracker for more than 180 min a day without GPS data loss to evaluate the authentication accuracy in more ideal conditions.

The FRR and FAR were calculated in each threshold λ from 0 to 0.99 in 0.01 steps. Also, only the days when $wcount_{all} \geq 3$ was satisfied were used for the authentication. The reason for this limitation of $wcount_{all}$ is that it would cause an unnecessary increase in the FAR if data with less than three walking periods per day were used.

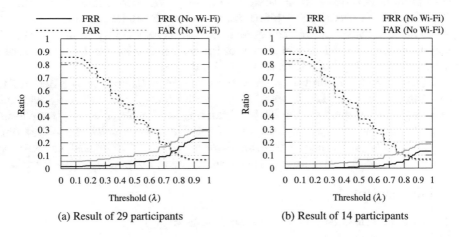

(a) Result of 29 participants   (b) Result of 14 participants

Fig. 1. Results of the experiment. In each result, "No Wi-Fi" means that only the GPS data (and activity history data) were used in the experiment.

Table 1. Result comparisons (Data Count shows the average number of preliminary data used per person. ACC denotes accuracy.)

Author	Method	Data Count	EER	FAR	FRR	ACC
Ours	Correlation	1728	8.7%	–	–	–
Gafurov et al. [2]	Template	–	6%	–	–	–
Kobayashi & Yamaguchi [5]	Template	8640	–	7.5%	9.9%	–
Sitová et al. [14]	Template	8000	7%	–	–	–
Muaaz et al. [9]	Template	3000	13%	–	–	–
Susuki & Yamaguchi [16]	Classification	–	–	–	–	89%
Lee et al. [7]	Classification	240000	–	7.5%	8.3%	–
Vhaduri & Poellabauer [18]	Classification	11250	–	–	–	93%
Fridman et al. [1]	Classification	1555200	2%	–	–	–
Monrose & Rubin [8]	Both	–	–	–	–	92%

4.3   Results

Figure 1(a) shows the authentication results of 29 participants. The black lines show the result of the proposed method, and the gray lines show the result without using Wi-Fi for comparison. We can see that the equal error rate (EER), i.e., the point where FRR and FAR are equal, is about 0.130, while the EER of the GPS-only method is about 0.172. When the GPS location information cannot be obtained correctly, our method, which also uses the nearby Wi-Fi access points, has a lower EER than the GPS-only method. Figure 1 also shows that using Wi-Fi access point information reduces FRR but increases FAR as a side effect. However, since the equal error rate is reduced by using the Wi-Fi data, it can be said that the accuracy of the authentication method as a whole is improved by using both GPS and Wi-Fi data.

Figure 1(b) shows the result with the ideal 14 participants (see Sect. 4.2). In an ideal case, our EER is about 0.087, while that of the GPS-only method is about 0.114. In general, the EER of existing studies for lifestyle/behavioral authentication is often around 0.1 (see Table 1). These results show that our method has the same level of authentication accuracy compared to existing research on lifestyle/behavioral authentication using templates or classification.

5   Discussion

5.1   Effectiveness of the Proposed Method

In our proposed method, authentication is based on the correlation between the GPS and Wi-Fi data of the user's smartphone and the activity tracker's data, implying the authentication accuracy may decrease if users wear only one of the devices or turn off the smartphone's GPS and Wi-Fi functions for some reason.

Still, based on the premise that the proposed method is intended to be one element of lifestyle authentication providing a convenient authentication method combined with other methods, and the final result of EER by our method are comparable to those of existing methods, it can safely be said that the proposed method has sufficient authentication accuracy.

5.2   Increasing Accuracy in Non-ideal Cases

In the case of users who often go out with only one device with them, it is not easy to improve the authentication accuracy of our method. On the other hand, even in the case of a user who wears both devices, it is still probable that the authentication accuracy will degrade if both GPS and Wi-Fi data cannot be acquired correctly for some reason.

In the dataset used in this study, accelerometer and gyroscope data, which can directly read the device's movement, were not included. However, there are existing studies on classification-based behavioral authentication using such data, and it is expected to be possible to perform a correlation-based authentication based on the characteristics of those data. By incorporating such data into our method, we can improve the authentication accuracy to some extent, even when GPS and Wi-Fi data cannot be obtained correctly.

5.3 Mitigation of Attacks

Our proposed method uses both the period when the activity tracker judges the user to be walking and the period when the activity tracker judges the user to be not walking. Since we lower the authentication score if the presented GPS or Wi-Fi data contradict the activity tracker's data, exploiting the authentication using simple fake data will be difficult. However, in our preliminary experiment, we found out that the FAR when fake data are given is still higher than when regular data are given. In this study, we tried to eliminate false data by imposing relatively simple penalties. We think that we can further eliminate false data by adding measures such as imposing larger penalties for longer discrepancies between activity tracker's data and GPS data. Also, since our method uses one day's worth of data for authentication, an attacker will be automatically authenticated if more than one day has passed since the attacker obtained both the smartphone and the activity tracker. This problem can be mitigated by using multi-factor authentication in combination with other lifestyle authentication methods.

6 Conclusions and Future Work

In this paper, we first described behavioral and lifestyle authentication as the fourth authentication method. We also pointed out that existing research on behavioral and lifestyle authentication requires a large amount of preliminary data, and few methods use correlations between authentication factors.

Based on these backgrounds, we proposed a new method for correlation-based lifestyle authentication using GPS/Wi-Fi data from smartphones and activity data from activity trackers. We conducted experiments using actual behavioral data of individuals collected in the MITHRA project, a demonstration experiment of lifestyle authentication. As a result, the equal error rates of the proposed method in both ideal and non-ideal cases were approximately 0.087 and 0.130, respectively. The results proved that the proposed method is applicable for a real-life lifestyle authentication.

As future work, we can consider combining other lifestyle authentication methods or acquiring accelerometer and gyroscope data from smartphones and utilizing them to further improve the authentication accuracy. For example, we can combine our method with existing lifestyle authentications using Wi-Fi information [5,6]. Although it is challenging to perform correlation-based authentication with these methods, it is expected that the FRR and FAR can be further improved by integrating the authentication scores of multiple authentication methods by score-level fusion and using them in the final authentication result.

It is also considered possible to use smartwatches equipped with more sensors to perform correlation-based authentication by combining other biometric data, such as heart rate, in addition to the amount and type of activity. In particular, the heart rate is considered to be closely related to an individual's behavioral pattern, and many existing studies use heart rate data obtained from smartwatches to identify individuals. We believe that combining these additional biometric data into the proposed method will also improve authentication accuracy.

References

1. Fridman, L., Weber, S., Greenstadt, R., Kam, M.: Active authentication on mobile devices via stylometry, application usage, web browsing, and GPS location. IEEE Syst. J. **11**(2), 513–521 (2017)
2. Gafurov, D., Helkala, K., Søndrol, T.: Biometric gait authentication using accelerometer sensor. J. Comput. **1**(7), 51–59 (2006)
3. Holst, A.: Smartwatch devices unit sales in the united states from 2016 to 2020. https://www.statista.com/statistics/381696/wearables-unit-sales-forecast-united-states-by-category/. Accessed 10 January 2021
4. Knoblauch, R.L., Pietrucha, M.T., Nitzburg, M.: Field studies of pedestrian walking speed and start-up time. Transp. Res. Rec. **1538**(1), 27–38 (1996)
5. Kobayashi, R., Yamaguchi, R.S.: A behavior authentication method using wi-fi BSSIDS around smartphone carried by a user. In: 2015 Third International Symposium on Computing and Networking (CANDAR), pp. 463–469 (2015)
6. Kobayashi, R., Yamaguchi, R.S.: One hour term authentication for wi-fi information captured by smartphone sensors. In: 2016 International Symposium on Information Theory and Its Applications (ISITA), pp. 330–334 (2016)
7. Lee, W.H., Lee, R.: Implicit sensor-based authentication of smartphone users with smartwatch. In: Hardware and Architectural Support for Security and Privacy 2016, HASP 2016, pp. 1–8. Association for Computing Machinery, New York (2016)
8. Monrose, F., Rubin, A.D.: Keystroke dynamics as a biometric for authentication. Future Gener. Comput. Syst. **16**(4), 351–359 (2000)
9. Muaaz, M., Mayrhofer, R.: Smartphone-based gait recognition: from authentication to imitation. IEEE Trans. Mobile Comput. **16**(11), 3209–3221 (2017)
10. Nakanishi, M., et al.: Estimating metabolic equivalents for activities in daily life using acceleration and heart rate in wearable devices. Biomed. Eng. Online **17**(1), 100 (2018)
11. Ohkawara, K., Oshima, Y., Hikihara, Y., Ishikawa-Takata, K., Tabata, I., Tanaka, S.: Real-time estimation of daily physical activity intensity by a triaxial accelerometer and a gravity-removal classification algorithm. Br. J. Nutr. **105**(11), 1681–1691 (2011)
12. Roh, J., Lee, S., Kim, S.: Keystroke dynamics for authentication in smartphone. In: 2016 International Conference on Information and Communication Technology Convergence (ICTC), pp. 1155–1159 (2016)
13. Shen, C., Li, Y., Chen, Y., Guan, X., Maxion, R.A.: Performance analysis of multi-motion sensor behavior for active smartphone authentication. IEEE Trans. Inf. Forensics Secur. **13**(1), 48–62 (2018)
14. Sitová, Z., Šeděnka, J., Yang, Q., Peng, G., Zhou, G., Gasti, P., Balagani, K.S.: HMOG: New behavioral biometric features for continuous authentication of smartphone users. IEEE Trans. Inf. Forensics Secur. **11**(5), 877–892 (2016)
15. Susuki, H., Kobayashi, R., Saji, N., Yamaguchi, R.S.: Lifestyle authentication social experiment -MITHRA project-. In: 2017 Symposium on Cryptography and Information Security, 4D2-1, pp. 1–8 (2017)
16. Susuki, H., Yamaguchi, R.S.: Cost-effective modeling for authentication and its application to activity tracker. In: Kim, H.W., Choi , D.(eds.) Information Security Applications, pp. 373–385. Springer, Cham (2016)

17. Thao, T.P., Irvan, M., Kobayashi, R., Yamaguchi, R.S., Nakata, T.: Self-enhancing GPS-based authentication using corresponding address. In: Singhal, A., Vaidya, J. (eds.) Data and Applications Security and Privacy XXXIV, pp. 333–344. Springer, Cham (2020)
18. Vhaduri, S., Poellabauer, C.: Wearable device user authentication using physiological and behavioral metrics. In: IEEE 28th Annual International Symposium on Personal, Indoor, and Mobile Radio Communications (PIMRC), pp. 1–6 (2017)
19. Yamaguchi, R.S., Nakata, T., Kobayashi, R.: Redefine and organize, 4th authentication factor, behavior. In: 2019 7th International Symposium on Computer and Networking Workshops (CANDARW), pp. 412–415 (2019)
20. Zhu, T., Qu, Z., Xu, H., Zhang, J., Shao, Z., Chen, Y., Prabhakar, S., Yang, J.: RiskCog: Unobtrusive real-time user authentication on mobile devices in the wild. IEEE Trans. Mobile Comput. **19**(2), 466–483 (2020)

Identify Encrypted Packets to Detect Stepping-Stone Intrusion

Jianhua Yang[✉], Lixin Wang, Suhev Shakya, and Michael Workman

TSYS School of Computer Science, Columbus State University, 4225 University Avenue, Columbus, GA 31907, USA
{yang_jianhua,wang_lixin,shakya_suhev,
workman_michael}@ColumbusState.edu

Abstract. Most attackers exploit stepping-stone to launch their attacks to avoid being captured. An encrypted TCP session established by attackers using ssh makes stepping-stone intrusion detection harder than non-encrypted sessions. Even though the contents of an encrypted packet are not readable, its header fields in different layers are not encrypted. In this paper, we propose a novel algorithm to detect stepping-stone intrusion based on IP address, port number, TCP packet flags, and the length of an encrypted packet. A preliminary experimental result in a local area network shows that the proposed algorithm cannot only detect stepping-stone intrusion, but also resist intruders' session manipulation.

1   Introduction

Most professional intruders launch their attacks by exploiting stepping-stones [1] because intruders can hide their identities in depth and escape from detection. The attacks by making use of stepping-stones are called stepping-stone intrusion. The approaches focusing on determining if stepping-stones are used to access remote hosts are called stepping-stone intrusion detection techniques. The idea to determine if a host is used as a stepping-stone is to compare the incoming and outgoing network traffic of a host. If the incoming network traffic of a host presents the similar behavior as the outgoing network traffic, the host is used as a stepping-stone and it is highly suspicious that the host is used by intruders to launch their attacks. The techniques to detect stepping-stone intrusion have been developed since 1995.

The first approach to detect stepping-stone intrusion was proposed by S. Staniford-Chen, and L. T. Heberlein [2] in 1995. They designed a thumbprint approach by hashing the contents of the packets captured in the connection of a host to describe the behavior of network traffic. By comparing the thumbprint of an incoming connection with the thumbprint of an outgoing connection of a host, it is trivial to decide if the host is used as a stepping-stone. The biggest disadvantage of this approach to detect stepping-stone intrusion is that it cannot apply to encrypted TCP sessions. Zhang and Paxson [1] proposed a time-based approach to detect stepping-stones or to trace back an intrusion even if the session is encrypted. However, there are three major problems in the time-based approach. First, it can be easily manipulated by intruders using chaff perturbation and/or

L. Barolli et al. (Eds.): AINA 2021, LNNS 226, pp. 536–547, 2021.
https://doi.org/10.1007/978-3-030-75075-6_43

time-jittering evasion. Second, the method requires that the packets of connections have precise and synchronized timestamps in order to correlate them properly. This makes it difficult or impractical to correlate the measurements those were taken at different points in a network. Third, Zhang and Paxson also were aware of the fact that a large number of legitimate stepping-stone users routinely traverse a network for a variety of reasons. Yoda and Etoh [3] proposed a deviation-based approach that is a network-based correlation scheme. This method is based on the observation that the deviation for two unrelated connections is large enough to be distinguished from the deviation of those connections within the same connection chain. In addition to the problems the time-based approach has, this method has other problems, such as not efficient and not applicable to compressed sessions and to padded payload.

Blum [4] proposed a packet number difference-based (PND-based) approach that detects stepping-stones by checking the difference of Send packet numbers between two connections. This method is based on the idea that if two connections are relayed, the difference between the two connections should be bounded; otherwise, it should not. This method can resist intruders' evasions such as time jittering and chaff perturbation. D. Donoho et al. [5] showed for the first time that there are theoretical limits on the ability of attackers to disguise their traffics using evasions during a long interactive session. The major problem with the PND-based approach is due to the fact that the upper bound on the number of packets required to monitor is large, while the lower bound on the amount of chaff an attacker needs to evade the detection is small. This fact makes Blum's method very weak in resisting to intruders' chaff evasion.

X. Wang, et al. [6–8] conceived an approach using watermarks to decide relayed connections. Injecting a watermark to a TCP/IP session may result in lots of computations, thus making their approach inefficient. Another issue is that the injected watermarks may be manipulated by intruders. T. He and L. Tong proposed an algorithm DBDC (DETECT-BOUNDED-MEMORY-CHAFF) [9] for detecting stepping-stone intrusion with bounded memory or bounded delay perturbation. It is stated that DBDC can deal with chaff evasion and tolerate a number of chaff packets proportional to the size of the attacking traffic. Their study shows that an intruder needs to insert at least $\frac{n}{1+\lambda\Delta}$ chaff packets in every n packets to evade DBDC detection if the packets delay is bounded by Δ. This tells us the chaff-rate of DBDC is $\frac{1}{1+\lambda\Delta}$, where λ is a parameter of a Poisson distribution which indicates the expected number of occurrences during a given time interval. It is obvious that a smaller λ and Δ would make DBDC tolerate more chaff, but would also make DBDC have a high false alarm probability for a wide range of normal traffic.

The common issue of the above approaches to detect stepping-stone intrusion is that their capability to resist intruders' manipulation is more or less limited. In this paper, in order to improve resistance performance to stepping-stone intrusion, we propose a novel approach by making use of the features of encrypted packets. After analyzing encrypted packets captured in an OpenSSH session, we observed that the contents of a packet at application layer are encrypted. The fields of a packet header in different layers are not encrypted. The most important observation is that the length of a single character in transport layer keeps the same regardless of the content of the character. Some header fields remains the same between two OpenSSH connections if the two connections are relayed which means the host is used as a stepping-stone. The length of an encrypted packet can also be used to detect stepping-stone since it also keeps the same regardless the encryption key and the character content. If we combine both the header fields and the length of an encrypted packet, we can not only detect stepping-stone intrusion, but also resist intruders' manipulation. Our experimental results in a local area network show this approach can improve resistance performance to intruders' manipulation.

2 Preliminaries

Most professional intruders make a connection chain using remote accessing tool, such as OpenSSH, to launch their attacks. Figure 1 shows such kind of connection chain where Host 0 is used by an attacker, Host N is used by the victim, and all other hosts in between are used as stepping-stones. We can detect an intrusion at any one of the stepping-stone hosts. We assume that the detection occurs at Host i. In the following, we will introduce some preliminaries needed in this paper.

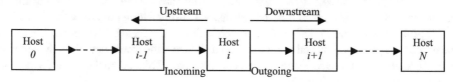

Fig. 1. An attacking connection chain

2.1 Relayed Sessions

As shown in Fig. 1, Host i is used as a stepping-stone of the connection chain. Detection program runs at Host i which is called a sensor. The connection from Host $i - 1$ to Host i is called the incoming connection of Host i. Similarly, the connection from Host i to Host $i + 1$ is called the outgoing connection of Host i. If an incoming connection and an outgoing connection of Host i belong to the same TCP connection chain, they are called relayed sessions. Detecting if a host is used as a stepping-stone is actually to find if there exist relayed sessions. We already knew that the way to determine a relayed session pair is to see if the same packet appears on both incoming and outgoing connections. For an encrypted session, the content of a packet cannot be observed. So comparing the header fields of the encrypted packets from incoming and outgoing connections respectively would be a possible way to help us to find a relayed session pair. In the following section, we will discuss the header fields of a packet in different layers to see which one could be used to detect stepping-stone intrusion.

2.2 Packet Header and its Fields

A packet captured in different layers can present different headers. Either the existing packet sniffing tools, such as Wireshark and TCPdump, or self-developed packet capturing programs using WinPcap (MS Windows related Operating System (OS)) or Libpcap

(Unix/Linux related OS) can only sniff packets in transport layer and below. So we do not consider any packet header fields in application layer for our detection algorithm.

The header fields in transport, network, and datalink layers are in our consideration to design an algorithm to detect stepping-stone intrusion. In transport layer, some header fields can be used to determine a relayed session pair since they remain unchanged from an incoming connection to an outgoing connection of the same host. These fields include destination port, source port, and TCP flags. Sequence number, acknowledgement number, offset, window size, checksum, urgent pointer and options cannot be used to detect stepping-stone intrusion since they are packet/session oriented. In network layer, none of the header fields of a packet can be considered since they remain either the same or packet/session oriented except the source IP address, destination IP address, and Total Length fields which will be discussed in Sect. 2.4. Similarly, in data link layer, even though there are many header fields such as destination MAC address, source MAC address, protocol type, or CRC code, none of them is considered in our algorithm design.

2.3 Packet Encryption

We normally use OpenSSH to access and operate a server remotely. Most intruders also use this tool to launch attacks to a remote server. The difference is that a regular user may access a remote server directly. However, a malicious user may indirectly access a remote server via a long connection chain. One biggest advantage is that OpenSSH can provide a secured communication. Each packet sent from the first host (intruder's host) of a connection chain to the last host (victim's host) is encrypted with different keys in different connections. In order to understand the proposed algorithm to detect stepping-stone intrusion, we first introduce how a packet is encrypted and delivered in a TCP session chain, which includes multi connections.

What shown in Fig. 1 is a session chain established using OpenSSH which is composed of N connections from Host 0 to Host N with each connection having its own encryption key. We assume Host 0 is used by an intruder, and Host N is the victim. When a connection is established, an encryption key is selected. For example, when the intruder makes a connection from Host 0 to Host 1, the system asks the user to input the password for authentication, and if it is the first time connecting to Host 1 from Host 0, the system also reminds the user to click "Yes" to accept a public key which is for a session key distribution. As long as the user connects to Host 1 successfully, an encryption key is generated and distributed from Host 0 to Host 1. Apparently, different connections have different encryption keys. For convenience, we assume the encryption keys for the connections Host 0 to Host 1, Host 1 to Host 2, ..., Host $i-1$ to Host i, Host i to Host $i+1$, ..., Host $N-1$ to Host N are $EnpK_{0,1}$, $EnpK_{1,2}$, ..., $EnpK_{i-1,i}$, $EnpK_{i,i+1}$, ..., $EnpK_{N-1,N}$ respectively. When attacks are launched, each packet is sent from Host 0 to Host 1 encrypted with $EnpK_{0,1}$, from Host 1 to Host 2 encrypted with $EnpK_{1,2}$, ..., form Host $N-1$ to Host N encrypted with $EnpK_{N-1,N}$. In each stepping-stone host, each packet is received from its incoming connection. Then the packet is decrypted and re-encrypted on its outgoing connection. We cannot obtain the content of each packet at each stepping-stone due to the secured design of OpenSSH.

OpenSSH is an application layer program which provides secured communication over an unsecured network channel. It lays in between application layer and transport layer. A packet in application layer can be encrypted first, then passed to transport layer and be encapsulated into a segment. Each segment contains a transport layer header and payload which is the encrypted application layer message. A transport layer segment can be passed to network layer and be encapsulated into a datagram which contains a network layer header and payload which is the transport layer segment. Similarly, a network layer datagram can be the payload of a datalink layer frame. A captured packet using a sniffing tool can show the three headers and the encrypted payload. Figure 2 shows the header and payload information of a captured packet using TCPdump in a computer with a Linux OS. We can see the IP addresses, port numbers, TCP flags, sequence and acknowledgement numbers, windows buffer size and the packet length in transport layer. In Fig. 2, it clearly shows the datalink header (yellow), IP header (red), and TCP header (pink). The application layer header and packet contents are not readable because they are encrypted.

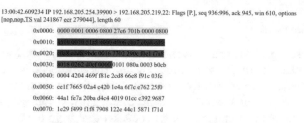

13:00:42.609234 IP 192.168.205.254.39900 > 192.168.205.219.22: Flags [P.], seq 936:996, ack 945, win 610, options [nop,nop,TS val 241867 ecr 279044], length 60

0x0000: 0000 0001 0006 0800 27e6 701b 0000 0800
0x0010:
0x0020: b8a8 edd9 9bde 0016 7702 290c f0e1 f7a8
0x0030: 8018 0262 d0ef 0000 0101 080a 0003 b0cb
0x0040: 0004 4204 469f f81e 2cd8 66e8 f91c 03fe
0x0050: ce1f 7665 02a4 c420 1e4a 6f7c e762 25f0
0x0060: 44a1 fe7a 20ba d4c4 4019 01cc c392 9687
0x0070: 1e29 f499 f1f8 7908 122e 44c1 5871 f71d

Fig. 2. A captured packet via TCPdump

2.4 The Length of Encrypted Packets

We are interested in the length of a TCP packet since this length remains the same under the same encryption algorithm regardless of the encryption keys and the characters. We found that for a certain encryption algorithm used in OpenSSH, the length of an encrypted string depends on not the content of the string, but the number of the characters in the string. This is significant because this feature can be used to detect stepping-stone intrusion. Before we go further to discuss our novel detection algorithm, we need to explain the length of encrypted packets in detail.

In order to simplify our discussion, we use a simplified model of stepping-stone intrusion as shown in Fig. 3 where the stepping-stone host has two connections: the incoming connection C_{in} and the outgoing connection C_{out}.

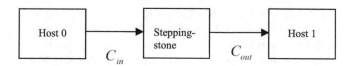

Fig. 3. A simplified model of stepping-stone

To verify the idea, we run OpenSSH to connect to the Stepping-stone host from Host 0 and connect out to Host 1 to make two connections as shown in Fig. 3. We type some characters to generate some packets at Host 0. The packets will be delivered to Stepping-stone via C_{in} and forwarded to Host 1 via C_{out}. An encryption algorithm is used in the two connections in which obviously encryption keys are different. All the three hosts run Ubuntu OS. We use TCPdump to capture all the packets coming from Host 0 at Stepping-stone, and also all the packets going to Host 1 from Stepping-stone. Our goal is to check the lengths of the packets captured at the incoming and outgoing connection of Stepping-stone, respectively. In order to make grouped characters into one packet, we use copy and paste. We typed 26 characters separately, as well as grouped characters, such as "ab", "abc", "abcd", and so on. The IP address, port number, flags, timestamp, and the total length of each packet are recorded. Table 1 shows the header fields of the packets captured at the incoming and outgoing connection of Stepping-stone, respectively. From the header information of the packets captured, we can get the IP addresses of Host 0, Stepping-stone, and Host 1 are 192.168.205.254, 192.168.205.219, and 192.168.205.236, respectively. Host 0 uses a port number 34652 sending packets to the SSH server in Stepping-stone. Stepping-stone uses a port number 55824 forwarding packets received from Host 0 to the SSH server of Host 1. If you check the timestamps of the packets from the incoming and outgoing connections, you would see the a little bit time lag.

Table 1. Captured packets comparison between the incoming and outgoing connection

Character(s)	Incoming connection	Outgoing connection
x	14:27:10.936370 IP 192.168.205.254.34652 > 192.168.205.219.22: Flags [P.], length 36	14:27:10.936712 IP 192.168.205.219.55824 > 192.168.205.236.22: Flags [P.], length 36
y	14:27:18.461314 IP 192.168.205.254.34652 > 192.168.205.219.22: Flags [P.], length 36	14:27:18.461411 IP 192.168.205.219.55824 > 192.168.205.236.22: Flags [P.], length 36
a	12:57:52.262636 IP 192.168.205.254.39900 > 192.168.205.219.22: Flags [P.], length 36	12:57:52.262809 IP 192.168.205.219.58534 > 192.168.205.236.22: Flags [P.], length 36
abc	12:58:15.767057 IP 192.168.205.254.39900 > 192.168.205.219.22: Flags [P.], length 44	12:58:15.767348 IP 192.168.205.219.58534 > 192.168.205.236.22: Flags [P.], length 44
abcdefghijk	12:59:00.740578 IP 192.168.205.254.39900 > 192.168.205.219.22: Flags [P.], length 52	12:59:00.740870 IP 192.168.205.219.58534 > 192.168.205.236.22: Flags [P.], length 52
abcdefghijklmnopqrs	13:00:37.931633 IP 192.168.205.254.39900 > 192.168.205.219.22: Flags [P.], length 60	13:00:37.932061 IP 192.168.205.219.58534 > 192.168.205.236.22: Flags [P.], length 60

From the results in Table 1, we conclude the following. 1) Different single characters have the same length of its encrypted packet. 2) Grouped characters (string) may have different lengths for its encrypted packet depending on the length of the string. 3) The same number of characters result in the same length of encrypted packet at the incoming and outgoing connections, respectively. Single characters 'x', 'y', and 'a' have the same length of encrypted packet 36. When the length of the string is in between 3 and 10, the length of the encrypted string is 44. It is easy to see from Table 1 that the encrypted string with the number of characters in between 11 and 18 has length of 52. When we increased the length of the string to be 19, the length of the encrypted packet becomes 60. We did not try more characters of a string, but we believe it would be changed as long as the number of the characters in a string reach a certain degree. We will discuss how this result will be used to detect stepping-stone intrusion in Sect. 4.

3   Reading Encrypted Packets

In order to design an algorithm and make a program to detect stepping-stone intrusion by inspecting each packet captured, it is necessary to understand each field of the header of a packet and know how to read the header information from an encrypted packet. In this section, we use an example of a captured packet to demonstrate what each binary number in the header exactly means. Before proceeding to the discussion, it is worth to mention that, when capturing packets using TCPdump, please specify the network interface. Otherwise, the destination and source MAC addresses would not be obtained because TCPdump captures packets in "Cooked" mode which has a different type of frame header from the standard.

In this experiment, we captured a packet from a computer host with MAC address 00:0c:29:3d:e7:e0 and IP address 192.168.1.115. The packet was sent from a computer host with MAC address dc:a6:32:98:0b:16 and IP address 192.168.1.103 via a ssh connection. In Fig. 4, it shows the captured packet in binary format. The first six bytes '000c293de7e0' from 0x0000 to 0x0005 represent the destination MAC address of the host we used to capture the packet. The second six bytes 'dca632980b16' represent the source MAC address of the host from which the packets were sent out. The next two bytes "0800" represent Ethernet type: Ethernet Type II.

```
0x0000:  000c 293d e7e0 dca6 3298 0b16 0800 4510
0x0010:  0058 648c 4000 4006 51d9 c0a8 0167 c0a8
0x0020:  0173 c3ce 0016 d2e2 e709 6086 c8ac 8018
0x0030:  01f5 edf4 0000 0101 080a 346e 1eab 9a5d
0x0040:  4cb2 0513 c261 54f7 2571 96cc 9e47 88ad
0x0050:  d637 912c 5302 3151 c750 65e3 85fb 8b3c
0x0060:  9a10 6b07 f273
```

Fig. 4. A captured packet in binary number format

The first number of the second last byte "4" indicates the IP version, and the second number "5" tells us the network IP header length = 5 * 32 bits = 160 bits = 20 bytes. The last byte "10" (in binary bits: 00010000) represent the Type of Service (ToS) with its first three bits "000" known as precedence bits, the next 4 bits (1000) indicate the ToS of "Minimize Delay", and the last bit is left unused.

The first two bytes of the second row starting from 0x0010 "0058" (Hex number) stand for the total length of the IP packet, which is 88 bytes in decimal. So the bytes from 0x000E (byte '45') to 0x0065 (byte '73') form the whole IP packet include the IP header and payload. Its first 20 bytes from 0x000E '45' to 0x0021 '73' denote the IP header fields. The second two bytes '648c' represent the identification number of the IP packet. If the packet is fragmented to smaller parts during transmission, all the fragments will share the same Identification number. The first byte of the third two bytes '40' (0100 0000) is the fragmentation flag and offset byte. The second bit '1' is set to represent "don't fragment", the third bit '0' shows no more fragment since the packet is not fragmented. The second byte of the third two bytes '00' is the fragmentation offset in case of a fragmented datagram. The first byte of the fourth two bytes '40' is the TTL field (Time To Live) indicating the number of the hops the packet is allowed to pass through in the network. The second byte of the fourth two bytes '06' represents the transport layer protocol "TCP" used. The next two bytes '51d9' is the header checksum field. The next four bytes 'c0a80167' represent the source IP address 192.168.1.103, and following four bytes 'c0a80173' tell us the destination IP address 192.168.1.115.

The rest part starting from 0x0022 'c3' to the last byte 0x0065 '73' is the TCP packet which is also the payload of the IP packet. The first two bytes 'c3ce' is the source port number, and the second two bytes '0016' is the destination port number 22 in decimal which is exactly the known ssh server port. The next four bytes "d2e2e709' show the sequence number, and the following four bytes "6086c8ac" represent the acknowledgement number. The first 4 bits of the next byte "80" (1000 0000) show the TCP header length and the remaining 4 bits are reserved. The TCP header length expresses a 32-bit word which must be multiplied by 4 to calculate the total byte value. So we can get 8 * 4 = 32 bytes TCP header length. The last byte '18" (Hex number, 00011000 in binary) is the TCP flag bits which represent "CWR, ECE, UR, ACK, PSH, RST, SYN and FIN" with each bit in order. The fourth and fifth bits are set to show this is a packet carrying data, as well as acknowledging the previously received packets. The first two bytes of the row 0x0030 '01f5' is the window size field. The second two bytes 'edf4' is the TCP checksum field, and the third two bytes '0000' represent the Urgent pointer. This field is 0 since the flag Urgent bit is not set.

We all know that the minimum TCP header length is 20 bytes. Therefore, we can calculate the TCP options length for this packet is $32 - 20 = 12$ bytes. The TCP packet payload is 12 bytes after the urgent pointer '0000' which is from the byte 0x0034 to the byte 0x0035. The bytes '0101 080a 34e6 1eab 9a5d 4cb2' are the TCP options. The payload bytes are from 0x0042 byte '05' to 0x0065 byte '73' (the last green piece in Fig. 4).

4 Detecting Stepping-Stone Intrusion

The way to determine if a host is used as a stepping-stone is to see if there is a relayed connection pair for the host. If we monitor an enterprise level server, we would see tons of incoming connections, as well as some outgoing connections. Most servers provide services to their clients. This means, in most cases, the connections to enterprise servers are not necessary to connect out of the servers. If we can find a ssh session connecting to a server also connects out to another host, it is highly suspicious that the server is used as a stepping-stone by an attacker to launch attacks to other hosts. So stepping-stone intrusion detection needs to check all the incoming connections of a server and all the outgoing connections of the server and compare them to see if there exists any relayed connections. In this paper, we only focus on a model as shown in Fig. 5 in which the server has only one incoming connection and one outgoing connection. The goal is to apply what we have discussed in the above to the two connections to determine if the host H1 is used as a stepping-stone.

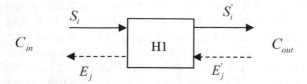

Fig. 5. A simplified model of stepping-stone with only one incoming and outgoing connection

As shown in Fig. 5, the host H1 has an incoming connection C_{in} and an outgoing connection C_{out}. We run TCPdump at H1 to capture the packets from the connections C_{in} and C_{out} respectively. When capturing packets, it is important to be aware of that the source IP and destination IP address cannot be used in TCPdump filter because before we start capturing, we actually do not know which host connects to H1 and to which host H1 will connect. We also do not know the source port number used to connect to H1, as well as the port number H1 uses to connect out to a remote host. However, we can use destination port number 22 since they are all ssh connection, as well as the IP address of H1 which can be used for the incoming connection, as well as the outgoing connection.

When we only use the lengths of captured packets to determine a stepping-stone, it is possible to introduce false-positive errors. The reason is that if two different users type the same command at their hosts respectively, it is hard to tell the packets captured at H1 are actually coming from two different hosts. In order to fix this issue, we identify the packets captured at H1 as two different categories: Send and Echo. When an intruder or a user types any Unix/Linux command, the packet of each character from each keystroke can arrive at the host H1 via its incoming session, and be forwarded to the host connected by the outgoing connection. Finally, each packet invoked from the intruder's host comes to the end host, a victim of the connection chain. We call this type of packet a Send packet. For each received Send packet, the victim host echoes the request, and the echoed response can go back to the intruder's host along the connection chain reversely. We call this type of packet an Echo packet. We have explained the concept of Send and Echo packets. But how can we identify them technically?

The way to identify if a captured packet is a Send or an Echo is to make use of both the packets flow direction and the TCP header flags. As shown in Fig. 5, either the incoming connection or the outgoing connection has Send and Echo packets. In the incoming connection of H1, a Send is defined as a packet that its destination IP is the IP address of H1, the destination port number is 22, and its TCP flags have PSH bit set up. An Echo is defined as a packet that the source IP address is the IP address of H1, the source port number is 22, and its TCP flags have PSH bit set up. In the outgoing connection of H1, a Send is defined as a packet that its source IP is the IP address of H1, the destination port number is 22, and its TCP flags have PSH bit set up. An Echo is defined as a packet that its destination IP address is the IP address of H1, the source port number is 22, and its TCP flags have PSH bit set up. Using these rules, it is trivial to write a TCPdump filter, or make our own program to capture not only the packets sent from H1, but also distinguish them to be Send and Echo packets. As shown in Fig. 5, in the incoming connection, the Send packet is denoted as S_i, and the Echo packet as E_j.

Similarly, in the outgoing connection, the Send packet is denoted as S_i', and the Echo packet as E_j'. The significance to introduce Send and Echo packets is that the length of each Send packet from different users may be the same, but as long a command is executed at a victim host, the length of each Echo packet should be different since they are echoed from different servers, which would result in different size of response packets.

We put the captured packets from the incoming connection of the host H1 into a packet stream $\{p_{in}\}$ with m packets, as well as the ones from the outgoing connection into $\{p_{out}\}$ with n packets. If the two connections are relayed, m is either equal or close to n. We check each packet in the input stream $\{p_{in}\}$, and decide not only the type of each packet (Send or Echo) but also its length. We obtain the following sequence $C_{in} = \{(S_i/E_j, \text{Len})\}$ for the input stream $\{p_{in}\}$, as well as a sequence $C_{out} = \{(S_i'/E_j', \text{Len})\}$ for the output stream $\{p_{out}\}$, here S_i/E_j means either S_i or E_j.

The next step is to compare the similarity between the two sequences C_{in} and C_{out}. We use the comparison algorithm proposed by S. Wu in 1990 [10] to compute the similarity between two sequences. We assume there are two sequences A with length M, B with length N, as well as $N \geq M$ without loss of generality. The way to compare the two sequences A and B is to either find the longest common subsequence (LCS) or find a shortest edit script (SES). Here it is supposed we are going to find SES. An edit script is to edit one sequence to another one via delete/insert actions in which "delete" action specifies which character in sequence A to delete and "insert" action specifies which character in sequence B to insert. A SES is an edit script whose length is the minimum among all possible edit scripts that edit sequence A to sequence B. The similarity between two sequences can be defined as the number of delete/insert actions needed to edit A to B. for the details of sequence comparison, please refer to the paper [10].

We summarize the proposed algorithm to detect stepping-stone intrusion as the following.

Input: Similarity threshold Δ
Step1: Capture packets using TCPdump and put them into a stream $\{p_{in}\}$ or $\{p_{out}\}$
Step2: Apply the header information of each packet to convert each stream into a sequence C_{in} or C_{out}
Step3: Call sequence comparison algorithm to obtain the similarity β between the two sequences
Step4: if $\beta > \Delta$, stepping-stone intrusion is detected

If a connection is manipulated by an intruder, such as chaff-perturbation, it will not affect the similarity between the two sequences because it does not change that sequence C_{in} is the part of C_{out}, or C_{out} is the part of C_{in}. Here we use a fact that is an intruder can only chaff a connection, but cannot remove any packet from a connection. So this detection algorithm can resist intruders' manipulation. In the following section, we will use some experimental results completed in a local area network to verify the performance of this algorithm.

5 Experimental Results and Analysis

We conducted some experiments in a local area network of our department collecting network traffic data to verify the performance of the proposed algorithm. Our goal is to justify if the algorithm not only detects stepping-stone intrusion, but also resists intruders' chaff evasion.

In the lab, there are 30 computers with each installed Windows 10 OS. Under Windows 10 OS, we installed virtual Ubuntu Linux OS under which we have ssh client and server installed. We pick three computers as an intruder's host (Intruder), a stepping-stone (Sensor), and a victim's host (Victim). We make a connection chain from Intruder to Sensor, then from Sensor to Victim by running command "ssh UserName@host-IP". After input the password for each host and pass the authentication, a connection chain is established. The follow Linux commands were typed at Intruder to make some packets: ls, date, mkdir test_folder, cd test_folder, vi test_file.txt, more test_file.txt, find –name test_file.txt, chmod –v u = rwx, g = rx, o = r test_file.txt, lsmod, ps, rm test_file.txt, cd .., rmdir test_folder, last –w.

We collect all the packets at Sensor using TCPdump, put them into two streams $\{p_{in}\}$, $\{p_{out}\}$ respectively. Each stream is converted to $\{(\text{Send (Echo)}, \text{Len})\}$ sequence. We obtain a sequence $\{(S_i/E_j, \text{Len})\}$ containing 380 elements for the incoming connection C_{in}, and $\{(S_i'/E_j', \text{Len})\}$ containing 380 elements for the outgoing connection C_{out}. If we compare the two sequences, it is obviously their similarity is 100% since we made Sensor as a stepping-stone host in our experiment. We inserted 10%, 20%, 30% and 40% packets into either C_{in} or C_{out}, and compare the two sequences. We still obtained 100% similarity since either sequence is the sub-sequence of the inserted sequence. The experiment results showed that the proposed algorithm cannot only detect stepping-stone intrusion, but also resist intruders' chaff perturbation manipulation.

6 Conclusion

In this paper, after analyzing packet encryption, and the header fields of a packet in different layers, we found the length of a packet can be used to detect stepping-stone intrusion since it remains the same from an incoming connection to an outgoing connection of a host. Based on this idea, we propose a novel algorithm to detect stepping-stone intrusion by making use of header fields of an encrypted packet in different layers. The experiment results show that the proposed algorithm not only detects stepping-stone intrusion, but also resists intruder's manipulation. Due to time limit, we did not conduct the experiments in the Internet to test the false-negative and false-positive errors of the algorithm since this needs at least 10 students to work together to make multiple connection chains. This will be our future work.

Acknowledgments. This research has been funded by NSA grant H98230-20-1-0293.

References

1. Zhang, Y., Paxson, V.: Detecting stepping-stones. In: Proceedings of the 9th USENIX Security Symposium, Denver, CO, pp. 67–81 (2000)

2. Staniford-Chen, S., Heberlein, L.T.: Holding intruders accountable on the internet. In: Proceedings of IEEE Symposium on Security and Privacy, Oakland, CA, pp. 39–49 (1995)
3. Yoda, K., Etoh, H.: Finding connection chain for tracing intruders. In: Proceedings of 6th European Symposium on Research in Computer Security, Toulouse, France, Lecture Notes in Computer Science, vol. 1985, pp. 31–42 (2000)
4. Blum, A., Song, D., Venkataraman, S.: Detection of interactive stepping-stones: algorithms and confidence bounds. In: Proceedings of International Symposium on Recent Advance in Intrusion Detection, Sophia Antipolis, France, pp. 20–35 (2004)
5. Donoho, D.L.: Detecting pairs of jittered interactive streams by exploiting maximum tolerable delay. In: Proceedings of the 5th International Symposium on Recent Advances in Intrusion Detection, Zurich, Switzerland, pp. 45–59 (2002)
6. Wang, X., Reeves, D.S., Wu, S.F., Yu, J.: Sleepy watermark tracing: an active network based intrusion response framework. In: Proceedings of the 16th International Conference on Information security, USA, pp. 369–384 (2001)
7. Wang, X., Reeves, D.S.: Robust correlation of encrypted attack traffic through stepping stones by manipulation of inter-packet delays. In: Proceedings of the 10th ACM Conference on Computer and Communications Security, USA, pp. 20–29 (2003)
8. Wang, X.: The loop fallacy and serialization in tracing intrusion connections through stepping stones. In: Proceedings of the 2004 ACM Symposium on Applied Computing, USA, pp. 404–411 (2004)
9. He, T., Tong, L.: Detecting encrypted stepping-stone connections. IEEE Trans. Signal Process. 55(5), 1612–1623 (2007)
10. Wu, S., Manber, U., Myers, G., Miller, W.: An O(NP) sequence comparison algorithm. Inf. Process. Lett. 35(6), 317–323 (1990)

Visual Paths in Creation of User-Oriented Security Protocols

Marek R. Ogiela[1]([✉]) and Lidia Ogiela[2]

[1] Cryptography and Cognitive Informatics Laboratory, AGH University
of Science and Technology, 30 Mickiewicza Ave., 30-059 Kraków, Poland
mogiela@agh.edu.pl
[2] Pedagogical University of Krakow, Podchorążych 2 St., 30-084 Kraków, Poland

Abstract. In modern security systems very often are used personal features, what allow to define user-oriented security protocols. Among such procedures we can consider approaches based on application of eye movements or blinking actions analyzed using eye tracking sensors. In this paper will be described such solutions dedicated for security purposes, and creation of personal security locks using visual paths. In particular new procedures will be defined, which will be based on visual cognitive skills, and possibilities of defining semantic commands during evaluation of visual patterns.

Keywords: Personalized cryptography · User-oriented security systems · Eye tracking technologies

1 Introduction

In advanced security technologies we can use cognitive approaches or user perception abilities in creation of security protocols. Many authentication procedures allow to involve different personal features or biometric parameters in creation user-oriented cryptographic application or personal security locks [1, 2]. Also, some multimedia interfaces can be applied as sensors, which register movements, gestures or personal characteristics required for security purposes. In our research we proposed cognitive-based security approaches, in which visual sequences called visual paths can be implemented. In such solution we must use cognitive systems, which allow to generate a thematic sequence in the form of visual patterns representing selected area of interest [3, 4]. Such sequences can be presented to the user, who according his knowledge and perception skills should quickly find the answer for several questions, connected with presented patterns. Having knowledge from this area he can find the proper selection of visual patterns and remotely select them using eye tracking devices.

2 Cognitive Systems in Cryptography

The main idea of cognitive information systems lays on application of cognitive resonance processes, which, allow to imitate the natural way of human thinking. Such

© The Author(s), under exclusive license to Springer Nature Switzerland AG 2021
L. Barolli et al. (Eds.): AINA 2021, LNNS 226, pp. 548–551, 2021.
https://doi.org/10.1007/978-3-030-75075-6_44

processes, and its application for pattern classification were described in [1]. Also, application of such systems for creation a new branch of modern cryptography called cognitive cryptography was presented in [5]. Now we can also try to apply such systems in creation of thematic-based visual paths oriented for user authentication. Application of such systems allow to create a very specific protocols, which consider users' expertise knowledge, thematic preferences, perception abilities, or external world details during security procedure.

Cognitive information systems try to imitate the natural way of human thinking [1]. In such processes there is a comparison of real feature or parameters registered by sensors with some expectations stored in databases. During this comparison hypothesis verification is performed, and decision if it is similar with expectation generated during resonance processes is made. Cognitive information systems are based on one of the models of human perception i.e. knowledge-based perception. According this model human mind cannot recognize properly any pattern or situation if we haven't any knowledge or previous experiences in recognizing such pattern. So, based on knowledge-based perception model we can implement several classes of cognitive information systems. Such systems connected with multimedia user interfaces like eye-tracking devices allow to construct efficient tools for moving attention analysis and measure perception skills.

3 Visual Paths in Security Protocols

In this section will be described solutions based on visual patterns, which for security purposes require specific information from selected area of interest. In such visual paths it is possible to check if user is a human (like in CAPTCHA codes) or if he belongs to trusted group [6, 7]. In such protocols it is necessary to quickly and correctly select the semantic combination of visual elements, from presented group of visual patterns. Selection should follow requested goals or fulfil requirements about semantic meaning. For example, we can select all or several of correct parts of visual patterns presenting specific information, and dependent on the asked questions. It is possible to specify very simple or more complex answers, which required basic or specific information about presented patterns. During visual analysis user should move his attention and concentrate on particular elements selected by blinking with his eyes [8, 9]. Such signal can be registered by eye-tracker sensors, what allow to trace the moving attention actions and create answers in the form of visual paths.

Verification of users can be done in two different manners. The first solution lays on checking user's information, which require selection in proper order of visual patterns depending on the verification questions. The second solution lays on following the question sequences connected with single pattern, containing several elements, which should be determined according semantic path. In this procedure on each stage a question connecting with different visual element should be asked. Questions can be understood only by users having knowledge related to the content of patterns and understanding relation between particular elements presented on the whole image.

In Fig. 1 is presented example presenting several block ciphers [10]. When user recognize them, can be asked about encryption keys length. In subsequent iterations he can be asked for grouping procedures, performing particular number of encryption iteration or based on Feistel schema etc. In next stage he can be asked to select procedures according time from the oldest one to the newest one.

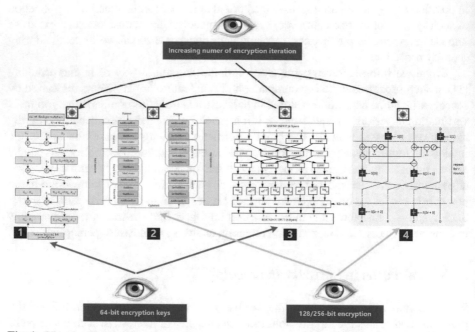

Fig. 1. Visual paths based on thematic sequence connected with encryption procedures. A set of ciphers are presented: 1-DES, 2-AES, 3-SAFER, 4-RC6. User oriented thematic paths can relate to selection of particular encryption procedures or sorting them according any features like key lengths (64-bits, 256-bits), number of iterations (8, 10, 12, 14, 16 etc.), complexity etc.

In such procedures it is possible to determine if user possess specific knowledge from particular areas of interest, but also verify if he can understand the questions, and is able to create right pattern sequences reflecting semantic meaning of the asked question.

4   Conclusions

In this paper we've described new security verification procedures in the form of visual paths. Such procedures can be implemented with application of cognitive information systems and eye tracking devices. Cognitive systems allow to generate thematic sequences of visual patterns, which can be recognized by users using eye tracking sensors. Presented procedures can be implemented in user-oriented security protocols, considering different levels of expertise knowledge possessed by particular users' group. During verification it also allows to check perception skills, by checking how users can understand visual pattern sequences [11, 12]. Application of eye tracking devices allow to introduce mobile security systems, working in different places and considering not only users knowledge but also external features observed by users in surrounding world.

Acknowledgments. This work has been supported by the AGH University of Science and Technology research Grant No 16.16.120.773. This work has been supported by the National Science Centre, Poland, under project number DEC-2016/23/B/HS4/00616.

References

1. Ogiela, M.R., Ogiela, L.: Cognitive keys in personalized cryptography. In: IEEE AINA 2017 - The 31st IEEE International Conference on Advanced Information Networking and Applications, Taipei, Taiwan, March 27–29, pp. 1050–1054 (2017)
2. Ogiela, M.R., Ogiela, L., Ogiela, U.: Biometric methods for advanced strategic data sharing protocols. In: 2015 9th International Conference on Innovative Mobile and Internet Services in Ubiquitous Computing IMIS 2015, pp. 179–183 (2015). https://doi.org/10.1109/imis.2015.29
3. Ogiela, U., Ogiela, L.: Linguistic techniques for cryptographic data sharing algorithms. Concurr. Comput. Pract. Exp. **30**(3), e4275 (2018). https://doi.org/10.1002/cpe.4275
4. Ogiela, M.R., Ogiela U.: Secure information splitting using grammar schemes. In: New Challenges in Computational Collective Intelligence, Studies in Computational Intelligence, vol. 244, pp. 327–336. Springer, Berlin (2009)
5. Ogiela, L., Ogiela, M.R., Ogiela, U.: Efficiency of strategic data sharing and management protocols. In: The 10th International Conference on Innovative Mobile and Internet Services in Ubiquitous Computing (IMIS-2016), July 6–8, Fukuoka, Japan, pp. 198–201 (2016). https://doi.org/10.1109/imis.2016.119
6. Alsuhibany, S.: Evaluating the usability of optimizing text-based CAPTCHA generation. Int. J. Adv. Comput. Sci. Appl. **7**(8), 164–169 (2016)
7. Osadchy, M., Hernandez-Castro, J., Gibson, S., Dunkelman, O., Perez-Cabo, D.: No bot expects the DeepCAPTCHA! introducing immutable adversarial examples, with applications to CAPTCHA generation. IEEE Trans. Inf. Forensics Secur. **12**(11), 2640–2653 (2017)
8. Ancheta, R.A., Reyes Jr., F.C., Caliwag, J.A., Castillo, R.E.: FEDSecurity: implementation of computer vision thru face and eye detection. Int. J. Mach. Learn. Comput. **8**, 619–624 (2018)
9. Guan, C., Mou, J., Jiang, Z.: Artificial intelligence innovation in education: a twenty-year data-driven historical analysis. Int. J. Innov. Stud. **4**(4), 134–147 (2020)
10. Menezes, A., van Oorschot, P., Vanstone, S.: Handbook of Applied Cryptography. CRC Press, Waterloo (2001)
11. Yang, S.J.H., Ogata, H., Matsui, T., Chen, N.-S.: Human-centered artificial intelligence in education: seeing the invisible through the visible. Comput. Educ. Artif. Intell. **2**, 100008 (2021)
12. Ogiela, L.: Cryptographic techniques of strategic data splitting and secure information management. Pervasive Mob. Comput. **29**, 130–141 (2016)

SeBeST: Security Behavior Stage Model and Its Application to OS Update

Ayane Sano$^{(\boxtimes)}$, Yukiko Sawaya, Akira Yamada, and Ayumu Kubota

KDDI Research, Inc., 2–1–15, Ohara, Fujimino-shi 356–8502, Saitama, Japan
{ay-sano,yu-sawaya,ai-yamada,kubota}@kddi-research.jp

Abstract. To protect computers from various types of cyberattack, users are required to learn appropriate security behaviors. Different persuasion techniques to encourage users to take security behaviors are required according to user attitude toward security. In this paper, we first propose a Security Behavior Stage Model (SeBeST) which classifies users into five stages in terms of attitude toward security measurements; having security awareness and taking security behaviors. In addition, we focus on OS updating behaviors as an example of security behaviors and evaluated effective OS update messages for users in each stage. We create message dialogs which can promote user OS updating behaviors. We conduct two online surveys; we analyze the validity of SeBeST in Survey1 and then evaluate effective messages for each stage in Survey2. We find that SeBeST has high validity and appropriate messages for the users in each stage differ from one another.

1 Introduction

Cyberattacks today are many and varied, and damages by these cyberattacks are increasing. To protect a computer from such attacks, users are required to learn appropriate security behavior as well as deploy technical solutions. Security notices are usually common to all users; however, awareness of security notices differ from user to user [1]. Therefore, different persuasion techniques to encourage the users to take security behaviors are required according to user attitude toward security. In the health field, a transtheoretical model of behavior change (TTM) [2] has been commonly used. This model classifies users into five stages in terms of attitude toward health. Many approaches [2, 3] using TTM have been proposed, and these are practical for behavior change. It is important to understand user awareness in each stage and approach each stage accordingly.

In this paper, we propose a Security Behavior Stage Model (SeBeST) that references TTM. This model classifies users into five stages (precontemplation, contemplation, preparation, action and maintenance) in terms of attitude toward security measurements, i.e., having security awareness and taking security behaviors. By approaching users in each stage accordingly, the number of users who take security behaviors by themselves is expected to increase. Additionally, we focus on OS updating behaviors as an example of security behaviors and create message dialogs which can promote user OS updating behaviors.

L. Barolli et al. (Eds.): AINA 2021, LNNS 226, pp. 552–566, 2021.
https://doi.org/10.1007/978-3-030-75075-6_45

We conducted two online surveys; we verified the validity of SeBeST in Survey1 and then evaluated effective OS update messages for each stage in Survey2. In Survey1, these questionnaires consisted of three items: SeBeST, awareness and attitude toward security behaviors (AASB) and status of taking each security behavior (SSB). We verified the validity of SeBeST by using correlation analysis between AASB, SSB and SeBeST. We evaluated the reliability of SeBeST by using Cronbach's α. As a result of Survey1, we classified participants into five stages by using SeBeST and verified the high validity and reliability of SeBeST. In Survey2, these questionnaires consisted of three items: SeBeST, decision making after viewing each message and impression of each message. We analyzed effective OS update message dialogs for users in each stage and the reason why they selected the messages by comparing the proposed message and the original message. As a result of Survey2, we found appropriate OS update messages for users in the stage of preparation, action and maintenance differ from one another. For example, the message indicating the ease of OS update is suitable for users in the preparation stage. The message indicating the advantage of OS update is not suitable for users in the action and maintenance stages.

2 Related Work

In this section, we describe related works on the transtheoretical model of behavior change (TTM), framing effect and security awareness.

2.1 Transtheoretical Model (TTM)

Prochaska et al. [2] proposed a transtheoretical model of behavior change (TTM). It has been commonly used in the health field, and consists of five stages (precontemplation, contemplation, preparation, action and maintenance). Precontemplation is the stage in which people do not intend to take action within the next six months. Contemplation is the stage in which people intend to change in the next six months. Preparation is the stage in which people intend to take action in the next month. Action is the stage in which people have changed their life styles within the last six months. Maintenance is the stage in which people maintain the same styles at more than six months after changing their life style. In general, people move up one stage at a time, but sometimes go back stages. Therefore, it is important to move up at the maintenance stage and stay at the maintenance stage. Additionally, there are many factors that affect behavior change. First, there are ten processes of behavior change, which differ at each stage. Second, there is decision making balance, which is affected advantages and disadvantages against oneself by taking action. Finally, there is self-efficacy [4] which users recognize that they can take action by themselves. The user who has high self-efficacy tends to promote action. Therefore, it is important to understand user awareness in each stage and approach each stage accordingly.

2.2 Framing Effect

Tversky et al. [5] proposed the framing effect that affects decision making. The framing effect has different outcomes from two messages: a loss framed message (LFM) and a gain framed message (GFM). LFM is the message that is disadvantageous and GFM is the message that is advantageous. Latimer et al. [6] showed LFM and GFM regarding physical activity for users; GFM is more appropriate than LFM for users. McNeil et al. [7] conducted an experiment on the framing effect. They asked the subjects to imagine that they had lung cancer and to choose between surgery and radiation therapy by viewing two messages (LFM, GFM). As a result, LFM was relatively less attractive than GFM regarding surgery. Takemura [8] conducted experiments regarding the influence of framing effects on a medical decision in two elaboration conditions (low-elaboration condition and high-elaboration condition). In the high-elaboration condition, subjects were asked to think about the justification of their decisions and were told that after making each decision they had to write down the content of the justification. As a result, in the low-elaboration condition, some subjects preferred the riskless option for GFM and some subjects preferred the risky option for LFM. However, in the high-elaboration condition, there is no significant difference between percentages of risky and riskless choices for LFM and GFM. Therefore, the framing effect differs in terms of the type of behavior and user awareness.

2.3 Security Awareness

Many researches have been conducted on security awareness which affects security behavior. Wash et al. [1] conducted a survey on the relationship of security beliefs on computer security. As a result of the survey, different demographic segments of populations are likely to respond differently to persuasive and educational messages. Metalidou et al. [9] showed the relationship of human factors on information security. Any human factors (lack of motivation and awareness, risky belief and so forth) are related to security behaviors. Glaspie et al. [10] surveyed many researches on human factors. They showed models of the relationship between human factors, such as "information security policy", "attitudes and involvement", "deterrence and incentives", "training and awareness", "management support" and "information security". Das et al. [11] proposed security sensitivity which are the awareness of, motivation to use and knowledge of how to use security privacy tools. The result of conducting the interviews is that "observed friends", "social sensemaking" and "prank/demonstration" affect behavior change. These factors show the influence of others, effectiveness against security behavior and experience of security damage. Inglesant et al. [12] conducted an experiment on password policies. As a result, users comply with policies in terms of cost. Therefore, these factors which are awareness, motivation, knowledge obtained from others and literatures, effectiveness and costs of security behaviors affect security behaviors.

3 Proposed Method

To encourage users to take secure behaviors, it is important to classify users into the stage model and approach users at each stage accordingly as described in related work

[2]. However, the stage model of security behaviors has not been proposed. In this paper, we propose a Security Behavior Stage Model (SeBeST) referred to as TTM. SeBeST classifies users into five stages (precontemplation, contemplation, preparation, action and maintenance) in terms of having security awareness and taking security behaviors. After verifying the validity of SeBeST, we focus on OS updating behaviors as an example of security behaviors and evaluate effective OS update messages for users in each stage. We create message dialogs which can promote user OS updating behaviors. These messages include the framing effect as described in Sect. 2.2 and security awareness as described in Sect. 2.3.

3.1 Security Behavior Stage Model (SeBeST)

We define a Security Behavior Stage Model (SeBeST). The same as TTM, SeBeST classifies users into five stages (precontemplation, contemplation, preparation, action and maintenance) in terms of attitude toward security measurements, i.e., having security awareness and taking security behaviors. In the case of security behavior, it is difficult for users to recognize a period that has been defined in TTM such as one month, six months, so we do not include the period. We define the stage of SeBeST as follows.

1. Precontemplation: people do not have interest in security behaviors, do not understand them.
2. Contemplation: people have interest in security behaviors (or/and users understand them), but do not want to take them
3. Preparation: people want to take security behaviors, but do not take them now.
4. Action: people sometimes take security behaviors, but not continuously.
5. Maintenance: people take security behaviors continuously.

3.2 Messages of OS Update Dialog

We focus on OS updating behaviors, and create message dialogs of OS update message. Figure 1 shows an example of original messages. We propose 20 types of messages in Table 1. These messages contain the original message, messages based on the framing effect (LFM, GFM), and security awareness. The original message is a message that we usually see when we update OS. LFM shows the disadvantages which occur by not updating OS (e.g., security damage and reduced usability). GFM shows the advantages which occur by updating OS (e.g., security effect and improved usability). Security awareness consists of five factors: "Increase Security Effect (ISE)", "Increase Security Interest (ISI)", "Receive Other Request (ROR)", "Recognize About Contribution (RAC)" and "Decrease Security Cost (DSC)". 20 types of messages have the same user interface as that in Fig. 1. In fact, these messages are shown in the Japanese language.

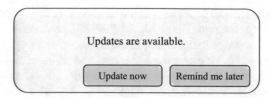

Fig. 1. An example of OS update message dialog. This dialog has two buttons ("Update now" and "Remind me later"). If the user pushes "Update now", OS will be updated now. On the other hand, if the user pushes "Remind me later", the user will show this dialog later.

Table 1. Category and contents of proposed messages. These messages include contents of the original message.

Category	Content
Original	Updates are available
LFM-1	Original+ "If you do not update, you may be infected by ransomware and not be able to access data in your computer."
LFM-2	Original+ "If you do not update, you may be infected by ransomware and need money to use your computer."
LFM-3	Original+ "If you do not update, others may access your computer dishonestly and leak your personal data."
LFM-4	Original+ "If you do not update, others may access your computer dishonestly and damage people around you."
LFM-5	Original+ "If you do not update, you can not use new functions of usability."
GFM-1	Original+ "If you update, you can solve security problems and improve security."
GFM-2	Original+ "If you update, you can use good user interfaces and improve usability."
GFM-3	Original+ "According to the NISC ^{a}, if you always update OS, your computer will be safe."
GFM-4	Original+ "If you always update, you will gain a discount of 10% off the license fee."
ISE-1	Original+ "If you update, you can defend yourself from various attacks."
ISE-2	Original+ "According to recent research, more than 60% of people have incurred security damages in the past one year."
ISI-1	Original+ "According to recent research, about 90% of people update now."
ISI-2	Original+ "It is important for you to update now because the attacks are sophisticated."
ROR-1	Original+ "According to the office or school, you need to update now."
ROR-2	Original+ "Your friend who is knowledgeable about security advises you to update now."
RAC-1	Original+ "If you update, you can defend not only yourself but also others from security damage."
RAC-2	Original+ "If you update, the rate of incurring security damage decreases by 50%."
DSC-1	Original+ "If you update once at a month, your computer will be safe."
DSC-2	Original+ "If you only reboot your computer, you can update in five minutes."

^{a}NISC is the abbreviation for National center of Incident readiness and Strategy for Cybersecurity.

4   Evaluation

In this section, we verify the validity of SeBeST and suitable messages for users in each stage. To do so, we conducted two online surveys by using the services of Macromill [13] from February 21st to February 26th 2020. Macromill is a Japanese research company which has a pool of about 10 million participants in Japan. In Survey1, we analyzed validity and reliability of SeBeST. In Survey2, we analyzed effective OS update messages for users in each stage. We analyzed the two surveys by using the HAD [14] for statistical analyses.

4.1 Survey1: Methodology

We recruited 1748 participants who were selected from registrants of Macromill. The method of selection is as follows. First, we extracted a pool of 15000 participants from age 15 to 69 and calculated the rate of population composition of computer user. Second, we extracted 2579 participants from this pool to correspond to the rate of population composition of computer user. Finally, we removed the participants who are taking security behavior by others and the participants who answered questions about user device differently between preliminary research and this research.

Table 2. The distribution of participants (1748 users) in Survey1.

Age	15–29	30–39	40–49	50–59	60–69
Male	131	163	235	260	372
Female	63	83	119	129	193

The distribution of participants is shown in Table 2. In this research, we did not collect any information which could identify the participants. The questions are consisted of three items as follows.

1. SeBeST as described in Sect. 3.1
2. Awareness and attitude toward security behaviors (AASB) on the scale of the degree of precontemplation, contemplation, and preparation
3. Status of taking each security behavior (SSB) on the scale of the degree of action and maintenance and created based on the Refined Security Behavior Intensions Scale (RSeBIS) [15]. RSeBIS examined the end-user's security behavior.

We used the answer of above questions as observed variables. To verify the validity of questionnaires, we used exploratory factor analysis of SeBeST and AASB which we newly created. We used observed variables which meet the recommended condition, 0.350 for factor loading and 0.160 for communality, respectively. As a result, we used all observed variables; SeBeST is consisted of one factor, and AASB is consisted of three factors (AASB-precontemplation, AASB-contemplation and AASB-preparation). Then, the average of observed variables in one factor is called the "factor average" in this paper. To verify the validity of SeBeST, we used correlation analysis to determine the relationship between AASB and SeBeST, and between SSB and SeBeST. To verify the reliability of SeBeST, we used Cronbach's α.

Moreover, we compared the factor average between users, and verified the user classification method of SeBeST accordingly.

4.2 Survey1: Results

We used exploratory factor analysis for four questions of SeBeST (Appendix: SeBeST-1~SeBeST-4) and the result is shown in Table 3. As shown in Table 3, four questions are consisted of one factor and meet the recommended condition, 0.350 for factor loading and 0.160 for communality, respectively. Cronbach's α is 0.808 and meet the recommended condition of more than 0.6 [16].

Table 3. The result of exploratory factor analysis for four questions of SeBeST.

SeBeST	Factor loading	Communality
SeBeST-1	.703	.494
SeBeST-2	.717	.514
SeBeST-3	.766	.587
SeBeST-4	.689	.475

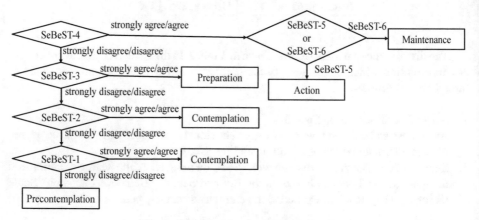

Fig. 2. Classification method of SeBeST.

The classification method of SeBeST is shown in Fig. 2. First, the users who answer "strongly agree/agree" at SeBeST-4 choose SeBeST-5 or SeBeST-6. We classify participants into two stages by using two questions (SeBeST-5, SeBeST-6). For example, in the case of the action stage, the user's answer is "SeBeST-5 is true". Next, we classify participants into four stages by using four questions (SeBeST-1~SeBeST-4). For example, in the case of the preparation stage, the user's answer is "SeBeST-4 is strongly disagree/disagree" and "SeBeST-3 is strongly agree/agree". As a result, we classified 1748 participants into five stages in Fig. 2; the precontemplation stage had 116 participants, the contemplation stage had 35 participants, the preparation stage had 138 participants, the action stage had 680 participants, the maintenance stage had 779 participants. Moreover, we set the classification of SeBeST (from precontemplation stage to maintenance stage) as 1~5 and used correlation analysis: between AASB-precontemplation and SeBeST is $r = -.440^{**}(p < .01)$, between AASB-contemplation and SeBeST is $r = .476^{**}(p < .01)$, between AASB-preparation and SeBeST is $r = .434^{**}(p < .01)$, between SSB and SeBeST is $r = .431^{**}(p < .01)$. Therefore, there are correlations between all items.

Table 4. The factor average between users. The vertical axis of the table shows each stage. The horizontal axis of the table shows AASB and SSB.

	AASB-precontemplation	AASB-contemplation	AASB-preparation	SSB
Precontemplation	5.60	9.91	8.34	69.55
Contemplation	5.74	12.31	9.91	85.57
Preparation	3.43	11.96	11.96	85.52
Action	2.91	14.92	12.62	97.18
Maintenance	1.50	17.00	14.08	108.76

Table 4 shows the factor average between users. In Table 4, it shows the higher the average, the greater the suitability of AASB or SSB. In the case of the precontemplation stage and the contemplation stage, AASB-precontemplation is the same level, but in terms of AASB-contemplation, the contemplation stage is higher than the precontemplation stage. This is because users in the contemplation stage have both AASB-precontemplation and AASB-contemplation, but the level of AASB-contemplation is higher than that of AASB-precontemplation. Moreover, we found that the higher the stage, the higher the level of AASB and SSB. We used t-test, and there are significant differences between stages expect for the following three cases. In the case of AASB-precontemplation between the precontemplation stage and the contemplation stage, in the case of AASB-contemplation between the contemplation stage and the preparation stage and in the case of SSB between the contemplation stage and the preparation stage, there are no significant differences. However, as most of the stages have significant differences between them, we verified the validity and reliability of the stage model.

4.3 Survey2: Methodology

We analyze effective OS update messages as described in Sect. 3.2. We recruited 2418 participants (only Windows PC users) who were selected from registrants of Macromill. The method of selection is as follows. First, we extracted a pool of 21000 participants from age 15 to 69 and calculated the rate of population composition of computer users. Second, we extracted 3611 users from this pool to correspond to the rate of population composition of computer users. Finally, we removed the users who take security behavior by others.

The distribution of participants is shown in Table 5. In this research, we did not collect any information which could identify the participants. The questions are consisted of three items as follows, shown in the Appendix.

1. SeBeST as described in Sect. 3.1
2. Decision making after viewing each message in Table 1
3. Impression of each message as shown in Table 1, i.e., importance, disadvantage, necessity, advantage, ease, effect.

Table 5. The distribution of participants (2418 users) in Survey2.

Age	15–29	30–39	40–49	50–59	60–69
Male	205	264	376	364	450
Female	113	114	178	162	192

We classified participants into five stages by using Fig. 2 to analyze effective OS update messages for users in each stage. We think it is important that users update now after viewing the message, so we compared the ratio of those who answered "I update OS now" in Question 2. Therefore, we calculated the ratio of those who answered "I update OS now" in Question 2 for each message and used t-test to compare 19 messages (excluding the original message) with the original message. If there is a significant difference between two messages, the message in which the ratio of those who answered "I update OS now" is higher than another message is an effective message. Moreover, we analyzed the reason why some people answered "I update OS now". We used chi-square test or Fisher's exact test (in the case that total frequency is less than 40 or expectation frequency is less than 5) as test for independence of each question in Question 3. If there is a significant difference between two messages, the message in which the ratio of those who answered each question is higher than another message is the reason why some people answered "I update OS now".

4.4 Survey2: Results

We classified 2418 participants into five stages as shown in Fig. 2. The precontemplation stage had 153 participants, the contemplation stage had 38 participants, the preparation stage had 172 participants, the action stage had 1112 participants, the maintenance stage had 943 participants. There were few participants in the contemplation stage, so we did not analyze the effective OS update message for users in that stage. Table 6 shows the result of t-test regarding the ratio of those who answered "I update OS now" in Question 2. In this paper, the level of significance is 5%. If the p-value is less than 0.05, the value is statistically significant. If the value is statistically significant and positive, the message for comparison is an effective message, but if the value is statistically significant and negative, the message for comparison is a backfire message which decreases people who update OS. In Table 6, these are no effective and backfire messages for users in the precontemplation stage, however, LFM-1~LFM-4, GFM-3, ISE-1, ISI-1, ISI-2, ROR-1, DSC-1 and DSC-2 are effective for users in the preparation stage. For users in the action stage, LFM-1~LFM-4, ISI-2 and DSC-2 are effective, but GFM-2, GFM-4 and ROR-2 are backfire. Moreover, for users in the maintenance stage, LFM-1~LFM-4, GFM-1, ISE-1 and ISI-2 are effective, but GFM-4, ROR-2 and RAC-1 are backfire. LFM-3 is the most effective messages for users in preparation, action and maintenance stages. We conducted a test for independence of each question in Question 3 after we extracted the users that answered "I update OS now" in Question 2. As a result, impression of these messages is almost nothing for users in the precontemplation stage. For users in the preparation stage, most of them answered "I recognize importance of security behaviors" for LFM-1~LFM-4, however, most of them did not answer "I can update OS easily" for effective messages (LFM-1~LFM-4, ISI-2, ROR-1). For users in the action stage, most of them answered "I can update OS easily" for DSC-2. Moreover, in the case of answering "I recognize importance of security behaviors", "I recognize disadvantages (security risks) which occur by not updating OS" and "I need to update OS", these messages are effective (LFM-1~LFM-4, ISI-2). However, in the case of answering "I recognize advantages which occur by updating OS", these messages are backfire (GFM-2, GFM-4). For users in the maintenance stage, in the case of answering one of the following four items: "I recognize importance of security behaviors", or "I recognize disadvantages (security risks) which occur by not updating OS" or "I need to update OS" or I recognize effect of security behaviors", these messages are effective (LFM-1~LFM-4, GFM-1, ISE-1, ISI-2). The same as users in the action stage, in the case of answering "I recognize advantages which occur by updating OS", these messages are backfire (GFM-4, ROR-2, RAC-1).

Table 6. The result of t-test regarding the ratio of those who answered "I update OS now" in Question 2. It shows average of the original message and difference between the message for comparison and the original message. The effective messages are shown in red; the backfire messages are shown in blue.

SeBeST	Precontempla-tion	Preparation	Action	Maintenance
original	0.085	0.029	0.108	0.146
comparison	difference	difference	difference	difference
LFM-1	-0.026	0.087	0.050	0.056
LFM-2	-0.007	0.081	0.055	0.037
LFM-3	0.000	0.128	0.074	0.060
LFM-4	-0.007	0.093	0.047	0.048
LFM-5	0.026	0.041	0.014	0.000
GFM-1	0.007	0.012	-0.001	0.029
GFM-2	0.000	0.023	-0.018	-0.019
GFM-3	0.007	0.058	0.004	-0.004
GFM-4	-0.033	0.012	-0.039	-0.058
ISE-1	-0.007	0.093	0.019	0.028
ISE-2	-0.020	0.047	-0.004	-0.022
ISI-1	-0.026	0.047	-0.017	-0.015
ISI-2	-0.013	0.116	0.050	0.054
ROR-1	-0.026	0.058	-0.001	-0.005
ROR-2	-0.013	-0.006	-0.043	-0.060
RAC-1	-0.013	0.029	-0.014	-0.023
RAC-2	-0.020	0.029	-0.001	-0.016
DSC-1	-0.013	0.076	-0.002	0.004
DSC-2	0.000	0.081	0.021	-0.001

5 Discussion

In this section, we discuss the validity of SeBeST, appropriate messages for users in each stage and future work of our research.

In Sect. 4.2, we described the validity and reliability of the stage model. SeBeST is an appropriate method for classifying users in terms of attitude toward security measurements. From the result of classification, maintenance users account for 44.6% in Survey1, and 39.0% in Survey2. Thus, we believe two approaches are required: advancing to next stage for users in the precontemplation, contemplation, preparation and action stages, and maintaining stage for users in the maintenance stage.

In Sect. 4.4, we stated that there are no suitable messages and impression of messages for users in the precontemplation stage. Therefore, as the approach of displaying messages is not suitable for them, other approaches are required. However, the approach of displaying messages is suitable for users in the preparation, action and maintenance stages. Our findings are that the rate of updating OS is higher by viewing appropriate messages for users in each stage as shown in Table 6. For example, LFM-1~LFM-4, ISI-2 are appropriate for users in the stages of preparation, action and maintenance. We find that LFM-1~LFM-4 are effective and LFM-5 is not effective, so it is important to teach users about security damage. In addition, GFM-3, ISI-1, ROR-1 and DSC-1 are suitable only for users in the preparation stage, GFM-1 is suitable only for users in the maintenance stage. Moreover, impressions of the same messages differ from user to user in each stage. Therefore, it is important to display appropriate messages for users in each stage.

Finally, we describe the future works of this research. We conducted two online surveys only for computer users. However, these related works [17, 18] state that user intention and behavior differ in terms of user device such as computer and smartphone. We believe the population ratio of SeBeST for smartphone users is different to the result in this paper. In survey2, we recruited only Windows PC users because Windows PCs have been commonallly used in Japan [19]. There is a possibility that the appropriate messages may differ from users who use different devices. Moreover, user interfaces affect decision making [20], so we will discuss the appropriate user interfaces of dialogs. Additionally, this research shows user intention, but not user behavior. We will conduct demonstration experiments of displaying OS update messages to participants and verify the rate of updating OS. After that, we will consider another application of SeBeST in addition to OS Update.

6 Conclusion and Future Research

We propose a Security Behavior Stage Model (SeBeST), which classifies users into five stages (precontemplation, contemplation, preparation, action and maintenance) in terms of having security awareness and taking security behaviors. We designed two online surveys to analyze the validity and reliability of SeBeST and effective OS update messages for users in each stage. As a result, we verified the high validity and reliability of SeBeST and found that appropriate OS update messages differ from one another. For example, LFM-1~LFM-4, GFM-3, ISE-1, ISI-1, ISI-2, ROR-1, DSC-1 and DSC-2 are suitable for users in the preparation stage. In future works, we will study the population ratio of SeBeST for users who use different devices with the same procedure in this paper. In addition, we will discuss appropriate user interfaces for users in each stage and conduct demonstration experiment.

Acknowledgments. The research results have been achieved by WarpDrive: Web-based Attack Response with Practical and Deployable Research InitiatiVE, the Commissioned Research of National Institute of Information and Communications Technology (NICT), JAPAN.

Appendix: User Survey Questionnaire

Survey1

SeBeST (4point Likert Scale: From "strongly agree" to "strongly disagree").

1. I am interested in security behaviors.
2. I understand security behaviors.
3. I want to take security behaviors.
4. I take security behaviors by myself.

SeBeST (in the case of 4/ "strongly agree" or "agree", 2point Likert Scale: "true" or "false")
5. I sometimes take security behaviors by myself.
6. I continuingly take security behaviors by myself.

AASB (5point Likert Scale: From "strongly disagree" to "strongly agree")
Precontemplation

1. I have nothing to do with security behavior.
2. Security threats are other people's concern.
3. I have no interest in security behavior.

Contemplation

1. I understand about security threats.
2. I have heard about security behaviors.
3. I understand the content of security behaviors.
4. I understand why taking security behaviors are required.
5. I understand the method of taking security behaviors.

Preparation

1. I need to take security behavior.
2. Bad things happen to me if I do not take security behavior.
3. There are advantages for me and the people around me if I take security behavior.
4. It is important to take security behavior.

SSB (5point Likert Scale: From "I have no interest in it" to "I always take it")

1. Lock my computer (password, fingerprint and so on)
2. Set a strong password when I create an account and set a password for the internet services if the website requirement is low
3. Use different passwords for different accounts that I have
4. Use password manager
5. Use two element authentication system if I can use it
6. Use one-time password authentication system if I can use it
7. Logout from the service and shutdown the browser
8. Set account lock policy on my computer
9. Set my computer screen to automatically lock if I do not use it for a prolonged period of time
10. Lock my computer screen when I step away from it
11. Know what website I'm visiting by looking at the URL bar, rather than by the website's look and feel
12. Verify that information will be sent securely (e.g., SSL, "https://", a lock icon) before I submit it to websites
13. When browsing websites, I mouseover links to see where they go, before clicking them
14. When using internet, I confirm whether it is fake warning or not in case of showing warning and alert
15. When using internet, I do not go next page to adjust warning and alert in case of showing them
16. Use privacy mode when using internet
17. Update anti-virus software of my computer
18. Update OS of my computer right away if I am prompted it
19. Update software of my computer right away if I am prompted it
20. Not to connect public Wi-fi
21. Confirm authorized software when I download it
22. Install high evaluation software
23. Take virus scan regularly
24. Back up information of my computer regularly
25. Remove software which I need not to use
26. Set to block malicious mail
27. Send to encrypt important file
28. Pay attention to SNS privacy (e.g., scope of opening information)
29. Use filter which prevents from others

Survey2

SeBeST uses the same questions as those for Survey1.

Decision making after viewing each message in Table 1(4point Likert Scale, From "I ignore the message." to "I update OS now.")

Impression of each message in Table 1 (2point Likert Scale: "true" or "false")

1. I recognize importance of security behaviors.
2. I recognize disadvantages (security risks) which occur by not updating OS.
3. I need to update OS.
4. I recognize advantages which occur by updating OS.
5. I can update OS easily.
6. I recognize effects of security behaviors.
7. I am not impressed by message.

References

1. Wash, R., Rader, E.: Too much knowledge? Security beliefs and protective behaviors among United States internet users. In: Proceedings of the Eleventh USENIX Conference on Usable Privacy and Security, SOUPS, pp. 309–325 (2015)
2. Prochaska, J.O., Velicer, W.F.: The transtheoretical model of health behavior. Am. J. Health Promot. **12**, 38–48 (1997)
3. Velicer, W.F., Prochaska, J.O., et al.: Using the transtheoretical model for population-based approaches to health promotion and disease prevention. Homeost. Health Dis. **40**, 174–195 (2012)
4. Bandura, A.: Self-efficacy: toward a unifying theory of behavior change. Psychol. Rev. **84**, 191–215 (1977)
5. Tversky, A., Kahneman, D.: The framing of decisions and the psychology of choice. Science **211**, 453–458 (1981)
6. Latimer, A., Rench, T., et al.: Promoting participation in physical activity using framed messages: an application of prospect theory. Br. J. Health. Psychol. **13**, 659–681 (2007)
7. McNeil, B.J., Pauker, S.G., et al.: On the elicitation of preferences for alternative therapies. N. Engl. J. Med. **306**, 1254–1262 (1982)
8. Takemura, K.: The influence of elaboration on the framing of decision. J. Psychol. **128**, 33–39 (1994)
9. Metalidou, E., Marinagi, C.C., et al.: The Human Factor of Information Security: Unintentional Damage Perspective, Social and Behavioral Sciences, pp. 424–428 (2014)
10. Glaspie, H., Karwowski, W.: Human factors in information security culture: a literature review. In: Proceedings of the Advances in Human Factors in Cybersecurity, AHFE, pp. 269–280 (2017)
11. Das, S., Kim, T.H., et al.: The effect of social influence on security sensitivity. In: Proceedings of the Tenth USENIX Conference on Usable Privacy and Security, SOUPS, pp. 143–157 (2014)
12. Inglesant, P.G., Sasse, M.A.: The true cost of unusable password policies: password use in the wild. In: Proceedings of the SIGCHI Conference on Human Factors in Computer Systems, CHI, pp. 383–392 (2010)
13. Macromill: https://www.macromill.com/service/. Accessed 21 Dec 2020
14. Shimizu, H.: An introduction to the statistical free software HAD: suggestions to improve teaching, learning and practice data analysis. J. Med. Inf. Commun. **1**, 59–73 (2016)
15. Sawaya, Y., Sharif, M., et al.: Self-confidence trumps knowledge: a cross-cultural study of security behaviour. In: Proceedings of the CHI Conference on Human Factors in Computer Systems, CHI, pp. 2202–2214 (2017)

16. Egleman, S., Peer, E.: Scaling the security wall: developing a security behavior intentions scale (SeBIS). In: Proceedings of the 33rd Annual ACM Conference on Human Factors in Computer, CHI, pp. 2873–2882 (2015)
17. Ndibwile, J.D., Luhanga, E.T., et al.: A demographic perspective of smartphone security and its redesigned notifications. J. Inf. Process. **27**, 773–786 (2019)
18. Mathur, A., Chetty, M.: Impact of user characteristics on attitudes towards automatic mobile application updates. In: Proceedings of the Thirteenth USENIX Conference on Usable Privacy and Security, SOUPS, pp. 175–193 (2017)
19. Statcounter Global Stats: https://gs.statcounter.com/os-market-share/desktop/japan. Accessed 23 Dec 2020
20. Johnson, J.: Designing with the Mind in Mind: Simple Guide to Understanding User Interface Design Guidelines Second Edition, Elsevier Inc., Impress Corporation (2014)

Investigation of Power Consumption Attack on Android Devices

Bin Cheng(✉), Tsubasa Kikuta, Yoshinao Toshimitsu, and Takamichi Saito

Meiji University, Tokyo, Japan
{bin_cheng,ce195013,ce205027}@meiji.ac.jp, saito@saitolab.org

Abstract. With the evolution of mobile technology, advanced features, such as location services and multimedia functions, have been developed for smartphones and tablets. However, the use of these functions consumes a large amount of electric power. A malicious application can exploit these functions and intentionally consume electric power to adversely affect the device. In this study, we investigated and evaluated a method of intentionally draining the device battery using multiple functions that consume a large amount of electric power on Android devices. As a result, we were able to reduce the usable time of the batteries of two Android devices to 3.1 h and 3.2 h, respectively. The results indicated that it is possible to execute an attack that intentionally increases the power consumption of Android devices and reduces the battery life.

1 Introduction

With the spread of mobile technology, smartphones are becoming increasingly accessible. According to the Ministry of Internal Affairs and Communications, their penetration rate in Japan in 2019 was 83.4% [14], and they are becoming a daily necessity. However, with the development of mobile technology, advanced functions, such as Global Positioning System (GPS) and cameras, which consume a large amount of power, have been introduced. If these functions are used for a long time, the usable time of the mobile device is reduced. In addition, a malicious mobile application may use these functions for a long time unnecessarily and intentionally consume power to adversely affect the device. This malicious behavior is hereinafter referred to as a power consumption attack.

In this study, among the standard functions installed on Android devices, various functions that consume a large amount of power, such as the central processing unit (CPU), display, wireless communication function, and media-related functions, were used to intentionally consume energy. A program was created to perform a power consumption attack, and the results indicated that a power consumption attack is possible. In addition, it was observed that this attack is difficult to handle because it does not use functions that require special privileges.

© The Author(s), under exclusive license to Springer Nature Switzerland AG 2021
L. Barolli et al. (Eds.): AINA 2021, LNNS 226, pp. 567–579, 2021.
https://doi.org/10.1007/978-3-030-75075-6_46

2 Related Research

Ogawa et al. [15] investigated the relationship between the performance of Android devices and the total power consumption required for processing. The results indicated that speed-up methods, such as Just-in-Time (JIT) and Vector Floating Point (VFP), and lowering the CPU clock frequency are effective in reducing the total power consumption.

Li et al. [12] conduct a comprehensive power evaluation under a predefined set of test cases, and identify a number of primary power-hungry modules, such as the screen display, GPS and Wi-Fi modules. Finally, an energy model for these modules is established.

In order to understand where and how is energy drain happening on users' phones under normal usage, for example, in a one-day cycle. Chen et al. [3] conduct the first extensive measurement and modeling of energy drain of 1520 smartphone in the wild.

Carroll et al. [2] develop a power model of the Freerunner device and analyse the energy usage and battery lifetime under a number of usage patterns.

Focusing on the power consumption of Android devices in the non-operating state, Nakamura et al. [13] investigated a method for identifying applications that consume power during non-operation based on the number of executions of WakeLock. Their results indicated that deleting applications that frequently execute WakeLock is effective in reducing power consumption.

3 Related Knowledge

3.1 Power-Consuming Hardware and Features

CPU

The CPU performs the main calculations on mobile devices, and power consumption increases by performing a large number of high-load calculations. The usage status can be verified by examining the CPU utilization rate. Generally, the higher the CPU utilization rate, the higher the power consumption.

Display

Liquid crystal displays (LCDs) and organic electroluminescent (EL) displays are the main displays used in mobile devices. Unlike organic EL, the liquid crystal molecules of LCDs cannot emit light for each pixel; thus, a backlight that illuminates the entire screen is required. In addition, since liquid crystal molecules rotate due to the action of the voltage applied to the LCD and display by partially blocking or transmitting light from the light source, power is also consumed when changing the screen display. Because the display is often used for a long time by the device and is the most important means of transmitting information to the user, it consumes a large amount of power.

Wireless communication function

In addition to mobile communication, wireless communication, such as Wi-Fi, Bluetooth, and GPS, are also causes of large power consumption. The power consumption of these services changes depending on the frequency of communication and the strength of radio waves during communication.

Media-related functions

Mobile devices may be equipped with media-related hardware, such as flash-lights, cameras, and speakers, to realize various functions. Although these functions are rarely used continuously, they can deplete the battery if used for a long time.

3.2 Power Saving Mode for Android Devices

The Android operating system (OS) is equipped with several power- saving modes to reduce the power consumption of a device. With these modes, it is possible to control the state of the application executed on the device and prevent unnecessary power consumption. In this study, to intentionally consume power, the following transitions to a power saving mode were all prevented by using WakeLock, which is described in Sect. 3.3.

App standby mode

App standby mode [4], which is a power-saving function added in Android 6.0, detects inactive applications that are temporarily not in use. It then limits the background network activity of these applications to reduce power consumption. Applications that are in use are not affected by this function.

Sleep mode (CPU idle)

Sleep mode is a power- saving mode at the Linux kernel level, and is generally called sleep mode. Based on process scheduling, when the CPU is idle, the system enters a low power consumption state, stopping the CPU or lowering its operating frequency. In addition to the CPU, some CPU-dependent functions and services are turned off or restricted. In other words, the CPU is idle to save power.

Doze mode (deep sleep mode)

Doze mode [4], which saves more power than sleep mode, has been added.which saves more power than sleep mode, was introduced in Android 6.0. The operation of doze mode is illustrated in Fig. 1.

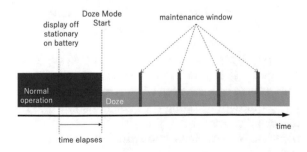

Fig. 1. Doze

When the Android device is not used for a long time, this mode limits high-load functions and services, such as the CPU and network, and reduces power consumption by deep sleep. The conditions for entering doze mode are that the mobile device is not connected to a power supply and that a certain time elapses while the screen remains off. The system also periodically pauses doze mode to continue the activities in applications for a period called the maintenance window.

3.3 WakeLock

WakeLock is a function provided by the Android OS to prevent the system from entering the power-saving mode. As long as an application uses WakeLock, the transition to the power-saving mode discussed in Sect. 3.2 can be avoided. Because WakeLock is a standard function, it can be used without special means, such as root authority.

WakeLock can be divided into two types: a timeout lock and normal lock. The timeout lock is automatically released when the preset timeout period elapses. In contrast, the normal lock must be released manually. The usage of WakeLock is described in Sect. 5.1.

4 Proposed Method

We selected several main functions that consume a large amount of power, as described in Sect. 3.1, and devised a method of intentionally consuming power using these functions. Furthermore, we created a program for consuming power using the devised method. This section describes these methods.

Avoiding power saving mode with WakeLock
To intentionally consume power, a device can be prevented from entering a power-saving mode, such as sleep mode. Therefore, WakeLock is used to prevent the device from entering the power-saving mode to keep it running at all times.
Power consumption by CPU
To increase power consumption, the CPU can be made to perform a large number of complex calculations to increase the CPU utilization rate. Here, hash value calculation is used as the calculation algorithm. Hash value calculation involves a large number of complex calculations and is used as a means to generate a load in many CPU performance evaluation tools.
Power consumption by display
In mobile device displays, the brightness of the LCD backlight is related to power consumption. Therefore, it is possible to consume a large amount of power by maximizing the brightness of the screen of an Android device. In addition, power can be consumed by quickly switching the screen display and rotating the liquid crystal molecules for all pixels.
Power consumption by wireless communication
A large amount of power is consumed when using wireless communication, such as Wi-Fi and Bluetooth. In Android OS, power consumption can be suppressed by reducing the number of times an application accesses the network through the power saving mode. Therefore, the power consumption is increased by intentionally increasing the number of wireless communications.

Power consumption by media-related functions

A large amount of power can be consumed by continuously operating the following media-related functions:

- Blinking or constantly turning on the flashlight
- Playing sound with the speaker always at maximum volume
- Using a vibration motor

We devised several other hardware and functional methods; however, in a preliminary test, their impact on power consumption was low. Therefore, we created a program that intentionally consumes power using only the functions introduced in this section.

5 Power Consumption Application

This section describes an Android application that intentionally consumes power (hereinafter referred to as the power consumption application) implemented based on the proposed method. The power consumption application consists of the following components, each of which can start a thread individually and turn it on and off without affecting other components.

5.1 Implementation of WakeLock

To use WakeLock [5], a new instance of the PowerManager.WakeLock class (new WakeLock) is created by calling android.os.PowerManager.newWakeLock(int flags, String tag). It is necessary to select the flag passed as the first argument (int flags) and specify the operation pattern of the backlight of the CPU, display, and keyboard. The flags that can be specified and the corresponding WakeLock operation patterns are displayed in Table 1.

Table 1. Flags to specify the type of WakeLock

Flag (int flags)	CPU	Display	Keyboard
PARTIAL_WAKE_LOCK	ON	OFF	OFF
SCREEN_DIM_WAKE_LOCK	ON	Dim	OFF
SCREEN_BRIGHT_WAKE_LOCK	ON	Brighten	OFF
FULL_WAKE_LOCK (deprecated from API level 17)	ON	Brighten	Brighten

WakeLock can be used by calling PowerManager.WakeLock.acquire(). By specifying the argument of acquire(), as in the Table 2, WakeLock can be used as a timeout lock or normal lock.

Table 2. Types of acquire()

Argument type	WakeLock type	Release method
acquire()	Normal lock	Manual release by calling release()
acquire(long timeout)	Time-out lock	Automatic cancellation after time elapses

In this study, two types of normal WakeLock are implemented by selecting SCREEN_BRIGHT_WAKE_LOCK, which keeps the display on, and PARTIAL_WAKE _LOCK, which holds only the CPU operation.

5.2   Implementation of Power Consumption Function

CPU

Makes the CPU perform a large number of calculations and generates a load. This function causes the CPU to perform a large number of calculations, thereby generating a load. In this study, we use the CPU load test tool ElephantStress published on Github [1]. This tool launches multiple threads at the same time and calculates MD5 values in parallel. As a result, high power consumption is achieved by keeping the CPU utilization rate high.

Display

In this study, the brightness of the display is set to the maximum to maximize the power consumption of the LCD backlight or the organic EL pixel In addition, the color of the screen displayed in full screen mode is inverted between black and white every 10 ms, and the liquid crystal molecules of the LCD are rotated at maximum angle and in high speed to consume power.

Wi-Fi

Wi-Fi scanning can be executed by calling WifiManager.startScan(), which is an application programming interface (API) [6] that detects the surrounding Wi-Fi networks. In this study, power consumption is realized by repeating execution at 5-s intervals and increasing the frequency of Wi-Fi scans.

Bluetooth

By calling BluetoothAdapter.startDiscovery(), which is an API [7] that detects surrounding Bluetooth devices, the system starts scanning for surrounding Bluetooth devices. In this study, as with Wi-Fi, power consumption is achieved by repeating execution at 5-s intervals and increasing the frequency of Bluetooth scans.

LED flashlight

The LED flashlight can be used by calling CameraManager.setTorchMode (String cameraId, boolean enabled), which is an API [8] for the camera-related function of the Android OS. Specifically, the LED flashlight can be turned on or off by specifying True or False for the second argument (boolean enabled). In this study, power consumption is realized by blinking the LED flashlight in a loop at 10 ms intervals.

Speaker

A beep sound can be generated and played using ToneGenerator.startTone (Tone Generator.TONE_PROP_BEEP), which is an API [9] for the media-related function of Android OS. The argument ToneGenerator.TONE_PROP_BEEP must be specified for the beep sound type [10] provided by the Android OS. In this study, power consumption is realized by continuously playing sound through the speaker.

Vibration function

Continuous vibration can be generated using VibrationEffect.createWaveform (long[] timings, int[] amplitudes, int repeat), which is an API [11] for the vibration function of Android OS. In this study, power consumption is achieved by continuously rotating the vibration motor.

Because the aforementioned functions are all created using the standard API for Android OS, this method does not require special privileges.

6 Experiment

This section describes an experiment that intentionally consumed power from an Android device.

6.1 Experimental Environment

The two Android devices used in the experiment are described in Table 3.

Table 3. Devices used in the experiment

Device	Device 1	Device 2
CPU	Qualcomm Snapdragon 660 (2.2 GHz)	MediaTek Helio P70 (2.1 GHz)
RAM	6 GB	6 GB
Display	2280 × 1080 (Organic EL)	2160 × 1080 (LCD)
OS	Android 9	Android 9
Battery	3,350 mAh (4.2 V)	4,220 mAh (4.2 V)

The power consumption of the Android device was measured directly by connecting a USB multi-meter to the device. This measurement method made it possible to obtain more accurate values than those obtained by calculating the average consumption from the remaining battery level using the Android API. Furthermore, because physical access to the battery was difficult, the USB multi-meter was used in this experiment. In the USB multi-meter used in this experiment, in addition to the instantaneous current (A) and voltage (V) at the device, the power (W) was also displayed and the values were recorded.

6.2   Experimental Procedure

In this study, we conducted the following experiments on two Android devices.

1. **Installation of power consumption application**
   We installed the power consumption application described in Sect. 5 on each device through the ADB.
2. **Fully charging the devices**
   To perform accurate measurements, we fully charged the battery before each experiment.
3. **USB multi-meter connection**
   We connected a USB multi-meter and external power supply to the experimental device. Because the USB multi-meter device is a mechanism to measure the power consumption from an external power supply, it was necessary to connect it to the external power supply. However, the device started charging automatically, and the power consumption due to charging was included in the measured value. To eliminate this component and measure the power consumption accurately, it was necessary to wait until charging stopped. Therefore, we fully charged the device and proceeded to the next step after confirming that charging stopped.
4. **Execution of power consumption application**
   After confirming that no applications were running, we ran the pre-installed power consumption application.
5. **Measurement of power consumption for functions other than the display**
   We executed each power consumption function of the power consumption application to measure the power consumption. First, measurement was performed for each function other than the display. Specifically, we referred to the items listed in Table 4, pressed the button to control the start/stop of each function, switched the operation pattern of the power consumption application, and displayed it on the USB circuit meter. We then recorded the value. For each measurement item, we performed the measurement three times: immediately after the start, 30 s later, and 1 min later, and recorded the average value.

Table 4. Measurement items other than display

Measurement item	Description
Standby	Display turned off, no applications running
CPU load test	CPU load
Wi-Fi scan	Searching for surrounding Wi-Fi network
Bluetooth scan	Searching for nearby Bluetooth devices
Flashlight lit	Flashlight lit continuously
Flashlight flashing	Flashlight flashing rapidly
Vibration function	Vibration motor continuously rotating
Speaker	Speaker playing sound at maximum volume

During measurement, to prevent the effects by other functions, the functions other than the measurement items are stopped and the display is turned off. Also, for hardware that cannot be stopped, such as CPU, memory, and sensors, try not to execute it with other apps or services as much as possible.

6. **Measurement of display power consumption**
   The power consumption of the display was measured by the same measurement method as in experiment 5. However, when measuring the power consumption of the display, we aimed to evaluate the effects of brightness adjustment and black-and-white inversion; thus, we measured each of the three display patterns listed in Table 5.

Table 5. Display measurement items

Measurement item	Description
Maximum screen brightness	Maximum display brightness, displaying a still white screen
Flashing screen (minimum brightness)	Minimum display brightness, high-speed black-and-white inversion of the entire screen
Flashing screen (maximum brightness)	Maximum display brightness, high-speed black-and-white inversion of the entire screen

In addition, we stopped all functions other than the display and avoided running other applications or services as much as possible.

7. **Measurement of power consumption during full-featured operation**
   For each of the two devices, we activated all the functions of the power consumption application and measured the total power consumption at that time. However, for the display items, we selected the state in which screen flashing at maximum brightness.

8. **Measurement of battery life during full-featured operation**
   We ensured that each of the two devices was fully charged. Then, with all the functions of the power consumption application running, we disconnected the external power supply and measured the time until the battery was depleted.

7   Experimental Results

Table 6 presents the results of measuring the power consumption of each item in the power consumption application on the two devices.

Table 6. Power consumption for each function

Model name	Device 1	Device 2
Standby	0.5 W	0.4 W
CPU load test	5.1 W	4.3 W
Wi-Fi scan	1.5 W	1.2 W
Bluetooth scan	0.9 W	1.1 W
Flash on	0.8 W	1.2 W
Flashing	1.0 W	1.5 W
Vibration function	0.8 W	1.1 W
Speaker	1.8 W	2.5 W
Screen brightness maximum	1.9 W	1.7 W
Blinking screen (minimum brightness)	0.9 W	1.0 W
Blinking screen (maximum brightness)	2.0 W	2.3 W
Full function operation	5.8 W	6.5 W

The results of the investigation revealed that the method for consuming the largest amount of power was to impose a load on the CPU. The results also demonstrated that when the brightness of the display was maximized, more than twice as much power was consumed as when the brightness was minimized.

Table 7 presents the battery life when all the functions of the power consumption application were operated on the two devices.

Table 7. Battery life when all features are used

Device	Device 1	Device 2
Time control	3.1 h	3.2 h
Battery capacity	3,350 mAh (4.2 V)	4,220 mAh (4.2 V)

It was determined that the battery life of both experimental devices was significantly shorter than the empire proximately 8–12 h, depending on the usage) in daily use.

8 Discussion

8.1 Battery Life When All Functions Are in Operation

For each device's battery power $W_p(Wh)$ and estimated time $T(h)$, the following equations hold:

$$W_p = Q \times U$$
$$T = W_p \div P$$

Here, $Q(Ah)$ represents the battery capacity, $U(V)$ represents the battery voltage, and $P(W)$ represents the power consumption during full-function operation. Using these formulas, the estimated battery life of each device was calculated.

For device 1, the following results were obtained:

$$W_{p1} = 3,350\,\text{mAh} \times 4.2\,\text{V} \div 1000 = 14.07\,\text{Wh}$$

$$T_1 = 14.07\,\text{Wh} \div 5.8\,\text{W} = 2.4\,\text{h}$$

For device 2, the results were as follows:]

$$W_{p2} = 4,220\,\text{mAh} \times 4.2\,\text{V} \div 1000 = 17.72\,\text{Wh}$$

$$T_2 = 17.72\,\text{Wh} \div 6.5\,\text{W} = 2.7\,\text{h}$$

Comparing these values with the results in Table 7 reveals that the actual measured battery life was longer than the estimated value. This discrepancy is discussed in Sect. 8.2.

8.2   Power Consumption by CPU

To clarify the reason for the discrepancy between the estimated time and measured time for each device, an additional experiment was conducted in which only the CPU was loaded. The results revealed that the CPU utilization rate and frequency increased immediately after the start of the measurement, but fluctuated significantly thereafter. In addition, it is likely that the temperature of the device rose significantly during the experiment, and the CPU performance deteriorated due to the heat limitation of the system, leading to differences between the measured time and estimated time.

8.3   Power Consumption by Display

Table 6 presents the difference in power consumption between organic EL (device 1) and LCD (device 2). Organic EL achieved power saving by partially controlling the light emission of the pixel itself. In the experiment involving blinking the screen at the maximum brightness, a black screen was displayed approximately half the time. Therefore, since all pixels of the organic EL did not emit light during that period, the power consumption during that period was lower than that of the LCD. However, while the white screen was displayed, the organic EL consumed more power than the LCD. Furthermore, the power consumed while the organic EL displayed a white screen was not very different from the power consumed by the LCD, even though the organic EL took approximately half the time than the LCD. In addition, when the brightness was at the minimum, the backlight of the LCD had low power; therefore, there was no difference from the organic EL display. The above results indicate that the method of consuming power by the display used in this study is effective for both organic EL and LCD.

8.4   Possibility of Attack

In the experiments in this study, the battery life, which is usually approximately 8–12 h, was reduced to 3.1 h for device 1 and 3.2 h for device 2. As the usable time was significantly reduced, the devised attack method can be considered effective. By using these methods and incorporating them into a normal application, it is possible to intentionally cause the user's device to consume more power than necessary. In addition, attacks using the proposed method are easy to execute because they do not require root privileges and are not based on specific vulnerabilities. However, because the main purpose of this experiment was to confirm the power consumption of each function of a mobile device, we also used functions that are easily noticed by users, such as blinking of the display and LED flashlight. From the point of view of an attack, it is possible to attack without being noticed, such as by loading the CPU or playing a continuous sound with a frequency that humans cannot hear.

8.5   Future Work

The experimental results demonstrate that the devised method is effective; however, several points can be improved. In this experiment, the power consumption by each function of the device was determined by measuring the accurate instantaneous power. However, due to performance degradation due to the heat limitation of the CPU, there was a discrepancy between the estimated battery life during full-function operation and the actual battery life. In the future, an important task is to consider a method for solving this problem and improve the accuracy. Another important task is to change the functions and implementation methods so that users will not notice when an attack occurs.

9   Research Ethics

In this paper, we created a program that intentionally consumes power, and conducted a series of experiments on mobile devices using the program. The purpose of these experiments was to evaluate the potential for power consumption attacks. The created power consumption application and its source code are not used for purposes other than research, and are strictly managed in the laboratory.

10   Conclusion

In this study, we investigated and evaluated a method of intentionally consuming power by using multiple functions that consume a large amount of power among the functions installed on an Android device. As a result, we were able to reduce the usable time of the batteries of two Android devices to 3.1 h and 3.2 h, respectively. In addition, the results indicated that the most effective means of consuming a large amount of power in mobile devices is to load the CPU. This study thus verifies that a power consumption attack is possible.

Acknowledgement. Part of this research was supported by JSPS Grant-in-Aid for Scientific Research JP18K11305.

References

1. https://github.com/SyncHack/ElephantStress
2. Carroll, A., Heiser, G.: An analysis of power consumption in a smartphone. In: Proceedings of the 2010 USENIX Conference on USENIX Annual Technical Conference, USENIXATC'10, p. 21, USA (2010). USENIX Association
3. Chen, X., Ding, N., Jindal, A., Hu,, Y.C., Gupta, M., Vannithamby, R.: Smartphone energy drain in the wild: analysis and implications. In: Proceedings of the 2015 ACM SIGMETRICS International Conference on Measurement and Modeling of Computer Systems, SIGMETRICS '15, pp. 151–164, New York, NY, USA (2015). Association for Computing Machinery
4. Google. Documentation for app developers. https://developer.android.com/training/monitoring-device-state/doze-standby?hl=ja
5. Google. Documentation for app developers. https://developer.android.com/training/scheduling/wakelock?hl=ja
6. Google. Documentation for app developers. https://developer.android.com/guide/topics/connectivity/wifi-scan
7. Google. Documentation for app developers. https://developer.android.com/guide/topics/connectivity/bluetooth
8. Google. Documentation for app developers. https://developer.android.com/reference/android/hardware/camera2/CameraManager#setTorchMode (java.lang.String,%20 boolean)
9. Google. Documentation for app developers. https://developer.android.com/reference/android/media/ToneGenerator#ToneGenerator (int,%20int)
10. Google. Documentation for app developers. https://developer.android.com/reference/android/media/ToneGenerator#constants_1
11. Google. Documentation for app developers. https://developer.android.com/reference/android/os/VibrationEffect
12. Li, X., Zhang, X., Chen, K., Feng, S.: Measurement and analysis of energy consumption on android smartphones. In: 2014 4th IEEE International Conference on Information Science and Technology, pp. 242–245 (2014)
13. Nakamura, Y., Hayakawa, A., Koyanagi, A., Nakarai, A., Takemori, K., Oguchi, M., Yamaguchi, S.: A study on power consumption by launching applications on android os. In: 77th IPSJ, vol. 2015, pp. 255–256 (2015)
14. Ministry of Internal Affairs and Communications. Reiwa 2nd year information and communication white paper (2020). https://www.soumu.go.jp/johotsusintokei/whitepaper/ja/r02/pdf/index.html
15. Ogawa, H., Yamaguchi, S.: Evaluation of computing performance and measurement of power consumption on android devices. 74th IPSJ **2012**(1), 145–146 (2012)

The Novel System of Attacks Detection in 5G

Maksim Iavich[1]([⊠]), Sergiy Gnatyuk[2], Roman Odarchenko[2], Razvan Bocu[3],
and Sergei Simonov[4]

[1] Caucasus University, Paata Saakadze st.1, Tbilisi, Georgia
miavich@cu.edu.ge
[2] National Aviation University, Liubomyra Huzara Ave, 1, Kiev, Ukraine
odarchenko.r.s@ukr.net
[3] Transilvania University of Brasov, Bulevardul Eroilor 29, Brașov 500036, Romania
razvan@bocu.ro
[4] Scientific Cyber Security Association, Lortkipanidze st.26, Tbilisi, Georgia
s_simonovi@cu.edu.ge

Abstract. The amount of traffic carried over wireless networks is growing rapidly
and is being driven by many factors. The telecommunications industry is undergoing a major transformation towards 5G networks in order to fulfill the needs
of existing and emerging use cases. The paper studies the existing vulnerabilities
of the 5G ecosystem. Considering this study, we propose a new cyber security
model that considers machine learning algorithms. The function contains Firewall
and IDS/IPS. We integrate the described model into an existing 5G architecture.
The methodology and the pseudo code of the algorithmic core is provided. The
paper also studies the efficiency of this approach. The tests are performed in a
test laboratory, which includes a server and 60 raspberry pi hardware systems that
are used in order to simulate attacks on the server. The tests show that the offered
approach identifies DOS/DDOS attack much better than methods described in the
related works. The paper also suggests the improvement strategy, which will be
implemented in the future versions of the system.

1 Introduction

The amount of traffic carried over wireless networks and the number of mobile devices
(including IoT) are growing rapidly and are being driven by many factors. The telecommunications industry is undergoing a major transformation towards 5G networks in order
to fulfill the needs of existing and emerging use cases. Therefore, the vision of 5G wireless networks lies in providing very high data rates and higher coverage through dense
base station deployment with increased capacity, significantly better Quality of Service
(QoS), and extremely low latency. The provision of the necessary services that are envi-
sioned by 5G requires that novel networking architectures, service deployment models,
storage and processing technologies are defined. These technologies should bring new
challenges for the 5G cybersecurity systems and their functionality.

The 5G data networks will connect critical infrastructures that will require more
security in order to ensure safety of not only the respective infrastructure, but the general
safety of the society as a whole. For example, a security breach in the online power

L. Barolli et al. (Eds.): AINA 2021, LNNS 226, pp. 580–591, 2021.
https://doi.org/10.1007/978-3-030-75075-6_47

supply systems can be catastrophic for all the electrical and electronic systems that the society depends upon. Therefore, it is highly important to investigate and highlight the important security challenges in 5G networks and create an overview related to the potential solutions that could lead to the design of secure 5G systems. The researchers and developers are working hard to make 5G systems secure. Furthermore, it is important to analyze the difference between 4G and 5G systems' security [1–4].

2 Literature Review

The authors of [1] discuss the changes which may be caused by the 5G technology. They discuss on the key challenges, which are needed for the future research and the activities for the 5G standardization. Furthermore, they realize the overview of the existing related literature. The authors of [2–4] analyze the differences that exist between the security of 4G and 5G systems. They discuss the novel architectural changes related to the radio access network (RAN) design, which includes antennas, air interfaces, smart, heterogeneous RAN and the cloud. The authors of [5, 6] describe the new authentication framework of 5G. The authors of [5], in addition, offer a reputation system-based lightweight message authentication framework and the protocol for vehicular networks with 5G enabled. The authors of [6] also propose an authentication framework, which is efficient, secure and service-oriented. It supports fog computing and network slicing for 5G-enabled IoT services. They assess the efficiency of this framework using simulations, in order to demonstrate its feasibility and efficiency. The authors of [7, 8] make the analysis of 5G network slicing and they offer a they analyze the existing works. They propose the framework. The authors of [9–12] analyze the security of 5G, and identify the existing vulnerabilities of it. They describe also the advantages of 5G's security model. Report on EU coordinated risk assessment of 5G is offered in [11]. This contribution reports the identification of several relevant security problems, which will appear or become much more relevant in 5G networks. Thus, it is emphasized that these vulnerabilities are much more relevant for 5G systems, then for the existing networks. The authors of [12] analyze the 5G security by means of combining the logical layer and the physical layer from the perspective of an automated attack and defense mechanism. They offer to use an automated solution framework for the assurance of the 5G security.

The authors of [13] illustrate three new classes of attacks, which exploit unprotected devices in 4G and 5G networks. In [14], the authors describe KDD99 dataset, they analyze 149 academic papers from 65 different journals. Finally, the authors of [15–17] offer different approaches of creating intrusion detection systems (IDS) by means of KDD99 dataset. They offer different approaches of training the intrusion detection system (IDS) by means of using the KDD99 dataset for 5G systems. The authors of [16] also use the ReliefF algorithm, in order to attach to every attribute of network in the KDD99 dataset the weight, which shows the relationship between attributes and the final class, in order to get the better classification results. They illustrate that the new IDS has a higher true positive rate, and it exhibits the lower false positive rate concerning the detection performance.

3 Remarks on the 5G Security Features

It must be emphasized that 5G and 4G security architectures are rather similar. The network nodes of LTE/4G and 5G, which provide the security and the communication links, which must be protected, are similar. In both cases, the security mechanisms can be grouped in two categories. The first one includes the security mechanisms for the network access. They are used to provide users with the secure access to different services using some devices and secure the system against the attacks on an air interface, which is located between the concrete device and a radio node. The second one includes different security mechanisms concerning the network access. They are used to exchange different signaling data and user's data between the nodes that are part of the radio and core network. The security architecture of the LTE/4G and 5G is illustrated in Fig. 1.

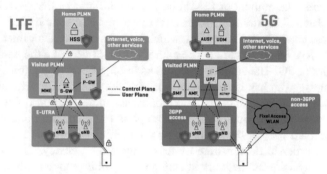

Fig. 1. Security architecture of the LTE/4G and 5G.

The standard 5G considers a new authentication framework. The access authentication is the basic security procedure in 3GPP networks and is considered as a basic authentication related to the security standards of the 3GPP 5G. This procedure is performed in the initial phase of the device's authentication. After the successful start of this procedure, the session keys must be established. The keys are used to secure the communication between the device and the network. This procedure is designed relative to the 3GPP 5G security in order to support an extensible authentication protocol (EAP). EAP is widely used in different IT areas. It gives us the opportunity to use different types of credentials, which are stored in the SIM cards, such as pre-shared keys, certificates, usernames and passwords [5, 6].

The authentication framework and the procedures of the key agreement gives the opportunity for the authentication to manifest between the user equipment (UE) and the network, which is needed for the communication between UE and the network in the future security scenarios. The keys, which are generated by the primary authentication framework and the procedure of key agreement form the anchor key, which is called (KSEAF). This key is used as the seed for the secure channel, which the subscriber and the serving networks (SEAF) need, after it is transferred by the Authentication Server Function (AUSF) of a home network to the serving network's SEAF.

Thus, KSEAF can be used to generate the keys for different security contexts, in which the new authentication process is not needed. For example, an authentication process is used in a 3GPP access network, which provides keys in order to establish the secure data channels between UE and a Non-3GPP Interworking Function (N3IWF). This function is used for the untrusted non-3GPP access.

In a 5G network, a globally unique Subscriber Permanent Identifier (SUPI) must be assigned to each subscriber. The Subscriber Permanent Identifier formats include the international mobile subscriber identity (IMSI) and Network Access Identifier (NAI). It is important not to disclose the SUPI, when a concrete mobile device connection has to be established. For example, 3G and 4G networks have this vulnerability, as in these networks, the IMSI is disclosed when a concrete device is passing an attach procedure, before this device can authenticate with a new network.

It must be mentioned that the Subscription Concealed Identifier (SUCI) must be used before the network and device may pass through an authentication process. Only after this, SUPI of the home network can be disclosed to the serving network. This procedure is used to prevent an international mobile subscriber identity catchers (IMSI) from steeling the identity of the subscriber. This is achieved by means of making a device either to be attached to the Rogue Base Station (RBS) or to be attached to the Base Station of operator while the unencrypted traffic over the air is sniffed.

The Subscription Identifier De-Concealing Function (SIDF) must be used to de-conceal the SUPI from the SUCI. The SIDF uses the private key part of the privacy-related home network public/private key pair that is securely stored by the home network of the operator. The de-concealment must take place at the UniFi Dream Machine (UDM). Access rights to SIDF must be defined, in the way that only a network element of the home network is allowed to request SIDF.

Security Edge Protection Proxy (SEPP) is implemented in the 5G architecture for the protection of messages that are sent over the N32 interface at the perimeter of the Public Land Mobile Network (PLMN). SEPP receives all the service layer messages from the Network Function (NF), and protects them before they are sent out of the network via the N32 interface. Additionally, it receives all messages on the N32 interface and after verifying its security data, messages are forwarded to the appropriate network function layer. The SEPP implements an application layer security for all the information, which is interchanged between two NFs across two different PLMNs.

The 5G architecture uses the concept of network slicing. By definition, a network slice is an independent end-to-end logical network that runs on a shared physical infrastructure, which is capable of providing a negotiated service quality level. This means that in a 5G network, the whole structure can be divided into components, without intersection between them. This concept allows us to divide the network and manage the related services in a secure fashion. This is beneficial from security perspective, because system will be more structured, secure and easily manageable [7, 8].

4   Security Problems of the 5G Standard and Machine Learning

The researches show that 5G still has security problems [9–12]. The analysis that is presented in this paper has found different security problems. These problems are the following:

1) A 5G network has a much consistent exposure to software attacks and it has much more entry points for the hackers, because the 5G networks are mostly based on software configurations of their logical architecture. The attacks target the different security flaws and bugs, which can impact the 5G network's operation.
2) As 5G network architecture provides new functionalities, some parts of the network equipment and functions became much more sensitive for the hacker attacks. The base stations and the network's key management functions can be the aim of the hackers' attacks.
3) The fact that mobile network operators rely on suppliers, can also lead to additional attack routes. It can also greatly increase the severity of such attacks' impact.
4) Many critical IT applications will use 5G networked structures, so the attacks concerning the availability and integrity of such type application can become a great security concern.
5) A lot of devices will be the part of the 5G network, which can provoke different types of denial of service (DoS) and distributed denial of service (DDoS) attacks.
6) Network slicing can also become a security problem, as hackers may try to force the concrete device to use the slice that was not intended for it.

In recent years, the researchers discovered vulnerabilities in 5G security systems that allow attackers to integrate malicious code into the system and conduct illegitimate actions [13].

These attacks are grouped into several categories.

- **MNmap** - A team of researchers sniffed information which was sent in plaintext and using it they recreated the map of devices connected to this network. They created the fake base station and recorded capabilities of connected devices. The scientists could identify the manufacturer, model, operating system, and the versions of connected devices as well as type of type of the devices.
- **MiTM** - Current 5G infrastructures also allow for Man-in-the-middle (MiTM) attacks to occur. Thus, through the implementation of the MiTM, bidding-down and battery drain attacks become possible. Attackers can remove Multiple-Input Multiple-Output (MIMO) enablement, which means that high speed in not achievable, and without that component it is the same as to be connected to a 3G/2G networks. The MIMO enablement is a feature that significantly boosts the data transfer speed over the 5G network.
- **Battery drain attack** - This attack is targeted to NB-IoT devices. Those are low-power sensors that rarely send little packets of information, which drain the battery power even in the energy saving mode. An attacker modifies PSM, such that devices has continuous activity and is searching for a network to connect to. Thus, the attacker can propose desirable networks to the victim, and then the equipment may be used maliciously.

Machine learning algorithms are widely used in different scientific areas. They are also very important for cybersecurity. Machine learning can identify cyber threats using pattern detection. It is very relevant to use these algorithms to identify the threat patterns mentioned above.

It is immediate to notice that it is necessary to define a new architecture for the 5G and future 6G networks in order to provide novel AI/ML based algorithms, which should give great opportunities to provide the highest cybersecurity level, and to ensure mobile subscribers, industry, government are adequately protected.

5   Description of the Approach

We offer to integrate a cyber security module on each 5G station as an additional server. Thus, this server will integrate the firewall and the IDS/IPS. The idea of this approach is illustrated in Fig. 2.

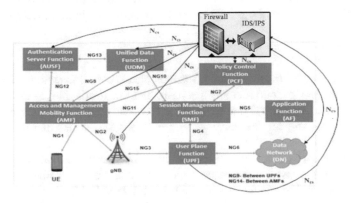

Fig. 2. The cyber security module

Considering the research that is presented in this paper it can be inferred that 5G cellular networks are vulnerable to Probe, Dos and to the software-based attacks. The initial version of our intrusion detection system (IDS) is designed using machine learning algorithms. It is relevant to note that the resistance against these types of attacks is ensured using different datasets for the training process. Thus, the first dataset is KDD99. It must be emphasized that it is very relevant and common to use KDD99 in academic research and in order to create prototypes of intrusion detection systems. Thus, KDD99 is the most common dataset that is considered for the anomaly detection calibration [14]. This particular dataset is based on the data captured in the framework of the DARPA'98 IDS evaluation program. It contains about four gigabytes of raw tcpdump data, which is generated by seven weeks of captured network traffic. The data that corresponds with the duration of two weeks includes about two million connection records. Furthermore, the entire dataset consists of about 5 million samples that are labeled as either normal or attack. All of the attacks are divided into four basic groups of attacks.

1) **Denial of Service (DoS) Attacks:** the attacker sends a lot of requests that overwhelm the computers' resources so that it can not handle the requests of the legitimate users.
2) **User to Root (U2R) Attacks:** this class of attacks supposes that the intruder gains access to legitimate user accounts data (by sniffing, social engineering, etc.), which

can be exploited at the level of their inner vulnerabilities in order to gain full access to the system.

3) **Remote to Local (R2L) Attacks:** the attacker does not have an account on the machine, but packets can be sent remotely in order to gain access as a user on the local machine.

4) **Probing Attacks:** they are user in order to gather information about networks in order to bypass their security mechanisms.

The entire KDD99 dataset is separated into training and test attack types, which include 24 and 14 entities, respectively. The features can be listed into three groups.

1) Basic Features: all the information that is gathered from the analysis of TCP/IP connections. These features lead to the introduction of implicit delays during the detection process.

2) The traffic features are divided into two groups.

2.1) "Same host" features: These features check only connections that occurred during the last two seconds that have the same destination as the current connection. Additionally, these features calculate statistics that are related to the protocol and service behavior, and possibly for entities that belong to other categories.

2.2) "Same service" features: These features check only connections that occurred during the last 2 s that have same service as current connection. There are some probing attacks that do not use the interval of 2 s but use a longer time period, as an example, a 1 min interval. The resolution of this problem involves that "same host" and "same service" features are recalculated for every 100 connection. These features are called "Connection-based traffic features".

3) Content features: DoS and Probing attacks require many connections to some host in a short period, but R2L and U2R attacks do not require this high amount of connections to be established. The R2L and U2R attacks are sent with data packets, which require only one connection. In this case, the proper detection of the traffic features involves that the monitor process considers some suspicious actions in that particular data portion, as an example, the number of failed login attempts.

It can be observed that the KDD99 dataset is divided into four main categories: DOS, R2L, U2R, and PROBE. Thus, the DOS category contains the following subcategories: APACHE2, PROCESSTABLE, UDPSTORM, BACK, LAND, NEPTUNE, POD, SMUR, MAILBOMB and TEARDROP, which belong to the category of DOS attacks. Moreover, the U2R category contains the following subcategories: BUFFER_OVERFLOW, PS, SQLATTACK, XTERM, PERL, LOADMODULE, and ROOTKIT. Furthermore, the R2L category contains the subcategories FTP_WRITE, GUESS_PASSWD, HTTPTUNNEL IMAP, MULTIHOP, NAMED, SENDMAIL, SNMPGETATTACK, SNMGUESS, WXLOCK, XSNOOP, PHF, SPY, WAREZCLIENT and WAREZMASTER, while the category PROBE contain the subcategories: IPSWEEP, NMAP, PORTSWEEP, NMA, MSCAN, SAINT and SATAN. The

R2L and U2R categories are related to software security attacks. Therefore, the training of the described IDS system using the KDD99 dataset is rather relevant in the context of the 5G attacks, as this dataset covers all the attacks categories, which can be filtered on a 5G network.

6 Methodology

We have trained our IDS using KDD and two DOS datasets containing DOS/DDOS attacks. Because of high data transfer speed, 5G clients are even more exposed to the reflected and volumetric DDoS. Therefore, it is very relevant to train the system additionally using DOS/DDOS modern dataset. The first dataset contains the information about the following DOS attacks: 'LDAP', 'MSSQL', 'NetBIOS', 'Syn', 'UDP', 'UDPLag' and its size is 380 MB, let us refer to it as DOS1. The second dataset contains only the information about the 'Portmap' attack and its size is 85 MB, let us refer to it as DOS2.

We partitioned the KDD99 dataset into test and training datasets. The training datasets contains 90% of the information, and the test dataset includes the remaining 10% of the information. The datasets DOS1 and DOS2 were also partitioned into two datasets, each of them. The training datasets contains 80% of the data, and correspondingly the test dataset contains 20% of the data. Both the DOS1 and DOS2 datasets are divided in the same way. This method that was chosen for the data partitioning generated the best accuracy results after the training process. We are training the model with each dataset separately. The accuracy of the model in the case of the KDD99 dataset is 0.9611049372916336, in the case of DOS1 is 0.9937894736842106 and in the case of DOS2 is 0.9998956703182055.

After the training process is completed, the system is waiting for input data from the network sniffer. First, the input is checked for the attacks that are contained in the KDD99 dataset. If an attack pattern is identified, it is transferred to the intrusion protection system. In the case when the attack not identified, the input is automatically checked relative to the attack patterns contained in the DOS1 dataset. If an attack pattern is determined, it is transferred to the intrusion protection system. The same procedure is applied relative to the DOS2 dataset, if an attack pattern is not determined. If an attack pattern is not identified, the IDS system reports that the analyzed traffic is not suspicious, and it moves forward to process the next input data block.

The algorithmic core of the IDS system is determined by the following pseudo code.

```
Class IDS():
    Private:
            X = None  # variable for training stage
            Y = None  # variable for training stage
        Model = None              #variable for NN model
        Def __init__(file_name, model_type):
        #Constructor for preprocess data.
                Preprocess data …
        Def create_model(model_type):
                Create NN model …
                Return model
        Def train_model(data_frame, model_type):
                Training model with data …
                Return model
        Def test_model(model, model_type):
                Testing and measure accuracy …
                Return accuracy_score
    Public:
        Def predict(x_data):
                Predicting with our model …
                Return predictions
        Def print_accuracy(model_type):
                Return IDS.test_model()
    IDS_concrette = IDS(df1)     # creating concrete dos
attack prediction model
    IDS_KDD = IDS(df2)      # creating KDD2020 attack
prediction model
    Waiting_procces(traffick) # process that catches
traffick
    IF IDS_KDD.predict(df) == 'KDD_attack'
        IPS.response_KDD()
    Elif IDS_concrette.predict(df) == 'DOS1':
        IPS.response_dos1()
    Elif IDS_concrette.predict(df) == 'DOS2':
        IPS.response_dos2()
    Else:
        Print "not vulnerable"
```

7   Experiments

The experimental assessment of the described IDS system involves the creation of the laboratory using 60 RASPBERRY PI devices, and a server where we have located our IPS software module. We have implemented the attack vectors, which are critical to the modern analyzed infrastructures.

During the attacks using the network sniffer, we have examined the generated traffic. We have parsed all the parameters that are relevant for our KDD99 and DOS samples, using the Python programming language. We have transformed this output to the format of the original datasets. After this, we have passed all this information to our IDS system and analyzed the result.

The attacks that are discovered pertain to the following types: BACK, LAND, NEPTUNE, POD, SMURF, TEARDROP, BUFFER_OVERFLOW, LOADMODULE, ROOTKIT FTP_WRITE, GUESS_PASSWD, IMAP, MULTIHOP, PHF, SPY, PROBE, IPSWEEP, NMAP, PORTSWEEP, SATAN, LDAP, MSSQL and Portmap (Table 1).

Table 1. Identified attacks by IDS.

Attack type	Number of attacks	Identified attacks
BACK	100	84
LAND	100	94
NEPTUNE	100	100
NetBIOS	100	98
SMURF	100	84
SYN	100	91
BUFFER_ OVERFLOW	100	76
FTP_WRITE	100	86
LOADMODULE	100	91
ROOTKIT	100	62
GUESS_PASSWD	100	100
MULTIHOP	100	91
SPY	100	51
PROBE	100	98
IPSWEEP	100	92
NMAP	100	81
PORTSWEEP	100	98
SATAN	100	82
LDAP	100	81
MSSQL	100	99
Portmap	100	98

These results prove that the IDS is rather useful and it can be used as the prototype version of the future real world IDS system.

8   Conclusion and Future Plans

The system that is described in this paper covers the majority of the attacks, which can break a real world 5G system. The described approach is very different from relevant similar contributions in the field [15–17]. The existing approaches usually train the model of the implemented intrusion detection system using the KDD99 dataset. Nevertheless, we consider the KDD99 dataset as one of the datasets, which is used for the training purposes. The experiments that have been conducted prove that the model can successfully identify the majority of the attacks that have been mentioned. The real time detection core of the IDS system will be improved. Furthermore, it is planned to create more precise and complex attack vectors. We shall create our own dataset, which will be used in order to thoroughly test, assess and optimize the intrusion detection system. Consequently, the system will be deployed in a real world 5G data network.

Acknowledgment. The work was financed by Shota Rustaveli National Science Foundation and Caucasus University in the frame of the [CARYS-19-121] grant and Caucasus University grant project.

References

1. Andrews, J.G., et al.: What will 5G be? IEEE J. Sel. Areas Commun. **32**(6), 1065–1082 (2014). https://doi.org/10.1109/jsac.2014.2328098
2. Osseiran, A., et al.: Scenarios for 5G mobile and wireless communications: the vision of the METIS project. IEEE Commun. Mag. **52**(5), 26–35 (2014). https://doi.org/10.1109/mcom. 2014.6815890
3. Shafi, M., et al.: 5G: a tutorial overview of standards, trials, challenges, deployment, and practice. IEEE J. Sel. Areas Commun. **35**(6), 1201–1221 (2017). https://doi.org/10.1109/ jsac.2017.2692307
4. Agiwal, M., Roy, A., Saxena, N.: Next generation 5G wireless networks: a comprehensive survey. In: IEEE Communications Surveys and Tutorials, vol. 18, no. 3, pp. 1617–1655, Thirdquarter (2016). https://doi.org/10.1109/comst.2016.2532458
5. Cui, J., Zhang, X., Zhong, H., Ying, Z., Liu, L.: RSMA: reputation system-based lightweight message authentication framework and protocol for 5G-enabled vehicular networks. IEEE Internet Things J. **6**(4), 6417–6428 (2019). https://doi.org/10.1109/jiot.2019.2895136
6. Ni, J., Lin, X., Shen, X.S.: Efficient and secure service-oriented authentication supporting network slicing for 5G-enabled IoT. IEEE J. Sel. Areas Commun. **36**(3), 644–657 (2018). https://doi.org/10.1109/jsac.2018.2815418
7. Foukas, X., Patounas, G., Elmokashfi, A., Marina, M.K.: Network slicing in 5G: survey and challenges. IEEE Commun. Mag. **55**(5), 94–100 (2017). https://doi.org/10.1109/mcom.2017. 1600951
8. Li, X., et al.: Network slicing for 5G: challenges and opportunities. IEEE Internet Comput. **21**(5), 20–27 (2017). https://doi.org/10.1109/mic.2017.3481355
9. Huawei 5G Security White Paper (2019). https://www-file.huawei.com/-/media/corporate/ pdf/trust-center/huawei-5g-security-white-paper-4th.pdf
10. 5G Americas: The evolution of Security in 5G (2019). https://www.5gamericas.org/files/ 4715/6450/22-67/5G_Security_White_Paper_07-26-19_FINAL.pdf

11. Report on EU coordinated risk assessment of 5G (2019). https://ec.europa.eu/comm-ission/presscorner/detail/en/IP_19_6049
12. Sun, Y., Tian, Z., Li, M., Zhu, C., Guizani, N.: Automated attack and defense framework toward 5G security. IEEE Netw. **34**(5), 247–253 (2020). https://doi.org/10.1109/mnet.011.1900635
13. Shaik, A., Borgaonkar, R.: New Vulnerabilities in 5G Networks. Black Hat USA Conference (2019)
14. Özgür, A., Erdem, H.: A review of KDD99 dataset usage in intrusion detection and machine learning between 2010 and 2015. PeerJ Preprints-4:e1954v1 (2010). https://doi.org/10.7287/peerj.preprints.1954v1
15. Li, J., Zhao, Z., Li, R.: Machine learning-based IDS for software-defined 5G network. IET Netw. **7**(2) (2017)
16. Wang, Y.: A Novel Intrusion Detection System Based on Advanced Naive Bayesian Classification. Lecture Notes of the Institute for Computer Sciences, Social Informatics and Telecommunications Engineering, vol 211. Springer, Cham. https://doi.org/10.1007/978-3-319-72823-0_53
17. Iwendi, C., Khan, S., Anajemba, J.H., Mittal, M., Alenezi, M., Alazab, M.: The use of ensemble models for multiple class and binary class classification for improving intrusion detection systems. Sensors **20**, 2559 (2020)

Risk Analysis for Worthless Crypto Asset Networks

Wataru Taguchi[1(✉)] and Kazumasa Omote[1,2]

[1] University of Tsukuba, Tennodai 1-1-1, Tsukuba 305-8572, Japan
s1920588@s.tsukuba.ac.jp, omote@risk.tsukuba.ac.jp
[2] National Institute of Information and Communications Technology,
4–2–1 Nukuikitamachi, Koganei, Tokyo 565–0456, Japan

Abstract. Blockchain technology has been used as the basis of crypto assets since the proposal of Bitcoin [8] by Satoshi Nakamoto, and new operation methods are being discussed every day. However, because of decentralization which is one of the characteristics of the blockchain network, such network cannot easily stop the service, and the blockchain network with high anonymity may be abused. Therefore, in this study, we investigated the Altcoin network of approximately 240 stocks was investigated and analyzed and considered the sustainability of the network, which has no asset value and is close to service outage, was analyzed. It was confirmed that there are active nodes in a worthless cryptoasset network, and that there are malicious nodes regardless of the worth of a cryptoasset network.

1 Introduction

Blockchain technology has emerged as the core technology of the Bitcoin [8], which was proposed by an anonymous person called Satoshi Nakamoto in 2008. Since then, blockchain technology has had a considerable influence in various fields, such as academic research and technology. As of December 2020, there are more than 8,000 types of crypto assets.[1]

Because a blockchain network communicates directly between nodes, if there are a sufficiently many nodes in the network, the system as a whole can continue to operate even if some nodes fail. With this structure, the service can be operated permanently and the information stored in the blockchain can be retained. This approach has been proposed as a solution to the risk of loss of reference domains such as educational backgrounds and qualifications [5].

As mentioned above, sharing a blockchain on a network has various advantages. However, the termination of services using blockchain has not been fully addressed, and the blockchain network used for services terminated for some reason may continue to operate and be abused. Previous studies have confirmed

---

[1] CoinMarketCap https://coinmarketcap.com/all/views/all/.

This work was supported by a Grant-in-Aid for Scientific Research (B) (19H04107).

that blockchains are actually targeted for poisoning, [6,9], and are diverted to a command and control(C&C) server [1,2]. Although many previous studies have been performed on high-profile stocks such as Bitcoin and Ethereum, there are few surveys have been conducted on Altcoin, which is low-profile. Therefore, it is extremely important to analyze the risk of the blockchain network, which is a system common to crypto assets including Altcoin.

Our paper makes the following two contributions:

- It clarifies that there are active nodes in a worthless crypto asset network.
- It empirically confirms that there are malicious nodes regardless of the value of a crypto asset network.

2 Background

2.1 Blockchain Network

A blockchain is a data structure that connects blocks that store transaction groups, and is operated as a basic technology for various crypto assets such as Bitcoin [3]. A peer-to-peer (P2P) network consisting of nodes running by users is called a blockchain network. Combining the blockchain structure and characteristics of a P2P network, a blockchain network has three characteristics: tamper resistance, non-stopping, and decentralization.

An attacker can tamper with a single non-decentralized blockchain by re-calculating the random numbers. As a countermeasure against this type of attack, the same blockchain is distributed to the nodes of the blockchain network. A legitimate blockchain in the network makes the rule that it is the longest in the network. The network mines legitimate blockchains faster than an attacker can tamper with them. A legitimate blockchain in a network grows faster than a tampered blockchain, so the blockchain with which an attacker has tampered is not treated as a legitimate blockchain. Thus, tampering with blockchains is virtually impossible.

A P2P network has no single point of failure; the network does not fail as long as there are nodes, and the service downtime converges to zero. Consequently, such networks are resistant to attacks aimed at network outage, such as distributed denial of service attacks by attackers, and it is possible to carry out transactions without delay, even if the number of transactions increases. However, the fact that network continues to operate forever means that when a service using a blockchain network is performed, there is a risk that data will remain forever as long as the node remains.

2.2 Altcoin

Altcoin is an abbreviation for Alternative Coin, which often refers to crypto assets other than Bitcoin. To be precise, it has not been determined conclusively whether to refer to all crypto assets other than Bitcoin as Altcoin or only coins other than Bitcoin that are intended to be used as a currency [4], but in this paper it is treated as a currency other than Bitcoin. Ethereum, Litecoin, and Monacoin are well-known brands, and as of December 2020, there are more than 8,000 types of Altcoin with various functions and operational principles.

3   Related Work

3.1   Poisoning of Bitcoin Blockchain

By using the instruction OP_RETURN, which is a Bitcoin opcode, it is possible to freely insert a byte character string freely into the 80 B area that exists after the instruction word. A study conducted by Matzutt et al. [6,7] found that illegal data that infringed privacy and copyrights, such as group photographs and Bitcoin papers, ware injected by dividing it into an 80 B character string space and embedding into the blockchain.

Using the Bitcoin contamination method using OP_RETURN, it is possible to share data freely by dividing all data into 80 B and embedding them into the Bitcoin block, and network may be used for illegal data sharing.

3.2   Diversion of Bitcoin Network to C&C Server

Ali et al.'s paper [1,2] stated that there is a possibility that a blockchain can be used as a C&C server by developing data sharing using blockchain and embedding C&C instructions into the 80 B area.

Attackers with C&C communication use the server they manage. This method entails computational resource and electricity burdens for server management, and there is a risk that attacker's identity will be revealed by information such as the Internet protocol (IP) address.

Ali et al. claimed that diverting a blockchain network to a C&C server can reduce its disadvantages while retaining its advantages. By embedding the instructions in the block and connecting it to the blockchain, the only monetary cost involved in sending the instructions is the transaction transmission cost. In addition, as the direct information of IP addresses is not disclosed, there is a great advantage over using a server managed by an attacker. Furthermore, because the blockchain is shared by the P2P network, service outage is unlikely to occur.

3.3   Risk of Diversion of the Above Research to Blockchain Other Than Bitcoin

The above research was focused on the Bitcoin blockchain, but in the research by Sato et al. [9], conducted a similar experiment was conducted on the Ethereum blockchain. Bitcoin has a byte string area of 80 B in which information can be freely embedded, but an area called the init data area of Ethereum can store hundreds of killobytes of data. The transmission fee is also greater than that of Bitcoin.

This study suggests that Bitcoin risks may exist in Altcoin as well.

3.4   Position of This Research

Previous articles have mentioned the risk of blockchains being exploited in attacks, and if a blockchain and blockchain network are exploited by an attacker, the attacker will realize considerable benefits. High-profile stocks such as Bitcoin and Ethereum have been researched relatively extensively and have relatively high risks among crypto assets, but there is little discussion of the common problems of Altcoin.

Many Altcoins have smaller networks than Bitcoin and Ethereum, and it is thought that the characteristics of blockchain networks are not fully revealed.

Therefore, in this study, we confirmed the operating status of blockchain networks of worthless stocks from the perspective of market capitalization of stocks and considered the risks of worthless crypto asset networks in which the characteristics of blockchain networks are not fully utilized.

4   Node Status Analysis of Worthless Altcoin Network

In this analysis, we investigated the number of nodes in the Altcoin network, which is considered to be near the end of service. If no node is found to join a worthless network, the reason for the participation of the blockchain network user depends on the market capitalization, and the Altcoin network naturally decreases as the market capitalization decreases.

4.1   Analysis Content

We will investigated 238 stocks that CryptoID[2] handled as of June 11, 2020; obtained the USD market capital and number of nodes; and checked if there were nodes in the worthless stocks. For data acquisition, we used the Web application programming interface (API) provided by CryptoID.

4.2   Results

Table 1 shows the numbers of stocks with value and with and without nodes, as well as the average, maximum, and minimum numbers of nodes. Figure 1 presents a graph of the number of nodes versus the market capital. Stocks with a market capital of 0 are arranged in the order of their stock symbols, and the boundary between stocks with and without market capital is indicated by the broken line on the right side of the graph. The data were acquired on 2020/6/11.

---

[2] CryptoID https://chainz.cryptoid.info/.

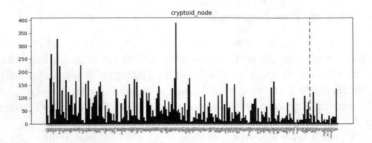

Fig. 1. Numbers of nodes of stocks

Table 1. Analysis result of the numbers of nodes

	Valuable	Not valuable
Total number of stocks	213	25
Number of stocks with nodes	213	24
Average number of nodes	63	31
Maximum number of nodes	390	134
Minimum number of nodes	1	0

Of the 238 stocks acquired from CryptoID, 24 (12%) do not have market prices. Of the 25 worthless stocks, 24 stocks (96%) have nodes, and almost all worthless stocks have nodes. Nodes exist in all 213 stocks with value. The average, minimum, and maximum numbers of nodes are greater for stocks with value than for those without value. It turns out that the USD market capital is not proportional to the number of nodes. No correlation can be seen between these quantities in Fig. 1.

5   Correlation Between USD Market Capital and Network Pollution Degree

In the previous section, we focused on market capital and confirmed that there are nodes in the network for which there is no valid reason for existence. The purpose of nodes that exist in the network without a valid reason is unclear, and these nodes pose a potential risk.

5.1   Purpose of This Analysis

This section identifies the motivation to join a network of nodes for which there is no reason to exist. Therefore, it addresses whether there is a correlation between high and low USD market capital of crypto assets and high and low network activity levels, the number of malignant judgments of nodes belonging to the network, the number of relationships with malignant identifiers, and the malignancy judgment ratio per node.

The number of malignant judgments refers to the number of pieces of antivirus software that have been judged to be malignant among all the VirusTotal antivirus software[3].

The number of relationships with malicious identifiers refers to the number of servers and pieces of software that have been judged to be malicious, with which the node to be scanned is communicating. The malignancy judgment ratio per node represents the number obtained by dividing the total number of malignancy judgments of the nodes belonging to the network by the number of nodes.

5.2   Blockchain Network Activity Level

The following discussion deals with an index called the network activity level, which is defined as the weighted standard deviation of the time interval of the blocks.

A decreased blockchain network activity level refers to a state in which the number of nodes participating in the network is decreasing, computational resources are exhausted, and distribution of transactions added to the block is decreasing.

5.3   Analysis Content

5.3.1   Calculation of Malignant Node Rate and Malignant Judgment Rate per Node

The analysis method is shown below.

1. The addition time of the latest block of each brand was obtained using index.data.dws of the CryptoID API.
2. Classification into three groups was performed: less than 1 min, less than 1 min and less than 3 min, and 3 min or more according to the time elapsed since the latest block was added.
3. The nodes of brands belonging to each group were scanned using the IP-address/report of VirusTotal API.
4. From the results, we calculated the percentage of nodes that received a malignant judgment and the percentage of pieces of malignant software detected for each number of malignant judgments.

5.3.2   Calculation of Network Activity Level

The calculation method was as follows.

1. The addition time of the latest 100 blocks was obtained using index.data.dws of the CryptoID API.

---

[3] VirusTotal https://www.virustotal.com/gui/home/.

2. Ninety-nine additional intervals were calculated from the additional time of 100 blocks.
3. The average and variance of the additional intervals were calculated.
4. Classification into four groups was performed: bottom 10 stocks with long average addition intervals, top 10 stocks with short average addition intervals, bottom 10 stocks with large variance, and top 10 stocks with small variance.
5. One node was randomly acquired from 10 stocks in each group (40 nodes were acquired in total).
6. Forty randomly acquired nodes were scanned using the IP-address/report of the VirusTotal API.
7. The average was calculated by performing 10 sets of 5 and 6.

Based on the above analysis method, the graph was calculated by sorting in descending order by the USD market capital of the target stocks. The data used in this analysis were current as of October 29, 2020.

5.4   Results

Figure 2 presents the network activity levels of each crypto asset network. Figure 3 depicts the total number of malicious judgments of nodes belonging to each network. Figure 4 illustrates the total number of connections with malicious servers and software on the Internet. Figure 5 shows the average number of malicious judgments per node belonging to each crypto asset network. For all graphs, the x-axis represents the capital of the crypto asset stock in the USD market (in descending order from left to right).

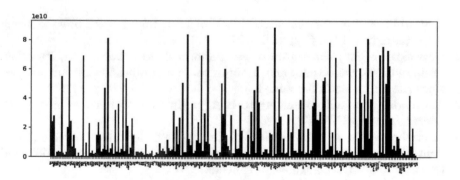

Fig. 2. Network activity level of each crypto asset network

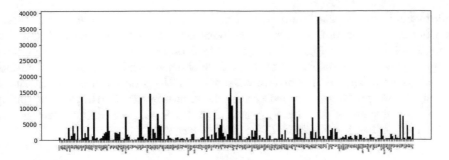

Fig. 3. Total number of malicious judgments of nodes belonging to each crypto asset network

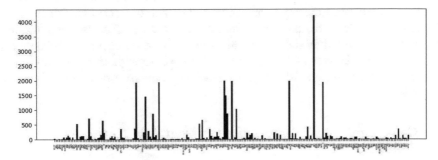

Fig. 4. Numbers of connections with malicious servers and pieces of software on the Internet

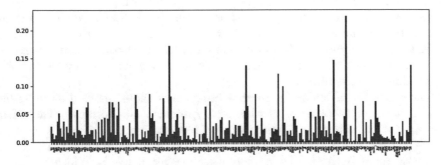

Fig. 5. Average number of malicious judgments per node belonging to each crypto asset network

Because the network activity level for each crypto asset network in Fig. 2 shows the number of digits of variance, the larger the bar in the graph, the lower the activity. Looking at the number of malignant judgments for each network in Fig. 3, it can be seen that the number of malignant judgments for many brands is 5000 or less. However, as Fig. 3 is the total of the results of scanning the nodes belonging to the network, it is considered that the number of malignant judgments increases as the number of nodes increases. The same applies to the number of relationships between the nodes belonging to the network and the malicious identifier in Fig. 4. Therefore, Fig. 5 depicts the number of malignant judgments per node, regardless of the scale of the network by dividing the number of malignant judgments in the entire network by the number of nodes. In Figs. 3 and 4, the protruding parts are almost the same, but in Fig. 5, the protruding parts do not match. In Figs. 2, 3 and 4, there is no correlation between the market capital of the stocks and the survey target.

6   Consideration of Network Non-disruption

In this study, in order to investigate the risk at the end of the service of the Altcoin network, we confirmed the status of the Altcoin network, which is considered to be difficult to maintain because the number of users is small due to lack of transaction value and decentralization cannot be ensured. We were able to confirm the existence of observable nodes even though they had no market value. Based on these results, we considered the non-stop nature of blockchain networks.

6.1   Consideration of Factors that Prevent the Networks from Stopping

As shown in Table 1, an average of 31 nodes was confirmed even in non-value currencies. It is unlikely that these nodes happened to come into contact with each other during the period and more likely they were operating for another purpose. This survey cannot determine the motivation for a node that is connected to the network to join the Altcoin network. However, in currencies with up to 134 connections, the existence of all nodes in one worthless Altcoin network seems to have some purpose other than legitimate purposes such as currency transfer and observation of transactions.

The existence of nodes for reasons other than those justified as described above is not preferable for users who use cryptocurrencies correctly. The reason that the service can be continued without terminating even if nodes exist for invalid reasons is that the majority is prioritized in determining the service continuity, as in Byzantine fault tolerance. However, it can be inferred that the decision to terminate the service may be due to the fact that some malicious opinions have become effective and the service cannot be terminated.

6.2   Benefits for Attackers Who Keep an Anonymous Altcoin Networks up and Running

Firstly, information sharing on blockchain has been be mentioned in previous research. Unfamous stocks with fewer users and low activity networks are maintained by some users. This network configuration is not healthy. It is thought that some users who make up such networks continue to operate the blockchain to enjoy the benefits that other general users do not have.

Well-known brands such as Bitcoin and Ethereum have websites operated by third parties that can obtain the information stored in blocks. On the other hand, unknown brands rarely operate such websites, so information can only be obtained by combination with a node of the blockchain network and sharing the blockchain. Furthermore, some Altcoin node software is not available, making it virtually impossible for new nodes to enter. When users who have server software run the server software with their peers, some users will be able to continue running the blockchain network when there are no general users. By using this approach, it is possible to share files without being seen by others, decentralizing information, and considering the risk of loss.

6.3   Correlation Between the Asset Value of Crypto Assets and Risk

In Analysis 2, we investigated whether there was a correlation between the market capitalization of approximately 240 crypto assets that were surveyed, number of malignant judgments, number of relationships with malignant identifiers, and network activity level. Crypto assets have monetary value as they are called assets, and high and low values are important, so stocks with monetary value, that is, high market capital, tend to attract attention. Therefore, it is thought that the number of nodes participating in the network will increase. Conversely, stocks with low market capital are thought to have fewer nodes, networks with few users are expected to be less active, and the number of abandoned nodes judged to be malignant increases. However, Figs. 2, 3, 4 and 5 reveal no correlation between the market capital and the number of malignant judgments, number of relationships with malignant identifiers, or network activity level.

The reason for the lack of correlation with the number of malignant judgments and malignant identifiers is that the merit of an attacker attacking a node exists in stocks with both high and low market capital. We expected that nodes belonging to stocks with low market capital would tend to be left unattended and that there would be potential risks, such as not updating server software. It is conceivable that an attacker will come up with the same idea and attack. On the other hand, for stocks with high market capital, an attack that invades a node and seizes the assets it holds is suitable.

As described above, it is considered that the number of malignant judgments and the relationship with the malignant identifier were distributed as a whole, and no correlation was found because the brand to be attacked changed depending on the objective of the attacker.

7   Conclusion

In this study, we confirmed that there are nodes in a worthless Altcoin network for which the reason for existence is unthinkable and investigated the relationships between USD market capital, network activity level, and malicious software. As a result, it was confirmed that there are active nodes in a worthless crypto asset network and that there are malicious nodes regardless of the worth of the network.

References

1. Ali, S.T., McCorry, P., Lee, P.H.-J., Hao, F.: Zombiecoin: powering next-generation botnets with bitcoin. In: International Conference on Financial Cryptography and Data Security, pp. 34–48, Springer (2015)
2. Ali, S.T., McCorry, P., Lee, P.H.-J., Hao, F.: Zombiecoin 2.0: managing next-generation botnets using bitcoin. Int. J. Inform. Secur. **17**(4), 411–422 (2018)
3. Binance academy. Difference Between Blockchain and Bitcoin. https://academy.binance.com/en/articles/difference-between-blockchain-and-bitcoin. Accessed on 6 Jan 2021
4. Bit flayer. What is an Altcoin. https://bitflyer.com/en-eu/faq/55-7. Accessed on 6 Jan 2021
5. Holotescu, C., et al.: Understanding blockchain opportunities and challenges. In: Conference proceedings of eLearning and Software for Education (eLSE), vol. 4, pp. 275–283. Editura Universitatii Nationale de Aparare , Carol I" (2018)
6. Matzutt, R., Henze, M., Ziegeldorf, J.H., Hiller, J., Wehrle, K.: Thwarting unwanted blockchain content insertion. In: 2018 IEEE International Conference on Cloud Engineering (IC2E), IEEE, pp. 364–370 (2018)
7. Matzutt, R., Hiller, J., Henze, M., Ziegeldorf, J.H., Mullmann, D., Hohlfeld, O., Wehrle, K.: A quantitative analysis of the impact of arbitrary blockchain content on bitcoin. In: Proceedings of the 22nd International Conference on Financial Cryptography and Data Security (FC), Springer (2018)
8. Nakamoto, S., et al.: Bitcoin: A peer-to-peer electronic cash system
9. Sato, T., Mitsuyoshi Imamura, K.O.: Threat analysis of poisoning attack against ethereum blockchain. In: IFIP International Conference on Information Security Theory and Practice (WISTP 2019), vol. 9 (2019)

Human Centered Protocols in Transformative Computing

Lidia Ogiela[1]([⊠]), Makoto Takizawa[2], and Urszula Ogiela[1]

[1] Pedagogical University of Krakow, Podchorążych 2 Street, 30-084 Kraków, Poland
[2] Research Center for Computing and Multimedia Studies,
Hosei University, 3-7-2, Kajino-cho, Koganei-shi,
Tokyo 184-8584, Japan
`makoto.takizawa@computer.org`

Abstract. This paper presents new protocols dedicated to transformative computing technology. The main idea is based on human centered solutions characteristic for individual human perceptual and decision-making processes. Proposed solutions will be based on perception, recording and description processes leading to the meaningful data analysis. Evaluation of proposed techniques will be presented, as well as possible application for such protocols in transformative computing areas.

Keywords: Human centered protocols · Data analysis · Transformative computing

1 Introduction

Due to their nature, transformative computing allow to processing of data with different notation, format and quantity. Their characteristic feature is that the data obtained in the recording process with the use of various sensors are processed and analyzed in real time in response to their modifications and changes [1–3, 6, 10].

Transformative computing allows to obtain data registered by a set of sensors while considering the assessment of the significance of the acquired data. Also, allow to the process of significant data analysis, which, based on the assessment of the degree of influence of individual information on the analyzed data set, allows to a deeper interpretation and evaluation. These processes were discussed in the [4, 6], where the main advantages of the presented solution were presented.

An extension of transformative methods with elements of semantic interpretation allowed to take into account changes that have a significant impact on the course of the entire analysis process [4–6, 9]. At the same time, it made it possible to properly assess the impact of individual pieces of information on the overall analysis process. The assessment of significance is carried out on the basis of a cognitive analysis based on the formalisms of data linguistic interpretation. The use of linguistic formalisms for interpretation processes allows to conclude what is the impact of individual data in the entire analyzed data sets.

© The Author(s), under exclusive license to Springer Nature Switzerland AG 2021
L. Barolli et al. (Eds.): AINA 2021, LNNS 226, pp. 603–607, 2021.
https://doi.org/10.1007/978-3-030-75075-6_49

The novelty of this work is the development of a solution dedicated to distributed service management processes based on the use of transformative computing methodology. Therefore, the discussed issue will concern the possibility of using these methods in the processes of data protection and security in distributed systems.

2 Distributed Service Management in Human-Centered Concept

Distributed service management concept has been proposed to improve information management processes. From this perspective, all services are understood as a kind of information that determines their type, form, offer, availability, supplier, concurrency, etc. The concept of a service in the context of the information it contains allows for its deeper analysis and proper understanding. Individual associations and preferences influence the choice of service management method.

The service management process contained in systems with a distributed structure, such as cloud or fog, allows for their more functional operation, provides easier access to them for both the provider and the recipient, and minimizes the costs of service management. It allows for quick modification and adaptation to the preferences of recipients. The assessment of recipients' preferences is carried out on the basis of the meaning interpretation of the services offered and it is possible due to the assessment of the degree of influence of individual preferences on the final demand for the service.

Distributed service management concept allows for the implementation of management processes both from the level of the organization offering specific services and from the superior levels of their protection are carried out. Distributed service management concept guarantees safe access to strategic and confidential information, only to persons authorized to have it.

The basic elements of building distributed service management idea are based on:

- selection of a distributed structure,
- the method of data distribution between the basic and distributed structures,
- choosing the method of safeguarding protected and confidential data.

An important element of the design of the appropriate protocol is the choice of the method of data protection. In such solutions, a useful method of securing data is the use of cryptographic threshold schemes, and particularly important are linguistic and biometric threshold schemes that guarantee information protection based on the assessment of the degree of importance of individual preferences and the characteristics of biometric features of all protocol participants. The main idea of the mentioned solutions in the field of data protection was presented in the works [6–8].

Distributed service management concept using linguistic and biometric threshold schemes are used to secure information both at the level of the basic structure – the service provider – and at the other management levels – fog and cloud. The choice of the management levels does not directly affect the possibility of using the preferred threshold schemes. They are universal and can be used at different of data management levels. The difference that may occur in the process of their adaptation is related to the possibility of data distribution between particular levels of the structure:

- In the case of a single-level structure, the process of distributing shadows generated in threshold diagrams in order to protect information, will be implemented only from and at the basic level of the structure.
- In the case of two-level structures (organization-fog, organization-cloud), the shadow generation and distribution process can be carried out at the structure level (sub-level), and the existence of a higher level provides the ability to read and manage the secret from the higher level.
- In the case of three-level structures (organization-fog-cloud), the process of generating and distributing shadows can be carried out at the level of the structure (lowest level) and the existence of higher levels (fog, cloud) provides the ability to read and manage the secret on each of them. At the same time, managing a secret from a higher level (fog) can be independent and different from the process of managing and recreating a secret on the highest level (cloud).

The horizontal structure of service management processes is also typical for obtaining data from distributed systems. These systems record data from a variety of readers, sensors and recorders, and their combination is characteristic of transformative methods and computing.

3   Transformative Computing for Proposed Protocols

Transformative computing carried out at the structure level are used to obtain various information recorded by specific sensors, on the basis of which it will be possible to perform a multifaceted analysis of the issue. The method of selecting the sources of data acquisition and recording depends on the technical, computational and financial capabilities of the organization.

Recorded data is subject to both preliminary analysis and their semantic interpretation, which is performed at the level of a given organization. The meaningful assessment of the obtained information allows for the extraction of the data that will have a significant impact on the further course of the entire process. Due to the acquisition of data in the process of continuous registration, their evaluation and analysis is also carried out in a continuous manner, taking into account possible modifications and changes.

The processed data is sent to higher management levels, where the processes of their protection are carried out. The use of linguistic and biometric protocols makes it possible to secure this information which, due to its nature, cannot be made public or disclosed. At the same time, the implementation of transformative computing methods allows to obtain new data registered at higher management levels, where they can be directly protected and classified. Then it will be possible to implement parallel data protection protocols using the same threshold schemes or completely independent solutions, guaranteeing full data protection.

The scheme of such solution presents Fig. 1.

The formal record of this protocol as is follows.

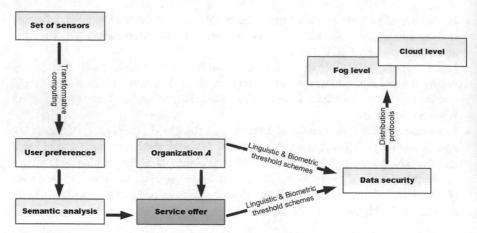

Fig. 1. The scheme of use transformative computing methodology in distributed service management protocols.

Protocol 1. Human-centered transformative protocol.

1. determination of the set of external sensors
2. data recording with the use of transformative computing techniques
3. recording of user preferences (human centered solutions)
4. semantic analysis and determination of the meaning of the obtained results
5. service offer (firm A)
6. application of linguistic and biometric threshold schemes to protect sensitive/confidential data (human centered solutions)
7. use of secret distribution protocols to secure confidential parts of information (shadows)
8. choice of service management levels (fog/cloud).

4   Conclusions

This paper proposes a new management and data protection solutions dedicated to the service area. The novelty of such solution is the possibility of using transformative computing in the process of obtaining information, its analysis as well as quick and efficient distribution.

The process of meaningful interpretation of the recorded data allows to clearly indicate the data that will be of key importance for the optimization of the service offer.

Additionally, the developed methodology was enriched with the possibility of incorporating linguistic and biometric data partitioning techniques in order to fully protect them.

It has also been shown that the proposed methodology can be effectively used in the process of distributing secret parts of a shared secret to different structures and at different levels of information management. The versatility of the developed solution results from the use of universal data protection techniques in the form of threshold schemes as well as transformative-cognitive techniques based on human-centered analysis, which together provide an effective tool supporting complex data management processes.

Acknowledgments. This work has been supported by the National Science Centre, Poland, under project number DEC-2016/23/B/HS4/00616.

References

1. Benlian, A., Kettinger, W.J., Sunyaev, A., Winkler, T.J.: The transformative value of cloud computing: a decoupling, platformization, and recombination theoretical framework. J. Manag. Inf. Syst. **35**, 719–739 (2018)
2. Gill, S.S., et al.: Transformative effects of IoT, blockchain and artificial intelligence on cloud computing: evolution, vision, trends and open challenges. Internet Things **8** (2019). Article Number 100118
3. Nakamura, S., Ogiela, L., Enokido, T., Takizawa, M.: Flexible synchronization protocol to prevent illegal information flow in peer-to-peer publish/subscribe systems. In: Barolli, L., Terzo, O. (eds.) Complex, Intelligent, and Software Intensive Systems. Advances in Intelligent Systems and Computing, vol. 611, pp. 82–93. Springer, Cham (2018)
4. Ogiela, L.: Cognitive informatics in automatic pattern understanding and cognitive information systems. In: 7th IEEE International Conference on Cognitive Informatics (ICCI 2008), Stanford University, Stanford, CA, 14–16 August 2008. Advances in Cognitive Informatics and Cognitive Computing, Studies in Computational Intelligence, vol. 323, pp. 209–226 (2010)
5. Ogiela, L.: Cryptographic techniques of strategic data splitting and secure information management. Pervasive Mob. Comput. **29**, 130–141 (2016)
6. Ogiela, L.: Transformative computing in advanced data analysis processes in the cloud. Inf. Process. Manag. **57**(5) (2020). Article Number 102260
7. Ogiela, M.R., Ogiela, U.: Secure information splitting using grammar schemes. In: Nguyen, N.T., Katarzyniak, R.P., Janiak, A. (eds.) New Challenges in Computational Collective Intelligence. Studies in Computational Intelligence, vol. 244, pp. 327–336. Springer, Berlin (2009)
8. Ogiela, M.R., Ogiela, U.: Linguistic cryptographic threshold schemes. Int. J. Future Gener. Commun. Netw. **2**(1), 33–40 (2009)
9. Ogiela, U., Takizawa, M., Ogiela, L.: Classification of cognitive service management systems in cloud computing. In: Barolli, L., Xhafa, F., Conesa, J. (eds.) Advances on Broad-Band Wireless Computing, Communication and Applications. BWCCA 2017. LNDECT, vol. 12, pp. 309–313. Springer, Cham (2008)
10. Yan, S.Y.: Computational Number Theory and Modern Cryptography. Wiley, Hoboken (2013)

Off-Chain Execution of IoT Smart Contracts

Diletta Cacciagrano, Flavio Corradini, Gianmarco Mazzante, Leonardo Mostarda[✉],
and Davide Sestili

University of Camerino, Camerino, Italy
{diletta.cacciagrano,flavio.corradini,gianmarco.mazzante,leonardo.mostarda,
davide.sestili}@unicam.it

Abstract. Modern blockchains allow the definition of smart contracts (SCs). An SC is a computer protocol designed to digitally ease, verify, or enforce the terms of agreement between users. SCs execution can require high fees when lots of computation is required or a high volume of data is stored. This is usually the case of Internet-of-Things (IoT) systems where a large amount of devices can produce a high volume of data. Off-chain contract execution is a viable solution to decrease the blockchain fees. Users can agree on an on-chain SC which is stored in the main chain. Computation can then be moved securely outside the chain to reduce fees. In this paper we propose DIVERSITY a novel approach that allows off-chain execution of SCs. DIVERSITY provides a novel model for defining on-chain contracts that can be securely executed by using a novel off-chain protocol. We have validate our approach on a novel IoT case study where fees have been greatly reduced.

1 Introduction

The success of Bitcoin and its increasing use has placed blockchain as a promising solution for enabling trust in a decentralised system. Bitcoin [10] and Ethereum [17] are but a few examples of cryptocurrencies that make use of blockchain in order to enable payments without a trusted party (e.g., a bank). Modern blockchains also allow the definition of complex smart contracts (SCs). These automatically execute, control or document legally important events and actions according to the contract terms or the agreement. SCs are run by miners which can result in slow execution time (e.g., 10 min to mine a Bitcoin block while 20 s to mine an Ethereum one) and expensive fees [12, 13]. These issues are exacerbated in Internet-of-Things (IoT) systems, where a large amount of devices can produce a high volume of data which require real-time elaboration. There is a voluminous literature about IoT scalability methods and consensus protocols for blockchains [14]. As the authors in [11] show, IoT devices can require high monetary cost for storing and elaborating data and the time required to mine and validate blocks can be too high.

Off-chain computation seems a viable solution to decrease the blockchain fees while reducing the response time [12, 13]. Off-chain approaches are not bound to the transactional speed and high fee limitations of the on-chain transactions since they move

This paper has been supported by the italian national project industry 4.0.

the computation outside the main chain. Although off-chain computation can take several forms, it usually requires two or more users to agree on an on-chain SC which is stored in the main chain. Computation can then be moved securely outside the chain. In [6,7] the authors explore different aspects about off-chaining techniques. In [7] they listed five different off-chaining patterns, such as performing a high number of microtransactions off-chain and then summing them up on a single on-chain transaction (e.g., Lightning Network [13]) or such as storing a large amount of data off-chain and putting only the reference on-chain (e.g., Swarm [16]). In [6] they analyse a couple of generic models for moving computation or storage off-chain. Then they drew up an overview of off-chain computation techniques and compared them with respect to scalability, privacy, security and programmability providing also some examples of implementations. In a different work [3] Eiberhart et al. propose an off-chain model in which SC computation is off-loaded to an external node to improve scalability and to enable private data to be used in SC execution. The main blockchain contains a verification SC which verifies the result of a computation received from the off-chain node along with a zkSNARK proof [9]. Their proposed model achieves greater scalability since on-chain verification is less computationally expensive than on-chain execution.

A noticeable contribution in the literature is the Lightning Network [13] which solves the scalability challenge of the Bitcoin blockchain. Lightning allows users to perform multiple microtransactions over an off-chain peer to peer network without any broadcasting on the main blockchain. Users can take back their balance on the main blockchain by providing a proof that their balance in the lighting network is the one claimed, if the proof is not challenged then they are allowed to take back their balance on the main chain. This protocol allows users to perform several microtransactions that would be otherwise prohibitively expensive in terms of fees if they were performed on the main blockchain. A similar proposal but for the Ethereum network is Plasma [12]. Its structure consists of a tree of blockchains. A root Ethereum blockchain can be connected to multiple Plasma child chains which can themselves have other child chains. The tree structure of the Plasma blockchain enables users to offload transactions and computation to child chains while enforcing correct behaviour of the child chains through fraud proofs. This can be filed to the parent blockchain in case malicious behaviour is detected, allowing them to roll back a block or retrieve their funds.

On the other hand, on-chain computation has its representatives in the literature. Since devices with limited storage capabilities cannot store ever-growing blockchains - such as Bitcoin - Richard Dennis et al. [4] proposed the concept of a rolling blockchain, in which only data stored for a predetermined period of time is stored in the chain and any data older than that is automatically discarded.

In [15], Shahid et al. present two interesting aggregation approaches enabling blockchain footprint to scale gracefully on low memory sensor nodes. The first approach is time-based, and exploits a time window concept. Sensors store blocks for a certain time window, then they aggregate data into that time window and they sum them up in an aggregated block. The second approach consists in dividing the blockchain in geographical cells, where each node is only required to store blocks about nodes in its cell, thus saving storage. Finally, they combine the two approaches in order to optimize the blockchain footprint to its limit. These approaches have two main flaws, unfortunately. The first one is that they require to delete and re-create a new blockchain for each aggregation step. The other one is that even though storage space is preserved, IoT nodes are still required to perform all the computation needed for the maintenance of a blockchain.

In [5] the authors propose a proxy-based approach to the off-chain computation. They implement a blockchain proxy since the use of a blockchain is infeasible for a low-power IoT nodes. A Certification Authority issues an identity certificate for each sensor node, which in turn uses it to sign its blockchain transactions. Then the sensor sends the signed transactions to the proxy, which commits them to the ledger. Also, they prove the potential of their approach proposing an interesting use-case scenario about IoT sensors for cold-chain monitoring. The use of a proxy allows IoT nodes to sign their own transactions without requiring them to store blocks or perform heavy computation. Nevertheless, the proxy is easily identifiable as a single point of failure, and it can create scalability issues with growing sensor networks.

Joshua Elul et al. [8] focus on IoT behaviour definition through blockchain smart contracts. They proposed a split virtual machine architecture to enable IoT device programming through the Ethereum Blockchain. In their proposed architecture an IoT device communicates with a special node in the blockchain which looks for smart contracts containing code to be executed on IoT devices. The code is then sent to the device that will execute it. IoT devices can also request data to the special node when needed. The behaviour of the IoT device is encoded in the contracts bytecodes deployed on the main blockchain, this allows for IoT application code to be developed for the specific smart contract without requiring manual updating of the end IoT devices.

1.1 Contribution of the Paper

This paper introduces DIVERSITY (Div), a novel second layer decentralised network that allows off-chain and secure execution of SCs. Div is suitable when SCs require a large amount of CPU computation or high volume of data needs to be processed. In this cases Div can lower the high fees required by the main chain and reduce the computation performed by the blockchain peers. The Div protocol uses an on-chain SC as a trust and arbitration means in order to securely perform an off-chain computation and notify any dispute. The on-chain structure also allow the definition of reactions when unanimous consensus is reached (e.g., the transfer of cryptocurrency). Div allows users to define on-chain SCs with a well defined structure. A contract specifies sensors, off-chain computation to be performed and the intermediate nodes that will execute it. We assume each user can use its intermediate nodes although delegation mechanisms can be used. The intermediate nodes execute a secure protocol that verifies unanimous agreement on the result of the off-chain computation. An honest node that performs a correct computation will be enough to detect any dishonest ones and open a dispute. While any blockchain which supports the execution of SCs can be enhanced with the Div novel approach, its current version supports Ethereum and the application case studies have been performed in the IoT area. In this context we assume there are some IoT devices (in the following refereed to as sensors) producing data plus two or more users that are interested in verifying and enforcing the terms of an agreement. Our contributions can be summarised as follows: (i) Div provides a novel combination of a unique model for defining on-chain contracts with a protocol for off-chain secure execution; (ii) Div

intermediate nodes can run an off-chain code (e.g., Java) that meet the demands of specific use cases, specifically those that are not feasible on the considered blockchain (e.g., Ethereum); (iii) Its peculiar computation model is based on streams of data thus is suitable to IoT data processing; (iv) We have validate our approach on a novel lock case study where fees have been greatly reduced.

2 Motivation Scenario

A smart lock is installed on different electric bicycles that are deployed in a smart city. A user can register on the service provider web site in order to buy a token k for renting the bicycles for a certain amount of time T_k. Every time a user u interacts with the locker of a bicycle b a message $m = [b,k,s,t]$ is received by the server where (i) s is equal to either u (i.e., lock unlocked) or l (lock locked); (ii) t is the time at which the operation was performed. Suppose that a token k is used to open a bicycle at time t_1 and the bicycle is closed at time t_2, then we denote with $t(k,t_1,t_2) = t_2 - t_1$ the time between a lock and an unlock operation (i.e., the time the bicycle was used). We denote with $T_d(k)$ the total amount of time the token k was used in a day. This can be defined as $T_d(k) = \sum_i t(k,t_i,t_{i+1})$ where all t_i belongs to the same day. An insurance company will bill the server provider monthly by considering the monthly bicycle rental time. This can be calculated as follows $T_m(k) = \sum_{k \in K_m} T_d(k)$ where K_m are all tokens that were used in the month m.

The first version of the system would store all lock and unlock operations at the service provider database. This would calculate the monthly bicycle rental time at the end of each month in order to pay the insurance company. This solution was immediately found inappropriate by the insurance that would rather prefer the use of the blockchain in order to ensure data immutability and avoid the need of a third trusted part. The main problem of the blockchain solution is that an high volume of data would be stored in the blockchain. This would cost additional money that would increase the cost of the service of the final user. A SC for calculating the monthly bicycle rental time had also to be run at the end of each month. This would transfer the money from the wallet of the service provider to the insurance one. This would cost money as well. $\mathcal{D}iv$ was used to greatly reduce the amount of money paid by allowing off-chain and secure execution of the terms of agreement between insurance and service provider.

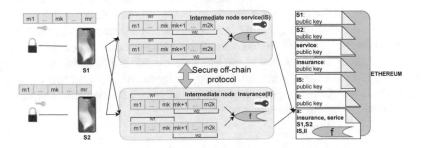

Fig. 1. $\mathcal{D}iv$ applied to the smart lock case study.

Figure 1 outlines the $\mathcal{D}iv$ architecture for the lock case study. The first step when using $\mathcal{D}iv$ is the definition of an on-chain SC. This specifies the following main parts: (i) the set of sensors that produce data; (ii) all users that are involved in the contract; (iii) the off-chain computation to be performed; (iv) a set of intermediate nodes that will perform the off-chain computation; (v) the quality of service. Users, sensors and intermediate nodes must be registered inside the main chain before the on-chain smart contract definition. Although they have different data, their registration always require a unique ID and a public key.

Figure 1 shows the lock case study where two locks (i.e., the sensors s_1 and s_2, the insurance and service intermediate nodes) have been registered. The on-chain SC a that contains the off-chain code f has also been registered. At run time, each sensor generates messages that contains its id, its type, some data and the sensor's signature. This is used in order to ensure integrity, authentication and non-repudiation of the messages. In our trust model we assume that sensors are trusted. Each sensor message is multi-cast to a set of intermediate nodes. Each intermediate node organises the sensor messages into windows that have a fixed-length size. A window is the basic computation unit that is used to compute the terms of agreement that is the off-chain code (in the following denoted with f). This has been agreed between two or more users that do not trust each other. The off-chain code can perform data aggregation and/or event generation and can output a value to be stored in the on-chain SC. $\mathcal{D}iv$ leverages on the use of two or more intermediate nodes for off-chain computation. Users can have their own intermediate nodes but they can also trust external ones. When external intermediate nodes are used fees can be transferred after their correct execution is validated. Intermediate nodes run a secure protocol that ensures the following properties: (i) when all intermediate nodes perform the same off-chain computation on the same window (i.e., unanimous consensus is reached) the on-chain contract is updated and the reaction is run; (ii) if unanimous consensus is not reached a dispute is open. In our lock case study the off-chain code computes the monthly bicycle rental time and stores it inside the related on-chain contract. Insurance and service providers have their own intermediate nodes. This will run the off-chain code which is expected to give the same output for the same windows. Locks are installed by a third-party company.

$\mathcal{D}iv$ provides the following three basic quality of service: (i) no-proof of storage; (ii) off-chain proof of storage; (iii) on-chain proof of storage. These quality of services always store on the on-chain SC the output of a non-null computation and log any dispute that emerges. The quality of services differ for the policy they use to store sensor messages. In the no-proof of storage service, window data are discarded after the unanimous off-chain computation is reached. When a dispute is detected data are locally stored at the intermediate nodes for dispute resolution. This solution minimises the amount of data that are stored in the intermediate nodes and in the main chain. In the off-chain proof storage service, all data are locally stored by the intermediate nodes. When on-chain proof of storage is used data are stored on the on-chain contract when computation is not null or a dispute emerges.

3 $\mathcal{D}iv$ System Model

In our system model we assume to have a set U of users that are interested in defining agreements. We denote with $u_1 \ldots u_n$ elements in U. We assume to have a set of sensors S that produce data. We denote with $s_1, \ldots, s_p$ elements in S. Each sensor s is declared into the main chain by using a declaration of the form $d_s = \{PU_s, c_s, type_1 \; x_1, \ldots, type_h \; x_h\}$ where PU_s is an asymmetric public KEY, c_s is the type of the sensor (such as temperature or light) while x_i is a variable of the type $type_i$. While running a sensor s must send messages that comply with its declaration. A sequence number is used to totally order the sensor messages. The q-th message is denote with $m_{s,q}$ and is of form $m_{s,q} = (PU_s||c_s||q||d_q) : s$ where q is the sequence number, $d_q = v_{x_1}, \ldots, v_{x_h}$ is the array of values assigned to variable $x_1, \ldots, x_h$ and $: s$ is the signature of the whole message performed by the sensor s. Each sensor s generates a local trace $Y_s = m_{s,0} \ldots m_{s,q} \ldots$. We assume to have a set of intermediate node I and we denote with $i_1, \ldots, i_m$ elements in I. An intermediate node i can receive messages from various sensors thus it can observe various sensor traces.

We use a window as a basic unit of computation. A window is a sub-string of a sensor trace Y_s. We recall that a sub-string is a contiguous sequence of characters within a string. We define with $w(s,q,z)$ a window of z messages $(m_{s,q}m_{s,q+1} \ldots m_{s,q+z-1})$ that belong to the sensor trace Y_s. We call q the starting point of the window. Our execution model is based on a sequence of windows. We denote with $W(s,q_0,sl,z)$ the sequence of windows $w(s,q_0,z), w(s,q_0+sl,z), w(s,q_0+2sl,z), \ldots$ where $w(s,q_0,z)$ is the starting window and sl is a positive integer referred to as sliding factor. Effectively, each window can be obtained by sliding the previous of sl messages. For the sake of presentation simplicity we denote with $w_s(t)$, with $t = 0$, the first window $w(s,q_0,z)$ while with $w_s(t)$ the window $w(s,q_0+t\times sl,z)$ (i.e., the window after t slidings). Effectively, the notation $w_s(t)$ omits the sliding factor sl and window size z. Windows from a set of different sensors $S = \{s_1 \ldots s_k\}$ can be organised into an array of windows. We denote with $w_S[t]$, with $t = 0$, the starting array $[w_{s_1,0}, \ldots, w_{s_k,0}]$ while with $w_S[t]$ the array $[w_{s_1}(t), \ldots, w_{s_k}(t)]$. We use $w[t]$ instead of $w_S[t]$ when the set S is clear from the context. We use $Q_0 = \{q_{s_1,0}, \ldots, q_{s_k,0}\}$ to denote the window starting points.

We assume to have a set A of on-chain smart contracts and we denote with $a_1, \ldots, a_n$ elements in A. Each smart contract a is a tuple $(U_a, S_a, Q_0, f_a, I_a, r_a) : U_a$ where: (i) $U_a = \{u_1 \ldots u_n\}$, with $n > 1$, is a set of users with $U_a \subseteq U$; (ii) $S_a = \{s_1 \ldots s_k\}$ is a set of sensors with $S_a \subseteq S$; (iii) $Q_0 = \{q_{s_1,0}, \ldots, q_{s_k,0}\}$ contains the starting points of all windows; (iv) f_a is the off-chain computation function that takes as an input the array $w_{S_a}[t]$ of window $[w_{s_1}(t), \ldots, w_{s_k}(t)]$ and outputs a result (i.e., a string) or null; (v) $I_a = \{(u_1, I_{u_1}) \ldots (u_n, I_{u_n})\}$ is the set of intermediate nodes where for each tuple (u_i, I_{u_i}) each u_i is a user in U_a and $I_{u_i} \subseteq I$ is a set of intermediate nodes trusted by u_i; (vi) r_a contains a reactions that is a piece of code that is executed when f_a is not null; (vii) $: U_a$ is the signature of all the users.

An SC a specifies a set of users $U_a = \{u_1 \ldots u_n\}$ that are interested in the agreement. The terms of agreement concerns the data produced by the sensors in S_a. Each user $u_i \in U_a$ can specify a set of intermediate nodes that trusts for the computation. This is done by adding the tuple (u_i, I_{u_i}) to the contract where I_{u_i} is a set of intermediate nodes trusted by u_i. Each intermediate node can run the function f_a that takes as an input the array of windows $w_{S_a}[t]$ (in the following denoted with $w[t]$). When $f_a(w[t])$ is *null* no data will be written in the on-chain contract a otherwise the result will be store inside a SC associative map map_a. The unique execution entry t will be used for storing a three-tuple $(f_a(w[t]), P_t, w[t])$, where $f_a(w[t])$ is the off chain computation of the windows $w[t]$ and P_t its proof of computations. This ensures that all intermediate nodes in I_a reached unanimous consensus. Otherwise a dispute will be logged. The next section describes the proof of computation P_t in details. When no-proof storage and off-chain proof quality of services are considered the $w[t]$ data will be empty, for the on-chain proof storage service $w[t]$ is stored on the map. The on-chain contract also defines a reaction r_a. This is a piece of code that can be executed when f_a is evaluated on a new window and all the intermediate nodes reach consensus on the result. For instance reactions can include exchange of cryptocurrency between accounts, auditing, or actuation of IoT devices.

Consensus and Security: Our main goal is to ensure that the intermediate nodes securely and correctly perform the off-chain SC execution. This means that the following two properties must be ensured: (P1) all nodes agree on the same computation otherwise a dispute is open; (P2) a node cannot learn and copy the computation of another node. This last property tries to avoid lazy nodes. Correct SC execution is a achieved by a secure computation protocol that, for each window, performs the following three phases: (i) proof of computation; (ii) computation agreement checking; (iii) SC update.

Proof of Computation: Each intermediate node i calculates the off-chain code $f(w_i[t])$ by considering its local window $w_i[t]$. Instead of revealing the computation $f(w_i[t])$ the intermediate node calculates the proof of computation $p_{t,i} = H(f(w_i[t])||S_{t,i}||ID_i)||t||a||H(w_i[t]))$ where H and $||$ denote a cryptographic hash function and the concatenation operation respectively; $S_{t,i}$ is a fresh random secret that uniquely identifies the t-th computation of f; ID_i is the identity of i; a and t the contract id and the window unique identifier; and $H(w_i[t])$ the hash of the window data. This is used by an intermediate node j to check that its local window has the same hash $H(w_j[t])$. When $H(w_i[t]) \neq H(w_j[t])$ j notifies a dispute on the on-chain contract. The intermediate node i signs and sends its proof of computation to all other nodes for signature. At the end of the proof of computation phase each intermediate node must have the signed proof of computation of all nodes. In the following we denote with $M : i$ the signature of the message M by the node i. Figure 2A shows a way to implement the proof of computation protocol. The proof of computation is performed by a node i_1 for an SC that includes three nodes that are i_1, i_2 and i_3. The node i_1 computes the off-chain code $f(w_{i_1}[t])$ and generates the prof of computation $p_{t,i_1} = H(f(w_{i_1}[t])||S_{t,i_1}||ID_{i_1}) ||a||t||H(w_{i_1}[t])$. This is signed by i_1 and is sent to i_2 for signature. This will append its signature and forwards the message to i_3. This signs the message and sends it back to i_1. At this point i_1 has its proof signed by all nodes. This is broadcast to all other nodes. The same steps will be performed by i_2 and i_3. Each node performs a ring based signature protocol that can be further optimised by using more efficient multi-signature approaches or by electing a leader.

Computation Agreement Checking and Smart Contract Update: Each intermediate node signs and reveals its secret plus its computation after receiving all proof of computations signed by all nodes. More precisely, a node i reveals its computation by broadcasting the message $c_{t,i} = [f(w_i[t])||S_{t,i}] : i$. This can be used by other intermediate nodes for verifying the off-chain code computation. More precisely, a node j that receives $c_{t,i} = f(w_i[t])||S_{t,i}$ can now verify the hash $H(f(w_i[t])||S_{t,i}||ID_i)$ sent by i and compare its computation with the one of i. When $f(w_j[t]) \neq f(w_i[t])$ a dispute is open by setting a notification on the on-chain contract. A node terminates with a successful on-chain contract update when its computation $f(w_i[t])$ its equal to all the other nodes. In this case the node sends its proof of computation $p_{t,i}$ and $c_{t,i}$ signed by all nodes. Figure 2A shows the message c_{t,i_1} sent by i_1 and the proof of computation sent by all nodes (denoted with P_t). This is register on the on-chain SC.

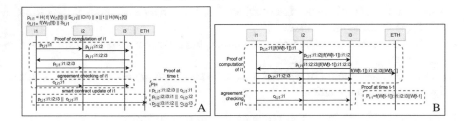

Fig. 2. A) Intermediate node unanimous consensus protocol, B) Fast protocol

We have designed and implemented an optimisation of the protocol (fast protocol). This decreases the amount of messages and reduces the size of the data that are stored in the on-chain SC during the SC update phase. While the proof of computation and agreement checking phases are the same, a leader node is entrusted to produce a compact proof of computation for the previous round of the protocol. More precisely, a leader node i en reaches its proof of computation message $p_{t,i}$ with a message $P_{t-1} = f(w[t-1])$ that contains the computation agreed in the previous run $t-1$. This is sent by the leader node and signed by all the nodes. When the leader gets all the signatures, the on-chain SC is updated. Figure 2B shows an example of the fast protocol. We can see that the leader node i_1 sends together with its proof of work the computation agreed in the previous round. In the example an on-chain proof of storage is implemented since the window data are also stored in the main chain. It is worth mentioning that the leader node can be statically designated during the contract definition or can dynamically change over the time. A dispute must be notified to the users that will try to analyse the window data and try to solve it. When the off-chain proof storage service is used the data will be available in the intermediate nodes. When the on-chain proof of storage is used the data will be available in the main chain.

Implementation: The blockchain stores a SC called *mainContract*. This SC allows Ethereum accounts to register sensors, intermediate nodes, users and on-chain contracts inside the blockchain. Each time a user registers an entity by calling the appropriate method of *mainContract* a new SC representing that entity is created. A reference to this new SC is kept in the main SC.

A sensor is represented by a SC containing the *publicKey* used for signing messages and its id. Intermediate nodes are represented as SC containing an Ethereum address and its ip address.

A *User* is represented as a SC containing an Ethereum address and an id with the relative getters. Only registered users can be referenced by an on-chain SC.

On-chain SC specifies the off-chain code that should be performed by intermediate nodes on the data provided by the specified sensors and the intermediate nodes and sensors interested in the contract. On-chain SC contain the following attributes:

- *intermediateNodes*: it is the list of the intermediate nodes allowed to partake in the execution of the agreement SC;
- *parties*: it is a list containing the users interested in the SC execution;
- sensors: a list of the sensors that provides data used in the execution of the off-chain SC by the intermediate nodes;
- *offChainCode*: it is a JSON string containing the off-chain code that should be executed by the intermediate nodes.
- *reward*: it is the amount of wei that should be rewarded to nodes when they agree on the result of a computation.

The agreement SC contains also methods that allows intermediate nodes to perform reactions (e.g., log data or to move funds from the SC to a *party*) if they are provided with the signature of all the intermediate nodes referenced by the on-chain SC.

4   Performance Evaluation

In this section we tested a simplified version of our proposed model on a smart lock case study. The main objective here is to compare three approaches in terms of fees spent. In the first one, the data produced by a smart lock is directly committed to the blockchain. In the second one, the data produced is first sent to an intermediate node that aggregates them and commits the result of the aggregation to the blockchain. In the third approach, both the result and the data used in the aggregation are sent to the blockchain.

Experimental Setup: The experimental setup consists of two virtual machines. One who runs Ubuntu 18.04.4 LTS and four validator nodes of a private Hyperledger Besu network [2]. The blockchain is set up with the IBTF 2.0 proof of authority consensus protocol. Another Ubuntu 18.04.4 LTS Virtual machine acts as an intermediate node and performs two tasks: (i) complex event processing through Flink [1] and (ii) blockchain interaction. The java application running flink reads data streams on an artificially generated dataset. When a Flink pattern is applied to the stream an aggregation function is executed. The combination of the flink pattern and the aggregation function acts as the off-chain smart contract. The result is then sent as a JSON object to a NodeJS app through a socket. The app prepares a transaction for the blockchain using the received JSON as argument and signs the transaction.

Smart Lock Case Study: This example is a simulation of an hypothetical scenario where a smart lock sends messages related to its opening and closing to an intermediate node that will aggregate the data received into a single transaction containing the amount of time the lock remained open in a month. The intermediate node reads the message data on 10 artificially generated datasets. The first dataset contains data for 10 opening and 10 closing messages in a month[1], each subsequent dataset contain 10 opening and closing messages more than the previous one, up to the last dataset containing 100 opening messages and 100 closing messages. Messages contain two attributes: *state* and *timestamp*. *state* is a boolean and determines if the message refers to the opening of the smart lock (true) or it is a closing message (false). *timestamp* is the value that represents the moment in time when the event occurred. The applications tested simulate a stream of events based on the messages of the datasets and applies to it a Flink pattern that matches whenever the difference between the timestamp of the first element of its trace and the last one is more than one month. A transaction is sent to a logging smart contract on the blockchain whenever the pattern is matched. The content of the transaction depends on the application tested:

- The session application (SA) sends only the value of the session. A session is an integer representing the amount of time a lock remained open in a month. The gas spent executing SA is equivalent to the amount of gas spent when the quality of service is set either to *no-proof of storage* or *off-chain proof of storage*
- The session application with trace (SAT) stores in the blockchain both the session and all the opening and closing events that triggered the pattern along with the SHA256 with ECDSA signature of the event. This corresponds to the quality of service *on-chain proof of storage*

An additional application called *raw data application* (RDA) has been tested for comparison reasons. The RDA application sends a transaction containing the data in a message each time an opening or closing message is generated.

Results: The results of the simulations are shown in Fig. 3 and Table 1. The Y-axis shows the amount of gas spent for storing data relative to one month of activity. The X-axis shows the amount of opening and closing events in the dataset taken into consideration. The amount of gas spent in the SA application is constant and always lower than the gas spent in the other applications. This is due to the fact that only a single transaction containing an integer representing the session is stored in the blockchain no matter how many events are in the pattern trace. The increase in gas spent in the SAT application is linear, this happens because each transaction stores every event matched in the pattern, this means that the higher the amount of events are in the dataset the higher the amount of data to be stored. In the RDA application the amount of gas spent increases linearly because it generates a transaction for each event. The reason why the amount of gas spent in the RDA application is significantly higher than the gas spent in the SAT application is because the constant cost of sending a transaction is higher than the cost of adding more data in a transaction, thus, although the amount of data sent by

---

[1] We denote with *OCM* the amount opening and closing in a month.

the two applications to the blockchain is almost the same, the amount of transactions performed are way higher in the RDA application than in the SAT application.

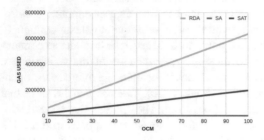

Fig. 3. Gas consumption chart for different approaches

Table 1. Gas consumption table for different approaches

OCM	RDA	SA	SAT
10	634876	23952	222607
20	1269688	23952	416244
30	1904628	23952	610379
40	2539952	23952	803924
50	3206367	23952	998992
60	3809896	23952	1193854
70	4444836	23952	1388638
80	5079584	23952	1584176
90	5714780	23952	1780853
100	6349208	23952	1976492

5   Conclusion and Future Work

In this paper we present *Diversity*, a second layer decentralised network for off-chain smart contract execution. *Diversity* allows users to define on-chain contracts that are automatically and securely executed off-chain by intermediate nodes. An honest intermediate node that performs a correct off-chain computation will be enough to detect any dishonest ones and open a dispute. This is achieved by *Diversity* novel protocol. Off-chain smart contract code is not bound the main chain contracting language and can be written in any language. We have validate our approach on a novel lock case study where fees have been greatly reduced.

References

1. Apache flink (2020). https://flink.apache.org/
2. Hyperledger besu (2020). https://besu.hyperledger.org/en/stable/
3. Cheng, R., Zhang, F., Kos, J., He, W., Hynes, N., Johnson, N., Juels, A., Miller, A., Song, D.: Ekiden: a platform for confidentiality-preserving, trustworthy, and performant smart contracts. In: 2019 IEEE European Symposium on Security and Privacy (EuroS P), pp. 185–200 (2019)
4. Dennis, R., Owenson, G., Aziz, B.: A temporal blockchain: a formal analysis. In: 2016 International Conference on Collaboration Technologies and Systems (CTS), pp. 430–437 (2016)
5. Dittmann, G., Jelitto, J.: A blockchain proxy for lightweight IoT devices. In: 2019 Crypto Valley Conference on Blockchain Technology (CVCBT), pp. 82–85 (2019)
6. Eberhardt , J., Heiss, J.: Off-chaining models and approaches to off-chain computations. In: Proceedings of the 2nd Workshop on Scalable and Resilient Infrastructures for Distributed Ledgers, SERIAL 2018, pp. 7–12. Association for Computing Machinery, New York (2018)

7. Eberhardt, J., Tai, S.: On or off the blockchain? Insights on off-chaining computation and data. In: De Paoli, F., Schulte, S., Broch Johnsen, E. (eds.) Service-Oriented and Cloud Computing, pp. 3–15. Springer International Publishing, Cham (2017)

8. Ellul, J., Pace, G.J., Alkylvm: a virtual machine for smart contract blockchain connected Internet of Things. In: 2018 9th IFIP International Conference on New Technologies, Mobility and Security (NTMS), pp. 1–4 (2018)

9. Gennaro, R., Gentry, C., Parno, B., Raykova, M.: Quadratic span programs and succinct NIZKs without PCPs. In: Johansson, T., Nguyen, P.Q. (eds.) Advances in Cryptology – EUROCRYPT 2013, pp. 626–645. Springer, Berlin, Heidelberg (2013)

10. Nakamoto, S.: Bitcoin: a peer-to-peer electronic cash system. Technical report, Manubot (2019)

11. Peker, Y.K., Rodriguez, X., Ericsson, J., Lee, S.J., Perez, A.J.: A cost analysis of internet of things sensor data storage on blockchain via smart contracts. Electronics 9(2), 244 (2020)

12. Poon, J., Buterin, V.: Plasma: Scalable autonomous smart contracts. White paper, pp. 1–47 (2017)

13. Poon, J., Dryja, T.: The bitcoin lightning network: scalable off-chain instant payments (2016)

14. Salimitari, M., Chatterjee, M.: A survey on consensus protocols in blockchain for IoT networks, arxiv, 1809.05613 (2019)

15. Shahid, A.R., Pissinou, N., Staier, C., Kwan, R.: Sensor-chain: a lightweight scalable blockchain framework for Internet of Things. In: 2019 International Conference on Internet of Things (iThings) and IEEE Green Computing and Communications (GreenCom) and IEEE Cyber, Physical and Social Computing (CPSCom) and IEEE Smart Data (SmartData), pp. 1154–1161 (2019)

16. Trón, V., Fischer, A., Nagy, D.A., Felföldi, Z., Johnson, N.: Swap, swear and swindle - incentive system for swarm (2016). https://ethersphere.github.io/swarm-home/ethersphere/orange-papers/1/sw%5E3.pdf

17. Wood, G., et al.: Ethereum: A secure decentralised generalised transaction ledger. In: Ethereum Project Yellow Paper, vol. 151, pp. 1–32 (2014)

Fast and Accurate Function Evaluation with LUT over Integer-Based Fully Homomorphic Encryption

Ruixiao Li[1](✉) and Hayato Yamana[2]

[1] Department of CSCE, Waseda University, Tokyo, Japan
liruixiao@yama.info.waseda.ac.jp
[2] Faculty of Science and Engineering, Waseda University, Tokyo, Japan
yamana@yama.info.waseda.ac.jp

Abstract. Fully homomorphic encryption (FHE), which is used to evaluate arbitrary functions in addition and multiplication operations via modular arithmetic (mod q) over ciphertext, can be applied in various privacy-preserving applications. However, big data is difficult to adopt owing to its high computational cost and the challenges associated with the efficient handling of complex functions such as $log(x)$. To address these problems, we propose a method for handling any multi-input function using a lookup table (LUT) to replace the original calculations with array indexing operations over integer-based FHE. In this study, we extend our LUT-based method to handle any input values, i.e., including non-matched element values in the LUT, to match with a near indexed value and return an approximated output over FHE. In addition, we propose a technique for splitting the table to handle large integers for improved accuracy with only a slight increase in the execution time. For the experiments, we use the Microsoft/SEAL library, and the results show that our proposed method can evaluate a 16-bit to 16-bit function in 2.110 s and a 16-bit to 32-bit function in 2.268 s, thereby outperforming previous methods implemented via bit-wise calculation over FHE.

1 Introduction

Cloud computing provides the ability to scale to large systems, such as big data analysis, making it convenient for providing services. However, this aspect causes privacy protection issues [1]. To protect users' sensitive data, traditional cryptography that promises security, such as the advanced encryption standard (AES) or data encryption standard (DES), can be used during data transfer and storage. However, the data must be decrypted when the cloud computes them, and such decryption increases security risks. Fully homomorphic encryption (FHE) allows a third party, i.e., cloud, to evaluate arbitrary functions without decryption, which makes it one of the promised solutions for achieving privacy-preservation.

Gentry [2] first proposed an FHE scheme with bootstrapping technique in 2009, enabling an arbitrary number of arithmetic operations; however, the major

L. Barolli et al. (Eds.): AINA 2021, LNNS 226, pp. 620–633, 2021.
https://doi.org/10.1007/978-3-030-75075-6_51

problems were its high computational cost and the limitation that it could only perform addition and multiplication. After his work, many studies, such as those carried out by [3–6], were conducted to improve FHE schemes that focus on reducing the noise that increases after multiplications to require bootstrapping. Apart from improving the FHE scheme, many researchers focus on adopting FHE to various applications in cooperation with private information retrieval (PIR) [7–9].

FHE supports both addition and multiplication via modular arithmetic (mod q). When the modular $q = 2$, e.g., the GSW scheme [5], numbers are encoded bit-by-bit so that any function can be evaluated by constructing logic circuits. However, such bit-wise operations incur a large computational cost. When the modular $q > 2$, e.g., the BFV scheme [3], numbers are encoded as integers. Therefore, we can process more data, i.e., multiple bits, in an operation. The remaining problem is that the functions that cannot be represented using addition and multiplication, e.g., $f(x) = \log x$ and the sigmoid function, are difficult to implement. To evaluate such functions using FHE, some studies have adopted a lookup table (LUT) [10,11].

Crawford et al. [10] implemented a low-precision approximation of arbitrary functions using a pre-computed LUT over binary representation. The lookup table $T_f[x] = f(x)$ returns bits of the output value $T_f[x]$ for an encrypted bit-representation of the input x, whose range is pre-defined. The homomorphic table lookup with encrypted indexes only needs a simple multiplexer (MUX). However, the execution time increases in proportionally to the number of bits. Chillotti et al. [11] proposed two techniques- horizontal and vertical packing, for table lookup. To evaluate arbitrary functions $f : \mathbb{B}^d \to \mathbb{T}^s$, they prepared an LUT that contained 2^d input values and corresponding output values for the number of s sub-functions. Note that the function is separated as follows: $f(x) = (f_0(x), ..., f_{s-1}(x))$, where the d-bit integer $x = \sum_{i=0}^{d-1} x_i 2^i$ [12,13]. These works adopted bit-wise encoding, which results in a large table size and inefficient evaluations because the table contains all possible inputs.

In our previous study [14], we introduced a protocol that adopts integer encoding to improve efficiency. By preparing an LUT matrix for each input, we can evaluate a multi-input function using a multi-threading operation. However, the table size is still huge, and it cannot handle the input that does not match in the LUT.

This study extends our previous protocol to enable the following:

- The approximation of any input value by matching it with the nearest indexed value over FHE. Consequently, the LUT size is reduced.
- We propose a method to separate a table into a set of tables to provide additional plaintext space for handling large integers. It is indispensable for handling accurate input and output values because the decimal point must be moved to convert an input/output value into an integer when calculating decimals, and this requires more plaintext space.

We expect that this method can be used in practical cloud applications, such as recommendation systems or some neural network algorithms, including CNNs, to improve calculation speed and protect sensitive data.

The notations and techniques used in this paper are presented in Sect. 2. Details of the extension for handling non-matched input values are presented in Sect. 3, followed by the proposal of the separation scheme of LUT to handle large plaintext space in Sect. 4. Security analysis is presented in Sect. 5. Section 6 presents the results of the experiments on 1) different table sizes and 2) two specific functions. Finally, the conclusion is presented in Sect. 7.

2   Preliminary

2.1   Notations

We implemented our protocol using the Microsoft's Simple Encrypted Arithmetic Library (SEAL) [15] and the BFV scheme [3]. We used $T_{in}(T_{out})$ to represent the LUT matrix of the input(output) values, x for the input values, and $f(x)$ for the corresponding outputs. By adopting the BFV scheme in SEAL [15], we packed l complex scalar values into one plaintext or ciphertext (hereafter referred to as the packing technique). c is a single ciphertext consisting of an encrypted vector whose all elements have the same value as that of a query from the client. We then represented $T_{in}(T_{out})$ as a matrix by adopting the packing technique such that each row of data contained l elements represented as a plaintext (a vector). Additionally, every encrypted l elements of the $T_{in}(T_{out})$, which is denoted as $Enc(T_{in})[i](Enc(T_{out})[i])$, where $i \in [0, k_{in}(k_{out})]$, is represented as a ciphertext. The number of rows in $T_{in}(T_{out})$ is expressed as $k_{in}(k_{out})$. The intermediate and final results are denoted as $\boldsymbol{res}$ and $\boldsymbol{r}$, respectively. The FHE slot-wise addition and multiplication over the ciphertext are represented as $\boxplus$ and $\boxdot$, respectively.

2.2   SIMD Operation over FHE

Smart and Vercauteren [16] proposed the polynomial-Chinese remainder theorem (polynomial-CRT) packing method, which supports SIMD-style operations. By adopting this method, l integers, which are denoted as a vector, can be packed into a CRT-represented ciphertext. The SIMD-style operation over the ciphertext is performed slot-wise and in parallel. For example, two vectors $\boldsymbol{x}$ and $\boldsymbol{y}$, each containing l integers, are encrypted as two ciphertexts $Enc(\boldsymbol{x})$ and $Enc(\boldsymbol{y})$. The slot-wise addition and multiplication are performed using the following equations.

$$Dec(Enc(\boldsymbol{x}) \boxplus Enc(\boldsymbol{y})) = [(x_0 + y_0), \cdots, (x_{l-1} + y_{l-1})] \tag{1}$$

$$Dec(Enc(\boldsymbol{x}) \boxdot Enc(\boldsymbol{y})) = [(x_0 \cdot y_0), \cdots, (x_{l-1} \cdot y_{l-1})] \tag{2}$$

2.3   Private Information Retrieval over FHE

Chor et al. [17] presented the idea of private information retrieval (PIR). Using PIR, a user can access the t-th entry in the database without revealing the entry, i.e., the access information, to the database as well as other information, except for the matched result in the database. Many studies implemented PIR using FHE. We introduced a solution based on the work of [18]. We assumed that there are n entries in a database D. If a user wants to privately access the t-th entry in the database D, where $t \in [0, n-1]$, the user generates a PIR query q whose elements are as follows:

$$q_i = \begin{cases} 1 (\text{if} \quad i = t) \\ 0 (\text{if} \quad i \neq t) \end{cases} \tag{3}$$

After encrypting q with FHE, which is then sent to the database, the database returns a ciphertext result by calculating $res = q \boxdot D$. Only the data of entry t remained in the result. This ensures that the entry t is not revealed to the database.

3   Approximate Search

In this section, we propose an approximate search on LUT over FHE.

3.1   Overview of System

Our proposed model consists of four parties: a computation server (CS), a trusted authority (TA), an LUT provider, and a client (or clients), as shown in Fig. 1. We assume all parties are honest-but-curious (semi-honest), i.e., they follow the protocol but try to learn some information, and do not collude with each other.

1) The TA generates and holds a pair of keys: a public key (PK) and a secret key (SK). It then shares the PK with the CS, the LUT provider, and the client. 2) The LUT provider generates the LUTs of a predefined function, encrypts them, and sends them to the CS. 3) The client(s) sends an encrypted query c, which is the input for the function, and a ciphertext r_{noise}, which is an encrypted vector all of whose elements are random values, to the CS. Note that the r_{noise} is used for hiding the final result from the TA. 4) The CS searches the index of the element that matches with the query c over the LUT and sends the intermediate result res to the TA (Sect. 3.3). 5) The TA generates encrypted PIR queries to send back to the CS (Sect. 3.4). 6) The CS selects an output value over the LUT to generate the result r' (Sect. 3.5). 7) The CS adds the r_{noise} to the result and sends it to the TA. 8) Finally, the TA decrypts and sends the result to the client.

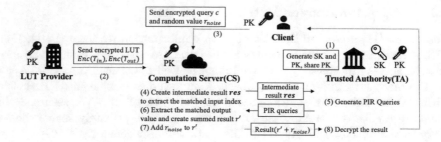

Fig. 1. System overview

3.2   Construction of LUT

Because the details of the lookup table matrix construction method are described in the works of [14,19], we only describe the different parts for approximate search herein.

All the element-values of the input x in T_{in} and the output $f(x)$ in T_{out} are integers. Therefore, the possible range of input is $x \in [-\frac{t}{2}, \frac{t}{2}] \cap \mathbb{Z}$, where $(t+1)$ represents a plaintext modulus(see [15] for more detail). Here, we can generate a small LUT containing a subset of the inputs arranged according to their values, to reduce the LUT size. The entries of T_{in} can be decided based on the nature of the function, e.g., fewer entries for stable outputs. The element-values in T_{in} and T_{out} are encrypted so that the elements are not revealed to the TA. Regenerating the LUT for each lookup can improve security.

3.3   Preparation of Intermediate Result

In this section, we describe the algorithm for preparing the intermediate result for the TA ((4) in Fig. 1).

The ciphertext c is an encrypted vector whose elements have the same value x as the query from the client. The CS computes the intermediate result **res** by Formula 4,

$$\mathbf{res}[i] = \begin{cases} c \boxplus -Enc(T_{in})[i + r_{val}] & \text{if} \quad (i + r_{val}) < k_{in} \\ c \boxplus -Enc(T_{in})[i + r_{val} - k_{in}] & \text{if} \quad (i + r_{val}) \geq k_{in} \end{cases}, \tag{4}$$

where $i \in [0, k_{in} - 1]$. The intermediate result **res** is an encrypted vector whose length is k_{in}. Then, the CS sends **res** to the TA at which the matched input index is chosen. To hide the matched index of T_{in} with the query from the TA, we right rotate the encrypted LUT matrix $Enc(T_{in})$ by $r_{val} \in [0, k_{in} - 1]$ elements, where r_{val} represents a random value created and owned by the CS.

An example of preparing the intermediate result **res** is shown in Fig. 2, where the query value $x = 10$ and $r_{val} = 2$. Note that the plaintext modulus is 31. After right rotating the $r_{val} = 2$ elements of $Enc(T_{in})$, the CS adds $-Enc(T_{in})$ to c and sends the intermediate result **res** to the TA.

3.4   Generation of PIR Query

In this section, we describe the algorithm for generating the PIR query ((5) in Fig. 1).

The TA decrypts the ciphertexts of the intermediate result **res** to a plaintext matrix. The input element value that matches or is closest to the query value is the smallest absolute value on the either side of the change from positive to negative in $Dec(\mathbf{res})$. Here, zero represents the exact match, and the smallest non-zero absolute value shows the closest element-value index in $Dec(\mathbf{res})$, which is denoted as (ind_{row}, ind_{col}). Then, the TA generates PIR queries using the PIR query generation method [19] (for the m-input function, see [14]). The algorithm is shown in Algorithm 1.

An example is shown in Fig. 2. After decrypting the intermediate result **res**, the element-value "1" is the smallest absolute value on either side of the change from positive to negative, such that the TA chooses the position of "1" as the closest input position to the query followed by the generation of PIR queries. Because the index of "1" in $Dec(\mathbf{res})$ is $(ind_{row}, ind_{col}) = (1, 2)$, the TA generates two PIR queries $\mathbf{q_0} = [0, 0, 1, 0, 0]$ and $\mathbf{q_1} = [0, 1, 0, 0, 0]$ by adopting the PIR query generation method [19]. The $\mathbf{q_0}$ is a vector showing that the ind_{col}-th element is one, and the others are zero. The $\mathbf{q_1}$ is a rotated vector representing left-rotated ind_{row} elements from $\mathbf{q_0}$. The TA then encrypts the PIR queries to send them to the CS.

Algorithm 1: Generation of PIR query

Input: Intermediate result **res**, the number of **res**'s rows k_{in}, the number of slots l

Output: PIR queries $Enc(\mathbf{q_0})$, $Enc(\mathbf{q_1})$

1 Decrypt the **res** to $Dec(\mathbf{res})$
2 **for** $row = 0$ to $k_{in} - 1$ **do**
3   **for** $col = 0$ to $l - 1$ **do**
4        $row_{next} \leftarrow row + \lfloor (col + 1)/l \rfloor$
5        $col_{next} \leftarrow (col + 1) \mod l$
6   **if** $Dec(\mathbf{res})[row][col] \geq 0$ and $Dec(\mathbf{res})[row_{next}][col_{next}] < 0$ **then**
7           $ind_{row}, ind_{col} \leftarrow$ element index of $min(Dec(\mathbf{res})[row][col], -Dec(\mathbf{res})[row_{next}][col_{next}])$
8   exit whole loops

9 Create plaintexts $\mathbf{q_0}$ and $\mathbf{q_1}$ whose length of l
10 $\mathbf{q_0} \leftarrow ind_{col}$-th element is one, and the others are zero
11 $\mathbf{q_1} \leftarrow \mathbf{q_0} << ind_{row}$ ▷ described in [14]
12 **return** $Enc(\mathbf{q_0})$, $Enc(\mathbf{q_1})$

3.5   Extraction of Matched Output Value

In this section, we describe the algorithm for extracting the matched output value ((6) in Fig. 1).

The CS generates a query vector q by using $Enc(q_0)$ and $Enc(q_1)$ [14]. The size of query vector q is the same as that of $Enc(T_{out})$, where only the queried element is 1 and the other elements are 0, as shown in Fig. 3. Then, q is then multiplied element-wise using $Enc(T'_{out})$ to retrieve the matched result, where T'_{out} represents the right-rotated T_{out} by r_{val} elements because T_{in} are also right rotated by r_{val} elements. The final vector r is calculated from Formula 5.

$$r[i] = \begin{cases} q[i] \boxdot Enc(T_{out})[i + r_{val}] & \text{if} \quad (i + r_{val}) < k_{out} \\ q[i] \boxdot Enc(T'_{out})[i + r_{val} - k_{out}] & \text{if} \quad (i + r_{val}) \geq k_{out} \end{cases} \quad (5)$$

After applying the summation for all elements of r, we obtain one cipher-text result r'. In the example shown in Fig. 3, we obtain the matched output $Enc(f(9))$.

4   Table Separation

In this section, we propose a method to split $T_{in}(T_{out})$ to store large integers when the initial plaintext space is not sufficient.

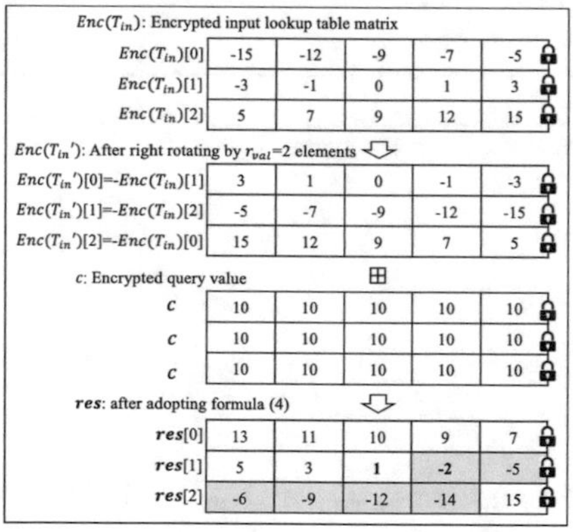

Fig. 2. Preparation of intermediate result

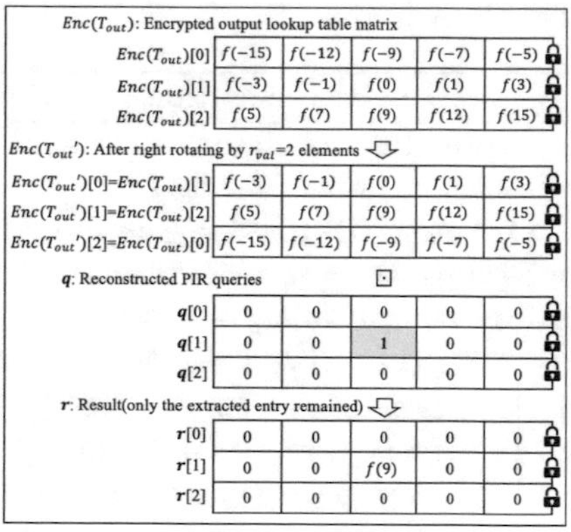

Fig. 3. Extraction of matched output value

Large plaintext space is needed when evaluating large integers or decimals. Since we adopt integer encoding, the decimal point must be moved to convert decimals into an integer when calculating. Thus, a large plaintext results in a better accuracy of the values in the lookup table. Further, we propose a technique to reduce the size of the separated input LUTs, in order to reduce the communication cost between the CS and TA.

4.1   Separation Algorithm

To split an LUT, we decompose the s-bit input x in the input LUT to separated m inputs of w bits each, where $s = m \times w$, or decompose the z-bit output $f(x)$ in the output LUT to separated p outputs of w bits each, where $z = p \times w$. By decomposing each element-value in the LUT, we construct separated LUTs.

$$x = x_0 + x_1 \times 2^w + ... + x_{m-1} \times 2^{(m-1)w}$$
$$f(x) = f(x)_0 + f(x)_1 \times 2^w + ... + f(x)_{p-1} \times 2^{(p-1)w} \tag{6}$$

By adopting Formula 6 to each element-value in $T_{in}(T_{out})$, we can separate it into m input(output) LUTs, each having x_i instead of x as the element value. Thus, when the original input LUT stores s-bit integers, we divide it into m input LUTs each of which stores w-bit integers. The input LUT separation is not directly related to the output LUT separation and can be performed separately as per the requirements.

An example of table separation is shown in Fig. 4. We decompose the 4-bit input integer x into two 2-bit integers x_0 and x_1, where each x is calculated as $x = x_0 + x_1 \times 2^2$, such that the original LUT is separated into two input LUTs. When $x = 6 = 2 + 1 \times 2^2$, we split it as two inputs $x_0 = 2$ and $x_1 = 1$ in $T_{in}^{x_0}$ and $T_{in}^{x_1}$. The corresponding outputs, $f(x)_0 = 2$ and $f(x)_1 = 2$ are selected from $T_{out}^{f(x)_0}$ and $T_{out}^{f(x)_1}$, because $f(6) = 10 = 2 + 2 \times 2^2$ (shadowed elements in Fig. 4).

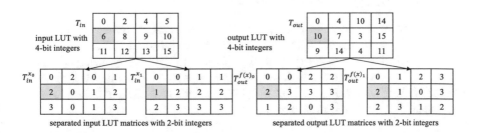

Fig. 4. Example of table separation

4.2   Size Reduction of Separated Input LUTs

In this section, we propose a technique for reducing the size of separated input LUTs. Note that this technique can only be applied to the exact search and it cannot handle an approximate search. Therefore, as mentioned before, we decompose each s-bit integer used in T_{in} into m w-bit integers, where $s = m \times w$, by adopting the Formula 6.

Instead of decomposing each s-bit integers in T_{in}, we make smaller input LUTs comprising all possible w-bit integers by only preparing the ranges of x_i using Formula 6. An example of this is shown in Fig. 5. In this example, the range of both x_0 and x_1 is calculated as from 0 to 3. By creating all combinations of

x_0 and x_1, we can reproduce all the element-values in T_{in}. Through the use of the separated input LUTs, we can decrease the communication cost between the CS and the TA. The total size of the separated input LUTs is $m \times w \times 2^w$ bits, which is the same as the size of the intermediate result. We can adopt the aforementioned method if and only if the total size of the separated input LUTs becomes smaller than the size of the input LUT for an approximate search.

The calculation of the output index in T_{out} using the index of input in separated input LUTs is described as follows. We first assume that all the separated input LUTs have the same size. Since the separated LUT matrices are generated from the original LUTs, which are denoted as T_{in} and T_{out} as the example shown in Fig. 5, each element x in the input LUT corresponds to the index of $f(x)$ in the output LUT, i.e., the index of x in input LUT is same as that of $f(x)$ in output LUT. Here, we define the number $index(ori)$ as shown in Formula 7, which shows the element number of one dimensional matrix converted from T_{in}(or T_{out}). By using $index(ori)$, the TA calculates the index of the matched output in T_{out}. We then define $index(x_i) = index(x_i^{row}) \times l + index(x_i^{col})$, where $(index(x_i^{row}), index(x_i^{col}))$ is the index of the matched x_i in $T_{in}^{x_i}$. The index of the matched output in T_{out} is represented as $(index(f(x)^{row}, index(f(x)^{col}))$. The formula is expressed as follows.

$$
\begin{aligned}
index(ori) &= index(x_0) + ... + index(x_{m-1}) \times (l \times k_{in})^{m-1} \\
index(f(x)^{row}) &= index(ori) \quad /l \\
index(f(x)^{col}) &= index(ori) \quad \mathrm{mod}\ l
\end{aligned}
\tag{7}
$$

In the example shown in Fig. 5, the number of slots(columns) l is four. When the input $x = 6$, $index(x_0^{row}, x_0^{col}) = (0, 2)$ and $index(x_1^{row}, x_1^{col}) = (0, 1)$. The TA calculates $index(x_0) = 2$ and $index(x_0) = 1$. By using Formula 7, $index(ori) = 6$, the output index of the separated output LUT matrix is $(index(f(x)^{row}), index(f(x)^{col})) = (1, 2)$, where $index(f(x)^{row}) = 6/4 = 1$ and $index(f(x)^{col}) = 6 \mod 4 = 2$.

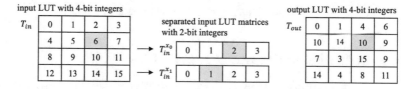

Fig. 5. Size reduction of separated input LUTs

5   Security Analysis

We describe the security analysis in this section. Our protocol has four parties: a CS, a TA, a LUT provider, and a client(or clients). All the parties are honest-but-curious (semi-honest), i.e., they follow the protocol but try to learn some information and do not collude with each other. The TA generates a pair of secret and public keys and keeps the secret key privately.

The client encrypts the input x while the LUT provider constructs a set of encrypted LUTs (input LUT and output LUT), followed by sending them to the CS as ciphertexts. Therefore, the CS cannot know neither the input x nor the output $f(x)$, because all computations in the CS are over ciphertexts. To hide the matched element-position in the input LUT from the TA, the CS randomly rotates the input LUT by r_{val} rows to calculate the intermediate result $\boldsymbol{res}$, as described in Sect. 3.3. The TA only knows the difference between input x and each element-value of the input LUT. Although the TA cannot obtain the matched value in the input LUT, the matched element-position in the input LUT can be revealed to the TA. Thus, the random rotation is required. Note that the difference is shown within the modulus. The TA then decrypts the $\boldsymbol{res}$ to generate the encrypted PIR queries to indicate the matched element-position in the $\boldsymbol{res}$, after which it sends them back to the CS, as described in Sect. 3.4.

The CS cannot obtain the matched element-position in the LUTs because of the encrypted PIR queries. After receiving the encrypted PIR queries, the CS calculates the final result $\boldsymbol{r}$ with the random value r_{noise} sent from the client. Then, the final result $\boldsymbol{r}$ is then sent to the TA for decryption. However, the TA cannot obtain the exact result because of the random value r_{noise}. After that, the decrypted $\boldsymbol{r}$ is sent back to the client, in which the final output value $f(x)$ is retrieved by using r_{noise}. Therefore, neither the CS nor the TA can obtain the input value x and the output value $f(x)$.

We show two different LUT constructions in Fig. 6. With the different r_{val} values, the intermediate results are different even with the same input value of 10.

Ex. 1 — $Enc(T_{in})$: Encrypted input lookup table matrix

$Enc(T_{in})[0]$	-15	-12	-9	-7	-5
$Enc(T_{in})[1]$	-3	-1	0	1	3
$Enc(T_{in})[2]$	5	7	9	12	15

$Enc(T_{in}')$: After right rotating by r_{val}=2 elements

$Enc(T_{in}')[0]=-Enc(T_{in})[1]$	3	1	0	-1	-3
$Enc(T_{in}')[1]=-Enc(T_{in})[2]$	-5	-7	-9	-12	-15
$Enc(T_{in}')[2]=-Enc(T_{in})[0]$	15	12	9	7	5

c: Encrypted query value

c	10	10	10	10	10
c	10	10	10	10	10
c	10	10	10	10	10

$\boldsymbol{res}$: after adopting formula (4)

$res[0]$	13	11	10	9	7
$res[1]$	5	3	1	-2	-5
$res[2]$	-6	-9	-12	-14	15

Ex. 2 — $Enc(T_{in})$: Encrypted input lookup table matrix

$Enc(T_{in})[0]$	-14	-10	-8	-6	-4
$Enc(T_{in})[1]$	-1	1	2	3	4
$Enc(T_{in})[2]$	7	10	11	12	15

$Enc(T_{in}')$: After right rotating by r_{val}=1 elements

$Enc(T_{in}')[0]=-Enc(T_{in})[2]$	-7	-10	-11	-12	-15
$Enc(T_{in}')[1]=-Enc(T_{in})[0]$	14	10	8	6	4
$Enc(T_{in}')[2]=-Enc(T_{in})[1]$	1	-1	-2	-3	-4

c: Encrypted query value

c	10	10	10	10	10
c	10	10	10	10	10
c	10	10	10	10	10

$\boldsymbol{res}$: after adopting formula (4)

$res[0]$	3	0	-1	-2	-5
$res[1]$	-7	-11	-13	-15	14
$res[2]$	11	9	8	7	6

Fig. 6. Examples of intermediate results for different LUT constructions

6   Experimental Evaluation

We implemented our proposed methods and experimented with Microsoft/SEAL 3.2.0[1]. We also adopted multi-thread operations with OpenMP[2]. The experiments were performed on a machine prepared for the operations of the four parties. The machine has four Intel Xeon E7-8880 v3 @2.3 GHz (Turbo Boost: 3.1 GHz) CPUs and 3 TB main memory.

The FHE parameters are shown in Table 1. In all the experiments, we set the input LUT, which holds all the inputs of w-bit integer $x_i \in [-\frac{t}{2}, \frac{t}{2}) \cap \mathbb{Z}$. The number of integers in the input LUT is $(t - 1)$ because $2^w + 1 = t$. The two divided output LUTs, each of which holds the $(t - 1)$ corresponding w-bit integers, and two tables express the $2w$-bit integer output. If we adopt a smaller LUT that does not contain all the possible inputs in the range but can still handle any $x \in [-\frac{t}{2}, \frac{t}{2}) \cap \mathbb{Z}$ via the approximate search, the runtime becomes shorter.

Table 1. FHE parameters

Scheme	Poly modulus degree	Coeff modulus size	Noise standard deviation
BFV	8,192	152bit	3.2

In Table 2, we show the runtime and RAM usage of the experiments in each step with different plaintext modulus, number of threads, and number of output LUTs. Additionally, the communication cost is shown as the transferred data size between the CS and TA in Table 2.

The result shows that we can evaluate an arbitrary 16-bit to 16-bit function in approximately 2.110 s (T_{in} and T_{out} each contain $2^{16} = 65536$ element-values) and an arbitrary 16-bit to 32-bit function in approximately 2.268 s (T_{in} and two T_{out}s each contain 2^{16} element-values) in one thread. Compared to the approach applied by [11], in which they used a core i7-4910MQ@2.90 GHz CPU that can evaluate a 16-bit to 8-bit function in 2.192 s, our approach is more practical because, theoretically, the approach applied by [11] takes a longer time when implementing 16-bit to 16-bit function evaluation. Additionally, we evaluated two specific functions: the sigmoid-function and the tanh-function, and we show the runtime and RAM usage in Table 3. The significant digits of the input values are three decimal places, whereas the output is accurate to 10 decimal places using two divided output LUTs. We set the plaintext modulus 557,057, which is larger than the required accuracy. The intermediate result file size is 1.9 MB for the sigmoid-function and 1.2 MB for the tanh-function. The PIR query file size is 769 KB.

---

[1] https://github.com/microsoft/SEAL.
[2] https://www.openmp.org/.

Table 2. Runtime and RAM experiment of different size of table (shown in seconds and MB)

Plaintext modulus		1-thread		2-thread		4-thread		8-thread	
		Runtime	RAM	Runtime	RAM	Runtime	RAM	Runtime	RAM
65,537(2^{16}) CStoTA:3.1 MB TAtoCS:769 KB	(a)	0.527	138	0.531	138	0.533	138	0.530	137
	(b)	0.489	21	0.483	22	0.474	23	0.473	25
	(c)	1.094	142	0.892	154	0.737	170	0.652	198
	(d)	1.252	151	1.035	165	0.813	178	0.697	211
	(a)+(b)+(c)	2.110	–	1.906	–	1.744	–	1.655	–
	(a)+(b)+(d)	2.268	–	2.049	–	1.820	–	1.700	–
557,057 CStoTA: 26 MB TAtoCS:769 KB	(a)	0.636	227	0.631	228	0.629	228	0.614	228
	(b)	0.657	53	0.620	53	0.558	54	0.531	56
	(c)	5.346	188	3.334	200	2.105	215	1.423	239
	(d)	7.111	239	4.292	251	2.609	270	1.711	303
	(a)+(b)+(c)	6.639	–	4.585	–	3.292	–	2.568	–
	(a)+(b)+(d)	8.404	–	5.543	–	3.796	–	2.856	–
1,032,193 CStoTA: 48 MB TAtoCS:769 KB	(a)	0.776	315	0.767	315	0.756	314	0.746	314
	(b)	0.813	82	0.730	82	0.654	83	0.603	86
	(c)	9.803	232	5.678	241	3.402	256	2.175	295
	(d)	13.210	327	7.352	339	4.314	357	2.694	389
	(a)+(b)+(c)	11.392	–	7.175	–	4.815	–	3.524	–
	(a)+(b)+(d)	14.799	–	8.849	–	5.727	–	4.043	–

(a): preparation of intermediate result, (b): generation of PIR query, (c): extraction of matched output value, (d): extraction of matched output value with two divided output LUTs, CStoTA: size of transfer file from CS to TA, TAtoCS: size of transfer file from TA to CS

Table 3. Runtime and RAM experiment of specific function (shown in seconds and MB)

		1-thread		2-thread		4-thread	
		Runtime	RAM	Runtime	RAM	Runtime	RAM
Sigmoid function $\frac{1}{1+e^{(-x)}}$ $x \in [-20, +20]$	(a)	0.538	134	0.526	133	0.530	133
	(b)	0.476	20	0.470	20	0.467	21
	(d)	0.998	144	0.810	159	0.722	172
	(a)+(b)+(d)	2.012	–	1.806	–	1.719	–
Tanh function $\frac{e^x - e^{-x}}{e^x + e^{-x}}$ $x \in [-10, +10]$	(a)	0.526	131	0.518	131	0.523	131
	(b)	0.465	19	0.464	19	0.465	19
	(d)	0.803	141	0.728	151	0.667	158
	(a)+(b)+(d)	1.794	–	1.710	–	1.655	–

7   Conclusion

In this study, we extended our previous work on handling the approximate search for an LUT and increased the plaintext space by separating LUTs over FHE with integer encoding. This enabled the approximation of the output for any input in a given range, even when the input did not match with any of the LUT entries. Our approach resulted in an improved accuracy when the decimals were evaluated. We also reduced the table size to decrease the communication cost between the computation server and the trusted authority. The experimental results showed that our proposed method is highly practical as a 16-bit to 16-bit function could be evaluated in 2.110 s and a 16-bit to 32-bit function could be evaluated in 2.268 s on a single thread even when the approximate search was adopted. Compared to the bit-wise implementations [10,11], our proposed method is more practical because the bit-wise implementations require a long execution time, which increases in proportionally to the number of bits used, especially when evaluating the large integers. However, our implementation adopted integer encoding, which results in a slight increase in the execution time even if the number of bits increases.

In our future studiesm we shall scale our work to cover additional applications such as smart grid system, or some neural network algorithms, including CNNs, to improve calculation speed and protect sensitive data.

Acknowledgement. This work was supported by JST CREST(Grant Number JPMJCR1503), and Japan and Japan–US Network Opportunity 2 by Commissioned Research of the National Institute of Information and Communications Technology (NICT), Japan.

References

1. Zhang, D.: Big data security and privacy protection. In: Proceedings of the 8th International Conference on Management and Computer Science (ICMCS 2018), pp. 275–278. Atlantis Press (2018)
2. Gentry, C.: Fully homomorphic encryption using ideal lattices. In: Proceedings of the Forty-First Annual ACM Symposium on Theory of Computing, pp. 169–178 (2009)
3. Fan, J., Vercauteren, F.: Somewhat practical fully homomorphic encryption. IACR Cryptol. ePrint Arch. **2012**, 144 (2012)
4. Brakerski, Z.: Fully homomorphic encryption without modulus switching from classical GapSVP. LNCS, vol. 7417, pp. 868–886 (2012)
5. Gentry, C., Sahai, A., Waters, B.: Homomorphic encryption from learning with errors: conceptually-simpler, asymptotically-faster, attribute-based. LNCS, vol. 8042, pp. 75–92 (2013)
6. Brakerski, Z., Gentry, C., Vaikuntanathan, V.: (Leveled) fully homomorphic encryption without bootstrapping. ACM Trans. Comput. Theory (TOCT) **6**(3), 1–36 (2014)
7. Boneh, D., Gentry, C., Halevi, S., et al.: Private database queries using somewhat homomorphic encryption. LNCS, vol. 7954, pp. 102–118 (2013)
8. Aguilar-Melchor, C., Barrier, J., Fousse, L., et al.: XPIR: private information retrieval for everyone. In: Proceedings on Privacy Enhancing Technologies, vol. 2, pp. 155–174 (2016)
9. Angel, S., Chen, H., Laine, K., et al.: PIR with compressed queries and amortized query. In: Proceedings of the 2018 IEEE Symposium on Security and Privacy (S&P), pp. 962–979. IEEE (2018)
10. Crawford, J.L.H., Gentry, C., Halevi, S., et al.: Doing real work with FHE: the case of logistic regression. In: Proceedings of the 6th Workshop on Encrypted Computing & Applied Homomorphic Cryptography, pp. 1–12 (2018)

11. Chillotti, I., Gama, N., Georgieva, M., et al.: TFHE: fast fully homomorphic encryption over the torus. J. Cryptol. **33**(1), 34–91 (2020)

12. Brakerski, Z., Vaikuntanathan, V.: Lattice-based FHE as secure as PKE. In: Proceedings of the 5th Conference on Innovations in Theoretical Computer Science, pp. 1–12 (2014)

13. Chillotti, I., Gama, N., Georgieva, M., et al.: Faster fully homomorphic encryption: bootstrapping in less than 0.1 seconds. LNCS, vol. 10031, pp. 3–33 (2016)

14. Li, R., Ishimaki, Y., Yamana, H.: Privacy preserving calculation in cloud using fully homomorphic encryption with table lookup. In: Proceedings of the 2020 5th IEEE International Conference on Big Data Analytics (ICBDA), pp. 315–322. IEEE (2020)

15. Microsoft Research, Redmond, WA, Microsoft SEAL (release 3.2) (2019). https://github.com/Microsoft/SEAL

16. Smart, N.P., Vercauteren, F.: Fully homomorphic SIMD operations. Des. Codes Crypt. **71**(1), 57–81 (2014)

17. Chor, B., Goldreich, O., Kushilevitz, E., et al.: Private information retrieval. In: Proceedings of IEEE 36th Annual Foundations of Computer Science, pp. 41–50. IEEE (1995)

18. Kushilevitz, E., Ostrovsky, R.: One-way trapdoor permutations are sufficient for non-trivial single-server private information retrieval. LNCS, vol. 1807, pp. 104–121 (2000)

19. Li, R., Ishimaki, Y., Yamana, H.: Fully homomorphic encryption with table lookup for privacy-preserving smart grid. In: Proceedings of the 2019 IEEE International Conference on Smart Computing (SMARTCOMP), pp. 19–24. IEEE (2019)

Reidentification Risk from Pseudonymized Customer Payment History

Hiroaki Kikuchi[(✉)]

Department of Frontier Media Science, School of Interdisciplinary Mathematical Sciences, Meiji University, 4-21-1 Nakano, Nakano Ku, Tokyo 164-8525, Japan
kikn@meiji.ac.jp

Abstract. To assess accurately the risk of being compromised, anonymized data requires a balance between utility and security. This paper studies the risk of reidentification for long-term transaction records aggregated by pseudonym (pid). Given a set of transaction records with pseudonyms, an adversary attempts to identify individuals by maximizing the similarity between sets of goods that have been purchased using the pseudonyms. Assuming a uniform probability for the choice of goods and applying Zipf's law to the behavior of the number of records per individual, we investigate the likelihood of an individual being correctly reidentified. Our model reveals that the risk of reidentification increases as the number of records associated with the pseudonym increases. A similar effect in terms of risk and the number of pseudonyms was found in a competition for data anonymization (PWS Cup 2017), where the Online Retail dataset comprising 40,000 records for 500 individuals was securely anonymized by each team and then reidentified each other. The competition results suggested that the reidentification rate increases the longer a pseudonym remains unchanged.

1 Introduction

Anonymization plays an important role in the safe sharing of data in the age of big data because it enables massive-scale data collection of human activities involving banks, stores, and social networking sites without revealing individual identities. Accumulating personally identifiable information (PII) is not only risky, but is also restricted by data protection legislation, such as the EU General Data Protection Regulation [1]. For this reason, data controllers charged with managing PII aim to prevent the sharing of data violating individuals' privacy. To achieve this, they carefully modify the collected data to reduce the disclosure risk.

The traditional approach to anonymization is based on a privacy model in which a privacy condition guarantees an upper bound on the risk of reidentification disclosure by an intruder. Examples of privacy models include k-anonymity [10], ℓ-diversity [11], t-closeness [12], β-likeness [13], ϵ-differential pri-

© The Author(s), under exclusive license to Springer Nature Switzerland AG 2021
L. Barolli et al. (Eds.): AINA 2021, LNNS 226, pp. 634–645, 2021.
https://doi.org/10.1007/978-3-030-75075-6_52

vacy [14], a maximum knowledge attacker model [7], a general confidentiality metrics [16], and the copula-based model [9].

All of these conventional approaches assume a *stable* data for the database in the sense that an individual is described by a fixed number of attributes per individual, e.g., demographic variables and survey responses. While, in practical use cases, we often need to deal with a *history* data in which an individual has multiple records. Examples of such features include the trajectory of locations, the statistics for the transactions, the URLs of websites that the target has visited, the titles of movies that the target has seen, the items that customer has been rated [8], the medical records that patient has been operated [6], and the list of goods that the target has purchased [4].

In practical anonymization, a data controller pays attention only to stable data but leaves history data unchanged, or just applies a naive anonymization, called *pseudonymization*, that simply replaces names by pseudonyms (pids). According to ISO/IEC 2018 [19], pseudonymization used alone does not reduce the risk that an individual data principal can be singled out. Despite this, very limited number of studies have been made. Xiao et al. studied a privacy preserving publication schema of fully-dynamic datasets [17]. To deal with dynamic datasets that are allowed to be altered through insertions and/or deletions of records, they propose a counterfeited generalization technique and *m*-invariance, which limits the risk of privacy disclosure in re-identification. Cummings studies an algorithm for answering privately to adaptive queries to growing datasets [18].

The risks associated with pseudonyms are hard to evaluate precisely because the anonymization of dynamic transaction data has to deal with the unpredictable dynamism of records across a diversity of privacy models. That is, the risk of reidentification increases if the fixed pseudonym per individual has been reused for so long that many useful features can be learned from the history of records for that pseudonym.

The goal of this study is to model an adversary who employs these features to identify individuals. We introduce a *profile* which is defined as a vector that indicates the goods an individual has purchased previously. Under some simple assumptions, we can quantify the likelihood of a pseudonym being reidentified correctly as an expected value of probability distribution for the number of records associated with the pseudonym. Our model enables quantification of the risk of reidentification without requiring empirical analysis and can suggest appropriate durations for pseudonyms to a data controller.

Our contributions in this work are as follows.

- We study a problem with the pseudonymized payment transaction records that are associated with individuals. It extends the typical setting of a stable table, whereby an individual is described by a single record comprising a fixed number of attribute values.
- We propose a mathematical model for the risk of pseudonymized records being reidentified correctly based on the number of records associated with a pseudonym. The proposed model will enable data controllers to maintain a sufficiently low probability of reidentification.

Table 1. Example customer master table M

$c_{.,1}$	$c_{.,2}$	$c_{.,3}$	$c_{.,4}$
Cust. ID	Sex	Birth	Country
12360	f	1950/1/1	Others
12361	m	1960/1/1	Germany
12362	m	1950/1/1	France
12363	f	1970/1/1	UK

Table 2. Example transaction table T

$t_{.,1}$	$t_{.,2}$	$t_{.,3}$	$t_{.,4}$	$t_{.,5}$	$t_{.,6}$	$t_{.,7}$
Cust. ID	Invoice	Date	Time	Stock ID	Price	qty
12361	0	2011/2/17	10:30	21913	3.75	4
12361	0	2011/2/17	10:30	22431	1.95	6
12361	0	2011/2/25	13:51	22630	1.95	12
12361	0	2011/2/25	13:51	22555	1.65	12
12362	0	2011/4/28	9:12	21866	1.25	12
12362	0	2011/4/28	9:12	20750	7.95	2
12362	0	2011/4/28	9:12	22908	0.85	12
12360	0	2011/5/23	9:43	21094	0.85	12
12360	0	2011/5/23	9:43	23007	14.95	6

– We analyze a past open competition for data anonymization using the Online Retail dataset [4] as an example of the anonymization of dynamic transactions. Using measures of mean numbers of pseudonyms and mean duration, we show the correlation between risk and the frequency of purchase orders.

The remainder of the paper is organized as follows. Section 2 provides a sample of transaction records and some characteristics observed from the data. In Sect. 3, after defining a model for the anonymization and the adversary from the history data, we formalize a feature of transaction records called the "profile" and make some assumptions. In Sect. 4, we analyze probability distributions for the profile that show the likelihood of reidentification. The characteristics of our model are found using the practical data competition results described in Sect. 4.2. We conclude our study in Sect. 5.

2 Preliminary

2.1 Payment History Data

Anonymization or deidentification is the process that removes the *association* between a set of data attributes and the data principal that they concern [5]. In our study, we model the association as a linkage, in terms of identity, between the records in an individual master dataset M and the records in a transaction dataset T. We assume that data principals are uniquely identified with the records of table M from certain external knowledge. This assumption is reasonable, given that each data principal is represented by a single record in table M. In contrast, the transaction table T stores multiple records for a data principal, making it relatively difficult to associate records with the correct data principals.

Tables 1 and 2 are examples of master M and transaction T table, respectively, taken from the Online Retail dataset in the UCI Machine Learning Repository[1]. The master M describes the customer ID, sex, day of birth, and country for $n = 4$ individuals, represented as a 4×4 matrix. The transaction T gives

---

[1] available at http://archive.ics.uci.edu/ml/datasets/Online+Retail/.

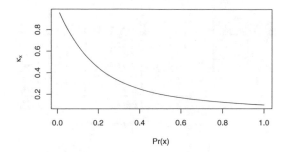

Fig. 1. Likelihood k_x of a record $\boldsymbol{x}$ being reidentified

purchase records, each comprising data, time, stock ID (good), unit price, and quantity for three individuals. We represent T as an $m \times 7$ transaction matrix, where $m = 9$ is the number of records in T. Each record is associated with a single individual via a customer ID. For example, the first four records are the purchase orders made by customer 12361 (female, born 1960, Germany) for two days (February 17 and 25). The number of records varies depending on the individual. For example, customer 12361 is a very frequent buyer, customer 12362 ordered three kinds of goods, but customer 12363 has never ordered anything.

The values of attributes in table M are *stable* in the sense that the values associated with a data principal do not change over the period of activity associated with the dataset. The number of records is identical to the number of customers (n), i.e., $n = m$. In contrast, the values in table T are *histories*. That is, customers buy as many arbitrary goods as they wish whenever they like. The number of records associated with a customer ID will vary and, typically, most records involve orders made by very few customers, known as a "power-law" or the Pareto distribution [15]. In general, the number of records is larger than the number of individuals, i.e., $n \leq m$.

2.2 Reidentification Probability

Rocher et al. [9] proposed a generative copula-based method for estimating the likelihood of an individual being correctly reidentified. They showed that their method could predict individual uniqueness ranging from 0.84 to 0.97. They proved that the likelihood κ_x of an individual's record $\boldsymbol{x}$ being correctly matched in a population of n individuals is

$$\kappa_i = E(\boldsymbol{x}_i \text{ correctly matched}) = \sum_{k=0}^{n-1} \frac{1}{k+1} Pr(\boldsymbol{x}_i \text{ equals } k \text{ signatures}) \quad (1)$$

$$= \sum_{k=0}^{n-1} \frac{1}{k+1} \binom{n-1}{k} Pr(\boldsymbol{x}_i)^k (1 - Pr(\boldsymbol{x}_i))^{n-1-k} \quad (2)$$

$$= \frac{1}{nPr(\boldsymbol{x}_i)}(1 - (1 - Pr(\boldsymbol{x}_i))^n), \quad (3)$$

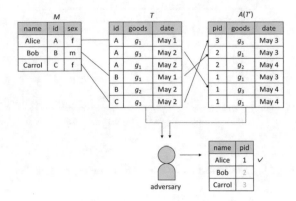

Fig. 2. Model of anonymized data

which is distributed with respect to $Pr(\boldsymbol{x})$ as shown in Fig. 1. The likelihood κ_x becomes $1/n$ when $Pr(\boldsymbol{x}) = 1$, i.e., all records have the same value and increases as $Pr(\boldsymbol{x})$ approaches 0. If we assume a uniform probability for the records, say, $Pr(\boldsymbol{x}_1) = \cdots = Pr(\boldsymbol{x}_n) = 1/4$, then $\kappa_x \approx 1/(n/4)$ by ignoring $1 - (3/4)^n \approx 1$ when n is large. This implies that the set of individuals is divided into four subsets and thereby obtains κ_x four times higher.

3   Anonymization of Payment History Records

3.1   Idea

We consider the problem of quantifying risk of reidentification for anonymized transaction data.

Figure 2 illustrates how transaction records are anonymized in a simplified model. Three customers ($n = 3$) purchase some goods from a set $\{g_1, g_2, g_3\}$ of $\ell = 3$ kinds. Let n and m be the number of individuals (customers) with specified attributes in table M and the number of records of orders specified with their date of payment in T, respectively. To reduce the risk of identification of individuals via this data, the database owner pseudonymizes the data by replacing customer names by *pseudonyms* (e.g., $1, 2, 3$), shuffling the order of records, suppressing nonstandard records, and perturbing the values (date) before publishing the anonymized transaction table $A(T')$.

Given T (M) and $A(T')$, an adversary attempts to identify the owners of the anonymized records. Even without an exact association of IDs with pseudonyms (pids), adversaries can compare $A(T')$ with T and may discover some features of individuals. For example, they may learn the frequency of records per pseudonym from their analyses and thereby guess that pid 1 is likely to belong to Alice. A set of goods associated with a pseudonym conveys a unique feature of the customer because it represents goods that the target customer is interested in. Based on features, an adversary can identify the mapping $\hat{f}$ from IDs to pseudonyms.

Suppose that an adversary estimate $f(A) = 1$ (correct), $f(B) = 3$, and $f(C) = 2$ (wrong). We say the reidentification rate is $1/3$, which is same risk to random guess.

This example shows that the accuracy of estimation depends on the frequency of orders. The more records per pseudonym, the more likely it is that the corresponding individual will be reidentified. Namely, identification of A is easier than that of B or C. Let m_i be the number of records in $A(T')$ that are associated with the i-th pid. That is, $m_1 = 3, m_2 = 2, m_3 = 1$, and $m_1 + m_2 + m_3 = m$. If m_i is sufficiently large that the individual can be identified uniquely, we should no longer use that pid and renew the assignment mapping f. This would prevent the i-th individual from being tracked via anonymized transaction records. This comes up with a simple question; *How much risky is if an individual with m_i records in dataset of order histories of n individuals?*

3.2 Profile Probability

For simplicity, we suppress redundant columns from a record and regard it as a time series of values in a single column. Let $D_i^{(g)}$ be a projection of dataset D_i to the column regarded as the goods that the i-th customer has purchased during the observation period. Let $G = \{g_1, \ldots, g_\ell\}$ be a set of goods that a shop carries, where the total number of goods is ℓ. Then, the probability of feature can be given in the following theorem.

Theorem 1 (profile probability). *[20] Let D_i be a sequence of m_i records of purchase for the i-th customer that are independent and identically distributed over G with $1/\ell$ probability. Then, the i-th profile x occurs with $\Pr[A(D_i) = x] = p_i^z(1 - p_i)^{\ell-z}$, where z is the Hamming weight of x and p_i is the probability that a certain bit of x is 1, computed as*

$$p_i = 1 - (1 - \frac{1}{\ell})^{m_i}. \tag{4}$$

Combined the profile probability with the Rocher's model, we have the likelihood of individual being correctly reidentified as follows.

Theorem 2. *Let x_i be the profile of i-th individual in ℓ-bit vector. Then, the likelihood κ_i of the i-th signature x_i being correctly reidentified from n individuals is*

$$\kappa_i = \frac{1}{n\tilde{p}^{z_i}(1 - \tilde{p})^{\ell-z_i}}(1 - (1 - \tilde{p}^{z_i}(1 - \tilde{p})^{\ell-z_i})^n),$$

where $\tilde{p}$ is the mean probability that a specific good has been purchased by any one of n individuals with m_i records and z_i is the Hamming weight of x_i.

Proof. Given the marginal probability, signature x_i is supposed as independent and identically distributed random variable drawn from $\{0, 1\}^\ell$ with probability

$$Pr(x_i) = \frac{1}{\binom{\ell}{z_i}} Pr(z_i) = \tilde{p}^{z_i}(1 - \tilde{p})^{\ell-z_i}.$$

Applying it to Eq.(3), the theorem is immediately given.

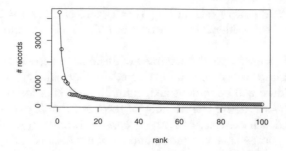

Fig. 3. Number of records in the Online Retail dataset [4] with respect to the rank ranging 1 to 100

4   Evaulation

4.1   Data Model

We model the number of records per individual as following Zipf's law [15], which is a power-law probability distribution observed widely in descriptions of social, scientific, and geophysical events, among many other kinds of events.

Definition 1 (Zipf's model). *The frequency of records associated with the i-th individual m_i follows Zipf's law, that is, $m_i = m_0/r_i^c$, where r_i is the rank of the i-th individual in frequency and $m_0 > 0$ and $c \in R$ are parameters.*

For example, Fig. 3 shows the distribution of the number of records m_i for the Online Retail dataset ($m = 45,047$ and $\ell = 3,090$) with respect to the rank r_i over the range 1 to 100. We see the power-law behavior from the fitted plot of nonlinear least-square estimation as

$$m_i = 4250/r_i^{0.908} + 14.06.$$

Figure 6 shows the probability distributions of $Pr(z_i)$, $Pr(x_i)$, and κ_i for z_i in $0, \ldots, \ell$ ($\ell = 10$, $n = 100$, $m_i = 70/r_i$). The mean of p_i for $m_i = 70, 35, 23, 18, \ldots$ is $\tilde{p}_i = 0.221$ and the mean of z_i is $\ell\tilde{p}_i = 2.21$, which matches the peak of the distribution $Pr(z_i)$. The corresponding likelihood $\kappa_z = 0.733$, which implies that individuals who have a signature with a Hamming weight of $z_i = 2$ are likely to be successfully reidentified from the anonymized transaction records $A(T')$ with a probability of 0.733.

If we set a cut-off likelihood of $\kappa_z^* = 0.733$, the corresponding rank is $r_i^* = 34$, for which the number of records $m_{34} = 2.05$ and the Hamming weight of $z_{34} = 1.95$. This implies that the top 34 individuals with respect to frequency of orders m_i should be suppressed or be replaced by more than 34 individuals with newly assigned pids. Figure 7 shows the change in the likelihood κ of being reidentified with respect to the number of records m_i for various $\ell = 10, 5, 3$.

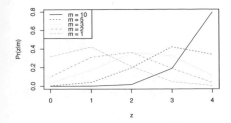

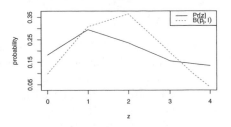

Fig. 4. Probability distribution $Pr(z|m_i)$

Fig. 5. Marginal probability distribution $Pr[z]$

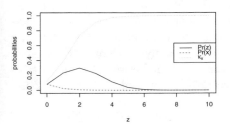

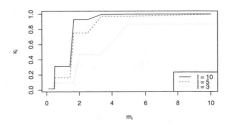

Fig. 6. Probability distributions $Pr(z)$, $Pr(x)$, and κ_x with respect to z

Fig. 7. Likelihood k_i for the i-th individual with m_i records being reidentified

4.2 Evaluation Using Competition Data

This work has been motivated by the open data-anonymization competition PWS Cup [2,3]. Before discussing the correlation between the risk of reidentification and the number of records per pseudonym, we briefly describe the rules of the competition and the results. (See Appendix A for more information)

In 2017, the target dataset was the Online Retail dataset [4] that comprised 45,000 records of purchase transactions from about 500 customers. Tables M and T in Table 1, 2 are examples of the data from this dataset.

Participants in the competition anonymized the transaction records so that no record could be reidentified correctly by other participants. The anonymization methods allowed for only a fixed set of processes, i.e., suppression of records, perturbation of dates and values, and replacement of customer identity by arbitrary pseudonyms. They were allowed to assign multiple monthly pseudonyms to a single identity. This meant that if a busy customer made purchases too frequently, which was evaluated as too risky, the participant could change the pseudonym of the records as if the purchase orders were made by other customers.

In the reidentification attempts, participants were free to choose any algorithm to identify pseudonyms, based on arbitrarily defined features of the transactions. In the 2017 competition, the dominant strategy was maximum similarity estimation, based on the Jaccard coefficient between a set of goods in the records. This approach, modeled in terms of the Jaccard signature, is used in this paper.

4.3 Results

On October 23th 2017, 14 teams participated in the competition to estimate the mapping between customer identities and pseudonyms. The winner balanced well the utility, defined as a set of error functions for the anonymized data, and its privacy, which was evaluated in terms of the maximum reidentification ratios submitted by other teams.

The history of purchases is one of the instance of our proposed model, where a bit of profile is distributed in Eq. (4). The Hamming weight of purchase records has many 1-bit as the number of records for the given customer increases. This provides a clear evidence that records are independent and identically distributed. As shown in Fig. 7, the likelihood of identification κ increases as the number of records m_i increases. Namely, the number of anonymized records must be more frequently identified than that without anonymization if Theorem 2 holds here. We are interested in whether the behavior between the number of records m_i and the frequency of identified records is observed in the competition data, or not.

The risk of reidentification of a pseudonym increases with the frequency of orders aggregated for the pseudonym. To quantify the degree of frequency, we denote P_i as the set of pseudonyms assigned to customer i. Let $|P|$ and $\mu(P)$ be the mean length and duration of pseudonyms for all customers, respectively. For example, consider three customers assigned monthly pseudonyms for three months: $P_1 = (10, 10, 10)$, $P_2 = (20, 20, 20)$, and $P_3 = (30, 30, 35)$. Here, the lengths are $|P_1| = |P_2| = 1$ and $|P_3| = 2$, and the durations are $\mu(P_1) = \mu(P_2) = 3$ month and $\mu(P_3) = (2+1)/2 = 1.5$. The mean length and the mean duration of the anonymized data are $|P| = 4/3$ and $\mu(P) = (3+3+1.5)/3 = 2.5$, respectively.

Figure 8 shows the number of correctly reidentified customers in a scatterplot for the mean duration $\mu(P)$ (X-axis) and the mean number of pseudonyms $|P|$. The winning team reidentified seven customers, assigning about two pseudonyms per customer for about a 1.7-month duration. Table 3 gives the results for an average team (M-OND-A), showing the counts of customers with respect to their successful reidentification. This indicates that the unidentified pseudonyms have a greater mean length and longer mean duration.

From the observation, we find that data with longer duration have many identified customers (e.g., 250 and 488). However, even if the numbers of records m_i are minimized, i.e., $m_i = 1.0$, the customers are not always safe against re-identification (e.g., 465). This shows the failure of Theorem 2 and we need further improvement of our model.

Table 3. Contingency table of the reidentification-success statistics of customers for team M-OND-A

Item	Unidentified	Reidentified
N	339	161
Mean length of pseudonyms	1.223	1.104
Mean duration of pseudonyms	3.378	2.491

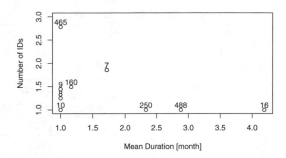

Fig. 8. Numbers of reidentified customers

5   Conclusions

We have studied the risk of anonymized data being reidentified from a feature of long-term transaction records aggregated by pseudonym. In our model, given the pseudonymized transaction records, an adversary attempts to identify individuals from the assigned pseudonyms by maximizing the similarity defined by the Jaccard coefficient between sets of goods that have been purchased by the pseudonym. Under the twin assumptions of a uniform probability of choice of goods and the Zipf's-law behavior of the number of records per individual, we investigated the likelihood of an individual being identified correctly from the transaction signature as a function of the number of records. Our model indicated that the risk of an individual being identified increases as the number of records associated with its pseudonym increases.

A similar effect, in terms of risk and the number of pseudonyms, was found for the data-anonymization competition PWS Cup 2017, in which the online retail dataset comprising 40,000 records for 500 individuals was securely anonymized and then reidentified where possible. The competition results suggested that the reidentification rate increases if the use of a particular pseudonym continues for too long.

A   Pwscup competition

PWS Cup [2,3] has been held since 2015 and supported by the Special Interest Group on Computer Security (CSEC) Working Group of the Information

Processing Society Japan (IPSJ)[2]. Table 4 gives the specifications for the competitions in 2015–2017.

Table 4. The PWS Cup competition editions

	2015	2016	2017
Date	Oct. 21,22	Oct. 11,12	Oct. 23,24
Venue	Nagasaki Brick Hall	Akita Castle Hotel	Yamagata Int. Hotel
# participants	20 members	42 members	43 members
	13 teams	15 teams	14 teams
Dataset	NSTAC	UCI Dataset Online Retail	
	Synthesized data	POS records	
# attributes	25	11 (4 customers + 7 attributes)	
# customers n	8,333	400	500
# records m	8,333	18,524	44,917
Duration T	N/A	1 year T	12-month $T^1, \ldots, T^{(12)}$
Adversary model	Maximum knowledge		Partial knowledge

References

1. General data protection regulation, regulation (EU) 2016/679. (https://gdpr-info. eu refereed in 2019)
2. Kikuchi, H., Yamaguchi, T., Hamada, K., Yamaoka, Y., Oguri, H., Sakuma, J.: Ice and fire: quantifying the risk of reidentification and utility in data anonymization. In: 2016 IEEE 30th International Conference on Advanced Information Networking and Applications (AINA), Crans-Montana, pp. 1035–1042 (2016)
3. Kikuchi, H., Yamaguchi, T., Hamada, K., Yamaoka, Y., Oguri, H., Sakuma, J.: A study from the data anonymization competition Pwscup 2015. In: Data Privacy Management and Security Assurance (DPM 2016), pp. 230–237. Springer (2016)
4. Chen, D., Sain, S.L., Guo, K.: Data mining for the online retail industry: a case study of RFM model-based customer segmentation using data mining. J. Database Market. Cust. Strategy Manage. **19**(3), 197–208 (2012)
5. Information Commissioner's Office (ICO), Anonymisation: managing data protection risk code of practice (2012)
6. El Emam, K., Arbuckle, L.: Anonymizing Health Data Case Studies and Methods to Get You Started. O'Reilly, Sebastopol (2013)
7. Domingo-Ferrer, J., Ricci, S., Soria-Comas, J.: Disclosure risk assessment via record linkage by a maximum-knowledge attacker. In: 2015 Thirteenth Annual Conference on Privacy. IEEE, Security and Trust (PST) (2015)
8. Linden, G., Smith, B., York, J.: Amazon.com recommendations: item-to-item collaborative filtering. IEEE Internet Comput. **7**(1), 76–80 (2003)

---

[2] http://www.iwsec.org/pws/pwscup/PWSCUP2017.html.

9. Rocher, L., Hendrickx, J.M., de Montjoye, Y.-A.: Estimating the success of re-identifications in incomplete datasets using generative models. Nat. Commun. **10**(3069), 1–9 (2019)
10. Samarati, P., Sweeney, L.: Protecting privacy when disclosing information: k-anonymity and its enforcement through generalization and suppression, SRI Int., Menlo Park, CA, USA, Technical report (1998)
11. Machanavajjhala, A., Gehrke, J., Kiefer, D., Venkatasubramanian, M.: L-diversity: privacy beyond k-anonymity. ACM Trans. Knowl. Discov. Data **1**(1), Article no. 3 (2007)
12. Li, N., Li, T., Venkatasubramanian, S.: t-Closeness: privacy beyond k-anonymity and ℓ-diversity. In: Proceedings of 23rd IEEE International Conference on Data Engineering (ICDE), pp. 106–115 (2007)
13. Cao, J., Karras, P.: Publishing microdata with a robust privacy guarantee. Proc. VLDB Endowment **5**(11), 1388–1399 (2012)
14. Dwork, C.: Differential privacy. In: Proceedings 33rd International Colloquium on Automata, Languages, and Programming (ICALP) (LNCS), vol. 4052, pp. 1–12. Springer, Berlin, Germany (2006)
15. Mitzenmacher, M., Upfal, E.: Zipf's law and other examples. In: Probability and Computing: Randomization and Probabilistic Techniques in Algorithms and Data Analysis 2nd Edition, Sect. 16.2.1, pp. 417–418, Cambridge University Press (2017)
16. Domingo-Ferrer, J., Muralidhar, K., Bras-Amorós, M.: General confidentiality and utility metrics for privacy-preserving data publishing based on the permutation model. IEEE Trans. Dependable Secure Comput. (2020). https://doi.org/10.1109/TDSC.2020.2968027
17. Xiaokui, X., Tao, Y.: M-invariance: towards privacy preserving re-publication of dynamic datasets. In: Proceedings of the 2007 ACM SIGMOD International Conference on Management of Data, pp. 689-700 (2007)
18. Cummings, R., Krehbiel, S., Lai, K.A., Tantipongpipat, U.: Differential privacy for growing databases. In: Advances in Neural Information Processing Systems, vol. 31, pp. 8864–8873 (2018)
19. ISO/IEC 20889, Privacy enhancing data de-identification terminology and classification of techniques (2018)
20. Kikuchi, H.: Differentially private profiling of anonymized customer purchase records. In: Data Privacy Management, Cryptocurrencies and Blockchain Technology, DPM 2020, CBT 2020, LNCS, vol. 12484, pp. 19–34. Springer (2020)

Trust-Based Detection Strategy Against Replication Attacks in IoT

Bacem Mbarek[1]($\boxtimes$), Mouzhi Ge[2], and Tomáš Pitner[1]

[1] Faculty of Informatics, Masaryk University, Brno, Czech Republic
bacem.mbarek@mail.muni.cz, tomp@fi.muni.cz
[2] Deggendorf Institute of Technology, Deggendorf, Germany
mouzhi.ge@th-deg.de

Abstract. The integration of 6LoWPAN standard in the Internet of Things (IoT) has been emerging and applied in many domains such as smart transportation and healthcare. However, given the resource constrained nature of nodes in the IoT, 6LoWPAN is vulnerable to a variety of attacks, among others, replication attack can be launched to consume the node's resources and degrade the network's performance. This paper therefore proposes a trust-based detection strategy against replication attacks in IoT, where a number of replica nodes are intentionally inserted into the network to test the reliability and response of witness nodes. We further assess the feasibility of the proposed detection strategy and compare with other two strategies of brute-force and first visited with a thorough simulation, taking into account the detection probability for compromised attacks and the execution time of transactions. The simulation results show that the proposed trust-based strategy can significantly increase the detection probability to 90% on average against replication attacks, while within the number of nodes up to 1000 the detection runtime on average keeps around 60 s.

Keywords: Internet of Things · Detection strategy · Trust in IoT · Replication attack · 6LoWPAN

1 Introduction

The Internet of Things (IoT) plays a key role in the development of modern wireless telecommunications that covers the integration of the Lossy and Low power Networks (LLNs). Internet Engineering Task Force suggested that IPv6 over Low -Power Wireless Personal Area Networks (6LoWPAN) as the communication protocol for LLNs [17]. 6LoWPAN is one of the most successful standards that defines the approach for routing IPv6 over low-power wireless networks. Thus, 6LoWPAN is considered as a promising IoT technology, and can be applied even to the smallest IoT devices. In 6LoWPAN, routing functionalities have been investigated under a dedicated routing protocol, namely RPL (Routing Protocol for Low Power and Losy Networks). RPL topology and 6LoWPAN mapper are based on a tree topology called Destination Oriented Directed Acyclic Graphs (DODAG) [5]. Each RPL instance is associated to an

L. Barolli et al. (Eds.): AINA 2021, LNNS 226, pp. 646–656, 2021.
https://doi.org/10.1007/978-3-030-75075-6_53

objective function that is responsible for calculating the best path depending on a set of metrics or constraints [14].

However, RPL-based 6LoWPAN network is usually vulnerable to various attacks given the resource constrained of IoT devices. One of the most severe cyber attacks is replication attack, where it creates replicated nodes in the IoT network. Once the replication attack compromises witness nodes, the adversary will make all the replicated nodes hidden as 6Mapper provides node ID and rank of each node. In a RPL network, when packets are delivered to one of the replicated nodes. These packets will be forwarded to one of the replicated nodes based on the routing metrics in the network, and the rest of the replicated nodes will be unreachable in the network [11]. Thus the detection of replication attacks is critical in IoT networks.

In this paper, we shed the light to the security vulnerabilities of RPL-based 6LoW-PAN network and their inability to tackle replication attack. We propose a trust-based strategy to detect and prevent node replication attack, specifically tailored to identify and counteract compromised witness nodes in the presence of testing replica nodes. Those testing replica nodes is deployed intentionally to the witness node and uses the reaction of the tested witness node to consolidate the detection of compromised witness nodes. Furthermore, we compare the proposed trust-based strategy with two baselines strategies, which are brute-force strategy and first visited strategy. To this end, we assess the feasibility of our proposed approach by evaluating the strategies and the impact of each strategy on detecting compromising attacks.

The remainder of the paper is organized as follows. Section 2 provides a brief overview of the replication attack and discusses the state-of-the-art solutions that are proposed to detect intrusions in the IoT networks. Section 3 proposes a trust-based solution that can improve the detection against replication attacks. In order to evaluate the proposed solution, Sect. 4 conducts a security analysis simulation by considering probability of detection and runtime performance. Finally, Sect. 5 concludes the paper and outlines future works for the paper.

2   Replication Attack

The replication attack in an IoT network is launched based on the duplication of sensor nodes by copying the same properties such node ID and node location [1,3]. In order to explain the replication attack in IoT network, we present in Fig. 1 an example of RPL protocol with the replication attack, where the network is represented by 6 Sensor Nodes (SNs), 2 witness nodes and 1 replica node which duplicates the properties of the sensor node $SN1$. Let's consider the *witness node* 1 as a compromised node, where the *replica node* is inserted next to the compromised witness node. The witness node is able to detect the replica node by using its intrusion detection system based on detection finding conflicting location.

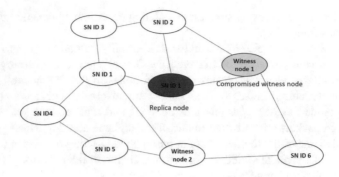

Fig. 1. Witness compromising attack

In order to intervene the security control from witness node, an attacker can compromise a witness node, and then make a replication attack to all of the neighbor nodes. As shown in Fig. 1, the *compromised witness node* can prevent nodes from *replica nodes* to be detected. Therefore, detection tolls will be vulnerable to witness compromising attacks [9].

2.1 Different Attack Detection in RPL

In order to detect the attacks in RPL-based 6LoWPAN network, there are a variety of solutions that are reviewed as follows. In [6], authors have presented the different types of attacks against the RPL network. Moreover, authors have classified the attacks according to the attacker's goal and means considering the specific properties of RPL networks. The attacks against resources based on implementing unnecessary processing in order to reduce the device lifetime and exhaust their resources. The attacks against the topology which able to isolate the node and disrupt its communication with their parents or with the root. Finally, attacks against network traffic by analysing and capturing a part of the traffic such as detecting root nodes and or spoofing the address of the DODAG root graph in RPL. Based on the attack classification, authors have discussed different techniques to avoid or prevent them. The classification of routing attacks against the RPL protocol is depicted in 1.

In [8], authors have studied and analyzed the RPL behavior in the presence of a sybil attack on the RPL protocol implemented in a WSN network with mobile nodes. In this attack, malicious mobile node takes several sybil identities to pollute the network with fake control messages from different locations. In [13], authors have designed an attack graph which can give an overview of different possible threats associated with rank property in the routing path. Based on the attack graph, authors analyzed the consequences of exploiting the rank property. In [7], have proposed a new intrusion detection system for IoT, called ENIDS. Their scheme considers replica detection algorithm in RPL-based 6LoWPAN network based on sending a replica test node to an arbitrarily selected witness node. Without a coherent strategy, ENIDS not sufficient to determine all legitimate nodes.

RPL-based 6LoWPAN network suffers from many routing security threats coming from both the external and internal attackers. In this paper, we focus on the replication attacks that are considered as the Attack against Topology. The replication attacks are classified as a sub-optimal path attack in IoT network (Table 1).

Table 1. Classification of attacks against RPL networks

Attacks against resources		Attacks against topology		Attacks against traffic	
Direct attacks	Indirect attacks	Sub-optimization	Isolation	Evesdropping	Misappropriation
• Flooding • Routing table overload	• Increased rank attack • DAG inconsistency • Version number modification	• Routing table flasification • Sinkhole • Wormhole • **Replication attack**	• Blackhole • DAO inconsistency	• Sniffing • Traffic analysis	• Decreased rank attack • Identity attack

3 Trust-Based Detection Against Replication Attacks

The rationale behind the trust-based detection solution is that we intentionally insert an arbitrary number of replica test nodes into the network to test the response of witness nodes. If the witness nodes do not detect these replica test nodes, the network will be considered as the case that there are compromised witness nodes used by the adversary who protects replicas from being detected. In this way, the deployed replica would be able to verify the compromised witness node. The replica node is inserted next to the witness node. Our new solution can test the reaction of the witness node to detect if it is compromised.

3.1 Replica Test

The replica test in our work is based on inserting a test replica node and uses the reaction of the witness node to decide if the witness node is compromised. Moreover, replica test is a procedure that consists in detecting replication attack with identity and location information. It enables the witness nodes in RPL-based 6LoWPAN network to detect replication nodes and determine the routing damage.

In this paper, we consider the detection of replication attacks in 6LoWPAN based on replica test strategy. We first create a duplication node using the same properties such as node ID and location. Second, we use trust-based detection strategy to insert the replica test node next to the selected witness node. Then, the replica test node will detect the reaction of the tested witness node against the replication attack. As such, if the witness node receives different locations for the same node ID, it should inform the original node about the detected replica node. The compromised witness node will be detected by not reporting to the original node.

In the node replication attack, an adversary physically captures a sensor node, extracts cryptography secrets from the original node, and distributes a large number of replication nodes of the captured node throughout the network and launch a variety of insider attacks. When the witness node realizes that two nodes (the original node and the replica) have the same ID, it will directly delete the replica node from the network. However, the adversary can launch a node compromising attack and compromises a witness node. The compromised witness nodes and replicated sensor nodes are entirely controlled by the adversary and can communicate with each other at any time. Then the compromised witness node can protect other replicated nodes.

3.2   Trust Factors in Trust-Based Strategy

The trust score depends on three factors: (1) Energy used to transmit a packet (ε), (2) Transmission delay (D_T) and (3) Packet delivery ratio (*PDR*). TrustScore can be for example scaled into a rating from 1 to 10. Each node in the network start with a score equal to 10. All three factors serve as important functions in determining if a node has the possibility to be hacked, loaded with malware, or with other cyber-attacks. For example, a node can be considered as possible malfunctions or under attack if it expends abnormal energy in the transmission of a packet or a large delay in the submission of a packet or a low propagation speed.

Energy Used to Transmit a Packet

To calculate the energy consumed during every packet transmission, authors in [16] proposed the equation below.

$$\varepsilon_{u,v}(x,r) = (P_t + \frac{P_{u,v}}{K})T \tag{1}$$

where ε is the power that is consumed by a sender node u to transmit a packet of length x bits to a receiver node v. (u,v) denotes the wireless link between u and v. Let r be the rate at which u transmits data to v over the physical link (u,v). Let P_t be the power required to run the receiving circuit of the transmitter, $P_{u,v}$ be the transmission power from u to v, and K be the efficiency of the power amplifier. T is the time required to transmit x bits with the rate rbps.

Transmission Delay

Transmission delay D_T is the time needed to transmit the entire packet (or frame, bit train), from first bit to last bit, over a communication link. The transmission delay D_T is calculated as follow.

$$D_T = \frac{N}{R} \tag{2}$$

Where D_T is the transmission delay in seconds, N is the number of bits, and R is the rate of transmission in bits per second.

Packet Delivery Ratio

Packet delivery ratio (PDR) can be measured as the ratio of number of packets delivered within the total number of packets sent from source node to destination node in the network [15]. The PDR is calculated as follow in Eq. 3.

$$PDR = \frac{Received\ packets}{Sent\ packets} \times 100 \tag{3}$$

3.3 Trust Score Assignment Algorithm

In the RPL-based 6LoWPAN network, each sensor node can be routed in a directed acyclic graph, and each node will connect with its parent to join a goal-oriented directed acyclic graph. This forms a channel for communications between the nodes and base station. In a predefined interval, the base station should analyze the behavior of each node by comparing their trust factors (ε, D_T, and PDR) with the average score of each factor. The Trust Scores Variation (TSV) of each trust factor is represented as follow: TSV, where i represents the trust factor ε or D_T or PDR. T_i will be calculated by Eqs. 1, 2, and 3 respectively, and $Ti_{average}$ will be calculated by the based station across all the nodes in the network. If the value of the trust factor $T_i > Ti_{average}$, then a malfunction is detected and TSV will be -1, otherwise it is 0. The base station detects the malfunction of nodes as depicted in Eq. 4.

$$TSV = \begin{cases} -1, & \text{if } T_i > Ti_{average} \\ 0, & \text{otherwise} \end{cases} \qquad (4)$$

Algorithm 1: Trust Score Assignment Algorithm

  Result: TrustScore
1 $TrustScore = 10$;
2 $NumOfRounds - 10$;
3 $StableThreshold = 2$;
4 $StableRound = 0$;
5 $Round = 0$;
6 **while** $0 \leq i \leq NumOfRounds$& **do**
7 | $TrustScore = TrustScore + (TVS_\varepsilon + TVS_{DT} + TVS_{PDR})$;
8 | **if** $StableRound = StableThreshold$ **then**
9 | | $TrustScore = TrustScore + 1$;
10 | | $StableRound = 0$;
11 | **end**
12 | $i = i + 1$;
13 | $Round = Round + 1$;
14 | **if** $TrustScore_{Round} = TrustScore_{Round-1}$ **then**
15 | | $StableRound + +$;
16 | **else**
17 | | $StableRound = 0$;
18 | **end**
19 **end**

Algorithm 1 describes the calculation of the trust score. Consider that we have a predefined number of replica testing rounds as $NumOfRounds$. In each round, the base station should select a number of nodes for replica test based on the existing trust score. For example, in the first deployment each node has an initial trust score $TrustScore$, which can be set to 10. In each replica test, the base station will update the trust score of each trust factor based on the Eq. 4. Then, the final trust score of each node is the sum of all the updated scores from the three trust factors. *StableRound* is calculated by comparing the $TrustScore_{Round}$ of the current and the $TrustScore_{Round-1}$ from last round. If it is the same, then we increase the number of the *StableRound*. When the number of *StableRound* is equal to a *StableThreshold*, in our example, we set 2 as *StableThreshold*, that means, the node trust score does not change in 3 consecutive testing rounds, then the Trust score in the coming round will increase its value by $+1$.

4   Evaluation

In order to evaluate the trust-based strategy, we used the Contiki network simulator COOJA [10] to analyze the performance of the trust-based strategy in terms of probability of detection of replication attacks and in terms of the runtime of detection. This embedded operating system, using the COOJA network simulator, allows a comprehensive of the new security scheme. It is written in the C language, and enables an easy deployment of the 6LoWPAN security extension.

During replication attacks in RPL-based 6LoWPAN networks, the attacker can use compromised nodes to send wrong information about their rank or one of their rank of neighbors to the intrusion detection modules [2, 12]. It is also possible to obtain an incorrect or an inconsistent view of the network because of the lossy links in the RPL-based 6LoWPAN network. It is therefore important to detect the inconsistencies, distinguish between valid and invalid consistencies, and correct the invalid information [4]. If an attacker created and implemented a cloned node in the RPL-based 6LoWPAN network by duplicating ID from an existing device, it will cause a series of security and privacy problem in routing information systems. Therefore, preventing replication attacks and detecting compromised replica node are critical for RPL-based 6LoWPAN network [4].

4.1   Baseline Detection Strategies

We focus on two typical strategies as baselines that can be used to choose the witness node, which are Brute-force Strategy, and First visit strategy. (1) Brute-force Strategy (BS): In Brute-force search, all the nodes are visited. We send the replication nodes to all the witness nodes at the same time. The replica tests will be executed on all the witness nodes. Running multiple tests at the same time against different. (2) First Visited Strategy (FVS): To visit all the nodes connected to the IoT network, we used a FIFO (First In First Out) order. Each time a new node is visited, the FIFO algorithm update the history-pointer by storing the initially visited node at the last array index of the queue. Both strategies can carry out the detection of the replication attacks with certain advantages.

4.2   Evaluation Results

We evaluated the *probability of detection* of replication attacks and the *runtime replication attack detection* of two main strategies: Brute-force Strategy (BS), Trust-based

Strategy (TBS) and First Visited Strategy (FVS). The probability of witness nodes P_{w-node} that hides the replica node is obtained by Eq. 5.

$$P_{w-node} = 1 - \frac{D_{replica}}{n_\alpha} \qquad (5)$$

where $D_{replica}$ is the detection probability of replica nodes, n_α is the number of replica test nodes into the network. We inject the replication attack in a compromised node which is a node never been tested, with higher trust score and in the N last positions of the queue. We come up with two interesting outcomes.

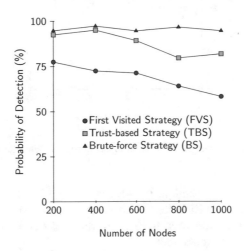

Fig. 2. Experimental results for comparing probability of detection

As shown in Fig. 2, the first outcome is that the TBS and BS approaches have higher probability of detection of replication attacks compared to FVS, when increasing the number of nodes. For example, when the number of nodes is equal to 600, TBS, and BS detect more than 90% of replication attacks; while FVS detects less than 75%. An explanation to this behavior is that in the case of inserting the replication attack in node never been tested, with higher trust score, then the first visit trust will not be convenient.

The second outcome is that the BS strategy requires a significant runtime for replication attack detection compared to TBS and FVS with a large number of nodes (as shown in Fig. 3). Also the simulation result shows that the approximate maximum number of nodes which can be considered as a small size of IoT network nodes is around 50 nodes. This scale of IoT network will TBS take around 40 s for replication attack detection. Even as the number of nodes increases to 1000, the runtime on average keeps around 60 s.

4.3 Discussions

In order to evaluate the detection strategies, it is important to maintain an efficiency balance between the runtime to detect a replication attack and the probability of

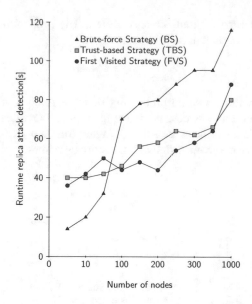

Fig. 3. Experimental results for comparing replica detection runtime

detection. In our proposed solution, trust-based strategy is used to detect replication attack in IoT, particularly in RPL-based 6LoWPAN network. Our test was to run one or many replica nodes in the network. In this way, the deployed replica would be able to verify the compromised witness node.

We consider that there is no one strategy that fits all the detection scenarios. For example, the results show that trust-based Detection Strategy can help to overall satisfaction in terms of runtime replication attack detection and the probability of detection, however there is still a lack on improving our solution in terms of adding more factors of trust and representing a very dynamic solution for gathering them. This suggests that we may invoke brute-force strategy when we notice the presence of a low number of nodes. In some detection scenario, the random strategy is also can be used. However, the arbitrary test selection of witness node is not able to construct a proper statistical inferences. Instead, brute-force strategy, the trust-based Strategy and the first visited strategy have a coherent mechanism.

Furthermore, strategy switching has the potential to improve the detection effectiveness. The strategy switching depends on the status of the network. We speculate that in this case the strategy switching plays a trade-off according to the number of nodes, network status and trust score. It can be seen that the random selection strategy can be used when all the network nodes haven't been tested and all nodes have the similar score of trust. In the case of different scores of trust, our replica test algorithm classify nodes in two categories: (1) Nodes have the similar score of trust, (2) Nodes with different scores of trust. For the first category, the replica test algorithm will apply the random selection strategy. For the second category, the replica test algorithm can apply the trust-based strategy. If the IoT network nodes have different scores of trust, the replica test may apply the FVS Strategy. When a part of network nodes has been and we reach a small size of network which has been not tested, the replica test can apply the brute-face strategy.

5 Conclusion

In this paper, we have proposed a new trust-based detection strategy and algorithm against replication attacks in IoT network. This strategy has been used to detect the replication attacks in the presence of compromised witness node against the RPL-based 6LoWPAN network. The simulation results have shown that the trust-based strategy can significantly improve the effectiveness of detection probability of replication attacks while being comparable to brute-force strategy and first visited strategy when the number of nodes is larger. To further evaluate the runtime performance, we found that the proposed trust-based detection strategy has a significantly lower delay than the other two strategies in the presence of higher number of nodes. In the future, we plan to generate a dynamic strategy for replica detection according to different network factors, such as number of nodes, network status, nodes' behavior, etc. In addition, we plan to apply different evaluation metrics to further investigate the robustness of the trust based detection strategy in various business domains such as connected transport and smart home.

Acknowledgements. This research was supported from ERDF/ESF "CyberSecurity, Cyber-Crime and Critical Information Infrastructures Center of Excellence" (No. CZ.02.1.01/0.0/0.0/ 16_019/ 0000822).

References

1. Benkhelifa, E., Welsh, T., Hamouda, W.: A critical review of practices and challenges in intrusion detection systems for IoT: toward universal and resilient systems. IEEE Commun. Surv. Tutor. **20**(4), 3496–3509 (2018)
2. Bogari, E.A., Zavarsky, P., Lindskog, D., Ruhl, R.: An analysis of security weaknesses in the evolution of RFID enabled passport. In: World Congress on Internet Security (WorldCIS-2012), pp. 158–166. IEEE (2012)
3. Buczak, A.L., Guven, E.: A survey of data mining and machine learning methods for cyber security intrusion detection. IEEE Commun. Surv. Tutor. **18**(2), 1153–1176 (2016)
4. Jaballah, W.B., Conti, M., Filè, G., Mosbah, M., Zemmari, A.: Whac-a-mole: smart node positioning in clone attack in wireless sensor networks. Comput. Commun. **119**, 66–82 (2018)
5. Kamble, A., Malemath, V.S., Patil, D.: Security attacks and secure routing protocols in RPL-based Internet of Things: survey. In: 2017 International Conference on Emerging Trends & Innovation in ICT (ICEI), pp. 33–39. IEEE (2017)
6. Mayzaud, A., Badonnel, R., Chrisment, I.: A taxonomy of attacks in RPL-based Internet of Things. Int. J. Netw. Secur. **18**(3), 459–473 (2016)
7. Mbarek, B., Ge, M., Pitner, T.: Enhanced network intrusion detection system protocol for Internet of Things. In: Proceedings of the 35th Annual ACM Symposium on Applied Computing, pp. 1156–1163 (2020)

8. Medjek, F., Tandjaoui, D., Romdhani, I., Djedjig, N.: Performance evaluation of RPL proto-col under mobile Sybil attacks. In: 2017 IEEE Trustcom/BigDataSE/ICESS, pp. 1049–1055. IEEE (2017)
9. Mishra, P., Varadharajan, V., Tupakula, U., Pilli, E.S.: A detailed investigation and analysis of using machine learning techniques for intrusion detection. IEEE Commun. Surv. Tutor. **2018**, 686–728 (2018)
10. Österlind, F., Dunkels, A., Eriksson, J., Finne, N., Voigt, T.: Cross-level sensor network sim-ulation with COOJA. In: First IEEE International Workshop on Practical Issues in Building Sensor Network Applications (SenseApp 2006) (2006)
11. Pongle, P., Chavan, G.: A survey: attacks on RPL and 6LoWPAN in IoT. In: 2015 Interna-tional Conference on Pervasive Computing (ICPC), pp. 1–6. IEEE (2015
12. Ray, B.R., Chowdhury, M.U., Abawajy, J.H.: Secure object tracking protocol for the Internet of Things. IEEE Internet Things J. **3**(4), 544–553 (2016)
13. Sahay, R., Geethakumari, G., Modugu, K.: Attack graph—based vulnerability assessment of rank property in RPL-6LOWPAN in IoT. In: 2018 IEEE 4th World Forum on Internet of Things (WF-IoT), pp. 308–313. IEEE (2018)
14. Shreenivas, D., Raza, S., Voigt, T.: Intrusion detection in the RPL-connected 6LoWPAN networks, pp. 31–38 (2017)
15. Singh, B., Srikanth, D., Suthikshn Kumar, C.R.: Mitigating effects of black hole attack in mobile Ad-Hoc NETworks: military perspective. In: 2016 IEEE International Conference on Engineering and Technology (ICETECH), pp. 810–814. IEEE (2016)
16. Vazifehdan, J., Prasad, R.V., Jacobsson, M., Niemegeers, I.: An analytical energy consump-tion model for packet transfer over wireless links. IEEE Commun. Lett. **16**(1), 30–33 (2011)
17. Verma, A., Ranga, V.: Mitigation of DIS flooding attacks in RPL-based 6LoWPAN networks. Trans. Emerg. Telecommun. Technol. **31**(2), e3802 (2020)

Reinforcement Learning Based Smart Data Agent for Location Privacy

Harkeerat Kaur[1][(✉)], Rohit Kumar[1], and Isao Echizen[2]

[1] Indian Institute of Technology Jammu, Jammu, India
{harkeerat.kaur, 2017ucs0054}@iitjammu.ac.in
[2] National Institute of Informatics Tokyo, Tokyo, Japan
iechizen@nii.ac.jp

Abstract. A "privacy by design" approach is presented for location management on a smartphone that is based on the concept of "smart data". Extensive use of location-based services has given rise to various privacy concerns such as processing and selling of personal information. The proposed work addresses these concerns by developing an intelligent agent that controls the release of the user's information in accordance with its preferences, context, nature of situation. The proposed model uses reinforcement learning to provide an on-the-go smart data agent that learns the user's privacy policy in a manner that is interactive and adaptive, enabling it to adjust itself to changes in user preferences over time. The agent aimed at mimicking the user, deciding when and where exact location information should be revealed and evolving over time to become the user's single trusted virtual proxy in cyberspace, managing its interactions with various applications in different environments.

1 Introduction

The digital world is now hosting a complete palate of IoT and web-based applications, such as smart homes, smart cars, smart wearables, smart locks, that are coupled with various online services for shopping, renting, education, insurance, etc. To avail themselves of these value-added services, users must often submit extensive personal information, including their preferences, behaviours, eating habits, and social status. Moreover, location information, another type of personal information is increasingly, inadvertently and sometimes covertly being captured by location-based services (LBS). Various mobile applications for route mapping, recommendation, weather and traffic prediction, and even non-LBS applications like music and entertainment continuously capture user location data, which is then used to monitor user movement patterns and places of interest. A present-day smartphone user often faces a choice like "allow ***
[app] to access your location even when you are not using it". Unless permission is granted, the user cannot use the app. There is thus a strong need for technological solutions that integrate "privacy by design" with IoT frameworks so that a user can enjoy the benefits of using such services while retaining control over the extent to which the information can be shared.

© The Author(s), under exclusive license to Springer Nature Switzerland AG 2021
L. Barolli et al. (Eds.): AINA 2021, LNNS 226, pp. 657–671, 2021.
https://doi.org/10.1007/978-3-030-75075-6_54

A model is proposed here using the concept of "*smart data*" i.e., data that can "think for itself" [5,19] to provide location data privacy. It requires development of an intelligent web-based agent that acts as a *virtual proxy* for the user in cyberspace, controlling the release of data in accordance with its instructions and/or preferences. Modelling the design of an agent that mimics the user's privacy behaviour is the most important and challenging aspect of this approach. This challenge is met by the development of reinforcement learning (RL) based adaptive and automated "smart data agent model". In comparison to supervised and unsupervised learning, RL is based on dynamic mapping and learning by interaction. The proposed agent model interacts with the environment in order to understand how to map a situation and to decide the appropriate action to be taken to maximize the total possible reward. The ability to learn dynamically in a human-like way has led to RL being used to develop adaptive software programs [13]. The proposed approach is the first step towards an "intelligent data agent" specific to each user, supporting the user's privacy preferences learned through reinforcement learning, to the best of our knowledge.

The work is organized as follows. The location privacy problem is described and existing approaches are reviewed in Sect. 2. The RL formulation of the proposed smart data agent model is discussed in Sect. 3. The simulation results are presented and discussed in Sect. 4. The key points are summarized in Sect. 5.

2   Location Privacy and Related Works

Location privacy essentially refers to a person's control over the storing, sharing, and dissipation of their location information [1]. Several types of mechanisms have been proposed for managing users' location information.

Centralized Third-Party Mechanism: In this approach, a centralized anonymizer or a third party acts as a trusted agent between the user and LBS. The anonymizer cloaks the actual location of the user within a set of locations known as a "cloaking region" [3]. In the basic K-anonymity approach, the anonymizer first encloses the actual location within a set of $K - 1$ similar locations and then forwards the query to the LBS [2]. Several improvements like K-anonymity and L-diversity [10], distributed K-anonymity [22], and (K,T)-anonymity [11] have been proposed. However, the major issues are reliability of the anonymizer and its probability of getting attacked. The need for maintaining a sufficient number of users may cause expansion of the cloaking regions which can degrade quality of service (QoS) and service time.

Collaborative Mechanism: In this approach a group of users collaborate to hide their location information [17,18]. It requires peer-to-peer connections of location-aware devices connected over a wireless infrastructure (Bluetooth, Wi-Fi) that can communicate with each other. Main working principle is that if a peer in the network has location-specific information fetched from LBS, it can share it with any peer who is seeking the same information without it having to submit its actual location. The devices can also use these locations for collaborative k-anonymization [18], obfuscation [6,8] etc. Major challenges with this approach are the presence of neighbours, their probability having similar queries, willingness to cooperate, and network viability for such operations [15]. Geographical location and population density play a very important role here and privacy cannot always be guaranteed.

User-Centric Mechanism: Here a user itself endeavours to protect its location instead of relying on peers or third parties. Location is distorted by cloaking, enlarging, or reducing the target region. A basic way to obfuscate is to distort on the basis of geometry [12]. Another way to protect actual location is to submit fake or dummy queries to the LBS directly rather than having a centralized anonymizer server generate them. However, if the locations for submitting the fake or dummy queries is not carefully selected, the attacker may be able to infer the actual location. Historical proximity and kernel transformation are other approaches in this direction [4,21].

Blockchain Mechanism: In this recently developed approach a framework is created that uses multiple private blockchains to collapse direct contact between users and an LBS. Transactions are used to record the user's location on a blockchain, and incentivized mining is used to query an LBS and return results to users [9,14].

While various approaches are proposed but the most important concern remains the same i.e., the restricted choice of using either original or distorted location. A map app, for example, gets permission to access exact location for navigation, which it may misuse to record user's location even when it is not in use. Furthermore, user's personality also affects their privacy preferences. A college going teenager would likely be fine with revealing its exact location to a social networking LBS while it would be matter of great concern to a medical practitioner, and unimaginable for a military person. The same teenager's behavior may change if he spends hours on that app and may not like getting its exact location recorded throughout.

Until now there is no provision of letting a user decide with *whom* they want to share information, *when*, and to *what extent*. This motivates the design of an intelligent model that first analyses the environment and then predicts the amount of privacy needed. It is cognizant of the user's needs and adapts to changes accordingly over time. We recently proposed a primitive model of such a data agent in [7]. It uses supervised learning on static input-output training data recorded to predict privacy behaviour over simple neural network architecture. However, it was cumbersome to analyse all possible states and give a scalable solution suitable for all possible scenarios. Our newly proposed model is an advanced, reconfigurable, self-evolving, and a more practical version that uses RL for interactive behaviour to mimic a proxy agent sensitive to the user's personal choices, app behaviours, and context.

3   Proposed Smart Data Agent RL Formulation

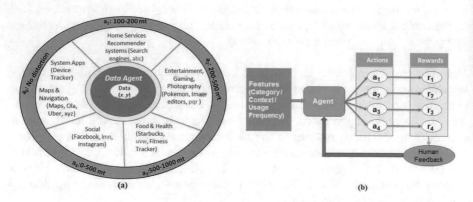

Fig. 1. (a) Smart data agent concept; (b) Depicting state, actions, and rewards in setup.

Smart Data: The smart data concept allows data to *"think for themselves"* by transforming themselves into different *"active forms"* in accordance with the user's preferences, content, context, and nature of the interacting environment with the help of an agent [5,19]. Data are instructed how to expose or transform themselves to the interacting environment. The proposed model transcends the above definition of the smart data concept to facilitate the creation of *smart location data*. As illustrated in Fig. 1(a) the original data, i.e., the coordinates of the user's actual location are at the center. Instead of directly revealing themselves (coordinates (x,y)) to the interacting environment (apps), the data pass through a transformation/cloaking layer. This process is supported by the agent, which senses the environment and interacting apps (e.g., Maps & Navigation, Social, Food & Health), and associates different actions (distortion ranges) in different situations.

Problem Formulation: The goal is to learn policies by testing different actions and automatically learn which one has the most rewarding outcome for a given situation [20]. This sets up the basis for *contextual bandit RL formalism*. In our contextual bandit formalism, we need to capture the user's preferences in our agent's decision making. Thus, there is a need for human input to evaluate the agent's behaviour. Therefore, the natural question to ask is how would a person evaluate an agent's behaviour? A natural way to capture such knowledge is from an object's group behaviour, an object's specific behaviour, an object's query goodness, and the agent's action. Figure 1(b) outlines the basic set up of the RL agent. The agent is a representative of *"user preferences"* and the object is an *"app querying for location"*. This notion of capturing user preference and evaluation guides us to form the elements of the problem, i.e., the *states*, *actions*, and *rewards*.

Table 1. Apps, Tags and Contexts

Apps	Tags	Context Words
Ola,Uber, Maps, Sygic	Maps & Navigation	Book, ride, go, look, find, way, map, guide, locate
SnapChat, Facebook, Instagram	Social	Check-in, live, tag, locate, share
Starbucks, Baskin Robins, Dominos	Food & Drinks	Shop, order, book, deliver, address, locate
Urban Clap	Home Services	Book, order, address, locate
Chrome	Communication	Find, search, nearby
Book My Show	Entertainment	Nearby, locate, maps
PicArt, Reading	Others	Locate, find

1. States: States are the input to the agent and should be mapped in accordance with the user's evaluation system. Three properties describe object's/app's state.

- Object's group behaviour : app tag (category)
- Object's specific behaviour : usage frequency level (low, medium, high)
- Object's query goodness : context (present or absent)

a) App Tags: The agent begins by crawling the web and/or app stores to obtain initial tags for the installed apps. An online tool such as the Google Play scrapper is available [16]. After scrapping the Google Play store, we identified seven major categories as shown in Table 1.

b) Usage Frequency: The usage frequency (UF) corresponds to the number of times location data is accessed within a certain interval of time. It is used to determine the level of interaction and is computed for each app as:

$$UF_{app} = \frac{\text{No. of times location accessed in } TIME\_BANDWIDTH}{TIME\_BANDWIDTH \text{ in seconds}}. \quad (1)$$

It is updated every $TIME\_BANDWIDTH$ (e.g. every minute or every hour) factor. Due to the tabular formation of our problem, we discretize the UF_{app} into three categories (*low*, *medium* and *high*) using some percentage threshold values.

$$UF_{app}^{category} = \begin{cases} low, & \text{if } UF_{app} > Low\_UF \\ medium, & \text{if } Low\_UF \leq UF_{app} < High\_UF; \\ high, & \text{if } UF_{app} \geq Low\_UF. \end{cases} \quad (2)$$

c) Context Words: Context words are obtained by inputting words for each app category that are relevant or may require the user's precise location (Table 1).

2. Actions: A set of valid location distortion ranges is defined as actions:

- $A_0 \Rightarrow$ no distortion
- $A_1 \Rightarrow$ distort by a random number in range (a_{1start}, a_{1end}) e.g. (0–250 mt)
- $A_2 \Rightarrow$ distort by a random number in range(a_{2start}, a_{2end}) e.g. (250–500 mt)
- $A_3 \Rightarrow$ distort by a random number in range(a_{3start}, a_{3end}) e.g. (>500 mt)

The number of actions and the range of distortion can be set by the user or defined by default, thus giving great flexibility.

3. Rewards: A non-stationary reward system is needed for the agent having reward $R \in \{0,1\}$, where 0 is negative feedback and 1 is positive feedback. It is non-stationary because human preferences may change over time.

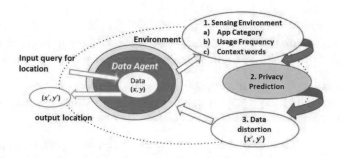

Fig. 2. Main operations performed by agent - sensing, prediction, and distortion.

The environment is specified by a tuple $M = (S; A; R)$, where S stands for state, A for action, and R for reward. The first step of the algorithm requires sensing the environment by analysing the state of the interacting object ("app querying for location"). The state is defined by extracting its app category (seven here), the level of its usage frequency (three here, $UF_{app}^{Category}$=low, medium, high), and query context present/absent (Table 1). Thus, the total number of possible states S is (app_categories $\times UF_{app}^{Category} \times$ context), which is 42 (7×3 × 2) here. Let there be four possible actions, A_0 to A_3. Thus, the total number of possible combinations of state-action pairs can be defined by a matrix with dimensions (no. states × no. actions), i.e. 42 × 4 here. This matrix is known as Q-matrix, and the agent maintains the value of each (state, action) tuple in the Q-matrix. This setup is flexible and can be similarly configured for different numbers of app categories, $UF_{app}^{Category}$, and actions. Figure 2 depicts main operations performed by the agent.

Algorithm 1 describes the basic flow of various steps that the model takes while learning to record data and make decisions. A reward table with dimensions similar to those of Q matrix records the user feedback for each state-action pair. Algorithm 2 initializes various important parameters, including the app categories, context words, UF categories, and the number of actions and their distortion range. One log file is maintained for each app to record its

Algorithm 1. PRIVACY_PREDICTION ()

1: INITIALISATION()
2: **for** each query q **do**
3:     q'= FORMULATE_STANDARD_QUERY(q)
4:     Action = AGENT_DECISION_MAKING (q')
5:     New_location = GET_NEW_LOCATION (Action)
6:     reward = GET_REWARD (q', action)
7:     UPDATE_AGENT (q', action , reward)
8:     UPDATE_LOGS (q) // *update logs*
9: **end for**
10: **return** New_location

behaviour and act as meta data for computing the current value of UF. LOG-FILE has data entry for each app and for the following keys $APP\_CATEGORY$, UF_COUNTS, $PREV\_UF$, UF_CAT. The UF_UPDATE() in Algorithm 3 is a considered to be a concurrent process that continuously update the log file after each $TIME\_BANDWIDTH$ unit. The UF is computed as a weighted average of the apps' current ($CURRENT\_UF$) and previous usage frequency ($PREV\_UF$). A weight factor α is used to ensure that the behaviour transitions are smooth.

Algorithm 2. INITIALISATION()

1: Initialise VALID_APP_CATEGORIES= ['maps&navigation', 'social', 'food&drinks', etc.] // *list of valid app categories as shown in Table 1*
2: Initialise VALID_CONTEXT_WORDS for each category
3: Initialise No_ACTIONS=4
4: Initialise VALID_ACTIONS = A_0 : (start = 0, end = 0), A_1 : (start = 0, end = 250mt), A_2 : (start = 250mt, end = 500mt), A_3 : (start = 500mt, end = 1000mt)
5: Initialise VALID_UF_CATS = ['low', 'mid', 'high'] // *a list of valid UF categories*
6: Initialise matrix Q of dimensions (app_categories, $UF_{app}^{Category}$, context, actions) with zeros
7: Initialise matrix REWARD_TABLE with same shape as matrix Q, each entry = NULL.
8: An empty LOGFILE // a log file maintains the logs necessary for calculation
9: OUR SMART DATA AGENT IS AWARE OF THE TRUE LOCATION.

Algorithm 3. UF_UPDATE( )

1: Initialise
   TIME_BANDWIDTH = 20 // *in seconds*
   $\alpha = 0.9$ // *contribution factor of current frequency in weighted average*
   THRESHOLD_LOW = 0.2 ,THRESHOLD_MID = 0.4 //*thresholds to discretise the UF*
2: **while** True **do**
3:     logs = readLogFile(LOGFILE)
4:     **for** each app's Entry in logs **do**
5:         get PREV_UF // *previous usage frequency if not present than it's 0*
6:         UPDATE PREV_UF with UF
7:         get UF_COUNTS // *frequency counts in previous TIME_BANDWIDTH period*
8:         CURRENT_UF = UF_COUNTS/ TIME_BANDWIDTH
9:         UF = α * CURRENT_UF + (1 - α) * PREV_UF
10:        UF_CAT = 0 if UF < THRESHOLD_LOW
11:        UF_CAT = 1 if THRESHOLD_LOW $\leq$ UF < THRESHOLD_MID
12:        UF_CAT = 2 if UF $\geq$ THRESHOLD_MID
13:     **end for**
14; **end while**

Algorithm 4. FORMULATE_STANDARD_QUERY (q)

Input: original query q
Output: formatted_query q'
    // template for input query q
      q = { appName : "string", appCategory : "string", contextWord :"string" }
    // template for output query q'
      q' = { appName : "string", appCategory : "numeric", contextWord :"binary", UF_CAT : "numeric" }
1: logs = read LOGFILE
2: q' [appName]= q[appName]
3: q' [appCategory] = get index of q [appCategory] in VALID_APP_CATEGORIES *// if app-Category not in VALID_APP_CATEGORIES, index of "other" is used*
4: q' [contextWord] = 1 if q [contextWord] in VALID_CONTEXT_WORDS else 0
5: q' [UF_CAT] = read logs for UF_CAT
6: **return** q'

Algorithm 5. AGENT_DECISION_MAKING (q')

Input: formatted query q'
1: Initialize $\epsilon = 0.1$
2: **return** random action with prob = ϵ
3: **return** greedy action w.r.t to Q values with prob = 1 - ϵ
    with ties broken randomly.

Algorithm 6. GET_REWARD(q', action)

Input: (q', action)
Output: reward
1: Initialize $\phi = 0.01$ *// probability with which to ask for human feedback*
2: **if** REWARD_TABLE does not have reward corresponding to (q', action): **then**
3:     fill REWARD_TABLE with human feedback
4:     **return** lookup REWARD_TABLE value corresponding to (q', action)
5: **else**
6:     ask human for feedback with probability ϕ
7:     **return** lookup in REWARD_TABLE with probability $(1-\phi)$
8: **end if**

Algorithm 7. UPDATE_AGENT(q', action, reward)

Input: q', action, reward
1: STEP_SIZE = 0.9
2: update Q [q', action] = Q[q', action] + STEP_SIZE * (reward - Q[q', action])

Algorithm 8. UPDATE_LOGS()

1: appName, appCategory = get from q
2: **if** no entry of appName in database: **then**
3:     SET default values of PREV_UF = 0 , UF_COUNTS = 0 , UF_CAT = 0, APP_CATEGORY
        = appCategory
4: **end if**
5: logs = read LOGFILE for data corresponding to appName
6: logs [UF_COUNTS] = logs[UF_COUNTS] + 1
7: logs [APP_CATEGORY] = appCategory
8: **return** write back logs

Originating query q is simple, carrying only the app name and the context words to make a LBS request. Algorithm 4 determines the goodness of a query (calling context word present or absent in user defined set) and its category of interaction using UF_CAT (low, medium, high). It outputs a formatted query q' as input to decision making Algorithm 5. Algorithm 5 follows the epsilon greedy approach to balance exploration with exploitation. It instructs the agent to take the random action with a probability ϵ (exploration) and the best action (one with highest reward value) suggested by the REWARD_TABLE and Q-matrix with probability (1-ϵ). It is followed by Algorithm 6, which collects user feedback for giving rewards or penalties. Algorithms 5 and 6 together enable continuous user feedback recording and behavioural adaptation.

Algorithm 7 is the agent update mechanism, which is based on the principle that $NewEstimate = OldEstimate + step\_size * (target - OldEstimate)$. Increasing or decreasing the $step\_size \in [0, 1]$ may affect the convergence. Since our problem has a non-stationary distribution, i.e., the distortion preferences and user feedback change over time, a constant step size is used. The logs need to be updated after each action as in Algorithm 8.

4   Simulation Results

Agent location distortion behaviour was simulated for two users (A and B) with different personalities and professional backgrounds.

1. **User A**   is a male university graduate who *is not concerned much* about location sharing. He is fine with sharing his exact location with any app as long as UF is low. He chooses the number of actions to be 4 with distortion ranges [A_0: 0–0 mt, A_1: 0–250 mt, A_2: 250–500 mt, A_3: 500–1000 mt]. The agent initially begins exploration and records a feedback whenever it encounters a state that does not have any value in Q matrix. Its interactions for two app categories is presented below.
   A. Interaction with app belonging to category "Maps & Navigation"
   Assume that the user identifies the following words belonging to this category as *contextwords* = [book, ride, go, look, find, way, map, guide, locate]. The user wants the original location to be shared in the presence of context and

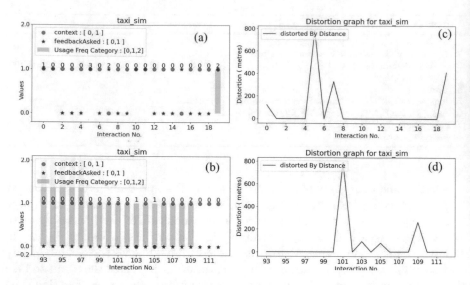

Fig. 3. User: A, app: 'taxi' (a)-(b) mapping of states and actions (c)-(d) corresponding distortion.

irrespective of UF. To simulate an app's behaviour we make the following settings:

Probability of non-contextual location access $= 0.1$

Probability of a contextual location access $= (\ 1 - 0.1\ )$

Then the state-action mapping as explored by the agent is given below

- **State** $=$ 'appCategory' : Maps & Navigation, 'contextWord' : present, 'UF' : any of [low mid high] $\Rightarrow$ preferred **Action** $= A_0$
- **State** $=$ 'appCategory' : Maps & Navigation, 'contextWord' : absent, 'UF' : [low] $\Rightarrow$ preferred **Action** $= A_0$
- **State** $=$ 'appCategory' : Maps & Navigation, 'contextWord' : absent, 'UF' : any of [mid high] $\Rightarrow$ preferred **Action** $= A_2/A_3$(medium to high distortion)

The agent's interaction with a 'taxi' app is visualized in Fig. 3 (a) and (b), where the x-axis depicts the ith interaction. The state at each interaction is depicted by a green dot and a pink bar, for which the value is shown on the y-axis. The green dot at value 1 indicates the presence of context and at value 0 indicates it absence. The UF category is depicted by the pink bar for which the value is set to 0 for 'low', 1 for 'medium', and 2 for 'high'. The user feedback is recorded using a red star: value 1 indicates when the agent receives feedback (exploration) and value 0 indicates when it does not (exploitation). The action (A_0, A_1, A_2, or A_3) taken at a particular state is numbered above the pink bar. The distortion corresponding to each ith interaction is plotted in Fig. 3 (c) and (d).

At interaction i_0, the user has first state, context present, and UF low, so the agent explores by taking a random action A_1 and records the feedback.

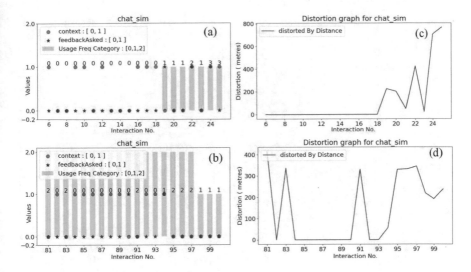

Fig. 4. User: A, app: 'chat' (a)-(b)mapping of states and actions (c)-(d) corresponding distortion

The feedback tells the agent that the action was wrong. Thus again when same state appears it takes another action A_0 and on recording feedback it finds to be appropriate. It exploits this choice in following iterations which have same state. However, with probability ϕ it attempts random actions to continue exploration at interactions i_5, i_{10}, and i_{15} as defined in algorithm 6. By interaction i_{100}, the agent has seen mid and high UF states and has learned the user choices for no distortion in the presence of context, otherwise distorted using A_2/A_3 in range 250–500/500–1000 mt. Since context is absent these distortions will give distorted locations if the app is frequently collecting location in the background.

B. Interaction with app belonging to category "Social"

Let the app name be 'chat', the *contextwords* be [check-in, live, tag, locate, share], and state-action mapping as explored for following settings:

Probability of non-contextual location access = 0.6

Probability of a contextual location access = (1–0.6)

- **State** = 'appCategory' : Social, 'contextWord' : present, 'UF' : any of [low mid high] $\Rightarrow$ preferred **Action** = A_0
- **State** = 'appCategory' : Social, 'contextWord' : absent, 'UF' : low $\Rightarrow$ preferred **Action** = A_0
- **State** = 'appCategory' : Social, 'contextWord' : absent, 'UF' : mid $\Rightarrow$ preferred **Action** = A_1 (slight distortion)
- **State** = 'appCategory' : Social, 'contextWord' : absent, 'UF' : high $\Rightarrow$ preferred **Action** = A_2 (high distortion)

Figure 4 (a) and (c) during initial interactions the agent learns not to distort (A_0) when UF is low whether context present or absent. Around i_{18} it learns if UF is mid and context absent distort by A_1 (0–250). Later around i_{80},

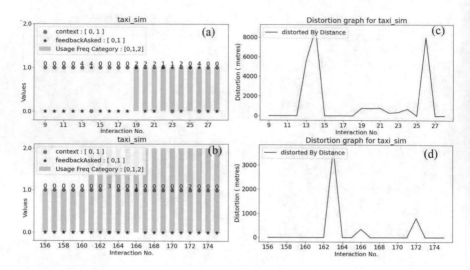

Fig. 5. User: B, app: 'taxi' (a)-(b) mapping of states and actions (c)-(d) corresponding distortion.

Fig. 4 (b) and (d) when UF is high then A_2 (250–500 mt) is learned while continuous exploration in process.

2. **User B** : She is a professional working in a government agency and highly concerned about revealing her location. She chooses the number of actions to be 5 with distortion ranges [A_0: 0–0 mt, A_1: 0–300 mt, A_2: 300–600 mt, A_3: 600–1000 mt, A_4: 1000–5000 mt, A_5: $\geq$ 5000mt]. She needs to use apps in the maps and navigation category and share her original location only in the presence of context otherwise, only the distorted location is shared even if the UF is low. She uses her social networking app to communicate with friends and family but without revealing her exact location even in the presence of context.

A. Interaction with app belonging to category "Maps & Navigation"

We assume that the context words and simulation probabilities are the same as for the category above. The state-action mapping as explored by the agent is

- **State** = 'appCategory' : Maps & Navigation, 'contextWord' : present, 'UF' : any of [low mid high] $\Rightarrow$ preferred **Action** = A_0
- **State** = 'appCategory' : Maps & Navigation, 'contextWord' : absent, 'UF' : [low] $\Rightarrow$ preferred **Action** = A_1 (slight distortion)
- **State** = 'appCategory' : Maps & Navigation, 'contextWord' : absent, 'UF' : any of [mid high] $\Rightarrow$ preferred **Action** = A_2/A_3 (higher distortion)

As shown in Fig. 5 (a) and (c), during the initial interactions, the UF is low and the agent takes random actions. From user feedback, it learns not to distort (A_0) when context is present. Even when context is absent, it continues exploring to learn the correct actions. At around iteration i_{160}, Fig. 5 (b) and

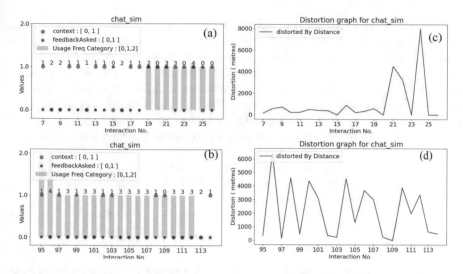

Fig. 6. User: B, app: 'chat' (a)-(b)mapping of states and actions (c)-(d) corresponding distortion.

(d), it learns that if UF is mid/high and context is absent, distort by 300–600 or 600–1000 mt (A_2/A_3).

B. Interaction with app belonging to category "Social"

The state-action mapping as explored by the agent is given below

- **State** = 'appCategory' : Social, 'contextWord' : present, 'UF' : any of [low mid high] ⇒ preferred **Action** = A_1 (slight distortion)
- **State** = 'appCategory' : Social, 'contextWord' : absent, 'UF' : any of [low mid high] ⇒ preferred **Action** = $A_2/A_3/A_4$ (high distortion)

Figure 6(a) and (c) shows initial interactions i_7 to i_{17} when UF is low and the actions learned are A_1 for context present and A_2 when context is absent. At around interaction i_{20}, UF is transitioning to mid-range and actions continue to be explored. At around interaction $i_{95} - i_{113}$, the agent has learned to take action A_1 in the presence of context for any UF and actions $A_2/A_3/A_4$ otherwise[1].

5   Conclusions

Agent location distortion behaviour was simulated for two users (A and B) with different personalities for two types of applications. The simulation demonstrated agent's adaption to different numbers of distortion actions, distortion ranges, and context words. After a few interactions, the agent was able to learn the desired policies through an interaction and reward mechanism. The epsilon greedy approach and continuous feedback estimation lets users change their policy settings

---

[1] A complete video demonstration of the above behaviours can be found at https://github.com/rohitdavas/Smart-data-2-An-RL-agent.

over time. Although simulations were run for only two users, the agent behaviours was found similar to those for other user settings as well. Overall, the proposed model behaves as a desired cyber proxy and is a one-stop solution for managing various apps on a user's smart phone. It enables the user to decide with whom to share the user's location and in which manner while enjoying the value-based services provided location-based services. The proposed model is practical application of privacy by design for handling location privacy and can be deployed on smartphone to make users aware and control their location revelation.

Acknowledgement. This work was partially supported by JSPS KAKENHI Grants JP16H06302 and JP18H04120, and by JST CREST Grants JPMJCR18A6 and JPMJCR20D3, Japan.

References

1. Bettini, C., Jajodia, S., Samarati, P., Wang, S.X.: Privacy in location-based applications: research issues and emerging trends, vol. 5599. Springer Science & Business Media (2009)
2. Gedik, B., Liu, L.: Protecting location privacy with personalized k-anonymity: architecture and algorithms. IEEE Trans. Mob. Comput. **7**(1), 1–18 (2007)
3. Gruteser, M., Grunwald, D.: Anonymous usage of location-based services through spatial and temporal cloaking. In: First International Conference on Mobile Systems, Applications and Services, pp. 31–42 (2003)
4. Guo, X., Wang, W., Huang, H., Li, Q., Malekian, R.: Location privacy-preserving method based on historical proximity location. Wirel. Commun. Mob. Comput. **2020** (2020)
5. Harvey, I., Cavoukian, A., Tomko, G., Borrett, D., Kwan, H., Hatzinakos, D.: Smartdata. Springer (2013)
6. Hashem, T., Kulik, L.: Safeguarding location privacy in wireless ad-hoc networks. In: International Conference on Ubiquitous Computing, pp. 372–390. Springer (2007)
7. Kaur, H., Echizen, I., Kumar, R.: Smart data agent for preserving location privacy. In: 2020 IEEE Symposium Series on Computational Intelligence (SSCI), pp. 2567–2575. IEEE (2020)
8. Le, T., Echizen, I.: Lightweight collaborative semantic scheme for generating an obfuscated region to ensure location privacy. In: IEEE International Conference on Systems, Man, and Cybernetics (SMC), pp. 2844–2849. IEEE (2018)
9. Li, B., Liang, R., Zhu, D., Chen, W., Lin, Q.: Blockchain-based trust management model for location privacy preserving in VANET. IEEE Trans. Intell. Transp. Syst. (2020)
10. Machanavajjhala, A., Kifer, D., Gehrke, J., Venkitasubramaniam, M.: l-diversity: privacy beyond k-anonymity. ACM Trans. Knowl. Discov. Data (TKDD) **1**(1), 3–es (2007)
11. Masoumzadeh, A., Joshi, J.: An alternative approach to k-anonymity for location-based services. Procedia Comput. Sci. **5**, 522–530 (2011)
12. Mokbel, M.F.: Privacy in location-based services: State-of-the-art and research directions. In: International Conference on Mobile Data Management, pp. 228–228. IEEE (2007)

13. Palm, A., Metzger, A., Pohl, K.: Online reinforcement learning for self-adaptive information systems. In: International Conference on Advanced Information Systems Engineering, pp. 169–184. Springer (2020)
14. Qiu, Y., Liu, Y., Li, X., Chen, J.: A novel location privacy-preserving approach based on blockchain. Sensors **20**(12), 3519 (2020)
15. Santos, F., Humbert, M., Shokri, R., Hubaux, J.P.: Collaborative location privacy with rational users. In: International Conference on Decision and Game Theory for Security, pp. 163–181. Springer (2011)
16. Scrapper. https://pypi.org/project/google-play-scraper/. Accessed 14 May 2020
17. Shokri, R., Papadimitratos, P., Theodorakopoulos, G., Hubaux, J.P.: Collaborative location privacy. In: 2011 IEEE Eighth International Conference on Mobile Ad-Hoc and Sensor Systems, pp. 500–509. IEEE (2011)
18. Takabi, H., Joshi, J.B., Karimi, H.A.: A collaborative k-anonymity approach for location privacy in location-based services. In: 2009 5th International Conference on Collaborative Computing: Networking, Applications and Worksharing, pp. 1–9. IEEE (2009)
19. Tomko, G.J., Borrett, D.S., Kwan, H.C., Steffan, G.: Smartdata: make the data think for itself. Identity Inf. Soc. **3**(2), 343–362 (2010)
20. Wiering, M., Van Otterlo, M.: Reinforcement learning, vol. 12. Springer (2012)
21. Zhang, L., Song, G., Zhu, D., Ren, W., Xiong, P.: Location privacy preservation through kernel transformation. In: Concurrency and Computation: Practice and Experience, p. e6014 (2020)
22. Zhong, G., Hengartner, U.: A distributed k-anonymity protocol for location privacy. In: IEEE International Conference on Pervasive Computing and Communications, pp. 1–10. IEEE (2009)

Comparison of Personal Security Protocols

Radosław Bułat and Marek R. Ogiela[(✉)]

Cryptography and Cognitive Informatics Laboratory, AGH University of Science
and Technology, 30 Mickiewicza Ave, 30-059 Kraków, Poland
{bulat,mogiela}@agh.edu.pl

Abstract. One of the most critical issues in modern society is the possibility of
efficient users' authentication in all the systems and items creating the IoT. There
is a need to determine persons' rights and privileges to use devices they operate in
their daily routines. This work aims to explore the most promising research avenues
that could prove the user identity based on their behavior, personal characteristics, or history, preferably on-the-fly. There is already a wide range of techniques
used, from simple biometrics, through the patterns describing the keystroke and
cadence of a given individual during any performed operation. All those methods
are considered for their usefulness in more contactless or decentralized systems
in a systematic review. That will form a factual research basis for a further avenue
of personalized cryptographic protocols. These would not determine what a person has or knows, but more likely – who such a person is and what makes them
unique. The issue of false positives and negatives should also balance the tight
line between strict metrics of peoples' behavior and privacy issues.

1 Introduction

As society became more connected in the Internet era and more and more processes of
our daily lives are now automated, streamed, or dependent on interconnected machines'
assistance, user identification is becoming more prevalent. As people are being surrounded by helper machines, often with minimal computational power, there is a risk of
granting privileges or services to individuals who are not entitled to them [1]. Furthermore, given that it is the most often the personal data at stake, the consequences of its
leak may be dire and most often irrevocable [2].

As of now, there are three main avenues of personal authentication of human beings
[3]. The first one is knowledge-based – asking about "something you know." Those
methods most often ask for the password, personal data, PINs, or any other sequence
of symbols known to the user. Though it does not mitigate the risk of a false positive
– there is no guarantee that the data has not been inadvertently, or by force, made public,
and the individual knowing the secret may not be entitled to it.

The second method, as opposed to knowledge, is more physical based – it works on
the principle of "something you have," which might be a tangible physical object (e.g.,
token or keycard) or a digital good (e.g., digital certificate). Still, even if there is only
a single instance of such an object, it does not protect from identity theft if lost to the
attacker.

L. Barolli et al. (Eds.): AINA 2021, LNNS 226, pp. 672–678, 2021.
https://doi.org/10.1007/978-3-030-75075-6_55

The third method – personal protocol, as opposed to those above, is not something separate from the individual who is being authenticated. It is "something you are" – which covers biometrics (physical features of a person, e.g., fingerprint or retina scan) but also other unique features (typing cadence, gait, subconscious movements or twitches, etc.). It can be freely combined with any other method (for example, typing the correct password with the correct rhythm inherent to the rights owner) and could be hard to replicate by an assailant.

The projected risks are in the finetuning the system – too strict criteria may lead to high FRR (False Rejection Rate) as even the rightful owner can sometimes vary in his habits. On the contrary, too relaxed ones (for user accessibility) may make the system vulnerable by driving up the FAR (False Acceptance Rate), defeating the system's whole purpose. Also, there is a need to consider the implications of storing the (often sensitive) medical or personal data (used for biometrics) in a system database. Lastly, if we consider the practical applications – many of the devices that make the Internet of Things (IoT) can not handle the processing power needed for complex computations or advanced image recognition needed for some of those methods [4].

In the following sections, we will describe the most popular personal characteristics, which could be considered in the creation of advanced security protocols.

2 Personal Security Protocols

In this section, selected methods used for personal authentication will be presented and compared according to their importance and potential as future research avenues [5].

2.1 Biometrics

From a security point of view, the most important biometric patterns which can be applied in the creation of security protocols are the following:

Fingerprints. Fingerprints are one of the oldest and most prevalent in consumer electronics forms of biometrics, based on unique characteristics of a person's finger ridges. It can be used to discriminate between individuals, even twins, with a high value of certainty [6]. Though it is a contact-based protocol, and in the case of some skin-altering diseases (e.g., psoriasis) or hard manual labor, it is prone to wear or being inconsistent.

Retinal Patterns. Such biometrics involve the recognition of the person's unique blood vessel pattern in the retina. Considered more secure than fingerprinting, it remains an invasive and much more costly procedure, limiting its usefulness in everyday applications [7].

Iris Recognition. A scan involving pattern recognition of a person's iris. Unlike the above, it is a contact-less procedure and can guarantee a similar degree of uniqueness. Like all image recognition methods, it remains vulnerable to low light conditions or distances. Sufficiently high-quality images or contact lenses can be used to spoof the detectors with a relatively high rate of success in case of the attacker obtaining the impersonated victim template [8].

2.2 Neuroscience Patterns

Besides standard biometric patterns, for security purposes, especially in creating user-oriented security protocols, we can also consider psychological and neural patterns. It seems that the most important are the following:

Subconsciously Imprinted Movements. There has been proposed a novel approach based on the subconscious mind. The main goal of such a method is the resistance to coercion – a password sequence is encoded into an individual brain (as a series of gestures/hand movements using a simple game) without the user's knowledge of the exact sequence, and thus, the inability to divulge the secret. Once the correct movements are ingrained into an individual's muscle memory, there is a noticeable change in reaction times during the gestures that form the sequence parts versus those that do not – which forms the basis of identity key generation. Such method, by definition, stays resistant against social engineering that might be employed against the user [9] – but may require constant usage, as to not lose the learned ability. In some ways, the method is similar to the cadence-tracking methods of authentication. In this case, the measure checks not an inherent metric of an individual but rather the presence of implicitly learned skill (which requires time and effort to master) [10].

Emotions. Staying on the topic of coercion, or physical abuse, used to obtain a key for an encrypted material, another approach has been proposed. In this approach, the correctness of the entered credential is measured as well as the mental, or emotional, state of an individual entering it. As has been proven, there is a distinct possibility of measuring the person's skin conductance (using electrodes simultaneously with fingerprint scan). Such a method would take a key from an individual (in the cited work, the key has been a voiceprint, but it is possible to use other keys) along with a fingerprint and a conductance scan. If a user's emotional state were deemed abnormal and under duress, even the correct user key would not work, and the whole authentication would fail. Still, this method raises some objections – the conductance scan needs to use some form of authentication too (in this case, a fingerprint) to eliminate a spoofing case, when the conductance measured would come from the attacker, and not the user (who would provide the password under duress). It is costly and impractical in common usage as it involves at least three authentication methods. There is also an ethical dilemma – if the user would fail to provide the passphrase being already under physical threat and would be left without any means to provide it (which sometimes would be the safest and advised course of action), it could lead to unnecessary danger or loss of life from the hands of the attacker, judging the user as being unhelpful on purpose [11].

2.3 Personal Characteristics

Also, other personal characteristics connected with some habits or preferences can be considered in the creation of personalized security protocols. The most typical are:

Typing Cadence. As of now, the methods above have been based on separate keys or values, sometimes used in unison (in case of emotions paired with voice key) in single or multiple factored authorization. Apart from the passphrase/image that should be produced, it would be prudent to involve something more to obfuscate the key. Such an attempt could involve not only the presentation of the correct key, but also another variable on "how" it was done, and the consistency with that individual (on the contrary to the emotion model which checks for a general emotional state, without regard to prior user mental state history). One example of such authentication protocol could be measuring a person's typing cadence during a mandatory password check. Authenticating user would store the template of his writing habits (by the use of various samples) – including the keypress strength, the regular cadence (or lack thereof) of writing, or some other similar characteristics like a tendency to press two keys at the same time (typing with both hands) or to finger-type. Such a combination would prove difficult to the brute-force attacks (having to match not only the correct passphrase but also the rhythm as a second factor). Also, in the case of identity theft – it might prove very difficult or impossible (depending on the number of features tracked in the pattern) for the attacker to mimic the victim's unique way of writing. This method is also easy to implement. It does not need any specialized tools apart from a keyboard (sometimes with keystroke strength checking). It is then suitable to be used over the internet to authenticate people at their workstations or personal machines. The success rate depends on the calibration of the recorded pattern and its details – too much can lead to high FRR and the risk of inability to access protected data. It has also been proven that the method accuracy varies according to the input text or password – the template recognition works best on a free text, which would need some revisions for a fixed text. Also, it is not a contactless protocol and requires the user's full interaction [12].

Eye Movements. Another distinguishable characteristics of an individual (without the use of touch-based devices) are eye movements. In this case, the authorization process is based on drawing a correct shape (similar to existing keycode-based touch devices) using the user's focused attention on the shown pattern elements. It retains the individual characteristics of motion, such as speed, squint, and order. It thus could differentiate the personae even presented with the same motion sequence (similar to the case above). Another security layer could be introduced when the proper sequence would not be static but rather tailored to the individual in question. For example, a grid of pictures involved with the user knowledge (or the resource being accessed) could be shown and needed to be recognized, categorized, and sequenced by the user. It would also involve intricate pattern and context recognition unavailable for simple impostor machines without cognitive ability. The method can be classified as secure, difficult to mimic, and relatively simple in its implementation. The drawbacks could include its short range of capture and dependence on the outside light (similar to iris scan). It would also need unique calibration to accept people with disabilities (to incorporate it in the pattern) [13].

Sitting Postures and ECG. Apart from the computer devices the user interacts with via keyboard or screen, some of the more prevalent smart devices are cars, requiring another authentication method. Thus, the devised one is based on a sitting posture (measured by pressure images of sensors in the driver's seat) and ECG data (collected by contactless electrodes in the same seat). The preliminary research shows promise, as there can be many pressure maps of the user (during casual driving, accelerating, braking, turning, changing gears, et caetera.), making continuous authentication possible. Also, an ECG reaction for a given situation or route may differ between individuals and be a factor in this method (similar to the neuroscientific ones). However, the ECG emotional reaction is not an individual characteristic that can be solely used for authentication, so the main factor still has to be the driver's sitting posture and lateral movement. Such a system can generally be viewed as an acceptable way to identify and continuously monitor an individual (as continuity is one of this approach's strengths). Still, it requires a particular setup and would not be useful in a general setting other than a transport device [14].

3   Cognitive Cryptography

As can be seen, most of the above methods are based on static "secret" known (as password) or had (as part of biometrics) by the individual as the means of authentication. However, if compromised or leaked, those methods can prove ineffective. According to Shannon's information theory, the best keys are random (or one-time pads) without superfluous information – which excludes a password or characteristics that is not changing with every use.

The solution might be cognitive cryptography – a model that uses the more secure approach of random key generation but infuses it with some data taken from the individual and held on file/database. The selected data pattern should be held secret. It would be obtained by semantically analyzing data (images, sound files, multimedia), selecting some unique characteristics from them, and inserting them as a part of a one-time key. Also, the nature and interpretation of acquired data would be analyzed by semantic mechanism as well – and according to the inferred context, different encryption algorithms and techniques could be used.

Such keys would not only be a standard cryptographic mechanism, but their pseudo-random nature would take an individual fingerprint or watermark on them. That could, in some cases, serve as a digital signature as well, since if a key could be undoubtedly assigned to an individual, that would make a certainty of non-repudiation on the encrypted material.

The cognitive approach is supposed to use significantly less computing power than is needed for traditional secure techniques, relying on the inferred contextual information instead, which would benefit the smaller systems. The drawback might be a need for a database feed with an individual's sensitive data, raising privacy concerns and requiring further study [15, 16].

4   Comparison of Personal Features Used in Security Protocols

As has been presented above, we can consider different personal features and parameters in the creation of new security protocols. All of them can be applied either as single or hybrid solutions. The development of security systems, which are based on such personal features, may depend on different factors (like sensors availability, complexity of analysis, uniqueness, available computational power, etc.) Also, there may be privacy laws, which would limit the usefulness of methods invading ones' privacy or accessing one biomedical image (sometimes a method can work even without those, albeit at a reduced personalization level). In Table 1 are compared selected features of previously mentioned personal characteristics.

Table 1. Comparison of features for selected personal patterns and their usefulness.

	Ease of use, sensors accessibility	Availability for minimal computational power	Uniqueness	No private data used	Contactless
Fingerprints	+	+	+	−	−
Retina	−	−	+	−	−
Iris	+	−	+	−	+
Imprinted patterns	−	+	−	+	−
Emotions	−	−	−	+	−
Typing cadence	+	+	−	+	−
Eye movement	+	+	−	±	+
Sitting posture	−	−	−	+	−
Cognitive cryptography	+	+	+	±	+

5 Conclusions

As can be seen, there are many methods to describe an individual, some of them being innate to the user in question. Thus it raises some interesting research problems – which avenue would be worth pursuing in the future of distributed small systems? What would make a man unique in the context of authentication? The need for specialized equipment or usability only in particular situations (e.g., driving or armed robbery) has to be taken into account and, as should in a post-COVID society, the protocol's contactlessness. As has been proven already, the concept of having some kind of key (which can be lost, stolen, or otherwise used in a malicious way by other individuals) can be inferior to "being" the key. As for the need for the key mutability – the cognitive techniques seem most promising, offering the personalization seen in the biometrics methods and adding another security layer on top of that. However, in some more distributed, or those that do not share a central user database, the cadence or eye movements tracking methods can work and take advantage of not needing any specialized equipment from the client-side other than a keyboard and a simple camera.

Acknowledgments. This work has been supported by the AGH University of Science and Technology Research Grant No 16.16.120.773. This work has been supported by the AGH Doctoral School Grant No 10.16.120.7999.

References

1. Sfar, A.R., Natalizio, E., Challal, Y., Chtourou, Z.: A roadmap for security challenges in the Internet of Things. Digit. Commun. Netw. **4**(2), 118–137 (2018)
2. Loukil, F., Ghedira, C., Aïcha-Nabila, B., Boukadi, K., Maamar, Z.: Privacy-aware in the IoT applications: a systematic literature review. In: International Conference on Cooperative Information Systems (CoopIS) 2017. Proceedings, Part I. LNCS, vol. 10573, pp. 552–569 (2017)
3. Menezes A., Van Oorschot P., Vanstone S.: Handbook of Applied Cryptography, CRC, New York (1997)
4. Liang, Y., Samtani, S., Guo, B., Yu, Z.: Behavioral biometrics for continuous authentication in the Internet-of-Things era: an artificial intelligence perspective. IEEE Internet Things J. **7**(9), 9128–9143 (2020)
5. Traore, I., Ahmed, E.A.: Continuous Authentication Using Biometrics: Data, Models, and Metrics, 1st edn. IGI Global, Harrisburg (2011)
6. Srihari, S., Srinivasan, H., Fang, G.: Discriminability of fingerprints of twins. J. Forensic Ident. **58**, 2008–109 (2008)
7. Mazumdar, J.: Retina based biometric authentication system: a review. Int. J. Adv. Res. Comput. Sci. **9**, 711–718 (2018). https://doi.org/10.26483/ijarcs.v9i1.5322
8. Galbally, J., Gomez-Barrero, M.: A review of iris anti-spoofing, pp. 1–6 (2016). https://doi.org/10.1109/iwbf.2016.7449676
9. Hadnagy, C.: Social Engineering: The Science of Human Hacking (2018)
10. Bojinov, H., Sanchez, D., Reber, P., Boneh, D., Lincoln P.: Neuroscience meets cryptography: designing crypto primitives secure against rubber hose attacks. In: Security Symposium, USENIX, pp. 129–141 (2012)
11. Gupta, P., Gao, D.: Fighting coercion attacks in key generation using skin conductance. In: USENIX Security Symposium (2010)
12. Panasiuk, P., Saeed, K.: A modified algorithm for user identification by his typing on the keyboard. In: Choraś, R.S. (ed.) Image Processing and Communications Challenges 2. Advances in Intelligent and Soft Computing, vol 84. Springer, Berlin, Heidelberg (2010)
13. Ogiela, M.R., Ogiela, L.: Cognitive keys in personalized cryptography. In: The 31st IEEE International Conference on Advanced Information Networking and Applications (AINA 2017), Taipei, Taiwan, 27–29 March 2017, pp. 1050–1054 (2017). https://doi.org/10.1109/aina.2017.164
14. Riener, A.: Sitting postures and electrocardiograms: a method for continuous and non-disruptive driver authentication. In: Continuous Authentication Using Biometrics: Data, Models, and Metrics, pp. 137–168 (2011)
15. Ogiela, M.R., Ogiela, L.: Cognitive cryptography techniques for intelligent information management. Int. J. Inf. Manage. **40**, 21–27 (2018)
16. Ogiela, L.: Cryptographic techniques of strategic data splitting and secure information management. Perv. Mob. Comput. **29**, 130–141 (2016)

Anonymous Broadcast Authentication for Securely Remote-Controlling IoT Devices

Yohei Watanabe[1,2(✉)], Naoto Yanai[2,3], and Junji Shikata[4,5]

[1] Graduate School of Informatics and Engineering, The University of Electro-Communications, Chofu, Japan
watanabe@uec.ac.jp
[2] Japan Datacom Co. Ltd., Tokyo, Japan
[3] Graduate School of Information Science and Technology, Osaka University, Suita, Japan
yanai@ist.osaka-u.ac.jp
[4] Graduate School of Information and Environments, Yokohama National University, Yokohama, Japan
[5] The Institute of Advanced Science, Yokohama National University, Yokohama, Japan
shikata-junji-rb@ynu.ac.jp

Abstract. In this paper, we present a basic system for controlling IoT devices in remote environments with the following requirements: (1) in a situation where an operation center broadcasts information to IoT devices, e.g., wireless environment, only the designated devices can identify operations sent from the center; (2) the devices can detect manipulation of the broadcast information and hence prevents maliciously generated operations from being executed. We formalize a model of the basic system and its essential requirements and propose *anonymous broadcast authentication (ABA)* as its core cryptographic primitive. We formally define the syntax and security notions for ABA and show provably-secure ABA constructions.

1 Introduction

Internet-of-Things (IoT) such as smart speakers and sensors have been widely deployed in our lives. There are several devices called edge AI whereby the devices managed by a user contain machine learning models, and hence more various devices will be introduced in the near future [27].

Meanwhile, one of the most important issues for IoT is cybersecurity. For example, the outbreak of an IoT malware named Mirai is significantly rapid, and some botnet that infects nearly 65,000 IoT devices in its first 20 h is also found [3]. However, implementing and deploying security functions for each specific security threat is not necessarily desirable for IoT devices, which are resource-constrained ones. In particular, an adversary may find new types of attacks to circumvent existing solution [1]. Therefore, we need to discuss a solution that considers various threats to realize IoT devices' security. Proposing such a versatile solution is challenging [6].

Our Contributions. Based on the background described above, in this paper, we focus on how we arbitrarily control devices infected with malware instead of protecting devices against malware. We aim to develop a framework to control IoT devices

© The Author(s), under exclusive license to Springer Nature Switzerland AG 2021
L. Barolli et al. (Eds.): AINA 2021, LNNS 226, pp. 679–690, 2021.
https://doi.org/10.1007/978-3-030-75075-6_56

remotely towards the deployment in existing IoT systems. Specifically, we consider a basic system where a central entity (e.g., an operation center) tries to simultaneously control many target entities (e.g., IoT devices). Note that we do not consider individual communication between the central entity and each target device due to numerous devices. Namely, the central entity broadcasts command, say, an abort instruction, to all IoT devices so that only designated devices can execute the command. Consequently, even when devices with vulnerability are attacked, their resultant damage can be reduced by controlling them. We believe that this simplified model of the remote control system captures characteristics of the IoT era as well as massive Machine Type Communications (mMTC) in the context of (Beyond) 5G. We describe the simple model for the above remote control system more concretely and discuss its essential requirements in Sect. 2.

We then propose anonymous broadcast authentication (ABA) as a core cryptographic primitive that fulfills the requirements for our remote control system. ABA allows the central entity to choose an arbitrary subset of target devices and create authenticated commands so that only the designated targets' verification algorithms accept them. We formalize the syntax and security notions for ABA in Sect. 3 and show provably-secure ABA constructions in Sect. 4. Our ABA schemes can be constructed from only message authentication codes (MACs) and pseudorandom functions (PRFs), thus providing efficient running time.

Related Work. To the best of our knowledge, there is no broadcast authentication scheme which enables a user to specify any chosen subset from a set of devices although there are several schemes targeting one-to-many communication [8,22,23] or many-to-many communication [25,26]. Besides, broadcast encryption [7,20] and anonymous broadcast encryption [11,16,18,19] have been proposed as a cryptographic scheme to specify any subset from multiple users. However, these are in the context of encryption, and no scheme is presented in the authentication.

To realize the IoT security, researchers have devoted effort from the firmware level [9,10] to the application [12,24]. Whereas the conventional approach is based on the data flow [13], the use of cryptography is discussed in the past years [2,17,21]. ABA is a framework to realize IoT security by leveraging cryptography.

Notations. For all natural number $n \in \mathbb{N}$, $\{1,\ldots,n\}$ is denoted by $[n]$. For a finite set $\mathcal{X}$, we denote by $|\mathcal{X}|$ the cardinality of $\mathcal{X}$. For finite sets $\mathcal{X}, \mathcal{Y}$, let $\mathcal{X} \triangle \mathcal{Y}$ be the symmetric difference $\mathcal{X} \triangle \mathcal{Y} := (\mathcal{X} \setminus \mathcal{Y}) \cup (\mathcal{Y} \setminus \mathcal{X})$. For any finite set $\mathcal{X}$ and any natural number $\ell \in \mathbb{N}$, let $2^{\mathcal{X}}_{\leq \ell} := \{\mathcal{Y} \subset \mathcal{X} \mid |\mathcal{Y}| \leq \ell\}$ be a family of subsets of $\mathcal{X}$ such that its cardinality is at most ℓ (i.e., a part of a power set of $\mathcal{X}$). Concatenation is denoted by $\|$. For any algorithm A, out $\leftarrow$ A(in) means that A takes in as input and outputs out. If we write A(in; r), r indicates an internal randomness that is chosen uniformly at random. Throughout the paper, we denote by κ a security parameter and consider probabilistic polynomial-time algorithms (PPTAs). We say a function $\mathsf{negl}(\cdot)$ is negligible if for any polynomial $\mathsf{poly}(\cdot)$, there exists some constant $\kappa_0 \in \mathbb{N}$ such that $\mathsf{negl}(\kappa) < 1/\mathsf{poly}(\kappa)$ for all $\kappa \geq \kappa_0$. We list other notations used in this paper in Appendix.

2   System Model

This section describes a basic model for the remote control system we consider throughout this paper. The system aim to have only designated devices securely execute commands requested by the operation center. Specifically, we consider the following basic system.

(1) The operation center decides a command and the corresponding devices that execute it.
(2) The operation center sends the command to *all* devices via a broadcast channel.
(3) Each device executes the command if it is designated, or does not otherwise.

As described above, we assume broadcast channels and that each device receives commands without errors. We would like to emphasize that each command is broadcasted and is common for all devices. Moreover, we do not consider upstream communication (i.e., from devices to the center). The operation center can designate devices with some identifiers associated with them. We assume that those identifiers are public information. We give essential requirements, which should be resolved by mechanism design, for the system below.

Requirement 1: To Accurately Control Which Devices Execute Commands.   We require the following correctness of the system.

- The devices designated by the center surely execute a command unless the command is (maliciously) modified.
- The non-designated devices never execute a command even if the command is (maliciously) modified.

Requirement 2: To Have Resilience Against Falsification of Commands.   Roughly speaking, each device should confirm the validity of transmitted commands, i.e., whether the commands are (maliciously) modified. We consider a situation where an adversary can obtain information about a certain number of IoT devices by infecting them with malware, (physically) stealing them, etc. We require that even if the adversary (maliciously) modifies commands transmitted over broadcast channels, one cannot have non-designated devices execute the (modified) commands. Note that we do not consider attacks whose aim to make commands undelivered such as jamming, and the falsification of verification mechanisms in IoT devices through their firmware updates.

Requirement 3: To Hide Which Devices Are Designated. The information on which devices are designated should not be leaked from broadcast information and operations. In general, cybersecurity-related information, such as vulnerable devices, should be treated as sensitive information [28]. More specifically, if an adversary observes that an abort instruction is sent to some device, they can know that the device is vulnerable through the instruction. Namely, naively sending a command that includes the information on designated devices may attract further attacks. Therefore, we require that it should be hidden.

Requirement 4: To Have Efficient Procedures and Compact Commands. As described above, commands are broadcasted to all devices. Therefore, taking into account practical situations such as the bandwidth of broadcast channels the center would use, the size of commands should be small enough. Furthermore, we assume IoT devices, and hence, their resources might be poor. Thus, the process executed by the devices should be efficient enough. Especially on the communication overhead, we intend that the communication complexity for an operation center is independent of the number of devices. Ideally, the system should be efficient so that even ARM Cortex-M3 can execute commands.

3   Anonymous Broadcast Authentication

In this section, we put forward anonymous broadcast authentication (ABA), which is a core cryptographic primitive for the remote control system described in the previous section. We will describe how to realize the system from ABA at the end of the next subsection. The requirement 1 follows from verification correctness below, and the requirements 2 and 3 follow from security notions of ABA. We also aim to satisfy the requirement 4 with our constructions in Sect. 4. We can regard ABA as the basic system in the previous section if we see it through the cryptographic lens.

3.1   Syntax

We introduce ABA $\Pi = (\mathsf{Setup}, \mathsf{Join}, \mathsf{Auth}, \mathsf{Vrfy})$ as follows. At the beginning of the protocol, a manager runs Setup to get an authentication key ak. Each device with an identifier id has its verification key $\mathsf{vk}_{\mathsf{id}}$ embedded inside, which is obtained by executing Join. Let $\mathscr{D}$ be an identifier set of activated devices (i.e., identifiers whose verification key has been generated by Join). We assume the maximum number of devices that join the protocol is predetermined, and we describe it as N. The manager can run Auth with ak to generate an authenticated command $\mathsf{cmd}_{\mathscr{S}}$ for a command m with a privileged set $\mathscr{S} \subset \mathscr{I}$ so that only $\mathsf{id} \in \mathscr{S}$ can check the validity of $\mathsf{cmd}_{\mathscr{S}}$. Note that $\mathscr{S}$ should satisfy $|\mathscr{S}| \leq \ell$, where ℓ is also predetermined value at the beginning of the protocol. Each device designated by the manager (i.e., each device whose identifier is included in $\mathscr{S}$) runs Vrfy with $\mathsf{vk}_{\mathsf{id}}$ to check the validity of $\mathsf{cmd}_{\mathscr{S}}$. If the device received $\mathsf{cmd}_{\mathscr{S}}$ as it is, Vrfy outputs m, which should be the same as what the manager sent. Otherwise, it outputs $\bot$, which indicates "reject." Non-designated devices (i.e., devices whose identifiers are included in $\mathscr{D} \setminus \mathscr{S}$) definitely output $\bot$ regardless of whether or not $\mathsf{cmd}_{\mathscr{S}}$ is (maliciously) modified.

1. $\mathsf{Setup}(1^\kappa, N, \ell) \to \mathsf{ak}$: a probabilistic algorithm for setup. It takes a security parameter 1^κ, the maximum number of devices $N \in \mathbb{N}$, the maximum number of devices ℓ designated at once as input, and outputs authentication key ak.
2. $\mathsf{Join}(\mathsf{ak}, \mathsf{id}) \to \mathsf{vk}_{\mathsf{id}}$: a verification-key generation algorithm. It takes ak and an identifier $\mathsf{id} \in \mathscr{I}$ as input, and outputs verification key $\mathsf{vk}_{\mathsf{id}}$ for i, where $\mathscr{I}$ is a set of all possible identifiers. We assume that $\mathscr{D}$ is simultaneously updated (i.e., $\mathscr{D} := \mathscr{D} \cup \{\mathsf{id}\}$).
3. $\mathsf{Auth}(\mathsf{ak}, \mathsf{m}, \mathscr{S}) \to \mathsf{cmd}_{\mathscr{S}}$: an algorithm for generating authenticated commands. It takes ak, a command $\mathsf{m} \in \mathscr{M}$, and a privileged set $\mathscr{S} \subset \mathscr{D}$ as input, and outputs

cmd$_{\mathscr{S}} \in \mathscr{T}$, where $\mathscr{M}$ and $\mathscr{T}$ are sets of commands and authenticated commands, respectively.

4. Vrfy(vk$_{\mathsf{id}}$, cmd$_{\mathscr{S}}$) $\rightarrow$ m / $\perp$: a deterministic algorithm for verification. It takes vk$_{\mathsf{id}}$ and cmd$_{\mathscr{S}}$ as input, and outputs m if it accepts cmd$_{\mathscr{S}}$. Otherwise (i.e., it outputs $\perp$), it rejects it.

Verification Correctness. For all $\kappa, N \in \mathbb{N}$, all ℓ such that $1 \le \ell \le N$, all ak $\leftarrow$ Setup($1^{\kappa}, N, \ell$), all $\mathscr{D} \subset \mathscr{I}$, all m $\in \mathscr{M}$, all $\mathscr{S} \subset \mathscr{D}$ such that $|\mathscr{S}| \le \ell$, the following holds with overwhelming probability:

$$\begin{cases} \mathsf{Vrfy}(\mathsf{Join}(\mathsf{ak},\mathsf{id}), \mathsf{Auth}(\mathsf{ak},\mathsf{m},\mathscr{S})) \rightarrow \mathsf{m} \text{ if } \mathsf{id} \in \mathscr{S}, \\ \mathsf{Vrfy}(\mathsf{Join}(\mathsf{ak},\mathsf{id}), \mathsf{Auth}(\mathsf{ak},\mathsf{m},\mathscr{S})) \rightarrow \perp \text{ if } \mathsf{id} \notin \mathscr{S}. \end{cases}$$

Remark 1 (On the Syntax). One may think that the verification algorithm should be Vrfy(vk$_i$, m, cmd$_{\mathscr{S}}$) $\rightarrow \top$ / $\perp$, instead of Vrfy(vk$_i$, cmd$_{\mathscr{S}}$) $\rightarrow$ m / $\perp$, as in traditional MAC/signatures. The reason why we adopt this syntax is that we will consider "confidentiality" of cmd$_{\mathscr{S}}$ in the full version, though we cannot include it due to page limitation. In such a case, the manager would not send m as it is, and it should be included in cmd$_{\mathscr{S}}$ (in an encrypted form).

We can realize a remote control system we want with ABA. According to which devices the operation center wants to allow to execute a command m, the center designates a subset $\mathscr{S}$ of all devices, and runs Auth with m and $\mathscr{S}$. The center broadcasts the resultant authenticated command cmd$_{\mathscr{S}}$ to all devices. Each device check the validity of cmd$_{\mathscr{S}}$, and executes the command m only if Vrfy outputs m.

3.2 Unforgeability

Unforgeability against chosen message attacks (UF-CMA), which is a fundamental security requirement for ABA, guarantees that even if an adversary A with polynomially many corrupted devices attempts to modify authenticated commands, it is difficult for A to derive another authenticated command such that at least one honest (i.e., uncorrupted) and non-designated device outputs m. Specifically, we consider the following experiment $\mathsf{Exp}_{\Pi,\mathsf{A}}^{\mathsf{CMA}}(\kappa, N, \ell)$ between a challenger C and any PPTA A.

$\boxed{\mathsf{Exp}_{\Pi,\mathsf{A}}^{\mathsf{CMA}}(\kappa, N, \ell)}$

C runs Setup($1^{\kappa}, N, \ell$) to get ak. Let $\mathscr{W}$, $\mathscr{M}_{\mathsf{a}}$, and $\mathscr{M}_{\mathsf{v}}$ be empty sets, and flag be a flag, where flag is initialized as 0. $\mathscr{W}$ indicates sets of identifiers of corrupted devices during the experiment, and $\mathscr{M}_{\mathsf{a}}$ and $\mathscr{M}_{\mathsf{v}}$ are sets of commands used for authentication queries and verification queries, respectively. A may adaptively issue the following queries to C.

Key-Generation Query. Upon a query id $\in \mathscr{I}$ from A, C adds id to $\mathscr{D}$ and generates vk$_{\mathsf{id}} \leftarrow$ Join(ak, id). Note that A obtains nothing, and that A is allowed to make this query at most N times.

Corruption Query. Upon a query id $\in \mathscr{D}$ from A, C adds id to $\mathscr{W}$, and returns vk$_{\mathsf{id}}$ to A.

Authentication Query. Upon a query $(m, \mathscr{S}) \in \mathscr{M} \times 2^{\mathscr{D}}_{\leq \ell}$ from A, if m has not used for any verification queries (i.e., $m \notin \mathscr{M}_v$), C returns $cmd_{\mathscr{S}} \leftarrow \mathsf{Auth}(ak, m, \mathscr{S})$ to A, and adds m to $\mathscr{M}_a$. Otherwise, it returns $\perp$.

Verification Query. Upon a query $(m, \mathscr{S}, cmd_{\mathscr{S}}) \in \mathscr{M} \times 2^{\mathscr{D}}_{\leq \ell} \times \mathscr{T}$ from A, C runs $\overline{\mathsf{Vrfy}}(vk_{id}, cmd_{\mathscr{S}})$, adds m to $\mathscr{M}_v$, and returns its output to A. If there exists at least one device's index $id \in \mathscr{S}$ such that the following conditions hold, C sets flag $:= 1$.

1. The output of Vrfy is the same as m. Namely, $\mathsf{Vrfy}(vk_{id}, cmd_{\mathscr{S}}) = m$ holds.
2. The device is not corrupted. Namely, $id \notin \mathscr{W}$ holds.
3. A has not created any authentication queries for m, i.e., $m \notin \mathscr{M}_a$.

A is restricted to issue this query only once.

At some point (right after some verification query without loss of generality), A terminates the experiment, and C sets flag as the output of $\mathsf{Exp}^{\mathsf{CMA}}_{\Pi,A}(\kappa, N, \ell)$. A's advantage is defined by $\mathsf{Adv}^{\mathsf{CMA}}_{\Pi,A}(\kappa, N, \ell) := \Pr[\mathsf{Exp}^{\mathsf{CMA}}_{\Pi,A}(\kappa, N, \ell) = 1]$.

Definition 1 (Unforgeability). Let Π be an ABA scheme. We say Π is UF-CMA secure if for any PPTA A, it holds that $\mathsf{Adv}^{\mathsf{CMA}}_{\Pi,A}(\kappa, N, \ell) < \mathsf{negl}(\kappa)$ for all sufficiently-large $\kappa \in \mathbb{N}$, all $N \in \mathbb{N}$, and all $\ell \ (\leq N)$.

Remark 2 (On the Restriction of the Number of Verification Queries). We can also consider unforgeability against chosen message and verification attacks (UF-CMVA), which allows the unlimited numbers of verification queries. It can be viewed as an analogy of a similar notion in traditional MAC [5], though we do not deal with it in this paper.

3.3 Anonymity

Anonymity is also a fundamental security notion in ABA. Specifically, we consider two kinds of anonymity notions, called ANO-CMA and ANO-eq-CMA security. Roughly speaking, UF-CMA-secure ABA is said to is ANO-CMA secure if an authenticated command $cmd_{\mathscr{S}}$ does not tell any information on the corresponding privileged set $\mathscr{S}$ except for corrupted devices. If the number of devices in $\mathscr{S}$ is further leaked (but their identifiers themselves are still hidden), ABA is said to is ANO-eq-CMA secure.

   Formally, we define them based on anonymous broadcast encryption [16, 19] and privacy-preserving MAC [4]. Specifically, for ANO-security, we consider an experiment $\mathsf{Exp}^{\mathsf{ANO}}_{\Pi,A}(\kappa, N, \ell)$ between a challenger C and any PPTA A.

$$\boxed{\mathsf{Exp}^{\mathsf{ANO}}_{\Pi,A}(\kappa, N, \ell)}$$

C runs $\mathsf{Setup}(1^{\kappa}, N, \ell)$ to get ak. Let $\mathscr{W}, \mathscr{M}_a$ be empty sets, ctr be a counter initialized as 0. $\mathscr{W}$ and $\mathscr{M}_a$ are the same as in unforgeability. A may adaptively issue the following queries to C.

Key-Generation Query and Corruption Query. Same as the unforgeability experiment $\mathsf{Exp}^{\mathsf{CMA}}_{\Pi,A}$.

Authentication Query. Upon a query $(\mathsf{m}, \mathscr{S}) \in \mathcal{M} \times 2^{\mathscr{D}}_{\leq \ell}$ from A, C returns $\mathsf{cmd}_{\mathscr{S}} \leftarrow \mathsf{Auth}(\mathsf{ak}, \mathsf{m}, \mathscr{S})$ to A, and adds m to $\mathcal{M}_{\mathsf{a}}$.

Challenge Query. Upon a query $(\mathsf{m}, \mathscr{S}_0, \mathscr{S}_1) \in \mathcal{M} \times (2^{\mathscr{D}}_{\leq \ell})^2$ from A, C randomly chooses $b \in \{0,1\}$ and returns $\mathsf{cmd}_{\mathscr{S}_b} \leftarrow \mathsf{Auth}(\mathsf{ak}, \mathsf{m}, \mathscr{S}_b)$ to A. A is allowed to make this query only once under the following restriction.

(i) $(\mathscr{S}_0 \triangle \mathscr{S}_1) \cap \mathscr{W} = \emptyset$.
(ii) A has not created any authentication queries for m, i.e., $\mathsf{m} \notin \mathcal{M}_{\mathsf{a}}$.

The former restriction is necessary to avoid trivial attacks: If there exists at least one $\mathsf{id}^\star \in ((\mathscr{S}_0 \triangle \mathscr{S}_1) \cap \mathscr{W})$, A can distinguish $\mathscr{S}_0$ and $\mathscr{S}_1$ with probability one depending on the output of $\mathsf{Vrfy}(\mathsf{vk}_{\mathsf{id}^\star}, \mathsf{cmd}_{\mathscr{S}_b})$. The latter restriction is the same as the one in the unforgeability experiment. Therefore, after the challenge query, A is not allowed to make any authentication query for m.

At some point, A outputs b'. If $b' = b$, C then sets 1 as the output of $\mathsf{Exp}^{\mathsf{ANO}}_{\Pi,\mathsf{A}}(\kappa, N, \ell)$. Otherwise, C then sets 0. C terminates the experiment.

We can also define ANO-eq-security with an experiment $\mathsf{Exp}^{\mathsf{ANO\text{-}eq}}_{\Pi,\mathsf{A}}(\kappa, N, \ell)$, which is the same as $\mathsf{Exp}^{\mathsf{ANO}}_{\Pi,\mathsf{A}}(\kappa, N, \ell)$ except for the following additional condition of the restriction during authentication queries: (iii) $|\mathscr{S}_0| = |\mathscr{S}_1|$.

Definition 2 (Anonymity). Let Π be a UF-CMA-secure ABA. We say Π is X-CMA secure $(X \in \{\mathsf{ANO}, \mathsf{ANO\text{-}eq}\})$ if for any PPTA A, it holds that $|\mathsf{Adv}^{\mathsf{X}}_{\Pi,\mathsf{A}}(\kappa, N, \ell) - 1/2| < \mathsf{negl}(\kappa)$ for all sufficiently-large $\kappa \in \mathbb{N}$, all $N \in \mathbb{N}$, and all $\ell\ (\leq N)$, where $\mathsf{Adv}^{\mathsf{X}}_{\Pi,\mathsf{A}}(\kappa, N, \ell) := |\mathsf{Pr}[\mathsf{Exp}^{\mathsf{X}}_{\Pi,\mathsf{A}}(\kappa, N, \ell) = 1] - 1/2|$.

4 Constructions

We proposed two concrete constructions of ABA schemes. Taking into account the requirement 4, we only use symmetric-key primitives for compactness of authenticated commands and efficiency of Vrfy.

Before going into the constructions, we would like to discuss achievable lower bounds on efficiency in ABA schemes. Kiayias and Sanari [16] showed that a lower bound on ciphertext sizes in ANO-secure anonymous broadcast encryption schemes is $\Omega(n \cdot \kappa)$-bits, and that in ANO-eq-secure schemes is $\Omega(|\mathscr{S}| \cdot \kappa)$-bits, where n is the number of all users. Therefore, we expect that ANO-secure and ANO-eq-secure ABA schemes have similar lower bounds on the size of authenticated commands. Indeed, our constructions below achieve $\Theta(n \cdot \kappa)$-bits and $\Theta(|\mathscr{S}| \cdot \kappa)$-bits for ANO-secure and ANO-eq-secure ABA schemes, respectively.

Building Blocks. Due to the page limitation, we briefly describe MACs and PRFs, which are building blocks of our constructions.

A MAC Π_{MAC} consists of the following three algorithms. A secret key K is generated by executing MAC.Gen. MAC.Auth takes K and a message $\mathsf{m} \in \mathcal{M}$ as input, and outputs an authentication tag τ. The tag τ can be verified by running MAC.Vrfy with K and m. If it is accepted, MAC.Vrfy outputs $\top$. Otherwise, it outputs $\bot$. We consider the following two security notions for MACs.

- **Unforgeability:** We say Π_{MAC} is UF-CMA-secure if no PPTA A can create a tag τ' (so that MAC.Vrfy outputs $\top$) for arbitrary message m' without the secret key K.
- **Confidentiality:** We say Π_{MAC} meets the confidentiality, or is privacy-preserving MAC (PP-MAC) if for any message m, no information on m is leaked from τ ($\leftarrow$ MAC.Auth(K, m)).

A PRF Π_{PRF} consists of PRF.Gen, which generates a secret key k, and PRF.Eval, which takes k and a value $x \in \mathscr{D}_{PRF}$ as input, and outputs $y \in \mathscr{R}_{PRF}$. Though we omit the formal definition, we say Π_{PRF} is a pseudorandom function (PRF) if no PPTA A can distinguish the output of PRF.Eval(k, x) and a random value $r \in \mathscr{R}_{PRF}$ for any $x \in \mathscr{D}_{PRF}$.

4.1 ABA from MACs and PRFs

We construct an ABA scheme $\Pi = (\text{Setup}, \text{Join}, \text{Auth}, \text{Vrfy})$ from any MAC Π_{MAC} and two PRFs Π_{PRF} and $\overline{\Pi}_{PRF}$ where:

- $\Pi_{MAC} = (\text{MAC.Gen}, \text{MAC.Auth}, \text{MAC.Vrfy})$;
- $\Pi_{PRF} = (\text{PRF.Gen}, \text{PRF.Eval})$ such that $\text{PRF.Eval}_k : \mathscr{D}_{PRF} \rightarrow \mathscr{R}_{PRF}$, where $k \leftarrow \text{PRF.Gen}(1^\kappa)$;
- $\overline{\Pi}_{PRF} = (\overline{\text{PRF.Gen}}, \overline{\text{PRF.Eval}})$ such that $\overline{\text{PRF.Eval}}_{\overline{k}} : \mathscr{R}_{PRF} \rightarrow \mathscr{R}_{PRF}$, where $\overline{k} \leftarrow \overline{\text{PRF.Gen}}(1^\kappa)$. Strictly speaking, $\overline{\Pi}_{PRF}$ is pseudorandom permutation.

$\underline{\text{Setup}(1^\kappa, N, N) \rightarrow \text{ak}}$. Run $k \leftarrow \text{PRF.Gen}(1^\kappa)$ and output ak := k.

$\underline{\text{Join}(\text{ak}, \text{id}) \rightarrow \text{vk}_{id}}$. Run $r_{id} \leftarrow \text{PRF.Eval}(k, 0\|\text{id})$ and $y_{id} \leftarrow \text{PRF.Eval}(k, 1\|\text{id})$. Compute $K_{id} \leftarrow \text{MAC.Gen}(1^\kappa; r_{id})$ and output $\text{vk}_{id} := (K_{id}, y_{id})$.

$\underline{\text{Auth}(\text{ak}, m, \mathscr{S}) \rightarrow \text{cmd}_{\mathscr{S}}}$. Let $n := |\mathscr{D}|$ be the number of devices currently involved in the scheme, and suppose that $\text{vk}_{id_1}, \ldots, \text{vk}_{id_n}$ are generated by Join. Compute $\overline{k} \leftarrow \overline{\text{PRF.Gen}}(1^\kappa)$ and a random permutation $\sigma : [n] \rightarrow [n]$. For all $j \in [n]$, run $\text{Join}(\text{ak}, \text{id}_j)$ to get (K_{id_j}, y_{id_j}), and then compute the following:

$$(1)\ \gamma_j \leftarrow \overline{\text{PRF.Eval}}(\overline{k}, y_{id_j}),$$

$$(2)\ \begin{cases} \tau_j \leftarrow \text{MAC.Auth}(K_{id_j}, 1\|m), & \text{if id}_j \in \mathscr{S}, \\ \tau_j \leftarrow \text{MAC.Auth}(K_{id_j}, 0\|m), & \text{if id}_j \notin \mathscr{S}. \end{cases}$$

Output $\text{cmd}_{\mathscr{S}} := (m, \overline{k}, \gamma_{\sigma(id_1)}\|\tau_{\sigma(id_1)}, \ldots, \gamma_{\sigma(id_n)}\|\tau_{\sigma(id_n)})$.

$\underline{\text{Vrfy}(\text{vk}_{id}, \text{cmd}_{\mathscr{S}}) \rightarrow m/\bot}$. Parse vk_{id} and $\text{cmd}_{\mathscr{S}}$ as (K_{id}, y_{id}) and $(m, \overline{k}, \gamma_1\|\tau_1, \ldots, \gamma_n\|\tau_n)$, respectively. Compute $\overline{\gamma}_{id} \leftarrow \overline{\text{PRF.Eval}}(\overline{k}, y_{id})$ and find j such that $\gamma_j = \overline{\gamma}_{id}$. Run $\text{MAC.Vrfy}(K_{id}, m, \tau_j)$ and output the output as it is.

Theorem 1. *If Π_{MAC} is a PP-MAC and Π_{PRF} and $\overline{\Pi}_{PRF}$ are PRFs, the construction of Π described above is* UF-CMA-*secure and* ANO-CMA-*secure.*

Proof (Sketch). Due to the page limitation, we here give a sketch. The underlying PRF guarantees that all MAC keys are correctly generated and are independent of each other. Therefore, even if an adversary A gets arbitrary devices $\mathscr{W} \subset \mathscr{I}$, A cannot forge MAC tags for honest devices $\mathscr{I} \setminus \mathscr{W}$ due to UF-CMA security of the underlying MAC. Thus, the proposed construction meets UF-CMA security. Furthermore, the confidentiality of the underlying MAC hides a message (i.e., $1\|m$ or $0\|m$) corresponding to each tag τ_{id_i} ($\forall i \in [n]$). Due to the underlying PRF, each γ_{id_i} leaks no information on y_{id_i} (and hence $1\|\mathrm{id}_i$). Therefore, A obtains no information which devices (except for corrupted ones $\mathscr{W}$) are designated. Hence, the proposed construction satisfies ANO-CMA security. $\qquad \square$

We next show an ANO-eq-secure ABA scheme from the same primitives by slightly modifying the above Auth algorithm as follows.

$\underline{\mathsf{Auth}(\mathsf{ak}, \mathsf{m}, \mathscr{S}) \to \mathsf{cmd}_{\mathscr{S}}.}$ Suppose $\mathscr{S} = \{\mathrm{id}_1, \ldots, \mathrm{id}_k\}$. Compute $\bar{\mathsf{k}} \leftarrow \overline{\mathsf{PRF.Gen}}(1^{\kappa})$ and a random permutation $\sigma : [|\mathscr{S}|] \to [|\mathscr{S}|]$. For all $j \in [k]$, run $\mathsf{Join}(\mathsf{ak}, \mathrm{id}_j)$ to get $(\mathsf{K}_{\mathrm{id}_j}, y_{\mathrm{id}_j})$, and then compute $\gamma_{\mathrm{id}_j} \leftarrow \overline{\mathsf{PRF.Eval}}(\bar{\mathsf{k}}, y_{\mathrm{id}_j})$ and $\tau_{\mathrm{id}_j} \leftarrow \mathsf{MAC.Auth}(\mathsf{K}_{\mathrm{id}_j}, \mathsf{m})$. Output $\mathsf{cmd}_{\mathscr{S}} := (\mathsf{m}, \bar{\mathsf{k}}, \gamma_{\mathrm{id}_{\sigma(1)}} \| \tau_{\mathrm{id}_{\sigma(1)}}, \ldots, \gamma_{\mathrm{id}_{\sigma(k)}} \| \tau_{\mathrm{id}_{\sigma(k)}})$.

Theorem 2. *If Π_{MAC} is UF-CMA secure and Π_{PRF} and $\overline{\Pi}_{\mathrm{PRF}}$ are PRFs, the construction of Π described above is UF-CMA-secure and ANO-eq-CMA-secure.*

Proof (Sketch). As in Theorem 1, we can prove that the proposed ABA scheme meets UF-CMA security. ANO-eq-CMA security can also be proved in a way similar to Theorem 1. Although each authenticated command $\mathsf{cmd}_{\mathscr{S}}$ leaks the number of designated devices $|\mathscr{S}|$, it does not violate the ANO-eq security due to the condition (iii) $|\mathscr{S}_0| = |\mathscr{S}_1|$ in $\mathsf{Exp}_{\Pi,\mathsf{A}}^{\mathrm{ANO\text{-}eq}}(\kappa, N, N)$. Note that we do not require the confidentiality of the underlying MAC since the underlying message (i.e., m) does not contain any information on $\mathscr{S}$. $\qquad \square$

Remark 3. It is known that (PP-)MACs are strictly weaker than PRFs [4]. Hence, we can also obtain UF-CMA-secure and ANO-secure ABA schemes by replacing Π_{MAC} with Π_{PRF} in the above constructions. Since PRFs and (PP-)MACs can be constructed from one-way functions [14, 15], the constructions can be viewed as ABA schemes from one-way functions. Furthermore, if we change Setup so that it runs MAC.Gen N times and outputs N MAC keys as ak, the construction turns to an ABA scheme from only (PP-)MACs.

5   Conclusion

In this paper, we considered a basic system for securely remote-controlling IoT devices, and discussed its requirements. We then put forward an anonymous broadcast authentication (ABA) scheme as its core cryptographic primitive. We formalized a model and security notions of ABA and showed provably-secure constructions. The proposed scheme consists of only symmetric-key cryptographic primitives. Throughput is thus expected to be fast since our ABA schemes can be implemented with only, say, HMAC or AES.

Research into implementing the ABA construction is in progress. In particular, we intend to evaluate the performance of ABA via a prototype implementation. Another future work is to design a real-world application of ABA (and hence the remote control system discussed in the paper) for intending to realize IoT security.

Acknowledgements. This research was conducted under a contract of "Research and development on IoT malware removal / make it non-functional technologies for effective use of the radio spectrum" among "Research and Development for Expansion of Radio Wave Resources (JPJ000254)", which was supported by the Ministry of Internal Affairs and Communications, Japan. We would like to thank Hirokazu Kobayashi for his useful comments on existing broadcast authentication protocols and Tatsuya Takehisa for his valuable comments on the system model.

Appendix

We provide lists for the notations regarding ABA and building blocks in Tables 1 and 2, respectively.

Table 1. Notation list for ABA.

Notations for ABA	Description
N	The maximum number of devices
ℓ	The maximum number of devices designated at once
id	Device identifier
$\mathscr{I}$	Identifier set
$\mathscr{D}\ (\subset \mathscr{I})$	Identifier set of activated devices
$\mathscr{S}\ (\subset \mathscr{D})$	Identifier set of designated devices
$\mathscr{W}\ (\subset \mathscr{D})$	Identifier set of corrupted devices
m	Command
$\mathscr{M}$	Set of commands
$\mathscr{M}_{\mathsf{a}}$	Set of commands used for authentication queries
$\mathscr{M}_{\mathsf{v}}$	Set of commands used for verification queries
$\mathsf{cmd}_{\mathscr{S}}$	Authenticated command for $\mathscr{S}$
$\mathscr{T}$	Set of authenticated commands
ak	Authentication key
$\mathsf{vk}_{\mathsf{id}}$	Verification key for id

Table 2. Notation list for MAC and PRF.

Notations for MAC	Description	Notations for PRF	Description
K	MAC key	k	PRF key
m	Message	$\mathscr{D}_{\mathsf{PRF}}$	Domain of PRF
τ	MAC tag	$\mathscr{R}_{\mathsf{PRF}}$	Range of PRF

References

1. Al-Garadi, M.A., Mohamed, A., Al-Ali, A., Du, X., Ali, I., Guizani, M.: A survey of machine and deep learning methods for internet of things (IoT) security. IEEE Commun. Surv. Tuts. **22**(3), 1646–1685 (2020)
2. Andersen, M.P., et al.: WAVE: a decentralized authorization framework with transitive delegation. In: Proceedings of USENIX Security 2019, pp. 1375–1392. USENIX Association (2019)
3. Antonakakis, M., et al.: Understanding the mirai botnet. In: Proceedings of USENIX Security 2017, pp. 1093–1110. USENIX Association (2017)
4. Bellare, M.: New proofs for NMAC and HMAC: security without collision-resistance. In: Proceedings of CRYPTO 2006. LNCS, vol. 4117, pp. 602–619. Springer, Berlin, Heidelberg (2006)
5. Bellare, M., Goldreich, O., Mityagin, A.: The power of verification queries in message authentication and authenticated encryption. Cryptology ePrint Archive, Report 2004/309 (2004)
6. Bertino, E., Islam, N.: Botnets and internet of things security. Computer **50**(2), 76–79 (2017)
7. Boneh, D., Gentry, C., Waters, B.: Collusion resistant broadcast encryption with short ciphertexts and private keys. In: Proceedings of CRYPTO 2005. LNCS, vol. 3621, pp. 258–275. Springer, Berlin, Heidelberg (2005)
8. Chan, H., Perrig, A.: Round-efficient broadcast authentication protocols for fixed topology classes. In: Proceedings of IEEE S&P 2010, pp. 257–272. IEEE (2010)
9. Costin, A., Zaddach, J., Francillon, A., Balzarotti, D.: A large-scale analysis of the security of embedded firmwares. In: Proceedings of USENIX Security 2014, pp. 95–110. USENIX Association (2014)
10. Costin, A., Zarras, A., Francillon, A.: Automated dynamic firmware analysis at scale: a case study on embedded web interfaces. In: Proceedings of ASIACCS 2016, pp. 437–448. ACM (2016)
11. Fazio, N., Perera, I.M.: Outsider-anonymous broadcast encryption with sublinear ciphertexts. In: Fischlin, M., Buchmann, J., Manulis, M. (eds.) Public Key Cryptography - PKC 2012, pp. 225–242. Springer, Berlin Heidelberg (2012)
12. Fernandes, E., Jung, J., Prakash, A.: Security analysis of emerging smart home applications. In: Proceedings of IEEE S&P 2016, pp. 636–654. IEEE (2016)
13. Fernandes, E., Paupore, J., Rahmati, A., Simionato, D., Conti, M., Prakash, A.: FlowFence: practical data protection for emerging iot application frameworks. In: Proceedings of USENIX Security 2016, pp. 531–548. USENIX Association (2016)
14. Goldreich, O., Goldwasser, S., Micali, S.: How to construct random functions. J. ACM **33**(4), 792–807 (1986)
15. HÅstad, J., Impagliazzo, R., Levin, L., Luby, M.: A pseudorandom generator from any one-way function. SIAM J. Comput. **28**(4), 1364–1396 (1999)
16. Kiayias, A., Samari, K.: Lower bounds for private broadcast encryption. In: Proceedings of IH 2013. LNCS, vol. 7692, pp. 176–190. Springer, Berlin, Heidelberg (2013)
17. Kumar, S., Hu, Y., Andersen, M.P., Popa, R.A., Culler, D.E.: JEDI: many-to-many end-to-end encryption and key delegation for IoT. In: Proceedings of USENIX Security 2019, pp. 1519–1536. USENIX Association (2019)
18. Li, J., Gong, J.: Improved anonymous broadcast encryptions. In: Proceedings of ACNS 2018. LNCS, vol. 10892, pp. 497–515. Springer (2018)
19. Libert, B., Paterson, K.G., Quaglia, E.A.: Anonymous broadcast encryption: adaptive security and efficient constructions in the standard model. In: Proceedings of PKC 2012. LNCS, vol. 7293, pp. 206–224. Springer, Berlin, Heidelberg (2012)

20. Naor, D., Naor, M., Lotspiech, J.: Revocation and tracing schemes for stateless receivers. In: Proceedings of CRYPTO 2001. LNCS, vol. 2139, pp. 41–62. Springer, Berlin, Heidelberg (2001)
21. Neto, A.L.M., et al.: AoT: authentication and access control for the entire iot device life-cycle. In: Proceedings of Sensys 2016, pp. 1–15. ACM (2016)
22. Perrig, A.: The biba one-time signature and broadcast authentication protocol. In: Proceedings of CCS 2001, pp. 28–37. ACM (2001)
23. Perrig, A., Canetti, R., Tygar, J.D., Song, D.: Efficient authentication and signing of multicast streams over lossy channels. In: Proceedings of IEEE S&P 2000, pp. 56–73. IEEE (2000)
24. Ronen, E., Shamir, A., Weingarten, A.O., Olynn, C.: IoT goes nuclear: creating a ZigBee chain reaction. In: Proceedings of IEEE S&P 2017, pp. 195–212. IEEE (2017)
25. Safavi-Naini, R., Wang, H.: Broadcast authentication for group communication. Theor. Comput. Sci. **269**(1), 1–21 (2001)
26. Shim, K.A.: Basis: a practical multi-user broadcast authentication scheme in wireless sensor networks. IEEE Trans. Inf. Forensics Secur. **12**(7), 1545–1554 (2017)
27. Wang, X., Han, Y., Leung, V.C., Niyato, D., Yan, X., Chen, X.: Convergence of edge computing and deep learning: a comprehensive survey. IEEE Commun. Surv. Tuts. **22**(2), 869–904 (2020)
28. Zhauniarovich, Y., Khalil, I., Yu, T., Dacier, M.: A survey on malicious domains detection through DNS data analysis. ACM Comput. Surv. **51**(4) (2018)

An Efficient and Privacy-Preserving Billing Protocol for Smart Metering

Rihem Ben Romdhane[1]($\boxtimes$), Hamza Hammami[2], Mohamed Hamdi[1], and Tai-Hoon Kim[3]

[1] Digital Security Research Lab, Higher School of Communications of Tunis, Sup'Com, Ariana, Tunisia
rihem.benromdhane@supcom.tn, mmh@supcom.rnu.tn
[2] Faculty of Sciences of Tunis, University of Tunis El Manar, LIPAH-LR11ES14, 2092 Tunis, Tunisia
hamza.hammami@aol.fr
[3] School of Economics and Management, Beijing Jiaotong University, Beijing 100044, China

Abstract. Due to the advancements in digital information and information technology, the automated meter reading has been replaced with advanced metering infrastructure which is based on smart metering. Smart metering enables the collection of fine-grained meter readings, which allows several features such as real time monitoring of electricity consumption, dynamic pricing and time-of-use billing. However, those fine-grained consumption measurements raise serious privacy concerns by inferring the consumers' personal habits and behaviors. In this paper, we propose an efficient privacy-preserving time-of-use billing protocol for smart metering. Our proposed solution uses the homomorphic property of the elliptic curve Pedersen commitment to perform the billing process without revealing the consumption readings to the utility provider. Security and performance analysis show that our suggested approach offers better privacy protection in comparison with existing solutions, in addition to providing efficient computational cost.

1 Introduction

In the recent years, the rapid growth of the Internet of Things has transformed the traditional power grids into smart grids by incorporating intelligent measuring devices called Smart Meters (SMs) in a large scale [1]. SMs collect fine-grained measurements and report them periodically (e.g. every 15 min) to the Utility Provider (UP). These real time meter readings enable the UP to monitor and manage the power flows efficiently. Hence, the detection of power outages, network leakage and fraud is becoming faster and more accurate. Another use is the electricity trading at different tariffs or the so-called SM billing. In fact, the regular fine-grained measurements allow the UP to bill customers accurately depending on pricing schemes based on the current offer and demand. However, the transmission and recording of customers' consumption measurements raise significant privacy concerns by enabling the creation of usage profiles related to specific energy consumers (persons, companies, houses…). By deeply analyzing these profiles, accurate conclusions about consumers' private lives can be deduced.

© The Author(s), under exclusive license to Springer Nature Switzerland AG 2021
L. Barolli et al. (Eds.): AINA 2021, LNNS 226, pp. 691–702, 2021.
https://doi.org/10.1007/978-3-030-75075-6_57

For billing, time-of-use tariffs and flexible pricing schemes are needed to accurately bill the grid users. Therefore, consumption information needs to be transmitted frequently to the UP each billing period, and so it may pose several privacy issues [2]. Given the harmful effect of disclosing related privacy data, legal measures have been taken in most countries to protect customers' privacy [3, 4]. However, these measures might not be enough. A stronger privacy guarantee can be given by technical approaches that rely on cryptographic protocols [5, 6].

In this paper, a technical approach to protect householders' privacy in smart metering systems is described: We put forward an efficient and privacy-preserving SM billing protocol. This approach provides a high degree of privacy by never reporting the fine-grained consumption readings outside the customer's home area network. The SM sends signed commitments to the UP accompanied with the result of the bill calculation. Using zero knowledge proofs, the UP can verify the correctness of the bill calculation.

The major contributions of our protocol can be summarized as follows:

- We propose a novel SM billing protocol that provides the privacy protection of customers' consumption measurements.
- We conduct performance analysis to show that our suggested protocol significantly reduces communication overhead in comparison with existing solutions and supports efficient computation cost imposed on both the SM and the UP.

The remainder of this paper is organized as follows: In Sect. 2, some related work is presented. Section 3 presents the system model, the privacy requirements and the design goals of the proposed approach. In Sect. 4, we review some preliminaries. Then, our suggested protocol is described in Sect. 5. The performance evaluation is given in Sect. 6. Finally, Sect. 7 concludes the paper and discusses future research directions.

2 Related Work

In the recent years, several privacy-preserving smart metering billing approaches have been proposed. Current research studies can be classified into three categories:

The first one is based on the use of a trusted entity (trusted third party or trusted platform module) [7–10]. In this approach, the calculation of the bill is isolated into the trusted entity. However, the householder has to trust the trusted entity with his/her sensitive consumption data and the right and exact computation of his/her invoice. On the other hand, the UP has to trust the trusted entity with the generation of bills. Therefore, this approach poses a major issue since the trust that both parties have to put into the trusted entity is very enormous. The second category is based on distorting the consumption readings. For example, the solution put forward in [11] relies on the use of additional consumption which is added as noise. This approach protects consumers' consumption readings from being revealed and provides differential privacy. However, the tariff of this additional consumption is expected to be very high. Similarly, the authors in [12] used a rechargeable battery that allowed the modification of the total amount of energy consumption by adding or subtracting noise. Other approaches relying on rechargeable batteries were given in [13, 14]. However, the feasibility of these approaches has been related to the possibility of powerful, economical and low-cost batteries. The third category employs cryptographic schemes to privacy, preserving the billing process in smart metering systems. In [15], the authors proposed two decentralized privacy-friendly protocols for billing and electricity trading. Their approach was based on an aggregation technique in addition to the use of a bidding algorithm based on secure multiparty computations. Nevertheless, in their approach, they considered that the communication channels between the different entities were secure and authentic. This assumption needs more clarification, particularly in the privacy-preserving context. The authors in [16] suggested a privacy-preserving aggregation and billing protocol for the smart grid based on Paillier's homomorphic encryption. Gope et al. [17] put forward lightweight privacy-preserving approach for dynamic pricing-based billing for the smart grid. Their proposed protocol uses the concept of single-pass authenticated encryption.

To offer ideal privacy, several solutions [18–20] have relied on homomorphic commitments and zero knowledge proofs to allow users to perform the bill calculation and to prove the correctness of their calculation without revealing their fine-grained consumption readings to external entities. However, even their approaches offer ideal privacy, they require high computation complexity. To solve this problem, we propose a lightweight and efficient privacy preserving billing protocol based on the elliptic curve Pedersen commitment. The main idea of our protocol relies on the fact that the bill calculation is performed without the need for plaintext consumption profiles to leave the household.

3   System Model, System Requirements and Design Goals

3.1   System Model

As shown in Fig. 1, our system model consists of three major entities in the SM billing system: the SM, the gateway and the UP.

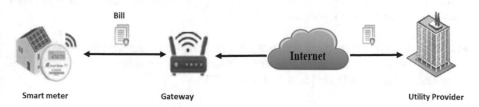

Fig. 1. System model under consideration

In our system model, the SM is responsible for measuring the electricity consumption of a household and for transmitting the meter readings to the UP at regular periods for billing purposes. The UP periodically collects readings from the SM, applies accurate tariffs and sends calculation to each household at regular periods. The gateway, in our model, works as a relay node and just forwards the exchanged messages between the SM and the UP. The communication between the different entities is through WIFI.

3.2 System Requirements

1) Households' sensitive consumption data should be protected from adversaries. Especially, these data should not be revealed to the UP. In this way, customers' consumption data can achieve the privacy-preserving requirement.
2) The UP must be able to verify the correctness of the received bill amount.
3) Due the limited resources of SMs, our protocol must require low communication and computation overhead.

3.3 Design Goals

Under the system model mentioned above and system requirements, our design goal is to develop an efficient privacy preserving the billing protocol for smart metering, which enables the SM to compute the householder's invoice and additionally allows the UP to ensure the correctness of the calculation without disclosing the household's consumption profile.

4 Preliminaries

The core of our proposed protocol relies on the elliptic curve Pedersen commitment [21]. In this section, we provide a brief description of this cryptographic tool.

Generally, a commitment scheme is a cryptographic tool based on two functions: *com* and *open*.

Function *com (m, r)* takes as input a value m to commit to and a random number r which provides hiding. As output it produces the commitment c such that $c = \text{com}(m, r)$.

The function *open(c, m, r)* takes as input commitment c and the value of m and r and returns true if c is a commitment to m and false if not. A commitment scheme provides the following security properties:

- *Hiding:* It means that, given commitment c, it is hard to determine the value of m.
- *Binding:* Given commitment c, m and a random number r, it is hard to compute m $\neq$ m' and r', such that Open (c, m', r') returns True.

Elliptic Curve Pedersen Commitment
The elliptic curve Pedersen commitment is a variant of the Pedersen commitment [22] based on elliptic curve cryptography. The elliptic curve Pedersen commitment is described as follows:

- Let F_p be the group of elliptic curve points where p is a large prime number.
- Let G be a random generator point in F_p and H be another random generator point such that $H = qG$ (q is a chosen secret).
- Let $r \in Z_p$ be a random value which represents the blinding factor.

The commitment to value m $\in Z_p$ with random value r $\in Z_p$ is then computed as: c = com(m, r)= mG + rH

To open the commitment we simply reveal the values m and r:

$$open\ (c,\ m,\ r) : c =\ mG + rH.$$

The elliptic curve Pedersen commitment is perfectly hiding, computationally binding and additively homomorphic. The homomorphic property allows providing the following equations: Given two messages m1 and m2, blinding factors r1 and r2 and scalar k, we obtain:

$$Com(m_1 + m_2,\ r_1 + r_2) = Com(m_1,\ r_1) + Com\ (m_2,\ r_2). \tag{1}$$

$$Com\ (k \cdot m,\ k \cdot r) = k \cdot Com\ (m,\ r). \tag{2}$$

5   Privacy-Preserving Billing Protocol

5.1   Assumptions

In our suggested approach, the following assumptions will be made:

a) The SM has a small memory to store data, to compute commitments and to produce signatures.
b) The SM is tamper-resistant and trusted by the UP.
c) All the communications between the SM and the UP are established over a secure channel.

5.2   Proposed Protocol

In this section, we put forward our privacy-preserving billing protocol. The main idea of our approach is as follows:

The SM generates an elliptic curve Pedersen commitment for each energy measurement. Then, it sends the vector of commitments and the random values to the energy provider instead of sending the real consumption measurements. Thus, the UP cannot obtain any insight into the householder's energy consumption. Next, the SM performs a calculation on the energy consumption values using the tariff information (received from the UP) to obtain the final invoice. The same calculation is then performed on the random numbers used for the initial commitments. After that, the SM sends the random values and the result of calculations to the UP. The latter, using the received commitments, calculates a new commitment related to the final invoice. Then, if the new commitment can be opened using the values received by the SM, the UP can then be sure that the bill is correctly calculated. Our protocol consists of three phases, including the initialization phase, the reporting and bill calculation phase and the verification phase. The parameters and notations are described in Table 1.

Table 1. Basic notations used in our protocol and their description.

Symbol	Description
SM	Smart meter
UP	Utility provider
K_{SIG}	Signature key between SM and UP
E	Vector of energy consumption values
e_j	Energy consumed during the round j
r_j	Random number related to e_j
c_j	Commitment related to e_j and r_j
COM	Vector of commitments
COM_{SIG}	Signature of COM using K_{SIG}
R'	Vector of random values { $r_0, r_1, ..., r_N$ }
T	Vector of tariffs
t_j	Tariff related to round j
P	Total price to be paid by household
R'	Result of the calculation made on random numbers
$COM_{invoice}$	Commitment related to total price of invoice

As the main operations are performed by the two principle entities (SM and UP), for convenience, we do not include the gateway in the protocol. However, it is implicitly included and considered as an intermediate node that is only responsible for forwarding the exchanged messages between the SM and the UP.

Initialization Phase

To protect the integrity of the exchanged data, the SM and the UP use a signature scheme using the same key denoted with K_{SIG}. (The signature and the key generation are beyond the scope of this paper).

Reporting and Bill Calculation Phase

Step 1. First, the SM generates a vector of energy consumption values. Each value corresponds to the energy consumed by the household during round j. E= {$e_0, e_1,......, e_N$}

Step 2. For each consumption value, the SM generates commitment C_j = curve com (e_j, r_j), where com (m, r) corresponds to the elliptic Pedersen commitment of value m, and r is a random number. This value will be stored in a vector of commitments COM = {$c_0, c_1,...., c_N$}.

The random values used for the generation of commitments are stored in the SM memory in vector R = {$r_0, r_1,......, r_N$}.

Step 3. The SM signs the vector COM using K_{SIG} to provide the data integrity and obtain the vector COM_{SIG} which will be sent to the UP accompanied by the vector of commitments COM.

$$SM \xrightarrow{\quad COM, COM_{SIG} \quad} UP$$

Step 4. After verifying the SM signature on COM_{SIG}, the UP sends a tariff vector that contains a list of price[s for each time frame (which corresponds to round j), T= $\{t_0, t_1,\ldots\ldots t_N\}$.

$$UP \xrightarrow{\text{T}} SM$$

Step 5. After obtaining the tariff information, the SM performs the following calculations:

1. $P = \sum_{j=0}^{N} ej * tj$ where P corresponds to the total invoice that must be paid by the household's SM.
2. $R' = \sum_{j=0}^{N} rj * tj$

Step 6. The SM sends P, R', COM and COM_{SIG} to the UP.

$$SM \xrightarrow{P, R', COM, COM_{SIG}} UP$$

Verification Phase
Step 1. First, the UP verifies the SM signature on COM, and then performs a calculation on the received commitments to obtain: $COM_{invoice} = \sum_{j=0}^{N} tj * cj$

Step 2. To check if the price sent by the SM is correct and trustfully calculated, $COM_{invoice}$ must open to True if the correct price P and the correct random number R' are used. Therefore, the UP should perform the following function: *Open (COM_{invoice}, P, R')*. The result of this function returns True if $COM_{invoice} = COM(P, R')$ (i.e. the bill sent by the SM is trustfully calculated), and false if not. The different steps of our protocol are presented in Fig. 2.

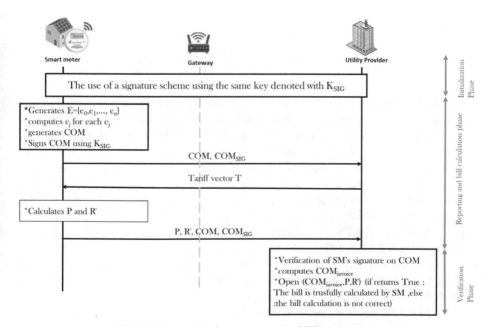

Fig. 2. Proposed privacy-preserving billing protocol

6 Performance Evaluation

As indicated in Sect. 2, some related work [19, 20] has relied on the Pedersen commitment to privacy preserving billing. However, their approaches require complex operations which are computationally expensive. To obtain better performances in terms of operation complexity and computation cost, we adopt the Pedersen commitment technique based on Elliptic Curve Cryptography (ECC). This can achieve lower complexity overhead since ECC offers the highest security per bit. It can achieve the same security level with much smaller key lengths.

To the best of our knowledge, we are the first to put forward a privacy preserving billing solution based on the elliptic curve Pedersen commitment algorithm. To evaluate the performance of our approach, we conduct simulation using the Python programming language on a computer with Intel Core i7-8565 2.0 GHz-processor, and 24.00 GB RAM. In our experiments, we use the recommended elliptic curve parameters sec256k1 presented in [23]. We assume that G is a point of the curve chosen using the following coordinates: Gx = 79BE667E F9DCBBAC 55A06295 CE870B07 029BFCDB 2DCE28D9 59F2815B 16F81798, Gy = 483ADA77 26A3C465 5DA4FBFC 0E1108A8 FD17B448 A6855419 9C47D08F FB10D4B8. The bit lengths related to the rest of parameters are presented in Table 2.

Table 2. Bit lengths of the used parameters

Parameter	Bit length (bit)
q (chosen secret)	256
p (the prime number)	256
e (energy consumption)	16
r (random number)	64
t (tariff)	16

To present the total computation overhead of our solution, we calculate the execution time of the different cryptographic operations involved in the proposal. To obtain accurate measures, we use the method proposed in [24]. The main idea is to first measure the time of one function call (the function corresponds to the operation we want to compute its execution time), next the time of two, after that the time of three, and so on. Then, through the measurements, we can fit a straight line. Hence, the overall execution time can be obtained by taking a slope from the straight line $f(x)$ such as $f(x) = a x + b$. For example, in Fig. 3, which represents the execution time of 1 commitment generation, the equation of the straight line is $f(x) = 8.56x + 2.84$, so the time required for generating 1 commitment is 8.56 ms. Similarly, in the case of 96 commitments, we can obtain the execution time related to the generation of commitments, the aggregation of randoms, the aggregation of commitments, the calculation of invoices and the verification of invoices, depicted respectively in Fig. 4, Fig. 5, Fig. 6, Fig. 7 and Fig. 8.

The results are summarized in Table 3 (in the case of generating 96 commitments).

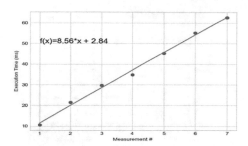

Fig. 3. Execution time of one commitment generation (8.56 ms)

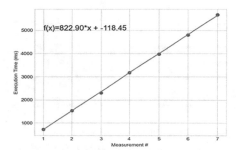

Fig. 4. Execution time of 96 commitments generation (822.9 ms).

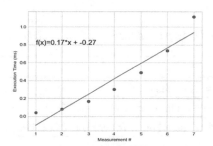

Fig. 5. Execution time of randoms aggregation (0.17 ms)

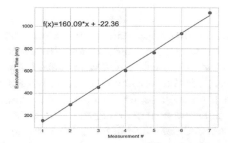

Fig. 6. Execution time of commitments aggregation (160.9 ms)

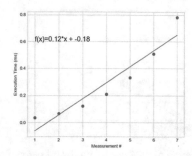

Fig. 7. Execution time of the invoice calculation (0.12 ms)

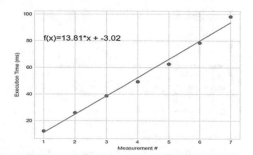

Fig. 8. Execution time of the invoice verification (13.81 ms)

Table 3. Execution time of the different operations.

Symbol	Operation	Time (ms)
GC	Generation of commitments(96 commitments)	822.9
AR	Aggregation of randoms (calculation of R')	0.17
AC	Aggregation of commitments(calculation of COMM$_{invoice}$)	160.09
CI	Calculation of invoices(calculation of P)	0.12
VI	Verification of invoices(Open function)	13.81

The results show that the different operations require low computation cost especially on the SM which has limited computation resources. To obtain the total computation overhead of our protocol, we suppose that the SM generates (as in the basic billing protocols) a meter reading every 15 min. As a consequence, the total number of commitments generated per day is 96. If we suppose, for example, that the billing period corresponds is one month, then the total number of commitments sent to UP is 96 * 30 accompanied by the aggregated random and the total bill price. If we suppose that T_{GC}, T_{AR}, T_{AC}, T_{CI}, TVI are respectively the execution time of GC, AR, AC, CI and VI, so the total computation overhead per month is $30T_{GC} + T_{AR} + T_{CI} + T_{AC} + T_{VI} = 24.86$ s.

7   Conclusion

In this paper, we have proposed an efficient and privacy preserving billing protocol for the smart grid based on the elliptic curve Pedersen commitment. Our approach offers ideal privacy by performing the bill calculation on the household's side without revealing the plaintext meter readings. Moreover, using the homomorphic property, the UP can verify the invoice correctness. The gateway, in our approach, works as a relay node and just forwards the exchanged messages between the two principal entities (SM and UP). The simulation results show that our suggested solution offers low computation overhead, which is affordable by the different entities, especially by the SM.

References

1. Saleem, Y., Crespi, N., Rehmani, M.H., Copeland, R.: Internet of things-aided smart grid: technologies, architectures, applications, prototypes, and future research directions. IEEE Access **7**, 62962–63003 (2019)
2. Asghar, M.R., Dán, G., Miorandi, D., Chlamtac, I.: Smart meter data privacy: a survey. IEEE Commun. Surv. Tuts. **19**(4), 2820–2835 (2017)
3. Papakonstantinou, V., Kloza, D.: Legal protection of personal data in smart grid and smart metering systems from the European perspective. In: Smart grid security, pp. 41–129. Springer, London (2015)
4. Mylrea, M.: Smart energy-internet-of-things opportunities require smart treatment of legal, privacy and cyber security challenges. J. World Energy Law Bus. **10**(2), 147–158 (2017)
5. Romdhane, R.B., Hammami, H., Hamdi, M., Kim, T.H.: At the cross roads of lattice-based and homomorphic encryption to secure data aggregation in smart grid. In: 2019 15th International Wireless Communications & Mobile Computing Conference (IWCMC), pp. 1067–1072. IEEE (2019)
6. Romdhane, R.B., Hammami, H., Hamdi, M., Kim, T.H.: A novel approach for privacy-preserving data aggregation in smart grid. In: 2019 15th International Wireless Communications & Mobile Computing Conference (IWCMC), pp. 1060–1066. IEEE (2019)
7. Ford, V., Siraj, A., Rahman, M.A.: Secure and efficient protection of consumer privacy in advanced metering infrastructure supporting fine-grained data analysis. J. Comput. Syst. Sci. **83**(1), 84–100 (2017)
8. Bohli, J.M., Sorge, C., Ugus, O.: A privacy model for smart metering. In: 2010 IEEE International Conference on Communications Workshops, pp. 1–5. IEEE (2010)
9. Zhao, J., Liu, J., Qin, Z., Ren, K.: Privacy protection scheme based on remote anonymous attestation for trusted smart meters. IEEE Trans. Smart Grid **9**(4), 3313–3320 (2016)
10. Eccles, T., Halak, B.: A secure and private billing protocol for smart metering. IACR Cryptol. ePrint Arch. **2017**, 654 (2017)
11. Danezis, G., Kohlweiss, M., Rial, A.: Differentially private billing with rebates. In: International Workshop on Information Hiding, pp. 148–162. Springer, Berlin, Heidelberg (2011)
12. Backes, M., Meiser, S.: Differentially private smart metering with battery recharging. In: Data Privacy Management and Autonomous Spontaneous Security, pp. 194–212. Springer, Berlin, Heidelberg (2013)
13. McLaughlin, S., McDaniel, P., Aiello, W.: Protecting consumer privacy from electric load monitoring. In: Proceedings of the 18th ACM Conference on Computer and Communications Security, pp. 87–98 (2011)

14. Yang, W., Li, N., Qi, Y., Qardaji, W., McLaughlin, S., McDaniel, P.: Minimizing private data disclosures in the smart grid. In: Proceedings of the 2012 ACM Conference on Computer and Communications Security, pp. 415–427 (2012)

15. Abidin, A., Aly, A., Cleemput, S., Mustafa, M.A.: Secure and privacy-friendly local electricity trading and billing in smart grid. arXiv preprint arXiv:1801.08354 (2018)

16. Wang, X.F., Mu, Y., Chen, R.M.: An efficient privacy-preserving aggregation and billing protocol for smart grid. Secur. Commun. Netw. **9**(17), 4536–4547 (2016)

17. Gope, P., Sikdar, B.: A lightweight and privacy-preserving data aggregation for dynamic pricing-based billing in smart grids. In: 2018 IEEE PES Innovative Smart Grid Technologies Conference Europe (ISGT-Europc), pp. 1–7. IEEE (2018)

18. Rial, A., Danezis, G.: Privacy-preserving smart metering. In: Proceedings of the 10th Annual ACM Workshop on Privacy in the Electronic Society, pp. 49–60 (2011)

19. Jawurek, M., Johns, M., Kerschbaum, F.: Plug-in privacy for smart metering billing. In: International Symposium on Privacy Enhancing Technologies Symposium, pp. 192–210. Springer, Berlin, Heidelberg (2011)

20. Eccles, T., Halak, B.: Performance analysis of secure and private billing protocols for smart metering. Cryptography **1**(3), 20 (2017)

21. França, B.F.: Homomorphic Mini-Blockchain Scheme (2015)

22. Pedersen, T.P.: Non-interactive and information-theoretic secure verifiable secret sharing. In: Annual International Cryptology Conference, pp. 129–140. Springer, Berlin, Heidelberg (1991)

23. http://www.secg.org/

24. Moreno, C., Fischmeister, S.: Accurate measurement of small execution times—getting around measurement errors. IEEE Embed. Syst. Lett. **9**(1), 17–20 (2017)

Intra Data Center MultiPath Optimization with Emulated Software Defined Networks

Lucio Agostinho Rocha[✉]

Departure of Computer Engineering, Federal University of Technology, Apucarana, PR, Brazil
luciorocha@utfpr.edu.br

Abstract. This paper investigates computational modeling for MultiPath Transmission Control Protocol (MPTCP) to optimize data paths in data centers. MPTCP is a protocol that distributes traffic over multiple paths and contributes to making more efficient use of the network capacity. But its automatic distribution of subflows generates suboptimal data traffic performance. So, the computational modeling of these data paths in the internal network topology contributes to sending data only in the best links. A testbed of emulated virtual switches is used to evaluates the performance of the MPTCP. These evaluations were done in network topologies with an OpenFlow controller. The results indicate that optimized data paths contribute to improving the overall goodput with balanced data transfer between computational resources.

1 Introduction

Cloud Computing is a well-consolidated solution for offering Internet pay-per-use computing, database storage, applications, and many other computational services on demand [1]. Some main key concepts that aim to understand its large adoption, as follows. Agility is about to offer computing services as soon as they are required with high availability and high coverage. This property reduces the costs of change and simplifies the experimentation of software solutions. Reliability is a metric about confidence and availability, i.e., the system is operational even in eventual failures. Resilience is a feature of the system about its ability to recover its functionalities over failures and/or minor shortage faults. Elasticity is auto-scale computing and database services over demand.

Recent studies [2] indicates that cloud computing models are evolving to distributed cloud models. A distributed cloud is a recent computing model that offers public cloud services to different physical locations with low-latency and near its customers. The responsibility of operation is given by the cloud provider. But even this new model needs to offer the best network performance inside the data center.

In this context, this paper investigates the performance of Multipath TCP protocol (MPTCP) [3] in an emulated software-defined network. Although MPTCP is not exclusive of enterprise networks, this research is particularly useful for data centers due to its high amount of data traffic, multiple low latency links, and high throughput. MPTCP is an ongoing effort from IETF Multipath TCP Working Group to allows TCP connection use multiple paths to increase redundancy and resilience of connections and optimize

L. Barolli et al. (Eds.): AINA 2021, LNNS 226, pp. 703–714, 2021.
https://doi.org/10.1007/978-3-030-75075-6_58

the bandwidth utilization. MPTCP is an extension to TCP and acts on the transport layer of TCP/IP stack. So, end-user applications are not affected. This protocol sends simultaneously many TCP subflows through distinct paths. Distributing the throughput through multiple paths potentially improves the performance of Ethernet networks. As a motivation, most network devices use one or more interfaces to communicate over Internet. But conventional TCP transmissions restrict network connections to a single data path, even if there are multiple paths to reach the destination. For example, if a device has cellular connectivity and Wi-Fi only one connection is used, even though both are available [4]. As an alternative to address this issue, MPTCP generates subflows which are multiple data paths of the same data flow, i.e., a set of TCP connections. So, this protocol offers concurrent multipath transport and also provides multi-homing and multipath transport. These features are useful for many web applications with intensive data traffic, such as big data and cloud computing [5].

Integration of MPTCP with software-defined networking (SDN) aims to pragmatically simplify the management of large data centers. By your centralized nature, SDN provides a global view of the entire network. SDN improves resource utilization when integrated with MPTCP [6]. SDN is a network paradigm that decouples the management control plane from the data plane [7–9]. SDN networks distribute forwarding rules to the network devices to handle flows. A central (or distributed) controller device manages the deployment of these rules on the managed devices [10]. Openflow [11] is the de facto protocol that establishes the communication between the controller and its compliance devices. Besides, experimentation with SDN can be done in emulated environments, giving an alternative for analysis of the network performance with a large set of configurations, avoiding the unavailability of production services of cloud providers during the tests.

In this paper, the motivation is to use mathematical modeling of operational research to define these data paths on-demand. The research problem is to distribute MPTCP subflows in emulated networks generated from computational modeling. The goal is to identify how computational modeling contributes to leverage the data transfer over multipath in data centers. The proposed solution is a methodology conceived in two folds: a) modeling of the data center traffic to generates the data paths according to traffic load demand; b) MPTCP kernel setup to leverage the MPTCP performance to forward subflows over the generated multipath links from the computational modeling. The focus of this paper is to evaluate the performance of MPTCP with multipath links from the computational modeling in emulated topologies. The results indicate that optimized data paths contribute to improving the goodput data transfers supported by kernel fine-tuning. As a consequence, this approach contributes to reducing the network costs to forward traffic of cloud services and aims to optimize the network energy consumption.

The remains of this paper are organized as follows. Section 2 is about related works; Sect. 3 relates the methodology; Sect. 4 describes the evaluation, analysis and results; finally, Sect. 5 does final considerations and comments about future works.

2   Related Works

A common approach in local area networks with multiple paths is to use a spanning tree protocol to avoid layer 2 loops. These protocols block redundant links and reduce the alternatives to forward traffic through multiple paths. These data paths are primarily tree-like paths, and alternate paths can not be used because they would form a loop in the topology [12]. This assumption implies that many links can be overwhelmed, reducing the total throughput. Another consideration is that these algorithms must know the intermediate nodes to converge and establish the data path before submitting the traffic. There are many protocols to enable multipath routing.

Multipath routing is done with ECMP, SCTP, MPLS [13] some other compliance layer 3 routing protocol, or performed at layer 2 with IEEE Shortest Path Bridging (SPB) [14] or IETF TRansparent Interconnection of Lots of Links (TRILL) [15]. However, all these protocols let the network performs the load balancing across the paths based on flows, calculating a hash based on Ethernet and IP addresses, and TCP/UDP ports of the packets. Packets from the same flow go through the same path to avoid unordered packets, and distinct flows are equally distributed through distinct paths, even if there are paths with different bandwidth, cost, or latency capacities [12].

On the other hand, the MPTCP [3] protocol does the load balancing in the end nodes as part of the TCP process. Several researchers [6] worked on improving the MPTCP performance in terms of path management and congestion control. MPTCP extends features of TCP and is defined on the transport layer. MPTCP handles different bandwidths because a congestion control mechanism acts over each subflow. This mechanism redistributes a subflow for a less congestion link according to demand [16]. On the sender's side, the MPTCP scheduling of packets splits the byte stream from the application in many segments. These segments are numbered and send through one of the subflows available. At the receiver's side, these segments are ordered, and the original byte stream is rebuilt [16]. MPTCP is an alternative to conventional TCP data traffic that goes through a single path to avoid performance problems related to packet reordering and lazy acks [17]. Besides, multiple interfaces are today low cost and very common, even in-home devices and/or enterprise data centers linked switches/routers from multiple vendors. In addition, Programming Protocol-Independent Packet Processors (P4) [18] is a high-level language for programming protocol-independent packet processors. P4 works in conjunction with SDN control protocols like OpenFlow and informs how packets are processed by the data plane of a programmable forwarding device. P4 provides an API for a controller populate tables and can implement a switch pipeline for any purpose. The motivation to use multiple paths is to transfer more traffic using the least congested paths, where a set of links behaves like a larger link with high bandwidth. A possibility is to do congestion control for each subflow, but this approach generates unfair traffic distribution when subflows share bottlenecks paths [19]. This occurs because the MPTCP data paths does not need to be entirely disjoint, i.e., these paths can share intermediary nodes. This protocol transparently identifies multiple paths by the presence of multiple network interfaces in the end-hosts.

MPTCP was designed for mobile clients with multiple network interfaces with intermittent connectivity. In the meantime, congestion control in data centers has distinct requirements than that in wide-area networks, mainly to solve the tension between 0-

RTT and packet losses, as applications strive for low latency. Under this aspect, MPTCP takes some time to react over link failures events compared with a managed solution provided by existing data center network operators.

3   Methodology

In this research, the MPTCP data paths are designed by mathematical modeling. The methodology of this evaluation is conceived in two folds: a) Modeling of the data center traffic: generates the data paths according to the traffic load demand; b) MPTCP kernel setup: adjust of kernel parameters of MPTCP hosts to forward subflows over the generated links from the latter computational modeling.

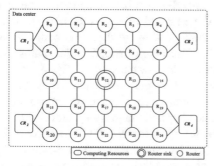

(a) Graph of the Modeled Data Center.

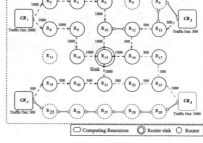

(b) Graph of the Resulting Data Paths.

Fig. 1. Data center modeling.

3.1   Modeling of the Data Center Traffic

Computing Resources (CR) are core components of a data center [20]. In this modeling, a Linear Programming (LP) model is defined, and the CR is a set of servers that provide processing to run low-cost applications. Two optimizations are done in one combined objective function: 1) minimize the cost of data transfer between CRs, and 2) minimize the cost to forward flows through the set of intermediary router nodes to reach a common sink node. This common sink is the endpoint to reach other geographic regions. Besides, the minimum cost path to this sink node guarantees an optimal common path to exchange a higher volume of traffic between CRs inside the data center.

In this strategy, the network between the data centers is modeled by directed graph $G(V,E)$ where the V is the set of vertices, i.e., network routers and CR, and $E \subseteq \{(i,j) \in V^2 | i \neq j\}$ is the set of edges, i.e., links between these nodes. A link is expressed by indexes $(i,j) \in E$. The Fig. (1a) shows the graph of the modeled data center. There are the following types of nodes:

- Source nodes: CR nodes generating network traffic in the data center;
- Destination nodes: CR nodes consuming network traffic in the data center.

Each CR eventually exchanges data with other neighbor CR in the same data center. Due to this issue, in this modeling, a CR node is a source and destination of data traffic. These nodes represent all traffic with origins and destinations between neighbor CRs. Table 1 shows the meaning of the main symbols in this modeling.

Table 1. Description of the symbols in the network data center modeling.

Symbols	Description
V	Vertices: set of routers and set of servers (CR)
E	Edges: set of links
$L_{i,j}$	Current traffic in the link between the nodes i and j
OUT_i	Maximum output supply from CR_i
IN_j	Minimum input demand to CR_j
$BW_{i,j}$	Bandwidth in the link between the nodes i and j, i.e., the maximum $Volume_{i,j}$ supported between the links i and j
$PoS_{i,j}$	Preference of Service in the link between the nodes i and j
$Cost_{i,j}$	Cost to send traffic between the node i and node j
$Volume_{i,j}$	Current traffic between the CR_i and CR_j

Equation (1) describes the cost to forward traffic in the links. In this equation, the bandwidth ($BW_{i,j}$) is the maximum data forwarded between i and j nodes. The cost is inversely proportional at BW, i.e., higher BW has a lower cost. Additionally, this equation includes the Preference of Service (PoS) of each link. This parameter guarantee preference to use links with higher PoS, even if all links have the same bandwidth (BW), i.e., links with higher PoS has a lower cost. The reason to include PoS is to offer an alternative to dynamically change the flows in links, but without affecting the bandwidth which is generally a static parameter.

$$Cost_{i,j} = \frac{1}{BW_{i,j} * PoS_{i,j}} \tag{1}$$

Equation (2) describes the amount of data traffic allowed to be forwarded through the link ($Volume_{i,j}$).

$$\sum Volume_{i,j} \geq 0 \tag{2}$$

Equation (3) is about the maximum supply of the source CR. It describes the sum of all volume j from the same source i must not bypass the maximum output throughput (OUT_i) of the source i.

$$\sum_j Volume_{i,j} \leq OUT_i \tag{3}$$

Equation (4) is about the minimum demand of the destination CR. It describes the sum of all volume i to a specific destination j must attend, at least, the minimum input demand (IN_j) of the destination j.

$$\sum_i Volume_{i,j} \geq IN_j \tag{4}$$

Until here, the modeling solves a simple transportation problem. But the communication between CRs deals with intermediary router nodes. So, the next steps complement the previous one, and add a minimum-cost flow modeling to the same model. The goal is to discover the traffic flows which satisfy the demand from a source CR to a destination CR, but without exceeding the link capacities.

Equation (5) is about the flow conservation. $L_{i,j}^{a,b}$ is the traffic volume from node $a \in V$ to node $b \in V$ routed through the link $(i, j) \in E$. This traffic volume should not bypass the bandwidth of this link $(BW_{i,j})$.

$$L_{i,j}^{a,b} \leq BW_{i,j} \tag{5}$$

Equation (6) indicates that the output traffic (OUT_i) is the sum of all traffic from servers j from CR_i.

$$\sum_j CR_{i,j} = OUT_i \tag{6}$$

Equation (7) describes the flow conservation of the router nodes. $L_{in,k}$ is the input traffic volume from router $k \in V$ routed through $(in, k) \in E$. The sum of all inputs (in) from a router k must be equal to the sum of outputs (out) to the same router. $L_{k,out}$ is the output traffic volume from router $k \in V$ routed through $(k, out) \in E$.

$$\sum_{in} L_{in,k} - \sum_{out} L_{k,out} = 0 \tag{7}$$

Equation (8) describes the flow conservation in the common sink router node. $L_{src,j}$ is the source of traffic volume from router $src \in V$ routed through $(src, j) \in E$, and $L_{k,dst}$ is the destination of traffic volume from router $k \in V$ routed through $(k, dst) \in E$. The sum of flows generated by the all CRs $(\sum_{src}(\sum_j L_{src,j}))$ must to be received by the domain sink router $(\sum_k L_{k,dst})$. This sink node is the Point-of-Presence (PoP) to reach other external sites. This equation is important because guarantee the existence of flows in the links between the intermediary router nodes.

$$\sum_{src}(\sum_j L_{src,j}) - \sum_k L_{k,dst} = 0 \tag{8}$$

Finally, the Eq. (9) describes the two optimizations combined in one objective function. The first optimization calculates the minimum cost of traffic $(Cost_{i,j})$ according to the volume of traffic from each CR $(Volume_{i,j})$. The second optimization is about the minimum cost $(Cost_{e,t})$ to forward the CR traffic in the intermediary router nodes $(Link_{e,t})$.

$$min\ E^{total} = \sum Cost_{i,j} * \sum Volume_{i,j} + \sum_e(\sum_t Cost_{e,t} * Link_{e,t}) \tag{9}$$

The Fig. 1b shows the resulting data paths obtained with this modeling written in the MathProg language and executed by the GLPK software. The feasible demands are set with 2000, 500, 500 and 1000 units for CR1 to CR4, respectively.

3.2 MPTCP Kernel Setup

The simple enabling of MPTCP in the operating system kernel is not sufficient to achieve better performance than conventional TCP transmission. The default scheduler of MPTCP Linux kernel always prefers to send data over the subflow with lesser round-trip-time (RTT) [21]. As an example, this happens when the client-side announces a receive window lesser than the server-side, e.g., cellular connections have the receive window lesser than the receive window of Wi-Fi connections [22]. In this scenario, most of subflows will be forwarded through the single path with a large receive window which has the lesser RTT.

Preliminary evaluations in this research revealed that MPTCP needs adjust kernel parameters to leverage its performance in congested links. This same conclusion was observed by some authors [23–25]. Sandri et al. [26] affirm that initiated more sub-flows than disjoint paths will not corroborate to improve the throughput. Even changes in tunning TCP [4,27], and/or changes in the links of topology will not generate bet-ter throughput if a properly set of schedulers, data path manager and congestion control are not well established to accomplish the many subflows. Mininet has some drawbacks to correctly emulate multiple paths, mainly due to the implementation of NetEm [28] and its queue management in the network stack [29]. In the MPTCP implementation, the default scheduler stops sending subflows in a path when a TCP small queue throt-tled occurs. TSQ [30] is an alternative to reduce queues in the network stack. Large queues imply higher latency due to queue delays. TSQ limits the amount of data that can be queued for transmission by sockets [30]. So, MPTCP avoids sending packets on subflows which will potentially have low throughput due to a path with TSQ throt-tled [29,31]. Mininet uses NetEm for delay emulation. However, delayed packets in NetEm are removed from the sending queue instead of is queued. As a consequence, MPTCP will never found a TSQ throttled subflow, even with delayed links in the source host, and a single path will be always used [29]. Define an emulated topology with pre-defined delays could artificially contribute to increasing the multipath utilization. So, the kernel setup must be considered to fine-tuning the MPTCP transmission with higher throughput. Disabling the `mptcp_checksum` is sufficient to enable goodput near the default TCP goodput, but it is not recommended to be done in production environ-ments. Checksum verifying is necessary to enabling only authorized subflows to join in the current MPTCP connection.

4   Evaluation of the MPTCP with Computational Modeling

In this section, the network topology resulting from the computational modeling of Sub-sect. 3.1 is implemented in MathProg language and executed by the GLPK software. The resulting data paths are sent to Mininet. This testbed has the OpenFlow controller ODL [32]. In ODL, the Model Driven-Service Abstraction Layer (MD-SAL) [33] is a middleware that offers a message bus to the intercommunication between applications. This middleware allows communication between the controller and applications using northbound (NB) interfaces and between the controller and switches using southbound (SB) interfaces. The implementation of the modeling and kernel parameters are avail-able in the Ref. [34].

The goal is to verify how the modeling contributes to increasing the goodput between the computing resources (CRs). This modeling gives the unused links and/or router nodes that can be turn off to reduce energy consumption. This scenario contemplates four end-points emulating CR servers with variable traffic output, and 25 OVS switch nodes. The following evaluations are done in the data paths presented in Fig. 1b with flow demands of 2000, 500, 500, and 1000 units of traffic. These values can be dynamically given to the testbed on demand.

It was performed five tests of 60 s each, using iperf3multi to simulate client/server MPTCP data transfer with until 8 subflows. Measures were done five times each to acquire the average of the data rate and bytes transferred. The performance was evaluated using the iperf3singleTCP script. This script runs iperf3 inside Mininet to transmit continuously between all hosts for 60 s each.

Figure 2 shows the average data transfer between the CRs. Figure 2a and b show the average data transfer between CRs without modeling and default MPTCP kernel setup [35]. This scenario reveals a non-balanced data transfer, and CR1 and CR2 have a preference to submit traffic. This occurs because Mininet has some drawbacks to correctly emulate multiple paths, mainly due to the implementation of NetEm and its queue management in the network stack. Besides, MPTCP always prefers the subflows with the smallest RTTs to send data and has a default scheduler that stops sending subflows in a path when a TCP small queue throttled occurs. The results of the proposed solution are shown in Fig. 2c and d. In this solution, the average data transfer between CRs is obtained with modeling and MPTCP kernel setup from the previous section. In this condition, the data transfer is balanced, and all CRs have a similar opportunity to schedule traffic. This occurs mainly due to non-congested data paths given by the modeling integrated with kernel fine-tuning for MPTCP subflows.

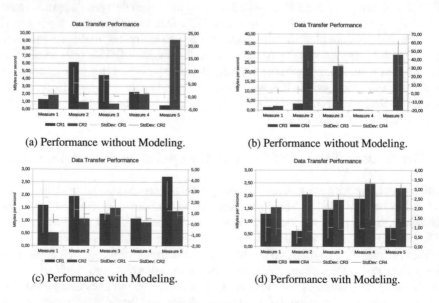

(a) Performance without Modeling.   (b) Performance without Modeling.

(c) Performance with Modeling.   (d) Performance with Modeling.

Fig. 2. MPTCP evaluation.

Table 2. MPTCP data transfer evaluation.

CR1	Scenario 1			Scenario 2		
	Mean (MB/s)	Std.Dev	CI	Mean (MB/s)	Std.Dev	CI
1	1,28	0,32	0,08	1,59	2,43	0,61
2	6,15	5,29	1,33	1,95	1,33	0,33
3	4,47	6,50	1,64	1,26	0,88	0,22
4	2,28	1,39	0,35	1,07	0,62	0,15
5	0,55	0,95	0,24	2,72	1,23	0,31
CR2	Mean (MB/s)	Std.Dev	CI	Mean (MB/s)	Std.Dev	CI
1	1,28	0,32	0,08	1,59	2,43	0,61
2	6,15	5,29	1,33	1,95	1,33	0,33
3	4,47	6,50	1,64	1,26	0,88	0,22
4	2,28	1,39	0,35	1,07	0,62	0,15
5	0,55	0,95	0,24	2,72	1,23	0,31
CR3	Mean (MB/s)	Std.Dev	CI	Mean (MB/s)	Std.Dev	CI
1	1,28	0,32	0,08	1,59	2,43	0,61
2	6,15	5,29	1,33	1,95	1,33	0,33
3	4,47	6,50	1,64	1,26	0,88	0,22
4	2,28	1,39	0,35	1,07	0,62	0,15
5	0,55	0,95	0,24	2,72	1,23	0,31
CR4	Mean (MB/s)	Std.Dev	CI	Mean (MB/s)	Std.Dev	CI
1	1,28	0,32	0,08	1,59	2,43	0,61
2	6,15	5,29	1,33	1,95	1,33	0,33
3	4,47	6,50	1,64	1,26	0,88	0,22
4	2,28	1,39	0,35	1,07	0,62	0,15
5	0,55	0,95	0,24	2,72	1,23	0,31

Table 2 shows the mean data rate (MB/s), standard deviation and confidence interval (CI) with confidence level of 95% for sixty samples. Scenario 1 runs without modeling and Scenario 2 runs with modeling integrated at kernel adjusts.

The results indicate that the proposed approach increases the overall throughput whilst keep the data transfer balanced between the hosts. The obtained standard deviation also gives balanced values for Scenario 2. The CI in Scenario 2 indicates that the difference between the parameters and the observed estimate is not statistically significant at the 5% level, i.e., the reliability of the estimation procedure gives values very closely at the mean data rate. Finally, packet losses were lesser than 5% and latency lesser than 100 ms for all CRs in Scenario 2. Scenario 1 also has low packet losses lesser than 5% for the CRs with higher precedence, but a latency higher than 500ms is given for the other non-privileged CRs generating low throughput for the other concurrent transmissions.

5   Conclusion

This article presents an evaluation of MPTCP with mathematical modeling of data paths for low-cost computing resources. The automatic topology discovery with MPTCP path-manager fullmesh generates suboptimal routes of data paths. This paper presents a possible solution with feasible computational modeling to obtain data paths with minimum cost in emulated data center topologies. As future works, it is been proposed emulation of larger topologies with Docker thin-clients, embed modeling in ODL, automatic setup of data paths with P4 language, evaluation of multipath priorities flags, such as MP_prio, and setup of multipath between distinct networks. Finally, many researching areas could use similar modeling to adaptative machine learning in assistive mobility and robotics [36,37] for path planning and obstacle avoidance.

Acknowledgment. The author gratefully acknowledges the contribution from UTFPR câmpus Apucarana in support of this research. The author also thanks Christoph Paach from the MPTCP team for his valuable contribution to this research.

References

1. Rad, B.B., Diaby, T., Rana, M.E.: Cloud computing adoption: a short review of issues and challenges. In: Proceedings of the 2017 International Conference on E-Commerce, E-Business and E-Government. ICEEG 2017. Association for Computing Machinery, New York, NY, USA, pp. 51–55 (2017). https://doi.org/10.1145/3108421.3108426
2. Buyya, R., Srirama, S.N., Casale, G., et al.: A manifesto for future generation cloud computing: Research directions for the next decade. ACM Comput. Surv. **51**, no. 5 (2018). https://doi.org/10.1145/3241737
3. Ford, A., Raiciu, C., Handley, M., et al.: TCP extensions for multipath operation with multiple addresses. In: IETF RFC 8684 (2020). https://tools.ietf.org/html/rfc8684
4. Zhou, F., Dreibholz, T., Zhou, X., et al.: The performance impact of buffer sizes for multipath TCP in internet setups. In: IEEE 31st International Conference on Advanced Information Networking and Applications (2017)
5. Wang, K., Dreibholz, T., Zhou, X., et al.: On the path management of multi-path TCP in internet scenarios based on the NorNet testbed. In: IEEE 31st International Conference on Advanced Information Networking and Applications (AINA) (2017)
6. Hussein, A., Elhajj, I.H., Chehab, A., et al.: SDN for MPTCP: an enhanced architecture for large data transfers in datacenters. In: IEEE International Conference on Communications (ICC), vol. 2017, pp. 1–7 (2017)
7. Feamster, N., Rexford, J., Zegura, E.: The road to SDN: an intellectual history of programmable networks. ACM SIGCOMM Comput. Commun. Rev. **44**(2), 87–98 (2014)
8. Kreutz, D., et al.: Software-defined networking: a comprehensive survey. Proc. IEEE **103**(1), 14–76 (2015)
9. Alvigzu, R., et al.: Comprehensive survey on T-SDN: software-defined networking for transport networks. IEEE Commun. Surv. Tuts. **19**(4), 2232–2283 (2017)

10. Bannour, F., Souihi, S., Mellouk, A.: Distributed SDN control: survey, taxonomy, and challenges. IEEE Communi. Surv. Tuts. **20**(1), 333–354 (2018)
11. Schaller, S., Hood, D.: Software defined networking architecture standardization. Comput. Stand. Interfaces **54**, 197–202 (2017)
12. van del Pol, R.: TRILL and IEEE 802.1aq Overview (2012)
13. Lim, C., Pahk, S., Kim, Y.: Model of transport SDN and MPLS-TP for T-SDN controller. In: 2016 18th International Conference on Advanced Communication Technology (ICACT), pp. 522–526 (2016)
14. Allan, D., Bragg, N.: IEEE 802.1aq in a Nutshell: antecedents and technology. In: 802.1aq Shortest Path Bridging Design and Evolution - The Architect's Perspective. Wiley (2012)
15. Li, Y., Umair, M., Banerjee, A., Hu, F.: Transparent interconnection of lots of links (TRILL): appointed forwarders. In: IETF RFC 8139 (2017). https://tools.ietf.org/html/rfc8139
16. van der Pol, R., Boele, S., Dijkstra, F., et al.: Multipathing with MPTCP and OpenFlow. In: Proceedings - 2012 SC Companion: High Performance Computing, Networking Storage and Analysis, SCC 2012, pp. 1617–1624 (2012)
17. Anjum, F.M.: Algorithms and multiple paths: considerations for the future of the internet. In: Information Systems Frontiers, vol. 6 (2004)
18. Bosshart, P., Daly, D., Gibb, G., et al.: P4: Programming Protocol-Independent Packet Processors (2014). https://arxiv.org/pdf/1312.1719.pdf
19. Raiciu, C., Handly, M., Wischik, D.: Coupled Congestion Control for Multipath Transport Protocols (2011). https://tools.ietf.org/html/rfc6356
20. Cisco.com: What is a Data Center? (2020). https://www.cisco.com/c/en/us/solutions/data-center-virtualization/what-is-a-data-center.html
21. Bonaventure, O.: Why is the Multipath TCP scheduler so important? (2018). http://blog.multipath-tcp.org/blog/html/2018/12/09/the_multipath_tcp_packet_scheduler.html
22. Paasch, C., Ferlin, S., Alay, O., Bonaventure, O.: Experimental evaluation of multipath TCP schedulers. In: ACM SIGCOMM Capacity Sharing Workshop (CSWS). ACM (2014)
23. Kukreja, N., Maier, G., Alvizu, R., et al.: SDN based automated testbed for evaluating multipath TCP. In: IEEE International Conference on Communications Workshops (ICC) (2016)
24. Nguyen, K., Golam, K.M., Ishizu, K., et al.: An approach to reinforce multipath TCP with path-aware information. In: Sensors (Basel), vol. 19, no. 3 (2019)
25. Ro, S., Van, D.N.: Performance evaluation of MPTCP over a shared bottleneck link. Int. J. Comput. Commun. Eng. **5**(3) (2016)
26. Sandri, M., Silva, A., Rocha, L.A., Verdi, F.L.: On the benefits of using multipath TCP and openflow in shared bottlenecks. In: 2015 IEEE 29th International Conference on Advanced Information Networking and Applications (2015)
27. Bonaventure, O.: Recommended Multipath TCP Configuration (2014). http://blog.multipath-tcp.org/blog/html/2014/09/16/recommended_multipath_tcp_configuration.html
28. Linux.org: NetEm - Network Emulator (2011). https://www.linux.org/docs/man8/tc-netem.html
29. Frömmgen, A.: Mininet/Netem Emulation Pitfalls - A Multipath TCP Scheduling Experience (2017). https://progmp.net/mininetPitfalls.html
30. Dumazet, E.: TCP: TCP Small Queues (2012). https://lwn.net/Articles/506237/
31. Paasch, C.: MPTCP: Do Not Schedule on TSQ-Throttled Subflows (2014). https://github.com/multipath-tcp/mptcp/commit/5c278893b37fe48c66ff226793607687b8482ba9
32. OpenDaylight Docummentation Magnesium (2020). https://docs.opendaylight.org/en/stable-magnesium/getting-started-guide/introduction.html
33. MD-SAL Overview (2020). https://docs.opendaylight.org/projects/controller/en/stable-magnesium/dev-guide.html#md-sal-overview
34. LPDC: Linear Programming for Data Center (2020). https://github.com/lpdc/lpdc

35. MultiPath TCP - Linux Kernel Implementation, "Configure MPTCP" (2020). https://multipath-tcp.org/pmwiki.php/Users/ConfigureMPTCP
36. Pinheiro, P.G., Pinheiro, C.G., Cardozo, E.: The wheelie — a facial expression controlled wheelchair using 3D technology. In: 2017 26th IEEE International Symposium on Robot and Human Interactive Communication (RO-MAN), pp. 271–276 (2017)
37. Rohmer, E., Pinheiro, P., Raizer, K., Olivi, L., Cardozo, E.: A novel platform supporting multiple control strategies for assistive robots. In: 24th IEEE International Symposium on Robot and Human Interactive Communication (RO-MAN), 2015, pp. 763–769 (2015)

Deploying Scalable and Stable XDP-Based Network Slices Through NASOR Framework for Low-Latency Applications

Rodrigo Moreira[1,2]([✉]), Pedro Frosi Rosa[1], Rui Luis Andrade Aguiar[3], and Flávio de Oliveira Silva[1]

[1] Faculty of Computing (FACOM), Federal University of Uberlândia, Uberlândia 38400-902, Brazil
{rodrigo.moreira,pfrosi,flavio}@ufu.br
[2] Institute of Exact and Technological Sciences (IEP), Federal University of Viçosa, Rio Paranaíba 38810-000, Brazil
rodrigo@ufv.br
[3] Telecommunications Institute (IT), University of Aveiro, Aveiro, Portugal
ruilaa@ua.pt

Abstract. The spreading and popularization of applications such as Artificial Intelligence (AI), the Internet of Things (IoT), and Virtual Reality (VR) have relentlessly changed the way we live and communicate. Besides, these applications, having specific requirements that transcend connectivity, have changed the way they have reserved the underlying resources. In this sense, to satisfy these applications agreeably, network slicing mechanisms are needed to make it possible to customize connectivity for applications that run on generic network hardware through a separate data and control plane. Besides, connectivity needs to be responsive to vertical business requirements, such as low-latency and high-reliability. This paper proposes the XCP Controller, a scalable and stable method of configuring data planes for network slices between multiple Internet domains for applications with low-latency requirements. Empirical results suggest that the combination of NASOR and XCP Controller enables to deploy network slices over early packet processing technologies to experience lower latencies than other baseline technologies.

1 Introduction

Due to the popularization and dissemination of novel applications that have impacted our daily lives, such as virtual reality, 3D media, artificial intelligence, and the Internet of Things (IoT), new connectivity models need to be considered [1]. These new applications running on the network have changed their needs and the way to acquire underlying resources [2]. Recent advances are wicked in mobile networks, especially with the deployment of the 5G network underway

© The Author(s), under exclusive license to Springer Nature Switzerland AG 2021
L. Barolli et al. (Eds.): AINA 2021, LNNS 226, pp. 715–726, 2021.
https://doi.org/10.1007/978-3-030-75075-6_59

worldwide. However, there is a consensus that the capabilities embedded in the design of this network do not deliver fully automatic connectivity or intelligence at the network level, especially at the core. In this sense, offering a network that provides human-centered connectivity, especially intelligently and reliably, raises challenges beyond the data rate [3,4].

Besides, technologies such as Software Defined Networking (SDN), Network Functions Virtualization (NFV), and Segment Routing [5] have changed how to deploy, deliver, and manage applications on network devices to satisfy the refined requirements for modern users and applications. In this sense, to provide personalized, flexible, scalable, end-to-end connectivity, the term network slicing was coined, recognized, and integrated as a technological component for 5G. In addition to technological impacts brings with them financial impacts [6,7].

State-of-the-art solutions that claim to enable network slicing do so considering predominantly the mobile network ecosystem [8]. However, delivering manageable, independent, and customizable networks to users and applications needs to occur at all network levels and domains. Technological advances are known that deliver resource slicing at the edge, at the radio interface, at the core of mobile networks on the network and backhaul [9], or even an end-to-end approach [10]. However, concise network slicing approaches at the core rather than VPN-based are little explored.

Previously, we proposed, implemented, and evaluated the proof of concept of a network slicing approach for multiple Autonomous Systems (ASs), called Network And Slice ORchestrator (NASOR) [11]. In this paper, considering the NASOR framework, we advanced the state-of-the-art by proposing and evaluating the XCP Controller combined with the NASOR to implement multi-domain network slices for applications with low-latency requirements. Among other findings, we conclude that early packet processing ensures connectivity stability in addition to ensuring better latency than fully-stacked approaches.

The main contributions of this paper are summarized as follows: (1) XCP Controller, a proactive control plane for early packet processing in low-layers of network hardware; (2) a multi-domain network slice deployment method for applications with low-latency requirements; (3) a performance evaluation of standard NASOR packet processing approach against XDP-based.

The remaining of this paper is organized as follows: Sect. 2 presents related works and highlights some gaps in which our method aims to sheds light. Section 3 describes the proposed solution with the technical details of the main components. Section 4 presents a method to experiment and measure our proof-of-concept described in Sect. 5. We analyze and discuss the results in Sect. 6 and we draw some conclusions in Sect. 7, pointing out future work.

2   Related Work

The work [12] describes the proposal and evaluation of a method for measuring packet processing time within VNFs. The proposal stands on the addition of information in the experimental header of a packet, allowing to measure the processing time considering both ingress and egress queues in each VNF. The authors evaluated the proposed method by implementing VNFs with specific functions, measuring the overhead of adding new fields to the packet and the processing time under some scenarios.

In Ricart-Sanchez, Ruben, *et al.* [13] we find a proposal based on the NetF-PGA SUME platform [14] for network slicing providing low latency and self-healing for Smart Grids. The solution bases on the early processing of packets, especially those containing specific power station messages, on programmable network hardware. Experimental results suggest that the network slicing proposal offers connectivity within the desirable limits such as latency, jitter, and packet loss. Unlike [13], our solution offers network slice deployment over multiple Internet domains, and for different use cases, providing low processing overhead on intermediate nodes, enabling through very early packet processing low-latency connectivity.

An integration of XDP with Service Function Chaining (SFC), as available at [15], allowed the authors to see a significant reduction in CPU consumption, latency and an increase in throughput through the early processing of packets in the XDP layer. The general idea is the proposal of hybrid VNFs, which perform simple tasks with functions implemented in the XDP layer and legacy VNFs with more complex tasks implemented in the userspace.

The work [16] contains a description, implementation, and evaluation of packet processing functions for Segment Routing Header (SRH) type packets. Among other findings, the authors realized that the proposition of early processing functions did not cause significant overhead in the transport of packets. Unlike the authors, leveraged by the NASOR framework, we use their distributed data plane combined with the XCP Controller to configure network slices for applications with low-latency requirements.

3   Our Proposal

We propose the eXpress Control Plane (XCP) Controller to perform low-latency at the network core and combined with the NASOR framework. NASOR offers network slice deployment over multiple Internet domains based on the routing data plane. As shown in Fig. 1, we shed light on our contribution through our conceptual scheme. It is possible to see different political and technological domains and the interaction that NASOR does between them to enable inter-domain connectivity through the exchange of domain capabilities.

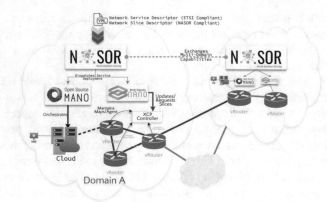

Fig. 1. Combining NASOR architecture with XCP controller

When NASOR receives a service deployment request, in the YAML standard, computing directs to the well-established MANO. However, the network part follows NASOR flow, which raises a Network and Orchestrator (NANO) (a NASOR component) to configure the routing parameters based on the source. However, our contribution adds the XCP Controller component between NANO and routers. The new component is responsible for maintaining the map in vRouters, creating virtual interfaces, and maintaining track of which XDP programs are currently running on each interface.

3.1   eXpress Data Path (XDP)

Proposed in 1993, the BSD Packet Filter (BPF) is a packet filtering method that uses the register buffering strategy to capture packets in the Kernel. With this, listening applications can receive a copy of the packets delivered by the network device drivers, enabling superior performance to approaches that use the Kernel network stack [17]. Subsequent advances, such as the insertion of new instructions and the addition of maps, culminated in the Extended BPF (eBPF) version, which today supports packet processing applications at various levels of the Kernel, such as Traffic Control (TC), Netfilter, and XDP.

The eXpress Data Path (XDP) is a programmable approach that allows the insertion of high-level codes such as C in an interface to provide packet processing instructions. XDP allows for cooperative integration with the operating system kernel. It allows the insertion of a hook in the RX path, allowing a program at the level supported by an eBPF machine to decide on processing the packet early. Before the memory allocation buffer, the hook places allow early processing or delivery packets to the conventional kernel stack [18].

3.2   eXpress Control Plane (XCP)

This paper proposes the eXpress Control Plane (XDP) concept, which represents a proactive mechanism for installing forwarding rules on the XDP Maps of the vRouters on which the multi-domain network slices are deployed. In Fig. 2, we propose an illustration of the interaction and behavior of the components of our mechanism. According to a bottom-up reading, immediately over the network hardware is the XDP layer, where we associate a program to each network interface.

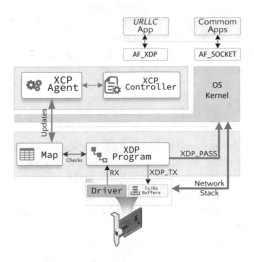

Fig. 2. XCP conceptual architecture.

The XDP Program contains the rules for matching Segment Routing Header (SRH) headers, a standard in the NASOR model for deploying network slices over the Internet. The packet processing is according to the *END* behavior of the Segment Routing technology. Step one is to parse the packet to identify whether it contains SRH fields. If the packet is not from a network slice, it will follow the standard forwarding flow of the Kernel. Step two, after parsing the SRH packet, the Hash Map is consulted, instructing which interface about forwarding behavior. If there is no forwarding rule installed on the Map, the packet will follow the kernel network stack flow. Step three consists of consulting the Forwarding Information Base (FIB) in the user space, built using the Internet routing algorithms. Finally, in step 4, the packet is forwarded to the appropriate interface.

The XCP Agent component is a user-space program responsible for updating the Kernel space map whenever a route is updated and notified by the XCP Controller. The XCP Controller component interacts with the NASOR through an API that can deploy network slices with low-latency requirements. Other applications or VNFs that do not have these connectivity requirements are deployed over the conventional stack.

The XDP Controller periodically checks the FIB built by the Internet routing algorithms. The XCP Agent updates a Hash Map that stores the most used matches. This method allows future incoming or passing packets to be processed early by an XDP program, avoiding processing by the network stack of Kernel.

The map update is performed by the XCP Agent, which runs as a daemon on virtual routers, Checks the FIB if there were any changes to the route. If there is a change, extract the IP-MAC component from the next-hop of SRH. It checks if the HASH Map contains this IP-MAC entry. If not, change the entry on the Map.

When using XDP-capable components as intermediate technology of a network slice, it is possible to establish logical connectivity at the core considering the low-latency requirement. The specification of this requirement is passed in the YAML configuration file. Our proof-of-concept current stage does not manage the capacity utilization of the interfaces.

4   Evaluation Method

Our evaluation method considers the performance baselines of the application verticals and performance KPIs [19] for network slicing. According to Ericsson [20], some performance baselines are desirable for some applications such as Autonomous vehicle control with a maximum latency of 5 ms, Factory cell automation below 1 ms, media on demand between 200 ms up to 5 s, High-speed Train 10 ms and others.

Regarding KPIs for network slicing, metrics including throughput, power consumption, and security are known [21]. In this paper, we consider KPI for network slices, as proposed in [22], the Integrity KPI that looks at end-to-end latency.

To evaluate the packet PT on each vRouter, we capture, using Wireshark, the timestamp, packets marks in the output queue and subtracts it from the packet timestamp input queue. In this way, we measure the time spent on processing. We certify that the machines do not exchange data related to any applications, so we set up Wireshark to use as reference time the arrival of the first packet in the entrance and egress queues. Besides, we defined packet capture accuracy as nanoseconds.

To evaluate the performance, stability, and scalability of network slicing considering early processing, we consider a partial factorial performance evaluation method. The method considers k factors with n_i levels for each factor i. We consider two factors for our experiment scenario: Slice Map Size and vRouter Flavor. The factors and their levels are summarized in Table 1.

Table 1. Performance evaluation setup.

Experiment	vRouter Flavor	Slice Map Size	vRouter Flavor Factor	Slice Map Size Factor
#1	2 vCPU 2 GB	Small (1 Slice)	-1	-1
#2	4 vCPU 4 GB	Small (1 Slice)	1	-1
#3	2 vCPU 2 GB	Large (100 k Slice)	-1	1
#4	4 vCPU 4 GB	Large (100 k Slice)	1	1

Using a regression model, we performed four experiments organized according to Table 1. The model for project 2^2 is given by equation: $y = q_0 + q_A x_A + q_B x_B + q_{AB} x_{AB}$. Replacing the four observations in the model, the values of q_0, q_A, q_B and q_{AB} are obtained according to the equations: $q_0 = \frac{1}{4} \times (y_1 + y_2 + y_3 + y_4)$, $q_A = \frac{1}{4} \times (-y_1 + y_2 - y_3 + y_4)$, $q_B = \frac{1}{4} \times (-y_1 - y_2 + y_3 + y_4)$, $q_{AB} = \frac{1}{4} \times (y_1 - y_2 - y_3 + y_4)$.

Thus, from the values q_0, q_A, q_B, q_{AB} it is possible to determine the Sum of the Squares (SS) of the factors based on the Sum of the Squares Total (SST) and the total variation of the response variables due to the influence and interaction between the factors. We ran each experiment of Table 1 30 times, consolidating an overall average for them.

5 Experimental Testbed

To validate the applicability of the NASOR integration with the XCP Controller, we propose a test environment on the topology, as shown in Fig. 3. Our testing environment disregarded the computing part of the Network Service Descriptor (NSD), which is usually handled by the OSM. A host machine with an Intel Core i7-7700HQ processor with 24 GB of RAM was used, hosting vRouters based on the Quagga Software Routing Suite [23], connected through an internal network.

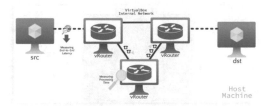

Fig. 3. Experimental testbed topology.

Besides, our test environment considered two VMs, source (*src*) and destination (*dst*) machines connected via vRouters running the Ping utility. After NASOR deployed a network slice considering this topology, we proposed some test cases to validate the proposed solution. We carried experiments to answer the following questions: (1) How much is the processing overhead of the packets considering the standard NASOR approach for segment routing (SREXT [24]) against XDP-based segment routing using REDIRECT and REDIRECT_MAP? (2) Which NASOR framework slice processing technology provides the best stability? (3) Considering the forwarding technique based on Maps, which factor is preponderant in the experienced *Latency (L)* and *Processing Time (PT)*?

6 Results and Discussion

To answer the guiding questions, we implemented network slices through the NASOR framework, as illustrated in Fig. 3. After starting the ping utility, 30 samples of 200 ICMPv6 encapsulated inside an SRH packet between *src* and *dst* were collected. After collected, we consolidated a sum of the average processing time of the packets considering the two approaches, the one based on the processing of SRH packets in the Kernel space and the XDP based approach.

As seen in the graph depicted in Fig. 4, the average *PT* of the packets using the SREXT approach was 490.133 µs. That is greater than the XDP considering REDIRECT_MAP routing, suggesting that REDIRECT_MAP provides less time overhead, therefore more suitable for network slices with low-latency requirements.

The processing time of REDIRECT was 490.418 µs, which in contrast to SREXT (482.525 µs) and considering a 95% confidence interval, are statistically equivalent. The graph depicted in Fig. 5 allows us to verify that the variability of processing times is less considering the REDIRECT_MAP technology than the SREXT and REDIRECT approaches in the XDP layer 75% of the measured processing times correspond to samples less than 469.523 µs.

As for the stability in the processing time, we found that, according to the graph in Fig. 5, the packets of the XDP-based network slices experienced a 21.211% less variability than the conventional NASOR approach, which is SREXT-based. That suggests that XDP-based slices may provide stability, according to Q3 and Q1 quartiles, in packet transport regarding *L* and *PT*.

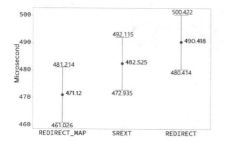

Fig. 4. Average *PT*

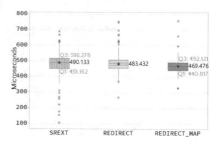

Fig. 5. Variability comparison.

This behavior was empirically expected because the processing of packets containing SRH headers in the SREXT approach occurs within the Kernel space. The packets travel through the Linux network stack, resulting in more significant overhead. On the other hand, considering the early processing that the XDP approach allows, there is less processing overhead. Among the reasons that justify this increase in performance of REDIRECT_MAP against REDIRECT include implementation enhancements and changing method of forwarding packets. Thus, the average processing time that the XDP approach with REDIRECT_MAP is 471.12 µs, and the SREXT is 482.52 µs, for a 95% confidence interval.

We validate that according to the experimental setup, the XDP REDIRECT_MAP forwarding approach exceeds the conventional REDIRECT approach through a third experiment. Hence, we found empirically that the REDIRECT_MAP approach spends 3.93% less processing time than REDIRECT.

In addition to providing better performance, the forwarding approach allows the addition of more sophisticated behaviors in the early processing of packets through maps. In this sense, considering this forwarding method, we proposed the XCP Controller integrated with the NASOR framework and installed rules for processing the forwarding for network slices with latency requirements.

To verify scalability and understand the reason for the stability of the approach to deploy XDP-based slices, we carried a performance evaluation to measure which aspects most influence the processing time and latency experienced by an application that runs on an XDP-based slice. For this, we consider the native approach of identification and processing of slice packets implanted by NASOR, SREXT approach, and the proposal in this paper, XDP-based.

After building a setup for performance evaluation, we plan and replicate experiments according to the methodology described in the Subsect. 4 section. We considered two response variables, PT and L, regarding the Flavor and Number of Network Slices, configured factors. We varied these factors on two levels according to the Partial Factor method. After carrying 30 experiments, we measure the final average according to Table 2.

Table 2. Performance evaluation: general averages of performance evaluation.

Experiment	Factors		Dependent variable	
	vRouter Flavor	Slice MAP Size	Latency (ms)	PT (seconds)
#1	−1	−1	0.5135	0.000471120
#2	1	−1	0.5013	0.000496615
#3	−1	1	1.4141	0.000469149
#4	1	1	0.5365	0.000162345

According to Table 3 it is possible to infer that the factor that empirically most influenced L is the Slice Map Size, with an influence of 36.24% on the response variable. The second factor that most influenced latency is the vRouter Flavor, with an influence of 32.76% on the response variable. The percentage variation between the two factors can be seen as significant since it is around 10.64%. It is also important to note that both factors combined represent a considerable percentage, with around 30.99%.

Table 3. Performance evaluation: Latency (L) and Processing Time (PT).

	Parameter				Sum of squares			
	q_0	q_A	q_B	q_{AB}	SS_A	SS_B	SS_{AB}	SST
Latency (L)	0.7413375	−0.2224425	0.2339625	−0.2163575	0.197922663	0.218953806	0.187242271	0.60411874
Influency (%)					**32.76%**	**36.24%**	**30.99%**	
Processing Time (PT)	0.000399807	−0.000070327	−0.000084060	−0.000083075	0.00000002	0.000000028	0.000000028	0.000000076
Influency (%)					**26.15%**	**37.36%**	**36.49%**	

Considering the PT, when analyzing the performance evaluation results, we found empirically, as available in Table 3, that the factor that most influenced the PT is also the Slice Map Size. It is worthy to point out that the interaction between the factors, vRouter Flavor and Slice Map Size represents the second-highest percentage of influence on the response variable, representing 36.49% of influence. Looking at the isolated vRouter Flavor factor, we infer that the variation of its levels influences PT by 26.15%.

Finally, it is possible to infer that the factor Slice Map Size isolated in contrast to the simultaneous interaction of the two factors is close to equidistant in the influence of PT, with a difference of only 2.38%. Besides, it is possible to note the stability due to its homogeneous influence of factors in the response variable, enabling network slicing in forwarding elements. Considering a scenario with about 100 K slices, it is possible to see the logical connectivity stability also showcasing scalability of our proposed method.

7   Concluding Remarks

We proposed the XDP Controller method that merges rules in the FIB of vRouters and adds them to the Kernel map to enable early packet processing. We showcased XCP Controller combined with NASOR provides scalability and stability in network slices deployment over early packet processing technology providing low-latency for vertical applications.

We carried a performance evaluation to understand which factors and levels most influence latency and processing time. In this way, we noticed factors such as the vRouter Flavor and Slice Map Size with homogeneous influence on the response variables, suggesting the stability of XDP-based slices despite the underlying technology.

As future work, we envision to compare the performance of slices XDP-based against approaches that use memory mapping as DPDK. Besides, we consider proposing and evaluating mechanisms for bandwidth reservation for XDP-based network slices.

Acknowledgements. This study was financed in part by the Coordenação de Aperfeiçoamento de Pessoal de Nível Superior - Brasil (CAPES) - Finance Code 001.

References

1. Afolabi, I., Taleb, T., Samdanis, K., Ksentini, A., Flinck, H.: Network slicing and softwarization: a survey on principles, enabling technologies, and solutions. IEEE Commun. Surv. Tutor. **20**(3), 2429–2453 (2018)
2. Kalør, A.E., Guillaume, R., Nielsen, J.J., Mueller, A., Popovski, P.: Network slicing in Industry 4.0 applications: abstraction methods and end-to-end analysis. IEEE Trans. Industr. Inf. **14**(12), 5419–5427 (2018)
3. Chowdhury, M.Z., Shahjalal, M., Ahmed, S., Jang, Y.M.: 6G wireless communication systems: applications, requirements, technologies, challenges, and research directions. IEEE Open J. Commun. Soc. **1**, 957–975 (2020)

4. Zhou, Y., Liu, L., Wang, L., Hui, N., Cui, X., Wu, J., Peng, Y., Qi, Y., Xing, C.: Service aware 6G: an intelligent and open network based on convergence of communication, computing and caching. Digit. Commun. Netw. **6**, 253–260 (2020)
5. Filsfils, C., Nainar, N.K., Pignataro, C., Cardona, J.C., Francois, P.: The segment routing architecture. In: 2015 IEEE Global Communications Conference (GLOBE-COM), pp. 1–6 (2015)
6. Khan, L.U., Yaqoob, I., Tran, N.H., Han, Z., Hong, C.S.: Network slicing: recent advances, taxonomy, requirements, and open research challenges. IEEE Access **8**, 36009–36028 (2020)
7. Markets and Markets: Network slicing market by component (solution and services (professional and managed)), end user (telecom operators and enterprises), application (manufacturing, government, automotive, media and entertainment), and region - global forecast to 2025
8. Zhang, S.: An overview of network slicing for 5G. IEEE Wirel. Commun. **26**(3), 111–117 (2019)
9. Cheng, X., Wu, Y., Min, G., Zomaya, A.Y., Fang, X.: Safeguard network slicing in 5G: a learning augmented optimization approach. IEEE J. Sel. Areas Commun. **38**(7), 1600–1613 (2020)
10. Taleb, T., Afolabi, I., Samdanis, K., Yousaf, F.Z.: On multi-domain network slicing orchestration architecture and federated resource control. IEEE Netw. **33**(5), 242–252 (2019)
11. Moreira, R., Rosa, P.F., Aguiar, R.L.A., de Oliveira Silva, F.: Enabling multi-domain and end-to-end slice orchestration for virtualization everything functions (VxFs). In: Barolli, L., Amato, F., Moscato, F., Enokido, T., Takizawa, M. (eds.) Advanced Information Networking and Applications, pp. 830–844. Springer, Cham (2020)
12. Van Tu, N., Yoo, J.H., Hong, J.W.K.: Measuring end-to-end packet processing time in service function chaining. In: 2020 16th International Conference on Network and Service Management (CNSM), pp. 1–9 (2020)
13. Ricart-Sanchez, R., Aleixo, A.C., Wang, Q., Alcaraz Calero, J.M.: Hardware-based network slicing for supporting smart grids self-healing over 5G networks. In: 2020 IEEE International Conference on Communications Workshops (ICC Workshops), pp. 1–6 (2020)
14. Zilberman, N., Audzevich, Y., Covington, G.A., Moore, A.W.: NetFPGA SUME: toward 100 Gbps as research commodity. IEEE Micro **34**(5), 32–41 (2014)
15. Van Tu, N., Yoo, J.H., Won-Ki Hong, J.: Accelerating virtual network functions with fast-slow path architecture using express data path. IEEE Trans. Netw. Serv. Manag. **17**(3), 1474–1486 (2020)
16. Xhonneux, M., Duchene, F., Bonaventure, O.: Leveraging eBPF for programmable network functions with IPv6 segment routing. In: Proceedings of the 14th International Conference on Emerging Networking EXperiments and Technologies, CoNEXT 2018, pp. 67–72. Association for Computing Machinery, New York (2018)
17. McCanne, S., Jacobson, V.: The BSD packet filter: a new architecture for user-level packet capture. In: USENIX Winter, vol. 46 (1993)
18. Høiland-Jørgensen, T., Brouer, J.D., Borkmann, D., Fastabend, J., Herbert, T., Ahern, D., Miller, D.: The express data path: fast programmable packet processing in the operating system kernel. In: Proceedings of the 14th International Conference on Emerging Networking EXperiments and Technologies, CoNEXT 2018, pp. 54–66. Association for Computing Machinery, New York (2018)
19. Kukliński, S., Tomaszewski, L.: Key performance indicators for 5G network slicing. In: 2019 IEEE Conference on Network Softwarization (NetSoft), pp. 464–471 (2019)

20. Ericsson: Ericsson white paper UEN 284 23-3251 rev B
21. Cunha, V.A., Corujo, D., Barraca, J.P., Aguiar, R.L.: MTD to set network slice security as a KPI. Internet Technol. Lett. **3**(6), e190 (2020). e190 ITL-20-0040.R1
22. 3GPP: Management and orchestration: 5G end to end key performance indicators (KPI) (2020)
23. Jakma, P., Lamparter, D.: Introduction to the quagga routing suite. IEEE Netw. **28**(2), 42–48 (2014)
24. Abdelsalam, A., Clad, F., Filsfils, C., Salsano, S., Siracusano, G., Veltri, L.: Implementation of virtual network function chaining through segment routing in a Linux-based NFV infrastructure. In: 2017 IEEE Conference on Network Softwarization (NetSoft), pp. 1–5 (2017)

A Monitoring Aware Strategy for 5G Core Slice Embedding

Oussama Soualah[1]([✉]), Omar Houidi[2], and Djamal Zeghlache[2]

[1] OS-Consulting, 79 Avenue François Mitterrand, Athis Mons, France
`oussama.soualah@os-c.fr`
[2] Telecom SudParis, Samovar-UMR 5157 CNRS, Institut Polytechnique de Paris,
Palaiseau, France
{`omar.houidi,djamal.zeghlache`}`@telecom-sudparis.eu`

Abstract. We address the question of how to deploy a monitored 5G Core network slice using virtual probes. We propose vProbe and vPacket-Broker (vPB) deployment schemes within the slice to meet the monitoring need of a network operator and to have full network visibility. The physical probes and packet brokers capture and monitor the traffic in the slices. In 5G and especially the 5G Core (5GC) network, network functions will be virtualized/containerized (VNFs/CNFs) and automatically provisioned in the 5GC slice. In this context, the deployment of vProbes and vPacket Brokers, should be also automated and simultaneously considered in 5GC slices instantiations and deployments. To realize this automated deployment, one of the challenges is how to capture the traffic between two given VNFs or CNFs without impacting network performance. We answer this question by describing potential architectures that fit such a virtual infrastructure and adapt and propose two embedding algorithms to optimize and automate the instantiation of a 5GC slice. Three different metrics are used to evaluate performance and compare with other existing schemes.

Keywords: 5G Core slice · Network traffic monitoring · vProbe · vPacket Broker · VNF-FG placement and chaining

1 Introduction

The increase in the number of applications, their diversification, and the improvement of the quality of mobile networks have led to an increase in demand, the emergence of new usages (connected objects, drones, etc.) and new users. 5G aims at addressing these new use cases and at responding better to this wide variety of needs and new demands through a unified technology that takes into account this diversity.

As stated by the Next Generation Mobile Networks (NGMN) alliance, 5G is an end-to-end ecosystem to enable a fully mobile and connected society. It empowers value creation towards customers and partners, through existing and

L. Barolli et al. (Eds.): AINA 2021, LNNS 226, pp. 727–744, 2021.
https://doi.org/10.1007/978-3-030-75075-6_60

emerging use cases, delivered with consistent experience, and enabled by sustainable business models.

To ensure a 5G service, the different network operators have the choice between (i) directly deploying an end-to-end 5G network a.k.a Standalone Architecture or (ii) start by deploying 5G new radio and rely on 4G Core network, a.k.a Non-Standalone Architecture. In the context of a 5G Core (5GC) network for a standalone architecture, the monitoring need remains mandatory while the network provisioning and deployment has drastically evolved. In fact, the evolution aims at replacing traditional legacy equipment by cloudified elements. The functionalities provided by the physical equipment should be logically transposed to the virtual environment.

We discuss and present the different architectures that can answer this need when dealing with a 5GC slice and highlight the related technical and operational limitations. Furthermore, we address the problem of how to embed (i.e., VNF placement and chaining problem) such kind of architectures within a given Network Function Virtualization Infrastructure (NFV-I). We provide a solution that enhances a given 5G Core slice by capturing and monitoring elements as well as embedding this new topology within the NFV-I. A simulation is used to evaluate the relevance and the performance of each architecture. We consequently provide some recommendations to help service providers in selecting a suitable architecture for their environments.

Our article is structured as follows. Section 2 of this paper presents related work. Section 3 describes traditional capturing and monitoring architecture in a network operator context while Sect. 4 introduces the considered capturing and monitoring solutions in a 5GC slice context and virtual environment. Section 5 presents the system model and the proposed algorithms. Section 6 reports the results of the performance evaluation of the different monitoring architectures based on the proposed algorithms.

2   Related Work

Over recent years, the integration of Network Function Virtualization (NFV) with other technologies, such as Software-Defined Networking (SDN), Cloud computing, and 5G [1,2] has attracted significant attention from both the academic research community and industry. The need to dynamically deploy virtualized network services on-demand, through VNF-FG or slice embedding, is identified as a core technology of 5G networks. As comprehensive surveys are already reported in [3] and [4], we only provide a short summary of VNF-FG embedding related works, as well as some works dealing with 5G slice network monitoring.

VNF-FG placement and chaining [5] was formulated and solved as an Integer Linear Programming (ILP) in [6–9]. The infrastructure providers can map the VNFs using multiple objectives: such as minimizing mapping costs [10], maximizing acceptance ratios and improving provider gains [11], and improving energy efficiency [12]. These methods are efficient in finding exact solutions but are subject to combinatorial explosion and do not scale polynomially with problem size.

There is an urgent demand for dynamic, elastic [13], and flexible [14] service chaining deployment model. Elastic orchestration of VNF is a key factor to achieve NFV goals. In fact, the resource dedicated or allocated to a VNF may have to scale due to dynamic traffic [15] variations. In order to overcome those challenges, some authors do address VNF scaling [16], migration [17], and extension [18] to ensure dynamic VNF chain placement for rising demand and traffic load.

Context-aware placement problems are also studied in [19] using VNF placement algorithms in virtual 5G network, with the goal of minimizing the path length and optimize the sessions' mobility. Few works address the joint monitoring and analytics for Service Assurance (SA) in 5G networks. Offering SA in network slicing implies full awareness and quick response to all frequent changes in topology, resources, function operation, service requests, and slice performance. It requires precise coordination between all network slicing components to facilitate efficient access, processing, and analysis of the information necessary to take the right actions. Service assurance is required to collaborate with orchestration and management to automate the 5G core slice provisioning process.

To address this open issue, authors of [20] study those new architectural components and propose a framework to realize network slicing SA enabled by NFV. The framework joins the two key components of SA, monitoring and analytics, in a hierarchical and distributed way. These components coordinate at various levels to optimize the tradeoff between monitoring cost and analytics accuracy.

In [21], authors propose a hierarchical, distributive, and modular SA architecture to assure the entire End-to-End (E2E) slice services, based on distributed modules assuring underlying services, functions, and resources. A closed-loop is also formed between assurance and orchestration to enable automation by cooperation between monitoring, analytics, and orchestration. Besides, the correlation across layers (infrastructure, network function, network service, and E2E slice) is considered to form a closed-loop through all layers to further improve efficiency.

3   Traditional Capturing and Monitoring Architecture in Telecommunication Networks

In a telecommunication operator context, probing is the process of acquiring user and control plane as well as signaling information in order to monitor the network. This kind of acquisition should be transparent to useful/user traffic. To ensure such probing without disturbing the network being monitored, some devices and functions should be introduced at minimum disruption to enhance and endow the initial topology. To do so we first describe the traditional, or legacy, capturing and monitoring architecture and then present potential architectures in virtualized environments where a 5G network slice is considered.

3.1   Traditional Architecture

Figure 1 depicts the traditional architecture of a captured and monitored traffic sent between a physical network function (PNF 1) and another function PNF 2. We highlight the introduction of an intermediate equipment such as a Test Access Point (TAP). The objective of this kind of equipment is to duplicate the transiting traffic between PNF 1 and PNF 2, and then to forward it to a Network Packet Broker (NPB). We briefly describe the main components in this architecture (Fig. 1) that enable the traffic monitoring.

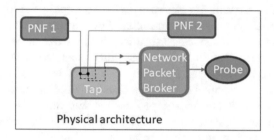

Fig. 1. Traditional capturing and monitoring architecture

3.2   Network Packet Broker

An NPB is a switch-like network device. An NPB may receive the traffic on one or more interfaces, perform some pre-defined function on that traffic, and output it to one or more interfaces according to the need. This is commonly referred to as "any to any", "many to any", and "any to many" port mapping. The functions that can be ensured by an NPB can be as simple as forwarding and/or dropping traffic and as complex as filtering on Layer 5+ information in order to identify a specific session. A packet broker is able to receive many flows and re-forward the traffic after applying some pre-defined functions to one equipment only, which can be a probe.

3.3   Probes

In the context of an operator network, the probes are mandatory to ensure the monitoring of the whole network by supervising the flows between its PNFs without any threat to the operation of the observed network. Traditionally the probes are bare-metal equipment that allow supervising and/or OSS teams to monitor the traffic in real time, to define some KPIs to prevent the network overload and bottlenecks. Besides, the probes can collect signaling information that enable the support teams to investigate and analyze problems detected in the network or raised by some clients. Furthermore, probes help operators to identify the nature/type of data traffic (4G for example). In this way, the operator can efficiently manage and size its Point of Presence (PoPs). These are some use cases of probes but each operator may customize the use of its probes to get extended visibility of its network.

3.4   Traffic Copy

In order to achieve a holistic view of all network traffic and fulfill visibility requirements, the network infrastructure must incorporate devices that copy the steering network traffic. To ensure the duplication of flows and the copy of traffic flows, taps and port **mirroring** can for example be used. **Tap** is a physical equipment that transmits sent and received data traffic simultaneously on separate dedicated channels. This ensures that all the data is forwarded to the target device (NPB in our case). **Port Mirroring** is realized by dedicated hardware that sends a copy of all network packets seen on one port to a target port. Switched Port Analyzer (SPAN) is an implementation of this technic. Tap and Port Mirroring present advantages and some weakness, but these shortcomings are out of the scope of this article and not discussed.

4   Capturing and Monitoring Architecture for 5G Core Slice

For the sake of clarity, we are representing a VNF as a VM to simplify the introduction of a monitoring architecture based on virtualization. But in the general case, a VNF is composed of multiple VNF Components (VNF-C). The main challenge in such environment is how to capture the traffic and send it to the vPacket-Broker and then to the vProbe in order to ensure the monitoring allowing a clear visibility of the network. That is why we consider this criterion (i.e., flow capturing mechanism) to distinguish between the different potential architectures in a virtual environment. Hereafter, we describe three architectures that present three different ways to handle the packets capturing.

4.1   Agent Based Architecture

In this architecture, each one of the VMs transiting a traffic that should be monitored, is enhanced by a light software called an agent that will re-send the traffic to the vPacket-Broker. We introduce Fig. 2 to explain and describe this architecture. The traffic in blue is the main one and each agent within its VM (for simplicity we make a one-to-one matching between a VM and VNF) is responsible to resend the traffic to the vPacket-Broker. This mirrored traffic is colored in orange. One of the major drawback and limitation of this architecture is that not all VNF providers/vendors are willing to have a third-party software plugged within their VMs that packages their proprietary systems. Besides, this architecture is generating the same amount of a given traffic. There is no optimization in the mirroring process. So the network can be rapidly overloaded by the mirrored traffic that will affect and degrade the performance of the network.

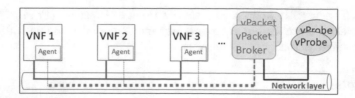

Fig. 2. Agent based architecture

4.2   Agent Free Architecture

vPort mirroring is a mechanism to replicate traffic supported by some Virtual Infrastructure Managers (VIMs) like Vmware (intuitively supported within the switch) and Openstack (Tap as a Service - TaaS module). Some SDN controllers are also proposing this kind of traffic copy. This mechanism is transparent to the VMs since it is the task of the VIM networking module to ensure port mirroring on the virtual switches and compute resources. Accordingly, it alleviates (offloads) the VMs from ensuring extra tasks rather than their main functionalities. Besides, it does not require any permission from the VNF provider to access its software and make some installation. We introduce Fig. 3 to describe the TaaS module.

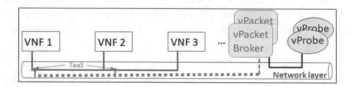

Fig. 3. Tap as a Service

4.3   vProbe Chaining (without vPB Scheme)

For this kind of architecture, note that a vPacket Broker is not used. In fact, a vProbe is inserted between each couple of VMs to monitor their exchanged traffic flows. This assumes that the traditional routines, performed by a Packet Broker, like filtering based on some criteria, will be ensured by the vProbe or the smartNICs. In this context, we are not interested in the aggregation capabilities of the vPB to be supported by the vProbe because there is no need to aggregate flows since we are considering only one flow. In this architecture, a new vProbe will be considered for each link to be monitored. The smartNICs are able to realize some filtering mechanisms so they can stop the reception of some packets and keep only the ones needed by the vProbe to be analyzed and monitored. In Fig. 4, we represent this vPB free architecture.

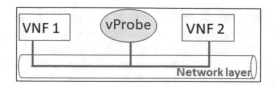

Fig. 4. Service chaining architecture

5 vProbe Aware Algorithm for 5GC Slice Embedding

5.1 NFV-I Model

The embedding of an enhanced 5GC slice by the capturing and monitoring services can be modeled as a VNF-FG placement and chaining problem within a physical infrastructure, the so-called NFV Infrastructure (NFV-I). In fact, the NFV-I represents a pool of virtualized servers thanks to a Virtual Infrastructure Management (VIM) software. Nowadays, modern servers include Non-Uniform Memory Access (NUMA) nodes that include a set of physical CPUs directly connected to a RAM. Two NUMA nodes within the same server are connected by a QuickPath Interconnect (QPI) bus. Hopefully, the scheduler of the current VIMs can distinguish between different NUMA nodes and consequently may specify and indicate in which NUMA node a VNF-C will be deployed. Without loss of generality, in Fig. 5, we describe a generic NFV-I. The nodes of this NFV-I are the NUMA nodes which match better to the current servers architecture. Nevertheless, they (NFV-I nodes) can be considered as traditional servers.

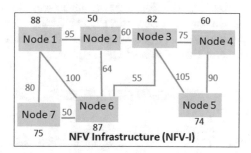

Fig. 5. NFV-I

5.2 Monitoring and VNF-FG Transformation

In Fig. 6, we introduce an example of a VNF-FG, representing a 5GC slice, with some monitoring needs/requirements. Indeed, a VNF is a logical element composed of a set (at least one) of VNF-Cs (VMs or Containers). In this figure, we assume that the network operator wants to monitor the link between the VNF-C2 and VNF-C3 as well as the link between VNF-C3 and VNF-C4. This monitoring need will impact the initial VNF-FG request in terms of required resources and virtual topology since some vProbes will be added as well as a potential vPacket Broker.

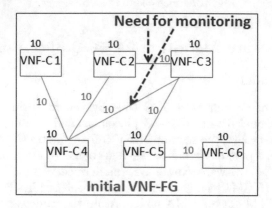

Fig. 6. VNF-FG and monitoring need

If the network operator opts for an **Agent Based Architecture (ABA)**, the initial VNF-FG can be updated accordingly as represented in Fig. 7. We can see that the initial topology is unchanged in links and/or nodes or network graph. In parallel, we notice a scaling-up of the related VNF-Cs. The scaling-up of VNF-Cs is justified by the additional resources needed by the plugged agent. The new added nodes are the vProbe and vPacket Broker. The new virtual links correspond to the connections between the related VNF-Cs and the vPB as well as the connection between the vPB and the vProbe.

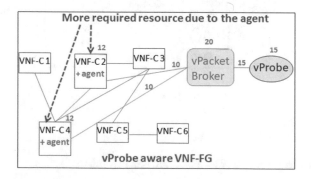

Fig. 7. VNF-FG and agent based architecture

In Fig. 8, we represent the **Mirroring Based Architecture (MBA)**. We model this by an added link between one of the two extremities of the monitored flow and the vPacket Broker. The vPacket Broker is automatically connected to the vProbe.

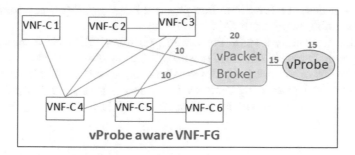

Fig. 8. VNF-FG and mirroring based architecture

Figure 9 describes the request model of the **Chaining Based Architecture (No-vPB)**. We notice that we introduce a vProbe between the two extremities of a monitored flow. Besides, we notice the absence of the vPacket Broker since each vProbe is capturing then handling the related flow.

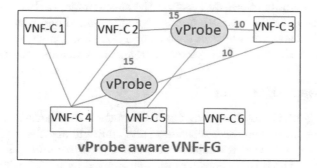

Fig. 9. VNF-FG and chaining based architecture

We highlight that the values over the vProbes and the vPacket Brokers, in Figs. 7 to 9, describe the required resources by these nodes. Besides, the values over the virtual links ending to a given vProbe and/or vPacket Broker is representing the required bandwidth for these new links.

5.3   Mathematical System Model

Figure 5 represents the substrate network, defined by the NFV Infrastructure (NFV-I). This NFV-I is modeled as an undirected weighted graph, denoted by

$G = (V(G), E(G))$ where $E(G)$ is the set of physical links and $V(G)$ is the set of physical nodes. Each physical node, $w \in V(G)$, is characterized by its (i) available resource denoted by $C(w)$ (representing the node remaining capacity in compute, memory or storage resources), and (ii) type $T(w)$: switch, server or Physical Network Function (PNF) [22]. A PNF is a dedicated hardware that implements a network function. The physical links $e \in E(G)$ are characterized by their available bandwidth $C(e)$.

The client requested VNF Forwarding Graph (VNF-FG) is modeled as a directed graph $D = (V(D), E(D))$ where $V(D)$ is the set of virtual nodes and $E(D)$ is the set of virtual links. Each virtual node, $v \in V(D)$, is characterized by its (i) required processing capacity $C(v)$ and (ii) its type $T(v)$: 5GC VNF, a vProbe or vPB. Each virtual link $d \in E(D)$ is described by its required bandwidth $C(d)$.

5.4   5GC Slice Embedding and Monitoring Problem

The VNF-FG embedding problem consists of placing a VNF within a host machine that fits the type constraint and having enough resources to serve the requirement of the related VNF. Note that in some cases it is preferable to share an already deployed VNF among different tenants and/or clients if some constraints can be satisfied. Once the VNFs of a given VNF-FG and/or 5GC slice are placed in hosted resources within the NFV-I, the service provider should ensure the chaining between every couple of VNFs that are neighbors in the VNF-FG request. This chaining allows and ensures the communication between the VNFs. A virtual link connecting two neighbors VNFs is mapped into a substrate path in the NFV-I. This selected physical path should have enough available bandwidth in order to meet the need of the virtual link and to guarantee the required QoS.

5.5   Proposed Algorithms

Based on the generic model already discussed and selected earlier in the previous section, we opt for our efficient placement and chaining algorithms [23] and [24]. We consequently describe the related ILP based strategy and Q-learning based algorithm. Since the initial client request represented by a 5GC slice is extended by vProbe(s) and/or vPB(s), these algorithms are especially relevant. The client request extended with the probes is handled and treated by our efficient algorithms to generate a suitable embedding.

The pseudo-code in Algorithm 1 summarizes the embedding strategy. We start by calling the $MonitorTrans$ procedure to make the VNF-FG transformation, based on the selected monitoring architecture, with the following 3 arguments. The first one D is the initial 5GC slice to be considered. The second is the set of links to be monitored $\{\mathcal{L}_\mathcal{M}\}$. The last argument is the selected monitoring architecture M_A. M_A can be the agent based architecture or mirroring based architecture or the vProbe chaining. The choice of M_A should be made by the service provider. Once the enhanced VNF-FG $\mathcal{D}_\mathcal{M}$ is generated we call our efficient R-ILP algorithm [23] in order to perform the placement and chaining

---

Algorithm 1: 5GC slice embedding and monitoring

---

1 Inputs: G, $\{\mathcal{L}_\mathcal{M}\}$, D, M_A
2 Output: $\mathcal{M}$
3 $\mathcal{D}_\mathcal{M} \leftarrow MonitorTrans(D, \{\mathcal{L}_\mathcal{M}\}, M_A)$
4 $\mathcal{M} \leftarrow$ R-ILP$(G, \mathcal{D}_\mathcal{M})$
5 return $\mathcal{M}$

---

step. It should be highlighted that any other embedding algorithm can be used without any limitation once the $\mathcal{D}_\mathcal{M}$ is already generated.

In the pseudo-code Algorithm 2, we summarize our Enhanced Q-Learning-based approach (EQL) for VNF-FG embedding to improve long term performance and reward using reinforcement learning combined with an expert knowledge system. The Q-learning algorithm, shown in Eq. (1), has a function that calculates the quality of a state-action combination:

$$
Q^{new}(s_t, a_t) \leftarrow (1-\alpha) \cdot \underbrace{Q(s_t, a_t)}_{\text{old value}}
$$

$$
+ \underbrace{\alpha}_{\text{learning rate}} \cdot \underbrace{\left(\underbrace{r_t}_{\text{reward}} + \underbrace{\gamma}_{\text{discount factor}} \cdot \underbrace{\max_a Q(s_{t+1}, a)}_{\text{estimate of optimal future value}} \right)}_{\text{learned value}}
\qquad (1)
$$

In Eq. (1), Q is a matrix that stores the recommended values of the executable actions in the current state. Depending on these values, the agent can decide which action to take next. In the Q matrix, t refers to the current time unit, $t + 1$: is the future time unit, max refers to the maximum value, s_t refers to the current state, which includes the resource requirements of VNFs and the remaining resources of the physical nodes, a_t to the action, s_{t+1} to the future state, and r_t is the reward value, which comes from the reward matrix R (r_t is the reward received when moving from the state s_t to the state s_{t+1}). The Q matrix is updated with reward r_t.

In the reward matrix R, we set the value of an element equal to the residual capacity in each physical node (increasing or decreasing the reward r_t value according to the conditions, based on CPU, of each physical node). Through this, the load of physical nodes will be intelligently distributed, so that each substrate node achieves optimal performance through the efficient use of resources.

The proposed EQL algorithm operates in 5 steps: (1) Compute the Q matrix using Eq. (1) and learn the rules during the training process; (2) Find a set of physical candidate nodes C related to a state s which are produced by the method *getPolicyFromState*: that decide to move to the state that has the maximum Q value based on all possible actions; and the method *getExpertKnowledgeCandidates*: used to select in priority the physical nodes with highest available residual capacity to favor load balancing. This method

---

Algorithm 2: EQL algorithm pseudo-code

---

1 Inputs: G, D, $maxTrainCycles$, $maxCand$
2 Output: SolutionMapping $SolMap$
3 initialize the Q matrix with all zero elements
4 initialize the R matrix with the residual capacity in each physical node
5 $queue \leftarrow \emptyset$
6 A set of candidate hosts $\mathcal{C} \leftarrow \emptyset$
7 **for** $n = 0$ to $maxTrainCycles$ **do**
8   Compute Q matrix using Eq. (1)
9   **if** $the\ Q\ matrix\ has\ basically\ converged$ **then**
10   Break; //return the Q matrix that can be used

11 **foreach** $virtual\ link\ d \in E\,(D)$ **do**
12     $s \leftarrow$ SelectInitialState(to host vnf)
13   **repeat**
14       $\mathcal{C} \leftarrow \mathcal{C} \cup$ getPolicyFromState(s) $\cup$ getExpertKnowledgeCandidates(s)
15   **until** $size(\mathcal{C})\ =\ maxCand$;
16     $stack \leftarrow$ Sort($\mathcal{C}$)
17   **while** $stack \neq \emptyset\ and\ solution\ =\ False$ **do**
18       $k \leftarrow$ Pop($stack$)
19   **if** $ComputeLinkMapping(s,\ k) = True\ and\ CheckLinks(s,\ k) = True$ **then**
20          $solution \leftarrow$ True
21          $SolMap \leftarrow SolMap +$ GetPathSolution($s,\ k$)
22   Update capacities on G
23   Update R matrix

24 return $SolMap$

---

accelerates the training process, and also, avoids random solutions at the beginning of the training process; (3) Compute the shortest path between s and candidate k using Dijkstra's algorithm based on available bandwidth; (4) Check link mapping: if links capacities are respected, confirm candidate k as a mapping solution for the VNF and add the solution path from s to k to the SolutionMapping $SolMap$ built so far; (5) Update capacities on G, update R matrix (that will impact and update the Q matrix based on network conditions);

Based on these two algorithms, we evaluate the relevance of the different monitoring architectures. In the next section, we present the associated performance study.

6   Performance Evaluation

This section focuses on assessing and evaluating the performance of the described monitoring architectures based on our efficient placement and chaining algorithms: R-ILP and EQL. Note that **our main focus in this paper is to qualify the different monitoring architectures and not the embedding**

algorithms. We first describe our simulation environment and then define our performance metrics used to evaluate the different architectures. Finally, we report and discuss the obtained results.

6.1 Simulation Environment

The simulations are performed on a virtual machine with 2.50 GHz Quad vCore and 6 GBytes of available vRAM. The VNF-FG requests are generated using a Poisson process with an average arrival rate of requests per 100 time units. The lifetime of each request follows an exponential distribution with a mean of 500 time units. We consider **a realistic network topology** to compare the performance of the different schemes and architectures. The **large Austrian core topology (TA2)** taken from [25], by Telekom Austria, is used for the performance assessment since it corresponds to the largest publicly available network topology in SNDlib [25] (with 65 nodes). The capacity of physical nodes and links are generated randomly in the [100, 150] interval.

For this realistic topology evaluation, 1000 VNF-FG requests are generated for each architecture scenario. The size of the initial VNF-FG requests is arbitrarily set to 6 nodes. The number of links we want to monitor is equal to 4. The ILP solver used in our experiments is Cplex [26]. We use our system simulator accepting both the NFV-I topologies (with their available resources for hosting) and the VNF-FG requests (with their resource requirements) as inputs to the ILP algorithm that search for a placement solution provided as an output of the Cplex solver integrated in a java-based simulator.

6.2 Performance Metrics

Acceptance Ratio: is the percentage of VNF-FG requests that have been successfully embedded within the NFV-I. Rejecting a request is justified by the unavailability of physical resources.

Acceptance Revenue: is the service provider realized (generated) revenue (benefits) when accepting VNF-FG client requests. This metric helps the network operator to make decisions based on the earned revenue. The acceptance revenue is formally expressed as:

$$\mathbb{R}(t) = \sum_{\mathcal{D} \in \mathcal{AR}_t} \mathbb{R}(\mathcal{D}) \tag{2}$$

where $\mathbb{R}(\mathcal{D})$ is the $\mathcal{D}$ request gain as it was defined in [27] and expressed as:

$$\mathbb{R}(\mathcal{D}) = \sum_{v \in V(\mathcal{D})} (C(v) \cdot U_c) + \sum_{d \in E(\mathcal{D})} (C(d) \cdot U_b) \tag{3}$$

where U_c is the gain from allocating one unit of CPU resources and U_b is the earned revenue from the allocation of one unit of bandwidth. Note that $\mathbb{R}$ is the total revenue taking into consideration all the incoming requests.

Physical Resource Usage: this indicator provides a comparison of the efficiency of available resource consumption. It gives an idea about a potential over-provisioning for some architectures. This metric reflects the possible "over-consumption" of the physical resources and gives an idea about the efficiency of the different architectures to deal with NFV-I resources.

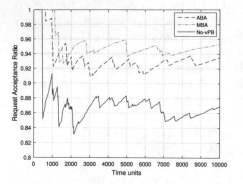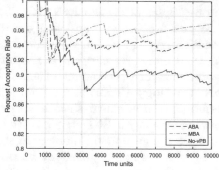

Fig. 10. Acceptance ratio of the R-ILP algorithm

Fig. 11. Acceptance ratio of the EQL algorithm

6.3   Evaluation Results

In Fig. 10, we present the acceptance ratio of the different architectures based on R-ILP algorithm. The No-vPB performs poorly compared with the two other architectures. In fact, the agent based architecture will not be adopted because of the need to add a third-party agent (external Software module) to the VNF.

In Fig. 11, we consider a second time the acceptance ratio metric but using our EQL algorithm in order to verify the previous results. Fortunately, the good results of MBA are also confirmed by the EQL algorithm since the architecture outperforms the other results. No-vPB architecture is still performing very poorly compared with the other architectures. We recall, this architecture may be adopted by a service provider with fewer constraints and requirements comparing with ABA. So the challenge for a service provider is to evaluate internally, based on his local environment, whether it is more suitable to adopt the ABA or MBA strategy. Obviously, the choice may be easily made if we consider only acceptance ratio and revenue metrics. Since the EQL algorithm outperforms the ILP based solution by accepting more requests, we will continue the rest of the performance evaluation with the best algorithm which is EQL.

In Fig. 12, we present the earned revenue from the acceptance of the incoming requests for the three architectures based on EQL. As expected from the results of Fig. 11, MBA achieves the best gain. This result is very motivating and important for service providers since the architecture (MBA) that represents less operational constraints leads to the best economical outcome.

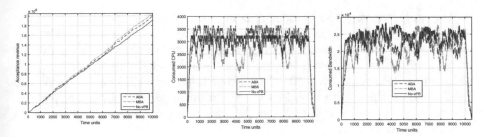

Fig. 12. Acceptance revenue of the EQL algorithm

Fig. 13. CPU usage for EQL algorithm

Fig. 14. Bandwidth usage for EQL algorithm

In Figs. 13 and 14, we give an idea about the status of the NFV-I when dealing with incoming requests for the three architectures. In order to analyze adequately the results plotted in these two Figs. 13 and 14, we have to keep in mind that MBA accepts more client requests comparing with the two other architectures. Indeed, accepting more clients is roughly equivalent to consuming more physical resources (CPU and bandwidth) in order to serve the accepted requests. The good and surprising result deduced from these two figures is that MBA is not always consuming more physical resources (CPU and bandwidth) comparing with the two other architectures even it is accepting more clients than the ABA and No-vPB architectures. This can be justified by the mirroring technics that ensure the traffic capturing without overloading the NFV-I resources. Besides, MBA does not require any scaling out/up of VNF-Cs like ABA. These results motivate the adoption of MBA instead of the two other competitor architectures. In Fig. 14, we notice that No-vPB architecture is consuming more bandwidth resources comparing with the two other architectures. This can be explained by the extra links introduced by integrating several instances of vProbes.

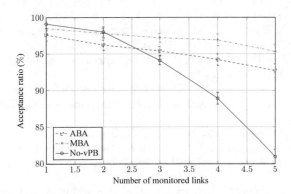

Fig. 15. Acceptance percentage w.r.t. monitored links variation

In Fig. 15, we vary the number of links we want to monitor (e.g., two links are monitored in Fig. 6), and we evaluate the acceptance rate metric. The performance of all architectures is roughly the same when the number of monitoring link is equal to one. However, the performance of No-vPB degrades when monitoring more links (e.g., five): 80.9% for No-vPB versus 95.3% for MBA. We can conclude from this figure that MBA is the most suitable architecture if scalability is required which is the case in practice where network/service providers monitor almost all links.

7 Conclusion

In this article, we discussed and presented different capturing and monitoring architectures when dealing with 5G Core slice embedding. Besides exposing the potential architectures, we adapt and propose two efficient algorithms: (i) R-ILP and (ii) Q-learning based approaches in order to address an end-to-end solution. The different monitoring architectures were evaluated based on three performance metrics: (i) acceptance ratio of client requests, (ii) earned revenue, and (iii) physical resources (CPU and bandwidth) usage rate. The obtained results show that the Mirroring Based Architecture (MBA) is the most suitable way to address the monitoring and capturing problem in the context of a 5G Core slice while respecting operational requirements.

References

1. Abdelwahab, S., Hamdaoui, B., Guizani, M., Znati, T.: Network function virtualization in 5G. IEEE Commun. Mag. **54**(4), 84–91 (2016)
2. Ordonez-Lucena, J., Ameigeiras, P., López, D., Ramos-Muñoz, J.J., Lorca, J., Folgueira, J.: Network slicing for 5G with SDN/NFV: concepts, architectures, and challenges. IEEE Commun. Mag. **55**(5), 80–87 (2017)
3. Yi, B., Wang, X., Li, K., Das, S.K., Huang, M.: A comprehensive survey of network function virtualization. Comput. Netw. **133**, 212–262 (2018)
4. Gil-Herrera, J., Botero, J.F.: Resource allocation in NFV: a comprehensive survey. IEEE Trans. Netw. Serv. Manag. **13**(3), 518–532 (2016)
5. Houidi, O.: Algorithms for Virtual Network Functions chaining. Ph.D. thesis, Institut polytechnique de Paris (2020)
6. Moens, H., De Turck, F.: VNF-P: a model for efficient placement of virtualized network functions. In: Network and Service Management (CNSM), pp. 418–423 (2014)
7. Gember, A., Krishnamurthy, A., John, S.S., Grandl, R., Gao, X., Anand, A., Benson, T., Akella, A., Sekar, V.: Stratos: a network-aware orchestration layer for middleboxes in the cloud. CoRR abs/1305.0209 (2013). http://arxiv.org/abs/1305.0209
8. Mehraghdam, S., Keller, M., Karl, H.: Specifying and placing chains of virtual network functions. In: 2014 IEEE 3rd International Conference on Cloud Networking (CloudNet), pp. 7–13 (2014)
9. Bari, M.F., Chowdhury, S.R., Ahmed, R., Boutaba, R.: On orchestrating virtual network functions in NFV. CoRR abs/1503.06377 (2015)

10. Papagianni, C.A., Leivadeas, A., Papavassiliou, S., Maglaris, V., Cervello-Pastor, C., Monje, Á.: On the optimal allocation of virtual resources in cloud computing networks. IEEE Trans. Comput. **62**(6), 1060–1071 (2013)
11. Soualah, O., Mechtri, M., Ghribi, C., Zeghlache, D.: An efficient algorithm for virtual network function placement and chaining. In: 14th IEEE Annual Consumer Communications and Networking Conference, CCNC 2017, Las Vegas, NV, USA, 8–11 January 2017, pp. 647–652 (2017)
12. Soualah, O., Mechtri, M., Ghribi, C., Zeghlache, D.: Energy efficient algorithm for VNF placement and chaining. In: Proceedings of the 17th IEEE/ACM International Symposium on Cluster, Cloud and Grid Computing, CCGRID 2017, Madrid, Spain, 14–17 May 2017, pp. 579–588 (2017)
13. Ghaznavi, M., Khan, A., Shahriar, N., Alsubhi, K., Ahmed, R., Boutaba, R.: Elastic virtual network function placement. In: 4th IEEE International Conference on Cloud Networking, CloudNet 2015, Niagara Falls, ON, Canada, 5–7 October 2015, pp. 255–260 (2015)
14. He, M., Alba, A.M., Basta, A., Blenk, A., Kellerer, W.: Flexibility in softwarized networks: classifications and research challenges. IEEE Commun. Surv. Tutor. **21**(3), 2600–2636 (2019)
15. Xie, Y., Liu, Z., Wang, S., Wang, Y.: Service function chaining resource allocation: a survey. CoRR abs/1608.00095 (2016)
16. Houidi, O., Soualah, O., Louati, W., Mechtri, M., Zeghlache, D., Kamoun, F.: An efficient algorithm for virtual network function scaling. In: 2017 IEEE Global Communications Conference, GLOBECOM 2017, Singapore, 4–8 December 2017, pp. 1–7 (2017)
17. Xia, J., Cai, Z., Xu, M.: Optimized virtual network functions migration for NFV. In: 22nd IEEE International Conference on Parallel and Distributed Systems, ICPADS 2016, Wuhan, China, December 13-16, 2016. pp. 340–346 (2016)
18. Houidi, O., Soualah, O., Louati, W., Zeghlache, D.: Dynamic VNF forwarding graph extension algorithms. IEEE Trans. Netw. Serv. Manag. **17**(3), 1389–1402 (2020)
19. Taleb, T., Bagaa, M., Ksentini, A.: User mobility-aware virtual network function placement for virtual 5G network infrastructure. In: 2015 IEEE International Conference on Communications, ICC 2015, London, United Kingdom, 8–12 June 2015, pp. 3879–3884 (2015)
20. Xie, M., Zhang, Q., Grønsund, P., Palacharla, P., Gonzalez, A.J.: Joint monitoring and analytics for service assurance of network slicing. In: IEEE Conference on Network Function Virtualization and Software Defined Networks, NFV-SDN 2018, Verona, Italy, 27–29 November 2018, pp. 1–6 (2018)
21. Xie, M., Zhang, Q., Gonzalez, A.J., Grønsund, P., Palacharla, P., Ikeuchi, T.: Service assurance in 5G networks: a study of joint monitoring and analytics. In: 30th IEEE Annual International Symposium on Personal, Indoor and Mobile Radio Communications, PIMRC 2019, Istanbul, Turkey, 8–11 September 2019, pp. 1–7 (2019)
22. ETSI GS NFV 003: Network Functions Virtualisation (NFV); Terminology for Main Concepts in NFV (2014)
23. Soualah, O., Mechtri, M., Ghribi, C., Zeghlache, D.: Online and batch algorithms for VNFs placement and chaining. Comput. Netw. **158**, 98–113 (2019)

24. Houidi, O., Soualah, O., Louati, W., Zeghlache, D.: An enhanced reinforcement learning approach for dynamic placement of virtual network functions. In: 31st IEEE Annual International Symposium on Personal, Indoor and Mobile Radio Communications, PIMRC 2020, London, United Kingdom, 31 August–3 September 2020, pp. 1–7 (2020)
25. Orlowski, S., Wessäly, R., Pióro, M., Tomaszewski, A.: SNDlib 1.0 - survivable network design library. Networks **55**(3), 276–286 (2010)
26. IBM ILOG CPLEX Optimization Studio. https://www-01.ibm.com/software/commerce/optimization/cplex-optimizer/
27. Zhu, Y., Ammar, M.: Algorithms for assigning substrate network resources to virtual network components. In: IEEE INFOCOM, pp. 1–12 (2006)

Performance Analysis of POX and OpenDayLight Controllers Based on QoS Parameters

Houda Hassen[✉], Soumaya Meherzi, and Safya Belghith

RISC Laboratory, National Engineering School of Tunis,
University of Tunis El Manar, Tunis, Tunisia
{houda.hassen,soumaya.meherzi,safya.belghith}@enit.utm.tn

Abstract. Software-defined networking (SDN) is an emerging network architecture that has been recently proposed to address the limitations of traditional networks. SDN consists in decoupling the forwarding plane and the control plane and moving this latter to a remote controller, which manages traffic routing rules in a centralized and dynamic fashion. Thus, the controller plays a crucial role in network management operations. A large diversity of open-source controllers, with different features, has been proposed for both industrial and research purposes. Extensive efforts are being deployed to provide a comparative study between available open-source controllers through performance evaluation of underlying SDN networks. In this paper, we provide an overarching performance evaluation of two popular SDN controllers namely, POX and OpenDayLight. Based on Mininet emulator, we assess the most relevant QoS parameters between end nodes in SDN networks namely, throughput, round trip time, jitter and packet loss. Moreover, we compare the behavior of the considered controllers within different network topologies such as, single, tree, linear and custom topologies.

Keywords: Software defined network · Controller · OpenFlow · OpenDayLight · POX · QoS

1 Introduction

Network Function virtualization (NFV) and software defined networking (SDN) are the two most prominent key enablers for next generation networks to overcome the limitations of traditional networks. Indeed, such technologies provide a high level of flexibility, programmability and scalability to network-management and -orchestration operations in a cost-effective way. From one hand, NFV performs the virtualization of network resources, functions and services, which offers an abstraction representation of the network components. From the other hand, SDN, which is based on the separation of the control plane and the forwarding plane, centralizes network management operations in a software-based controller.

L. Barolli et al. (Eds.): AINA 2021, LNNS 226, pp. 745–756, 2021.
https://doi.org/10.1007/978-3-030-75075-6_61

Based on the network state, the controller determines the optimal routing rules for data flows according to specific application requirements.

SDN is a three-layered architecture: application plane, control plane and data plane. Application or management plane is composed of a set of business- and service-oriented applications such as, traffic monitoring, load balancing, security, etc. Based on an abstraction representation of the network and instantaneous information about its state and topology, the application layer makes appropriate decisions about routing policies to be forwarded to the control plane. This latter applies the so dictated policies by programming the paths to be followed by data flows. The data or infrastructure plane is composed of switching nodes which simply forward incoming traffic flows according to the controller's instructions. Figure 1 illustrates a simplified scheme of SDN architecture.

The controller communicates with top and bottom layers by means of multiple Application Program Interfaces (APIs) according to specific protocols. For the northbound interface (NBI), i.e. between control and application layers, APIs are configured by protocols such as, REST, RPC, etc. For the southbound interface communications, i.e. between the controller and the forwarding devices, the most widely used protocol is OpenFlow.

Therefore, the controller plays a crucial role in SDN-based networks, which have been basically conceived to manage the network resources in order to meet a wide spectrum of QoS requirements stemming from multi-tenant applications. A large variety of open-source SDN controllers are available, for instance, POX [1], RYU [2], OpenDayLight (ODL) [3], Floodlight [4], Open Networking Operating System (ONOS) [5], etc. It has been shown in recent studies, that such controllers have not only different features and architectures, but also different performance behaviors in underlying SDN networks.

While interesting, most of these contributions however focused on few Quality of Service (QoS) parameters namely, latency and throughput/bandwidth. Based on Mininet [6] emulation platform, we propose in this work a comparative study between POX and ODL controllers, using more relevant parameters namely, throughput, round trip time, jitter and packet loss. We address, as well, the effect of network topology on performance behavior by considering different topology types such as, single, linear, tree and custom.

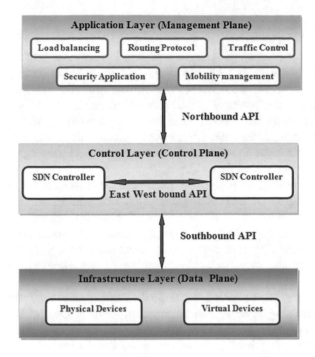

Fig. 1. SDN architecture

This paper is organized as follows. Section 2 presents a thorough overview of literature. In Sect. 3 we describe methodology and implementation tools that have been used in this paper. Then, we illustrate and discuss simulation results in Sect. 4. Last, some concluding remarks are drawn in Sect. 5.

Table 1. Features of POX and OpenDayLight controller [7,8]

Feature	POX	OpenDayLight
Programming language	Python	Java
Architecture	Centralized	Distributed
Southbound API	OpenFlow 1.0	OpenFlow 1.0,1.3
NorthBound API	REST API	REST API
EastWestbound API	–	Akka, Raft
Multithreading	No	Yes
Interface	CLI, GUI	CLI, Web UI
Application domain	Campus	Data center

2   Related Works

In paper [9], Badotra et al. provided a comparative evaluation of two most powerful and popularly known SDN controllers, ODL and ONOS. Using Mininet emulator, authors have evaluated burst rate, throughput, Round Trip Time (RTT) and bandwidth, as performance parameters within a tree-topology type network. TCP and UDP flows have been considered for bandwidth measurement using iperf and real-time packets have been captured and analyzed using Wireshark. Simulation results have shown that ODL controller performs better than ONOS for all considered criteria (Table 1).

In [10], Arahunashi et al. carried out a performance analysis of various SDN controllers like Ryu, ODL, Floodlight and ONOS, using two parameters, average delay and throughput. Authors addressed as well, the topology type and the complexity of the network in terms of Hosts' number. It has been concluded that Floodlight achieved the best performance levels in terms of delay and throughput. Hence, it is the most suitable controller for applications with such QoS requirements.

In [11] Islam et al. provided a thorough performance analysis of RYU controller, by addressing several QoS parameters such as, bandwidth, throughput, round-trip time, jitter and packet loss. Authors have used Mininet emulator with *iperf3* and *ping* tools. They have considered only single topology with three nodes. It has been concluded that RYU is an excellent choice for small businesses and research applications, for Python programmers.

In [12] Bholebawa et al. mainly focused on two controllers, POX and Floodlight. Based on Mininet emulator, a comparative study has been performed over different network topology types namely, single, linear, tree and custom. According to throughput and round-trip delay, as performance criteria, Floodlight performed better than POX for all topologies.

In paper [13] Eljack et al. focused on two popular SDN controllers, Floodlight and ONOS. A performance comparison has been carried out by addressing TCP and UDP traffic flows over different network topologies (single, linear and tree). The analysis was based on some metrics such as transfer, Delay, bandwidth and jitter. The ONOS has shown better performance than Floodlight in both TCP and UDP traffic in network.

In [14] Pushpa et al. discussed some of the SDN Controllers like POX, NOX, Floodlight, ODL and Open Contrail with their features. Authors have addressed centralized and distributed controllers' comparison through Floodlight and ODL controllers. Analysis has been performed for varying topology and according to throughput, rate of data transfer and RTT, as performance criteria. Simulation results have shown that ODL performed better than Floodlight in single topology for varying number of nodes but failed in linear and tree topology as structure enlarged. Floodlight outperformed both in linear and tree topology. Authors suggested centralized controller rather than distributed one for complex structure.

In [15] Mamushiane et al. evaluated and compared the performance of some SDN controllers (Ryu, Floodlight, ONOS and ODL) based on throughput and latency. They used an open source benchmarking tool Cbench with two operational modes, latency mode and throughput mode. Authors recommended the adoption of ODL. While ONOS exhibited the best throughput results, Ryu achieved the best latency levels.

In paper [16] Priyadarsini et al. presented the study of number of parameters that affect the performance of SDN controllers. They also stated the working procedure of various controllers like NOX, NOX-MT, POX, Rosemary, Floodlight, ODL, Beacon and Maestro. They evaluated the performance of NOX, POX and Beacon controller in terms of throughput, latency, jitter using *iperf* and *Cbench* tools. While POX has been shown to outperform NOX, Beacon provided higher response time and throughput than NOX and POX. According to authors, NOX and POX worked efficiently for homogeneous traffic while, for heterogeneous traffic, Beacon has been shown to be the best choice.

In paper [17] Li et al. compared the performance of RYU and Floodlight controllers under different topologies namely single, minimal, linear, tree, reversed and custom using Mininet emulator. The comparison was based on two metrics such as bandwidth and delay. Results show that Floodlight outperforms RYU in terms of delay and bandwidth in all topologies.

While interesting, the aforementioned works have focused on few metrics for performance evaluation of some selected SDN controllers. In addition, the effect of network topology on performance behavior of SDN controllers has not been deeply addressed. In this paper, we propose an overarching analysis by considering several performance parameters according to QoS criteria namely, round-trip time, throughput, jitter and packet loss. Moreover, we consider different network topology types such as, single, tree, linear and custom. We have selected POX and OpenDayLight, as the most two popular SDN controllers having different features, to carry out performance analysis using Mininet emulator.

3   Methodology

In this section, we present the methodology, tools and steps we have used to conduct our performance analysis. We have chosen to evaluate two different SDN controllers, POX and OpenDayLight, as two well-known examples of OpenFlow controllers. While POX is a centralized controller which is written in Python, OpenDaylight is a distributed one which is based on Java language. We have used Mininet, as a network emulator, to implement and test the controllers. Mininet is a popular open-source tool which is widely used to emulate SDN-based networks. Using Mininet, we can create networks of virtual hosts, switches, links and SDN controllers. There are some predefined topologies in Mininet such as minimal, single, reversed, linear and tree. We can also create some other custom topologies using Python code. In this work, four topologies are considered: single, linear, tree and custom topology.

- **Single topology:** It is a simple topology with one OpenFlow-enabled switch and a given number of hosts. We have considered 16 hosts, which are connected with the remote controller via one switch.

- **Linear topology:** In this topology there is as many switches as there are hosts. We have chosen 16 hosts, each of which is connected to the remote controller through one switch.
- **Tree topology:** It is a topology where all switches and hosts are linked to each other in a hierarchical mode. This topology has two parameters: depth and fanout. Depth is the number of hierarchical level of switches and fanout is the number of connections (hosts) at each level. We have considered $depth = 2$ and $fanout = 4$. For the same number of hosts as the previous topologies, we have chosen 4 switches that were connected the remote controller through one other switch.
- **Custom topology:** We have built a custom network topology with 16 hosts and 4 switches using Python script. A network was designed so that each switch would have an equal number of hosts connected (i.e. 4 hosts for each switch) and switches would be connected to each other and to the remote controller. The description of the created custom topology connected to ODL controller is shown in Fig. 2.

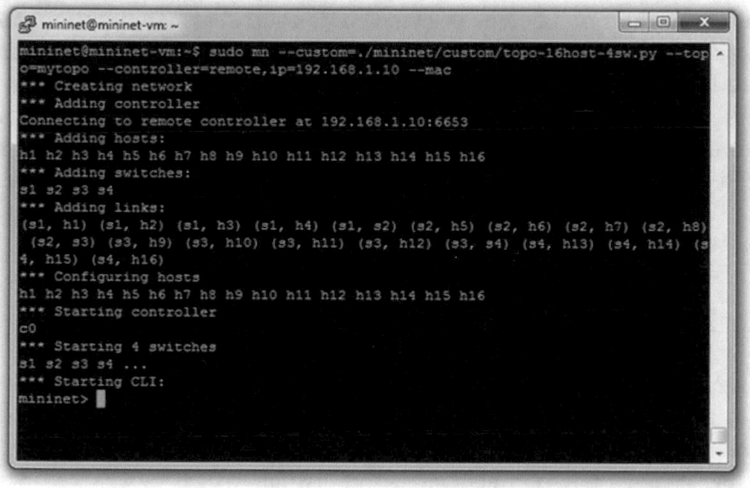

Fig. 2. Custom topology in Mininet VM

Firstly, the POX controller was made active on remote IP address (127.0.0.1) and it was programmed as $l3$ learning switch. In Mininet command-line interface (CLI), to create topology, for example a custom one, and connect it to POX controller we write the following command lines:

$sudo mn −custom=./mininet/custom/topo-16host-4sw.py −mac −topo= mytopo −controller=remote,ip=127.0.0.1

In a second time, OpenDaylight controller was made active on remote IP address (192.168.1.10). Also, we used the same command line to create a custom topology but we just need to change the remote IP address with the IP address of ODL. For the other topologies, we used the same command line with specification of parameters of the topology that we want to create and the remote IP address of each controller. For all tests, the number of end hosts for all topologies was fixed at 16 hosts.

To compare the performance of POX and ODL controllers, different metrics are considered such as throughput, round-trip time (RTT), jitter and packet loss. This comparison is realized by using of ping and iperf commands between two end hosts for each topology. Delay in terms of average RTT was calculated using *ping* command. *Iperf* command has been used in TCP mode to calculate throughput, and in UDP mode to estimate jitter and packet loss. Mininet, *ping*, and *iperf* tools run on the same machine.

The machine used to perform the experiments is an Intel ® Core™ $i3 - 4005U$ M CPU @1.70GHz and 4 GB RAM, running the Operating System Windows 7 $(64 - bit)$ and VirtualBox workstation. In this machine, we had installed Mininet VM based on Ubuntu server version 14.04. POX was automatically installed in Mininet VM. The ODL controller was installed in another virtual machine based on Ubuntu server version 16.04 with VirtualBox.

4   Simulation Results and Discussion

We present and discuss in this section the simulation results we have obtained. For each criterion, we compare performance behavior of POX and ODL for all topologies.

4.1   Round-Trip Time

In the first test, RTT parameter has been evaluated using ping command between end hosts. For each topology type, 20 ICMP packets have been transmitted from host $h1$ to host $h16$. Resulting average RTT is shown in Fig. 3.

We can clearly notice that ODL achieves lower RTT levels than POX for all topology types. The highest value over all topology types is achieved with POX controller for linear topology, which is 30 times higher than the ODL level. Thus, we can conclude that ODL performs better than POX, in terms of average RTT. Hence, it is more suitable for latency-sensitive applications. This result can be explained by the fact that POX is slower because it is written in Python, whereas ODL is written in Java.

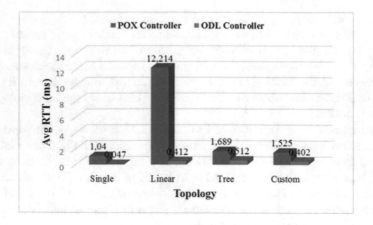

Fig. 3. Average round-trip time between end hosts

4.2 Throughput

This test aims to evaluate the throughput performance of POX and ODL controllers. TCP packets are transmitted between end hosts $h1$ and $h16$, which have been designed as client and server, respectively. Then, TCP throughput is measured by executing *Iperf* command. Results, which have been obtained by averaging over 10 realizations for different topologies, are illustrated on Fig. 4.

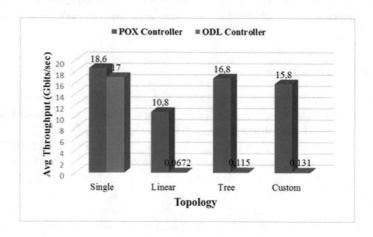

Fig. 4. Average TCP throughput between end hosts

The figure shows that POX controller can achieve higher throughput levels than ODL for all topologies. While for single topology throughput levels are comparable, for linear, tree and custom topologies, there is a large gap between the two controllers (more than 140 times). With regard to topology type, both controllers achieve the lowest levels in linear topology and the highest levels in single topology.

4.3 Jitter and Packet Loss

In the last performance test, we have measured jitter and packet loss using $Iperf$ command for each topology. UDP packets have been transmitted between end hosts $h1$ and $h16$, where $h1$ acted as a UDP client and $h16$ as a server. The test has been driven during 20 seconds and for different bandwidth values. The evolution of jitter versus bandwidth for each topology type is shown in Figs. 5, 6, 7 and 8, respectively.

We can clearly notice that POX controller achieves lower jitter values than ODL, for all topology types. Hence, POX controller performs better than ODL controller in terms of jitter. With regard to topology type, the lowest levels are achieved in single topology and the highest ones in linear topology. This can be explained by the fact that linear topology contains more switching devices than single topology.

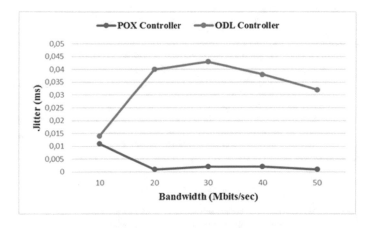

Fig. 5. Jitter of single topology

Concerning the packet loss, the simulation results of each controller is tabulated in Tables 2 and 3, respectively. It is clear that POX controller presents more packet loss rates than ODL for all topology types and all bandwidth values. In particular, POX presents the worst levels in linear topology which achieve 40%, whereas they do not exceed 1.7% for all other topologies. On the other hand, packet-loss rates of ODL are null in single topology and do not exceed 2.5% for other topologies. Thus, we can conclude, that ODL controller performs better than POX in terms of packet loss for all topologies. This can be explained by the ODL controller's multithreading feature, which enables better performance than a single-threaded controller such as POX.

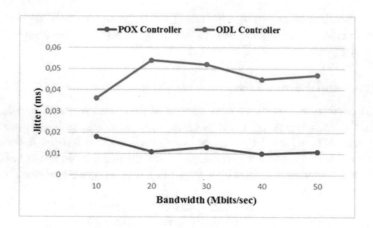

Fig. 6. Jitter of linear topology

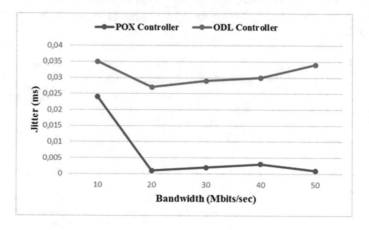

Fig. 7. Jitter of tree topology

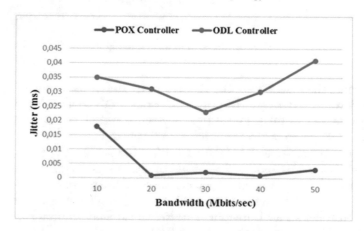

Fig. 8. Jitter of custom topology

Table 2. Packet Loss (%) between end hosts in each topology with POX controller

Bandwidth (Mbits/s)	Single topology	Linear topology	Tree topology	Custom topology
10	0.035	21	0.61	1.7
20	0.082	25	0.49	0.065
30	0.0098	36	1.1	0.26
40	0.056	31	1.2	1.6
50	0.031	40	1.3	0.67

Table 3. Packet Loss (%) between end hosts in each topology with ODL controller

Bandwidth (Mbits/s)	Single topology	Linear topology	Tree topology	Custom topology
10	0	0	0	0
20	0	0	0	0
30	0	0.29	0	0
40	0	2.5	0	0
50	0	1.3	0.0071	0.0024

5 Conclusion

In this paper, we have presented a comparative performance evaluation of two popular SDN controllers, POX and OpenDayLight. We have considered the most relevant QoS parameters such as, RTT, throughput, jitter and packet loss, as performance criteria. Furthermore, we have explored a large set of topology types like single, linear, tree and customized topologies. Based on Mininet emulator, we have shown that for all topology types, ODL controller performs better than POX controller in terms of RTT and packet loss, whereas POX outperforms ODL in terms of throughput and jitter. With regard to topology type, we have noticed a disparate performance behavior and, that the worst levels were achieved with linear topology for all parameters. We can conclude that the choice of the controller depends on the QoS requirements of the desired application in addition to other feature considerations such as, programming language or architecture. For a future work, we can explore the performance behavior of controllers with increasing complexity and varying network topology. This work will include the impact analysis of network scalability on the performance of various SDN controllers in different network topologies.

References

1. POX Wiki. https://openflow.stanford.edu/display/ONL/POX+Wiki.html
2. RYU SDN Framework. https://osrg.github.io/ryu-book/en/html/

3. The OpenDayLight Platform. https://www.opendaylight.org/
4. Floodlight controller. https://floodlight.atlassian.net/wiki/spaces/floodlightcontroller/overview
5. Open Network Operating System (ONOS) Wiki. https://wiki.onosproject.org/
6. Mininet. http://mininet.org/
7. Zhu, L., Karim, M. M., Sharif, K., Li, F., Du, X., Guizani, M.: SDN controllers: benchmarking & performance evaluation. arXiv preprint arXiv:1902.04491 (2019)
8. Lakhani, G., Kothari, A.: Fault administration by load balancing in distributed SDN controller: a review. Wireless Pers. Commun. 1–33 (2020)
9. Badotra, S., Panda, S.N.: Evaluation and comparison of OpenDayLight and open networking operating system in software-defined networking. Cluster Comput. 1–11 (2019)
10. Arahunashi, A.K., Neethu, S., Aradhya, H.R.: Performance analysis of various SDN controllers in Mininet emulator. In: 2019 4th International Conference on Recent Trends on Electronics, Information, Communication and Technology (RTEICT), pp. 752–756. IEEE, May 2019
11. Islam, M.T., Islam, N., Al Refat, M.: Node to node performance evaluation through RYU SDN controller. Wireless Pers. Commun., 1–16 (2020)
12. Bholebawa, I.Z., Dalal, U.D.: Performance analysis of SDN/OpenFlow controllers: POX versus floodlight. Wireless Pers. Commun. 98(2), 1679–1699 (2018)
13. Eljack, A.H., Hassan, A.H.M., Elamin, H.H.: Performance analysis of ONOS and Floodlight SDN controllers based on TCP and UDP Traffic. In: 2019 International Conference on Computer, Control, Electrical, and Electronics Engineering (ICCCEEE), pp. 1-6. IEEE, September 2019
14. Research, P.J., Raj, P.: Topology-based analysis of performance evaluation of centralized vs. distributed SDN controller. In: 2018 IEEE International Conference on Current Trends in Advanced Computing (ICCTAC), pp. 1–8. IEEE, February 2018
15. Mamushiane, L., Lysko, A., Dlamini, S.: A comparative evaluation of the performance of popular SDN controllers. In: 2018 Wireless Days (WD), pp. 54–59. IEEE, April 2018
16. Priyadarsini, M., Bera, P., Bampal, R.: Performance analysis of software defined network controller architecture—a simulation based survey. In: 2017 International Conference on Wireless Communications, Signal Processing and Networking (WiSPNET), pp. 1929–1935. IEEE, March 2017
17. Li, Y., Guo, X., Pang, X., Peng, B., Li, X., Zhang, P.: Performance analysis of Floodlight and Ryu SDN controllers under Mininet simulator. In: 2020 IEEE/CIC International Conference on Communications in China (ICCC Workshops), pp. 85–90. IEEE, August 2020

DoSSec: A Reputation-Based DoS Mitigation Mechanism on SDN

Ranyelson N. Carvalho$^{(\boxtimes)}$, Lucas R. Costa, Jacir L. Bordim, and Eduardo Alchieri

Department of Computer Science, University of Brasilia, Brasilia, Brazil
ranyelson.carvalho@aluno.unb.br, lucas.costa@gigacandanga.net.br,
{bordim,alchieri}@unb.br

Abstract. Denial-of-service (DoS) attacks bring many challenges in software-defined networks (SDN), mainly due to the vulnerabilities present in the communication between the control and data planes. Mitigation mechanisms have been proposed to alleviate these problems with the side effect of blocking legitimate flows. This work presents a DoS attack mitigation mechanism, named DoSSec, which encompasses a flow-based reputation strategy to improve detection of spurious traffic. The results show that DoSSec is able to prioritize and preserve an average of 95% legitimate traffic compared to state-of-the-art solutions, reducing impacts caused by SYN flood DoS attacks.

1   Introduction

Software Defined Networks (SDN) represents a network architecture with flexible management. The SDN architecture splits the network into three planes. The network applications are running in the application plane, the control plane manages the network and the data plane forwards the data packets. Main characteristic of an SDN is the decoupling the data and control planes. OpenFlow is the most widely used protocol for communication between the two planes. It is characterized as a generic method of communication between the controller and the switches. Despite the numerous benefits provided by this architecture, network security is still a matter of concern once the decoupling between planes increases the attack surface. In particular, denial-of-service (DoS) attacks challenge the SDN network in many different ways since the vulnerabilities in the communication between the two planes could be explored by such attacks. DoS attacks in an SDN architecture aim to exhaust the controller's processing capacity and reduce the application bandwidth, through a high volume of traffic in a short period of time [13].

DoS attacks detection is a well known and studied topic in traditional networks. Most of the proposed works employ methods based on traffic observation. Detection methods can be classified into four categories, according [2]: (*i*) statistical; (*ii*) sequential change-point detection; (*iii*) wavelets analysis; and (*iv*) neural networks. In statistical, the profile of normal traffic flow is provided by the information in the packet header, which is compared to spurious traffic. The sequential change-point detection analyzes traffic as a time series, while wavelet analysis performs an investigation of the network's spectral signals. Neural networks propose the use of machine learning algorithms that can make decisions automatically, through the analysis of traffic attributes [2].

© The Author(s), under exclusive license to Springer Nature Switzerland AG 2021
L. Barolli et al. (Eds.): AINA 2021, LNNS 226, pp. 757–770, 2021.
https://doi.org/10.1007/978-3-030-75075-6_62

Detecting a DoS attack is not enough to overcome security problems in SDN. It is necessary also to minimize the effects caused by these attacks through mitigation mechanisms. Mitigation strategies can be categorized into three classes: (*i*) blocking; (*ii*) control; and (*iii*) resource management [8]. The first class blocks the ports on which the malicious hosts are connected or immediately discards spurious traffic. The second class delays spurious traffic by reducing the bandwidth or the number of messages sent to the controller. Finally, the third class configures and manages the network through additional resources present in the infrastructure.

The literature has presented several mitigation strategics for DoS attacks on SDN [8], however, the proposed solutions require continuous communication between the control and the data plane to obtain information from the network. Moreover, many of the proposed solutions do not support any type of flow re-validation. In such case, the solution becomes vulnerable to a simple attacker which initially behave as a legitimate user and latter promote an attack. One of the main types of DoS attacks that efficiently exploit the aforementioned vulnerability is the SYN flood attacks. These attacks are a form of DoS attacks that send a flood of SYN requests to any destination (victim), in an attempt to exhaust the server's available resources, leaving it unavailable to respond to legitimate traffic. Existing solutions end up penalizing legitimate flows when trying to mitigate the attack. To reduce this problem, we introduce DoSSec. The solution uses traffic reputation mechanisms to avoid blocking legitimate traffic by distinguishing it from spurious traffic. In a nutshell, the main contributions of this paper are:

- It presents the design of a solution, called DoSSec, for mitigating DoS attacks in SDNs that validates legitimate flows, which also works directly in the data plane.
- It discusses a set of experiments to show the advantages of the proposed solution by a comparison with other approaches in the literature. The experimental results show that DoSSec is able to prioritize and preserve an average of 95% legitimate traffic, reducing the congestion and the impacts caused by SYN flood attacks.

The remainder of this paper is organized as follows. Section 2 presents related works. Section 3 presents the proposed solution. Section 4 shows the experimental results. Finally, Sect. 5 concludes the paper.

2   Related Works

In this section, we present the main works for mitigating SYN flood attacks on SDN networks, in addition some related to the concept of reputation on SDN networks.

Table 1. Overview of the literature on mitigating SYN flood attacks in SDN.

Solution	Detection technique	Plane	Mitigation technique	Protection area			
				Other hosts	Controller	Switch	Target
AVANT-GUARD [17]	Switch Proxy	Data	Blocking		✓		✓
NUGRAHA [16]	Statistical	Control	Resource Management		✓		✓
OPERETTA [7]	Controller Proxy	Control	Blocking		✓		✓
LINESWITCH [1]	Switch Proxy	Data	Blocking		✓		✓
PIU-DOS [3]	Statistical	Control	Control		✓		✓
FLOWSEC [10]	Statistical	Control	Control		✓		✓
SDN-GUARD [6]	Statistical with IDS	Control	Resource Management		✓	✓	✓
SLICOTS [14]	Statistic with temporary rules	Control	Blocking		✓		✓
SDN-SCORE [9]	Statistical with reputation	Control	Control	✓	✓		✓
SAFETY [11]	Statistical	Control	Blocking		✓	✓	✓
DoSSec (this work)	Statistical Proxy with reputation	Data	Control and Blocking	✓	✓	✓	✓

In [17], the authors proposed Avant-Guard, which transforms the switch into a proxy. The proposal intercepts TCP connection information in one session and forwards only complete requests to the controller. As a side effect, the solution increases the delay caused in establishing a connection for legitimate requests due to the proxy, in addition to allowing the attack on hosts with already validated IPs. The authors in [16] proposed a solution based on statistical data that consists of the cumulative sum of packets, classifying the attack when that sum reaches a certain threshold. The controller applies a blocking rule to the switch to reduce the impacts of the network attack. However, the solution does not show how legitimate and spurious flows are distinguished, which generates a large number of false positives. In [7], similarly to [17], the solution turns the controller into a proxy, which always rejects the client's first connection. The solution maintains a SYN request counter, and if the number of SYN requests exceeds the limit, the controller performs the attacker's blocking action. Although the solution is efficient, if the attacker completes the first TCP handshake, then he/she has no further obstacles to attack the victim. In the same line as above, in [1] the first connection is handled by the switch while the others are handled with a certain probability. Connections that do not perform the TCP handshake are discarded. However, the solution is vulnerable to attacks on IPs that have already been validated. In [3], the authors proposed a detection algorithm that uses cumulative sum to detect the SYN flood attack. Mitigation instructs the switch to send fewer packets so that it can delay some flows without overloading the controller. Likewise, the authors in [10] proposed a solution that applies a dynamic limit on the rate of the number of packets sent to the controller, when this limit is exceeded, the switch then applies the same mitigation mechanism as [3]. Although both solutions are effective, they do not distinguish the type of delayed flow (legitimate or spurious). The solution proposed in [6] current together with an IDS (Intrusion Detection System). Upon detecting an attack, the solution forwards the attacker to a less used link, ensuring that it does not compromise legitimate flows. However, the solution requires continuous communication with the IDS, which may incur in long detection delays. In [14] the authors developed a solution that monitors the connections between client and server and adds temporary short-term rules. When the number of half-open connections exceeds a predefined limit, the solution installs a blocking rule. The installation of rules is a problem for a large volume of hosts, which causes fast saturation of the routing devices. The authors at [9] proposed a solution that score the packets of each network flow, classifying it as legitimate or spurious. When a flow is considered spurious, it is discarded by the solution. Continuous sending of data between the switch and the controller for score control can overload the communication channel. The authors in [11] proposed SAFETY. Solution uses entropy as a detection mechanism. The analysis is based the destination IP of the packets to detect anomalies and identify the attack source. Mitigation involves discovering the malicious source and blocking it on the switch port. Although the solution is effective, blocking the switch port can penalize all legitimate flows linked to that port.

The authors in [19] analyzed the reliability of SDN applications, through reputation. The reputation is made through a beta distribution, granting a credit. Main factor for attributing the reputation is the feedback from the controller in relation to the customer's request. In [15] the authors used reputation to evaluate external applications. Solution takes historical data into account for establishing reputation. Reputable flows have higher priority to be served. The solutions presented do not explore the risks related to DoS attacks, but present how the concept of reputation can be used in an SDN network.

Table 1 shows an overview of the solutions discussed in this Section. We classify these solutions according to their detection and functioning. Most of the solutions are based on statistical information. While others use proxy techniques to validate connections. Regarding the action plane, 80% of the solutions act on the control plane, which affects the processing time due to excessive communication with the data plane. Solutions that act on the data plane is a challenge because it is necessary to change the routing devices to reduce communication and overload to controller. The mitigation techniques are based on the taxonomy presented in Sect. 1. Half of the solutions are based on blocking which can harm the percentage of false positives. A hybrid model would be most appropriate. Finally, we classify the protection area. In general, protection focuses on the controller and the target of the attack. The ideal would be the complete protection of the SDN infrastructure.

3   Proposed Solution

This section presents DoSSec (Denial of Service Security), our proposal to reduce the penalty/blocking of legitimate flows in SDN, due to the mitigation techniques applied by the solutions present in the state-of-the-art. DoSSec uses traffic reputation mechanisms to avoid blocking legitimate traffic and thereby reduce the impacts caused by SYN flood attacks. Solution consists of two components: detection and mitigation, as shown in Fig. 1. Detection module collects flow statistics and outputs changes in traffic volume that indicate a possible attack. We use the detection mechanism developed by [5], which uses entropy to analyze the randomness of traffic. Mitigation module consists in applying a reputation-based method, which uses priority queues to denote the level of confidence of the flows, establishing trust between hosts and the forwarding devices.

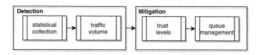

Fig. 1. Proposed SDN architecture.

Figure 2 shows the transition process between the types of classification, based on the behavior of the flow. New flows are treated as *unclassified*. These have no priority and are forwarding through the switch queue standard management. Priority is only given when the flow completes its first TCP connection, making up the *graylist*. Depending on its behavior, a flow can be promoted to the *whitelist* or downgraded to the *blacklist*, when its behavior is characterized as irregular. Also, a flow may lose its initial classification (graylist), when considered suspicious. A whitelist flow may lose the trust it has been given and may be characterized as unclassified or be blocked. Finally, the flow transferred to the blacklist is blocked and may return to the status of Unclassified after a certain time.

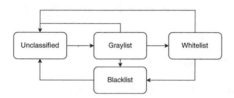

Fig. 2. Transition flows between classification levels.

DoSSec uses the aforementioned classification types to prioritize flows. Upon receiving a TCP connection request, DoSSec checks the prioritization level for this request. Each request is observed through connection monitoring, responsible for classifying connection. When DoSSec detects an attack, new flows first attempt to establish a connection will be redefined as a preventive measure. When redefined the connection, a record of the flow will be added to a list of pending connections, which contains the unfinished connections for the flows. Upon returning, these flow are removed from the list and has its connection forward.

Table 2. List of Variables

Notation	Description
f_i	Flow
Flags.SYN	Connection opening request
Flags.ACK	Response to the completion of the connection
connectionFault	Counter for the number of failure connections
connectionSucc	Counter for the number of successful connections
stats	Flow statistics
remove_queue	Counter for the number of queue removals
enqueue	Queuing time
α	Threshold of failure connections
β	Threshold queue packet rate
γ	Threshold for the number of queue removals
δ	Threshold for the reputation level of the queue
θ	Threshold for queuing time
μ	Threshold for the stay time in queue

3.1 Connection Monitoring

Connection monitoring register information and monitors connection requests, categorizing them as success or failure. The classification is done through the monitoring of TCP flags, exchanged between source and destination. Table 2 lists all symbols and variables that will be used in the algorithms presented below.

Connection monitoring details are shown in Algorithm 1. Upon receiving a connection establishment request (SYN), a timeout for flow f_i, $i > 0$, starts. This time is set based on historical data (previous TCP connection timeouts), considering an environment without the presence of spurious flows. More precisely, it considers the waiting time from the point when the switch receives the first SYN packet until the point where it receives the first ACK packet. Such information is used to define the average waiting time to conclude whether the connection was in fact established or not. Upon receiving an ACK packet, it indicates connection establishment and the successful connection counter is updated. It will be managed according to the method adopted for the flow in its respective queue. If the flow does not send the ACK packet within the expected time limit, the connection fails and its respective counter is increased and compared to the defined threshold. If the number of failed connections surpasses the threshold (α) the flow will be managed by the blacklist and then discarded. For each flow present in the queue, there is a method of manipulation. New flows have a set of statistics registered, such as queuing time, number of connections made (success or failure), volume of packets and the number of removals, etc. For existing flows, such information is maintained and updated.

Algorithm 1. Connection Monitoring for flow f_i

Upon receipt of f_i.Flags.SYN
Start f_i.timeout

Upon receipt of f_i.Flags.ACK
f_i.connectionSucc ++
if $(f_i \in Whitelist)$ **then**
| call handle_queue(f_i,Whitelist)
else
| call handle_queue(f_i,Graylist)
end

Upon f_i.timeout
f_i.connectionFault ++
if $(f_i.connectionFault > \alpha)$ **then**
| call handle_queue(f_i,Blacklist)
end
Discards connection f_i

Procedure handle_queue(f_i,Q)
if $f_i \notin Q$ **then**
| $Q = Q \cup f_i$
| Register(Q.stats.f_i)
else
| Update(Q.stats.f_i)
end

3.2 Queue Processing

As stated in the previous subsection, connection monitoring directs the incoming flow to corresponding priority queue. Queue processing monitors these queues so that there is no overload and it is able to provide the appropriate service. The flows present in graylist and whitelist are monitored, through the packet rate, which aims to detect when a flow uses the available resources of the queue in an excessive way. Packet rate (T_Q) is calculated at each time slot t_j, $j > 0$, by the following equation:

$$T_Q = \frac{\sum_{i=1}^{m} T_{p_i}}{m}, \tag{1}$$

where m is the number of flows in the queue and T_p is the total volume of packets in the queue. The flows present in the graylist have another approach, the reputation level, which allows assessing the behavior of the flow over the time it has been present, allowing to raise its level of classification. The reputation level (R_G) is calculated at each time slot t_j, $j > 0$ by the following equation:

$$R_G = \frac{\sum_{i=1}^{m} T_{c_i}}{m}, \tag{2}$$

where m is the number of flows in the queue and T_{c_i} is total volume of successful connections in the queue.

Algorithm 2. Queue processing (Graylist, Whitelist and Blacklist)

Upon Graylist.timeout
call priority_queue(Graylist)

Upon Whitelist.timeout
call priority_queue(Whitelist)

Upon Blacklist.timeout
call update_blacklist(Blacklist)

Procedure priority_queue(Q)
for $f_i \in Q$ **do**
 if *(f_i.rate > β)* **then**
  $Q = Q \setminus f_i$
  Q.stats.f_i.remove_queue ++
  if *(Q.stats.f_i.remove_queue > γ)* **then**
   | call handle_queue(f_i,Blacklist)
  end
 else
  if *($f_i \in$ Graylist and Q.stats.f_i.connectionSucc > δ and Q.stats.f_i.enqueue > θ)*
  then
   | Whitelist = Whitelist $\cup$ f_i
  end
 end
end

Procedure update_blacklist(Q)
for $f_i \in Q$ **do**
 if *(Q.stats.f_i.enqueue > μ)* **then**
  $Q = Q \setminus f_i$
  Update(Q.stats.f_i)
 end
end

Queuing processing details are shown in Algorithm 2. For each time slot t_j, $j > 0$, the packet rate (Eq. 1) and reputation levels (Eq. 2) are computed. This time slot will provide a check of the queues. Time slot that are too long or too short can cause heavy overloading of the solution. For this reason, graylist and whitelist management is performed by the *priority_queue* method, separated from the blacklist and unassigned flows. If the packet rate of the flow is above the maximum allowable packet rate (β), it is removed and loses its assigned rating and can be added to the blacklist if it exceeds the threshold (γ) for the number of allowed removals. However, if the flow belongs to graylist and is below the packet rate, with a volume of connections (success) above the reputation level (δ) and has the longest standing in the queue (θ), the flow is added to whitelist. The flows present in the blacklist have their information updated by the *update_blacklist* method. Thus, when there is a recurrence of the flow, it is verified its time of permanence in the queue, which is compared to the threshold (μ). The flow is removed from the blacklist and its information is updated after (μ) time slots. Otherwise, the flow remains blocked.

4   Performance Evaluation

In this section we present DoSSec. The solution used the P4 language, which allows adding new headers and specifying how packets should be analyzed and processed by the switches, turning it into a programmable switch [4]. However, solutions that adopt this language need to deal with a restricted set of functionalities for traffic analysis, making the scope of security solutions limited [12, 18]. One of the limitations is that it does not support mathematical operations that result in floating point numbers. To overcome these restrictions, we proposed the use of an auxiliary component next to the switch that returns information regarding flows to switch P4 in real time. In this work we used the Apache Thrift, which allows communication between different programming languages via RPC (Remote Procedure Call).

The experiments were performed on a computer with an operating system Ubuntu 16.04 LTS, with a processor $i7$ 3.0 GHz CPU and 4Gb of RAM. The network emulator was Mininet. The network topology consists of 5 switches, 11 hosts and 1 controller, as shown in Fig. 3. This was defined as allowing the distribution of traffic in order to evaluate the solution. The bandwidth of each link is 100 Mbps. In experiments, the attacker is assumed to control one of the hosts. The victim is a web server and is located on the $S2$ switch and the attacker initiates the attack from a host connected to the $S5$ switch, which does not run the OpenFlow protocol. The objective is to show the behavior in a heterogeneous network.

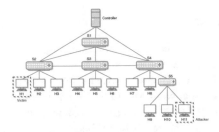

Fig. 3. Network topology.

To assess the performance of DoSSec, we compare it with SAFETY [11], shown in Table 1, as it is the most recent solution in the scope of SYN flood mitigation. Evaluation is performed in the following scenarios:

- **Scenario 1:** the attacker sends a large number of SYN packets with spoofed IP addresses to the server. This scenario is commonly used in the literature while evaluating the performance of attack detection and mitigation strategies.
- **Scenario 2:** the attacker completes some TCP connections and then starts the attack on the server, exploring validated connections.

Two types of traffic are generated: legitimate traffic and spurious traffic. The first is created using the Scapy tool. All client hosts generate traffic towards the server. The second is generated using the Hping tool. Legitimate traffic is used to calculate entropy under normal conditions and thus to define the detection threshold. Spurious traffic is injected to analyze the behavior of the solution. Legitimate traffic runs for 60 s. The attack starts after that and runs for 20 s. The solution assessment covers the following parameters:

- **Average detection time:** is the time difference between the instant an attack is detected and the attacker's first SYN packet.
- **Average packet volume for detection:** number of packets needed for each solution to identify that an attack is occurring.
- **Precision average detection:** precision was measured using the F1 score. This metric is defined as a harmonic average between precision and recall. Precision represents the proportion of positive identifications that have been correctly identified. The recall is the proportion of positives that were correctly identified, measuring the completeness of the mechanism. High F1 values indicate that the detection technique has higher precision.
- **Number of legitimate packets discarded:** number of legitimate packets sent to the server during the attack.

The simulations were performed under different traffic loads based on the total network capacity. The overall load of the network was kept constant in the experiments, while the load of spurious traffic ranges from 20%, 40%, 60% and 80% in relation to legitimate traffic. This configuration provides 4 workloads.

4.1 Experimental Results

The average detection time of the attack, can be seen in Fig. 4. As the spurious traffic increases, the detection time decreases, whereas we will have a more dense malicious traffic. In both scenarios, DoSSec was able to detect spurious traffic faster than SAFETY. Recall that DoSSec, contrarily to SAFETY, is located at the data plane, providing detection closer to the attacker. DoSSec is on average 40% faster than SAFETY to detect the attack. In the second scenario, we had a 12% increase in DoSSec detection time with a 20% load compared to the first scenario, due to the nature of the attack.

The volume of packets for detection can be seen in Fig. 5. In both scenarios, as the spurious traffic load increases, the number of spurious packets that make up the observation window becomes more present, facilitating the detection mechanism. The fact that DoSScc is closer to the attacker, contributes to a smaller number of packets to be analyzed and, consequently, less time to detect any change in traffic. SAFETY, in turn, needs more packets and more time to detect any changes, since it is located in the center of the network. In the second scenario, we have a greater presence of legitimate flows in the observation windows, a direct consequence of the type of attack carried out.

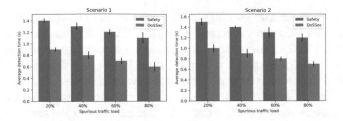

Fig. 4. Average detection time for each scenario, respectively.

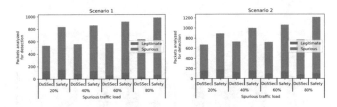

Fig. 5. Packets analyzed for detection for each scenario, respectively.

Figure 6 shows the result regarding the precision to detect the attack by F1 score. In both scenarios, as spurious traffic increases, precision reaches higher values. This precision is influenced by the variation in traffic, which becomes more concentrated and, consequently, easier to detect. This facility is intrinsic to the detection of anomaly-based attacks and is exacerbated when considering smaller proportions of traffic, which results in less precision estimates of statistical analysis techniques. SAFETY performed well, reaching 95% for the load of 80% in both scenarios, as it receives a wider range of packets to be analyzed, improving the quality of the result. SAFETY precision was about 3% higher than the precision of DoSSec in both scenarios. On the other hand, SAFETY requires a higher volume of traffic to be analyzed, consuming more time to detect an attack.

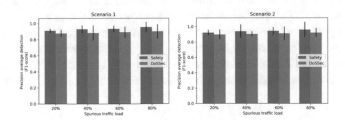

Fig. 6. Precision average detection for each scenario, respectively.

Figure 7 shows the number of legitimate packets discarded. The strategy adopted by SAFETY in blocking the port connected to the attacker, can become a problem. In the topology used, we have a switch that does not run the OpenFlow protocol, so the information collected related to the hosts connected to that switch is captured by the switch immediately above. In this way, the hosts belonging to the switch (S5) pass through a single port. Thus, the mitigation adopted by SAFETY ends up penalizing legitimate hosts connected to the same port as the attacker. The consequence of this is seen in the number of legitimate packets discarded, much higher than that of DoSSec. This, in turn, performs the prioritization of legitimate flows and the redefinition of malicious connections. In the first scenario, the loss is minimal. In the second scenario, there is a small increase in the number of discarded packets, due to the characteristic of the attack. However, the rate of packets in the queue removes the attacker and guarantees priority service to other flows.

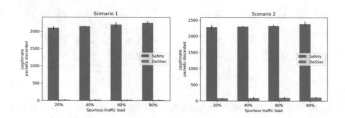

Fig. 7. Legitimate packets discarded for each scenario, respectively.

Finally, Fig. 8 shows the number of packets in the queues. Due to space limitation, loads of 20% and 80% are shown, as others loads follow a similar pattern. In the first scenario, the attacker is unable to access any of the priority queues, so the number of packets traversed comprises only the legitimate packets, as DoSSec prevents the attacker from flooding the queues, redefining the connections coming from the attack. There is only a decrease in the number of packets covered in the Graylist, a result of the transition of flows to Whitelist, as they reach the reputation level, generating an increase. In the second scenario, there is an increase in the volume of packets in the Graylist, due to the attacker completing some TCP connections and starting the attack from there, causing the discard of some legitimate packets. However, the packet rate, responsible for analyzing the behavior of the flows proved to be effective, identified the anomaly and performed the removals of the attacker's packets and added it to the Blacklist. As spurious traffic increases, the packet rate (Eq. 1) is more effective, preserving a larger number of legitimate packets. For example, with a load of 80%, the removal of the attacker occurs more faster, being four times faster compared to the load of 20%.

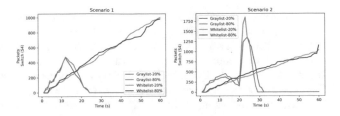

Fig. 8. Queue behavior for each scenario, respectively.

5 Conclusion

This paper present and analyzed DoSSec, an attack mitigation strategy aimed to reduce the impacts caused by SYN flood attacks in SDN networks. DoSSec uses reputation mechanisms to prioritize flows, through the management of queues on forwarding devices, with the purpose of ensuring that legitimate flows are served while restrained suspicious and spurious traffic. The proposed solution was analyzed in two scenarios and experimental results showed that DoSSec is 40% faster than similar alternatives in the literature. DoSSec was also able to preserve legitimate traffic when performing the mitigation process. Moreover, it is able to manage priority levels, providing a faster service to legitimate flows.

Acknowledgements. This work is partially supported by the MCTIC/RNP/CTIC (Brazil) through the project P4Sec.

References

1. Ambrosin, M., Conti, M., Gaspari, F.D., Poovendran, R.: LineSwitch: tackling control plane saturation attacks in SDN. IEEE/ACM Trans. Networking **25** (2017)
2. Beitollahi, H., Deconinck, G.: Analyzing well-known countermeasures against distributed denial of service attacks. Comput. Commun. **35**(11), 1312–1332 (2012)
3. Bera, P., Saha, A., Setua, S.K.: DoS in SDN. In: 5th International CCSNT, pp. 497–501, December 2016
4. Bosshart, P., Daly, D., Gibb, G., Izzard, M., McKeown, N., Rexford, J., Schlesinger, C., Talayco, D., Vahdat, A., Varghese, G., Walker, D.: P4: programming protocol-independent packet processors. SIGCOMM Rev. **44**(3), 87–95 (2014)
5. Carvalho, R.N., Costa, L.R., Bordim, J.L., Alchieri, E.A.P.: New programmable data plane architecture based on P4 OpenFlow agent. In: Barolli, L., Amato, F., Moscato, F., Enokido, T., Takizawa, M. (eds.) AINA. pp, pp. 1355–1367. Springer, Cham (2020)
6. Dridi, L., Zhani, M.F.: SDN-guard: DoS attacks mitigation in SDN networks. In: 5th International Conference Cloudnet, pp. 212–217, October 2016
7. Fichera, S., Galluccio, L., Grancagnolo, S.C., Morabito, G.: OPERETTA: an OPEnFlow-based REmedy to mitigate TCP SYNFLOOD Attacks against web servers. Comput. Netw. **92**, 89–100 (2015)
8. Imran, M., Durad, H., Khan, F., Derhab, A.: Toward an optimal solution against DoS attacks in software defined networks. Future Gener. Comput. Syst. **92**, 09 (2018)

9. Kalkan, K., Gür, G., Alagöz, F.: SDNScore: a statistical defense mechanism against DDoS attacks in SDN environment. In 2017 ISCC, pp. 669–675 (2017)
10. Kuerban, M., Tian, Y., Yang, Q., Jia, Y., Huebert, B.: FlowSec: DoS attack mitigation strategy on SDN controller. In: International CNAS, pp. 1–2, August 2016
11. Kumar, P., Tripathi, M., Nehra, A., Conti, M., Lal, C.: Safety: early detection and mitigation of TCP SYN flood utilizing entropy in SDN. IEEE Trans. Netw. Serv. Manag. (2018)
12. Lapolli, A.C., Marques, J.A., Gaspary, L.P.: Offloading real-time DDoS attack detection to programmable data planes. In: IFIP International Symposium on Integrated Network Management (2019)
13. Lau, F., Rubin, S.H., Smith, M.H., Trajkovic, L.: DDoS attacks, vol. 3, pp. 2275–2280, February 2000
14. Mohammadi, R., Javidan, R., Conti, M.: SLICOTS: an SDN-based lightweight countermeasure for SYN flooding attacks. IEEE Trans. Netw. Serv. Manag. **14**, 487–497 (2017)
15. Niemiec, M., Jaglarz, P., Jekot, M., Chołda, P., Boryło, P.: Risk assessment approach to secure northbound interface of SDN networks. In: 2019 ICNC, pp. 164–169 (2019)
16. Nugraha, M., Paramita, I., Choi, D., Cho, B.: Utilizing OpenFlow and sFlow to detect and mitigate SYN flooding attack. J. Korea Multimedia Soc. **8**(8) (2014)
17. Shin, S., Yegneswaran, V., Porras, P., Gu, G.: AVANT-GUARD: scalable and vigilant switch flow management in SDN. In: ACM SIGSAC Conference on Computer, pp. 413–424 (2013)
18. Sviridov, G., Bonola, M., Giaccone, P., Bianchi, G.: LODGE: LOcal Decisions on Global statEs in prograñanaable data planes. In: 4th CNSW, pp. 257–261 (2018)
19. Wang, Y., Liu, Y., Hu, J., Zhang, M., Wang, X.: Reputation and incentive mechanism for SDN applications. In: 14th International CMASN, pp. 152–157 (2018)

A Comparative Study Between Containerization and Full-Virtualization of Virtualized Everything Functions in Edge Computing

Hugo Gustavo Valin Oliveira da Cunha[1]([envelope]), Rodrigo Moreira[1,2], and Flávio de Oliveira Silva[1]

[1] Faculty of Computing (FACOM), Federal University of Uberlândia, Uberlândia 38400-902, Brazil
{hugo.cunha,rodrigo.moreira,flavio}@ufu.br
[2] Institute of Exact and Technological Sciences (IEP), Federal University of Viçosa, Rio Paranaíba 38810-000, Brazil
rodrigo@ufv.br

Abstract. The astronomical amount of data is generated today. The currents computer network architectures have trouble managing high bandwidth, dispersion, low geographical latency, and privacy. Edge Computing (EC) aims to solve these problems by allowing data processing to be done close to end users. Network Function Virtualization promises to simplify the network service creation workflow by turning traditional network functions. This work presents a detailed comparative study between containers and virtual machines. To this end, we present an architecture and implementation of a management and orchestration platform (MANO) for Virtualized Everything Functions (VxFs), unified across bare-metal servers as well as low-cost devices. A detailed comparison of the performance of VxFs based on containers and virtual machines is made, represented respectively by the LXD and KVM platforms on a Raspberry Pi 4. The results show that containerization is not an adequate solution for all cases. Instead, it can be thought of as a complementary solution to be used with full-virtualization making the most of these approaches.

1 Introduction

Data production is growing exponentially in today's communication networks [1]. This irreversible trend is driven by the increase in end-users and the widespread adoption of new mobile devices (smartphones, wearables, sensors, etc.). The currents computer network architectures have trouble managing high bandwidth, dispersion, low geographical latency, and privacy. Edge Computing (EC) was proposed to allow data processing close to end-users to solve these problems. Thus, latency can be minimized, high data rates achieved locally, and real-time information can be exploited to develop high-value services. To consolidate network elements and cutting edge applications into the same virtualization infrastructure, network operators intend to combine EC with Network Function Virtualization (NFV). Combined with the recent advances in EC, NFV can enable fast service provisioning, scaling, and migration.

However, even though NFV has received significant attention, there are still several open challenges [2], that need to be addressed before NFV can be effectively implemented in Edge Computing environments. This includes management and orchestration of network services, supporting highly heterogeneous access and processing technologies, from commodity servers to single-board computers.

This work aims to present a detailed comparative study between containers and virtual machines. To this end, we present and use an ETSI Management and Orchestration Platform (MANO) compliant architecture for Virtualized Everything Functions (VxFs) [3] in the edge. This architecture is capable of deploying services across bare-metal servers and low-cost devices in a seamless way. Using this architecture, we conduct a fair comparison of the performance of VxFs based on containers and virtual machines, represented respectively by the LXD and KVM platforms on a Raspberry Pi 4. Besides, a unified monitoring platform for computing resources is proposed, implemented. This monitoring platform produces all the data for our analysis.

The main contributions of this work compared to state of art are: 1) an architecture that enables seamless use of VxFs on bare metal and edge devices; 2) a monitoring infrastructure for seamless monitoring of core, fog, and edge resources and 3) a detailed comparison of the performance of VxFs based on LXD and KVM on low-cost devices (Raspberry Pi 4) using the same infrastructure.

The rest of the paper is structured as follows. Section 2 presents the works related to machine virtualization, some of which are applied to the Edge Computing environment. Section 3 describes the general NFV architecture in edge computing, highlighting its logical components. Section 4 presents the experimental scenario for implementing VxFs and their monitoring. Section 5 provides a detailed discussion of the results obtained and their reflexes in edge computing environments. Finally, Sect. 6 concludes this work and present future research directions.

2   Related Work

Table 1. State-of-the-art proposals.

	Virtualization	Infrastructure	Monitoring	Metrics
Richards et al. [4]	KVM	Raspberry Pi 3	Psutil tool	Percentage of CPU. Memory response. Buffer swap memory. Temperature of CPU. Boot time machines
Ramalho and Neto [5]	Docker and KVM	Cubieboard2	Benchmarks: NBench, SysBench, Bonnie++, LINPACK, STREAM	Schedule thread, floating-point operations, I/O operantions, memory bandwidth
Ahmad et al. [6]	LXD	Raspberry Pi 2 Model B	OpenStack Cloud Computing	Container batch launch times for operating systems Ubuntu 14.04 Trusty Tahr, Ubuntu 16.04 Xenial Xerus and CirrOS 0.3.4
Cziva and Pezaros [7]	XEN, ClickOS, KVM, Container	Commodity server	Virtual Infrastructure Manager (VIM) and OpenDaylight	Delay of idle round-trip time (RTT). Time to manage (create, start, and stop) virtual machines and containers. Idle memory consumption
Li et al. [8]	Docker and VMWare	Commodity server	Benchmarks: Iperf, HardInfo, STREAM, Bonnie++	Data throughput in communication. Latency in computation. Data throughput in memory. Transaction speed, Data throughput in disk
Riggio et al. [9]	Docker and VirtualBox	Commodity server	Command Linux: top and stat	Round Trip Time in send a web page request. Amount time required to boot application. CPU and memory utilization
This work	LXD and KVM	Raspeberry Pi 4 Model B and Commodity Server	Node exporter, Prometheus and VictoriaMetrics	CPU seconds on user/kernel mode. Memory allocation. Adjacent input operations. Network packets received

There are several research lines on machine virtualization, and some are applied to the Edge Computing environment. Besides, there is research related to Containerized Network Functions (CNF). Table 1 highlights a comparison between the different solutions and the present project proposal. In each approach, four aspects were analyzed. First, the virtualization platform, be it LXD, Docker, KVM, Xen, and others. Second, the type of infrastructure used to make virtual machines or containers run is on a single board computer or a commodity server. Third, the way used to collect metrics on the infrastructure was verified. Fourth and last, the metrics collected from the infrastructure itself.

Richards et al. [4] proposes management and orchestration (MANO) of virtual network functions (VNFs) in Edge Computing using standard solutions such as Open Source MANO (OSM) and OpenStack. To demonstrate its viability, an experimental evaluation using a Raspberry Pi 3 infrastructure is demonstrated. The assessment showed the feasibility of using low-cost devices, such as Raspberry Pi 3, with standard management solutions used in the central cloud.

Ramalho and Neto [5] conduct and evaluate the performance of the container-based approach compared to hypervisor-based virtualization when running on devices typically used at the network edge. The study was performed by executing several synthetic benchmarks providing an insight into the performance overhead introduced by Docker containers (lightweight-virtualization) and KVM VMs (hypervisor-virtualization) running at network edge devices.

Ahmad et al. [6] describes a virtualized test environment for the Internet of Things (vIoT). They argue in favor of infrastructure as an IoT service as a possible model for implementing future IoT services. The vIoT testbed was built from open source, more precisely from OpenStack, LXD, and Raspberry Pi.

Cziva and Pezaros [7] identifies virtualization opportunities at the edge of the network by presenting Glasgow Network Functions (GNF), a container-based NFV platform that performs and orchestrates VxFs. Finally, there is an economy of such an approach in using the central network, providing less latency to requests in networks.

Li et al. [8] investigates the performance overhead of a Docker container compared to a virtual machine (VM). It is shown that the performance of virtualization can vary from resource to resource. Furthermore, it is shown that the hypervisor-based solution does not come with higher performance overhead in all cases, such as QoS, in terms of transaction and storage speed.

Riggio et al. [9] discusses the fundamental challenges of deploying NFV in dispersed environments and then introduces the LightMANO framework, a multi-access network operating system that converges Software Defined Network (SDN) and NFV into a single lightweight platform for managing and orchestrating network services in an infrastructure distributed NFV.

3   Edge NFV Architecture

The Edge Computing NFV architecture is the infrastructure for comparison in this study. Figure 1 shows the general NFV architecture in Edge Computing, highlighting its logical components. Such architecture has several components, each developed as a module with independent submodules.

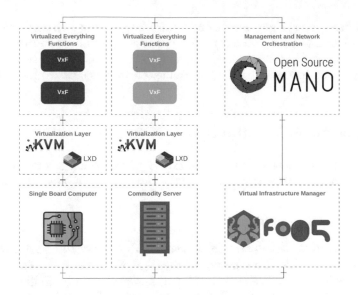

Fig. 1. Edge NFV architecture.

Open Source MANO (OSM) is an ETSI-hosted open-source stack for NFV, capable of consuming openly published information models, available to everyone, suitable for all VxFs, operationally significant, and VIM-independent.

Eclipse Fog05 is a component that allows the end-to-end management of computing, storage, networking, and I/O fabric in the Edge and Fog Environment. It is based on a decentralized architecture that allows users to manage and deploy different types of applications, packaged as containers, VMs, binaries, and so on. All of this can be deployed from big servers to micro-controllers.

A Single Board Computer (SBC), represented by a Raspberry Pi 4 in this work, is a complete computer built on a single circuit board, with microprocessor(s), memory, Input/Output (I/O), and other features required of a functional computer. A commodity server is a dedicated computer to running server programs and carrying out associated tasks.

Virtualization is the technology used to run multiple operating systems (OSs) or applications on top of a single physical infrastructure by providing each of them an abstract view of the hardware. It enables these applications or OSs to run in isolation while sharing the same hardware resources. Two virtualization approaches will be used for the proposed project, namely LXD (Linux Containers) and KVM. LXD is a next-generation system container manager. It offers a user experience similar to virtual machines but using Linux containers instead. KVM is a full virtualization solution containing virtualization extensions (Intel VT or AMD-V). Using KVM, a single node computer can run multiple virtual machines. Each one has private virtualized hardware: a network card, disk, graphics adapter, etc.

Virtualized Everything Functions [3] are one or more virtual machines (or containers like LXD, for example) that implement specialized software that runs with resources from standard high capacity servers, networked computers, and storage systems, instead of being implemented in specialized physical systems devices.

4   Experimental Evaluation

To evaluate the proposed solution, an experimental scenario was used in a real test environment. This section presents the tests executed, as well as its results.

4.1   Test Environment

The architecture of the experiment includes a Raspberry Pi 4 and a bare metal desktop computer that are together at the edge of the network. The bare metal desktop runs Virtual Infrastructure Manager, represented by Eclipse Fog05. On the other hand, the Raspberry Pi 4 represents Single Board Computing, capable of providing computational resources to support VxFs. It is noteworthy that any ARM64 (or AArch64) device with virtualization capability can be used in place of the Raspberry Pi 4, so the experimental setup here replicated using several other platforms commonly used in the edge.

The Raspberry Pi 4 is a Model B with a quad-core Cortex-A72 (ARMv8) 64-bit 1.5 GHz processor, 4 gigabytes of RAM. The VIM bare-metal computer is an Intel Core 2 Quad with 8 gigabytes of RAM. The Raspberry Pi 4 has an Ubuntu 18.04.5 LTS arm64 server, with support for hardware virtualization. The bare-metal desktop computer uses a Ubuntu 18.04.5 LTS 64-bit (AMD64) desktop image.

All the hardware was interconnected using a local physical network. The network's router was configured to support the communication between the Virtual Infrastructure Manager, the VxFs on Raspberry Pi 4, and the external world.

Two Network Service Descriptor (NSD) was uploaded to OSM for the testbed. These service descriptors indicated the creation of Virtual Machines (VMs) on the Raspberry Pi 4 located in the Edge. OSM interacts with the Virtual Infrastructure Manager (VIM), which is represented by the Eclipse Fog05. For the creation of the VxFs, the Eclipse Fog05 interacts with the Raspberry Pi 4, which in turn, will run the VxFs over the LXD or KVM.

4.2   Test Monitoring Enviroment

Figure 2 shows all the components necessary for monitoring and collecting information on infrastructure behavior for the present project.

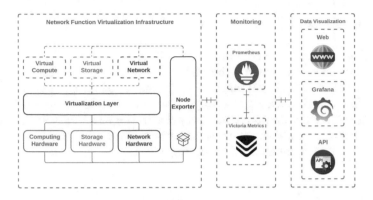

Fig. 2. Edge NFV monitoring architecture.

Network Function Virtualization Infrastructure is represented by Raspberry Pi 4. The infrastructure components, such as computing, storage, and networking, support the software needed to run network applications. This software is represented by the Virtualization Layer and can be a hypervisor like KVM or a container management platform like LXD.

Node Exporter is software for server level and OS level metrics with configurable metric collectors. Node Exporter can measure various server resources such as RAM, disk space, and CPU utilization.

Prometheus is an open-source system monitoring and alerting toolkit. VictoriaMetrics is a fast, cost-effective, and scalable monitoring solution and time-series database. Prometheus was configured to send all data collected to VictoriaMetrics.

All data can be accessed for viewing, such as through a Web Browser, third-party API, or even through a dashboard previously configured in Grafana.

4.3   Experiment Description

The Network Service(NS) used for testing on the container management LXD was composed of a single virtual machine using the Alpine Edge aarch64 operating system [10]. As for the test on the KVM hypervisor, the NS was also composed of a virtual machine, using the CirrOS 0.5.0 aarch64 operating system [11]. In both cases, it was defined that such a VM would have at its disposal a single virtual CPU, 256 megabytes of RAM, and 1 gigabyte of a disk.

To analyze the infrastructure's behavior, several Network Services were launched, as much as possible, so that such infrastructure was operational and could respond to new requests. Thus, the launch of any Network Service would not compromise the functioning of an existing one. Therefore, it was always established that, for both cases, the launch of a new Network Service was only possible after the previous service was in an operational state. In total, 8 VxFs were launched on the KVM platform and 27 VxFs on the LXD platform. These limits were experimentally obtained before running the experiments detailed in this section.

5   Analysis and Discussion

This section discusses the results obtained in experiments detailed in Sect. 4.3. For this discussion, we selected the highly descriptive metrics relevant for comparing containers and virtual machines. We considered CPU, memory, disk, and network. Thus, having the opportunity to get a better and deeper understanding of the Edge Compute environment's behavior to support VxFs. Figures 3, Fig. 4, Fig. 5 and 6 show each of the metrics in the form of time series, with each vertical line representing the moment of launching a VxF on a container or a VM.

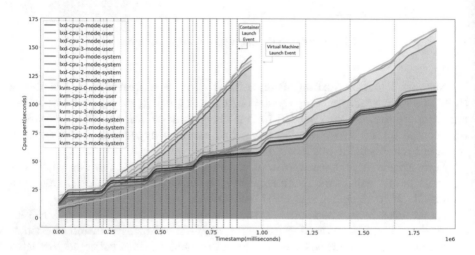

Fig. 3. Seconds the cpus spent in each mode.

CPU: the present work measured how many seconds are spent on the CPU with the operating system's operating modes, that is, kernel mode and user mode. Figure 3 shows the time spent in user mode and kernel mode for each platform. In both cases (KVM and LXD), the time spent in user mode is greater than the time spent in kernel mode. However, this difference is most noticeable with the LXD container manager. It can also be noted that the time spent for KVM is longer in both user mode and kernel mode. This can be considered an improvement of KVM over LXD since more time spent in one mode reduces the changes in the workspace exchange, reducing the overhead that exists in the exchange between user mode and kernel mode. Another important factor is how each processor core is worked. It can see that for LXD, regardless of the mode (kernel or user), the time spent on each processor core is similarly distributed. That is, they present similar behavior throughout the experiment. On the other hand, KVM has a similarity in the kernel mode, in contrast to the user mode. Such mode presents an asymmetry to the cores, probably due to the specialization in its architecture for VxFs components' virtualization.

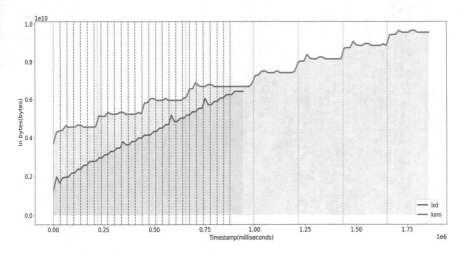

Fig. 4. The amount of memory presently allocated on the system.

Memory: for the present project, the amount of memory currently allocated in the system was measured. Figure 4 show the committed memory for each platform. Committed memory is the sum of all memory that has been allocated by the processes, even if it has not yet been "used" by them. A process that allocates 1 gigabyte of memory (using malloc or similar), but uses only 300 megabytes of that memory, will appear using only 300 megabytes of memory, even if you have the address space allocated for the entire 1 gigabyte. It appears that in both cases the whereas new NFVs are launched, more memory is consumed. However, it is noted that Network Services were established with the same address space, that is, 1 gigabyte, less consumption for the LXD and the KVM. Therefore, it is concluded that there is an optimization of the memory on the part of the LXD concerning the KVM, thus supporting more VxFs in an Edge Computing environment. Besides, it is noteworthy that such optimization can cause only information more relevant to VxF to be present in memory, creating low rate swap operations, thereby reducing I/O operations.

Disk: for the present project, the number of adjacent writings that can be combined for efficiency was measured. Thus, two 4 kilobyte recordings can become an 8-kilobyte recording before it is finally delivered to the disk, and, therefore, they will be counted (and queued) as just an I/O operation. This metric is critical. Figure 5 clearly shows that LXD creates highly performing writing operations concerning KVM. This performance can be a milestone for VxFs with a high storage rate in Edge Computing, which it can do, certain restrictions are not met. Or even more, writing congestion can cause the CPU to be idle, and the processes do not make progress, degrading the system's entire performance.

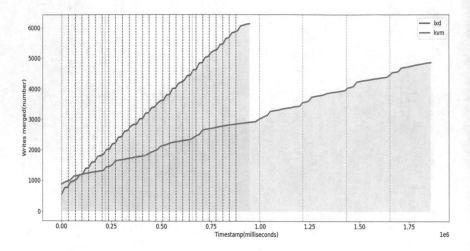

Fig. 5. The number of writes merged.

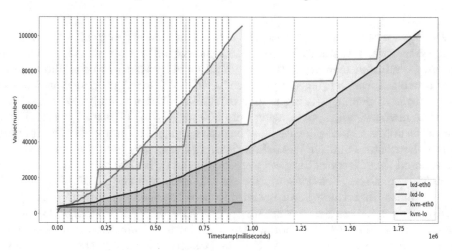

Fig. 6. Network received packets.

Network: Figure 6 shows the number of packets received by each interface. We can see that for both platforms (KVM and LXD), the lo interface (Loopback) has grown over time. Although for LXD, the growth is dizzyingly greater. This is due to the intensive use of data for control and management of EclipseFog05 and Zenoh (software responsible for the communication between the parts of EclipseFog05) present in the Raspberry Pi. Besides, the number of VxFs in LXD is strictly higher than in KVM, leading to the conclusion that more information must be exchanged between EclipseFog05 and VxFs for control management. All communication between VxFs and the outside world is done through the eth0 interface on a virtual network. It can be noted that for KVM, there is an increasing number of packets received as new VxFs are launched. The network metrics are the cumulative counter type. There is no receipt of new packages as soon as each of the VxFs is launched in the KVM. The same is true for LXD, although the number of packets received over the network is much smaller. This concludes that for KVM, data packets are received, perhaps by OSM MANO, for the communication, management, and execution of NS primitives.

6 Conclusion

Most NFV platforms have been targeted toward exploiting traditional or specialized VMs for hosting VxFs typically found in remote, overprovisioned data centers. However, as a programmable network edge is gaining traction, there is a need for lightweight NFV technologies that can exploit the benefits the edge offers (e.g., localized, high-throughput, low-latency network connectivity) and bring NFV to the network edge. Service providers have the ability to run customized, high-performance network services while reducing the increasing cost of core network management and operations.

This work launched several VxFs based on two different virtualization technologies on a Raspberry Pi 4. The results of the KVM-based solution showed significant inferiority to the management of LXD containers. LXD containers have proven to be more productive, optimized, and of high performance. Recent studies [12] have shown that, despite all the container benefits, some problems are still present, such as poor isolation, lack of support for cross-platform and portability limitations. On the other hand, VMs provide strong isolation, have a variety of reach tools to provide cross-platform support and management, offer better portability than containers. However, they take up more memory and have a slower startup.

Therefore, the future of Virtualized Everything Functions (VxF) in edge computing lies in the combination of containerization and full virtualization, making the most of such approaches. In this way, the basis for new applications in different scenarios is created. We can highlight data flow processing throughout the network, efficient data delivery with secure deduplication, low latency video analysis, and artificial intelligence.

Acknowledgements. This study was financed in part by the Coordenação de Aperfeiçoamento de Pessoal de Nível Superior - Brasil (CAPES) - Finance Code 001.

References

1. IDC, IDC's Global DataSphere Forecast Shows Continued Steady Growth in the Creation and Consumption of Data. https://www.idc.com/getdoc.jsp?containerId=prUS46286020
2. N. W. Paper, Network functions virtualisation: an introduction, benefits, enablers, challenges & call for action. issue 1, October 2012

3. Silva, A.P., Tranoris, C., Denazis, S., Sargento, S., Pereira, J., Luís, M., Moreira, R., Silva, F., Vidal, I., Nogales, B., Nejabati, R., Simeonidou, D.: 5ginfire: an end-to-end open5g vertical network function ecosystem. Ad Hoc Netw. **93**, 101895 (2019). http://www.sciencedirect.com/science/article/pii/S1570870518309387

4. Richards, V.M., Moreira, R., de Oliveira Silva, F.: Enabling the management and orchestration of virtual networking functions on the edge. In: CLOSER, pp. 338–346 (2020)

5. Ramalho, F., Neto, A.: Virtualization at the network edge: a performance comparison. In: 2016 IEEE 17th International Symposium on A World of Wireless, Mobile and Multimedia Networks (WoWMoM), pp. 1–6 (2016)

6. Ahmad, M., Alowibdi, J.S., Ilyas, M.U.: vIoT: a first step towards a shared, multi-tenant IoT infrastructure architecture. In: IEEE International Conference on Communications Workshops (ICC Workshops), pp. 308–313 (2017)

7. Cziva, R., Pezaros, D.P.: Container network functions: bringing NFY to the network edge. IEEE Commun. Mag. **55**(6), 24–31 (2017)

8. Li, Z., Kihl, M., Lu, Q., Andersson, J.A.: Performance overhead comparison between hypervisor and container based virtualization. In: 2017 IEEE 31st International Conference on Advanced Information Networking and Applications (AINA), pp. 955–962 (2017)

9. Riggio, R., Khan, S.N., Subramanya, T., Yahia, I.G.B., Lopez, D.: Lightmano: converging NFV and SDN at the edges of the network. In: NOMS 2018 - 2018 IEEE/IFIP Network Operations and Management Symposium, pp. 1–9 (2018)

10. Braun, P.J., Pandi, S., Schmoll, R., Fitzek, F.H.P.: On the study and deployment of mobile edge cloud for tactile internet using a 5g gaming application. In: 2017 14th IEEE Annual Consumer Communications Networking Conference (CCNC), pp. 154–159 (2017)

11. Shahzadi, S., Iqbal, M., Wang, X., Ubakanma, G., Dagiuklas, T., Tchernykh, A.: Lightweight computation to robust cloud infrastructure for future technologies (workshop paper). In: Wang, X., Gao, H., Iqbal, M., Min, G. (eds.) Collaborative Computing: Networking, Applications and Worksharing, pp. 3–11. Springer International Publishing, Cham (2019)

12. Watada, J., Roy, A., Kadikar, R., Pham, H., Xu, B.: Emerging trends, techniques and open issues of containerization: a review. IEEE Access **7**, 152-443–152-472 (2019)

A Proposal for Application-to-Application Network Addressing in Clean-Slate Architectures

Alisson O. Chaves[✉], Pedro Frosi Rosa, and Flávio O. Silva

Faculty of Computing (FACOM), Federal University of Uberlândia (UFU),
Uberlândia 38400-902, Brazil
{alisson,pfrosi,flavio}@ufu.br

Abstract. The addressing model presented in current networks has been around for 40 years, has become a standard, and follows the OSI reference model. In this sense, we have a different type of addressing at each layer. The addressing model is part of some research on new internet architectures of the future. These models allow the unified use of different addresses used, reducing computer networks' complexities and improving support for new applications' communication needs on these networks. This work aims to contribute to a simplification of the addressing process in application-to-application communication. With an experimental evaluation, we showcase how our work can benefit clean-slate network architectures.

Keywords: Future Internet · Addressing · Network computing

1   Introduction

The beginning of the concept of addressing in computer networks occurred through the human needs to store information in its memory. In the 70's we already had client computers and a few servers offering services to a company's network. The task of decorating the IP address of these servers was somewhat trivial.

As time passed, the global computerization process systematized across computing solutions that required creating more of these network servers. Thus, the task of memorizing IP addresses started to become more complex, requiring a simpler solution.

Proposals for developing new future internet architectures have been published, demonstrating a good definition of the control plan and data plan's roles. In which devices it influences together with these proposals, the addressing model has been redesigned to support new applications and reduce addressing complexity.

The use of headers in the application layer, among others, increases a large amount of non-useful data in network traffic, increases the use of computational resources for unpacking the package, and making the decision to forward the operating system socket.

This work assumes that it is possible to further reduce the addressing complexity by creating a specific Virtual Network Interface Card (vNIC) for each communication channel between clients and server applications - a service known as a socket. The practical applications of this solution are reflected even in the form of data sent by the applications.

This work contributes to supporting the development of agnostic applications to the network, reducing network traffic consumption, simplification in addressing applications, and an increase in the number of hosts available on the internet. The solution natively provides interoperability between different architectures and applications developed for possible heterogeneous environments. The remainder of this paper is organized as the following. Section 2 presents a brief history of the use of names in the addressing representation process and the concept of ETArch architecture. Section 3 discusses related works. Section 4 describes the application-to-application addressing proposal for this work. Section 5 presents the details of our experimental evaluation. Section 6 analyzes the results obtained with the tests. Finally, Sect. 7 presents concluding remarks.

2   Background

In this section, we will talk about the concepts and architectures used as principles for developing this project.

2.1   Domain Name Service

Dated 1982, a solution was introduced using the file *hosts.txt*, a text file that contains a list correlating an IP address to any name. This simple solution was implemented on the internet for all connected devices, and synchronization was done by transferring the file hosted on the SRI Network Information Center (SRI-NIC). The solution could also be used on local networks [9].

The concept of addressing is widely used in information technology, from memory locations, process queues to store information on a disk. After the solution was defined, in addition to the continuous growth in the number of services provided by the network, there was a great growth in the types of objects to be addressed in a network, such as switches and routers, it would literally be possible to have at least one name for each one of the 4,294,967,296 IP addresses available worldwide.

Managing, maintaining, organizing, and distributing this list has now become a complex and costly process for networks. The transfer of this file could cause degradation in the internet backbones and the large workload applied to the SRI-NIC server, and the number of changes to the *hosts.txt* file [9].

Works published in the late 1970s and early 1980s originated RFC 882 and 883, which define the concepts, specifications, and model for implementing domain names, which do not replace the use of the *hosts.txt* file, but rather define a global pattern of use [9].

From these definitions today, the DNS has undergone several improvements. However, for this work, we will describe some features.

2.1.1 Language

The language adopted for digital representations of information is ASCII, restricted to characters globally used in the vast majority of languages and some characters' reserves for special functions. There is no size limitation on domain names.

2.1.2 Hierarchy

In the hierarchical and tree operating mode, which in addition to being able to segment management into branches, also allows us to consult directly with branches, which reduces the need to consult the entire list of name-domain relationships, this increases the name resolution performance and decreases the load on domain servers and network backbone.

The branches are organized using the dot character [.]. For example: *mehar.facom.ufu.br*, where *mehar* is a branch of *facom* that is part of the *ufu* branch that participates in the trunk *br*.

2.1.3 Uniqueness

The principle of exclusivity was necessary for the DNS service to guarantee a single, impartial and secure access. Root, trunks, branches, sub-branches, and all their entries must be unique and exclusive throughout the internet. For the registration process of these names to be organized, the IANA - Internet Assigned Numbers Authority was created, which, among several functions, performs the registration of IP addresses on the internet and manages the assignment of domain name root servers, this way, professional, fair and neutral [5].

2.2 ETArch Architecture

The ETArch architecture project uses the concept of assigning titles between the components present in communication. The following are defined: the entities that are participants in the communication, the title, which is the unique form of identification independent of the topology, and the workspace, which is the means of interconnection for communication [3, 13].

Technically, the implementation of ETArch would be based on addressing the entity and workspace. By addressing fields available in the link layer of the OSI reference model, the entities and workspace are represented. An abstraction of the network, transport, session, and presentation layers is carried out, simplifying the traffic on the network bringing gains concerning performance and addressing. Figure 1 presents the basis of ETArch architecture and its components.

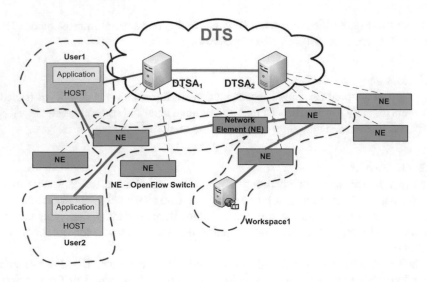

Fig. 1. ETArch architecture [12].

Using the implementation of this project, contributions were made to the definition of an addressing model.

2.3 Addressing

IP addressing solves the problem of finding a device presence on the internet. It represents using decimal numbers the addressing in the network layer of the OSI reference model. Its version 4 allows approximately 4 billion addresses, and since 2014 it has been running out, so after the creation of version 6, we have more than 340 undecillion of possible addresses. However, the use of a new protocol is only possible by updating all devices and operating systems present on the internet, making the process slow. Even today, several networks do not have IPv6 addresses.

With a size of 65536 possibilities, the port is known to also represent, through decimal numbers, the context of an application within a device. It is considered an application layer address, but the transport layer also uses it. They form the socket with the IP address, which is responsible for allowing the communication of different applications between one or several client/server devices simultaneously.

The MAC address, which has more than 281 trillion addresses, represents the address in the link layer of the OSI model through hexadecimal numbers and, by definition, has its use controlled for registered manufactures of network devices, where a unique MAC address is installed to each Network Interface Card manufactured in the world.

3 Related Work

There are different research related to the Future Internet. Several works have proposed a model of addressing networks.

The work from Seoul [2], proposes a division in the semantics of network locators and identifiers, which today is part of the addressing of the network layer, bringing improvements in support of mobility multi-homing.

Geoff Houston [4] presents the properties uniqueness, consistency, persistence, trust, and robustness, which are essential for the network operation. He also describes that the characteristic of the combination of functions present in the IP address of *who*, *where*, and *how* not only makes the use of the network more efficient, they are also the cause of the complexity existing on the internet today. Problems such as mobility, the addressing system's granularity, and the lack of identification of complete paths through the network are research challenges.

The NDN Project [14] proposes using identifiers hierarchically structured in all communication elements of the network. Instead, the names assume the role of identification and can reference the data more simply. In one of its principles, end-to-end, it allows the development of more robust applications in network failures. Some characteristics are unique and hierarchical names of variable length and addressing by names.

Researchers at MIT [10] present the principle of the end-to-end argument that suggests that the functions applied to the network can be redundant or costly, since in many services available on the internet today, they perform these functions directly at the application layer. Functions like Delivery Guarantees, Secure Transmission of Data, Duplicate Message Suppression, and point identification are examples of possible unnecessary use of resources that we use today.

Proposals for the development of new *clean-slate* architectures such as Mobiliy-First [11], NEBULA [1], and ETArch [12], have been published, demonstrating a good definition of the roles of the control plan and data plan and which devices it influences.

However, when we look at everyone involved in the communication, these proposals are focused only on network devices, not considering the operating systems of the final devices, thus requiring applications that consume network services to know the structures of the network and also to have greater interaction with the OS at the level of identifications in the application layer.

4 Proposal

Future Internet projects reduced the number of headers to only the one present in the link layer. Because it does not interfere with the application layer protocols, the headers must still exist for traffic to be redirected to the device's correct application.

Aiming to make usability in new internet technologies of the future simpler, in this work, we contemplate an effective communication between the new network architectures and the operating systems to provide transparency in the application data traffic.

Figure 2 presents an overview of the component of our solution. In this proposal, we separate the components into layers with their respective functions. In the application component layer, we have several applications developed for different types of architectures. Identification of these applications' communication is sent to the converter component layer, where it will be converted into a MAC address.

This MAC address is now used to create a Virtual Network Interface Card of the *macvlan* type in bridge mode for each application communication. Now the data is sent on the physical interface available on the host using its own identification on the network.

Without hurdling the functionalities described in this paper about the DNS, it is proposed to create a virtualized NIC for each application socket within the device's operating system. Thus, when an application invokes a socket to use through an application, this vNIC is created and used solely and exclusively for that application.

As sockets' concept was abstracted, the traffic of an application through this vNIC can be in raw mode, that is, send only the useful data related to the context of the application.

For example, after accessing the application socket and creating the vNIC, a chat application can traffic only the bits referring to the message's text. The management and use of the operating system's network sockets would be simpler, and the volume of data for the same application compared to today's networks would be less, which provides faster access to the services available in these architectures.

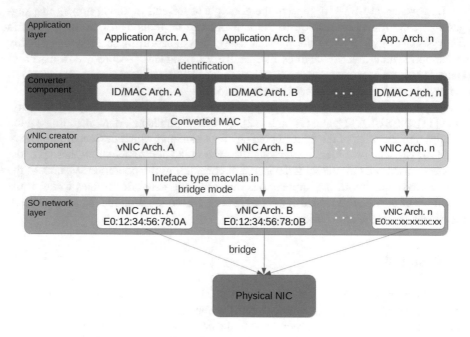

Fig. 2. Solution components.

In an application, a message exchange module and a video conference module belonging to the same server must have unique and hierarchical identifiers as a name, for example: *msg.chat.mehar.facom.ufu.br* and *vid.chat.mehar.facom.ufu.br*.

5 Experimental Evaluation

This section will demonstrate that the solution proposed in this work can work all the clean-slate network architecture cited in Sect. 2. To do the experimental evaluation of our proposal, we selected the ETArch network architecture that our research group proposed.

The addressing fields available in the ETArch architecture are the Destination MAC Address and Source MAC Address fields, which have a size of 48 bits. Of these 48 bits, the 7 bit is reserved to identify whether the address is globally unique or locally administered, and bit 8 is used to check whether the address is unicast or multicast.

Using the Linux OpenSuSE Leap 15.0 64bit operating system, the use of the network driver virtualization library called *macvlan* was defined. Macvlan was developed to create a communication channel between the physical NIC and virtual machine [7].

In this work, the use of the concept of domain names for use in titles and workspaces with all their language, hierarchy, and exclusivity functions were defined. A configuration file indicates the device's title and on which physical network interface the vNIC will be attacked.

The software created reads this information and uses the MAC address conversion library's domain name to create the vNIC corresponding to the device.

The library reserves the first 8 bits as *1110 0000* which will result in hexadecimal *E0* and converts the domain name to hexadecimal using the default unsecured hash function *CRC32 CRC16* [8] with 40-bit output, thus totaling the 48 bits present in the link-layer addressing fields.

After creating the workspace, the application must make a call informing the name of the workspace to the conversion software that will have "<device_title>+ <workspace>" input and 48-bit address output.

At this point, the software developed creates a vNIC *macvlan* type in bridge mode, and now the application can open a socket and send data in raw mode through this interface.

5.1 Test Environment

The proposal's test environment has all the necessary participants for the complete functioning of an ETArch network. DTSA is hosted in the MEHAR research data center. In another network connected via the internet, 2 x wireless TPLink TL-WR1043ND routers will be used, which had the operating system modified to OpenWRT 18.06, with the installation of the OpenVSwitch package and OpenFlow 1.0 protocol enabled. Connected to each of these wireless routers, a computer runs the Linux OpenSuSE Leap 15.0 64bit operating system where the developed applications will be implemented.

5.2 Experiment Description

Figure 3 shows the proposal for raw data sending on an application-to-application socket. In the host *alisson.ufu.br* we have two applications, and each application has two different services, and for each service, a corresponding vNIC with unique addressing would be created. These services communicate with the client host. We would have

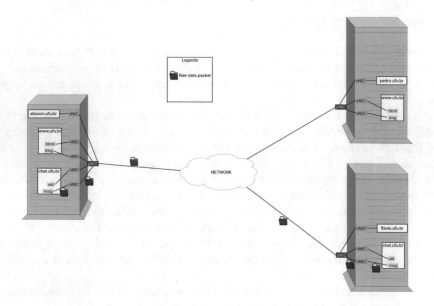

Fig. 3. Raw data sending on application-to-application socket.

access to the respective applications and services provided. Thus, a data package has its destination, the unique identification of the service delivered by the application.

For the tests, a simple ETArch based chat application was used, which sends through the interface intended that application socket in raw socket mode only the pure text to be sent in the message.

To better illustrate the proposal, a 162.2 MByte file was sent through the test structure described above in three network configuration scenarios: 1500 Bytes MTU, 500 Bytes MTU and 130 Bytes MTU. The different MTU sizes simulate sending packets of large and small sizes, and as a means of comparison with a TCP/IP application, we carry out the transmission of the same file through the simple File Transfer Protocol (FTP).

6   Results and Analysis

To perform the tests, we created the vNIC Creator and Converter components. We also used an ETArch chat application that performs the sending of pure data were used. Figure 4 shows the execution of the software developed in two hosts. It checks the configuration file, creates and verifies the interface related to the host identification. It also creates and verifies a workspace and carries out the exchange of messages in the chat application.

Figure 5 and 6 demonstrate the capture of packets sent by the chat application from both hosts, as seen, not being necessary the process of padding the message until completing 46 bytes in the data field of According to the standards of the Ethernet protocol, [6] this allows a reduction in the data sent over the network.

```
alisson:~ # cat /etc/etarch/etarch.conf
p7p1 alisson.ufu.br
alisson:~ # etarch-int start
alisson:~ # ifconfig alisson.ufu.br
alisson.ufu.br: flags=4163<UP,BROADCAST,RUNNING,MULTICAST>  mtu 1500
        inet6 fe80::e233:aaff:fe1e:894b  prefixlen 64  scopeid 0x20<link>
        ether e0:33:aa:1e:89:4b  txqueuelen 1000  (Ethernet)
        RX packets 9  bytes 2546 (2.4 KiB)
        RX errors 0  dropped 0  overruns 0  frame 0
        TX packets 25  bytes 4524 (4.4 KiB)
        TX errors 0  dropped 0 overruns 0  carrier 0  collisions 0

alisson:~ # etarch-ws start chat.ufu.br
alisson:~ # ifconfig chat.ufu.br
chat.ufu.br: flags=4163<UP,BROADCAST,RUNNING,MULTICAST>  mtu 1500
        inet6 fe80::e22d:3eff:fec4:a5c9  prefixlen 64  scopeid 0x20<link>
        ether e0:2d:3e:c4:a5:c9  txqueuelen 1000  (Ethernet)
        RX packets 15  bytes 3615 (3.5 KiB)
        RX errors 0  dropped 0  overruns 0  frame 0
        TX packets 24  bytes 4429 (4.3 KiB)
        TX errors 0  dropped 0 overruns 0  carrier 0  collisions 0

alisson:~ # etarch-chat chat.ufu.br
RTNETLINK answers: File exists
Send a message:
First msg

Received Message:
Second msg

Send a message:
Third msg

Received Message:
Fourth msg

Send a message:
```

```
hp:~ # cat /etc/etarch/etarch.conf
eth0 flavio.ufu.br
hp:~ # etarch-int start
hp:~ # ifconfig flavio.ufu.br
flavio.ufu.br: flags=4163<UP,BROADCAST,RUNNING,MULTICAST>  mtu 1500
        inet6 fe80::e2d1:25ff:fe68:2a9a  prefixlen 64  scopeid 0x20<link>
        ether e0:d1:25:68:2a:9a  txqueuelen 1000  (Ethernet)
        RX packets 7  bytes 1610 (1.5 KiB)
        RX errors 0  dropped 0  overruns 0  frame 0
        TX packets 19  bytes 3402 (3.3 KiB)
        TX errors 0  dropped 0 overruns 0  carrier 0  collisions 0

hp:~ # etarch-ws start chat.ufu.br
hp:~ # ifconfig chat.ufu.br
chat.ufu.br: flags=4163<UP,BROADCAST,RUNNING,MULTICAST>  mtu 1500
        inet6 fe80::e265:2dff:fe0b:9efb  prefixlen 64  scopeid 0x20<link>
        ether e0:65:2d:0b:9e:fb  txqueuelen 1000  (Ethernet)
        RX packets 7  bytes 1930 (1.8 KiB)
        RX errors 0  dropped 0  overruns 0  frame 0
        TX packets 18  bytes 3240 (3.1 KiB)
        TX errors 0  dropped 0 overruns 0  carrier 0  collisions 0

hp:~ # etarch-chat chat.ufu.br
RTNETLINK answers: File exists
Send a message:
Second msg

Received Message:
Third msg

Send a message:
Fourth msg
```

Fig. 4. Terminal configuration, execution and chat test.

No.	Time	Source	Destination	Protocol	Length	Info
39	52.212904404	e0:2d:3e:c4:a5:c9	e0:6e:71:f3:7b:ce	0x0880	24	Ethernet II

```
▶ Frame 39: 24 bytes on wire (192 bits), 24 bytes captured (192 bits) on interface chat.ufu.br, id 0
▶ Ethernet II, Src: e0:2d:3e:c4:a5:c9 (e0:2d:3e:c4:a5:c9), Dst: e0:6e:71:f3:7b:ce (e0:6e:71:f3:7b:ce)
▼ Data (10 bytes)
     Data: 5468697264206d73670a
     [Length: 10]
```

```
0000  11100000 01101110 01100001 11110011 01111011 11001110 11100000 00101101   ·nq·{··-
0008  00111110 11000100 10100101 11001001 00001000 10000000 01010100 01101000   >····Th
0010  01101001 01110010 01100100 00100000 01101101 01110011 01100111 00001010   ird msg·
```

Fig. 5. Send packet capture on first computer (alisson).

No.	Time	Source	Destination	Protocol	Length	Info
4	58.197214025	e0:65:2d:0b:9e:fb	e0:6e:71:f3:7b:ce	0x0880	25	Ethernet II

```
▶ Frame 4: 25 bytes on wire (200 bits), 25 bytes captured (200 bits) on interface chat.ufu.br, id 0
▶ Ethernet II, Src: e0:65:2d:0b:9e:fb (e0:65:2d:0b:9e:fb), Dst: e0:6e:71:f3:7b:ce (e0:6e:71:f3:7b:ce)
▼ Data (11 bytes)
     Data: 466f75727468206d73670a
     [Length: 11]
```

```
0000  e0 6e 71 f3 7b ce e0 65  2d 0b 9e fb 08 80 46 6f   ·nq·{··e····Fo
0010  75 72 74 68 20 6d 73 67  0a                        urth msg·
```

Fig. 6. Send packet capture on second computer (hp).

Figure 7 compares the amount of data sent over the network using the FTP protocol and our proposal concerning the original information contained in the file. The amount of protocol control data plus the size of the PDU layers of the TCP/IP model's layers represent the total send data in the network: 6.40%, 18.08%, and 55.88%, respectively in the MTU sizes of 1500, 500, and 130. In the proposed model, we do not have protocol control data, and the fields used as the combined PDU represent the total send data in the network: 0.92%, 2.72%, and 9.72% respectively in the MTU sizes of 1500, 500, and 130.

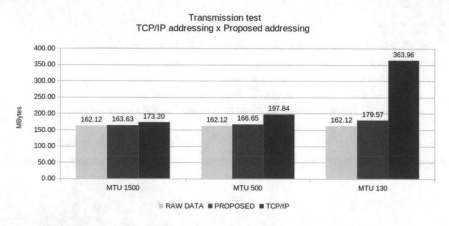

Fig. 7. Transmission test. TCP/IP addessing x Proposed addressing.

In the proposed model, we do not have protocol control data and the fields used as the combined PDU represent the total send data in the network: 0.92%, 2.72% and 9.72% respectively in the MTU sizes of 1500, 500 and 130.

7   Conclusion

Future Internet research projects create a new world of possibilities, which must be studied in depth. The ETArch architecture defines a title model that, among several gains, we can mention the abstraction of layers 3 and 4 of the TCP/IP reference model.

As a contribution, a unique addressing model was defined for the entire TCP/IP stack through the fields currently available in an ethernet protocol gained with the reduction in the minimum size of an ethernet frame.

As an objective, we have also defined a model for developing new applications where it is only necessary to send data through the network interface related to the application socket's name. Using the developed application, we created a channel that allows the creation and maintenance of these interfaces.

A simple sample application of a clean-slate network architecture was used to test this work's assumptions and achieved the expected objectives.

Future work must adapt the interface maintenance system as a module of the Linux operating system. The development of more complex applications in client / server architectures would be a powerful object of study.

Another future objective is to apply this model to other clean-slate network architectures, thus ensuring interoperability between the projects' concepts. Another possibility would be tested on legacy networks and the current internet to verify the applicability of the concepts defined here in the historical load of internet development.

References

1. Anderson, T. et al.: The nebula future internet architecture. In: SPRINGER. The Future Internet Assembly, pp. 16–26 (2013)
2. Choi, J., et al.: Addressing in future internet: problems, issues, and approaches (2008)
3. Corujo, D., et al.: Enabling network mobility by using IEEE 802.21 integrated with the entity title architecture (2013)
4. Huston, G.: Addressing the Future Internet. The ISP Column (2007)
5. IANA - About us. https://www.iana.org/about/. Accessed 12 Jan 2021
6. Kurose, J., Ross, K.: Computer networking: a top-down approach. CERN Document Server (2017). https://cds.cern.ch/record/2252697. Accessed 12 Jan 2021
7. Liu, H.: Introduction to Linux interfaces for virtual networking (2018). https://developers.redhat.com/blog/2018/10/22/introduction-to-linux-interfaces-for-virtual-networking/. Accessed 12 Jan 2021
8. Miller, F.P., et al.: Cyclic redundancy check: computation of CRC, mathematics of CRC, error detection and correction, cyclic code, list of hash functions, parity bit, information ... Cksum, Adler- 32, Fletcher's Checksum. Alpha Press (2009)
9. Mockapetris, P., Dunlap, K.J.: Development of the domain name system. In: Symposium Proceedings on Communications Architectures and Protocols, Association for Computing Machinery, pp. 123–133 (1988). https://doi.org/10.1145/52324.52338
10. Saltzer, J.H., et al.: End-to-end arguments in system design. ACM Trans. Comput. Syst. 2(4), 277–288 (1984). https://doi.org/10.1145/357401.357402
11. Seskar, I., et al.: Mobilityfirst future internet architecture project. In: ACM. Proceedings of the 7th Asian Internet Engineering Conference, pp. 1–3 (2011)
12. Silva, F.: Endereçamento por título: uma forma de encaminhamento multicast para a próxima geração de redes de computadores. Tese (Doutorado) — Universidade de São Paulo (2013)
13. Silva, F., et al.: Entity title architecture extensions towards advanced quality-oriented mobility control capabilities. In: 2014 IEEE Symposium on Computers and Communications (ISCC), pp. 1–6 (2014). https://doi.org/10.1109/ISCC.2014.6912459
14. Zhang, L., Estrin, D., Burke, J., Jacobson, V., Thornton, J., Smetters, D., Zhang, B., Tsudik, G., Massey, D., Papadopoulos, C., et al.: Named data networking (NDN) project. Technical report NDN-0001, Xerox Palo Alto Research Center-PARC 2010; 157:158 (2010)

Author Index

Printed in the United States
by Baker & Taylor Publisher Services